©1990 Tom Van Sant, Inc. / The GeoSphere Project
Santa Monica, California

# www.brookscole.com

*www.brookscole.com* is the World Wide Web site for Thomson Brooks/Cole and is your direct source to dozens of online resources.

At *www.brookscole.com* you can find out about supplements, demonstration software, and student resources. You can also send e-mail to many of our authors and preview new publications and exciting new technologies.

**www.brookscole.com**
Changing the way the world learns®

# Essentials of Physical Geography Eighth Edition

**Robert E. Gabler**
*Emeritus, Western Illinois University*

**James F. Petersen**
*Texas State University—San Marcos*

**L. Michael Trapasso**
*Western Kentucky University*

With contributions by

**Dorothy Sack**
*Ohio University, Athens*

**Robert J. Sager**
*Pierce College, Washington*

**Daniel L. Wise**
*Western Illinois University*

Australia • Brazil • Canada • Mexico • Singapore • Spain
United Kingdom • United States

### *Essentials of Physical Geography,* Eighth Edition
Robert E. Gabler, James F. Petersen, L. Michael Trapasso

Earth Sciences Editor: Keith Dodson
Assistant Editor: Carol Benedict
Editorial Assistants: Anna Jarzab, Megan Asmus
Technology Project Manager: Ericka Yeoman-Saler
Marketing Manager: Mark Santee
Marketing Assistant: Sylvia Krick
Marketing Communications Manager: Bryan Vann
Project Manager, Editorial Production: Harold Humphrey
Art Director: Vernon Boes
Print Buyer: Doreen Suruki
Permissions Editor: Joohee Lee
Project Coordination and Composition: Pre-Press Company
Text Designer: Jeanne Calabrese

Photo Researcher: Terri Wright
Illustrator: Accurate Art, Precision Graphics, Rolin Graphics
Cover Designer: Larry Didona
Cover Images: Top: The Kaokoveld, Namibia (Africa) © Eddi Böhnke/zefa/Corbis; Wave: Oahu North Shore, Hawaii © Brain Sytnyk/Masterfile; Death Valley, CA sand dune: Photo 24/Getty Images: Hurricane Katrina: NASA.
Image on p.7 by R.B. Husar, Washington University; the land layer from the Sea WiFs Project; fire maps from the European Space Agency; the sea surface temperture from the Naval Oceanographic Office's Visualization laboratory; and cloud layer from SSEC, University of Wisconsin
Text and Cover Printer: Transcontinental Printing/Interglobe

© 2007 Thomson Brooks/Cole, a part of The Thomson Corporation. Thomson, the Star logo, and Brooks/Cole are trademarks used herein under license.

ALL RIGHTS RESERVED. No part of this work covered by the copyright hereon may be reproduced or used in any form or by any means—graphic, electronic, or mechanical, including photocopying, recording, taping, web distribution, information storage and retrieval systems, or in any other manner—without the written permission of the publisher.

ExamView® and ExamView Pro® are registered trademarks of FSCreations, Inc. Windows is a registered trademark of the Microsoft Corporation used herein under license. Macintosh and Power Macintosh are registered trademarks of Apple Computer, Inc. Used herein under license.

Printed in Canada
1  2  3  4  5  6  7  09  08  07  06

Library of Congress Control Number: 2005938850

Student Edition: ISBN 0-495-01194-0

**Thomson Higher Education**
**10 Davis Drive**
**Belmont, CA 94002-3098**
**USA**

For more information about our products, contact us at:
**Thomson Learning Academic Resource Center**
**1-800-423-0563**
For permission to use material from this text or product, submit a request online at **http://www.thomsonrights.com**.
Any additional questions about permissions can be submitted by e-mail to **thomsonrights@thomson.com**.

# Preface

Earth, with its life-support system for humankind, is a wondrous and complex place, yet it is also enigmatic. In many ways our planet is robust and adaptable to environmental change, but in others it is fragile and threatened. In our modern day-to-day lives, our use of technology tends to insulate us from fully experiencing our environment, so we can become lulled into forgetting about our direct dependence on Earth's natural resources. Sometimes it is hard to imagine that as you read this you are also moving through space on a life-giving planetary oasis surrounded by the vastness of space—one that is empty and, as far as we know, devoid of life. Yet having knowledge of our planet, its environments, and how they operate is as critical as it has ever been for human society.

For as long as humans have existed, the resources provided by their physical environments have been the key to survival. Preindustrial societies, such as those dependent on hunting and gathering, or small-scale agriculture, tended to have small populations and therefore exerted relatively little impact on their natural surroundings. In contrast, today's industrialized societies have large populations, demand huge quantities of natural resources, and can influence or cause environmental change, not only on a local scale, but also on a global one.

As human population has increased, so have the scales, degrees, and cumulative effects of its impacts on the environment. Unfortunately, many of these impacts are detrimental. We have polluted the air and water. We have used up tremendous amounts of nonrenewable resources and have altered many natural landscapes without fully assessing the potential consequences. Too often, we have failed to respect the power of Earth's natural forces when constructing our homes and cities or while pursuing our economic activities. In the twenty-first century, it has become evident that if we continue to fail to comprehend Earth's potential and respect its limitations as a human habitat, we may be putting ourselves and future generations at risk. Despite the many differences between our current lifestyles and those of early humans, one aspect has remained the same: The way we use our physical environment is key to our survival.

Today, this important message is getting through to many people. We understand that Earth does not offer boundless natural resources. Broadcast and print media have expanded their coverage of the environment and human/environment relationships. Politicians are drafting legislation designed to address environmental problems. Scientists and statesmen from around the world meet to discuss environmental issues that increasingly cross international boundaries. Humanitarian organizations, funded by governments as well as private citizens, struggle to alleviate the suffering that results from natural disasters but also from mistreatment of the environment. The more we know about our environment, the more we can continue to strive to preserve it.

- The goal of studying physical geography is to develop a thorough knowledge, understanding, and appreciation of our Earth. In approaching this goal, this textbook focuses on three perspectives that physical geography employs for examining Earth and its environmental features and processes. Through a spatial science perspective, physical geography focuses on the location, distribution, and spatial interactions of natural phenomena on Earth.
- Through a physical science perspective, physical geography incorporates the techniques of models and systems analysis, as well as the accumulated knowledge of the companion natural and physical sciences, to examine Earth processes and patterns.
- Through an environmental science perspective, physical geography plays a special role because geographers consider the influence and interaction of both components in human-environment relationships.

Geography is a highly regarded subject in most nations of the world, and in recent years it has undergone a renaissance in the United States. More and more Americans are realizing the importance of knowing and respecting the people, cultural contributions, and environmental resources of other nations in addition to those of the United States. National education standards that include physical geography ensure that a high quality geography curriculum is offered in United States elementary, secondary, and postsecondary schools. Employers across the nation are recognizing the value and importance of geographic knowledge, skills, and techniques in the workplace. Physical geography as an applied field makes use of computer-assisted and space-age technologies, such as geographic information systems (GIS), computer-assisted map-making (cartography), the global positioning system, and satellite image interpretation. At the collegiate level, physical geography offers an introduction to the concerns, ideas, knowledge, and tools that are necessary for further study of our planet. More than ever before, physical geography is being recognized as an ideal science course for general-education students—students who will make decisions that weigh human needs and desires against environmental limits and possibilities. It is for these students that *Essentials of Physical Geography* has been written.

## Features

**Comprehensive View of the Earth System** *Essentials of Physical Geography* introduces all major aspects of the Earth system, identifying physical phenomena and natural processes and stressing their characteristics, relationships, interactions, and distributions. The textbook covers a wide range of topics, including the atmosphere, the solid Earth, oceans and other water bodies, and the biotic environments of our planet.

**Clear Explanation** The text uses an easily understandable, narrative style to explain the origins, development, significance, and distribution of processes, physical features, and events that occur within, on, or above the surface of Earth. The writing style is targeted toward rapid reader comprehension and toward making the study of physical geography an enjoyable undertaking.

**Introduction to the Geographer's Tools** Space-age and computer technologies have revolutionized the ways that we can study our planet, its features, its environmental aspects, and its natural processes. A full chapter is devoted to maps and other forms of spatial imagery and data used by geographers. Illustrations throughout the book include images gathered from space, accompanied by interpretations of the environmental aspects that the scenes illustrate. Also included are introductory discussions of techniques currently used by geographers to analyze or display location and environmental aspects of Earth, including remote sensing, geographic information systems, computer-assisted cartography, and the global positioning system.

**Focus on Student Interaction** The text uses numerous methods to encourage continuous interaction between students, the textbook, and the subject matter of physical geography. The Consider and Respond activities at the end of each chapter are designed to invite students to apply their newly acquired knowledge in different situations. Questions following the captions of many illustrations prompt students to think beyond the map, graph, diagram, or photograph and give further consideration to the topic.

**Map Interpretation Series** Learning map-interpretation skills is a priority in a physical geography course. To meet the needs of students who do not have access to a laboratory setting, this text includes map activities with accompanying explanations, illustrations, and interpretation questions. These maps give students an opportunity to learn and practice valuable map skills. In courses that have a lab section, the map activities are a supplementary link between class lectures, the textbook, and lab work.

## Objectives

Since the first edition, the authors have sought to accomplish four major objectives:

**To Meet the Academic Needs of the Student** Instructors familiar with the style and content of *Essentials of Physical Geography* know that this textbook is written specifically for the student. It has been designed to satisfy the major purposes of a liberal education by providing students with the knowledge and understanding to make informed decisions involving the physical environments that they will interact with throughout their lives. The text assumes little or no background in physical geography or other Earth sciences. Numerous examples from throughout the world are included to illustrate important concepts and help the non-science major bridge the gap between scientific theory and its practical application.

**To Strongly Integrate the Illustrations with the Written Text** Numerous photographs, maps, satellite images, scientific visualizations, block diagrams, graphs, and line drawings have been carefully chosen to clearly illustrate each of the complex concepts in physical geography addressed by the text. The written text discussions of these concepts often contain repeated references to the illustrations, so students are able to examine in model form, as well as mentally visualize, the physical processes and phenomena involved. Some examples of complex topics that are clearly explained through the integration of visuals and text include the seasons (Chapter 3), heat energy budget (Chapter 4), surface wind systems (Chapter 5), atmospheric disturbances (Chapter 7), soils (Chapter 12), plate tectonics (Chapter 13), rivers (Chapter 17), and glaciers (Chapter 19).

**To Communicate the Nature of Geography** The nature of geography and three major perspectives of physical geography (spatial science, physical science, and environmental science) are discussed at length in Chapter 1. In subsequent chapters, important topics of geography involving the spatial science perspective are emphasized, along with those relating to the physical and environmental sciences. For example, location is a dominant topic in Chapter 2 and remains an important theme throughout the text. Spatial distributions are stressed as the climatic elements are discussed in Chapters 4 through 6. The changing Earth system is a central focus in Chapter 8. Characteristics of places constitute Chapters 9 and 10. Spatial interactions are demonstrated in discussions of weather systems (Chapter 7), soils (Chapter 12), and tectonic activity (Chapters 13 and 14).

**To Fulfill the Major Requirements of Introductory Physical Science College Courses** *Essentials of Physical Geography* offers a full chapter on the important tools and the methodologies of physical geography. Throughout the book, the Earth as a system, and the physical processes that are responsible for the location, distribution, and spatial relationships of physical phenomena beneath, at, and above Earth's surface are examined in detail. Scientific method, hypothesis, theory, and explanation are continually stressed. In addition, questions that involve understanding and interpreting graphs of environmental data (or graphing data for analysis), quantitative transformation or calculation of environmental variables, and/or hands-on map-analysis directly support science learning and are included in sections at the end of many chapters. Models and systems are frequently cited in the discussion of important concepts and scientific classification is presented in several chapters—some of these topics include air masses (Chapter 7), climates (Chapters 8 and 9), biomes (Chapter 11), soils (Chapter 12), rivers (Chapter 17), and coasts (Chapter 21).

## Eighth Edition Revision

Revising *Essentials of Physical Geography* for an eighth edition involved considering the thoughtful input from many reviewers with varied opinions. Several reviewers recommended a reduction in the number of chapters and a few organizational changes. As authors we seek to expand or include coverage on physical geographic topics that will spark student interest. Recent natural disasters such as the extraordinary 2005 hurricane season, particularly Hurricane Katrina, and the tragic South Asia tsunami merited inclusion in this new edition. These events are addressed as examples of Earth processes and human-environmental interactions. In addition, we thoroughly revised the text; prepared new graphs, maps, and diagrams; integrated approximately 200 new photographs; and updated information on numerous worldwide environmental events. What follows is a brief review of major changes made to this eighth edition.

**New Contributing Author** The authors are privileged to welcome Dorothy Sack of Ohio University as a new contributing author. Dorothy is a geomorphologist with a broad background in physical geography and a strong interest in coastal environments. Her expertise and fresh outlook on geomorphology, particularly in arid, fluvial, and Quaternary environments, as well as her commitment to geographic education, have been a great asset to this edition.

**Chapter Reorganization** The number of chapters has been reduced to 21, though the content itself has not been changed. The most important aspects of the previous edition's summary chapter have been streamlined and placed in other chapters, particularly Chapter 1, allowing us to strengthen discussions without lengthening the book. The chapter on temperature and the heat budget was reorganized to allow students to follow a sequence of worldwide patterns and variations of heat, temperatures, and controls of temperature. The global climates and climate change chapter received major revision. An introduction to climate, climate classification systems including Köppen's, climographs, and climate change are addressed in a single, expanded chapter. Climate classification has moved from an appendix and been modified into a student activity involving climate data and climographs. The chapter on rivers and fluvial geomorphology has been strengthened with more information and a more strongly hydrology-oriented approach to rivers. Arid landforms related to running water and eolian processes are now united in a single chapter that precedes a chapter on glacial systems, and both chapters have been improved with new illustrations. The map and graph interpretation exercises have been moved to the ends of chapters to avoid interrupting the flow of text discussion. These changes provide increased course flexibility without significantly altering the sequence of topics or compelling instructors to make major changes in syllabi.

**New and Revised Text** New material has been added on a variety of topics. Hurricane Katrina and its impact has been included, as well as a section on the South Asia tsunami. Earth systems approaches are reinforced with additional content, illustrations, and examples. A new map interpretation activity deals with weather associated with frontal systems. New feature boxes have been added. The concept of spatial scale in atmospheric processes has been given a stronger emphasis. Sections on the greenhouse effect and global warming have been expanded. A graph interpretation activity is included that involves the analysis and classification of climatic data and characteristics through use of climographs. The chapter on soils has been extensively revised to include new soil profile images for each of the twelve NCRS soil orders. The chapter on Earth materials and plate tectonics has been revised to strongly focus on evidence for plate/continental motions, and a new approach to understanding rocks and minerals has been added. The groundwater and karst chapter highlights aquifers as a source of fresh water and considers the concern for their protection and their limitations as a resource. Many other sections of the book contain new material, new line art, new photographs, and new feature boxes. These include geography's spatial, physical, and environmental science perspectives (Chapter 1), GPS, GIS, and remote sensing (Chapter 2), Earth-Sun relationships (Chapter 3), upper air circulation (Chapter 5), weather systems (Chapter 7), soil-forming processes and soil classification (Chapter 12), evidence for plate tectonics (Chapter 13), fluvial systems (Chapter 17), arid landforms and processes (Chapter 18), and glacial systems (Chapter 19).

**Enhanced Program of Illustrations** The illustration program has undergone substantial revision. The new and expanded topics required many new figures and updates to others, including numerous photographs, satellite images, and maps. More than 200 figures have been replaced by new photographs and new or revised line drawings.

**An Increased Focus on Geography as a Discipline**  The undergraduate students of today include the professional geographers of tomorrow. Several changes in the text have been introduced to provide students with a better appreciation of geography as a discipline worthy of continued study and meriting serious consideration as a career choice, beginning with the definition of geography, the discipline's tools and methodologies, selected topics to illustrate the role of geography as a spatial science, and the practical applications of the discipline, all topics found in Chapter 1. In addition, there are few better ways to understand a discipline than to learn firsthand from its practitioners. The Career Vision essays provide a close look at geographers in the workplace, how they earn their living, and what educational programs prepared them for their occupations. These first-person reports describe some of the exciting career opportunities available to students who choose physical geography as their field of study.

The eighth edition also updates a popular feature illustrating the three major geographic perspectives among the sciences, with two article-boxes for each chapter. Articles titled "Geography's Spatial Science Perspective" demonstrate a major emphasis of geography, in comparison to other sciences. The nature of geography mandates a study of how phenomena are distributed in space. To this end, the spatial point of view becomes that of the geographer. Articles in the "Geography's Physical Science Perspective" series emphasize geography as a synthesis and applied representative of all the major sciences. In these articles, physics, chemistry, biology, geology, and mathematics are used to highlight selected aspects or questions derived from associated chapter text, but from the point of view of supporting a geographic analysis. "Geography's Environmental Science Perspective" reminds students that our physical environment comprises Earth, water, vegetation, and the atmosphere, all of which are components of physical geography essential to life on Earth. Thus, geography plays a central role in understanding environmental issues, human-environmental interactions, and in the solution of environmental problems. Spreading the message about the importance and relevance of geography in today's world is essential to the viability and strength of geography in schools and universities. *Essentials of Physical Geography,* Eighth Edition, seeks to reinforce that message to our students through appropriate examples in the text.

# Ancillaries

Instructors and students alike will greatly benefit from the comprehensive ancillary package that accompanies this text.

# For the Instructor

## Class Preparation and Assessment Support

**Instructor's Manual with Test Bank and Lab Pack**  The downloadable manual contains suggestions concerning teaching methodology as well as evaluation resources including course syllabi, listings of main concepts, chapter outlines and notes, answers to review questions, recommended readings and a complete test item file. The Instructor's Manual also includes answers for the accompanying *Lab Pack.* Available exclusively for download from our password-protected instructor's website: http://now.brookscole.com/gabler8

**ExamView® Computerized Testing**  Create, deliver, and customize tests and study guides (both print and online) in minutes with this easy-to-use assessment and tutorial system. Preloaded with the *Essentials of Physical Geography* test bank, *ExamView* offers both a *Quick Test Wizard* and an *Online Test Wizard*. You can build tests of up to 250 questions using as many as 12 question types. *ExamView's* complete word-processing capabilities also allows you to enter an unlimited number of new questions or edit existing questions.

## Dynamic Lecture Support

**Multimedia Manager: A Microsoft® PowerPoint® Link Tool**  This one-stop lecture tool makes it easy for you to assemble, edit, publish, and present custom lectures for your course, using Microsoft PowerPoint™. Enhanced with Living Lecture Tools (including all of the text illustrations, photos, and Active Figure animations), the *Multimedia Manager* quickly and easily lets you build media-enhanced lecture presentations to suit your course.

**JoinIn™ on TurningPoint® for Response Systems**  Using or considering an audience response system? Book-specific JoinIn™ content for Response Systems tailored to physical geography allows you to transform your classroom and assess your students' progress with instant in-class quizzes and polls. A Thomson exclusive agreement to offer TurningPoint® software lets you pose book-specific questions and display students' answers seamlessly within the Microsoft® PowerPoint® slides of your own lecture, in conjunction with the infrared or radio frequency "clicker" hardware of your choice. Enhance how your students interact with you, your lecture, and each other. For college and university adopters only. Contact your local Thomson Brooks/Cole representative to learn more about JoinIn, as well as our infrared and radio-frequency transmitter solutions.

**Transparency Acetates**  An updated collection of 100 acetates that include full-color images from the text.

## WebCT and Blackboard Support

**Physical GeographyNow, integrated with Web CT and Blackboard**  Now present the interactive testing and tutorial features of Physical GeographyNow with all the sophisticated functionality of a WebCT or Blackboard product. Contact your Thomson Brooks/Cole representative for more information on our solutions for your WebCT or Blackboard-enhanced courses.

### Laboratory and GIS Support

**Lab Pack**   ISBN: 0495011916
The perfect lab complement to the text, this *Lab Pack* contains over 50 exercises, varying in length and difficulty, designed to help students achieve a greater understanding and appreciation of physical geography.

**GIS Investigations**   Michelle K. Hall-Wallace, C. Scott Walker, Larry P. Kendall, Christian J. Schaller, and Robert F. Butler of the University of Arizona, Tucson.
The perfect accompaniment to any physical geography course, these four groundbreaking guides tap the power of *ArcView*® GIS to explore, manipulate, and analyze large data sets. The guides emphasize the visualization, analysis, and multimedia integration capabilities inherent to GIS and enable students to "learn by doing" with a full complement of GIS capabilities. The guides contain all the software and data sets needed to complete the exercises.

**Exploring the Dynamic Earth:**
**GIS Investigations for the Earth Sciences**
ISBN: 0-534-39138-9

**Exploring Tropical Cyclones:**
**GIS Investigations for the Earth Sciences**
ISBN: 0-534-39147-8

**Exploring Water Resources:**
**GIS Investigations for the Earth Sciences**
ISBN: 0-534-39156-7

**Exploring The Ocean Environment:**
**GIS Investigations for the Earth Sciences**
ISBN 0-534-42350-7

### For the Student

**Physical GeographyNow**   What do you need to learn NOW? Physical GeographyNow™ is a powerful, personalized online learning companion designed to help you gauge your own learning needs and identify the concepts on which you most need to focus your study time. How does it work? After you read the chapter, log on using the 1Pass™ access code packaged with most versions of this text and take the diagnostic Pre-Test ("What Do I Know?") for a quick assessment of how well you already understand the material. Your Personalized Learning Plan ("What Do I Need to Learn?") outlines the interactive animations, tutorials, and exercises you need to review. The Post-Test ("What Have I Learned?") helps you measure how well you have mastered the core concepts from the chapter. Study smarter—and make every minute count!

## Acknowledgments

This edition of *Essentials of Physical Geography* would not have been possible without the encouragement and assistance of editors, friends, and colleagues from throughout the country. Great appreciation is extended to Sarah Gabler and Martha, Emily, and Hannah Petersen for their patience, support, and understanding.

Special thanks go to the splendid freelancers and staff members of Brooks/Cole Thomson Learning. These include Keith Dodson, Earth Sciences Editor; Carol Benedict, Assistant Editor; Ericka Yeoman-Saler, Technology Project Manager; Mark Santee, Marketing Manager; Hal Humphrey, Production Project Manager; Patrick Devine, Designer; Terri Wright, Photo Researcher; illustrators Precision Graphics, Accurate Art, and Rolin Graphics; Sam Blake, Pre-Press Company, Inc. Production Coordinator; and Anna Jarzab, Editorial Assistant.

Colleagues who reviewed the plans and manuscript for this and previous editions include: Brock Brown, Texas State University; Leland R. Dexter, Northern Arizona University; Roberto Garza, San Antonio College; Perry Hardin, Brigham Young University; David Helgren, San Jose State University; Jeffrey Lee, Texas Tech University; John Lyman, Bakersfield College; Fritz C. Kessler, Frostburg State University; Michael E. Lewis, University of North Carolina, Greensboro; Charles Martin, Kansas State University; James R. Powers, Pasadena City College; Joyce Quinn, California State University, Fresno; Colin Thorn, University of Illinois at Urbana-Champaign; Dorothy Sack, Ohio University; George A. Schnell, State University of New York, New Paltz; David L. Weide, University of Nevada, Las Vegas; Thomas Wikle, Oklahoma State University, Stillwater; Amy Wyman, University of Nevada, Las Vegas; Craig ZumBrunnen, University of Washington, Redmond; Joanna Curran, Richard Earl, and Mark Fonstad, all of Texas State University; Percy "Doc" Dougherty, Kutztown State University; Ray Sumner, Long Beach City College; Ben Dattilo, University of Nevada, Las Vegas; Greg Gaston, University of North Alabama; Tom Feldman, Joliet Junior College; Andrew Oliphant, San Francisco State University; and Michael Talbot, Pima Community College.

Photos courtesy of: Bill Case and Chris Wilkerson, Utah Geological Survey; Center for Cave and Karst Studies, Western Kentucky University; Hari Eswaran, USDA Natural Resources Conservation Service; Richard Hackney, Western Kentucky University; David Hansen, University of Minnesota; Susan Jones, Nashville, Tennessee; Bob Jorstad, Eastern Illinois University; Carter Keairns, Texas State University; Parris Lyew-Ayee, Oxford University, UK; L. Elliot

Jones, U.S. Geological Survey; Dorothy Sack, Ohio University; Anthony G. Taranto Jr., Palisades Interstate Park–New Jersey Section; Justin Wilkinson, Earth Sciences, NASA Johnson Space Center.

The detailed comments and suggestions of all of the above individuals have been instrumental in bringing about the many changes and improvements incorporated in this latest revision of the text. Countless others, both known and unknown, deserve heartfelt thanks for their interest and support over the years.

Despite the painstaking efforts of all reviewers, there will always be questions of content, approach, and opinion associated with the text. The authors wish to make it clear that they accept full responsibility for all that is included in the eighth edition of *Essentials*.

*Robert E. Gabler*
*James F. Petersen*
*L. Michael Trapasso*

# Foreword to the Student

## Why Study Geography?

In this global age, the study of geography is absolutely essential to an educated citizenry of a nation whose influence extends throughout the world. Geography deals with location, and a good sense of where things are, especially in relation to other things in the world, is an invaluable asset whether you are traveling, conducting international business, or sitting at home reading the newspaper.

Geography examines the characteristics of all the various places on Earth and their relationships. Most important in this regard, geography provides special insights into the relationships between humans and their environments. If all the world's people could have one goal in common, it should be to better understand the physical environment and protect it for the generations to come.

Geography provides essential information about the distribution of things and the interconnections of places. The distribution pattern of Earth's volcanoes, for example, provides an excellent indication of where Earth's great crustal plates come in contact with one another; and the violent thunderstorms that plague Illinois on a given day may be directly associated with the low pressure system spawned in Texas two days before. Geography, through a study of regions, provides a focus and a level of generalization that allows people to examine and understand the immensely varied characteristics of Earth.

As you will note when reading Chapter 1, there are many approaches to the study of geography. Some courses are regional in nature; they may include an examination of one or all of the world's political, cultural, economic, or physical regions. Some courses are topical or systematic in nature, dealing with human geography, physical geography, or one of the major subfields of the two.

The great advantage to the study of a general course in physical geography is the permanence of the knowledge learned. Although change is constant and is often sudden and dramatic in the human aspects of geography, alterations of the physical environment on a global scale are exceedingly slow when not influenced by human intervention. Theories and explanations may differ, but the broad patterns of atmospheric and oceanic circulation and of world climates, landforms, soils, natural vegetation, and physical landscapes will be the same tomorrow as they are today.

## Keys to Successful Study

Good study habits are essential if you are to master science courses such as physical geography, where the topics, explanations, and terminology are often complex and unfamiliar. To help you succeed in the course in which you are currently enrolled, we offer the following suggestions.

### Reading Assignments

- Read the assignments before the material contained therein is covered in class by the instructor.
- Compare what you have read with the instructor's presentation in class. Pay particular attention if the instructor introduces new examples or course content not included in the reading assignment.
- Do not be afraid to ask questions in class and seek a full understanding of material that may have been a problem during your first reading of the assignment.
- Reread the assignment as soon after class as possible, concentrating on those areas that were emphasized in class. Highlight only those items or phrases that you now consider to be important, and skim those sections already mastered.
- Add to your class notes important terms, your own comments, and summarized information from each reading assignment.

### Understanding Vocabulary

Mastery of the basic vocabulary often becomes a critical issue in the success or failure of the student in a beginning science course.

- Focus on the terms that appear in boldface type in your reading assignments. Do not overlook any additional terms that the instructor may introduce in class.
- Develop your own definition of each term or phrase and associate it with other terms in physical geography.

- Identify any physical processes associated with the term. Knowing the process helps to define the term.
- Whenever possible, associate terms with location.
- Consider the significance to humans of terms you are defining. Recognizing the significance of terms and phrases can make them relevant and easier to recall.

## Learning Earth Locations

A good knowledge of place names and of the relative locations of physical and cultural phenomena on Earth is fundamental to the study of geography.

- Take personal responsibility for learning locations on Earth. Your instructor may identify important physical features and place names, but you must learn their locations for yourself.
- Thoroughly understand latitude, longitude, and the Earth grid. They are fundamental to location on maps as well as on a globe. Practice locating features by their latitude and longitude until you are entirely comfortable using the system.
- Develop a general knowledge of the world political map. The most common way of expressing the location of physical features is by identifying the political unit (state, country, or region) in which it can be found.
- Make liberal use of outline maps. They are the key to learning the names of states and countries and they can be used to learn the locations of specific physical features. Personally placing features correctly on an outline map is often the best way to learn location.
- Cultivate the atlas habit. The atlas does for the individual who encounters place names or the features they represent what the dictionary does for the individual who encounters a new vocabulary word.

## Utilizing Textbook Illustrations

The secret to making good use of maps, diagrams, and photographs lies in understanding why the illustration has been included in the text or incorporated as part of your instructor's presentation.

- Concentrate on the instructor's discussion. Taking notes on slides, overhead transparencies, and illustrations will allow you to follow the same line of thought at a later date.
- Study all textbook illustrations on your own. Be sure to note which were the focus of considerable classroom attention. Do not quit your examination of an illustration until it makes sense to you, until you can read the map or graph, or until you can recognize what a diagram or photograph has been selected to explain.
- Hand-copy important diagrams and graphs. Few of us are graphic artists, but you might be surprised at how much better you understand a graph or line drawing after you reproduce it yourself.
- Read the captions of photos and illustrations thoroughly and thoughtfully. If the information is included, be certain to note where a photograph was taken and in what way it is representative. What does it tell you about the region or site being illustrated?
- Attempt to place the principle being illustrated in new situations. Seek other opportunities to test your skills at interpreting similar maps, graphs, and photographs and think of other examples that support the text being illustrated.
- Remember that all illustrations are reference tools, particularly tables, graphs, and diagrams. Refer to them as often as you need to.

## Taking Class Notes

The password to a good set of class notes is selectivity. You simply cannot and, indeed, you should not try to write down every word uttered by your classroom instructor.

- Learn to paraphrase. With the exception of specific quotations or definitions, put the instructor's ideas, explanations, and comments into your own words. You will understand them better when you read them over at a later time.
- Be succinct. Never use a sentence when a phrase will do, and never use a phrase when a word will do. Start your recall process with your note-taking by forcing yourself to rebuild an image, an explanation, or a concept from a few words.
- Outline where possible. Preparing an outline helps you to discern the logical organization of information. As you take notes, organize them under main headings and subheadings.
- Take the instructor at his or her word. If the instructor takes the time to make a list, then you should do so too. If he or she writes something on the board, it should be in your notes. If the instructor's voice indicates special concern, take special notes.
- Come to class and take your own notes. Notes trigger the memory, but only if they are your notes.

## Doing Well on Tests

Follow these important study techniques to make the most of your time and effort preparing for tests.

- Practice distillation. Do not try to reread but skim the assignments carefully, taking notes in your own words that record as economically as possible the important definitions, descriptions, and explanations. Do the same with any supplementary readings, handouts, and laboratory exercises. It takes practice to use this technique, but

it is a lot easier to remember a few key phrases that lead to ever increasing amounts of organized information than it is to memorize all of your notes. And the act of distillation in itself is a splendid memory device.
- Combine and reorganize. Merge all your notes into a coherent study outline.
- Become familiar with the type of questions that will be asked. Knowing whether the questions will be objective, short-answer, essay, or related to diagrams and other illustrations can help in your preparation. Some instructors place old tests on file where you can examine them or will forewarn you of their evaluation styles if you inquire. If not, then turn to former students; there are usually some around the department or residence halls who have already experienced the instructor's tests.
- Anticipate the actual question that will likely be on the test. The really successful students almost seem to be able to predict the test items before they appear. Take your educated guesses and turn them into real questions.
- Try cooperative study. This can best be described as role playing and consists very simply of serving temporarily as the instructor. So go ahead and teach. If you can demonstrate a technique, illustrate an idea, or explain a process or theory to another student so that he or she can understand it, there is little doubt that you can answer test questions over the same material.
- Avoid the "all-nighter." Use the early evening hours the night before the test for a final unhurried review of your study outline. Then get a good night's sleep.

## The Importance of Maps

Like graphs, tables, and diagrams, maps are an excellent reference tool. Familiarize yourself with the maps in your textbook in order to better judge when it is appropriate to seek information from these important sources.

Maps are especially useful for comparison purposes and to illustrate relationships or possible associations of things. But the map reader must beware. Only a small portion of the apparent associations of phenomena in space (areal associations) are actually cause-and-effect relationships. In some instances the similarities in distribution are a result of a third factor that has not been mapped. For instance, a map of worldwide volcano distribution is almost exactly congruent with one of incidence of earthquakes, yet volcanoes are not the cause of earthquakes, nor is the obverse true. A third factor, the location of tectonic plate boundaries, explains the first two phenomena.

Finally, remember that the map is the most important statement of the professional geographer. It is useful to all natural and social scientists, engineers, politicians, military planners, road builders, farmers, and countless others, but it is the essential expression of the geographer's primary concern with location, distribution, and spatial interaction.

## About Your Textbook

This textbook has been written for you, the student. It has been written so that the text can be read and understood easily. Explanations are as clear, concise, and uncomplicated as possible. Illustrations have been designed to complement the text and to help you visualize the processes, places, and phenomena being discussed. In addition, the authors do not believe it is sufficient to offer you a textbook that simply provides information to pass a course. We urge you to think critically about what you read in the textbook and hear in class.

As you learn about the physical aspects of Earth environments, ask yourself what they mean to you and to your fellow human beings throughout the world. Make an honest attempt to consider how what you are learning in your course relates to the problems and issues of today and tomorrow. Practice using your geographic skills and knowledge in new situations so that you will continue to use them in the years ahead. Your textbook includes several special features that will encourage you to go beyond memorization and reason geographically.

**Consider and Respond**  At the end of each chapter, Consider and Respond questions require you to go well beyond routine chapter review. The questions are designed specifically so that you may apply your knowledge of physical geography and on occasion personally respond to critical issues in society today. Check with your instructor for answers to the problems.

**Caption Questions**  With almost every illustration and photo in your textbook a caption links the image with the chapter text it supports. Read each caption carefully because it explains the illustration and may also contain new information. Wherever appropriate, questions at the ends of captions have been designed to help you seize the opportunity to consider your own personal reaction to the subject under consideration.

**Map Interpretation Series**  It is a major goal of your textbook to help you become an adept map reader, and the Map Interpretation Series in your text has been designed to help you reach that goal.

**Environmental Systems Diagrams**  Viewing Earth as a system comprising many subsystems is a fundamental concept in physical geography for researchers and instructors alike. The concept is introduced in Chapter 1 and reappears frequently throughout your textbook. The interrelationships and dependencies among the variables or components of Earth systems are so important that a series of special diagrams (see, for example, Figure 7.8) have been included with the text to help you visualize how the systems work. Each diagram depicts the system and its variables and also demonstrates their interdependence and the movement or exchanges that occur within each system. The diagrams are designed to help you understand how human activity can affect the delicate balance that exists within many Earth systems.

**Physical GeographyNow** Most new editions of this text come with access to Physical GeographyNow, an online learning tool that helps you create a personalized learning plan for exam review. Each chapter on the Physical GeographyNow website allows you to review for exams by using the Pre-Test web quizzes. Your responses on these quizzes are then used to create a personalized learning plan. As part of this learning plan, you can explore Active Figures from the text (each of these animated figures also contains additional review questions). Finally, a Chapter Post-Test allows you to further check your understanding of core concepts after you have worked through your personalized learning plan.

As authors of your textbook, we wish you well in your studies. It is our fond hope that you will become better informed about Earth and its varied environments and that you will enjoy the study of physical geography.

# Brief Contents

1 Physical Geography: Earth Environments and Systems  2
2 Representations of Earth  30
3 Earth in Space and Solar Energy  66
4 The Atmosphere, Temperature, and the Heat Budget  86
5 Atmospheric Pressure, Winds, and Circulation Patterns  114
6 Moisture, Condensation, and Precipitation  144
7 Air Masses and Weather Systems  174
8 Global Climates and Climate Change  204
9 Low-Latitude and Arid Climate Regions  234
10 Middle-Latitude, Polar, and Highland Climate Regions  256
11 Biogeography  292
12 Soils and Soil Development  330
13 Earth Materials, the Lithosphere, and Plate Tectonics  360
14 Volcanoes, Earthquakes, and Tectonic Landforms  392
15 Gradation, Weathering, and Mass Wasting  422
16 Underground Water and Karst Landforms  446
17 Fluvial Processes and Landforms  468
18 Arid Landforms and Eolian Processes  502
19 Glacial Systems and Landforms  534
20 The Global Ocean  566
21 Coastal Landforms  592

# Contents

## 1 Physical Geography: Earth Environments and Systems  2

The Study of Geography  4
  Physical Geography  5
  Technology, Tools, and Methods  6
Major Perspectives in Physical Geography  7
**Geography's Spatial Science Perspective: The Regional Concept: Natural and Environmental Regions  8**
  The Spatial Science Perspective  8
**Geography's Physical Science Perspective: The Scientific Method  12**
  The Physical Science Perspective  13
  The Environmental Science Perspective  14
**Geography's Environmental Science Perspective: Human–Environment Interactions  16**
Models and Systems  21
  Systems Theory  22
  How Systems Work  22
  Equilibrium in Earth Systems  24
Physical Geography and You  27
Define & Recall  28
Discuss & Review  28
Consider & Respond  29

## 2 Representations of Earth  30

Location on Earth  32
  Maps and Mapmaking  32
  Size and Shape of Earth  32
  Globes and Great Circles  33
  Latitude and Longitude  34
The Geographic Grid  36
  Parallels and Meridians  36
  Longitude and Time  36
  The International Date Line  37
  The U.S. Public Lands Survey System  38

The Global Positioning System  39
Maps and Map Projections  39
  Advantages of Maps  40
  Limitations of Maps  41
  Properties of Map Projections  41
  Examples of Map Projections  43
  Map Basics  45
Career Vision: Robert Iantria  48
Displaying Spatial Data and Information on Maps  49
  Discrete and Continuous Data  49
  Topographic Maps  50
Modern Mapping Technology  51
**Geography's Spatial Science Perspective: Using Vertical Exaggeration to Portray Topography  52**
  Digital Mapmaking  52
  Geographic Information Systems  53
Remote Sensing of the Environment  56
  Aerial Photography  57
  Specialized Remote Sensing Techniques  57
  Multispectral Remote Sensing Applications  59
**Geography's Physical Science Perspective: Polar versus Geostationary Satellite Orbits  60**
Define & Recall  62
Discuss & Review  62
Consider & Respond  63
**MAP INTERPRETATION: TOPOGRAPHIC MAPS  64**

## 3 Earth in Space and Solar Energy  66

The Solar System and Beyond  68
  The Solar System  68
**Geography's Environmental Science Perspective: Solar Energy  69**
  The Sun and Its Energy  69
  The Planets  71
  Life beyond Earth?  72

The Earth–Sun System  74
  Solar Energy and Atmospheric Dynamics  75
  Movements of Earth  76
Sun Angle, Duration, and Insolation  78
  The Seasons  80
**Geography's Physical Science Perspective: Using the Sun's Rays to Measure the Spherical Earth—2200 Years Ago  81**
  Lines on Earth Delimiting Solar Energy  82
  The Analemma  83
  Variations of Insolation with Latitude  83
Define & Recall  85
Discuss & Review  85
Consider & Respond  85

## 4 The Atmosphere, Temperature, and the Heat Budget  86

Characteristics of the Atmosphere  88
  Composition of the Atmosphere  88
  Atmospheric Environmental Issues  89
**Geography's Spatial Science Perspective: Why Are Ozone Holes Located at the Poles?  92**
  Vertical Layers of the Atmosphere  92
  Effects of the Atmosphere on Solar Radiation  95
  Water as Heat Energy  95
**Geography's Physical Science Perspective: Colors of the Atmosphere  97**
Heating the Atmosphere  97
  Processes of Heat Energy Transfer  98
  The Heat Energy Budget  98
Air Temperature  99
  Temperature and Heat  99
  Temperature Scales  99
  Short-Term Variations in Temperature  100
  Vertical Distribution of Temperature  102
  Controls of Earth's Surface Temperatures  104
  Temperature Distribution at Earth's Surface  108
  Annual March of Temperature  109
Weather and Climate  110

Complexity of Earth's Energy Systems  111
Define & Recall  112
Discuss & Review  112
Consider & Respond  113

## 5 Atmospheric Pressure, Winds, and Circulation Patterns  114

Variations in Atmospheric Pressure  117
  Vertical Variations in Pressure  117
  Horizontal Variations in Pressure  117
Basic Pressure Systems  117
  Convergent and Divergent Circulation  117
  Mapping Pressure Distribution  118
Wind  118
  Pressure Gradients and Winds  118
**Geography's Environmental Science Perspective: Harnessing the Wind  119**
  The Coriolis Effect and Wind  120
  Friction and Wind  121
  Wind Terminology  121
  Cyclones, Anticyclones, and Winds  122
Subglobal Surface Wind Systems  122
  Local Winds  122
  Monsoon Winds  124
**Geography's Spatial Science Perspective: The Santa Ana Winds and Fire  125**
Global Pressure Belts  126
  Idealized Global Pressure Belts  126
  The Global Pattern of Atmospheric Pressure  127
  Seasonal Variations in the Pattern  127
Global Surface Wind Systems  129
  Idealized Model of Atmospheric Circulation  130
  Conditions within Latitudinal Zones  131
**Career Vision: Sandra Diaz  132**
  The Effects of Seasonal Migration  133
  Longitudinal Differences in Winds  134
Upper Air Winds  134
  Ocean Currents  136
El Niño  139
  El Niño and the Southern Oscillation  139
  El Niño and Global Weather  141

Define & Recall 142
Discuss & Review 142
Consider & Respond 143

## 6 Moisture, Condensation, and Precipitation 144

The Hydrologic Cycle 146
Water in the Atmosphere 147
  The Water Budget and Its Relation to the Heat Budget 147
  Saturation and Dew Point 148
  Humidity 149
Sources of Atmospheric Moisture 151
  Rate of Evaporation 151
  Potential Evapotranspiration 152
Condensation 152
  Condensation Nuclei 153
  Fog 153
  Other Minor Forms of Condensation 155
  Clouds 156
**Geography's Environmental Science Perspective: Cumulus Congestus: An Aspiring Thunderstorm 157**
Precipitation Processes 161
  Major Forms of Precipitation 162
  Factors Necessary for Precipitation 163
**Geography's Physical Science Perspective: The Lifting Condensation Level (LCL) 164**
  Distribution of Precipitation 166
Variability of Precipitation 170
Define & Recall 172
Discuss & Review 172
Consider & Respond 173

## 7 Air Masses and Weather Systems 174

Air Masses 176
  Modification and Stability of Air Masses 176
  North American Air Masses 176
Fronts 179
  Cold Front 179
  Warm Front 179
  Stationary and Occluded Fronts 180
Atmospheric Disturbances 180
  Cyclones and Anticyclones 181
**Career Vision: Thomas G. Novak 185**
  Thunderstorms 188
  Snowstorms and Blizzards 189
  Tornadoes 190
  Weak Tropical Disturbances 193
**Geography's Physical Science Perspective: Storm Chasers and Storm Spotters 194**
  Hurricanes 196
**Geography's Spatial Science Perspective: Hurricane Paths and Landfall Probability Maps 198**
Define & Recall 200
Discuss & Review 201
Consider & Respond 201
**MAP INTERPRETATION: WEATHER MAPS 202**

## 8 Global Climates and Climate Change 204

Classifying Climates 206
  The Thornthwaite System 206
  The Köppen System 208
  Climate Regions 215
  Climographs 216
  Scale and Climate 216
Climates of the Past 217
  The Ice Ages 217
  Modern Research 217
Rates of Climate Change 219
**Geography's Physical Science Perspective: Determining Past Climates 220**
Causes of Climate Change 220
  Orbital Variations 221
  Changes in Earth's Atmosphere 222
  Changes in the Ocean 223
  Changes in Landmasses 223
**Geography's Spatial Science Perspective: Climate Change and Its Impact on Coastlines 225**
Future Climates 226

Define & Recall 228
Discuss & Review 228
Consider & Respond 229
GRAPH INTERPRETATION 230

## 9 Low-Latitude and Arid Climate Regions 234

Humid Tropical Climate Regions 236
  Tropical Rainforest Climate 238
  Geography's Environmental Science Perspective: The Amazon Rainforest 241
  Tropical Monsoon Climate 242
  Tropical Savanna Climate 244
Arid Climate Regions 246
  Desert Climates 248
  Geography's Spatial Science Perspective: Desertification 250
  Steppe Climates 252
Define & Recall 254
Discuss & Review 254
Consider & Respond 255

## 10 Middle-Latitude, Polar, and Highland Climate Regions 256

Middle-Latitude Climates 258
Humid Mesothermal Climate Regions 258
  Mediterranean Climate 258
  Humid Subtropical Climate 262
  Marine West Coast Climate 264
Humid Microthermal Climate Regions 268
  Humid Continental, Hot Summer Climate 270
  Humid Continental, Mild Summer Climate 273
  Subarctic Climate 275
Geography's Physical Science Perspective: Effective Temperatures 276
Polar Climate Regions 279
  Tundra Climate 280
  Ice-Sheet Climate 281
  Human Activity in Polar Regions 283
Highland Climate Regions 284
Geography's Environmental Science Perspective: The Effects of Altitude on the Human Body 286
  Peculiarities of Mountain Climates 287
  Highland Climates and Human Activity 288
Define & Recall 289
Discuss & Review 289
Consider & Respond 290

## 11 Biogeography 292

Organization within Ecosystems 294
  Major Components 294
  Trophic Structure 295
  Nutrient Cycles 295
  Energy Flow 296
  Productivity 298
  Ecological Niche 300
Succession and Climax Communities 300
  Succession 301
Geography's Environmental Science Perspective: The Theory of Island Biogeography 302
  The Climax Community 304
Environmental Controls 306
  Climatic Factors 306
  Soil and Topography 308
  Natural Catastrophes 309
  Biotic Factors 309
  Human Impact on Ecosystems 310
Geography's Spatial Science Perspective: Introduction and Diffusion of an Exotic Species: Fire Ants 311
Classification of Terrestrial Ecosystems 312
Forest Biomes 312
  Tropical Forests 316
Career Vision: Roxie Anderson 317
  Middle-Latitude Forests 318
Grassland Biomes 321
  Tropical Savanna Grasslands 321
  Middle-Latitude Grasslands 322

Desert 323
Arctic and Alpine Tundra 324
Marine Ecosystems 325
Define & Recall 328
Discuss & Review 328
Consider & Respond 329

## 12 Soils and Soil Development 330

Major Soil Components 332
  Inorganic Materials 333
  Soil Water 333
  Soil Air 334
  Organic Matter 335
Characteristics of Soil 335
  Color 335
**Geography's Physical Science Perspective: Basic Soil Analysis 336**
  Texture 336
  Structure 337
  Acidity and Alkalinity 338
Development of Soil Horizons 339
  Soil Horizons 340
Factors Affecting Soil Formation 340
  Parent Material 341
  Organic Activity 341
  Climate 342
**Career Vision: Amy Jo Steffen 343**
  Land Surface Configuration 344
  Time 344
Soil-Forming Regimes 347
  Laterization 347
  Podzolization 347
  Calcification 347
**Geography's Environmental Science Perspective: Slash-and-Burn Agriculture Adds Nutrients to Tropical Soils 348**
  Regimes of Local Importance 349
Soil Classification 350
  The NRCS Soil Classification System 350
  NRCS Soil Orders 351

Soils as a Critical Natural Resource 356
Define & Recall 358
Discuss & Review 358
Consider & Respond 359

## 13 Earth Materials, the Lithosphere, and Plate Tectonics 360

Minerals and Rocks 362
  Minerals 362
  Rocks 363
Earth's Planetary Structure 371
  Earth's Core 371
  Earth's Mantle 372
  Earth's Crust 373
Continents in Motion: The Search for a Unifying Theory 374
  Continental Drift 374
  Supporting Evidence for Continental Drift 375
**Geography's Spatial Science Perspective: Paleomagnetism: Evidence of Earth's Ancient Geography 376**
Plate Tectonics 377
  Seafloor Spreading and Convection Currents 378
  Tectonic Plate Movement 379
  Hot Spots in the Mantle 384
Growth of Continents 384
**Geography's Physical Science Perspective: Isostasy: Balancing the Earth's Lithosphere 385**
Paleogeography 386
Define & Recall 390
Discuss & Review 390
Consider & Respond 391

## 14 Volcanoes, Earthquakes, and Tectonic Landforms 392

Landforms and Geomorphology 394
Igneous Processes and Landforms 396
  Volcanoes and Eruptions 396

CONTENTS  xxi

Lava Flows  397
Types of Volcanoes  397
Intrusive Activity and Intrusions  401
Tectonic Processes, Rock Structure, and Landforms  404
Warping of Earth's Crust  405
**Geography's Spatial Science Perspective: Spatial Relationships between Plate Boundaries, Volcanoes, and Earthquakes  406**
Folding  406
Faulting  408
Earthquakes  412
Measuring the Size of Earthquakes  413
Earthquake Hazards  413
**Geography's Environmental Science Perspective: Mapping the Distribution of Earthquake Intensity  414**
Relationships between Rock Structure and Topography  417
Define & Recall  418
Discuss & Review  419
Consider & Respond  419
**MAP INTERPRETATION: VOLCANIC LANDFORMS  420**

## 15 Gradation, Weathering, and Mass Wasting  422

Gradation and Tectonism  424
Weathering  425
Physical Weathering  425
Chemical Weathering  431
**Geography's Physical Science Perspective: Expanding and Contracting Soils  432**
Differential Weathering and Erosion  434
Topography Related to Differential Weathering and Erosion  435
The Importance of Weathering  436
Mass Wasting  436
Classification of Mass Wasting Processes  437
Rapid Mass Movement  438
Slow Mass Movement  441
**Geography's Environmental Science Perspective: Hazard Potential in the Cascades  442**

Weathering, Mass Wasting, and the Landscape  444
Define & Recall  445
Discuss & Review  445
Consider & Respond  445

## 16 Underground Water and Karst Landforms  446

Occurrence and Supply of Groundwater  448
Groundwater Zones and the Water Table  448
Factors Affecting the Distribution of Groundwater  450
**Career Vision: Carmen Yelle  451**
Availability of Groundwater  452
Wells  453
**Geography's Physical Science Perspective: Monitoring Groundwater Flow  454**
Artesian Wells  454
Landform Development by Solution  455
Karst Landscapes and Landforms  455
**Geography's Environmental Science Perspective: Sinkhole Formation  458**
Limestone Caverns and Cave Features  460
Geothermal Water  463
Define & Recall  464
Discuss & Review  464
Consider & Respond  465
**MAP INTERPRETATION: KARST TOPOGRAPHY  466**

## 17 Fluvial Processes and Landforms  468

Surface Runoff  470
The Stream System  471
Drainage Basins  473
**Geography's Spatial Science Perspective: Watersheds as Critical Natural Regions  474**
Drainage Density and Pattern  475
Stream Discharge  477
Stream Energy  479
**Career Vision: Andy Wohlsperger  481**
Fluvial Gradational Processes  482

Erosion by Streams **482**
    Stream Transportation **483**
    Stream Deposition **484**
**Channel Patterns 485**
**Land Sculpture by Streams 486**
    Features of the Upper Course **487**
    Features of the Middle Course **487**
    Features of the Lower Course **488**
**Deltas 490**
**Base Level Changes and Tectonism 492**
**Stream Hazards 492**
*Geography's Environmental Science Perspective: Restoring the Everglades* **494**
**The Importance of Surface Waters 495**
    Streams **495**
    Lakes **496**
**Quantitative Fluvial Geomorphology 497**
**Define & Recall 498**
**Discuss & Review 498**
**Consider & Respond 499**
**MAP INTERPRETATION: FLUVIAL LANDFORMS 500**

## 18 Arid Landforms and Eolian Processes 502

**Surface Runoff in the Desert 504**
**Water as a Gradational Agent in Arid Lands 506**
    Arid Region Landforms of Fluvial Erosion **507**
    Arid Region Landforms of Fluvial Deposition **510**
**Wind as a Gradational Agent 514**
    Wind Erosion and Transportation **515**
    Wind Deposition **517**
*Geography's Enviromental Science Perspective: Invasions by Sand Dunes* **519**
**Landscape Development in Deserts 524**
**Define & Recall 528**
**Discuss & Review 529**
**Consider & Respond 529**
**MAP INTERPRETATION: DESERT LANDFORMS 530**
**MAP INTERPRETATION: EOLIAN LANDFORMS 532**

## 19 Glacial Systems and Landforms 534

**Glacier Formation and the Hydrologic Cycle 536**
*Geography's Physical Science Perspective: Glacial Ice Is Blue!* **537**
**Types of Glaciers 538**
**Features of an Alpine Glacier 539**
    Equilibrium and the Glacial Budget **540**
    How Does a Glacier Flow? **541**
    Glaciers as Agents of Gradation **543**
    Erosional Landforms Produced by Alpine Glaciation **544**
    Depositional Landforms Produced by Alpine Glaciation **546**
**Continental Ice Sheets 548**
    Existing Ice Sheets **548**
    Pleistocene Ice Sheets **550**
**Movement of Continental Ice Sheets 551**
*Geography's Spatial Science Perspective: The Driftless Area—A Natural Region* **552**
    Ice Sheets and Erosional Landforms **553**
    Ice Sheets and Depositional Landforms **554**
    Glacial Lakes and Periglacial Landscapes **558**
**Define & Recall 560**
**Discuss & Review 561**
**Consider & Respond 561**
**MAP INTERPRETATION: ALPINE GLACIATION 562**
**MAP INTERPRETATION: CONTINENTAL GLACIATION 564**

## 20 The Global Ocean 566

**Introduction to the Oceans 568**
    Characteristics of Ocean Waters **568**
    Surface Variations **569**
    Variations with Depth **570**
**The Ocean Floor 570**
    Features of the Continental Margins **571**
    Features of the Deep-Ocean Floor **575**
    Islands and Coral Reefs **578**
**Tides and Waves 579**
    Tides **579**
    Waves **584**

Geography's Environmental Science Perspective: Rogue Waves  585

Geography's Physical Science Perspective: Tsunami Forecasts and Warnings  586

  Catastrophic Waves  587

Define & Recall  590

Discuss & Review  590

Consider & Respond  591

## 21 Coastal Landforms  592

Coastal Landforms and Processes  594

  The Breaking of Waves  594

  Wave Refraction  595

Career Vision: Aaron Young  596

Erosional Coastal Landforms  596

Beaches and Coastal Deposition  599

Geography's Environmental Science Perspective: Beach Building  600

  Coastal Currents  601

  Depositional Coastal Landforms  603

Geography's Physical Science Perspective: Beach Erosion and Protection  605

Types of Coasts  607

Define & Recall  610

Discuss & Review  610

Consider & Respond  611

**MAP INTERPRETATION: ACTIVE-MARGIN COASTLINES  612**

**MAP INTERPRETATION: PASSIVE-MARGIN COASTLINE  614**

Appendix A SI Units  617
Appendix B Topographic Maps  619
Appendix C Understanding and Recognizing Some Common Rocks  621

Glossary  625
Index  643

# List of Major Maps

## The World

| Figure | Description | |
|---|---|---|
| 2.8 | Time Zones | 37 |
| 4.27 | Average sea-level temperatures, January | 108 |
| 4.28 | Average sea-level temperatures, July | 109 |
| 5.14a | Average sea-level pressure, January | 128 |
| 5.14b | Average sea-level pressure, July | 128 |
| 5.24 | Major ocean currents | 138 |
| 5.25 | Satellite images, development of an El Niño episode | 140 |
| 6.22 | Map of average annual precipitation | 168–169 |
| 6.24 | Precipitation variability | 171 |
| 7.23 | Hurricane alleys | 199 |
| 8.6 | Climates, modified Köppen classification system | 212–213 |
| 9.1 | Humid tropical climates | 236 |
| 9.14 | Arid lands | 247 |
| 10.1 | Humid mesothermal climates | 258 |
| 10.14 | Humid microthermal climates | 269 |
| 11.6 | Satellite image, vegetation patterns | 299 |
| 11.21 | Natural vegetation | 314–315 |
| 12.24 | Distribution of NRCS soil orders | 352–353 |
| 13.21 | Wegener's continental drift hypothesis | 374 |
| 13.25 | Major tectonic plates | 378 |
| 13.37 | Last 250 million years of Earth history | 389 |
| p. 407a | Earthquake epicenters | 407 |
| p. 407b | Major volcanic regions | 407 |
| 18.38 | Major loess regions | 526 |
| 19.27 | Extent of Pleistocene glaciation | 553 |
| | Physical Map of the World | Front endpapers |
| | Population Density | Back endpapers |

## The Ocean

| Figure | Description | |
|---|---|---|
| 11.34 | Distribution of chlorophyll-producing marine plankton | 327 |
| 13.26 | Oceanic ridges and the age of the sea floor | 379 |
| 20.1 | Extent of Pacific Ocean | 568 |
| 20.8 | Relief, Floor of the ocean | 572–573 |

## The Contiguous United States

| Figure | Description | |
|---|---|---|
| 1.6 | Satellite image, population distribution | 10 |
| 2.10 | Baselines and principal meridians | 038 |
| 6.6 | Potential evapotranspiration | 152 |
| 7.6 | Common storm tracks | 182 |
| 7.16 | Average number of tornadoes | 192–193 |
| 8.2 | Thornthwaite climate regions | 208 |
| 10.15 | Average annual snow cover | 270 |
| 12.11 | Alkaline and acidic soils | 339 |
| 14.35 | Earthquake hazard distribution | 417 |
| 16.10 | Limestone (karst) regions | 456 |
| 18.14 | Pleistocene lakes, Western region | 512 |
| 19.32 | Glacial deposits, Great Lakes region | 556 |

## North America

| Figure | Description | |
|---|---|---|
| 2.24 | Magnetic declination | 49 |
| 7.1 | Air mass source regions | 178 |
| 8.17 | Pleistocene ice sheets | 218 |
| 13.36 | Rock ages | 387 |

## Environmental Systems Series

| Figure | Title | |
|---|---|---|
| 1.8 | Earth's Major Subsystems | 14 |
| 4.10 | Earth's Radiation Budget | 96 |
| 6.3 | The Hydrologic System—The Water Cycle | 148 |
| 7.8 | Middle-Latitude Cyclonic Systems | 184 |
| 11.3 | Middle-Latitude Ecosystem | 296 |
| 13.29 | Plate Tectonic Movement | 381 |
| 16.2 | The Groundwater System | 449 |
| 17.3 | The Hydrologic System—Streams | 472 |
| 19.6 | The Hydrologic System—Glaciers | 540 |

# Author Biographies

**Robert E. Gabler** During his nearly five decades of professional experience, Professor Gabler has taught geography at Hunter College, City of New York, Columbia University, and Western Illinois University, in addition to five years in public elementary and secondary schools. At times in his career at Western he served as Chairperson of the Geography and Geology Department, Chairperson of the Geography Department, and University Director of International Programs. He received three University Presidential Citations for Teaching Excellence and University Service, served two terms as Chairperson of the Faculty Senate, edited the *Bulletin of the Illinois Geographical Society*, and authored numerous articles in state and national periodicals. He is a Past President of the Illinois Geographical Society, former Director of Coordinators and Past President of the National Council for Geographic Education, and the recipient of the NCGE George J. Miller Distinguished Service Award.

**James F. Petersen** James F. Petersen is Professor of Geography at Texas State University, in San Marcos, Texas. He is a broadly trained physical geographer with strong interests in geomorphology and Earth Science education. He enjoys writing about topics relating to physical geography for the public, particularly environmental interpretation, and has written a landform guidebook for Enchanted Rock State Natural Area in central Texas and a number of field guides. He is a strong supporter of geographic education, having served as President of the NCGE in 2000 after more than 15 years of service to that organization. He has also written or served as a senior consultant for nationally published educational materials at levels from middle school through university, and has done many workshops for geography teachers. Recently, he contributed the opening chapter in an environmental history of San Antonio that explains the physical geographic setting of central Texas.

**L. Michael Trapasso** L. Michael Trapasso is Professor of Geography at Western Kentucky University, the Director of the College Heights Weather Station and a Research Associate with the Kentucky Climate Center. His research interests include human biometeorology, forensic meteorology, and environmental perception. He has received the Ogden College Faculty Excellence Award, and has also received Fulbright and Malone Fellowships to conduct research in various countries. His explorations have extended to all seven continents and he has written and lectured extensively on these travels. He has also contributed to a lab manual, a climatology textbook, and topical encyclopedias; and he has written and narrated educational television programming concerning weather and climate for the Kentucky Educational Television (KET) Network and WKYU-TV 24, Western Kentucky University Television.

# CHAPTER 1

The big picture: Earth's incredible diversity as viewed from space  NASA

# Physical Geography: Earth Environments and Systems

**PHYSICAL Geography Now™**  This icon, appearing throughout the book, indicates an opportunity to explore interactive tutorials, animations, or practice problems at now.brookscole.com/gabler8

## CHAPTER PREVIEW

**A major role of physical geography among the sciences is to investigate and attempt to explain the spatial aspects of Earth's physical phenomena.**

- Why is geography often called the spatial science?
- In what ways are locations and places fundamental in the science of geography?
- Why are the topics of spatial interaction and change so important in physical geography?

**Although physical geography is closely related to many other sciences, it is unique both in its focus and in its subjects of study.**

- How do physical geographers differ from other physical scientists in the topics or phenomena they study?
- What three major perspectives describe physical geography?
- What distinguishes physical geography from human or regional geography?

**The use of models and the analysis of various Earth systems are important research and educational techniques used by scientists who emphasize geography as a physical science.**

- What is a system?
- What kinds of models may be used to portray Earth and its physical processes?
- In what ways can systems analysis lead to an understanding of complex environments?

**Unlike some other physical sciences, physical geography places a special emphasis on human–environment relationships.**

- How does the nature of geography explain the role of geographers as environmental scientists?
- Why is this role of geography so important in the study of the environmental sciences today?
- Why does the study of ecosystems provide such an excellent opportunity for physical geographers to observe the interactions between humans and the natural environment?

**Every physical environment offers an array of advantages as well as challenges or hazards to the human residents of that location.**

- What adaptations are necessary for humans to live in your area?
- What environmental advantages do you enjoy where you live?
- What impacts do humans have on the environment where you live?

Viewed from space, far enough away to see an entire hemisphere, Earth is both beautiful and intriguing. From here we can begin to appreciate "the big picture," a global view of our planet's physical geography through its display of environmental diversity. Characteristics of the oceans, the atmosphere, the landmasses, and evidence of life, as presented by vegetated regions, are all apparent. Looking carefully, we can recognize geographic patterns in these features, shaped by the processes that make our world dynamic and ever changing. Except for the external addition of energy from the sun, our planet is a self-contained system that has all the requirements to sustain life. From the perspective of humans living on its surface, Earth may seem immense and almost limitless. In contrast, viewing the "big picture" reveals its conspicuous limits and fragility—a spherical island of life surrounded by the vast, dark emptiness of space. These distant images of Earth display the basic aspects of our planet that make our existence possible, but they only hint at the complexity of our planet. From a vantage point in space, we cannot comprehend the details of how air, water, land, and living things combine to create a diverse array of landscapes and environmental conditions on Earth's surface. This generalized global view, "the big picture," must be bolstered by a more detailed understanding of how Earth's features and processes interact to develop the extraordinary variety of environments that exist on our planet. Developing this understanding is the goal of a course in physical geography.

## The Study of Geography

*Geography* is a word that comes from two Greek roots. *Geo-* refers to "Earth," and *-graphy* means "picture or writing." Geography is distinctive among the sciences by virtue of its definition and central purpose. The primary objective of geography is the examination, description, and explanation of Earth—its variation from place to place and how places and features change over time. Geography is often called the **spatial science** because it includes recognizing, analyzing, and explaining the variations, similarities, or differences in phenomena located (or distributed) on Earth's surface (through Earth space). The major geography organizations in the United States have provided us with a good description of geography.

Where is something located? Why is it there? How did it get there? How does it interact with other things? Geography is not a collection of arcane information. Rather it is the study of spatial aspects of human existence.

People everywhere need to know about the nature of their world and their place in it. Geography has much more to do with asking questions and solving problems than it does with rote memorization of facts.

So what exactly is geography? It is an integrative discipline that brings together the physical and human dimensions of the world in the study of people, places, and environments. Its subject matter is the Earth's surface and the processes that shape it, the relationships between people and environments, and the connections between people and places.

Geography Education Standards Project, 1994
*Geography for Life*

Geography is distinctive among the sciences by virtue of its definition and central purpose. Unlike most scientists in related disciplines (for example, biologists, geologists, chemists, economists), who are bounded by the phenomena they study, geographers may focus their research on nearly any topic or subject related to the scientific analysis of human or natural processes on Earth ( Fig. 1.1). Geographers generally consider and examine *all* phenomena that are relevant to a given problem or issue; in other words, they often take a **holistic approach** to research.

Geographers study the physical and/or human characteristics of places, attempting to identify and explain the aspects that two or more locations may have in common as well as why places vary in their geographic characteristics. Geographers gather, organize, and analyze many kinds of geographic information, yet a unifying factor among them is a focus on explaining spatial locations, distributions, and relationships. They apply a variety of skills, techniques, and tools to the task of answering geographic questions. Geographers also study the processes that influenced Earth's landscapes in the past, how they continue to affect them today, how a landscape may change in the future, and the significance or impact of these changes.

Because geography embraces the study of virtually any global phenomena, it is not surprising that the subject has many subdivisions and it is common for geographers to specialize in one or more subfields of the discipline. Geography is also subdivided along academic lines; some geographers are social scientists, some are natural scientists, but most are involved in studying the human or natural processes that affect our planet, and the interactions among these processes. The main subdivision that deals with human

**FIGURE 1.1**
When conducting research or examining one of society's many problems, geographers are prepared to consider any information or aspect of a topic that relates to their study.
**What advantage might a geographer have when working with other physical scientists seeking a solution to a problem?**

**FIGURE 1.2**
Vancouver, British Columbia's magnificent physical setting strongly influences the human geography of this urban area. Settlement patterns, economic activities, recreational opportunities, and many aspects of human activities are a function of interactions among geographic factors, both human and physical.
**What human geographic characteristics can you interpret from this scene?**

activities and the results of these activities is called cultural or **human geography.** Human geographers are concerned with such subjects as population distributions, cultural patterns, cities and urbanization, industrial and commercial location, natural resource utilization, and transportation networks ( ▌ Fig. 1.2). Geographers are interested in how to divide and synthesize areas into meaningful divisions called **regions**, which are areas identified by certain characteristics they contain that make them distinctive and separates them from surrounding areas. Geographic study that concentrates on the general physical and human characteristics of a region, such as Canada, the Great Plains, the Caribbean, or the Sahara, is termed **regional geography.**

## Physical Geography

**Physical geography,** the focus of this textbook, encompasses the processes and features that make up Earth, including human activities where they interface with the physical environment. In fact, physical geographers are concerned with nearly all aspects of Earth and can be considered generalists because they are trained to view a natural environment in its entirety, functioning as a unit ( ▌ Fig. 1.3). However, after completing a broad education in basic physical geography, most physical geographers focus their expertise on advanced study in one or two specialties. For example, *meteorologists* and *climatologists* consider the atmospheric components that interact to influence weather and climate. Meteorologists are interested in the atmospheric processes that affect daily weather, and they use current data to forecast weather conditions. Climatologists are interested in the averages and extremes of long-term weather data, regional classification of climates, monitoring and understanding climatic change and climatic hazards, and the long-range impact of atmospheric conditions on human activities and the environment.

The study of the nature, development, and modification of landforms is a specialty called *geomorphology,* a major subfield of physical geography. Geomorphologists are interested in understanding and explaining variation in landforms, the processes that produce physical landscapes, and the nature and geometry of Earth's surface features. The factors involved

**FIGURE 1.3**
The Colorado Rocky Mountains in autumn. Physical geographers study the elements and processes that affect natural environments. These include rock structures, landforms, soils, vegetation, climate, weather, and human impacts.
**What physical geography characteristics can you interpret from this scene?**

in landform development are as varied as the environments on Earth and include, for example, gravity, running water in streams, stresses in the Earth's crust, the flow of ice in glaciers, volcanic activity, and the erosion or deposition of Earth's surface materials. *Biogeographers* examine natural and human-modified environments and the ecological processes that influence their nature and distribution, including vegetation change over time. They also study the ranges and patterns of vegetation and animal species, seeking to discover the environmental factors that limit or facilitate their distributions. Many *soil scientists* are geographers, and these individuals are often involved in the mapping and analysis of soil types, determining the suitability of a soil for certain uses, such as agriculture, and working to conserve soil as a natural resource. Finally, because of the critical importance of water to the existence of life on Earth, geographers are widely involved in the study of water bodies and their processes, movements, impact, quality, and other characteristics. They may serve as *hydrologists, oceanographers,* or *glaciologists.*

Many geographers involved with water studies also function as *water resource managers* to ensure that lakes, watersheds, springs, and groundwater sources are suitable for human use, provide an adequate supply of water, and are as free of pollution as possible.

## Technology, Tools, and Methods

The technologies used by physical geographers in their efforts to learn more about Earth are rapidly changing. The abilities of computer systems to capture, process, model, and display spatial data—functions that can be performed on a personal computer—were only a dream 30 years ago. Today the Internet provides access to information and images on virtually any topic. The amounts of data, information, and imagery available for studying Earth and its environments have exploded. Graphic displays of environmental data and information are becoming more vivid and striking as a result of sophisticated methods of data processing and visual representation. Increased computer power allows the presentation of high-resolution images, three-dimensional scenes, and animated images of Earth features, changes, and processes (Fig. 1.4).

Continuous satellite imaging of Earth has been ongoing for more than 30 years, which has given us a better perspective on environmental changes as they occur. Using satellite imagery it is possible to monitor changes in a single place over time or to compare two different places at a point in time. Using various energy sources to produce images from space, we are able to see, measure, monitor, and map processes and the effects of certain processes that are invisible to the naked eye. Satellite technology is being used to determine the precise location of a positioning receiver on Earth's surface, a capability that has many useful applications for geography and mapping. Today, mapmaking (*cartography*) and the analysis of maps are typically digital, computer-assisted operations.

Making observations and gathering data in the field are important skills for most physical geographers, but they must also keep up with new technologies that support and facilitate traditional fieldwork. Technology may provide maps, images, and data, but a person who is knowledgeable about the subject being studied is essential to the processes of analysis and problem solving (Fig. 1.5). Many geographers are gainfully employed in positions that apply technology to the problems of understanding our planet and its environments, and their numbers are certain to increase in the future.

**FIGURE 1.4**
Complex computer-generated model of Earth, based on data gathered from satellites.
**How does this image compare to the Earth image in the chapter opening?**

## Major Perspectives in Physical Geography

Your textbook has been designed to demonstrate three major perspectives that physical geography emphasizes: spatial science, physical science, and environmental science. Although the emphasis on each of these perspectives may vary from chapter to chapter, the contributions of all three perspectives to scientific study will be apparent throughout the book. As you read through the remainder of this chapter, take note of how directly each perspective in the sciences relates to the unique nature of the geographic discipline.

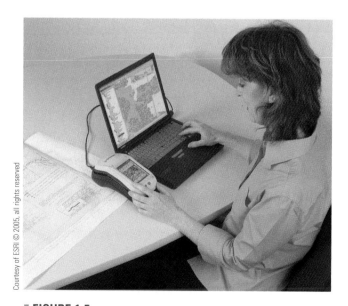

**FIGURE 1.5**
A geographer uses computer technology to analyze maps and imagery.

# Geography's Spatial Science Perspective

## The Regional Concept: Natural and Environmental Regions

The term *region* is familiar to us all, but it has a precise meaning and special significance to geographers. Simply stated, a region is an area that shares a certain characteristic (or a set of characteristics) within its boundaries. Regions are spatial models, just as systems are operational models. Systems help us understand how things work, and regions help us make spatial sense of our world. The concept of a region is a tool for thinking about and analyzing logical divisions of areas based on their geographic characteristics. Just as it helps us to understand Earth by considering smaller parts of its overall system, dividing space into coherent regions helps us understand the arrangement and nature of areas on our planet. Regions can be described based on either human or natural characteristics, or a combination of the two.

Regions can also be divided into subregions. For example, North America is a region, but it can be subdivided into many subregions. Examples of subregions based on natural characteristics include the Atlantic Coastal Plain (similarity of landforms, geology, and locality), the Prairies (ecological type), the Sonoran Desert (climate type and locality), the Pacific Northwest (general locality), and Tornado Alley (region of high potential for these storms).

Physical geographers are mainly interested in natural regions and human–environmental regions. The term *natural*, as used here, means primarily related to natural processes and features involved in the four major spheres of the Earth system. However, we recognize that today human activities have an impact on virtually every natural process, and human–environmental regions offer significant opportunities for geographic analysis. Geographers not only study and explain regions, their locations and characteristics but also strive to delimit them—to outline their boundaries on a map. An unlimited number of regions can be derived for each of the four major Earth subsystems.

There are three important points to remember about natural and environmental regions. Each of these points has endless applications and adds considerably to the questions that the process of defining regions based on spatial characteristics seeks to answer.

- **Natural regions can change in size and shape over time in response to environmental changes.** These changes can be fast enough to observe as they occur, or so gradual that they require intensive study to detect. An example is the change in the Amazonian rainforest, a natural region defined by vegetation associations, caused by deforestation. Using images from space, we can see and monitor changes in the area covered by the rainforests, as well as other natural regions.
- **Boundaries separating different natural or environmental regions tend to be indistinct or transitional, rather than**

Change in the Amazonian rainforest due to logging operations.

## The Spatial Science Perspective

The central role of geography among the sciences is best illustrated by its definition as the **spatial science** (the science of Earth space). No other discipline has the specific responsibility for investigating and attempting to explain the spatial aspects of Earth phenomena. Even though physical geographers may have many divergent interests, they share a common goal of understanding and explaining the spatial variation displayed on Earth's surface.

**sharp.** For example, on a climate map, lines separating desert from nondesert regions do not imply that extremely arid conditions instantly appear when the line is crossed; rather, if we travel to a desert, it is likely to get progressively more arid as we approach our destination.

- **Regions are spatial models, devised by humans, for geographic analysis, study, and understanding.** Natural or environmental regions, like all regions, are conceptual models that are specifically designed to help us comprehend and organize spatial relationships and geographic distributions. Learning geography is an invitation to think spatially, and regions provide an essential, extremely useful, conceptual framework in that process.

The rainforests form one of the most biologically diverse regions on Earth and provide excellent examples of natural regions. Over the last 30 years, Brazil, home to the largest rainforest region, has lost between 12 and 15% of its forest cover through deforestation. Since the late 1970s, an average of 5.6 million acres (8800 mi$^2$ or 22,800 km$^2$) of rainforest have been destroyed—each year. These are not global statistics, however, because they describe what has been happening to only the Brazilian rainforest region. A wide variety of processes can result in the change in the size of a region, and in this case, human activities are the responsible factor. Yet, the rainforest regions are important for many reasons including their great biodiversity and the interactions that these vast forests have with Earth's atmosphere and climate.

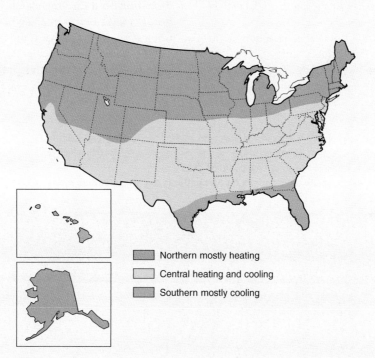

This human–environmental map divides the United States into three regions based on annual heating and cooling needs. Red means that heating is required more often than cooling. The pink region represents roughly equal heating and cooling needs. Blue represents a stronger or greater demand for cooling than for heating. The map is clearly related to climate regions. Do you think that the boundaries between these regions are as sharply defined in reality as they are on this map? Can you recognize the spatial patterns that you see? Do the shapes of these regions, and the ways that they are related to each other, seem spatially logical?

How do physical geographers examine Earth from a spatial point of view? What are the spatial questions that physical geographers raise, and what are some of the problems they seek to understand and solve? From among the almost unlimited number of topics available to physical geographers, we have chosen five that clearly illustrate the role of geography as the spatial science. In keeping with the quote from *Geography for Life*, that geography is about asking questions and solving problems, common study questions have been included for each topic.

**Location** Geographic knowledge and studies often begin with locational information. The location of a feature is usually expressed by one of two methods: **absolute location,** expressed by a coordinate system (or address), or **relative location,** which identifies where something is in relation to something else, usually a fairly well-known location. For example, Pikes Peak, in the Rocky Mountains of Colorado, with an elevation of 4301 meters (14,110 feet), has a location of latitude 38°51′ north and longitude 105°03′ west. A global address like this is an absolute location. However, another way to report its location would be to state that it is 36 kilometers (22 miles) west of Colorado Springs. This is an example of relative location (relative to Colorado Springs). Typical spatial questions involving location include the following: *Where is a certain type of Earth feature found, and where is it not found? Why is a certain feature located where it is? What methods can we use to locate a feature on Earth? How can we describe its location? What is the most likely or least likely location for a certain Earth feature?*

**Characteristics of Places** Physical geographers are interested in the environmental features and processes that combine to make a place unique, and they are also interested in shared characteristics between places. For example, what physical geographic features make the Rocky Mountains appear as they do? Further, how are the Appalachian Mountains different from the Rockies, and what characteristics are common to both of these mountain ranges? Another aspect of the characteristics of places is analyzing the environmental advantages and challenges that exist in a place. Other examples might include: *How does an Australian desert compare to the Sonoran desert of the southwestern United States? How do the grasslands of the Great Plains of the United States compare to the grasslands of Argentina? What are the environmental conditions at a particular site? How do places on Earth vary in their environments, and why? In what ways are places unique, and in what ways do they share similar characteristics with other places?*

**Spatial Distributions and Spatial Patterns** When studying how features are arranged in space, geographers are usually interested in two spatial factors ( ❙ Fig. 1.6). **Spatial distribution** means the extent of the area or areas where a feature exists. For example, where on Earth do we find the tropical rainforests? What is the distribution of rainfall in the United States on a particular day? Where on Earth do major earthquakes occur? **Spatial pattern** refers to the arrangement of features in space—are they regular or random,

❙ **FIGURE 1.6**
This view of North America by night provides several good illustrations of distribution and pattern. Spatial distribution is where features are located (or, perhaps, where they are absent), and spatial pattern refers to their arrangement. Geographers attempt to explain these spatial relationships.
**Can you locate and propose possible explanations for two patterns and two distributions in this scene?**

clustered together or widely spaced? From the window seat of an airplane on a clear day, it is obvious that the population distribution can be either dense or sparse. The spatial pattern of earthquakes may be linear on a map because earthquake faults display similar linear patterns. *Where are certain features abundant, and where are they rare? How are particular factors or elements of physical geography arranged in space, and what spatial patterns exist, if any? What processes are responsible for these patterns? If a pattern exists, what does it signify?*

**Spatial Interaction** Few processes on Earth operate in isolation; areas on our planet are interconnected, which means linked to conditions elsewhere on Earth. A condition, an occurrence, or a process in one place generally has an impact on other places. Unfortunately, the exact nature of this **spatial interaction** is often difficult to establish with certainty except after years of study. A cause–effect relationship can often only be suspected because a direct relationship is often difficult to prove. It is much easier to observe that changes seem to be associated with each other, without knowing if one event causes the other or if this result is coincidental.

For example, the presence of abnormally warm ocean waters off South America's west coast, a condition called El Niño, seems to be related to unusual weather in other parts of the world. Clearing the tropical rainforest may have an impact on world climates. Interconnections are one reason for considering Earth as a whole. *What are the relationships among places and features on Earth? How do they affect one another? What important interconnections link the oceans to the atmosphere and the atmosphere to the land surface?*

**Ever-Changing Earth** Earth's features and landscapes are continuously changing in a spatial context. Weather maps show where and how weather elements change from day to day, over the seasons, and from year to year. A future Hawaiian island is now forming beneath the waters of the Pacific Ocean. Storms, earthquakes, landslides, and stream processes modify the landscape ( ▌Fig. 1.7). Coastlines may change position because of storm waves, tsunamis, or changes in sea level. Areas that were once forested have been clear-cut, changing the nature of the environment there. Vegetation and wildlife are becoming reestablished in areas that were devastated by recent volcanic eruptions or wildfires. Desertlike conditions seem to be expanding in many arid regions of the world.

World climates have changed throughout Earth's history, with attendant shifts in the distributions of plant and animal life. Earth and its environments are always changing, but not all of this change can be directly monitored. *How are Earth features changing in ways that can be recorded in a spatial sense? What processes contribute to the change? What is the rate of change? Does change occur in a cycle? Can humans witness this change as it is taking place, or is a long-term study required to recognize the change? Do all places on Earth experience the same levels of change, or is there spatial variation?*

▌**FIGURE 1.7**
Natural changes in a landscape can be either beneficial or detrimental to human existence. An undersea earthquake triggered a tsunami wave that hit coastal areas along the Indian Ocean with great destructive force. The effect on the people living there was devastating. The better we understand an environment's characteristics and the processes that affect it, the more able we are to avoid situations such as this scene in Thailand: before the tsunami (left), and after the tsunami (right).

# Geography's Physical Science Perspective

## The Scientific Method

Science . . . is the systematic and organized inquiry into the natural world and its phenomena. Science is about gaining a deeper and often useful understanding of the world.

*Multicultural History of Science web page, Vanderbilt University*

The real purpose of the scientific method is to make sure nature hasn't misled you into thinking you know something you don't actually know.

Robert M. Pirsig,
*Zen and the Art of Motorcycle Maintenance*

Physical geography is a science that focuses on the Earth system, how its components and processes interact, and how and why aspects that affect Earth's surface are spatially arranged, as well as how humans and their environments are interrelated.

To wonder about your environment and attempt to understand it is a fundamental basis of human life. Increasing our awareness, satisfying our curiosity, learning how our world works, and determining how we can best function within it are all parts of a satisfying but never-ending quest for understanding. Without curiosity about the world, supported by making observations, noting relationships and patterns, and applying the knowledge discovered, humans would not have survived beyond their earliest beginnings. Science gives us a method for answering questions and testing ideas by examining evidence, drawing conclusions, and making new discoveries.

The sciences search for new knowledge using a strategy that minimizes the possibility of erroneous conclusions. This highly adaptable process is called the *scientific method*. The scientific method is a general framework for research, but it can accommodate an infinite number of topics and strategies for deriving conclusions.

Scientific method generally involves the following steps:

1. **Making an observation that requires an explanation.** We may wonder if the observation represents a general pattern or is a "fluke" occurrence. For example, on a trip to the mountains, you notice that it gets colder as you go up in elevation. Is that just a result of conditions on the day you were there, or just the conditions at the location where you were, or is it a relationship that generally occurs everywhere?

2. **Restating the observation as a hypothesis.** Here is an example: As we go higher in elevation, the temperature gets cooler (or, as a question, Does it get cooler as we go up in elevation?). The answer may seem obvious, yet it is generally but not always true, depending on environmental conditions that will be discussed in later chapters. Many scientists recommend a strategy called *multiple working hypotheses*, which means that we consider and test many possible hypotheses to discover which one best answers the questions while eliminating other possibilities.

3. **Determining a technique for testing the hypothesis and collecting necessary data.** The next step is finding a technique for evaluating data (numerical information) and or facts that concern that hypothesis. In our example, we would gather temperature and elevation data (taken at about the same time for all data points) for the area we are studying.

4. **Applying the technique or strategy to test the validity of the hypothesis.** Here we discover if the hypothesis is supported by adequate evidence, collected under similar conditions to minimize bias. The technique will recommend either acceptance or rejection of the hypothesis.

If the hypothesis is rejected, we can test an alternate hypothesis or modify our existing one and try again, until all of the conceivable hypotheses are rejected or we discover one that is supported by the data. If the test supports the hypothesis, we have confirmed our observation, at least for the location and environmental conditions in which our data and information were gathered.

After similar tests are conducted and the hypothesis is supported in many places and under other conditions, then the hypothesis may become a theory. Theories are well-tested concepts or relationships that, given specified circumstances, can be used to explain and predict outcomes.

The processes of asking questions, seeking answers, and finding solutions through the scientific method have contributed greatly to human existence, our technologies, and our quality of life. Obviously, there are many more questions to be answered and problems yet to be solved. In fact, new findings typically yield new questions.

Human curiosity, along with an intrinsic need for knowledge through observation and experience, has formed the basis for scientific method, an objective, structured approach that leads us toward the primary goal of physical geography—understanding how our world works.

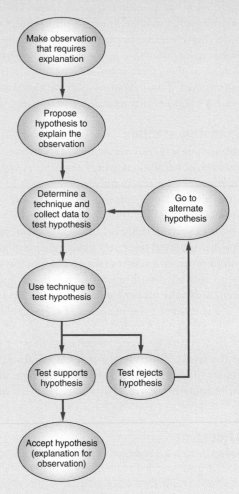

Steps in the scientific method

The previous five topics illustrate geography's strong emphasis on the spatial perspective. Learning what questions to ask is the first step toward finding answers and explanations, and it is a major objective of your physical geography course.

## The Physical Science Perspective

As geographers apply their expertise to the study of Earth, it is clear that physical geographers will be observing phenomena, compiling data, and seeking solutions to problems that are also of interest to researchers in one or more of the other physical sciences. Physical geographers, as climatologists, share ideas and information with atmospheric physicists. Soil geographers study the same elements and compounds analyzed by chemists. Biogeographers are concerned about the environments of the same plants and animals that are classified by biologists. However, to whatever question is raised and whichever problem requires a solution, the physical geographer brings a unique point of view—a spatial perspective and a mandate to carefully consider all Earth phenomena that may be involved. This is why an introductory course in physical geography and your textbook are so often focused on a global scale, although local and regional processes are important as well. Physical geographers are concerned with the processes that affect the Earth's physical environments. By examining the factors, features, and processes that influence the environment and learning how these elements work together, we can better understand the ever-changing physical geography of our planet. We can also appreciate the importance of viewing Earth in its entirety as a constantly functioning system—as a whole greater than but dependent upon each of its parts.

**The Earth System** Examining Earth as a system made up of an almost unlimited number of smaller subsystems constitutes a major contribution to the physical sciences. A **system** may be thought of as a set of parts or components that are interrelated, and the analysis of systems provides physical geographers with ideal opportunities to study these relationships as they affect Earth's features and environments. The individual components of a system, termed **variables,** are studied or grouped together because these variables interact with one another as parts of a functioning unit. Earth certainly fits this definition because many continuously changing variables combine to make our home planet, the **Earth system,** function the way that it does. Virtually no part of a system operates in isolation from another.

A change in one aspect of the Earth system affects other parts, and the impact of these changes can be significant enough to appear in worldwide patterns, clearly demonstrating the interconnections among these variables. For example, the presence of mountains influences the distribution of rainfall, and variations in rainfall affect the density, type, and variety of vegetation. Plants, moisture, and the underlying rock affect the kind of soil that forms in an area. Characteristics of vegetation and soils influence the runoff of water from the land, leading to completion of the circle, because the amount of runoff is a major factor in stream erosion, which eventually can reduce the height of mountains. Many cycles such as this operate to change our planet, but the Earth system is complex, and these cycles and processes operate at widely varying rates and over widely varying time spans.

**Earth's Major Subsystems** Four major divisions of our planet comprise the Earth system. The **atmosphere** is the gaseous blanket of air that envelops, shields, and insulates Earth. The movements and processes of the atmosphere create the changing conditions that we know as weather and climate. The solid Earth—landforms, rocks, soils, and minerals—makes up the **lithosphere.** The waters of the Earth system—oceans, lakes, rivers, and glaciers—constitute the **hydrosphere.** The fourth major division, the **biosphere,** is composed of all living things: people, other animals, and plants.

It is the nature of these four major subsystems and the interactions among them that create and nurture the conditions necessary for life on Earth ( ▌ Fig. 1.8). For example, the hydrosphere serves as the water supply for all life, including humans, and provides a home environment for many types of aquatic plants and animals. The hydrosphere directly affects the lithosphere as the moving water in streams, waves, and currents shapes landforms. It also influences the atmosphere through evaporation, condensation, and the effects of ocean temperatures on climate. The impact or intensity of interactions among Earth's subsystems is not identical everywhere on the planet, and it is this variation that leads to the geographic patterns of environmental diversity.

Many other examples of overlap exist among the four divisions. Soil can be examined as part of the biosphere, the hydrosphere, or the lithosphere. The water stored in plants and animals is part of both the biosphere and the hydrosphere, and the water in clouds is a component of the atmosphere as well as the hydrosphere. The fact that we cannot draw sharp boundaries between these divisions underscores the interrelatedness among various parts of the Earth system. However, like a machine, a computer, or the human body, planet Earth is a system that functions well only when all of its parts (and its subsystems) work together harmoniously.

We do know that the Earth system as well as its four major subsystems are *dynamic* (ever changing) and that we can directly observe some of these changes—the seasons, the ocean tides, earthquakes, floods, volcanic eruptions. Other aspects of our planet may take years, or even more than a lifetime, to accumulate enough total change to be directly noticed by humans. Such long-term changes in our planet are often difficult to understand or predict with certainty, so they must be carefully and scientifically studied to determine what is really happening and what the consequences might be. Changes of this type include shifts in world climates, drought cycles, the spread of deserts, worldwide rise or fall in sea level, erosion of coastlines, and major changes in river systems. Yet understanding

**FIGURE 1.8**
Earth's major subsystems. Studying Earth as a system is central to understanding changes in our planet's environments and adjusting to or dealing with these changes. Earth consists of many interconnected subsystems.
**How do these systems overlap? For example, how does the atmosphere overlap with the hydrosphere, or the biosphere?**

ecology, or ecosystem? In the broadest sense, our **environment** can be defined as our surroundings; it is made up of all physical, social, and cultural aspects of our world that affect our growth, our health, and our way of living ( Fig. 1.9).

Environments are also systems because they function through the interrelationships among many variables. Environmental understanding involves giving consideration to a wide variety of factors, characteristics, and processes involving weather, climate, soils, rocks, terrain, plants, animals, water, humans, and how they interact with each other to produce an environment. The holistic approach of physical geography is an advantage in this understanding, as the potential influence of each of these factors must not only be considered individually, but also as they affect one another in an environmental system.

Physical geographers are keenly interested in environmental processes and interactions, and they pay special attention to the relationships that involve humans and their activities. Much of human existence throughout time has been a product of the adaptations that various cultures have made to and the modifications they have imposed on their natural surroundings. Primitive skills and technology generally require people to make greater adjustments in adapting to their environment. The more sophisticated a culture's technology is, the greater the amount of environmental modification. Thus, human–environment interaction is a two-way relationship, with the environment influencing human behavior and humans impacting the environment.

changes in our planet is critical to human existence. We are, after all, a part of the Earth system. Changes in the system may be naturally caused or human induced, or they may result from a combination of these factors. To understand our planet, therefore, we must learn about its components and the processes that operate to change or regulate the Earth system. Such knowledge is in the best interest of not only humankind but also Earth, as a habitat for all living things.

## The Environmental Science Perspective

Today, we regularly hear talk about the environment and ecology and worry about damage to ecosystems caused by human activity. We also hear news reports of disasters caused by humans being exposed to such violent natural processes as earthquakes, floods, tornadoes, or the South Asia tsunami of 2004 with its terrible consequences. Newspapers and magazines often devote entire sections to discussions of these and other environmental issues. But what are we really talking about when we use words like *environment*,

Just as humans interact with their environment, so do other living things. The study of relationships between organisms, whether animal or plant, and their environments is a science known as **ecology.** Ecological relationships are complex but naturally balanced "webs of life." Disrupting the natural ecology of a community of organisms may have negative results (although this is not always so). For example, filling in or polluting coastal marshlands may disrupt the natural ecology of those wetlands. As a result, fish spawning grounds may be destroyed, and the food supply of some marine animals and migratory birds could be depleted. The end product may be the destruction of valuable plant and animal life.

The word *ecosystem* is a contraction of *ecological system*. An **ecosystem** is a community of organisms and the relationships of those organisms to their environment ( Fig. 1.10). An ecosystem is dynamic in that its various parts are always in flux. For instance, plants grow, rain falls, animals eat, and soil matures—all changing the environment of a particular ecosystem. Because each member of the ecosystem belongs to the environment of every other part of that system, a change in one alters the environment for the others. As those components react to the alteration, they in turn continue to transform the environment for the others. A change in the weather, from sunshine to rain, affects plants, soils, and animals. Heavy rain may carry away soils and plant nutrients so that plants may not be able to grow as well and animals, in turn, may not have as much to eat. In contrast, the addition of moisture to the soil may help some plants grow, increasing the amount of shade beneath them and thus keeping other plants from growing.

The ecosystem concept (like other systems models) can be applied on almost any scale from local to global, in a wide variety of geographic locations, and under all environmental conditions in which life is possible. Hence, your backyard, a farm pond, a grass-covered field, a marsh, a forest, or a portion of a desert can be viewed as an ecosystem. Ecosystems are found wherever there is an exchange of materials among living organisms and where there are functional relationships between the organisms and their natural surroundings. Ecosystems are open systems, as both energy and material move across their boundaries. Although some ecosystems, such as a small lake or a desert oasis, have clear-cut boundaries, the limits of many others are not as precisely defined.

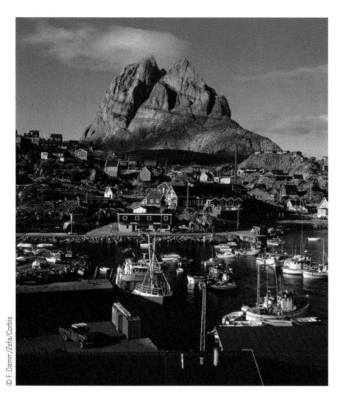

**FIGURE 1.9**
The physical and cultural attributes of a site combine to form a unique geographic environment.
**What can you learn about human–environment relationships by studying this photograph?**

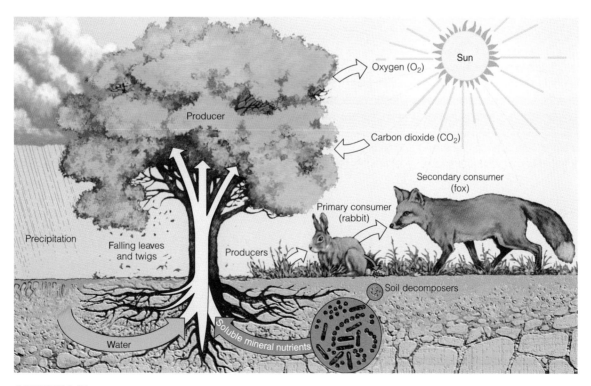

**FIGURE 1.10**
Ecosystems are an important aspect of natural environments.
**How do ecosystems illustrate the interactions in the environment?**

## Geography's Environmental Science Perspective
### Human–Environment Interactions

As the world population has grown, the effects of human activities on the environment, as well as the impacts of environmental processes on humans, have become topics of increasing concern. There are many circumstances where human–environment relationships have been mutually beneficial, yet two negative aspects of those interactions have gained serious attention in recent years. Certain environmental processes, with little or no warning, can become dangerous to human life and property, and certain human activities threaten to cause major, and possibly irrevocable, damage to Earth environments.

The environment becomes a hazard to humans and other life-forms when relatively uncommon and extraordinary natural events occur that are associated most directly with the atmosphere, hydrosphere, or lithosphere. Living under the conditions provided by these three subsystems, it is elements of the biosphere, including humans, which suffer the damaging consequences of sporadic natural events of extraordinary intensity. The routine processes of these three subsystems become a problem and spawn environmental hazards for two reasons. First, on occasion and often unpredictably, they operate in an unusually intense or violent fashion. Summer showers may become torrential rains that occur repeatedly for days or even weeks. Ordinary tropical storms gain momentum as they travel over warm ocean waters, and they reach coastlines as full-blown hurricanes. Molten rock and associated gases from deep beneath Earth move slowly toward the surface and suddenly trigger massive eruptions that literally blow apart volcanic mountains.

Human activities affect the natural environment. Pollution of water resources and our atmosphere at an industrial complex.

U.S. Environmental Protection Agency

The 2004 tsunami wave that devastated coastal areas along the Indian Ocean provided an example of the potential for the occasional occurrences of natural processes that far exceed our expectable "norm."

Each of these examples of Earth systems operating in sudden or extraordinary fashion is a noteworthy environmental event, but it does not become an environmental hazard unless people or their property are affected. Thus, the second reason environmental hazards exist is because people live where potentially catastrophic environmental events may occur. The greater the number of people and the greater the value of the property involved, the greater the catastrophe.

Why do people live where environmental hazards pose a major threat? Actually, there are many reasons. Some people have no choice. The land they live on is their land by birthright; it was their family's land for generations. Especially in densely populated developing nations, there may be no other place to go. Other people choose to live in hazardous areas because they believe the advantages outweigh the potential for natural disaster. They are attracted by productive farmland, the natural beauty of a region or building site, or by the

---

Often the change from one ecosystem to another is obscure and transitional, occurring gradually over distance.

**A Life-Support System** Certainly the most important attribute of Earth is that it is a **life-support system.** Like space vehicles that support astronauts, the Earth system provides the necessary environmental constituents and conditions to permit life, as we know it to exist ( Fig. 1.11). If a critical part of a life-support system is significantly changed or fails to operate properly, living organisms may no longer be able to survive. For instance, if all the oxygen in a spacecraft is used up, the crew inside will die. If a spacecraft cannot control the proper temperature range, its occupants may burn or freeze. If food supplies run out, the astronauts will starve. On Earth, natural processes must provide an adequate supply of oxygen; the sun must interact with the atmosphere, oceans, and land to maintain tolerable temperatures; and photosynthesis or other continuous economic possibilities available at a location. In addition, few populated areas of the world are not associated with an environmental hazard or perhaps several hazards. Forested regions are subject to fire; earthquake, landslide, and volcanic activities plague mountain regions; violent storms threaten interior plains; and many coastal regions experience periodic hurricanes or typhoons (the Pacific Ocean equivalent).

Just as the environment can pose an ever-present danger to humans, through their activities, humans can constitute a serious threat to the environment. Issues such as global warming, acid precipitation, deforestation and the extinction of biological species in tropical areas, damage to the ozone layer of the atmosphere, and desertification have risen to the top of agendas when world leaders meet and international conferences are held. Environmental concerns are recurring subjects of magazine and newspaper articles, books, and television programs.

Much environmental damage has resulted from atmospheric pollution associated with industrialization, particularly in support of the wealthy, developed nations. But as population pressures mount and developing nations struggle to industrialize, human activities are exacting an increasing toll on the soils, forests, air, and waters of the developing world as well. Environmental deterioration is a problem of worldwide concern, and solutions must involve international cooperation in order to be successful. As citizens of the world's wealthiest nation, Americans must seriously consider what steps can be taken to counter major environmental threats related to human activities. What are the causes of these threats? Are the threats real and well documented? What can I personally do to help solve environmental problems?

Natural processes affect the human environment. Destruction of property by a tornado that struck near Tuscaloosa, Alabama.

Examining environmental issues from the physical geographer's perspective requires that characteristics of both the environment and the humans involved in those issues be given strong consideration. As will become apparent in this study of geography, physical environments are changing constantly, and all too frequently, human activities result in negative environmental consequences. In addition, throughout Earth, humans live in constant threat from various and spatially distributed environmental hazards such as earthquake, fire, flood, and storm. The natural processes involved are directly related to the physical environment, but causes and solutions are imbedded in human–environmental interactions that include the economic, political, and social characteristics of the cultures involved. The recognition that geography is a holistic discipline—that it includes the study of all phenomena on Earth—requires that physical geographers play a major role in the environmental sciences.

cycles of creation must provide new food supplies for living things.

Earth, then, is made up of a set of interrelated components that are vital and necessary for the existence of all living creatures. About 40 years ago, Buckminster Fuller, a distinguished scientist, philosopher, and inventor, coined the notion of Spaceship Earth—the idea that our planet is a life-support system, transporting us through space. Fuller also thought that knowing how Earth works is important—indeed this knowledge may be required for human survival—but that humans are only slowly learning the processes involved. He compared this information to an operating manual, like the owner's manual for an automobile.

> One of the most interesting things to me about our spaceship is that it is a mechanical vehicle, just as is an automobile. If you own an automobile, you realize that you must put oil and gas into it, and you must put water in the radiator and take care of the car as a whole. You know that you are going to have to

**FIGURE 1.11**
Earth and space vehicles operate as life-support systems. Astronauts orbit in the lunar lander above the moon's lifeless surface, with a rising Earth above the horizon.

keep the machine in good order or it's going to be in trouble and fail to function.

We have not been seeing our Spaceship Earth as an integrally designed machine which to be persistently successful must be comprehended and serviced in total . . . there is one outstandingly important fact regarding Spaceship Earth, and that is that no instruction book came with it.

R. Buckminster Fuller
*Operating Manual for Spaceship Earth*

Today, we realize that critical parts of our life-support system, **natural resources**, can be abused, wasted, or exhausted, potentially threatening the function of planet Earth as a human life-support system. One such abuse is **pollution,** an undesirable or unhealthy contamination in an environment resulting from human activities ( Fig. 1.12). We are aware that some of Earth's resources, such as air and water, can be polluted to the point where they become unusable or even lethal to some life-forms. By polluting the oceans, we may be killing off important fish species, perhaps allowing less desirable species to increase in number. Acid rain, caused by atmospheric pollutants from industries and power plants, is damaging forests and killing fish in freshwater lakes. Air pollution has become a serious environmental problem for urban centers throughout the world ( Fig. 1.13). What some people do not realize, however, is that pollutants are often transported by winds and waterways hundreds or even thousands of kilometers from their source. Lead from automobile exhausts has been found in the snow of Antarctica, as

**FIGURE 1.12**
Toxic chemicals, such as the ones discovered in this solid-waste dump, pose a serious health hazard and threaten local water supplies.

**What pollutants form the major threat to the air and water supply in your community?**

MAJOR PERSPECTIVES IN PHYSICAL GEOGRAPHY   **19**

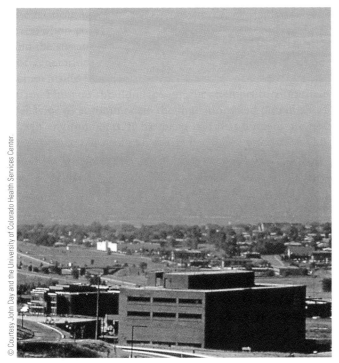

**▌FIGURE 1.13**
(a) Denver, Colorado on a clear day, with the Rocky Mountains visible in the background. (b) On a smoggy day from the same location, even the downtown buildings are not visible.
**If you were choosing whether to live in a small town or a major city, would pollution affect your decision?**

has the insecticide DDT. Pollution is a worldwide problem that does not stop at political boundaries.

Another concern is that humans may be rapidly depleting critical natural resources, especially those needed for fuel. Many natural resources on our planet are nonrenewable, meaning that nature will not replace them once they are exhausted. Coal and oil are nonrenewable resources. When nonrenewable resources such as these mineral fuels are gone, the alternative resources may be less desirable or more expensive.

We are learning that, much like life on a spaceship, there are limits to the suitable living space on Earth, and we must use our lands wisely. In our search for livable space, we occasionally construct buildings in locations that are not environmentally safe. Also, we sometimes plant crops in areas that are ill suited to agriculture while at the same time paving over prime farmland for other uses.

In modern times, the ability of humans to alter the landscape has been increasing. For example, a century ago the interconnected Kissimmee River–Lake Okeechobee–Everglades ecosystem constituted one of the most productive wetland regions on Earth. But sawgrass marsh and slow-moving water stood in the way of urban and agricultural development. Intricate systems of ditches and canals were built, and since 1900, half of the original 4 million acres (1.6 million hectares) of the Everglades has disappeared (▌ Fig. 1.14). The Kissimmee River was channelized into an arrow-straight ditch, and wetlands along the river were drained. Levees have prevented water in Lake Okeechobee from contributing water flow to the Everglades, and highway construction further disrupted the natural drainage patterns.

Fires have been more frequent and destructive, and entire biotic communities have been eliminated by lowered water levels. During excessively wet periods, portions of the Everglades are deliberately flooded to prevent drainage canals from overflowing. As a result, animals drown and birds cannot rest and reproduce. South Florida's wading bird population has decreased by 95% in the last hundred years. Without the natural purifying effects of wetland systems, water quality in south Florida has deteriorated; with lower water levels, saltwater encroachment is a serious problem in coastal areas.

Today, backed by government agencies, scientists are struggling to restore south Florida's ailing ecosystems. There are extensive plans to allow the Kissimmee River to flow naturally across its former flood plain, to return agricultural land to wetlands, and to restore water-flow patterns through the Everglades. The problems of south Florida should serve as a useful lesson. Alterations of the natural environment should not be undertaken without serious consideration of all consequences.

**The Human–Environment Equation** Despite the wealth of resources available in the air, water, soil, minerals, vegetation, and animal life on Earth, the capacity of our planet to support the growing numbers of humans may have an ultimate limit, a threshold population. Dangerous signs indicate that such a limit may someday be reached. The world population has passed the 6.5 billion mark, and United Nations' estimates indicate more than 9 billion people by 2050 if current growth rates continue. Today, more than half the world's people must tolerate subject. Geography is a way of looking at the world and of observing its features. It involves asking questions about the nature of those features and appreciating their beauty and complexity. It encourages you to seek explanations, gather information, and use geographic skills, tools, and knowledge to solve problems. Even if you forget many of the facts discussed in this book, you will have been shown new ways to consider, see, and evaluate the world around you. Just as you see a painting differently after an art course, so too will you see sunsets, waves, storms, deserts, rivers, forests, prairies, and mountains with an "educated eye." You should retain knowledge of geography for life. You will see greater variety in the landscape, not because there is any more variety there but because you will have been trained to observe Earth differently and with deeper understanding.

## Define & Recall

geography
spatial science
holistic approach
human geography
region
regional geography
physical geography
absolute location
relative location
spatial distribution
spatial pattern
spatial interaction
system
variable
Earth system
atmosphere
lithosphere
hydrosphere
biosphere
environment
ecology
ecosystem
life-support system
natural resource
pollution
model
physical model
pictorial/graphic model
mathematical/statistical model
conceptual model
mental map
systems theory
subsystem
input
output
closed system
open system
equilibrium
dynamic equilibrium
feedback
negative feedback
positive feedback
threshold
feedback loop

## Discuss & Review

1. What does a holistic approach mean in terms of thinking about an environmental problem?
2. Why can geography be considered both a physical and a social science? What are some of the subfields of physical geography, and what do geographers study in those areas of specialization?
3. How do geography's three major perspectives make it unique among the sciences?
4. Why is geography known as the spatial science? What are some topics that illustrate the role of geography as the spatial science?
5. What are the four major divisions of the Earth system, and how do the divisions interact with one another?
6. How does the study of systems relate to the role of geography as a physical science?
7. How does the examination of human–environment relationships in ecosystems serve to illustrate the role of geography as an environmental science?
8. What is meant by the human–environment equation? Why is the equation falling further out of balance?
9. How do open and closed systems differ? How does feedback affect the dynamic equilibrium of a system?
10. How does negative feedback maintain a tendency toward balance in a system? What is a threshold in a system?

## Consider & Respond

1. Give examples from your local area that demonstrate each of the five topics listed concerning spatial science.
2. How have various kinds of pollution affected your life? List some potential sources of pollution in your city or town.
3. Give one example of an ecosystem in your local area that has been affected by human activity. In your opinion, was the change good or bad? What values are you using in making such a judgment?
4. There are advantages and disadvantages to the use of models and the study of systems by scientists. List and compare the advantages and disadvantages from the point of view of a physical geographer.
5. How can a knowledge of physical geography be of value to you now and in the future? What steps should you take if you wish to seek employment as a physical geographer? What advantages might you have when applying for a job?

# CHAPTER 2

The physical environment of Cape Town, South Africa. Satellite imagery and data were computer enhanced to produce this scene. NASA Jet Propulsion Laboratory

# Representations of Earth

**PHYSICAL Geography Now™** This icon, appearing throughout the book, indicates an opportunity to explore interactive tutorials, animations, or practice problems at now.brookscole.com/gabler8

Knowing where something is located and being able to convey that information to others is essential to geographers as they describe and analyze aspects of the Earth system. Although many basic principles used in dealing with locational problems have been known for centuries, the technologies applied to these tasks are ever changing. Today, the tools that geographers and other scientists use to find, describe, record, and display locations are becoming more precise and accurate through the application of computers and space-age technologies.

# CHAPTER PREVIEW

**Maps and other graphic representations of Earth are essential to understanding geography.**
- In what ways are maps important in daily life or in the workplace?
- Why are geographers major users of maps?
- Why are maps an important means of communication?

**The Earth's shape is generally referred to as spherical.**
- How does the shape of Earth deviate from a sphere? Why and by how much?
- Why is a world globe the only representation of the entire Earth with little or no distortion?
- What does this statement imply about maps?

**The geographic grid is an arbitrary coordinate system for describing locations on Earth.**
- In what ways is it an arbitrary system of location?
- How was the geographic grid developed?
- How is the Earth grid associated with navigation and time zones?
- How is the Public Lands Survey System different from and similar to the geographic grid?

**All maps, projections, and images of Earth have beneficial and detrimental properties that affect the accuracy of the spatial or environmental information they portray.**
- What does this mean?
- What does this suggest in relation to the development of a "perfect" map?
- Why should representations of Earth be carefully selected to fit an intended purpose?

**Maps, global positioning systems, and remotely sensed images are useful tools for the physical geographer.**
- Why? What are they?
- How can maps, aerial photographs, and remotely sensed images provide complementary information about a place?

**Geographic information systems (GIS) allow the direct comparison and combination of many map information layers.**
- How can GIS help us understand spatial relationships in complex environmental systems?
- Why is it useful to be able to simultaneously compare the locations and distributions of two or more environmental variables?

# Location on Earth

Perhaps as soon as people began to communicate with language, they also began to develop a language of location, using landscape features as directional cues. Today, we still use familiar landmarks to help us find our way. When ancient peoples began to sail the ocean, they recognized the need for ways of finding directions and describing locations. Long before the first compass was developed, they discovered that positions of the sun and stars—rising, setting, or circling in the heavens—could provide accurate directions. Observing relationships between the sun and the stars to a position on Earth is a basic skill in **navigation,** the science of location and wayfinding. Navigation has been called the process of getting from where you are to where you want to go.

## Maps and Mapmaking

No one knows where or when the first map was made because its origin is lost in antiquity. Early humans certainly drew locational diagrams on rock surfaces and in the soil. Some of the earliest known maps were constructed of sticks or drawn on clay tablets, stone slabs, metal plates, papyrus, linen, or silk. Ancient maps were fundamental to the beginnings of geography because they helped humans communicate spatial thinking and were useful in finding directions ( Fig. 2.1).

Maps and globes fulfill the same functions today by conveying spatial information through graphic symbols, a "language of location," that must be understood to comprehend the rich store of information that they display. Although we typically think of maps as being visual representations of Earth or a part of its surface, maps and globes have now been made to show extraterrestrial features such as the surface of the moon or Mars.

**Cartography** is the science and profession of mapmaking. Geographers who specialize in cartography supervise the development of maps and globes to ensure that mapped information and data are accurate and effectively presented. Most cartographers would agree that the primary purpose of a map is to communicate spatial information. In recent years, computer technology has revolutionized cartography.

> The changes in map data collection and display that have occurred in the 20th century are comparable to the change from pedestrian to astronaut. Information that used to be collected little by little from ground observations can now be collected instantly by satellites hurtling through space, and recorded data can be flashed back to Earth at the speed of light.
>
> Cartographers can now gather spatial data and make maps faster than ever before—within hours—and the accuracy of these maps is excellent. Moreover, digital mapping enables mapmakers to experiment with a map's basic characteristics (for example, scale or projections), to combine and manipulate map data, to transmit entire maps electronically, and to produce unique maps on demand.
>
> United States Geological Survey (USGS)
> *Exploring Maps,* page 1

Maps are everywhere. We can all think of applications in navigation, political science, community planning, surveying, history, meteorology, and geology, in which maps are vital. We experience maps in our everyday lives, through education, travel, television, recreation, and reading. The maps in your daily newspaper provide excellent examples. How do they contribute to your understanding of the news? How many are there? How many would that equal in a year (365 daily papers)?

## Size and Shape of Earth

Although it was not until the 1960s that we were able to image Earth's shape from space, as early as in 540 B.C. ancient Greeks theorized that our planet was a sphere. In 200 B.C. a philosopher-geographer named Eratosthenes estimated the circumference of Earth within a few hundred miles of its actual size. Earth can generally be considered a sphere, with an equatorial circumference of 39,840 kilometers (24,900 mi), but the forces associated with Earth *rotation* bulge

**FIGURE 2.1**
When did humans make the first map? Cave paintings by Cro-Magnon people in France depict the animals that they were hunting, sometime between 17,000 and 35,000 years ago. Although this view shows detail of stags crossing a river, experts suggest that some of the artwork represents a rudimentary map. The paintings include lines that apparently represent migration routes, and other marks appear to represent locational information. If so, this is the earliest known example of humans recording their spatial knowledge.
**Why would these prehistoric humans want to record locational information?**

the equatorial region outward, and slightly flatten the polar regions. Earth's shape is basically an **oblate spheroid,** yet its deviations from being a true sphere are relatively minor. Earth's diameter at the equator is 12,758 kilometers (7927 mi), while from pole to pole it is 12,714 kilometers (7900 mi). On a globe with a 12-inch diameter (30.5 cm), this difference of 44 kilometers (27 mi) would be about as thick as the wire in a paperclip. This deviation is less than one third of 1 percent; and while viewing Earth from space it would not be noticeable to the unaided eye ( ▌Fig. 2.2a). Nevertheless, people working in navigation, surveying, aeronautics, and cartography require precise calculations of Earth's deviations from a perfect sphere.

Landforms also cause deviations from true sphericity. Mount Everest in the Himalayas is the highest point on Earth at 8850 meters (29,035 ft) above sea level. The lowest point is the Challenger Deep, in the Mariana Trench of the Pacific Ocean southwest of Guam, at 11,033 meters (36,200 ft) below sea level. The difference between these two elevations, 19,883 meters, or just over 12 miles (19.2 km), would also be insignificant when reduced in scale on a 12-inch (30.5 cm) globe.

## Globes and Great Circles

A world globe is a nearly perfect model of our planet ( ▌Fig. 2.2b). It shows Earth's spherical shape and accurately displays spatial relationships between landforms and water bodies, comparative distances between locations, relative sizes and shapes of Earth's features, and true compass directions. Having essentially the same shape as Earth, a globe represents geographic features and relationships associated with our planet virtually without distortion.

Yet, globes also have limitations. A world globe would not help us find our way on a hiking trail. It would be awkward to carry, and our location would appear as a tiny pinpoint, with little, if any, local information. We would need a *map* that clearly showed elevations, trails, and rivers and that could be folded to carry in a pocket or pack. However, if we want to view the entire world, a globe provides the most accurate representation, despite its limitations. Being familiar with the characteristics of a globe helps us understand maps and how they are constructed.

An imaginary circle drawn in any direction around its surface and whose plane passes through the center of Earth is called a **great circle** ( ▌Fig. 2.3a, b). It is "great" because this is the largest circle that can be drawn around Earth through two particular points. Great circles have several useful characteristics: (1) Every great circle divides Earth into equal halves called **hemispheres;** (2) every great circle is a circumference of Earth; and, perhaps most important, (3) great circles mark the shortest travel routes between locations on Earth's surface. An important example of a great circle is the *circle of illumination,* which divides Earth into light and dark halves—a day hemisphere and a night hemisphere. Circles whose planes do not pass through the center of Earth are called **small circles** ( ▌Fig. 2.3c). Any circle on Earth's surface that does not divide the planet into equal halves is a small circle.

▌**FIGURE 2.2**
(a) Earth as photographed from space by Apollo astronauts, showing most of Africa, the surrounding oceans, storm systems in the Southern Hemisphere, and the relative thinness of the atmosphere.
(b) A world globe oriented to represent the same general viewpoint.

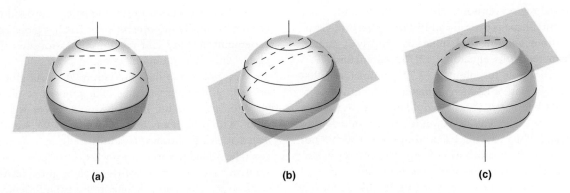

**FIGURE 2.3**
Geometric planes slicing through a globe's center (a) at the equator and (b) obliquely. In either case, the globe is divided into equal halves, and the line where the plane intersects the surface of the globe is a great circle with the same circumference as the globe itself. In (c), the plane slices the globe into unequal parts. The line of intersection of such a plane with the globe is a small circle.

The shortest route between two places can be located by finding the great circle that connects them. Put a rubber band (or string) around a globe to visualize this spatial relationship. Connect any two cities, such as Moscow and New York, San Francisco and Tokyo, New Orleans and Paris, or Kansas City and Singapore, by stretching a rubber band around the globe so that it touches both cities and divides the globe in half. The rubber band then marks the shortest route between these two cities. Navigators chart *great circle routes* for aircraft and ships because traveling the shortest distance saves time and fuel. The farther away two points are on Earth, the larger the distance savings will be by following the great circle route that connects them.

## Latitude and Longitude

Imagine you are traveling by car and you want to visit the Football Hall of Fame in Canton, Ohio. Using the Ohio road map, you look up Canton in the map index and find that it is located at "G-6." The letter G and the number 6 meet in a box marked on the map. Scanning the area within box G-6, you locate Canton (Fig. 2.4). What you have used is a **coordinate system** of intersecting lines, a system of *grid cells* on the map. Without a locational coordinate system, it would be difficult to describe a location. The problem is deciding where the starting points should be for a grid system on a sphere. Without points of reference, either natural or arbitrary, a sphere is a geometric form that looks the same from any direction.

**Measuring Latitude** The **North Pole** and the **South Pole** provide two natural reference points because they mark the opposite positions of Earth's axis, around which it turns in 24 hours. The **equator,** halfway between the poles, forms a great circle that divides the planet into the Northern and Southern Hemispheres. The equator is designated as 0° latitude, forming the reference line for measuring **latitude**

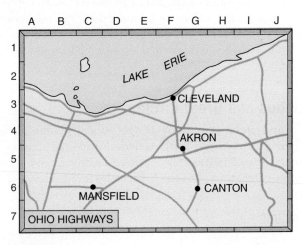

**FIGURE 2.4**
Using a simple rectangular coordinate system to locate a position. This map employs an alphanumeric location system, similar to that used on many road maps.
**What are the rectangular coordinates of Mansfield? What is at location F-3?**

in degrees north or degrees south. North or south of the equator, the angles and their arcs increase until we reach the North or South Pole at the maximum latitudes of 90° north or 90° south.

To locate the latitude of Los Angeles, imagine two lines that go outward from the center of Earth. One goes straight to Los Angeles and the other goes to a point on the equator directly south of the city. These two lines form an angle that is the latitudinal distance (in degrees) that Los Angeles lies north of the equator (Fig. 2.5a). The angle made by these two imaginary lines is just over 34°—so the latitude of Los Angeles is about 34°N (north of the equator). Because Earth's circumference is approximately 40,000 kilometers (25,000 mi) and there are 360 degrees in a circle, we can

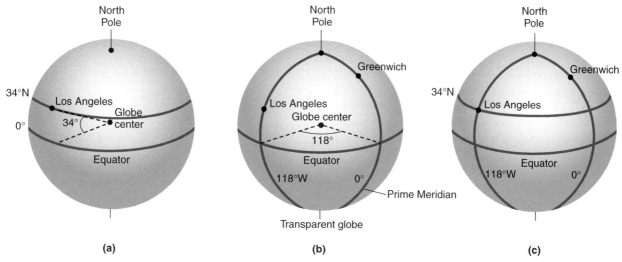

**FIGURE 2.5**
Finding a location by latitude and longitude. (a) The geometric basis for the latitude of Los Angeles, California. Latitude is the angular distance in degrees either north or south of the equator. (b) The geometric basis for the longitude of Los Angeles. Longitude is the angular distance in degrees either east or west of the prime meridian, which passes through Greenwich, England. (c) The location of Los Angeles is 34°N, 118°W.
**What is the latitude of the North Pole?**

divide (40,000 km/360°) to find that 1° of latitude equals about 111 kilometers (69 mi).

A single degree of latitude covers a relatively large distance, so degrees are further divided into minutes ( ' ) and seconds ( " ) of arc. There are 60 minutes of arc in a degree. Actually, Los Angeles is located at 34°03'N (34 degrees, 3 minutes north latitude). We can get even more precise: 1 minute is equal to 60 seconds of arc. We could locate a different position at latitude 23°34'12"S, which we would read as 23 degrees, 34 minutes, 12 seconds south latitude. A minute of latitude equals 1.85 kilometers (1.15 mi), and a second is about 31 meters (102 ft).

A **sextant** can be used to determine latitude by celestial navigation ( Fig. 2.6). This instrument measures the angle between our *horizon,* the visual boundary line between the sky and Earth, and a celestial body such as the noonday sun or the North Star (Polaris). The latitude of a location, however, is only half of its global address. Los Angeles is located approximately 34° north of the equator, but an infinite number of points exist on the same line of latitude.

**Measuring Longitude** To accurately describe the location of Los Angeles, we must also determine where it is situated along the line of 34°N latitude. However, to describe an east or west position, we must have a starting line, just as the equator provides our reference line for latitude. Actually, any half of a great circle, running from pole to pole, could serve as 0° longitude. The global position of the 0° east–west reference line for longitude is arbitrary, and in the past some countries designated their own zero line, passing through that country's

**FIGURE 2.6**
Finding latitude by celestial navigation. A traditional way to determine latitude is by measuring the angle between the horizon and a celestial body with a sextant. Today, most air and sea navigation (as well as certain land travel) is supported by a satellite-assisted technology called the global positioning system (GPS).
**With high-tech location systems like GPS available, why might understanding how to use a sextant still be important?**

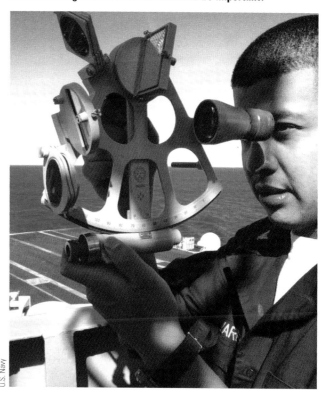

capital. This practice caused confusion because the longitudes on maps could vary with a map's country of origin. This difficulty was resolved by international agreement in 1884, when the longitude line passing through Greenwich, England (near London), was accepted as the **prime meridian,** or 0° longitude. **Longitude** is the angular distance east or west of the prime meridian.

Like latitude, longitude is also measured in degrees, minutes, and seconds. Imagine a line drawn from the center of Earth to the point where the north–south running line of longitude that passes through Los Angeles crosses the equator. A second imaginary line will go from the center of Earth to the point where the prime meridian crosses the equator (this location is 0°E or W and 0°N or S). Figure 2.5b shows that these two lines drawn from Earth's center define an angle, the arc of which is the angular distance that Los Angeles lies west of the prime meridian (118°W longitude). Figure 2.5c provides the global address of Los Angeles by latitude and longitude.

As we go farther east or west from 0° at the prime meridian, our longitude increases. Traveling eastward from the prime meridian, we will eventually be halfway around the world from Greenwich, in the middle of the Pacific Ocean at 180°E. This line is also 180°W. Longitude is measured in degrees up to a maximum of 180° east or west of the prime meridian.

# The Geographic Grid

Any point on Earth's surface can be located by its latitude north or south of the equator, measured in degrees, and its longitude east or west of the prime meridian, measured in degrees. Lines that run east and west around the globe to mark latitude and lines that run north and south from pole to pole to indicate longitude form the **geographic grid** ( Fig. 2.7).

## Parallels and Meridians

The east–west lines marking latitude circle the globe, are evenly spaced, and are parallel to the equator and each other. Hence, they are known as **parallels.** The equator is the only parallel that is a great circle; all other lines of latitude are small circles. One degree of latitude equals about 111 kilometers (69 mi) anywhere on Earth.

Lines of longitude, called **meridians,** run north and south, converge at the poles, and measure longitudinal distances east or west of the prime meridian. Each meridian of longitude, when joined with its mate on the opposite side of Earth, forms a great circle. Meridians at any given latitude are evenly spaced, although meridians get closer together as they move poleward from the equator. At the equator, meridians separated by 1° of longitude are about 111 kilometers (69 mi) apart, but at 60°N or 60°S latitude, they are only half that distance apart, about 56 kilometers (35 mi).

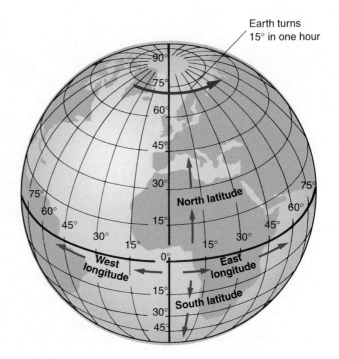

**FIGURE 2.7**
A globelike representation of Earth, which shows the geographic grid with parallels and meridians at 15° intervals.
**How do parallels and meridians differ?**

## Longitude and Time

**Time zones** were established based on the relationship between longitude and time. Until about 125 years ago, each town or area used what was known as *local time.* **Solar noon** was determined by the precise moment in a day when a vertical stake cast its shortest shadow. This meant that the sun had reached its highest angle in the sky for that day at that location—noon—and local clocks were set to that time. Because of Earth's *rotation* on its axis, noon in a town toward the east occurred earlier, and towns to the west experienced noon later.

By the late 1800s, advances in travel and communication made the use of local time by each community impractical. In 1884 the International Meridian Conference in Washington, D.C., set standardized time zones and established the longitude passing through Greenwich as the prime meridian (0° longitude). Earth was divided into 24 time zones, one for each hour of the day, because Earth turns 15° of longitude in an hour (360° divided by 24 hours). Ideally, each time zone spans 15° of longitude. The prime meridian is the *central meridian* of its time zone, and the time when solar noon occurs at the prime meridian was established as noon for all places between 7.5°E and 7.5°W of that meridian. The same pattern was followed around Earth. Every line of longitude evenly divisible by 15° is the central meridian for a time zone of 7.5° of longitude on either side. However, as shown in

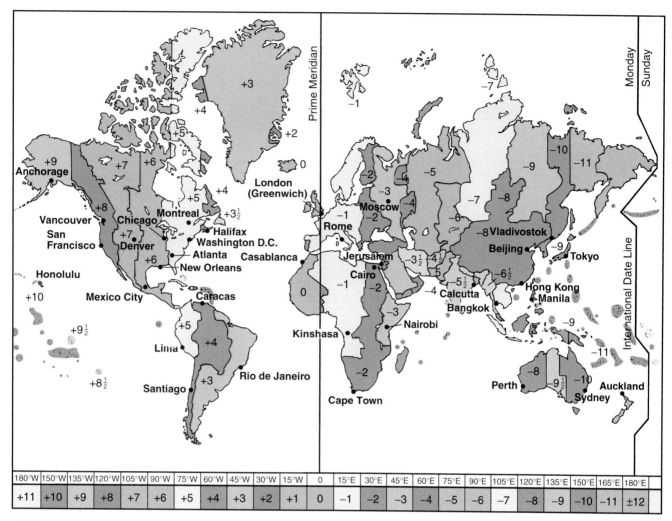

**FIGURE 2.8**
World time zones reflect the fact that Earth turns through 15° of longitude each hour. Thus, time zones are approximately 15° wide. Political boundaries usually prevent the time zones from following a meridian perfectly.
**What is the time difference between the time zone where you live and Greenwich, England?**

Figure 2.8, time zone boundaries do not follow meridians exactly. In the United States, time zone boundaries commonly follow state lines. It would be very inconvenient to have a time zone boundary dividing a city or town into two time zones, so jogs in the lines were established to avoid most of these problems. Imagine the confusion that would result if a city was divided into two time zones.

The time of day at the prime meridian, known as *Greenwich Mean Time* (sometimes called GMT, Universal Time, UTC, or Zulu Time), is used as a worldwide reference. Times to the east or west can be easily determined by comparing them to GMT. A place 90°E of the prime meridian would be 6 hours later (90°/15° per hour) while in the Pacific Time Zone of the United States and Canada, whose central meridian is 120°W, the time would be 8 hours earlier than GMT.

For navigation, longitude can be determined with a *chronometer*, an extremely accurate clock. Two chronometers are used, one set on Greenwich time, and the other on local time. The number of hours between them, earlier or later, determines longitude (1 hour equals 15° of longitude). Until the advent of electronic navigation by ground- and satellite-based systems, the sextant and chronometer were a navigator's basic tools for determining location.

## The International Date Line

On the opposite side of Earth from the prime meridian is the **International Date Line.** It is a line that generally follows the 180th meridian, except for jogs to separate Alaska and Siberia and to skirt some Pacific island groups

**FIGURE 2.9**
The International Date Line. The new day officially begins at the International Date Line (IDL) and then sweeps westward around the Earth to disappear when it again reaches the IDL. Thus, west of the line is always a day later than east of the line. Maps and globes often have either "Monday | Sunday" or "M | S" shown on opposite sides of the line to indicate the direction of the day change.
**Why does the International Date Line deviate from the 180° meridian in some places?**

substitute the current day for Monday or Sunday, and use the same relationship.

The International Date Line was not established officially until the 1880s, but the need for such a line on Earth to adjust the day was inadvertently discovered by Magellan's crew who, from 1519 to 1521, first circumnavigated Earth. Sailing westward from Spain around the world, the crew noticed when they returned that one day had apparently been missed in the ship's log. What had actually happened is that in going around the world in a westward direction, the crew had experienced one less sunset and one less sunrise than had occurred in Spain during their absence.

## The U.S. Public Lands Survey System

The geographic grid of longitude and latitude was designed to locate the *points* where these lines intersect. A different system is used in much of the United States to define and locate land *areas*. This is the **U.S. Public Lands Survey System,** or the *Township and Range System,* developed for parceling public lands west of Pennsylvania. The Township and Range System divides land areas into parcels based on north–south lines called **principal meridians** and east–west lines called **base lines** ( Fig. 2.10). Base lines were surveyed along parallels of latitude. The north–south meridians, though perpendicular to the base lines, had to be adjusted (jogged) along their length to counteract Earth's curvature. If adjustments were not made, the north–south lines would tend to converge and land parcels defined by this system would be smaller in northern regions of the United States.

The Township and Range System forms a grid of nearly square parcels called townships laid out in horizontal *tiers* north and south of the base lines and in vertical *columns*

( Fig. 2.9). At the International Date Line, we turn our calendar back a full day if we are traveling east and a full day forward if we are traveling west. Thus, if we are going east from Tokyo to San Francisco and it is 4:30 p.m. Monday just before we cross the International Date Line, it will be 4:30 p.m. Sunday on the other side. If we are traveling west from Alaska to Siberia and it is 10:00 a.m. Wednesday when we reach the International Date Line, it will be 10:00 a.m. Thursday once we cross it. As a way of remembering this relationship, many world maps and globes have Monday and Sunday (M | S) labeled in that order on the opposite sides of the International Date Line. To find the correct day, you just

**FIGURE 2.10**
Principal meridians and base lines of the U.S. Public Lands Survey System (Township and Range System).
**Why wasn't the Township and Range System applied throughout the eastern United States?**

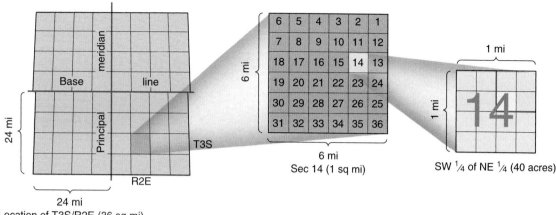

**FIGURE 2.11**
Method of location according to the Public Lands Survey System.
**How would you describe the extreme southeastern 40 acres of section 20 in the middle diagram?**

ranging east and west of the principal meridians. A **township** is a square plot 6 miles on a side (36 sq mi, or 93 sq km). As illustrated in Figure 2.11, townships are first labeled by their position north or south of a base line; thus, a township in the third tier south of a base line will be labeled Township 3 South, which is abbreviated T3S. However, we must also name a township according to its *range*—its location east or west of the principal meridian for the survey area. Thus, if Township 3 South is in the second range east of the principal meridian, its full location can be given as T3S/R2E (Range 2 East).

The Public Lands Survey System divides townships into 36 **sections** of 1 square mile, or 640 acres (2.6 sq km, or 259 ha). Sections are designated by numbers from 1 to 36 beginning in the northeasternmost section with section 1, snaking back and forth across the township, and ending in the southeast corner with section 36. Sections are divided into four *quarter sections*, named by their location within the section—northeast, northwest, southeast, and southwest, each with 160 acres (65 ha). Quarter sections are also subdivided into four *quarter-quarter sections*, sometimes known as *forties*, each with an area of 40 acres (16.25 ha). These quarter quarter-sections, or 40-acre plots, are named after their position in the quarter: the northeast, northwest, southeast, and southwest forties. Thus, we can describe the location of the 40-acre tract that is shaded in Figure 2.11 as being in the SW ¼ of the NE ¼ of Sec. 14, T3S/R2E, which we can find if we locate the principal meridian and the base line. The order is consistent from smaller division to larger, and township location is always listed before range (T3S/R2E).

The Township and Range system has exerted an enormous influence on landscapes in many areas of the United States and gives most of the Midwest and West a checkerboard appearance from the air or from space ( Fig. 2.12). Road maps in states that use this survey system strongly reflect its grid, and many roads follow the regular and angular boundaries between square parcels of land.

## The Global Positioning System

A modern technology for determining a location employs the **global positioning system (GPS).** The GPS uses a network of satellites orbiting 17,700 kilometers (11,000 mi) above Earth. This high-tech locational system was originally created for military applications but also has a host of civilian uses, from surveying to navigation. GPS can measure precise distances from a location on Earth to several orbiting satellites. The satellites transmit radio signals that are detected by a ground receiver. GPS receivers vary in size, and handheld units are common ( Fig. 2.13). The distance from receiver to satellite is calculated by measuring the time it takes for radio signals, traveling at the speed of light, to arrive. A GPS receiver performs these calculations and displays a locational readout in latitude, longitude, and elevation. Many receivers also display a digital map that marks the position of the receiver. GPS is based on the principle of triangulation. In other words, knowing the distance to our location, measured from three or more different satellites will determine our position ( Fig. 2.14). With sophisticated GPS equipment and techniques, it is possible to find locational coordinates within small fractions of a meter.

# Maps and Map Projections

It has been said that if a picture is worth a thousand words, then a map is worth a million. Maps can be reproduced easily, can depict the entire Earth or show a small area in great detail, are easy to handle and transport, and can be displayed on a computer monitor. There are many different varieties of

**FIGURE 2.12**
Rectangular field patterns resulting from the Public Lands Survey System in the western United States. Note the slight jog in the field pattern to the right of the farm buildings in the center of the photo.
**How do you know this photo was not taken in the Midwestern United States?**

**FIGURE 2.13**
Global positioning system (GPS). A GPS receiver provides a readout of its latitudinal and longitudinal position based on signals from a satellite network. Handheld units provide an accuracy that is acceptable for many uses. Precise land surveying, however, requires equipment and techniques that are more sophisticated.
**What uses can you think of for a small unit like this that displays its longitude, latitude, and elevation as it moves from place to place?**

maps, and they all have qualities that can be either advantageous or problematic, depending on the application. No map fits all possible uses, and a key to effectively using them is knowing some basic concepts concerning maps and cartography. This knowledge can help us select the right map for a particular task.

## Advantages of Maps

Maps are useful to people in every walk of life. They are valuable to tourists, political scientists, historians, geologists, pilots, soldiers, sailors, and hikers, among many others. As visual representations of Earth, maps supply an enormous amount of information that would take pages and pages to describe—probably less successfully—in words. Because they are graphic representations and use symbolic language, maps show spatial relationships with great efficiency. Imagine trying to verbally explain all the information that a map of your city, county,

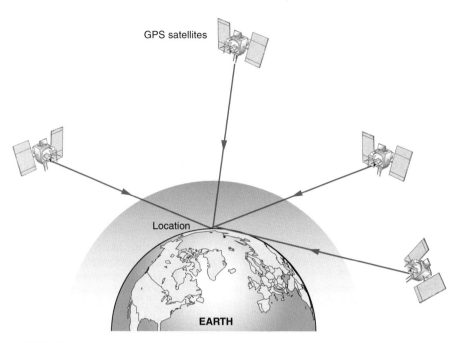

**FIGURE 2.14**
The global positioning system uses signals from a network of satellites to determine a position on Earth. The GPS receiver calculates the distances from several satellites (a minimum of three) to find its location by longitude, latitude, and elevation. With the distance from three satellites, a position can be located within meters; with more satellite signals and sophisticated GPS equipment, the position can be located very precisely.

state, or campus provides: sizes, areas, distances, directions, street patterns, railroads, bus routes, hospitals, schools, libraries, museums, highway routes, business districts, residential areas, population centers, and so forth. Maps can show true courses for navigation and true shapes of Earth features. They can be used to measure areas or distances, and they can show the best route from one place to another. The potential applications of maps are practically infinite, even "out of this world," because our space programs have produced maps of the moon, Mars, and other extraterrestrial features.

Geographers can produce maps to illustrate almost any relationship in the environment. For many reasons, whether it is presented on paper, on a computer screen, or in the mind, the map is the geographer's most important tool.

## Limitations of Maps

On a globe, we can compare the size, shape, and area of Earth features, and we can measure distance, direction, shortest routes, and true directions. Yet, because of the distortion inherent in maps, we can never compare or measure all these properties on a single map. It is impossible to present a spherical planet on a flat (two-dimensional) surface and accurately maintain all of its geometric properties. This process has been likened to trying to flatten out an eggshell. There is no such thing as a perfect map, at least not the perfect map for all uses.

Distortion is an unavoidable problem of representing a sphere on a flat map, but when a map depicts only a small area, the distortion should be negligible. If we use a map of a state park for hiking, this distortion will be too small to affect us. On maps that show large regions or the world, Earth's curvature causes apparent and pronounced distortion. To be skilled map users, we must know which properties a certain map depicts accurately, which features it distorts, and for what purpose a map is best suited. If we are aware of these map characteristics, we can make accurate comparisons and measurements on maps and better understand the information that the map conveys.

## Properties of Map Projections

The geometry of the geographic grid has four important properties: (1) Parallels of latitude are always parallel, (2) parallels are evenly spaced, (3) meridians of longitude converge at the poles, and (4) meridians and parallels always cross at right angles. There are thousands of ways to transfer a spherical grid onto a flat surface to make a **map projection,** but no map projection can maintain all four of these properties at once. Because it is impossible to have all these properties on the same map, cartographers must decide which properties to preserve at the expense of others. Closely examining a map's grid system for how these four properties are affected will help us discover areas of greatest and least distortion.

Although maps are not actually made this way, certain projections can be demonstrated by putting a light inside a transparent globe so that the grid lines are projected onto a plane or flat surface (*planar projection*), a cylinder (*cylindrical projection*), or a cone (*conic projection*), geometric forms that can be cut and then flattened out ( ▮ Fig. 2.15a–c). Today, map projections are developed mathematically, using computers to fit the geographic grid to a surface. The distortions in the geographic grid that are required to make a map can affect the geometry of several characteristics of the areas and features that a map portrays.

**Shape** Flat maps cannot depict large Earth features without distorting either their shape or their area. However, areas—continents, regions, mountain ranges, lakes, islands, and bays—can be depicted in their true shape by using the proper map projection. Maps that maintain the true shapes of areas are known as **conformal maps.** To preserve the shapes of Earth features on a conformal map, meridians and parallels always cross at right angles just as they do on the globe.

(a) Planar projection

(b) Cylindrical projection

(c) Conical projection

**▮ FIGURE 2.15**
The theory behind the development of (a) planar, (b) cylindrical, and (c) conic projections. Although projections are not actually produced this way, they can be demonstrated by projecting light from a transparent globe.
**Why do we use different map projections?**

generations of students to believe incorrectly that Greenland is as large as South America. On Mercator's projection, Greenland is shown as being about equal in size to South America (see again Fig. 2.16), but South America is actually about eight times larger.

**Area** Cartographers are able to create a world map that maintains true-area relationships; that is, areas on the map have the same proportions to each other as they have in reality. Thus, if we cover any two parts of the map with, say, a nickel, no matter where the nickel is placed it will cover equivalent areas on Earth. Maps drawn with this property, called **equal-area maps,** should be used if comparisons are being made between two or more areas shown on the map. The property of equal area is also essential when examining spatial distributions. As long as the map displays equal area and a symbol represents the same quantity throughout the map,

**▮ FIGURE 2.16**
The Mercator projection, often misused as a general-purpose world map, was designed for navigation. Its most useful property is that lines of constant compass heading, called rhumb lines, are straight lines. The Mercator is developed from a cylindrical projection.
**Compare the sizes of Greenland and South America on this map to their proportional sizes on a globe. Is the distortion great or small?**

Most of us are familiar with the **Mercator projection** (▮ Fig. 2.16), commonly used in schools and textbooks, although less so in recent years. The Mercator projection presents the correct shape of areas, so it is a conformal map, but areas away from the equator are exaggerated in size. Because of its widespread use, the Mercator projection's distortions led

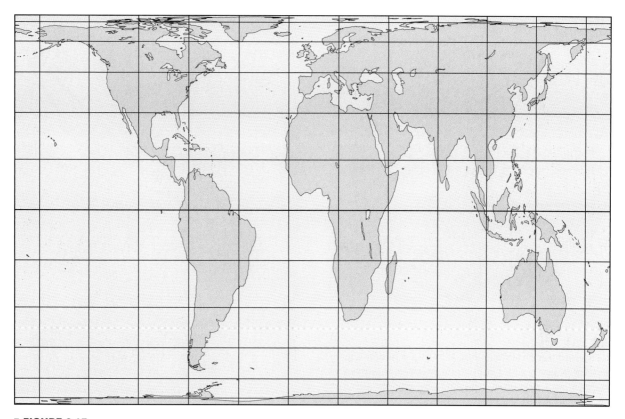

**FIGURE 2.17**
An equal-area world projection map. This map preserves area relationships but distorts the shape of landmasses.

we can get a good idea of the distribution of any feature—for example, people, churches, cornfields, hog farms, or volcanoes. However, equal-area maps distort the shapes of map features ( Fig. 2.17) because it is impossible to show both equal areas and correct shapes on the same map.

**Distance** No map can maintain a constant distance scale over Earth's entire surface. The scale on a map that depicts a large area cannot be applied equally everywhere on that map. On maps of small areas, however, distance distortions will be minor, and the accuracy will usually be sufficient for most purposes. It is possible for a map to have the property of **equidistance** in specific instances. That is, on a world map, the equator may have equidistance (a constant scale) along its length, or all meridians, but not the parallels, may have equidistance. On another map, all straight lines drawn from the center may have equidistance, but the scale will not be constant unless lines are drawn from the center.

**Direction** Because the compass directions on Earth curve around the sphere, not all flat maps can show true compass directions as straight lines. A given map may be able to show true north, south, east, and west, but the directions between those points may not be accurate in terms of the angle between them. So, if we are sailing toward an island, its location may be shown correctly according to its longitude and latitude, but the direction in which we must sail to get there may not be accurately displayed, and we may pass right by it. Maps that show true directions as straight lines are called **azimuthal map** projections. These are drawn with a central focus, and all straight lines drawn from that center are true compass directions ( Fig. 2.18).

## Examples of Map Projections

All map projections maintain one aspect of Earth—the property of location. Every place shown on a map must be in its proper location with respect to latitude and longitude. No matter how the arrangement of the global grid is changed by projecting it onto a flat surface, all places must still have their proper location—they must display their accurate latitude and longitude.

**The Mercator Projection** As previously mentioned, one of the best-known world maps is the Mercator projection, a mathematically adjusted cylindrical projection (see again Fig. 2.15b). Meridians appear as parallel lines instead of converging at the poles. Obviously, there is enormous east–west distortion of the high latitudes because the distances between meridians are stretched to the same width that they are at the

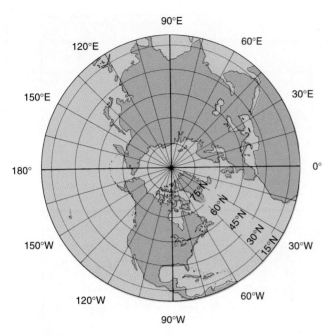

**FIGURE 2.18**
Azimuthal map centered on the North Pole. Although a polar view is the conventional orientation of such a map, it could be centered anywhere on Earth. Azimuthal maps show true directions between all points, and can only show half of Earth on a single map.

equator (see again Fig. 2.16). The spacing of parallels on a Mercator projection is also not equal, in contrast to their arrangement on a globe. The resulting grid consists of rectangles that become larger toward the poles. Obviously, this projection does not display equal area, and size distortion increases toward the poles.

Gerhardus Mercator, who devised this map in 1569, developed it to provide a property that no other world projection has. A straight line drawn anywhere on a Mercator projection is a true compass heading. A line of constant direction, called a **rhumb line,** has great value to navigators (see again Fig. 2.16). By connecting their location to where they wanted to go with a straight line on Mercator's map, navigators could follow a constant compass direction to get to their destination.

**Gnomonic Projections** **Gnomonic projections** are planar projections, made by projecting the grid lines onto a plane, or flat surface (see again Fig. 2.15a). If we put a flat sheet of paper tangent to (touching) the globe at the equator, the grid will be projected with great distortion. Despite their distortion, gnomonic projections (Figure 2.19) have a valuable characteristic: they are the only maps that display all arcs of great circles as straight lines. Navigators can draw a straight line between their location and where they want to go, and this line will be a great circle route—the shortest route between the two places.

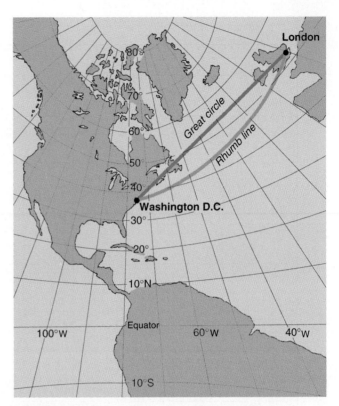

**FIGURE 2.19**
The gnomonic projection produces extreme distortion of distances, shapes, and areas. Yet it is valuable for navigation because it is the only projection that shows all great circles as straight lines. It is developed from a planar projection.
**Compare this figure with Figure 2.16. How do these two projections differ?**

An interesting relationship exists between gnomonic and Mercator projections. Great circles on the Mercator projection appear as curved lines, and rhumb lines appear straight. On the gnomonic projection the situation is reversed—great circles appear as straight lines, and rhumb lines are curves.

**Conic Projections** Conic projections are commonly used to map middle-latitude regions, such as the United States (other than Alaska and Hawaii), because they portray these latitudes with minimal distortion. In a simple conic projection, a cone is fitted over the globe with its pointed top centered over a pole (see again Fig. 2.15c). Parallels of latitude on a conic projection are concentric arcs that become smaller toward the pole, and meridians appear as straight lines radiating from the pole.

**Compromise Projections** In developing a world map, one cartographic strategy is to compromise by creating a map that shows both area and shape fairly well but is not really correct for either property. These world maps are **compromise projections** that are neither conformal nor equal area, but an effort is made to balance distortion to produce an

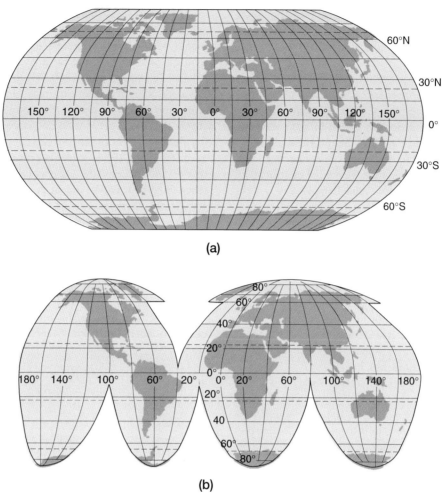

**FIGURE 2.20**
The Robinson projection (a) is considered a compromise projection because it departs from equal area to better depict the shape of the continents. Distortion in projections can be also reduced by interruption (b)—that is, by having a central meridian for each segment of the map.
**Compare the distortion of these maps with the Mercator projection (Fig. 2.16). What is a disadvantage of (b) in terms of usage?**

"accurate looking" global map (Fig. 2.20a). An *interrupted projection* can also be used to reduce the distortion of landmasses (Fig. 2.20b) by moving much of the distortion to the oceanic regions. If our interest was centered on the world ocean, however, the projection could be interrupted in the continental areas to minimize distortion of the ocean basins.

## Map Basics

Maps not only contain spatial information and data that the map was designed to illustrate but also display essential information about the map itself. This information and certain graphic features (often in the margins) are intended to facilitate understanding and using the map. Among these items are title, date, legend, scale, and direction.

**Title** A map should have a title that tells what area is depicted and what subject the map concerns. For example, a hiking and camping map for Yellowstone National Park should have a title like "Yellowstone National Park: Trails and Camp Sites." Most maps should also indicate when they were published or the date to which its information applies. For instance, a population map of the United States should tell when the census was taken, to let us know if the map information is current, or outdated, or whether the map is intended to show historical data.

**Legend** A map should also have a **legend**—a key to symbols used on the map. For example, if one dot represents 1000 people or the symbol of a pine tree represents a roadside park, the legend should explain this information. If color shading is used on the map to represent elevations, different climatic regions, or other factors, then a key to the color coding should be provided. Map symbols can be designed to represent virtually any feature (see Appendix B).

**Scale** Obviously, maps depict features smaller than they actually are. To measure size or distance, we need to know the map scale that we are using (Fig. 2.21). The **scale** of a map is an expression of the relationship between a distance on Earth and the same distance as it appears on the map. Most maps indicate the scale and to what parts of the map that scale is applicable. Knowing the map scale is essential to accurately measure distances and determine areas. Map scales can be conveyed in three basic ways.

A **verbal scale** is a statement on the map that says something like "1 inch to 1 mile" (1 inch on the map represents 1 mile on the ground), or "1 centimeter to 100 kilometers" (1 cm represents 100 km). Making a statement that includes a verbal scale tends to be the way that most of us would refer to a map scale in conversation. A verbal scale will no longer be correct if the original map is reduced or enlarged, but with verbal scales it is acceptable to use different map units (centimeters, inches) to represent another measure of true length on Earth (kilometers, miles).

A **representative fraction (RF)** scale is a ratio between a unit of distance on the map to the distance that unit represents in reality (expressed in the same units). Because a ratio is also a fraction, units of measure, being the same in the numerator and denominator, cancel each other out. An RF scale is therefore free of units of measurement and can be used with any unit of linear measurement—feet, inches, meters, or centimeters—as long as the same unit is used on both sides of the ratio. As an example, a map may have an RF scale of 1:63,360, which can also be expressed 1/63,360. This RF scale can mean that 1 inch on the map represents 63,360 inches on the ground. It also means that 1 cm on the map represents 63,360 cm on the ground. Knowing that 1 inch on the map represents 63,360 inches on the ground may be difficult to conceptualize unless we realize that 63,360 inches is the same as 1 mile. Thus, the representative fraction 1:63,360 means the map has the same scale as a map with a verbal scale of 1 inch to 1 mile.

A **graphic scale** (or **bar scale**) is useful for making distance measurements on a map. A graphic scale is a graduated line (or bar) marked with map distances that are proportional to distances on Earth. To use a graphic scale, take a straight edge, such as the edge of a piece of paper, and mark the distance between any two points on the map. Then use the graphic scale to measure the equivalent distance on Earth's surface. Graphic scales have two major advantages:

1. It is easy to determine true distances on the map, because the graphic scale can be used like a ruler to make measurements.
2. They are applicable even if the map is reduced or enlarged, because the graphic scale will also change proportionally in size. This is particularly useful because of the ease with which maps can be reproduced or copied in a reduced or enlarged scale using computers or photocopiers.

Maps are often described as being of small, medium, or large scale (Fig. 2.22). *Small-scale* maps show large areas in a relatively small size, include little detail, and have large denominators in their representative fractions. *Large-scale* maps show small areas of Earth's surface in greater detail and have smaller denominators in their representative fractions. To avoid confusion, remember that ½ is a *larger* fraction than 1/100. Maps with representative fractions larger than 1:25,000 are large scale. Medium-scale maps have representative fractions between 1:25,000 and 1:250,000. Small-scale maps have representative fractions less than 1:250,000. This classification follows the guidelines of the U.S Geological Survey (USGS), publisher of many maps for the federal government and the public.

**Direction** The orientation and geometry of the geographic grid give us an indication of direction because parallels of latitude are east–west lines and meridians of longitude run directly north-south. Many maps also have an arrow pointing to north as displayed on the map. A north arrow may indicate either *true north* or *magnetic north*—or two north arrows may be given, one for true north and one for magnetic north.

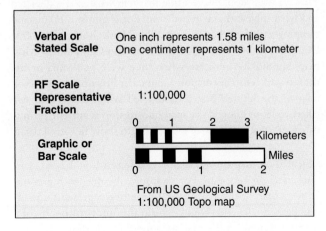

**FIGURE 2.21**
Kinds of map scales. A *verbal scale* is a statement of the relationship between a measurement on a map and the corresponding distance that it represents on the map. Verbal scales generally mix units (centimeter/kilometer or inches/mile). A *representative fraction (RF)* scale is a ratio between a distance as represented on a map (1 unit) and its actual length on the ground (here, 100,000 units). An RF scale requires that the measurements must be in the same units of measure on the map and on the ground. A *graphic scale* is a device used for making distance measurements on the map in terms of distances on the ground.

MAPS AND MAP PROJECTIONS 47

(a) 1:24,000 large-scale map

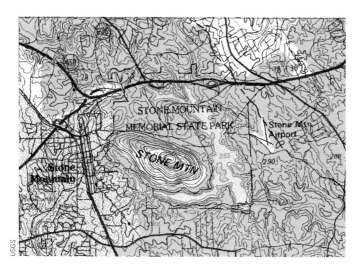

(b) 1:100,000 small-scale map

**FIGURE 2.22**
Map scales: Larger versus smaller. The designations *small scale* and *large scale* are related to a map's *representative fraction scale* (RF). Which of these maps of Stone Mountain has the smaller scale, (a) at 1:24,000 or (b) at 1:100,000? It is important to remember that an RF scale is a *fraction* that represents the proportion between a length on the map and the true distance it represents on the ground. One centimeter on the map would equal the number of centimeters in the denominator of the RF on the ground. **Which number is smaller 1/24,000 or 1/100,000? Which scale map shows more land area—the larger-scale map or the smaller-scale map?**

# Career Vision

**Robert Iantria, Physical Geographer, Terra Remote Sensing, University of Toronto, Canada**

After 2 years off to travel, I completed a community college degree and then earned a bachelor of science (BS) degree in Physical Geography from the University of Toronto. In addition, I have taken CAD, MS Project, and management courses. I liked math, that geography could be both science and humanities, and fieldwork. I love the outdoors, and opportunity to travel was very attractive. I chose geography because it was all encompassing in geosciences, and physical geography offered opportunities for fieldwork.

Two months before I graduated, I got a position with a company that was building an 80,000-mile global fiber-optic network. I was brought on board to oversee the GPS data collection, in order to locate the conduit, splice locations, and cable routing on a section of this network in southern Ontario.

I now work for Terra Remote Sensing. We specialize in marine geophysics, hydrography, photogrammetry, LIDAR, and GIS. I market technology to economic sectors that still use old ways of gathering geo-referenced data. I also research new ways for applying our technology. I travel extensively, doing presentations at conferences and to industries. People skills, presentation skills, and written and oral communication skills have helped me succeed in this position.

We use remote-sensing technology and high-resolution imagery, marketing them to utility companies. The mapping data that we provide enable these companies to run their businesses more effectively. I'm in charge of railroads, electrical transmission corridors, and floodplains. My knowledge of geomorphology, GIS, surveying, and statistics helps me provide solutions, documented via a GIS.

Things that I learned in school that helped me succeed include

- Research skills
- Writing and communication skills, especially presentations
- Statistics
- GIS and spatial theories
- Surveying theory

Statistics has everyday applications. Students interested in this field should take a GIS course and as many field courses as possible. I was able to use communication, research, and GIS skills to provide solutions that were cost effective, accurate, detailed, and met everyone's needs.

Physical geography helps people think on a broad scale and encourages them to examine the forces with which they interact, whether spatially or in business. It helped me look at processes, as opposed to cause and effect. Physical geography is the study of all physical processes on Earth and their interactions. We study features at various scales, from a sand grain, to a mountain, to an entire mountain range, to figure out why and how they have developed into what we see today and what we might see in the future. Physical geography helps us look at the big and small picture and recognize how different processes throughout time influence what we see today. In life, these skills of looking for and identifying processes and recognizing their interactions can be applied to many everyday social and business problems. The processes and the interactions may be different, but researching and studying them are the same.

My experience has been that a master's or a doctorate degree may put you outside the trainable scope that interests most employers. With a BS, you're more trainable in the employer's eyes. You're more adaptable. If you do some follow-up—specialized education and training with a BS— you're very marketable in a business environment.

Earth has a magnetic field that makes the planet act like a giant bar magnet, with a magnetic north pole and a magnetic south pole, each with opposite charges. Although the magnetic poles shift position slightly over time, they are located in the Arctic and Antarctic regions and do not coincide with the geographic poles. Aligning itself with Earth's magnetic field, the north-seeking end of a compass needle points toward the magnetic north pole. If we know the **magnetic declination,** the angular difference between magnetic north and true geographic north, we can compensate for this difference (Fig. 2.23). Thus, if our compass points north and we know that the magnetic declination for our location is 20°E, we can adjust our course knowing that our compass is pointing 20°E of true north. To do this, we should turn 20°W from the direction indicated by our compass in order to face true north. Magnetic declination varies from place to place and also changes through time. For this reason, magnetic declination maps are revised periodically, and using a recent map is very important. A map of magnetic declination is called an *isogonic map* (Fig. 2.24), and *isogonic lines* connect locations that have equal declination.

**FIGURE 2.23**
Map symbol showing true north, symbolized by a star representing Polaris (the North Star), and magnetic north, symbolized by an arrow. The example indicates 20°E magnetic declination.
**In what circumstances would we need to know the magnetic declination of our location?**

**FIGURE 2.24**
Isogonic map of North America, showing the magnetic declination that must be added (west declination) or subtracted (east declination) from a compass reading to determine true directions.
**What is the magnetic declination of your hometown to the nearest degree?**

Compass directions can be given by either the azimuth system or the bearing system. In the **azimuth** system, direction is given in degrees of a full circle (360°) clockwise from north. That is, if we imagine the 360° of a circle with north at 0° (and at 360°) and read the degrees clockwise, we can describe a direction by its number of degrees away from north. For instance, straight east would have an azimuth of 90°, and due south would be 180°. The **bearing** system divides compass directions into four quadrants of 90° (N, E, S, W), each numbered by directions in degrees away from either north or south. Using this system, an azimuth of 20° would be north, 20° east (20° east of due north), and an azimuth of 210° would be south, 30° west (30° west of due south). Both azimuths and bearings are used for mapping, surveying, and navigation for both military and civilian purposes (see Appendix B).

# Displaying Spatial Data and Information on Maps

Any information or data that vary spatially can be shown on a map. **Thematic maps** are designed to focus attention on the distribution of one feature (or a few related ones). Examples include maps of climate, vegetation, soils, earthquake epicenters, or tornadoes.

## Discrete and Continuous Data

There are two major types of spatial data, discrete and continuous (Fig. 2.25). **Discrete data** means that the phenomenon is either located at a particular place or it is not—for example, hot springs, tropical rainforests, rivers, or earthquake faults. Discrete data are represented on maps by point, area, or line symbols to show their locations and distributions (Fig. 2.25a–c). *Regions,* discrete areas that exhibit a common characteristic or set of characteristics within their boundaries, are represented by different colors or shading to differentiate one from another. This kind of representation is used to show distribution of soil, climate, and vegetation regions (see the world maps throughout this book).

**Continuous data** means that a measurable numerical value for a certain characteristic exists everywhere on Earth (or within the area of interest displayed); for example, every location on Earth has an elevation (or air pressure value or population density). The distribution of continuous data is often shown using **isolines**—lines on a map that connect points with the same numerical value (Fig. 2.25d). Isolines that we will be using later on include *isotherms,* which connect points of equal temperature; *isobars,* which connect points of equal barometric pressure; *isobaths* (also called bathymetric contours), which connect points with equal water depth; and *isohyets,* which connect points receiving equal amounts of precipitation.

### FIGURE 2.25
Discrete and continuous spatial data (variables). *Discrete variables* represent features that are present at certain locations but not found everywhere. The locations, distributions, and patterns of discrete features are of great interest in understanding spatial relationships. Discrete variables can be (a) *points* as shown by locations of large earthquakes in Hawaii (or places where lightning has struck or locations of water-pollution sources), (b) *lines* as in the path taken by Hurricane Andrew (or river channels or fault lines), (c) *areas* like the lands burned by wildfire (or clear-cuts in a forest or the area where an earthquake was felt). *Continuous* variables mean that every location possesses a measurable characteristic; for example, everywhere on Earth has an elevation, even if it is zero (at sea level) or below (a negative value). Changes in a continuous variable over an area can be represented in many ways, but isolines, shading, and colors can be used. The map (d) shows the continuous distribution of temperature variation for a day in part of eastern North America.
**Can you name other environmental examples of discrete and continuous variables?**

## Topographic Maps

**Topographic contour lines** are isolines connecting points on a map that are at the same elevation above mean sea level (or below sea level such as in Death Valley, California). For example, if we walk around a hill along the 1200-foot contour line shown on the map, we would be 1200 feet above sea level at all times, walking on a level line, and maintaining a constant elevation. Contour lines are an excellent means for showing the land surface configuration on a map (see Appendix B). The arrangement, spacing, and shapes of the contours give a map reader a mental image of what the topography (the "lay of the land") is like ( Fig. 2.26).

Figure 2.27 illustrates how contour lines show the character of the land surface. The bottom portion of the diagram is a simple contour map of an asymmetrical hill. Note that the elevation difference between adjacent contour lines on this map is 20 feet. The constant difference in elevation between adjacent contour lines is called the **contour interval.**

**FIGURE 2.26**
(Top) A view of a river valley and surrounding hills, shown on a shaded-relief diagram. Note that a river flows into a bay partly enclosed by a sand spit. The hill on the right has a rounded surface form, but the one on the left forms a cliff along the edge of an inclined but flat upland. (Bottom) The same features represented on a contour map.
**If you had only a contour map, could you visualize the terrain shown in the shaded-relief diagram?**

distance). When studying a contour map, we should understand that the slope in between contours almost always changes gradually, and it is unlikely that the land drops off in steps downslope as the contour lines might suggest.

Topographic maps show other features in addition to elevations (see Appendix B)—for instance, water bodies such as streams, lakes, rivers, and oceans or cultural features such as towns, cities, bridges, and railroads. The USGS produces topographic maps of the United States at several different scales. Some of these maps—1:24,000, 1:62,500, and 1:250,000—use English units for their contour intervals. Many recent maps are produced at scales of 1:25,000 and 1:100,000 and use metric units. Contour maps that show undersea topography are called *bathymetric charts*. In the United States, they are produced by the National Ocean Service.

## Modern Mapping Technology

Cartography has undergone a technological revolution, from slow and sometimes tedious manual methods to an automated process, using computer systems to store, analyze, and retrieve data and to draw the final map. For most mapping projects, computer systems are faster, more

What kind of terrain would we cover by hiking from point A to point B? We start from point A at sea level and immediately begin to climb. We soon cross the 20-foot contour line, then the 40-foot, the 60-foot, and, near the top of our hill, the 80-foot level. After walking over a relatively broad summit that is above 80 feet but not as high as 100 feet (or we would cross another contour line), we once again cross the 80-foot contour line, which means we must be starting down. During our descent, we cross each lower level in turn until we arrive back at sea level (point B).

In the top portion of Figure 2.27, a **profile** (side view) helps us to visualize the topography we covered in our walk. We can see why the trip up the mountain was more difficult than the trip down. Closely spaced contour lines near point A represent a steeper slope than the more widely spaced contour lines near point B. Actually, we have discovered something that is true of all isoline maps: The closer together the lines are on the map, the steeper the **gradient** (the greater the rate of change per unit of horizontal

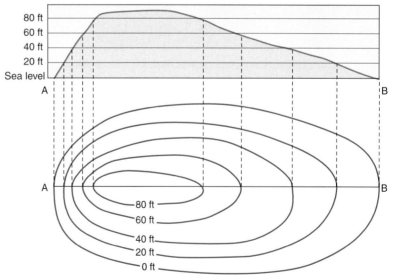

**FIGURE 2.27**
A topographic profile and contour map. Topographic contours connect points of equal elevation relative to mean sea level. The upper part of the figure shows the topographic profile (side view) of an island. Horizontal lines mark 20-foot intervals of elevation above sea level. The lower part of the figure shows how these contour lines look in map view.
**Study this figure and the maps in Figure 2.22. What is the relationship between the spacing of contour lines and steepness of slope?**

## Geography's Spatial Science Perspective
### Using Vertical Exaggeration to Portray Topography

Most maps present a landscape as if viewed from directly overhead, looking straight down. This perspective is sometimes referred to as a *map view* or *plan view* (like architectural house plans). Measurements of length and distance are accurate, as long as the area depicted is not so large that Earth's curvature becomes a major factor. Topographic maps, for example, show spatial relationships in two dimensions (length and width on the map, called $x$ and $y$ coordinates in mathematical Cartesian terms). Illustrating terrain, as represented by differences in elevation, requires some sort of symbol to display elevational data on the map. Topographic maps use contour lines, which can also be enhanced by relief shading (see the Map Interpretation, Volcanic Landforms, in Chapter 14 for an example).

For many purposes, though, a side view or an oblique view of what the terrain looks like (also called *perspective*) helps us visualize the landscape. Block diagrams, 3-D models of Earth's surface, are very useful for showing the general layout of topography from a perspective view. They provide a perspective with which most of us are familiar, similar to looking out an airplane window or from a high vantage point. Block diagrams are excellent for illustrating 3-D relationships in a landscape scene, and information about the subsurface can be included. But such diagrams are not intended for making accurate measurements, and many block diagrams represent hypothetical or stylized, rather than actual, landscapes.

A topographic *profile* illustrates the shape of a land surface as if viewed directly from the side. It is basically a graph of elevation changes over distance along a transect line. Elevation and distance information collected from a topographic map or from other elevation data in spatial form can be used to draw a topographic profile. Topographic profiles show the terrain. If the geology of the subsurface is represented as well, such profiles are called *geologic cross sections*.

Block diagrams, profiles, and cross sections are typically drawn in a manner that stretches the vertical presentation of the features being depicted. This makes mountains appear taller than they are in comparison to the landscape, the valleys deeper, the terrain more rugged, and the slopes steeper. The main reason why vertical exaggeration is used is that it helps make subtle changes in the terrain more noticeable. In addition, land surfaces are really much flatter than most people think they are. In fact, cartographers have worked with psychologists to determine what degree of vertical exaggeration makes a profile or block diagram appear most "natural" to people viewing a presentation of elevation differences in a landscape. For technical applications, most profiles and block diagrams will indicate how much the vertical presentation has been stretched, so that there is no misunderstanding. Two times vertical exaggeration means that the feature is presented two times higher than it really is, but the horizontal scale is correct. Note that the image of Capetown, in the chapter-opening photo, has two times vertical exaggeration; that is, the mountains appear to be twice as steep as they really are.

To illustrate why vertical exaggeration is used, look at the three profiles of a volcano in the Hawaiian Islands. Which do you think shows the true, natural-scale profile of this volcanic mountain? Which one "looks" the most natural to you? What is the true shape of this volcano? After making a guess, check below for the answer and the degree of vertical exaggeration in each of the three profiles. Note that this is a huge volcano—the profile extends horizontally for 100 kilometers.

---

precise, more efficient, and less expensive than the manual cartographic techniques they have replaced. However, it is still important to understand basic cartographic principles to make a good map. A computer mapping system will draw only what an operator instructs it to draw.

## Digital Mapmaking

A great advantage that computers offer is that maps can be revised without manually redrawing them. Map data displayed on a computer screen can be corrected, changed, or improved until the cartographer decides that the final map is ready to be produced. Hundreds of millions of data bits, representing elevations, depths, temperatures, or populations, can be stored in a digital database. A typical USGS topographic map has more than 100 million bits of information to be stored and thousands of bits of data to be plotted on the finished map. A database for a map may include information on coastlines, political boundaries, city locations, river systems, map projections, and coordinate systems.

The cartographer may tile together adjacent maps to cover a large area or zoom in to a small area of the map. In addition to scale changes, computer map revision allows easy metric conversion as well as changes in projections, contour intervals, symbols, colors, and directions of view (orientation). Computer map revision is essential for rapidly changing phenomena such as weather systems, air pollution, ocean currents, volcanic eruptions, and forest fires. Weather maps require continual revision.

Of particular interest to physical geographers, geologists, and engineers are **digital terrain models,** three-dimensional (3-D) views of topography ( Fig. 2.28). A 3-D model is particularly useful for visualizing topography and land surface features. Digital terrain models may be designed to show

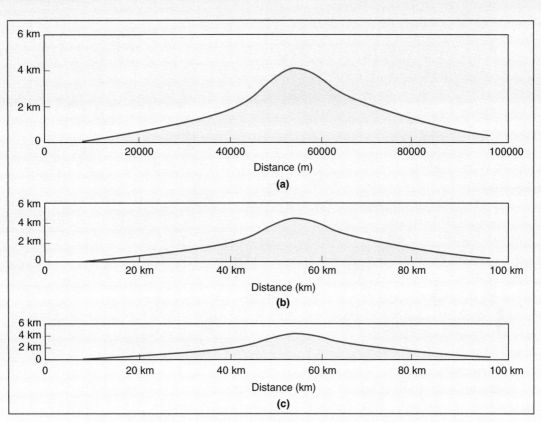

(a) 4X vertical exaggeration (b) 2X vertical exaggeration (c) natural-scale profile—no vertical exaggeration

Profiles of Mauna Kea, Hawaii (data from NASA)

**vertical exaggeration** by stretching the vertical part of the display to enhance the relief of an area. Digital elevation data can be translated into many types of terrain displays and maps, including color-scaled contour maps (in which areas between assigned contours are a certain color), conventional contour maps, and shaded relief maps.

Any factor represented by continuous data can be displayed either as a two-dimensional contour map or as a 3-D surface to enhance the visibility of its spatial variation (❚ Fig. 2.29). **Visualization** refers to the products of computer techniques used to generate three-dimensional, sometimes animated, image-models designed for illustrating and explaining complex processes and features. For example, the Earth image shown in the first chapter (Fig. 1.4) and the image at the beginning of this chapter are visualization models that present several components of the Earth system in stunning 3-D views, based on environmental data and remotely sensed images. Visualization models help us understand and conceptualize many environmental processes and features. Although the products and techniques of cartography are now very different from their beginnings and continue to be improved, the goal of making a representation of Earth remains the same—to effectively communicate geographic and spatial knowledge in a visual format.

## Geographic Information Systems

A versatile innovation in the handling of maps and spatial data is called a **geographic information system (GIS)**. A GIS is a computer-based technology that assists the user in the entry, analysis, manipulation, and display of geographic

**FIGURE 2.28**
Colorized elevation model of the Grand Canyon in Arizona. A digital terrain model is a computer-generated, 3-D view of a land surface. Many different kinds of terrain displays can be generated from digital elevation data, and the models can be rotated on a computer screen to be viewed from any angle or direction.

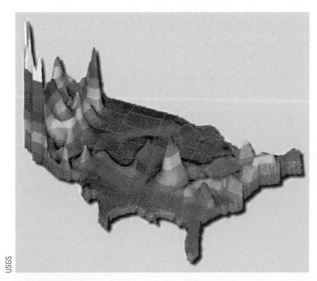

**FIGURE 2.29**
Earthquake hazard in the conterminous United States: a continuous variable displayed in 3-D perspective. Here it is easy to develop a mental map of how potential earthquake danger varies across this part of the United States.
**Other than where you might expect to find a high level of earthquake hazard, are there locations with a level of this characteristic you find surprising?**

information derived from map layers (Fig. 2.30). A GIS can also make the scale and map projection of these layers identical, thus allowing several or all layers to be integrated into new and more meaningful composite maps. GIS is especially useful to geographers because they often address problems that require large amounts of spatial data from a variety of sources.

**What a GIS Does** Imagine you are in a giant map library with thousands of maps, all of the same area but each showing a different aspect of the same place: one map shows roads, another highways, another trails, another rivers (or soils, or vegetation, or slopes, or rainfall, and on almost to infinity). Each map is at a different scale, and there are many different projections (some that do not even preserve shape or area). These factors will make it very difficult to compare the information among these different maps. You also have digital terrain models and satellite images that you would like to compare to the maps. Further, you want to be able to put some of this information together because few aspects of the environment involve only one factor or exist in spatial isolation. You have a geographic problem, and to solve that problem, you need a way to make these representations of a part of Earth directly comparable. What you need is a GIS and the knowledge of how to use this system.

**FIGURE 2.30**
Four basic subsystems of a geographic information system (GIS). (1) Input systems provide spatial data in digital form through scanning of maps and images, manual digitizing, and access of digital data files. The global positioning system can input locational coordinates directly from the field. (2) Data storage systems include disks, tapes, and hard drives. (3) Hardware and software systems for managing, accessing, analyzing, and updating the database. (4) Output systems include plotters for generating hard copies of maps and images. Access to many kinds of spatial data (maps, photographs, satellite imagery, digital elevations, and any data that can be mapped) is important in using a GIS. The technology is powerful, but people are the most important part of a GIS.
**How can computer technology help us to store, retrieve, and update maps?**

**Data and Attribute Entry** The first step in solving this problem is to enter the map and image data into a computer system. Each map is digitally entered and stored as a separate data file that represents an individual thematic map with a separate layer of information for each theme (Fig. 2.31). Another step is **geocoding,** locating all spatial data and information in relation to grid coordinates such as latitude and longitude. Further, a list of *attributes* (information) about each mapped feature can be stored and easily accessed.

**Registration and Display** Each map layer is geometrically registered to the same set of geographic coordinates. The GIS can now display any layer that you wish, stretched to fit any map projection that you specify, at a scale that you specify. The maps, images, and data sets can now be directly compared at the same size and on the same configuration of grid coordinates. It is also possible to combine any set of thematic layers that you wish. If you want to see the locations of *homes* on a *river floodplain,* a GIS can overlay these two thematic layers, giving you an instant map by retrieving, combining, and displaying the *home* and *floodplain* map layers. If you want to see *earthquake faults* and *artificially filled areas* in relation to locations of *fire stations* and *police stations,* that composite map will require four layers, but this is no problem for a GIS to display.

**FIGURE 2.31**
Geographic information systems store different information and data as individual map layers. GIS technology is widely used in geographic and environmental studies in which several different variables need to be assessed and compared spatially to solve a problem.
**Can you think of other applications for geographic information systems?**

**GIS in the Workplace** Suppose you are a geographer working for the Natural Resources Conservation Service. Your current problem is to control erosion along the banks of a newly constructed reservoir. You know that erosion is a function of many environmental variables, including soil types, slopes, vegetation characteristics, and others. Using a GIS, you would enter map data for each of these variables as a separate layer. These variables could be analyzed individually; however, by integrating information from individual layers (soils, slope, vegetation, and so on), you could identify the locations most susceptible to erosion. Your resources and personnel could then be directed toward controlling erosion in those target areas.

Many geographers are employed in career fields that apply GIS technology. The capacity of a GIS to integrate and analyze a wide variety of geographic information, from census data to landform characteristics, makes it useful to both human and physical geographers. With nearly unlimited applications in geography and other disciplines, GIS is now and will continue to be an important tool for spatial analysis.

# Remote Sensing of the Environment

**Remote sensing** is the collection of information and data about distant objects or environments. Remote sensing involves the gathering and interpretation of aerial and space imagery, images that have many maplike qualities. Using remote sensing systems, we can generate images of objects and scenes that would not be visible to humans and can display these scenes as images that we can visually interpret.

Remote sensing is commonly divided into photographic techniques (using film and cameras to record images of light) and digital imaging (which can record images from light or many other forms of energy). Photographs are made by using cameras to record a picture on film. Digital techniques use image scanners or digital cameras to produce a **digital image**—an image derived from numerical data (Fig. 2.32). Digital images consist of **pixels,** a term that is of computer-assisted data processing, image enhancement, and interpretation. A key factor in digital images is **resolution,** expressed as how small an area (on the Earth) a pixel represents—for example, 15–30 meters for a satellite image of a city or small region. Satellites that image large continental areas or an entire hemisphere at once use resolution that is much more coarse, to produce a more generalized scene (see again Fig. 2.32). Digital cameras for personal use express resolution in megapixels, or how many million pixels make an image. The more megapixels a camera or digital scanner can image, the better the resolution and the sharper the image will be (compare Fig. 2.32 to Fig 2.38).

## Aerial Photography

Aerial photographs have long provided us with important perspectives on our environment ( ▌ Fig. 2.33). Air photos and digital images may be *vertical* (looking straight down) or *oblique* (taken at an acute angle to Earth's surface). Image interpreters use aerial photographs and other kinds of imagery to examine and describe relationships among objects on Earth's surface. A device called a *stereoscope* allows overlapped pairs of images (typically aerial photos) taken from different positions to show features in three dimensions.

**Near-infrared (NIR)** energy, basically light energy at wavelengths that are too long for our eyes to see, cuts though atmospheric haze better than visual light does. Photographs and digital images that use NIR tend to provide very clear images when taken from high altitude or space. Color NIR photographs and images are sometimes referred to as "false color" pictures because, on NIR, healthy grasses, trees, and most plants will show up as bright red, rather than green ( ▌ Fig. 2.34). A widely held but incorrect notion of NIR techniques is that they image heat.

## Specialized Remote Sensing Techniques

A variety of remote sensing systems are designed for specific imaging applications. Remote sensing may use UV light, visible light, NIR light, thermal infrared energy (heat), and microwaves (radar) to produce images.

▌ **FIGURE 2.32**
Digital images beamed from satellites allow us to monitor environmental changes over time and as they occur. The spatial resolution of a digital image refers to how much area each pixel represents. In general, the smaller the area we are imaging, the greater the resolution we require. Coarse resolution is typically used in regional or global imaging (pixel inset). These images show changes in the cover of growing vegetation that occurs between (a) winter (early January) and (b) spring (late May–early June).
**What other environmental change can you see by comparing these images?**

short for "picture element," the smallest area resolved in a digital picture. A digital image is similar to a mosaic, made up of grid cells with varying colors or tones that form a picture. Each cell (pixel) has a locational address within the grid and a value that represents the brightness of the picture area that the pixel represents. The digital values in the array of grid cells are translated into an image.

Most images returned from space are digital because digital data can be easily broadcast back to Earth. Digital imagery taken by a camera or scanner also offers the advantage

(a) (b)

**FIGURE 2.33**
(a) Oblique photos provide a "natural view," like looking out of an airplane window. This oblique aerial photograph in natural color shows farmland, countryside, and forest. (b) Vertical photos provide a maplike view that is more useful for mapping and making measurements (as in this view of Tampa Bay, Florida).
**What are the benefits of an oblique view, compared to a vertical view?**

(a) Natural color           (b) Near-infrared (red tones)

**FIGURE 2.34**
A comparison of a natural color photograph (a) to the same scene in false color near-infrared (b). Red tones indicate vegetation; dark blue—clear, deep water; and light blue—shallow or muddy water. This is a wetlands area on the coast of Louisiana.
**If you were asked to make a map of vegetation or water features, which image would you prefer to use and why?**

**Thermal infrared (TIR)** images show patterns of heat energy and can be taken either day or night by TIR sensors. Spatially recorded heat data are digitally converted into a visual image. TIR images record contrasts in temperature, whether they are cold or hot. How well an object will show up on a thermal image depends on how different its temperature is from that of its surroundings. Hot objects show up in light tones, and cool objects will be dark, but a computer may be used to colorize the gray tones to emphasize heat differences. Some applications include finding leaks in pipelines, detecting thermal pollution, finding volcanic hot spots and geothermal sites, finding leaks in building insulation, and locating forest fires through dense smoke.

Weather satellites also use TIR to image atmospheric conditions. We have all seen these images on television when the meteorologist says, "Let's see what the weather satellite shows." Clouds are depicted in black on the original thermal image because they are colder than their background, the surface of Earth below. Because we don't like to see black clouds, the image tones are reversed, like a photo negative, so that the clouds appear white. These images may also be colorized to show cloud heights (Fig. 2.35).

**Radar** (*RAdio Detection And Ranging*) transmits radio waves and produces an image by reading the energy signals that are reflected back. Radar systems can operate day or night and can see through clouds.

# REMOTE SENSING OF THE ENVIRONMENT

**Sonar** (*SO*und *NA*vigation and *R*anging) uses the reflection of emitted sound waves to probe the ocean depths. Much of our understanding of sea floor topography, and mapping of the sea floor, has been a result of sonar applications.

## Multispectral Remote Sensing Applications

**Multispectral remote sensing** means using and comparing more than one type of image of the same place, whether taken from space or not (for example, radar and TIR images, or NIR and normal photos). Common on satellites, *multispectral scanners* produce digital images by sensing many kinds of energy simultaneously but in separate data files that are relayed to receiving stations.

■ **FIGURE 2.35**
Thermal infrared images such as this one show heat differences with gray tones (darker meaning warmer) or with colors. These images are like a "map" of heat (or cold) patterns. This is North America as beamed back from a U.S. weather satellite called GOES (Geostationary Operational Environmental Satellite). The whiter the cloud tops, the colder (and higher) they are, but here the highest cloud tops have been colorized for emphasis (the darker the colors, the higher the cloud tops).
**What other details can you see in this heat image? The season is spring (April 16).**

There are several kinds of **imaging radar** systems that sense the ground (topography, rock, water, ice, sand dunes, and so forth) by converting radar reflections into a maplike image. **Side-Looking Airborne Radar (SLAR)** was designed to image areas located to the side of an aircraft. Other imaging radar systems are mounted on satellites and the space shuttle. Imaging radar generally does not "see" trees (depending on the system), so it makes an image of the land surface rather than a crown of trees. SLAR is excellent for mapping terrain and water features (■ Fig. 2.36) and is used most often to map remote, inhospitable, inaccessible, cloudy, or heavily forested regions.

Radar is also used to monitor and track thunderstorms, hurricanes, and tornadoes (■ Fig. 2.37). **Weather radar** systems produce maplike images of precipitation. Radar penetrates clouds (day or night) but reflects back off of raindrops and other precipitation, producing a signal on the radar screen. Precipitation patterns are typically the kind of weather radar image that we see on television. The latest systems include *Doppler radar*, which can determine precipitation patterns, direction of movement, and how fast a storm is approaching (much as police radar measures vehicle speed).

■ **FIGURE 2.36**
Imaging radar reflections produce an image of a landscape. Radar reflections are affected by many factors, particularly the surface materials, as well as steepness and orientation of the terrain. This radar scene taken from Earth orbit, shows the topography near Sunbury, Pennsylvania, where the West Branch River flows into the Susquehanna River. Parallel ridges, separated by linear valleys, form the Appalachian Mountains in Pennsylvania. River bridges provide a sense of scale, and north is at the top of the image.

# Geography's Physical Science Perspective
## Polar versus Geostationary Satellite Orbits

Satellite systems that return images from orbit are designed to produce many different kinds of Earth imagery. Some of these differences are related to the type of orbit the satellite system is using while scanning the surface. There are two distinctively different types of orbits: the *polar orbit* and the *geostationary orbit* (sometimes called a *geosynchronous orbit*), each with a different purpose.

The polar orbit was developed first; as its name implies, the satellite orbits Earth from pole to pole. This orbit has some distinct advantages. It is typically a low orbit for a satellite, usually varying in altitude from 700 kilometers (435 mi) to 900 kilometers (560 mi). At this height, but also depending on the equipment used, a polar-orbiting satellite can produce clear, close-up images of Earth. However, at this distance the satellite must move at a fast orbital velocity to overcome the gravitational pull of Earth. This velocity can vary, but for polar orbiters it averages around 27,400 kilometers/hour (17,000 mph), traveling completely around Earth in about 90 minutes. While the satellite orbits from pole to pole, Earth rotates on its axis below, so each orbit views a different path along the surface. Thus, polar orbits will at times cover the dark side of the planet. To adjust for this, a slightly modified polar orbit was developed, called a *sun synchronous orbit*. If the polar orbit is tilted a few degrees off the vertical, then it can remain over the sunlit part of the globe at all times. Most modern polar-orbiting satellites are sun synchronous (a near-polar orbit).

The geostationary orbit, developed later, offered some innovations in satellite image gathering. A geostationary orbit must have three characteristics: (1) It must move in the same direction as Earth's rotation. (2) It must orbit exactly over the equator. (3) The orbit must be perfectly circular. The altitude of the orbit must be also exact, at 35,900 kilometers (22,300 mi). At this greater height, the orbital velocity is less than that for a polar orbit—11,120 kilometers/hour (6900 mph). When these conditions are met, the satellite's orbit is perfectly synchronized with Earth's rotation, and the satellite is always located over the same spot above Earth. This orbit offers some advantages. First, at its great distance, a geostationary satellite can view an entire Earth hemisphere in one image (that is, the half it is always facing—a companion satellite images the other hemisphere). Another great advantage is that geostationary satellites can send back a continuous stream of images for monitoring changes in our atmosphere and oceans. A film loop of successive geostationary images is what we see on TV weather broadcasts when we see motion in the atmosphere. Geostationary satellite images give us broad regional presentations of an entire hemisphere at once. Near polar–orbiting satellites take image after image in a swath and rely on Earth's rotation to cover much of the planet, over a timespan of about a week and a half.

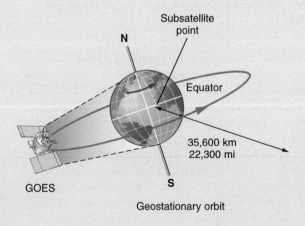

*Polar orbits* circle Earth approximately from pole to pole and use the movement of Earth as it turns on its axis to image small areas (perhaps 100 X 100 kilometers) to gain good detail of the surface. This orbital technique yields nearly full Earth coverage in a mosaic of images, and the satellite travels over the same region every few days, always at the same local time. (Not to scale.)

*Geostationary orbits* are used with satellites orbiting above the equator at a speed that is synchronized with Earth rotation so that the satellite can image the same location continuously. Many weather satellites use this orbit at a height that will permit imaging an entire hemisphere of Earth. (Not to scale.)

### FIGURE 2.37
NEXRAD radar image of thunderstorms associated with a storm front shows detail of the storm (124-nautical-mile circle). Blue represents clouds with no precipitation (except in the very center, where blue indicates stray reflections called ground clutter). Other colors show rainfall intensity: green—light rainfall, yellow—moderate, and orange-red—heavy. The storm is moving to the east-southeast.

**How are weather and imaging radar scenes different in terms of what they record about the environment?**

of the Earth system. Remote sensing provides rapid, worldwide coverage of environmental conditions. Through continuous monitoring of the Earth system, global, regional, and even local changes can be detected and mapped. Geographic information systems offer the ability to match and combine thematic layers of any sort, instantly accessing any combination of these that we need to solve complex spatial problems.

Maps and various kinds of representations of Earth continue to be essential tools for geographers and other Earth scientists whether they are on paper, displayed on a computer monitor, or stored as a mental image. Digital mapping, GPS, GIS, and remote sensing have revolutionized the field of geography, but the principles concerning maps and cartography remain basically unchanged.

Each part of the energy spectrum yields different information about aspects of the environment.

Many types of images can be generated from multispectral data, but the most familiar is the **color composite image** (Fig. 2.38). Blending three images of the same location, by overlaying pixel data from three different wavelengths, creates a digital color composite. The most common color composite image resembles a false-color NIR photograph, with the same color assignment as color NIR photos. On a standard NIR color composite, red is healthy vegetation; barren areas show up as white or brown; clear, deep water bodies are dark blue; and muddy water appears light blue. Clouds and snow are bright white. Urbanized areas are blue-gray. Although the colors are visually important, the greatest benefit of digital multispectral imagery is that computers can be used to identify, classify, and map (in a first approximation) these kinds of areas automatically. These digital images can be input as thematic layers for integration into a GIS, and geographic information systems will play a strong role in this kind of analysis.

The use of digital technologies in mapping and imaging our planet and its features continues to provide us with data and information that contribute to our understanding

### FIGURE 2.38
This satellite scene of San Francisco Bay and environs is a near-infrared color composite image. Digital data, beamed back from an orbiting satellite, were computer processed to produce the image. The enlarged inset shows the city's airport with pixel details.

**What features can you recognize on this false color image?**

## Define & Recall

navigation
cartography
oblate spheroid
great circle
hemisphere
small circle
coordinate system
North Pole
South Pole
equator
latitude
sextant
prime meridian
longitude
geographic grid
parallel
meridian
time zone
solar noon
International Date Line
U.S Public Lands Survey System
principal meridian
base line
township
section
global positioning system (GPS)
map projection
conformal map
Mercator projection
equal-area map
equidistance
azimuthal map
rhumb line
gnomonic projection
compromise projection
legend
scale
verbal scale
representative fraction (RF scale)
graphic (bar) scale
magnetic declination
azimuth
bearing
thematic map
discrete data
continuous data
isoline
topographic contour line
contour interval
profile
gradient
digital terrain model
vertical exaggeration
visualization
geographic information system (GIS)
geocoding
remote sensing
digital image
pixel
resolution
near-infrared (NIR)
thermal infrared (TIR)
radar
imaging radar
side-looking airborne radar
weather radar
sonar
multispectral remote sensing
color composite image

## Discuss & Review

1. Why is a great circle useful for navigation?
2. What great circle determines the zero point for latitude?
3. What are the latitude and longitude coordinates of your city?
4. Approximately how precise in meters could you be if you tried to locate a building in your city to the nearest second of latitude and longitude? Using a GPS?
5. What time zone are you in? What is the time difference between Greenwich time and your time zone?
6. If it is 2:00 a.m. Tuesday in New York (EST), what time and day is it in California (PST)? What time is it in London (GMT)?
7. If you fly across the Pacific Ocean from the United States to Japan, how will the International Date Line affect you?
8. How has the use of the Public Lands Survey System affected the landscape of the United States? Has your local area been affected by its use? How?
9. Why can't maps give an accurate representation of Earth's surface? What is the difference between a conformal map and an equal-area map?
10. What is the difference between an RF and a verbal map scale?
11. Why is the scale of a map so important? What does a small-scale map show in comparison with a large-scale map?
12. Why is the date of publication of a map important? List several reasons, including one related to Earth's magnetic field.
13. Why are topographic maps so important to physical geographers? What specific information can be obtained from a contour map that cannot be obtained from other maps?
14. What does the concept of *thematic map layers* mean in a geographic information system?
15. How have computers revolutionized cartography and the handling of spatial data? What specific advantages do computers offer to the mapmaking process?
16. What is the difference between a photograph and a digital image?
17. In what ways would a standard color composite image taken from a satellite resemble an NIR photograph taken of the same place from the same position in orbit? How do they differ?
18. What does a weather radar image show in order to help us understand weather patterns?

# Consider & Respond

1. Use an atlas and globe to determine what cities are located at the following grid coordinates: 40°N, 75°W; 34°S, 151°E; and 41°N, 112°W.
2. What are the grid coordinates of the following cities: Portland, Oregon; Rio de Janeiro, Brazil; your hometown?
3. You are located at 10°S latitude, 10°E longitude; you travel 30° north and 30° east. What are your new geographic coordinates?
4. You are located at 40°N latitude, 90°W longitude. You travel due north 40°, then due east 60°. What are your new geographic coordinates?
5. Using an atlas and Figure 2.24, identify a city in North America with each of the following magnetic declinations: 20°E; 20°W; 25°E; 0°.
6. Select a place within the United States that you would most like to visit for a vacation. You have with you a highway map, a USGS topographic map, and a satellite image of the area. What kinds of information could you get from one of these sources that is not displayed on the other two? What spatial information do they share (visible on all three)?
7. If you were an applied geographer and wanted to use a geographic information system to build an information database about the environment of a park (pick a state or national park near you), what are the five most important layers of mapped information that you would want to have? What combinations of two or more layers would be particularly important to your purpose?
8. GIS has been called a "power tool" for spatial (geographic) analysis. Why are people so important to its effective application?

# Map Interpretation
Topographic Maps

## The Map

A topographic map is the most widely used tool for graphically depicting elevational variation within an area. A contour line connects points of equal elevation above some reference datum, usually mean sea level. A vast storehouse of information about the relief and the terrain can be interpreted from these maps by understanding the spacing and configuration of contours. For example, elevations of mountains and valleys, steepness of slopes, and the direction of stream flow can be determined by studying a topographic map. In addition to contour lines, many standard symbols are used on topographic maps to represent mapped features, data, and information (a guide to these symbols is in Appendix B).

The elevation difference represented by adjacent contour lines depends on the map scale and the relief in the mapped area, and is called the contour interval. Contour intervals on topographic maps are typically in elevation measurements divisible by ten. In mountainous areas wider intervals are needed to keep the contours from crowding and visually merging together. A flatter locality may require a smaller contour interval to display subtle relief features. It is good practice to note both the map scale and the contour interval when first examining a topographic map.

Keep in mind several important rules when interpreting contours:

- Closely spaced contours indicate a steep slope, and widely spaced contours indicate a gentle slope.
- Evenly spaced contours indicate a uniform slope.
- Closed contour lines represent a hill or a depression.
- Contour lines never cross but may converge along a vertical cliff.
- A contour line will bend upstream when it crosses a valley.

## Interpreting the Map

1. What is the contour interval on this map?
2. The map scale is 1:24,000. One inch on the map represents how many feet on the Earth's surface?
3. What is the highest elevation on the map? Where is it located?
4. What is the lowest elevation on the map? Where is it located?
5. Note the mountain ridge between Boat and Emerald Canyons (C-4). Is it steeper on its east side or its west side? What led you to your conclusion?
6. In what direction does the stream in Boat Canyon flow? What led you to your conclusion?
7. The aerial photograph at left depicts a portion of the topographic map on the opposite page. What area of the air photo does the map depict? How well do the contours represent the physical features seen on the air photo?
8. Identify some cultural features on the map. Describe the symbols used to depict these features. The map shown is older than the aerial photograph. Can you identify some cultural features on the aerial photograph not depicted on the contour map?

Aerial photograph of the coast at Laguna Beach, California

Opposite:
Laguna Beach, Calfornia
Scale 1:24,000
Contour interval = 20 feet
U.S. Geological Survey

# CHAPTER 3

The sun affects Earth systems in myriad ways.  Courtesy of NASA/Marshall Space Flight Center

# Earth in Space and Solar Energy

**PHYSICAL Geography◉Now™** This icon, appearing throughout the book, indicates an opportunity to explore interactive tutorials, animations, or practice problems at now.brookscole.com/gabler8

With a radius 110 times that of Earth and a mass 330,000 times greater, the sun reigns as the center of our solar system. The gravitational pull of this fierce, stormy ball of gas holds Earth in orbit, and its emissions power the Earth–atmosphere systems on which our lives depend. As the source of almost all the energy in our world, it holds the key to many of our questions about Earth and sky.

Everyone has wondered about environmental changes that take place throughout the year and from place to place over Earth's surface. Perhaps when you were young, you wondered why it got so much warmer in summer than in winter and why some days were long whereas those in other seasons were much shorter. These questions and many like them are probably as old as the earliest human thoughts, and the answers to them help provide us with an understanding of the physical geography of our world.

Physical geographers' concerns take them beyond planet Earth to a consideration of the sun and Earth's position in the solar system. Geographers examine the relationship between the sun and Earth to explain such earthly phenomena as

# CHAPTER PREVIEW

**The universe is made up of billions of galaxies, each containing so many stars that the mathematical possibility of other planets containing life as we know it seems unlimited.**
- What is a galaxy, and what is its relationship to the universe?
- How is distance involved in the determination of whether life beyond our solar system does exist in the universe?

**Earth is one of nine planets that, together with the sun, comprise the major components of our solar system.**
- In what ways is the sun of most importance to life on Earth?
- What are the chief characteristics of the other planets?
- What are the other components of our solar system?

**The regular movements of Earth, termed *rotation* and *revolution*, are the fundamental elements of Earth–sun relationships, which initially control the dynamics of our atmosphere and the phenomena related to it.**
- Why is this one of the most important understandings in physical geography?
- What other understandings follow from this concept?

**The sun is the original and ultimate source of the energy that drives the various components of the Earth system.**
- How does the sun's energy reach Earth?
- How does this energy affect the Earth system?

**The types of energy emitted by the sun are represented in the electromagnetic spectrum.**
- What bands of this spectrum control heating and cooling in the Earth energy system?
- What bands of this spectrum affect humans directly?

**The relationship of Earth's axis to the plane of Earth's orbit is the key to an explanation of seasons on Earth.**
- What is the relationship?
- How does it operate in conjunction with Earth's revolution to produce seasons?
- How does it influence variations in the amounts of insolation reaching different portions of Earth's surface?

the alternating periods of light and dark that we know as day and night. Other relationships between Earth and sun also help explain seasonal variations in climate. Although the universe and solar system are not strictly within the province of physical geography, an acquaintance with each can be helpful in an examination of Earth as an environment for life as we know it.

## The Solar System and Beyond

If you look at the sky on a clear night, all the stars that you see are part of a single collection of stars called the Milky Way Galaxy. A **galaxy** ( Fig. 3.1) is an enormous island in the universe—an almost incomprehensible cluster of stars, dust, and gases. Our sun is one of hundreds of billions of stars that compose the Milky Way Galaxy. In turn, the observable universe appears to contain billions of other galaxies.

Distances within the universe are so vast that it is necessary to use a large unit of measure termed a **light-year**—the distance that light travels in 1 year. A light-year is equal to 6 trillion miles. Light travels at the amazing speed of 298,000 kilometers per second (186,000 mi/sec).

Thus, in 1 second, light could travel seven times around the circumference of Earth. Although that may seem like a great distance, the *closest* star to Earth, other than the sun, is 4.3 light-years away, and the *closest* galaxy to our galaxy is 75,000 light-years away.

### The Solar System

The sun is the center of our solar system. A **solar system** can be defined as all the heavenly bodies surrounding a particular star because of the star's dominant mass and gravitational attraction ( Fig. 3.2). The principal celestial bodies

**FIGURE 3.1**
Looking at one of the billions of galaxies that make up our universe.

**FIGURE 3.2**
The solar system, showing the sun and planets in their proper order. The approximate size relationships between the individual planets is shown. However, the planetary orbits are much condensed, and the scale of the sun and planets is greatly exaggerated. The planets would be much too small to be visible at the scale of the orbits shown.

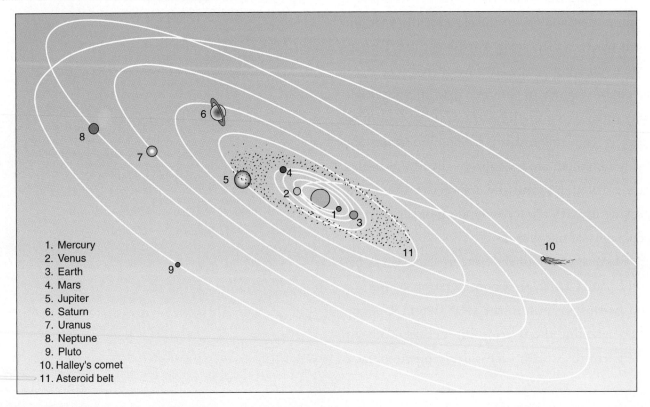

1. Mercury
2. Venus
3. Earth
4. Mars
5. Jupiter
6. Saturn
7. Uranus
8. Neptune
9. Pluto
10. Halley's comet
11. Asteroid belt

## Geography's Environmental Science Perspective
### Solar Energy

In the United States today, a growing number of dedicated scientists, environmentalists, government administrators, politicians, and business executives are convinced that solar energy provides the best alternative to fossil fuels as a power source. Power from solar radiation is nonpolluting and limitless. Earth receives, in just 2 weeks' time, an amount of solar energy that would equal our entire known global supply of fossil fuels.

However, a number of serious obstacles stand in the way of those who see solar energy as the power source of the future. Solar radiation is intermittent (no radiation at night, cloud cover), so a means of storing the power produced or of linking the power to an existing grid system that can accommodate large swings in production must be identified. In addition, while the cost of producing energy from solar radiation is reduced by new technologies, additional incentives must be provided for a switch to solar energy.

Two major types of solar technology are being actively developed in various parts of the world, especially in the United States, France, Israel, and Australia. One type involves solar thermal towers, in which racks of tracking mirrors (a heliostat field) follow the sun and focus its heat on a steam boiler perched on a high tower. Temperatures in the boiler may be raised more than 500°C (900°F). No different in principle from our youthful experiments with a magnifying glass to set fire to paper, this device is already operating at several sites. In California, Solar One in the Mojave Desert is the world's largest solar thermal electric power plant of this kind. However, development is slow because additional solar thermal towers must be considered in terms of economic feasibility.

The other major type of solar technology depends on photovoltaic (PV) cells that convert sunlight directly into electrical power (not unlike a camera's light meter). Although with PV technology the production of power is limited to daylight hours, this type of solar energy conversion is being given serious consideration as a supplementary source of energy for overtaxed power grids in high-pollution areas. For example, in Las Vegas, Nevada, there are plans to construct a major PV facility on the roof of the Las Vegas Convention Center. The solar PV system will cost between $.17 and $.25 per killowatt-hour, which is two to three times as much as the Convention Center currently pays for electricity but about the same as a customer pays in Marin County in energy-starved California. Although the cost will be high, there is a special reason for considering the project. Nevada law requires that, starting in 2003, a certain percentage of all energy sold in the state must come from solar energy. Nevada is not alone. Twelve other states have enacted such provisions in their laws. Solar energy may never fully replace fossil fuel as a source of power, but clearly the process has already begun.

One of three solar energy complexes in the Mojave Desert of California that together produce 90% of the world's grid-connected solar energy.

in our sun's system are the nine major **planets** (celestial bodies that revolve around a star and reflect the star's light rather than producing their own). Our solar system also includes no less than 130 **satellites** (like Earth's moon, these bodies orbit the planets) and numerous **asteroids** (very small planets, usually with a diameter of less than 500 miles), as well as **comets** and **meteors**. A comet is made up of a head—a collection of solid fragments held together by ice—and a tail, sometimes millions of miles long, composed of gases (Fig. 3.3). Meteors are small, stonelike or metallic bodies that, when entering Earth's atmosphere, burn and often appear as a streak of light, or "shooting star." A meteor that survives the fall through the atmosphere and strikes Earth's surface is called a **meteorite** (Fig 3.4).

## The Sun and Its Energy

The sun, like all other stars in the universe, is a self-luminous sphere of gases that emits radiant energy. A slightly less than average-sized star, our sun is the only self-luminous body in our solar system and is the source of almost all the light and heat for the surfaces of the various celestial bodies in our planetary system. It has an estimated surface temperature of between 5500°C and 6100°C (10,000°F and 11,000°F). The energy emitted by the sun comes from **fusion (thermonuclear) reactions** that take place in its interior. There, under high pressure, two hydrogen atoms fuse together to form one helium atom in a process with impact similar to that of a hydrogen bomb explosion (Fig. 3.5). This nuclear reaction releases

**FIGURE 3.3**
The comet Hale–Bopp shows a split tail because two different types of icy material are emitting different jets of gasses.

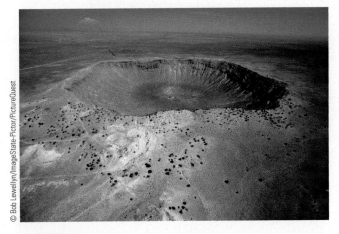

**FIGURE 3.4**
Barringer (meteor) Crater in Arizona displays some of the same characteristics as those found on the moon and other planets.

tremendous amounts of energy that radiate out from the sun in all directions at the speed of light.

Earth receives about 1/2,000,000,000 (one-two billionth) of the radiation given off by the sun, but even this tiny amount affects the biological and physical characteristics of Earth's surface. Other bodies in the solar system receive some of the remainder of the sun's radiant energy, but the vast proportion of it travels out through space unimpeded.

From the outermost layer of the sun, ionized gases acquire enough velocity to escape the gravitational pull of the sun and become **solar wind.** Solar wind is produced by streams of extremely hot protons and electrons that travel out at the speed of light. Occasionally, these solar winds approach Earth but are prevented from reaching the surface by Earth's magnetic field and are confined to the atmosphere. During these times, they can disrupt radio and television communication, may disable orbiting satellites, and intensify the **auroras.** The Aurora Borealis, known as the *northern lights,* and the Aurora Australis, called the *southern lights,* are caused by the interaction between incoming solar wind and the ions in our upper atmosphere ( Fig 3.6).

The intensity of solar winds is influenced by the best-known solar feature, **sunspots.** These dark regions are about 1500°C–2000°C cooler than the surrounding surface temperature. Galileo began recording sunspots back in the 1600s, and for many years they have been used to indicate solar activity. Sunspots seem to observe an 11-year cycle from one maximum (where 100 or so may be visible) to the next. An individual sunspot may last from a few days to a few months. Just how sunspots might affect Earth's atmosphere is still a matter of controversy. Proving the exact connections between sunspot numbers and weather and climate is difficult ( Fig. 3.7).

The sun's energy is the most important factor determining environmental conditions on Earth. With the exception of geothermal heat sources (such as volcanic eruptions and geyser springs) and heat emitted by radioactive minerals, the sun remains the source of all the energy for Earth and atmospheric systems.

The intimate and life-producing relationship between Earth and sun is the result of the amount and distribution of radiant energy received from the sun. Such factors as our planet's size, its distance from the sun, its atmosphere, the movement of Earth around the sun, and the planet's rotation on an axis all affect the amount of radiant energy that Earth receives. Though some processes of our physical environment

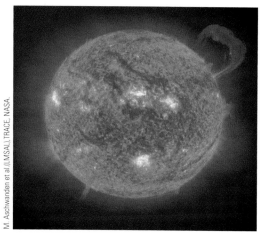

**FIGURE 3.5**
The fireball explosion of a hydrogen bomb is created by thermonuclear fusion. This same reaction powers the sun.

**FIGURE 3.6**
(a) Solar wind and the ions in Earth's atmosphere interact to produce the Aurora Borealis. (b) The record-setting solar activity of November 2004 causes the Aurora Borealis to be seen as far south as Houston, Texas. This photo was taken near Bowling Green, Kentucky, at 37°N latitude.

result from Earth forces not related to the sun, these processes would have little relevance were it not for the life-giving, life-sustaining energy of the sun.

Earth revolves around the sun at an average distance of 150 million kilometers (93 million mi). The sun's size and its distance from us challenge our comprehension. About 130 million Earths could fit inside the sun, and a plane flying at 500 miles per hour would take 21 years to reach the sun.

As far as we know with certainty, within our solar system, only on Earth has the energy from the sun been used to create life—to create something that can grow, develop, reproduce, and eventually die. Yet there remains a possibility of life, or at least the basic organic building blocks, on Mars and perhaps even on one or two of the moons of Saturn or Jupiter. What fascinates scientists, geographers, and philosophers alike, however, is the likelihood that millions of planets like Earth in the universe may have developed life-forms that might be more sophisticated than humans.

## The Planets

The four planets closest to the sun (Mercury, Venus, Earth, and Mars) are called the **terrestrial planets.** They are relatively small, warmed by their proximity to the sun, and composed of rock and metal. They all have solid surfaces that exhibit records of geological forces in the form of craters, mountains, and volcanoes. The next four planets (Jupiter, Saturn, Uranus, and Neptune) are much larger and composed primarily of lighter ices, liquids, and gases. These planets are termed the **giant planets,** or **gas planets.** Although they have solid cores at their centers, they are more like huge balls of gas and liquid with no solid surface

**FIGURE 3.7**
Sunspots as they appear on the solar surface. Inset shows relative size of Earth.

on which to walk. Finally, at the outer edge of the solar system is Pluto, which is neither a terrestrial nor a giant planet (Fig. 3.8). Through the years, controversy has arisen as to whether Pluto is officially one of our planets or a large body captured from the Kuiper Belt (a disk-shaped region containing small icy bodies that lies past the orbit of Neptune) by the gravitational pull of the sun. The question has been, Is Pluto a small outermost planet or a large captured member of the Kuiper Belt? For now it has been decided to keep Pluto as one of our family of planets.

The nine major planets that are known to revolve around the sun have several phenomena in common. From a point far out in space above the sun's "north pole," they would all appear to move around the sun in the same counterclockwise direction. Their orbits follow an elliptical, almost circular, path. All planets also rotate, or spin, on their own axes. With the exception of Venus and Uranus, all rotate in the same direction. All but Pluto lie close to the same plane (the plane of the ecliptic) passing through the sun's equator. All planets have an atmospheric layer of gases with the exception of Pluto and Mercury, which are not dense or heavy enough to hold any appreciable amounts of gases.

## Life beyond Earth?

Although the planets are similar in many ways, they differ in a number of vital characteristics, and the existence of life, as we know it, beyond Earth in our solar system seems highly unlikely (Table 3.1). Mercury, the smallest terrestrial planet, has a diameter less than half that of Earth, and its **mass** (the total amount of material in the planet) is only 5% of Earth's. This means that Mercury does not have sufficient mass and gravitational pull to hold onto a meaningful atmosphere. (**Gravitation** is the attractive force one body has for another. The greater the mass or amount of matter a body has, the greater the gravitational pull it will exert on other bodies.)

Also, partly because of the extremely thin atmosphere around Mercury, surface temperatures on that planet reach over 425°C (800°F) on the side facing the sun. As we will see later, Earth's atmosphere protects our planet's surface from much of the sun's radiation, which can be dangerous to life, and keeps temperatures on Earth within a livable range.

Further, because Mercury rotates on its axis only once every 59 days, the same side of the planet is exposed to the sun for long periods, while the opposite side receives none of the sun's energy. This situation compounds the problem of Mercury's surface temperatures. The side facing the sun is much too hot, and the opposite side is much too cold −183°C (−297°F) to support life.

Uranus, Neptune, and Pluto, the three planets farthest from the sun, are too distant to receive enough solar energy to have surface temperatures conducive to life, as we know it. In addition, though Uranus and Neptune are large enough to have atmospheres, theirs are made up primarily of hydrogen and helium; they apparently have no free oxygen, a factor necessary for the development of life like that on Earth.

Jupiter and Saturn are the two largest planets; yet, because significant parts of their masses are gaseous, both have very low densities. For instance, Saturn's density is less than that of water, whereas Earth's density is more than five times as great. The atmospheres of Jupiter and Saturn also have a high proportion of hydrogen and helium and no free oxygen. Even though Jupiter and Saturn are closer to the sun than Uranus, Neptune, and Pluto, they still do not receive enough solar energy to produce livable surface temperatures; their temperatures are down around −95°C to −150°C (−200°F to −300°F). It is unlikely that life exists on the two largest planets, and scientists have begun to look instead at the moons of Saturn and Jupiter for that possibility. The spacecraft *Cassini* launched by the United States in 1997 began orbiting Saturn in July 2004. In January 2005, it released its piggybacked Huygens probe for descent through the thick atmosphere of Titan, one of Saturn's moons. Planetary scientists are now receiving the data from Titan.

Venus and Mars, in the orbits closest to that of Earth, are the planets most similar to our own. Venus, in fact, has been called Earth's twin because it is most like Earth in size, density, and mass. However, we cannot see the surface of Venus because it is hidden by a permanent layer of thick clouds. Through information gathered by the orbiting spacecraft *Magellan* (from 1990 until it was destroyed in 1995), we have a much better understanding of surface conditions on Venus. The surface consists primarily of lowland

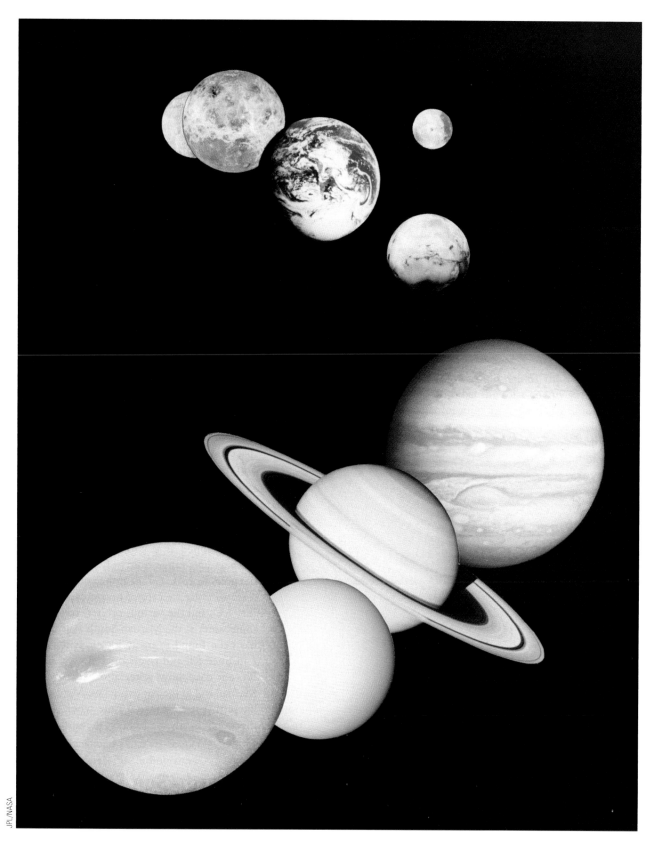

**FIGURE 3.8**
The planets in our solar system: Mercury, Venus, Earth, and Mars (the terrestrial planets); Jupiter, Saturn, Uranus, and Neptune (the giant gas planets). Pluto is not shown here.

farthest from the sun in July and the solar constant is lowest because of the distance from the sun, the Northern Hemisphere is in the midst of a summer with temperatures that are not significantly different from those in the Southern Hemisphere 6 months later. The solar constant also varies slightly with changes in activity on the sun; during intense sunspot or solar storm activity, for example, the solar constant will be slightly higher than usual. However, these variations are not even as great as those caused by Earth's elliptical orbit.

## Movements of Earth

Earth has three basic movements: **galactic movement, rotation,** and **revolution.** The first of these is the movement of Earth with the sun and the rest of the solar system in an orbit around the center of the Milky Way Galaxy. This movement has limited effect on the changing environments of Earth and is generally the concern of astronomers rather than geographers. The other two movements of Earth, rotation on its axis and revolution around the sun, are of vital interest to the physical geographer. The consequences of these movements are the phenomena of day and night, variations in the length of day, and the changing seasons.

**Rotation** Rotation refers to the spin of Earth on its axis, an imaginary line extending from the North Pole to the South Pole. Earth rotates on its axis at a uniform rate, making one complete turn with respect to the sun in 24 hours.

Earth turns in an eastward direction ( Fig. 3.11). The sun "rises" in the east and appears to move westward across the sky, but it is actually Earth, not the sun, that is moving, rotating toward the morning sun (that is, toward the east).

Earth, then, rotates in a direction opposite to the apparent movement of the sun, moon, and stars across the sky. If we look down on a globe from above the North Pole, the direction of rotation is counterclockwise. This eastward direction of rotation not only defines the movement of the zone of daylight on Earth's surface but also helps define the circulatory movements of the atmosphere and oceans.

The velocity of rotation at the Earth's surface varies with the distance of a given place from the equator (the imaginary circle around Earth halfway between the two poles). All points on the globe take 24 hours to make one complete rotation (360°). Thus, the *angular velocity* for all locations on Earth's surface is the same—360° per 24 hours, or 15° per hour. However, the *linear velocity* depends on the distance (not the angle) covered in that 24 hours. The linear velocity at the poles is zero. You can see this by spinning a globe with a postage stamp affixed to the North Pole. The stamp rotates 360° but covers no distance and therefore has no linear velocity. If you place the stamp anywhere between the North and South Poles, however, it will cover a measurable distance during one rotation of the globe. The greatest linear velocity is found at the equator, where the distance traveled by a point in 24 hours is largest. At Kampala, Uganda, near the equator, the velocity is about 460 meters

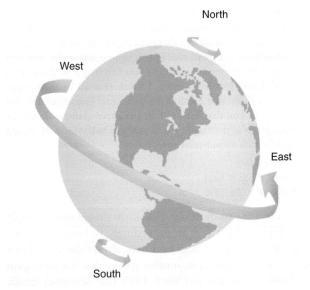

**FIGURE 3.11**
Earth spins around a tilted axis as it follows its orbit around the sun. Earth's rotation is from west to east, making the stationary sun appear to rise in the east and set in the west.

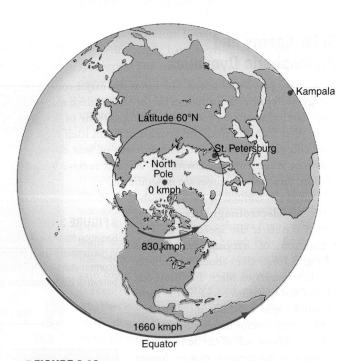

**FIGURE 3.12**
The speed of rotation of Earth varies with the distance from the equator.
**How much faster does a point on the equator move than a point at 60°N latitude?**

(1500 ft) per second, or approximately 1660 kilometers (1038 mi) per hour ( Fig. 3.12). In comparison, at St. Petersburg, Russia (60°N latitude), where the distance traveled during one complete rotation of Earth is about half that at the equator, Earth rotates about 830 kilometers per hour.

We are unaware of the speed of rotation because (1) the angular velocity is constant for each place on Earth's surface, (2) the atmosphere rotates with Earth, and (3) there are no nearby objects, either stationary or moving at a different rate with respect to Earth, to which we can compare Earth's movement. Without such references, we cannot perceive the speed of rotation.

Rotation accounts for our alternating days and nights. This can be demonstrated by shining a light at a globe while rotating the globe slowly toward the east. You can see that half the sphere is always illuminated while the other half is not and that new points are continually moving into the illuminated section of the globe while others are moving into the darkened sector. This corresponds to Earth's rotation and the sun's energy striking Earth. While one half of Earth receives the light and energy of solar radiation, the other half is in darkness. As noted in Chapter 2, the great circle separating day from night is known as the **circle of illumination** (Fig. 3.13).

**Revolution** While Earth rotates on its axis, it also revolves around the sun in a slightly elliptical orbit at an average distance from the sun of about 150 million kilometers (93 million mi) (Fig. 3.14). On about January 3, Earth is closest to the sun and is said to be at **perihelion** (from Greek: *peri,* close to; *helios,* sun); its distance from the sun then is approximately 147 million kilometers. At around July 4, Earth is about 152 million kilometers from the sun. It is then that Earth has reached its farthest point from the sun and is said to be at **aphelion** (Greek: *ap,* away; *helios,* sun). Five million kilometers is insignificant in space, and these varying distances from Earth to the sun only minimally (about 3.5%) affect the receipt of energy on Earth. Hence, they have no relationship to the seasons.

The period of time that Earth takes to make one revolution around the sun determines the length of 1 year. Earth makes 365¼ rotations on its axis during the time it takes to complete one revolution of the sun; therefore, a year is said to have 365¼ days. Because of the difficulty of dealing with a fraction of a day, it was decided that a year would have 365 days, and every fourth year, called *leap year,* an extra day would be added as February 29.

**Plane of the Ecliptic, Inclination, and Parallelism** In its orbit around the sun, Earth moves in a constant plane, known as the **plane of the ecliptic.** Earth's equator is tilted at an angle of 23½° from the plane of the ecliptic, causing Earth's axis to be tilted 23½° from a line perpendicular to the plane (Fig 3.15). In addition to this constant **angle of inclination,** Earth's axis maintains another characteristic called **parallelism.** As Earth revolves around the sun, Earth's axis remains parallel

**FIGURE 3.13**
The circle of illumination, which separates day from night, is clearly seen on this image of Earth.

**FIGURE 3.14**
Oblique view of the elliptical orbit of Earth around the sun. Earth is closest to the sun at perihelion and farthest away at aphelion. Note that in the Northern Hemisphere summer (July), Earth is farther from the sun than at any other time of the year.

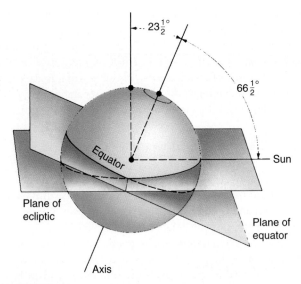

**FIGURE 3.15**
The plane of the ecliptic is defined by the orbit of Earth around the sun. The 23½° inclination of Earth's rotational axis causes the plane of the equator to cut across the plane of the ecliptic.

to its former positions. That is, at every position in Earth's orbit, the axis remains pointed toward the same spot in the sky. For the North Pole, that spot is close to the star that we call the North Star, or Polaris. Thus, Earth's axis is fixed with respect to the stars outside our solar system but not with respect to the sun (see the axis representation in Fig. 3.11).

Before continuing, it should be noted that, although the patterns of Earth rotation and revolution are considered constant in our current discussion, the two movements are subject to change. Earth's axis wobbles through time and will not always remain at an angle of exactly 23½° from perpendicular to the plane of the ecliptic. Moreover, Earth's orbit around the sun will change from more circular to more elliptical through periods that can be accurately determined. These and other cyclical changes were calculated and compared by Milutin Milankovitch, a Serbian astronomer during the 1940s, as a possible explanation for the ice ages. Since then the *Milankovitch Cycles* have often been used when climatologists attempt to explain climatic variations. These variations will be discussed in more detail along with other theories of climatic change in Chapter 8.

## Sun Angle, Duration, and Insolation

Understanding Earth's relationships with the sun leads us directly into a discussion of how the intensity of the sun's rays varies from place to place throughout the year and into an examination of the seasonal changes on Earth. Solar radiation received by the Earth system, known as **insolation** (for *in*coming *sol*ar radi*ation*), is the main source of energy on our planet. The seasonal variations in temperature that we experience are due primarily to fluctuations in insolation.

What causes these variations in insolation and brings about seasonal changes? It is true that Earth's atmosphere affects the amount of insolation received. Heavy cloud cover, for instance, will keep more solar radiation from reaching Earth's surface than will a clear sky. However, cloud cover is irregular and unpredictable, and it affects total insolation to only a minor degree over long periods of time.

The real answer to the question of what causes variations in insolation lies with two major phenomena that vary regularly for a given position on Earth as our planet rotates on its axis and revolves around the sun: the duration of daylight and the angle of the solar rays. The amount of daylight controls the duration of solar radiation, and the angle of the sun's rays directly affects the intensity of the solar radiation received. Together, the intensity and the duration of radiation are the major factors that affect the amount of insolation available at any location on Earth's surface (Table 3.2).

Therefore, a location on Earth will receive more insolation if (1) the sun shines more directly, (2) the sun shines longer, or (3) both. The intensity of solar radiation received at any one time varies from place to place because Earth presents a spherical surface to insolation. Therefore, only one line on the Earth's rotating surface can receive radiation at right angles while the rest receive varying oblique (sharp) angles ( Fig. 3.16a). As we can see from Figure 3.16b and c, solar energy that strikes Earth at a nearly vertical angle covers less area than an equal amount striking Earth at an oblique angle.

The intensity of insolation received at any given latitude can be found using *Lambert's Law,* named for Johann Lambert, an 18th-century German scientist. Lambert developed a formula by which the intensity of insolation can be calculated using the sun's zenith angle (that is, the sun angle deviating from 90° directly overhead). Using Lambert's Law, one can identify, based on latitude, where greater or lesser solar radiation is received on Earth's surface. Figure 3.17 shows the intensity of total solar energy received at various latitudes, when the most direct radiation (from 90° angle rays) strikes directly on the equator.

In addition, the atmospheric gases act to diminish, to some extent, the amount of insolation that reaches Earth's surface. Because oblique rays must pass through a greater distance of atmosphere than vertical rays, more insolation will be lost in the process.

Since no insolation is received at night, the duration of solar energy is related to the length of daylight received at a particular point on Earth (Table 3.3). Obviously, the longer the period of daylight, the greater the amount of solar radiation that will be received at that location. As we will see in our next section, periods of daylight vary in length through the seasons of the year as well as from place to place on Earth's surface.

## TABLE 3.2
**Radiation Intensity for Certain Solar Angles Expressed as a Percentage of the Maximum Possible (perpendicular beam)**

| Solar angle (degrees above the horizon) | 0.0 | 10.0 | 20.0 | 23.5 | 26.5 | 30.0 | 40.0 |
|---|---|---|---|---|---|---|---|
| Percent of Maximum | 00.0 | 17.4 | 34.2 | 39.9 | 44.6 | 50.0 | 64.3 |
| Solar Angle | 50.0 | 60.0 | 63.5 | 66.5 | 70.0 | 80.0 | 90.0 |
| Percent of Maximum | 76.6 | 86.6 | 89.5 | 91.7 | 94.0 | 98.5 | 100.0 |

(a)

(b)

(c)

**PHYSICAL Geography Now™ ■ ACTIVE FIGURE 3.16**

Watch this Active Figure at http://now.brookscole.com/gabler8.

(a) The angle at which the sun's rays strike Earth determines the amount of solar energy received per unit of surface area. This amount in turn controls the seasons. The diagram represents the June condition, in which solar radiation strikes the surface perpendicularly only in the Northern Hemisphere, creating summer conditions there. In the Southern Hemisphere, oblique rays are spread over large areas, producing less receipt of energy per unit of area and making this the winter hemisphere. **How would this figure differ in December?** The sun's rays in summer (b) and winter (c). In summer the sun appears high in the sky, and its rays hit Earth more directly, spreading out less. In winter the sun is low in the sky, and its rays spread out over a much wider area, becoming less effective at heating the ground.

## TABLE 3.3
### Duration of Daylight for Certain Latitudes

| | Length of Day (Northern Hemisphere) (read down) | | |
|---|---|---|---|
| LATITUDE (IN DEGREES) | MAR. 21/SEPT. 22 | JUNE 21 | DEC. 21 |
| 0.0 | 12 hr | 12 hr 0 min | 12 hr 0 min |
| 10.0 | 12 hr | 12 hr 35 min | 11 hr 25 min |
| 20.0 | 12 hr | 13 hr 12 min | 10 hr 48 min |
| 23.5 | 12 hr | 13 hr 35 min | 10 hr 41 min |
| 30.0 | 12 hr | 13 hr 56 min | 10 hr 4 min |
| 40.0 | 12 hr | 14 hr 52 min | 9 hr 8 min |
| 50.0 | 12 hr | 16 hr 18 min | 7 hr 42 min |
| 60.0 | 12 hr | 18 hr 27 min | 5 hr 33 min |
| 66.5 | 12 hr | 24 hr | 0 hr |
| 70.0 | 12 hr | 24 hr | 0 hr |
| 80.0 | 12 hr | 24 hr | 0 hr |
| 90.0 | 12 hr | 24 hr | 0 hr |
| LATITUDE | MAR. 21/SEPT. 22 | DEC. 21 | JUNE 21 |
| | Length of Day (Southern Hemisphere) (read up) | | |

## The Seasons

Many people assume that the seasons must be caused by the changing distance between Earth and the sun during Earth's yearly revolution. As noted earlier, the change in this distance is very small. Further, for people in the Northern Hemisphere, Earth is actually closest to the sun in January and farthest away in July (see again Fig. 3.14). This is exactly opposite of that hemisphere's seasonal variations. As we will see, seasons are caused by the 23½° tilt of Earth's equator to the plane of the ecliptic (see Fig. 3.15) and the parallelism of the axis that is maintained as Earth orbits the sun. About June 21, Earth is in a position in its orbit so that the northern tip of its axis is inclined toward the sun at an angle of 23½°. In other words, the plane of the ecliptic (the 90° sun angle) is directly on 23½°N latitude. This time in Earth's orbit is called the summer **solstice** (from Latin: *sol*, sun; *sistere*, to stand) in the Northern Hemisphere. We can best see what is happening if we refer to Figure 3.18, position A. In that diagram, we can see that the Northern and Southern Hemispheres receive unequal amounts of light from the sun. That is, as we imagine rotating Earth under these conditions, a larger portion of the Northern Hemisphere than the Southern Hemisphere remains in daylight. Conversely, a larger portion of the Southern Hemisphere than

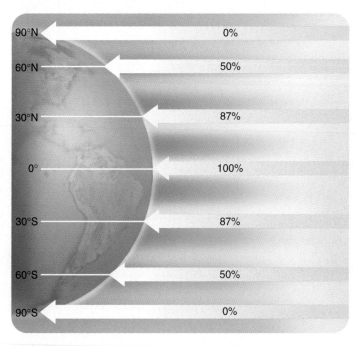

**FIGURE 3.17**
The percent of incoming solar radiation (insolation) striking various latitudes during an equinox date according to Lambert's Law.
**How much less solar energy is received at 60° latitude than that received at the equator?**

## Geography's Physical Science Perspective
### Using the Sun's Rays to Measure the Spherical Earth—2200 Years Ago

About 240 B.C. in Egypt, Eratosthenes, a Greek philosopher and geographer, observed that the noonday sun's angle above the horizon changed along with the seasons. Knowing that our planet was spherical, he used geometry and solar observations to make a remarkably accurate estimate of Earth's circumference.

A librarian in Alexandria, he read an account of a water well in Syene (today Aswan, Egypt), located to the south about 800 kilometers (500 mi) on the Nile River. On June 21 (summer solstice), this account stated, the sun's rays reached the bottom of the well and illuminated the water. Because the well was vertical, this meant that the sun was directly overhead on that day. Syene was also located very near the Tropic of Cancer, the latitude of the subsolar point on that date.

Eratosthenes had made many observations of the sun's angle over the year, so he knew that the sun's rays were never vertical in Alexandria, and at noon on that day in June a vertical column near the library formed a shadow. Measuring the angle between the column and a line from the column top to the shadow's edge, he found that the sun's angle was 7.2° away from vertical.

Assuming that the sun's rays strike Earth's spherical surface in a parallel fashion, Erathosthenes knew that Alexandria was located 7.2° north of Syene. Dividing the number of degrees in a circle (360°) by 7.2° he calculated that the two cities were separated by 1/50 of Earth's circumference.

The distance between Syene and Alexandria was 5000 stades, with a stade being the distance around the running track at a stadium. Therefore, 5000 stades times 50 meant that Earth must be 250,000 stades in circumference.

Unfortunately in ancient times, stades of different lengths were being used in different regions, and it is not certain which distance Eratosthenes used. A commonly cited stade length is about 0.157 kilometers (515 ft), and in using this measure, the resulting distance estimate would be 39,250 kilometers (24,388 mi). This distance is very close to the actual circumference of the great circle that would connect Alexandria and Syene.

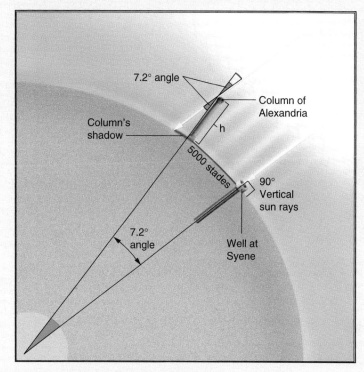

By observing the noon sun angle cast by a column where he lived and knowing that no shadow was cast on that same day in Syene to the south, Eratosthenes used geometry to estimate the Earth's circumference. On a spherical Earth, a 7.2° difference in angle also meant that Syene was 7.2° south in latitude from Alexandria, or 1/50 of Earth's circumference.

---

the Northern Hemisphere remains in darkness. Thus, a person living at Repulse Bay, Canada, north of the Arctic Circle, experiences a full 24 hours of daylight at the June solstice. On the same day, someone living in New York City will experience a longer period of daylight than of darkness. However, someone living in Buenos Aires, Argentina, will have a longer period of darkness than daylight on that day. This day is called the winter solstice in the Southern Hemisphere. Thus, June 21 is the longest day of the year in the Northern Hemisphere and the shortest day of the year in the Southern Hemisphere.

Now let's imagine the movement of Earth from its position at the June solstice toward a position a quarter of a year later, in September. As Earth moves toward that new position, we can imagine the changes that will be taking place in our three cities. In Repulse Bay, there will be an increasing amount of darkness through July, August, and September. In New York, sunset will be arriving earlier. In Buenos Aires, the situation will be reversed; as Earth moves toward its position in September, the periods of daylight in the Southern Hemisphere will begin to get longer, the nights shorter.

Finally, on or about September 22, Earth will reach a position known as an **equinox** (Latin: *aequus*, equal; *nox*, night). On this date (the autumnal equinox in the Northern Hemisphere), day and night will be of equal length at all locations on Earth. Thus, on the equinox, conditions are

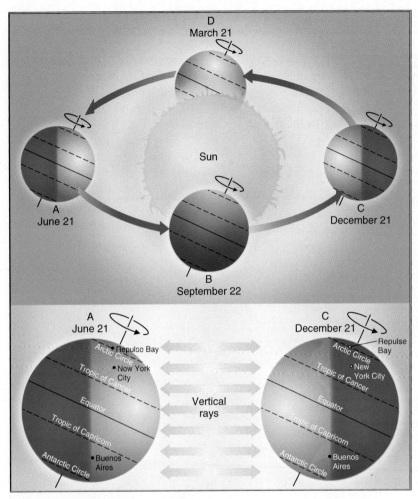

**ACTIVE FIGURE 3.18**
Watch this Active Figure at http://now.brookscole.com/gabler8.
Geometric relationships between Earth and the sun at the solstices. Note the differing day lengths at the summer and winter solstices in the Northern and Southern Hemispheres.

identical for both hemispheres. As you can see in Figure 3.19, position B, Earth's axis points neither toward nor away from the sun (imagine the axis is pointed at the reader); the circle of illumination passes through both poles, and it cuts Earth in half along its axis.

Imagine again the revolution and rotation of Earth while moving from September 22 toward a new position another quarter of a year later in December. We can see that in Repulse Bay the nights will be getting longer until, on the winter solstice, which occurs on or about December 21, this northern town will experience 24 hours of darkness (Fig. 3.18, position C). The only natural light at all in Repulse Bay will be a faint glow at noon refracted from the sun below the horizon. In New York, too, the days will get shorter, and the sun will set earlier. Again, we can see that in Buenos Aires the situation is reversed. On December 21, that city will experience its summer solstice; conditions will be much as they were in New York City in June.

Moving from late December through another quarter of a year to late March, Repulse Bay will have longer periods of daylight, as will New York, while in Buenos Aires the nights will be getting longer. Then, on or about March 21, Earth will again be in an equinox position (the vernal equinox in the Northern Hemisphere) similar to the one in September (Fig. 3.19, position D). Again, days and nights will be equal all over Earth (12 hours each). Finally, moving through another quarter of the year toward the June solstice where we began, Repulse Bay and New York City are both experiencing longer periods of daylight than darkness. The sun is setting earlier in Buenos Aires until, on or about June 21, Repulse Bay and New York City will have their longest day of the year and Buenos Aires its shortest. Further, we can see that on June 21, a point on the Antarctic Circle in the Southern Hemisphere will experience a winter solstice similar to that which Repulse Bay had on December 21 (Fig. 3.18). There will be no daylight in 24 hours, except what appears at noon as a glow of twilight in the sky.

## Lines on Earth Delimiting Solar Energy

Looking at the diagrams of Earth in its various positions as it revolves around the sun, we can see that the angle of inclination is important. On June 21, the plane of the ecliptic is directly on 23½°N latitude. The sun's rays can reach 23½° beyond the North Pole, bathing it in sunlight. The **Arctic Circle,** an imaginary line drawn around Earth 23½° from the North Pole (or 66½° north of the equator) marks this limit. We can see from the diagram that all points on or north of the Arctic Circle will experience no darkness on the June solstice and that all points south of the Arctic Circle will have some darkness on that day. The **Antarctic Circle** in the Southern Hemisphere (23½° north of the South Pole, or 66½° south of the equator) marks a similar limit.

Furthermore, it can be seen from the diagrams that the sun's **vertical (direct) rays** (rays that strike Earth's surface at right angles) also shift position in relation to the poles and the equator as Earth revolves around the sun. At the time of the June solstice, the sun's rays are vertical, or directly overhead, at noon at 23½° *north* of the equator. This imaginary line around Earth marks the northernmost position at which the solar rays will ever be directly overhead

SUN ANGLE, DURATION, AND INSOLATION    83

## The Analemma

The latitude at which the noon sun is directly overhead is also known as the sun's **declination.** Thus, if the sun appears directly overhead at 18°S latitude, the sun's declination is 18°S. A figure called an **analemma,** which is often drawn on globes as a big-bottomed "figure 8," shows the declination of the sun throughout the year. A modified analemma is presented in Figure 3.20. Thus, if you would like to know where the sun will be directly overhead on April 25, you can look on the analemma and see that it will be at 13°N. The analemma actually charts the passage of the direct rays of the sun over the 47° of latitude that they cover during a year.

## Variations of Insolation with Latitude

Neglecting for the moment the influence of the atmosphere on variations in insolation during a 24-hour period, the amount of energy received by the surface begins after daybreak and increases as Earth rotates toward the time of solar noon. A place will receive its greatest insolation at solar noon when the sun has reached its zenith, or highest point in the sky, for that day. The amount of insolation then decreases as the sun angle lowers toward the next period of darkness. Obviously, at any location, no insolation is received during the darkness hours.

We also know that the amount of daily insolation received at any one location on Earth varies with latitude (see again Fig. 3.17). The seasonal limits of the most direct insolation are used to determine recognizable zones on Earth. Three distinct patterns occur in the distribution of the seasonal receipt of solar energy in each hemisphere. These patterns serve as the basis for recognizing six latitudinal zones, or bands, of insolation and temperature that circle Earth (■ Fig. 3.21).

If we look first at the Northern Hemisphere, we may take the Tropic of Cancer and the Arctic Circle as the dividing lines for three of these distinctive zones. The area between the equator and the Tropic of Cancer can be called the north *tropical zone*. Here, insolation is always high but is greatest at the time of the year that the sun is directly overhead at noon. This occurs twice a year, and these dates vary according to latitude (see again Fig. 3.20). The north *middle-latitude zone* is the wide band between the Tropic of Cancer and the Arctic Circle. In this belt, insolation is greatest on the June solstice when the sun reaches its highest noon

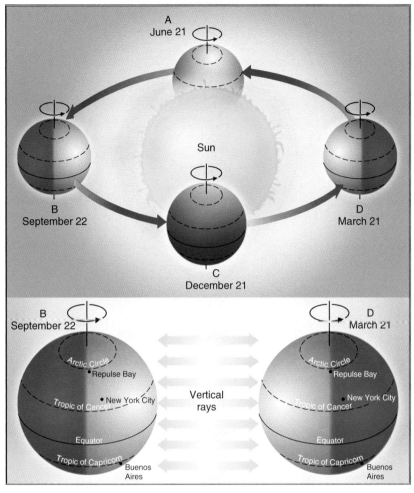

**ACTIVE FIGURE 3.19**

Watch this Active Figure at http://now.brookscole.com/gabler8.
Geometric relationships between Earth and the sun at the equinoxes. Day length is 12 hours everywhere because the circle of illumination crosses the equator at right angles and cuts through both poles.
**If Earth were not inclined on its axis, would there still be latitudinal temperature variations? Would there be seasons?**

during a full revolution of our planet around the sun. The imaginary line marking this limit is called the **Tropic of Cancer.** Six months later, at the time of the December solstice, the solar rays are vertical, and the noon sun is directly overhead 23½° *south* of the equator. The imaginary line marking this limit is known as the **Tropic of Capricorn.** At the times of the March and September equinoxes, the vertical solar rays will strike directly at the equator; the noon sun is directly overhead at all points on that line.

Note also that on any day of the year the sun's rays will strike Earth at a 90° angle at only one position, either on or between the two tropics. All other positions that day will receive the sun's rays at an angle of less than 90° (or will receive no sunlight at all).

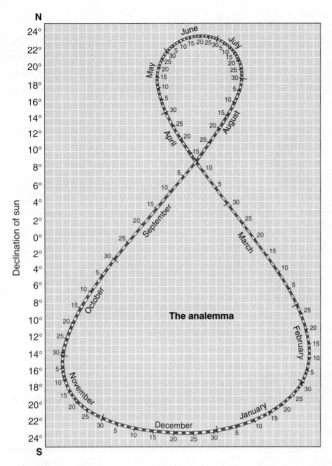

**FIGURE 3.20**
An analemma shows the location of the vertical noon sun throughout the year.

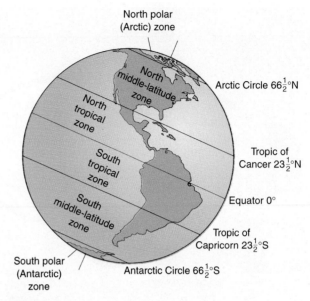

**FIGURE 3.21**
The line of the equator, the Tropics of Cancer and Capricorn, and the Arctic and Antarctic Circles define six latitudinal zones that have distinctive insolation characteristics.
**Which zone(s) would have the least annual variation in insolation? Why?**

angle and the period of daylight is longest. Insolation is least at the December solstice when the sun is lowest in the sky and the period of daylight the shortest. The north *polar zone,* or *Arctic zone,* extends from the Arctic Circle to the pole. In this region, as in the middle-latitude zone, insolation is greatest at the June solstice, but it ceases during the period that the sun's rays are blocked entirely by the tilt of Earth's axis. This period lasts for 6 months at the North Pole but is as short as 1 day directly on the Arctic Circle.

Similarly, there is a south tropical zone, a south middle-latitude zone, and a south polar zone, or *Antarctic zone,* all separated by the Tropic of Capricorn and the Antarctic Circle in the Southern Hemisphere. These areas get their greatest amounts of insolation at opposite times of the year from the northern zones.

Despite various patterns in the amount of insolation received in these zones, we can make some generalizations. For example, total annual insolation at the top of the atmosphere over a particular latitude remains nearly constant from year to year (the solar constant). Furthermore, annual insolation tends to decrease from lower latitudes to higher latitudes (Lambert's Law). The closer to the poles a place is located, the greater will be its seasonal variations caused by fluctuations in insolation.

The amount of insolation received by Earth systems is an important concept in understanding atmospheric dynamics and the distribution of climate and vegetation. Such climatic elements as temperature, precipitation, and winds are controlled in part by the amount of insolation received by Earth. People depend on certain levels of insolation for physical comfort, and plant life is especially sensitive to the amount of available insolation. You may have noticed plants that have wilted in too much sunlight or that have grown brown in a dark corner away from a window. Over a longer period of time, deciduous plants have an annual cycle of budding, flowering, leafing, and losing their leaves. This cycle is apparently determined by the fluctuations of increasing and decreasing solar radiation that mark the changing seasons. Even animals respond to seasonal changes. Some animals hibernate; many North American birds fly south toward warmer weather as winter approaches; and many animals breed at such a time that their offspring will be born in the spring, when warm weather is approaching.

## Define & Recall

galaxy
light-year
solar system
planet
satellite
asteroid
comet
meteor
meteorite
fusion (thermonuclear) reaction
solar wind
aurora
sunspot
terrestrial planet

giant planet (gas planet)
mass
gravitation
electromagnetic energy
shortwave radiation
calorie
solar constant
galactic movement
rotation
revolution
circle of illumination
perihelion
aphelion
plane of the ecliptic

angle of inclination
parallelism
insolation
solstice
equinox
Arctic Circle
Antarctic Circle
vertical (direct) rays
Tropic of Cancer
Tropic of Capricorn
declination
analemma

## Discuss & Review

1. What is a solar system? What bodies constitute our solar system?
2. How is the energy emitted from the sun produced?
3. Name the terrestrial planets. What do they have in common? Name the giant planets. What do they have in common?
4. Give a unique characteristic for each of the nine planets.
5. The electromagnetic spectrum displays various types of energy by their wavelengths. Where is the division between longwave and shortwave energy? In what ways do humans use electromagnetic energy?
6. Is the amount of solar energy reaching Earth's outer atmosphere constant? What might make it change?
7. Describe briefly how Earth's rotation and revolution affect life on Earth.
8. If the sun is closest to Earth on January 3, why isn't winter in the Northern Hemisphere warmer than winter in the Southern Hemisphere?
9. Identify the two major factors that cause regular variation in insolation throughout the year. How do they combine to cause the seasons?

## Consider & Respond

1. Do you think that we have discovered all the galaxies in the universe? Why?
2. Given what you know of the sun's relation to life on Earth, explain why the solstices and equinoxes have been so important to cultures all over the world. What are some of the major festivals associated with these times of the year?
3. Use the discussion of solar angle, including Figure 3.16, to explain why we can look directly at the sun at sunrise and sunset but not at the noon hour.
4. Imagine you are at the equator on March 21. The noon sun would be directly overhead. However, for every degree of latitude that you travel to the north or south, the noon solar angle would decrease by the same amount. For example, if you travel to 40°N latitude, the solar angle would be 50°. Can you explain this relationship? Can you develop a formula or set of instructions to generalize this relationship? What would be the solar angle at 40°N on June 21? On December 21?
5. Describe in your own words the relationship between insolation and latitude.
6. Use the analemma presented in Figure 3.20 to determine the latitude where the noon sun will be directly overhead on February 12, July 30, November 2, December 30.

# CHAPTER 4

In reality, the atmosphere is a very thin layer of gases held to the surface by Earth's gravity. NASA

# The Atmosphere, Temperature, and the Heat Budget

**PHYSICAL Geography Now™** This icon, appearing throughout the book, indicates an opportunity to explore interactive tutorials, animations, or practice problems at now.brookscole.com/gabler8

## CHAPTER PREVIEW

**Our planet's atmosphere is essential to life as we know it here on Earth.**

- How is this true?
- What is the significance of this statement for humans?
- How should this fact affect human behavior?

**Earth has an energy budget with a multiplicity of inputs and outputs (exchanges) that ultimately remains in balance despite recurring deficits and surpluses from time to time and from place to place.**

- How is the budget concept useful to an understanding of atmospheric heating and cooling?
- How can we tell that the budget remains in balance?

**Water plays a very important role in the exchanges of energy that fuel atmospheric dynamics.**

- What characteristics of water are responsible for its importance in energy exchange?
- In what ways is water involved in the heating of the atmosphere?

**As a direct result of differences in insolation and the mechanics of atmospheric heating, air temperature varies over time and both horizontally and vertically through space.**

- What are the most obvious variations?
- Why do they occur?
- How do temperatures stay within the ranges suitable for life if there are such great differences in the amounts of insolation received?

**Atmospheric elements are affected by atmospheric controls to produce weather and climate.**

- How do the elements differ from the controls?
- How does weather differ from climate?

Water and oxygen are vital for animals and humans to survive. Plant life requires carbon dioxide as well as a sufficient water supply. Most living things we know cannot survive extreme temperatures, nor can they live long if exposed to large doses of harmful radiation. It is the atmosphere, the envelope of air that surrounds Earth, that supplies most of the oxygen and carbon dioxide and that helps maintain a constant level of water and radiation in the Earth system.

Though actually a thin film of air, the atmosphere serves as an insulator, maintaining the viable temperatures we find on Earth. Without the atmosphere, Earth would experience temperature extremes of as much as 260°C (500°F) between day and night. The atmosphere also serves as a shield, blocking out much of the sun's ultraviolet (UV) radiation and protecting us from meteor showers. The atmosphere is also described as an ocean of air surrounding Earth. This description reminds us of the currents and circulation of the atmosphere—its dynamic movements—which create the changing conditions on Earth that we know as weather.

For comparison, we can look at our moon—a celestial body with virtually no atmosphere—in order to see the importance of our own atmosphere. Most obviously, a person standing on the moon without a space suit would immediately die for lack of oxygen. Our lunar astronauts recorded temperatures of up to 204°C (400°F) on the hot, sunlit side of the moon and, on the dark side, temperatures approaching −121°C (−250°F). These temperature extremes would certainly kill an unprotected human.

The next thing any astronaut on the moon would notice is the "unearthly" silence. On Earth, we hear sounds because sound waves move by vibrating the molecules in the air. Because the moon has no atmosphere and no molecules to carry the sound waves, the lunar visitor cannot hear any sounds; only radio communications are possible. Also, because there is no atmosphere, an astronaut cannot fly aircraft or helicopters, and it would be fatal to try to use a parachute. In addition, lack of atmosphere means no protection from the bombardment of meteors that fly through space and collide with the moon. Nearing Earth, most meteors burn up before reaching the surface because of friction within the atmosphere. Without an atmosphere for protection, the ultraviolet rays of the sun would also burn a visitor to the moon. On Earth, we are protected to a large degree from UV radiation because the ozone layer of the upper atmosphere absorbs the major portion of this harmful radiation.

We can see that, in contrast to our stark, lifeless moon, Earth presents a hospitable environment for life almost solely because of its atmosphere. All living things are adapted to its presence. For example, many plants reproduce by pollen and spores that are carried by winds. Birds can fly only because of the air. The water cycle of Earth is maintained through the atmosphere, as are the heat and radiation "budgets." The atmosphere diffuses sunlight as well, giving us our blue skies and the fantastic reds, pinks, oranges, and purples of sunrise and sunset. Without this diffusion, the sky would appear black, as it does from the moon ( Fig. 4.1)

Further, the atmosphere provides a means by which the systems of Earth attempt to reach equilibrium. Changes in weather are ultimately the result of the atmospheric effects that equalize temperature and pressure differences on Earth's surface by transferring heat and moisture through atmospheric and oceanic circulation systems.

# Characteristics of the Atmosphere

The atmosphere extends to approximately 480 kilometers (300 mi) above Earth's surface. Its density decreases rapidly with altitude; in fact, 97% of the air is concentrated in the first 25 kilometers (16 mi) or so. Because air has mass, the atmosphere exerts pressure on Earth's surface. At sea level, this pressure is about 1034 grams per square centimeter (14.7 lb/sq in.), but the higher the elevation, the lower is

**FIGURE 4.1**
Without an atmosphere, the moon's environment would be deadly to an unprotected astronaut.
**How do astronauts communicate with each other on the moon?**

the atmospheric pressure. In Chapter 5, we will examine the relationship between atmospheric pressure and elevation in more detail.

## Composition of the Atmosphere

The atmosphere is composed of numerous gases (Table 4.1). Most of these gases remain in the same proportions regardless of the density of the atmosphere. A bit more than 78% of the atmosphere's volume is made up of nitrogen, and nearly 21% consists of oxygen. Argon comprises most of the remaining 1%. The percentage of carbon dioxide in the atmosphere has risen through time, but is a little less than 0.04% by volume.

There are traces of other gases as well: ozone, hydrogen, neon, xenon, helium, methane, nitrous oxide, krypton, and others.

### Nitrogen, Oxygen, Argon, and Carbon Dioxide
Of these four most abundant gases that make up the atmosphere, nitrogen gas ($N_2$) makes up the largest proportion of air. It is a very important element supporting plant growth and will be discussed in more detail in Chapter 11. In addition, some of the other gases in the atmosphere are vital to the development and maintenance of life on Earth. One of the most important of the atmospheric gases is of course oxygen ($O_2$), which humans and all other animals use to breathe and oxidize (burn) the food that they eat. *Oxidation*, which is technically the chemical combination of oxygen with other materials to create new products, occurs in situations outside animal life as well. Rapid oxidation takes place, for instance, when we burn fossil fuels or wood and thus release large amounts of heat energy. The decay of certain rocks or organic debris and the development of rust are examples of

## TABLE 4.1
### Composition of the Atmosphere Near the Earth's Surface

| Permanent Gases | | | Variable Gases | | | |
|---|---|---|---|---|---|---|
| GAS | SYMBOL | PERCENT (BY VOLUME) DRY AIR | GAS (AND PARTICLES) | SYMBOL | PERCENT (BY VOLUME) | PARTS PER MILLION (PPM)* |
| Nitrogen | $N_2$ | 78.08 | Water vapor | $H_2O$ | 0 to 4 | |
| Oxygen | $O_2$ | 20.95 | Carbon dioxide | $CO_2$ | 0.037 | 368* |
| Argon | Ar | 0.93 | Methane | $CH_4$ | 0.00017 | 1.7 |
| Neon | Ne | 0.0018 | Nitrous oxide | $N_2O$ | 0.00003 | 0.3 |
| Helium | He | 0.0005 | Ozone | $O_3$ | 0.000004 | 0.04† |
| Hydrogen | $H_2$ | 0.00006 | Particles (dust, soot, etc.) | | 0.000001 | 0.01–0.15 |
| Xenon | Xe | 0.000009 | Chlorofluorocarbons (CFCs) | | 0.00000002 | 0.0002 |

*For $CO_2$, 368 parts per million means that out of every million air molecules, 368 are $CO_2$ molecules.
†Stratospheric values at altitudes between 11 km and 50 km are about 5 to 12 ppm.

### FIGURE 4.2
The equation of photosynthesis shows how solar energy (light) is used by plants to manufacture sugars and starches from atmospheric carbon dioxide and water, liberating oxygen in the process. The stored food energy is then eaten by animals, which also breathe the oxygen released by photosynthesis.

slow oxidation. All these processes depend on the presence of oxygen in the atmosphere. The third most abundant gas in our atmosphere is Argon (Ar). It is not a chemically active gas and therefore neither helps nor hinders life on Earth.

Carbon dioxide ($CO_2$), the fourth most abundant gas, is involved in the system known as the *carbon cycle*. Plants, through a process known as **photosynthesis**, use carbon dioxide and water to produce carbohydrates (sugars and starches), in which amounts of energy, derived originally from the sun, are stored and used by vegetation (Fig. 4.2). Oxygen is given off as a by-product. Animals then use the oxygen to oxidize the carbohydrates, releasing the stored energy. A by-product of this process in animals is the release of carbon dioxide, which completes the cycle when it is in turn used by plants in photosynthesis.

### Water Vapor, Liquids, and Particulates
Water vapor is always mixed in some proportion with the dry air of the lower part of the atmosphere; it is the most variable of the atmospheric gases and can range from 0.02% by volume in a cold, dry climate to more than 4% in the humid tropics. The percentage of water vapor in the air will be discussed later under the broad topic of humidity, but it is important to note here that the variations in this percentage over time and place are an important consideration in the examination and comparison of climates.

Water vapor also absorbs heat in the lower atmosphere and so prevents its rapid escape from Earth. Thus, like carbon dioxide, water vapor plays a large role in the insulating action of the atmosphere. In addition to gaseous water vapor, liquid water also exists in the atmosphere as rain and as fine droplets in clouds, mist, and fog. Solid water is found in the atmosphere in the form of ice crystals, snow, sleet, and hail.

Particulates are solids suspended in the atmosphere. These can be pollutants from transportation and industry, but the majority are natural particles that have always existed in our atmosphere (Fig. 4.3). Particles such as dust, smoke, pollen and spores, volcanic ash, bacteria, and salts from ocean spray can all play an important role in absorption of energy and in the formation of raindrops.

## Atmospheric Environmental Issues

Two gases in our atmosphere play significant roles in important environmental issues. One is carbon dioxide, a gas that is directly involved in an apparent slow but steady rise in global temperatures. The other is ozone, which comprises a layer in the upper atmosphere, that protects Earth from excessive UV

**FIGURE 4.3**
Volcanic eruptions add a variety of gases, particulates, and water vapor into our atmosphere.
**What other ways are particles added to the atmosphere?**

radiation but is endangered by other gases associated with industrialization.

## Carbon Dioxide and the Greenhouse Effect

We are all familiar with what happens to the inside of a parked car on a sunny street if all the windows are left closed. Shortwave radiation (mainly visible light) from the sun can penetrate the glass windows with ease (Fig. 4.4). When the insolation strikes the interior of the car, it is absorbed and heats the exposed surfaces. Energy, emitted from the surfaces as longwave radiation (mainly heat), cannot escape through the glass as freely. The result is that the interior of the vehicle gets hotter throughout the day. Many drivers recognize this as a blast of hot air as the car door opens. In extreme cases, windows in some cars have cracked under the heat as a result of *thermal expansion* (the property of all materials to expand when heated). What is more serious, temperatures have become so great in automobiles with closed windows as to pose a deadly threat to small children and pets left behind.

A similar phenomenon also occurs in the atmosphere. Like glass, carbon dioxide and water vapor (and other so-called greenhouse gases) can impede the escape of longwave radiation by absorbing it and then radiating it back to Earth. For example, carbon dioxide emits about half of its absorbed heat energy back to Earth's surface. Termed the **greenhouse effect,** this is the primary reason for the moderate temperatures observed on Earth. A greenhouse (plant beds surrounded by glass) will behave like the closed vehicle parked in the sun. Heat from the sun can flow through the glass roof and walls of the greenhouse, but the heat cannot escape as rapidly as insolation comes in and the greenhouse becomes warmer. The greenhouse effect in Earth's atmosphere is not necessarily a bad thing, for, without any green-

**FIGURE 4.4**
(a) Greenhouse gasses in our atmosphere allow short wavelength solar radiation (sunlight) to penetrate Earth's atmosphere relatively unhampered, while some of the long wavelength radiation (heat) is kept from escaping into outer space. (b) The same sort of heat buildup occurs in a closed car. The penetration of shortwave radiation through the car windows is plentiful, but the glass prevents some of the longwave radiation to escape.
**How do you handle it when your car interior becomes so hot?**

house gases in the atmosphere, Earth's surface would be too cold to sustain human life. The greenhouse process helps maintain the warmth of the planet and is a factor in Earth's *heat energy budget* (discussed later in this chapter). However, a serious environmental issue arises when increasing concentrations of greenhouse gases cause measurable increases in worldwide temperatures.

Since the Industrial Revolution, human beings have been adding more and more carbon dioxide to the atmosphere through their burning of fossil (carbon) fuels. At the same time, Earth has undergone massive *deforestation* (the removal of forests and other vegetation for agriculture and urban development). Vegetation uses large amounts of carbon dioxide in *photosynthesis* (see again Fig. 4.2), and removal of

**FIGURE 4.5**
Since 1958 these measurements of carbon dioxide taken at Mauna Loa, Hawaii, have shown an upward trend.
**Why do you suppose the line does a zigzag each year?**

the vegetation permits more carbon dioxide to remain in the atmosphere. Figure 4.5 shows how these two human activities have worked together to increase the amount of carbon dioxide in the atmosphere through time. Because carbon dioxide absorbs the longwave radiation from Earth's surface, restricting its escape to space, the rising amounts of carbon dioxide in the atmosphere increase the greenhouse effect and produce a global rise in temperatures.

The effects of greenhouse warming on Earth environments are many and varied. It is feared that such warming will raise sea levels around the world in at least two ways. First, ocean water will expand with increased temperature because of *thermal expansion*. Second, the higher average temperatures will cause the melting of major ice sheets and mountain glaciers worldwide, and the meltwater will add to the rising sea levels. Rising sea levels are of great concern to coastal regions around the globe and even more so to areas that lie below sea level, like the coastal Netherlands and such cities as New Orleans. The shifting temperature structure of the atmosphere can cause shifts in vegetation regions as plants begin to grow in new locations that previously lacked suitable growing conditions. With shifts in temperatures, scientists also expect variation in migration patterns of birds and other animals and the movement of insect-borne diseases to new areas. Last, but not least, with shifts in temperatures, it is anticipated that there will be a parallel shift in weather systems. Temperature and rainfall patterns may change throughout the globe.

At the present time, numerous researchers are closely monitoring the trends and amounts of change associated with Earth's temperatures and are watching for physical manifestations of greenhouse warming. This issue will be discussed in more detail in Chapter 8.

### The Ozone Layer
Another vital gas in Earth's atmosphere is **ozone.** The ozone molecule ($O_3$) is related to the oxygen molecule ($O_2$), except it is made up of three oxygen atoms whereas oxygen gas consists of only two. Ozone is formed in the upper atmosphere when an oxygen molecule is split into two oxygen atoms ($O^-$) by shortwave solar radiation. Then the free unstable atoms join two other oxygen molecules to form two molecules of ozone gas consisting of three oxygen atoms each:

$$2O_2 + 2O^- \rightarrow 2O_3$$

In the lower atmosphere, ozone is formed by electrical discharges (like high-tension power lines and lightning strikes) as well as incoming shortwave solar radiation. It is a toxic pollutant and a major component of urban smog, which can cause sore and watery eyes, soreness in the throat and sinuses, and difficulty in breathing. Near the surface of Earth, ozone is a menace and can only hurt life-forms. However, in the upper atmosphere, ozone is vital to both terrestrial and marine life. Ozone is vital to life-forms because it is capable of absorbing large amounts of the sun's UV radiation.

In the upper atmosphere, UV radiation is consumed as it breaks the chemical bonds of ozone ($O_3$) to form oxygen gas ($O_2$) and an oxygen ion ($O^-$). Then more UV radiation is consumed to recombine the oxygen gas and the oxygen ion back into ozone. This process is repeated over and over again, thereby involving large amounts of UV energy that would otherwise reach Earth's surface.

Without the ozone layer of the upper atmosphere, excessive UV radiation reaching Earth would severely burn human skin, increase the incidence of skin cancer and optical cataracts, destroy certain microscopic forms of marine life, and damage plants. Throughout the globe, UV radiation is responsible for painful sunburns or sensible suntans, depending on individual skin tolerance and exposure time.

For many years, there has been concern that human activity, especially the addition of chlorofluorocarbons (CFCs) and nitrogen oxides ($NO_x$) to the atmosphere, may permanently damage Earth's fragile ozone layer. CFCs, known commercially as Freon, have been used extensively in refrigeration and air conditioning. As refrigeration has become common in most parts of the world, CFCs have entered the upper atmosphere and, by removing ozone through chemical combination with the gas, have become a serious threat to our natural UV filter. Nitrogen oxide compounds, emitted along with automobile and jet engine exhaust, also have the ability to enter and destroy our ozone shield. Figure 4.6 shows an area of ozone depletion over the South Polar region, where the ozone layer is weakest.

The small proportion of UV radiation that the ozone layer allows to reach Earth does serve useful purposes. For instance, it is important in the production of certain vitamins (especially Vitamin D), it can help treat certain types of skin disorders, and it helps the growth of some beneficial viruses and bacteria. It also has a vital function in the process of photosynthesis. However, increasing amounts of UV radiation reaching Earth can become a serious problem for Earth environments, and the ozone layer must be protected from the industrial pollutants that threaten its existence.

## Geography's Spatial Science Perspective
### Why Are Ozone Holes Located at the Poles?

Many questions exist concerning the depleted region of our ozone layer, dubbed the "ozone hole." The first being, Is there indeed a hole in our protective ozone layer? The correct answer is no. From Earth's surface to the outer reaches of the atmosphere, there is always some ozone present (keeping in mind, of course, that ozone is only a trace gas). There have been years when all the ozone was missing from specific levels above the ground, but there is always some ozone above us. If all the ozone in our atmosphere were forced down to sea level, the atmospheric pressure there would compress the ozone into a worldwide layer of about 3–4 millimeters in thickness. So the ozone hole in reality is an area where the amount of ozone is considerably less than it should be.

When we see a satellite image of the infamous ozone hole, we see an elliptical area centered on the polar regions (especially the South Pole). The color pattern of the images corresponds to the amount of ozone in the atmosphere using Dobson units (du). Dobson units in the 300-du to 400-du range indicates a sufficient amount of ozone to prevent damage to Earth life-forms. G.M.B. Dobson developed these units, along with the instrument (spectrophotometer) used to measure atmospheric ozone, in the late 1920s. To understand these units better, consider that 100 Dobson units translate to 1 millimeter of thickness that the ozone would have at sea level. Ozone measured inside the hole has dropped as low as 95 Dobson units in recent years, and the areal extent of this ozone-deficient region (a more accurate description) has exceeded that of the North American continent.

People have also asked why ozone holes form at the South Pole (and to a lesser extent, the North Pole). Ozone in the upper atmosphere is destroyed by chemical reaction to chlorofluorocarbons (CFCs) and other gases that are products of pollution by industrial nations. With the exception of scientific research bases, no one lives near the South Pole. There are certainly no big cities or any widespread industrial activity present. Why should the hole appear there instead of over major urbanized countries?

Research through the decades has shown three factors necessary for ozone destruction and the creation of the "holes." First, there are circulation patterns in the stratosphere, which move from the tropics to the poles and back again. This slow, conveyor-belt-like movement transports pollution from the tropics, through populated middle latitudes, and on to the poles. Therefore, it is quite possible to bring pollutants to polar regions.

Second, during the *austral winter* (June, July, and August), the South Pole experiences total darkness and becomes the coldest place on Earth. A closed circulation pattern of extremely cold air, called the *circumpolar vortex*, forms above the dark pole. Ozone is trapped, along with destructive pollutants (like CFCs and $NO_x$), within this vortex. During the *austral spring* (September, October, and November), the incoming solar radiation starts to dissolve the vortex and powers the ozone-destroying process. Usually, the ozone hole reaches its greatest extent each year within the first 2 weeks of October.

The third factor involves stratospheric ice clouds (SIC) found many miles above the poles. These clouds, located in the stratosphere, do not affect the weather, but they act as the perfect laboratory for the ozone-destructive reaction to take place.

$$Cl + O_3 \rightarrow ClO + O_2$$

This ozone-destructive reaction takes place most efficiently on an ice crystal surface.

In light of these three factors, it becomes clear that if an ozone hole were going to appear in our atmosphere, it would necessarily appear first at the poles.

### FIGURE 4.6
For decades, satellite sensors have produced images of the ozone hole (shaded in purple) over Antarctica.
**Are there any populated areas within the area of the ozone hole?**

Sep 22 2004

## Vertical Layers of the Atmosphere

Though people function primarily in the lowest level of the atmosphere, there are times, such as when we fly in aircraft or climb a mountain, when we leave our normal altitude. The thinner atmosphere at these higher altitudes may affect us if we are not accustomed to it. Visitors to Inca ruins in the Andes or high-altitude Himalayan climbers may experience altitude sickness, and even skiers in the Rockies near mile-high Denver may need time to adjust. The air at these levels is much *thinner* than most of us are used to; by this we mean there is more empty space between air molecules and thus less oxygen and other gases in each breath of air.

As one travels from the surface to outer space, the atmosphere undergoes various changes, and it is necessary to look at the vertical layers that exist with Earth's atmospheric envelope. There are several systems used to divide the atmosphere into vertical layers. One system uses temperature and rates of temperature changes. Another uses the changes in

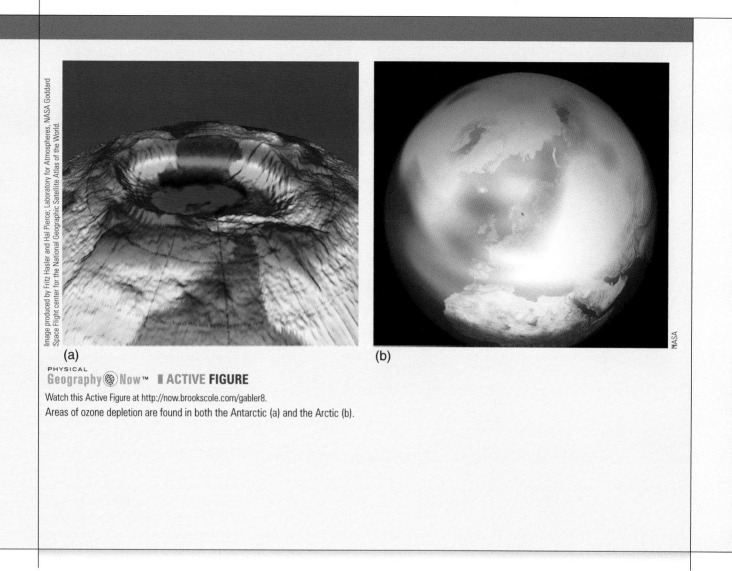

**PHYSICAL Geography Now™ ACTIVE FIGURE**
Watch this Active Figure at http://now.brookscole.com/gabler8.
Areas of ozone depletion are found in both the Antarctic (a) and the Arctic (b).

the content of the gases in the atmosphere, and yet a third deals with the functions of these various layers.

### System of Layering by Temperature Characteristics

The atmosphere can be divided into four layers according to differences in temperature and rates of temperature change ( Fig. 4.7). The first of these layers, lying closest to Earth's surface, is the **troposphere** (from Greek: *tropo,* turn—the turning or mixing zone), which extends about 8–16 kilometers (5–10 mi) above Earth. Its thickness, which tends to vary seasonally, is least at the poles and greatest at the equator. It is within the troposphere that people live and work, plants grow, and virtually all Earth's weather and climate take place.

The troposphere has two distinct characteristics that differentiate it from other layers of the atmosphere. One is that the water vapor and particulates of the atmosphere are concentrated in this one layer; they are rarely found in the atmospheric layers above the troposphere. The other characteristic of this layer is that temperature normally decreases with increased altitude. The average rate at which temperatures within the troposphere decrease with altitude is called the **normal lapse rate** (or the environmental lapse rate); it amounts to 6.5°C per 1000 meters (3.6°F/1000 ft).

The altitude at which the temperature ceases to drop with increased altitude is called the *tropopause*. It is the boundary that separates the troposphere from the **stratosphere**—the second layer of the atmosphere. The temperature of the lower part of the stratosphere remains fairly constant (about −57°C, or −70°F) to an altitude of about 32 kilometers (20 mi). It is in the stratosphere that we find the ozone layer that does so much to protect life on Earth from the sun's UV radiation. As the ozone layer absorbs UV radiation, this absorbed energy results in the release of heat, and thus temperatures increase in the upper parts of the stratosphere. Some water is available in the stratosphere, but it appears as stratospheric ice clouds. These thin veils of ice clouds have no effect on weather as we

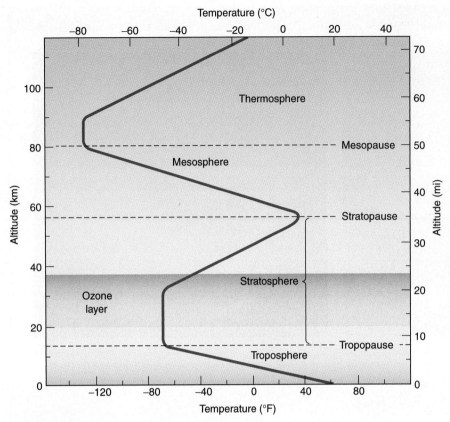

**FIGURE 4.7**
Vertical temperature changes in Earth's atmosphere are the basis for its subdivision into the troposphere, stratosphere, mesosphere, and thermosphere.
**At what altitude is our atmosphere the coldest?**

experience it. Temperatures at the *stratopause* (another boundary), which is about 50 kilometers (30 mi) above Earth, are about the same as temperatures found on Earth's surface, although little of that heat can be transferred because the air is so thin.

Above the stratopause is the **mesosphere,** in which temperatures tend to drop with increased altitude; the *mesopause* (the last boundary) separates the mesosphere from the **thermosphere,** where temperatures increase until they approach 1100°C (2000°F) at noon. Again, the air is so thin at this altitude that there is practically a vacuum and little heat can be transferred.

### System of Layering by Functional Characteristics

Astronomers, geographers, and communications experts sometimes use a different method of layering the atmosphere, one based on the protective function these layers provide. In this system, the atmosphere is divided into two distinct layers, the lowest of which is the **ozonosphere.** This layer lies approximately between 15 and 50 kilometers (10–30 mi) above the surface. The ozonosphere is another name for the ozone layer mentioned previously. Here, ozone effectively filters the UV energy from the sun and gives off heat energy instead. As we have noted, although ozone is a toxic pollutant at Earth's surface, aloft, it serves a vital function for Earth's life systems.

From about 60–400 kilometers (40–250 mi) above the surface lies the layer known as the **ionosphere.** This name denotes the ionization of molecules and atoms that occurs in this layer, mostly as a result of UV rays, X-rays, and gamma radiation. (Ionization refers to the process whereby atoms are changed to ions through the removal or addition of electrons, giving them an electrical charge.) The ionosphere in turn helps shield Earth from the harmful shortwave forms of radiation. This electrically charged layer also aids in transmitting communication and broadcast signals to distant regions of Earth. It is in the ionosphere that the auroras (defined in Chapter 3) occur. The ionosphere gradually gives way to interplanetary space ( ▌Fig. 4.8).

### System of Layering by Chemical Composition

Atmospheric chemists and physicists are at times concerned with the actual chemical makeup of the atmosphere. To this end, there is one more system to divide the atmosphere into two vertical layers. The first is termed the **homosphere** (from Greek: *homo,* same throughout). This layer begins on the surface and extends to an altitude of 80 kilometers (50 mi). In this layer, the gases in our atmosphere maintain the same percent by volume as those listed in Table 4.1. There are a few areas of concentration of specific gases, like the water vapor near Earth's surface and the ozone layer aloft, but for the most part the mixture is homogeneous. Other than rapid decreases in pressure and density while ascending through this layer, this is essentially the same air that we breathe on the surface. From an altitude of about 80 kilometers (50 mi) and reaching into the vacuum of outer space lies the **heterosphere** (from Greek: *hetero,* different). In this layer, atmospheric gases are no longer evenly mixed but begin to separate out into distinct sublayers of concentration. This separation of gases is caused by Earth's gravity in which heavier gases are pulled closer to the surface and the lighter gases drift farther outward. The regions of concentration and their corresponding gases occur in the following order: nitrogen gas ($N_2$) is the heaviest and therefore the lowermost, followed by atomic oxygen (O), then by helium (He), and finally atomic hydrogen (H)—the lightest element that concentrates at the outermost region ( ▌Fig. 4.9).

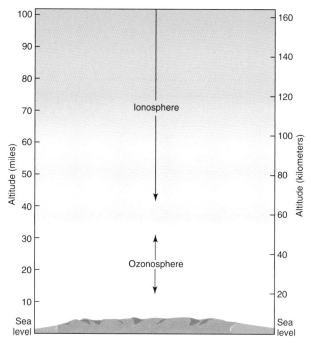

**FIGURE 4.8**
Vertical changes in Earth's atmosphere based on functions of the gases cause the atmosphere to be subdivided into the ozonosphere and the ionosphere.
**How do these layers protect life on Earth?**

**FIGURE 4.9**
The changes in chemical composition of the gases allow for the subdivision of the atmosphere into the homosphere and the heterosphere.
**What gas in the atmosphere is the last to reach out into outer space?**

It is interesting to note that, when watching television news reports from the International Space Station or NASA shuttle missions, the astronauts who we see are still in the atmosphere. The background appears black, making it look like interplanetary space, but in reality these missions still take place within the realm of the outer atmosphere. One should also keep in mind that these different layering systems can focus on the same regions. For example, note that the thermosphere, the ionosphere, and the heterosphere all occupy the same altitudes above Earth—that is, from 80 kilometers (50 mi) and outward. The names are different because of the criteria used in the differing systems.

## Effects of the Atmosphere on Solar Radiation

As the sun's energy passes through Earth's atmosphere, more than half of its intensity is lost through various processes. In addition, as discussed in Chapter 3, the amount of insolation actually received at a particular location depends on not only the processes involved but also latitude, time of day, and time of year (all of which are related to the angle at which the sun's rays strike Earth). The transparency of the atmosphere (or the amount of cloud cover, moisture, carbon dioxide, and solid particles in the air) also plays a vital role.

When the sun's energy passes through the atmosphere, several things happen to it (the following figures represent approximate averages for entire Earth; at any one location or time, they may differ): (1) 26% of the energy is reflected directly back to space by clouds and the ground; (2) 8% is *scattered* by minute atmospheric particles and returned to space as diffuse radiation; (3) 19% is *absorbed* by the ozone layer and water vapor in the clouds of the atmosphere; (4) 20% reaches Earth's surface as diffuse radiation after being scattered; (5) 27% reaches Earth's surface as direct radiation ( ■ Fig. 4.10). In other words, on a worldwide average, 47% of the incoming solar radiation eventually reaches the surface, 19% is retained in the atmosphere, and 34% is returned to space. Because Earth's energy budget is in equilibrium, the 47% received at the surface is ultimately returned to the atmosphere by processes that we now examine.

## Water as Heat Energy

As it penetrates our atmosphere, some of the incoming solar radiation is involved in energy exchanges. One such exchange involves how water in the Earth system is altered from one state to another. Water is the only substance that can exist in all three states of matter—as a solid, a liquid, and a gas—within the normal temperature range of Earth's surface. In the atmosphere, water exists as a clear, odorless gas called *water vapor*. It is also a liquid in the atmosphere, in the oceans, and in other water bodies on and beneath the surface of Earth. Water is also found within vegetation and animals. Water is a solid in snow and ice in the atmosphere, as well as on and under the surface of the colder parts of Earth.

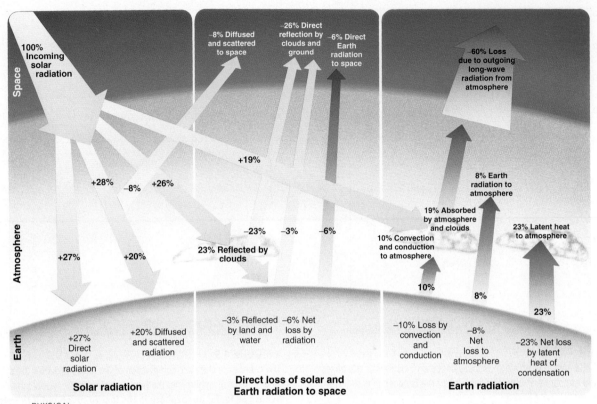

**PHYSICAL Geography Now™ ■ ACTIVE FIGURE 4.10**
Watch this Active Figure at http://now.brookscole.com/gabler8.

**Environmental Systems: Earth's Radiation Budget** From one year to the next, Earth's overall average temperature varies very little. This fact indicates that a long-term global balance, or equilibrium, must exist between the energy received and emitted by the Earth system. Note in the diagram that only 47% of the incoming solar energy reaches and is absorbed by Earth's surface. Eventually, all the energy gained by the atmosphere is lost to space. However, the radiation budget is a dynamic one. In other words, alterations to one element affect the other elements. As a result, there is growing concern that one of the elements, human activity, will cause the atmosphere to absorb more Earth-emitted energy, thus raising global temperatures.

**■ FIGURE 4.11**
The three physical states of water and the energy exchanges between them.
**Why do you suppose that some of the energy in these exchanges is referred to as "Latent Heat?"**

Not only does water exist in all three states of matter, but it can change from one state to another, as illustrated in Figure 4.11. In doing so, it is involved in the heat energy system of Earth. The molecules of a gas move faster than do those of a liquid. Thus, during the process of *condensation,* when water vapor changes to liquid water, its molecules slow down and some of their energy is released (about 590 cal/g). The molecules of a solid move even more slowly than those of a liquid, so during the process of *freezing,* when water changes to ice, additional energy is released (80 cal/g). When the process is reversed, heat must be added. Thus, *melting* requires the addition of 80 calories per gram, and *evaporation* requires the addition of 590 calories per gram. This added energy is stored in the water as *latent* (or hidden) *heat.*

## Geography's Physical Science Perspective
### Colors of the Atmosphere

When we look to the skies, we are used to seeing certain colors: the brilliant blue of a clear day, the vibrant reds and oranges of a sunset, clouds ranging from pure white to ominous gray-black, and sometimes even a colorful rainbow. These are wondrous colors, and we sometimes take them for granted. Why they exist, however, can be explained by the concept of *atmospheric scattering*. This process is explained in the text, but let's examine how it affects the colors of our atmosphere.

It is important to first realize that as sunlight is scattered in our atmosphere, certain wavelengths are scattered more than others. Each wavelength of visible light corresponds to one of the colors of the visible light spectrum. For example, a wavelength of 0.45–0.50 micrometers (1 micrometer = 1/1000 of a millimeter) corresponds to the color blue, and 0.66–0.70 micrometers to the color red. The wavelengths that are scattered, then enter the human eye, are transmitted to the human brain and translated into the colors. In other words, the wavelengths that are scattered are the colors that we see.

In 1871, Lord Rayleigh, a radiation scientist, explained that our atmospheric gases tend to scatter the shorter wavelengths of visible light. In other words, *Rayleigh scattering* makes a clear sky appear blue. In 1908, another radiation specialist, Gustav Mie, explained that oblique (or sharp) sun angles coming through the atmosphere tend to cause the scattering of the longer wavelengths of visible light. In other words, *Mie scattering* explains why sunsets bring about a red sky. Another form of scattering, called *indiscriminant scattering*, is caused by tiny water droplets and ice crystals in our atmosphere that tend to scatter all the wavelengths of visible light with equal intensity. When all the wavelengths are scattered equally, white light emerges, and this explains the white clouds we see. (Keep in mind that all clouds are white. When we see dark gray to black storm clouds, we are actually seeing white clouds plus shadows cast by the height and thickness of these massive clouds.)

Rainbows are much more complicated, and their full explanation involves a number of trigonometric functions. Rainbows are caused by water droplets suspended in the atmosphere. The droplets act like tiny glass prisms, which take the incoming white sunlight and project the colors of the spectrum. Key elements necessary to observe a rainbow involve the sun angle, water droplet density and orientation, and the observer angle. When you see a rainbow from a moving vehicle, it is only a matter of time until the angles change and the rainbow disappears.

Incidentally, the colors described above are those distinguishable by the human eye in regard to Earth's atmosphere. Humans visiting another planet with an atmosphere may never see the color blue at all. Furthermore, animals (those that are not color-blind) may see colors other than those described here.

Sunset over the Antarctic Ocean

Rainbow

Some of these energy exchanges can be easily demonstrated. For example, if you hold an ice cube in your hand, your hand feels cold because it is giving off the heat needed to melt the ice. We are cooled by evaporating perspiration from our skin because heat must be absorbed both from our skin and from the remaining perspiration, thereby lowering the temperature of both.

## Heating the Atmosphere

The 19% of direct solar radiation that is retained by the atmosphere is "locked up" in the clouds and the ozone layer and thus is not available to heat the troposphere. Other sources must be found to explain the creation of atmospheric warmth. The explanation lies in the 47% of incoming solar energy reaching Earth's surface (on both land and water) and in the transfer of heat energy from Earth back to the atmosphere. This is accomplished through such physical processes as (1) radiation, (2) conduction, (3) convection (along with the related phenomenon, advection), and (4) the latent heat of condensation.

## Processes of Heat Energy Transfer

**Radiation** The process by which electromagnetic energy is transferred from the sun to Earth is called **radiation.** We should be aware that all objects emit electromagnetic radiation. The characteristics of that radiation depend on the temperature of the radiating body. In general, the warmer the object, the more energy it will emit, and the shorter are the wavelengths of peak emission. Because the sun's absolute temperature is 20 times that of Earth, we can predict that the sun will emit more energy, and at shorter wavelengths, than Earth. This is borne out by the facts: The energy output per square meter by the sun is approximately 160,000 times that of Earth! Further, the majority of solar energy is emitted at wavelengths shorter than 4.0 micrometers, whereas most of Earth's energy is radiated at wavelengths much longer than 4.0 micrometers (see again Fig. 3.10). Thus, *shortwave radiation* from the sun reaches Earth and heats its surface, which, being cooler than the sun, gives off energy in the form of longwaves. It is this *longwave radiation* from Earth's surface that heats the lower layers of the atmosphere and accounts for the heat of the day.

**Conduction** The means by which heat is transferred from one part of a body to another or between two touching objects is called **conduction.** Heat flows from the warmer to the cooler part of a body in order to equalize temperature. Conduction actually occurs as heat is passed from one molecule to another in chainlike fashion. It is conduction that makes the bottom of your soup bowl too hot to touch.

Atmospheric conduction occurs at the interface of (zone of contact between) the atmosphere and Earth's surface. However, it is actually a minor method of heat transfer in terms of warming the atmosphere because it affects only the layers of air closest to the surface. This is because air is a poor conductor of heat. In fact, air is just the opposite of a good conductor; it is a good insulator. This property of air is why a layer of air is sometimes put between two panes of glass to help insulate the window. Air is also used as a layer of insulation in sleeping bags and ski parkas. In fact, if air were a good conductor of heat, our kitchens would become unbearable every time we turned on the stove or oven.

**Convection** In the atmosphere, as pockets of air near the surface are heated, they expand in volume, become less dense than the surrounding air, and therefore rise. This vertical transfer of heat through the atmosphere is called **convection;** it is the same type of process by which boiling water circulates in a pot on the stove. The water near the bottom is heated first, becoming lighter and less dense as it is heated. As this water tends to rise, colder, denser surface water flows down to replace it. As this new water is warmed, it too flows upward while additional colder water moves downward. This movement within the fluid is called a *convective current*. These currents set into motion by the heating of a fluid (liquid or gas) make up a *convectional system*. Such systems account for much of the vertical transfer of heat within the atmosphere and the oceans and are a major cause of clouds and precipitation.

**Advection** **Advection** is the term applied to horizontal heat transfer. There are two major advection agents within the Earth–atmosphere system: winds and ocean currents. Both agents help transfer energy horizontally between the equatorial and polar regions, thus maintaining the energy balance in the Earth–atmosphere system.

**Latent Heat of Condensation** As we have seen, when water evaporates, a significant amount of energy is stored in the water vapor as latent heat (see again Fig. 4.11). This water vapor is then transported by advection or convection to new locations where condensation takes place and the stored energy is released. This process plays a major role in the transfer of energy within the Earth system: The **latent heat of evaporation** helps cool the atmosphere while the **latent heat of condensation** helps warm the atmosphere and, in addition, is a source of energy for storm activity. The power of all severe weather is supplied by the latent heat of condensation.

## The Heat Energy Budget

**The Heat Energy Budget at Earth's Surface** Now that we know the various means of heat transfer, we are in a position to examine what happens to the 47% of solar energy that reaches Earth's surface (see again Fig. 4.10). Approximately 14% of this energy is emitted by the Earth in the form of longwave radiation. This 14% includes a net loss of 6% (of the total originally received by the atmosphere) directly to outer space; the other 8% is captured by the atmosphere. In addition, there is a net transfer back to the atmosphere (by conduction and convection) of 10 of the 47% that reached Earth. The remaining 23% returns to the atmosphere through the release of latent heat of condensation. Thus, 47% of the sun's original insolation that reached Earth's surface is all returned to other segments of the system. There has been no long-term gain or loss. Therefore, at Earth's surface, the heat energy budget is in balance.

Examination of the heat energy budget of Earth's surface helps us understand the *open energy system* that is involved in the heating of the atmosphere. The *input* in the system is the incoming shortwave solar radiation that reaches Earth's surface; this is balanced by the *output* of longwave terrestrial (Earth's) radiation back to the atmosphere and to space. As these functions adjust to remain in balance, we can say that the overall heat budget of Earth's surface is in a state of *dynamic equilibrium*.

Of course, it should be noted that the percentages mentioned earlier represent an oversimplification in that they refer to *net* losses that occur over a long period of time. In the shorter term, heat may be passed from Earth to the atmosphere and then back to Earth in a chain of cycles before it is finally released into space.

**The Heat Energy Budget in the Atmosphere** At one time or another, about 60% of the solar energy intercepted by the Earth system is temporarily retained by the atmosphere. This includes 19% of direct solar radiation absorbed by the clouds and the ozone layer, 8% emitted by *longwave radiation* from the Earth's surface, 10% transferred from the surface by *conduction* and *convection,* and 23% released by the *latent heat of condensation.* Some of this energy is recycled back to the surface for short periods of time, but eventually all of it is lost into outer space as more solar energy is received. Hence, just as was the case at Earth's surface, the heat energy budget in the atmosphere is in balance over long periods of time—a dynamically stable system. However, many scientists believe that an imbalance in the heat energy budget, with possible negative effects, could develop due to greenhouse warming.

**Variations in the Heat Energy Budget** Remember that the figures we have seen for the heat energy budget are averages for the whole Earth over many years. For any *particular* location, the heat energy budget is most likely not balanced. Some places have a surplus of incoming solar energy over outgoing energy loss, and others have a deficit. The main causes of these variations are differences in latitude and seasonal fluctuations.

As we have noted previously, the amount of insolation received is directly related to latitude ( Fig. 4.12). In the tropical zones, where insolation is high throughout the year, more solar energy is received at Earth's surface and in the atmosphere than can be emitted back into space. In the Arctic and Antarctic zones, on the other hand, there is so little insolation during the winter, when Earth is still emitting longwave radiation, that there is a large deficit for the year. Locations in the middle-latitude zones have lower deficits or surpluses, but only at about latitude 38° is the budget balanced. If it were not for the heat transfers within the atmosphere and the oceans, the tropical zones would get hotter and the polar zones would get colder through time.

At any location, the heat energy budget varies throughout the year according to the seasons, with a tendency toward a surplus in the summer or high-sun season and a tendency toward a deficit 6 months later. Seasonal differences may be small near the equator, but they are great in the middle-latitude and polar zones.

# Air Temperature
## Temperature and Heat

Although heat and temperature are highly related, they are not the same. **Heat** is a form of energy—the total kinetic energy of all the atoms that make up a substance. All substances are made up of molecules that are constantly in motion (vibrating and colliding) and therefore possess kinetic energy—the energy of motion. This energy is manifested as heat. **Temperature,** on the other hand, is the average kinetic energy of the individual molecules of a substance. When something is heated, its atoms vibrate faster, and its temperature increases. It is important to remember that the amount of heat energy depends on the mass of the substance under discussion, whereas the temperature refers to the energy of individual molecules. Thus, a burning match has a high temperature but minimal heat energy; the oceans have moderate temperatures but high–heat energy content.

## Temperature Scales

Three different scales are generally used in measuring temperature. The one with which Americans are most familiar is the **Fahrenheit scale,** devised in 1714 by Daniel Fahrenheit, a German scientist. On this scale, the temperature at which water boils at sea level is 212°F, and the temperature at which water freezes is 32°F. This scale is used in the English system of measurements.

The **Celsius scale** (also called the **centigrade scale**) was devised in 1742 by Anders Celsius, a Swedish astronomer. It is part of the metric system. The temperature at which water freezes at sea level on this scale was arbitrarily set at 0°C, and the temperature at which water boils was designated as 100°C.

The Celsius scale is used nearly everywhere but in the United States, and even in the United States, the Celsius scale is used by the majority of the scientific community. By this time, you have undoubtedly noticed that throughout this book comparable figures in both the Celsius (centigrade) and Fahrenheit scales are given side by side for temperatures. Similarly, whenever important figures for

■ **FIGURE 4.12**
Latitudinal variation in the energy budget. Low latitudes receive more insolation than they lose by reradiation and have an energy surplus. High latitudes receive less energy than they lose outward and therefore have an energy deficit.
**How do you think the surplus energy in the low latitudes is transferred to higher latitudes?**

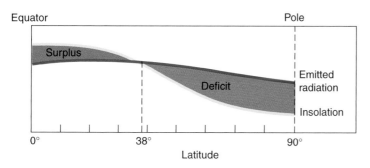

# FIGURE 4.13

The Fahrenheit and Celsius temperature scales. The scales are aligned to permit direct conversion of readings from one to the other.
**When it is 70°F, what is the temperature in Celsius degrees?**

Fahrenheit scale

Celsius scale

distance, area, weight, or speed are given, we use the metric system followed by the English system. Appendix A at the back of your text may be used for comparison and conversion between the two systems.

Figure 4.13 can help you compare the Fahrenheit and Celsius systems as you encounter temperature figures outside this book. In addition, the following formulas can be used for conversion from Fahrenheit to Celsius or vice versa:

$$C = (F - 32) \div 1.8$$
$$F = (C \times 1.8) + 32$$

The third temperature scale, used primarily by scientists, is the **Kelvin scale.** Lord Kelvin, a British radiation scientist, felt that negative temperatures were not proper and should not be used. In his mind, no temperature should ever go below zero. This scale is based on the fact that the temperature of a gas is related to the molecular movement within the gas. As the temperature of a gas is reduced, the molecular motion within the gas slows. There is a temperature at which all molecular motion stops and no further cooling is possible. This temperature, approximately −273°C, is termed *absolute zero*. The Kelvin scale uses absolute zero as its starting point.

Thus, 0°K equals −273°C. Conversion of Celsius to Kelvin is expressed by the following formula:

$$K = C + 273$$

## Short-Term Variations in Temperature

Local changes in atmospheric temperature can have a number of causes. These are related to the mechanics of the receipt and dissipation of energy from the sun and to various properties of Earth's surface and the atmosphere.

### The Daily Effects of Insolation

As we noted earlier, the amount of insolation at any particular location varies both throughout the year (annually) and throughout the day (diurnally). Annual fluctuations are associated with the sun's changing declination and hence with the seasons. Diurnal changes are related to the rotation of Earth about its axis. Each day, insolation receipt begins at sunrise, reaches its maximum at noon (local solar time), and returns to zero at sunset.

Although insolation is greatest at noon, you may have noticed that temperatures usually do not reach their maximum until 2–4 p.m. ( Fig. 4.14). This is because the insolation received by Earth from shortly after sunrise until the afternoon hours exceeds the energy

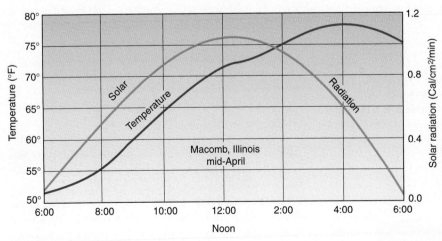

# FIGURE 4.14

Diurnal changes in insolation and temperature in mid-April for Macomb, Illinois.
**Why does temperature rise even after solar energy declines?**

being lost through Earth radiation. Hence, during that period, as Earth and atmosphere continue to gain energy, temperatures normally show a gradual increase. Sometime around 3–4 p.m., when outgoing Earth radiation begins to exceed insolation, temperatures start to fall. The daily lag of Earth radiation and temperature behind insolation is accounted for by the time it takes for Earth's surface to be heated to its maximum and for this energy to be radiated to the atmosphere.

Insolation receipt ends at sunset, but energy that has been stored in Earth's surface layer during the day continues to be lost throughout the night and the ability to heat the atmosphere decreases. The lowest temperatures occur just before dawn, when the maximum amount of energy has been emitted and before replenishment from the sun can occur. Thus, if we disregard other factors for the moment, we can see that there is a predictable hourly change in temperature called the **daily march of temperature.** There is a gentle decline from midafternoon until dawn and a rapid increase in the 8 hours or so from dawn until the next maximum is reached.

**Cloud Cover** The extent of cloud cover is another factor that affects the temperature of Earth's surface and the atmosphere ( Fig. 4.15). Weather satellites have shown that, at any time, about 50% of Earth is covered by clouds ( Fig. 4.16). This is important because a heavy cloud cover can reduce the amount of insolation a place receives, thereby causing daytime temperatures to be lower on a cloudy day. On the other hand, we also have the greenhouse effect, in which clouds, composed in large part of water droplets, are capable of absorbing heat energy radiating from Earth. Clouds therefore keep temperatures near Earth's surface warmer than they would otherwise be, especially at night. The general effect of cloud cover, then, is to moderate temperature by lowering the potential maximum and raising the potential minimum temperatures. In other words, cloud cover makes for cooler days and warmer nights.

**Differential Heating of Land and Water** For reasons we will later explain in detail, bodies of water heat and cool more slowly than the land. The air above Earth's surface is heated or cooled in part by what is beneath it. Therefore, temperatures over bodies of water or on land subjected to ocean winds (**maritime** locations) tend to be more moderate than those of land-locked places at the same latitude. Thus, the greater the **continentality** of a location (the distance removed from a large body of water), the less its temperature pattern will be modified.

**Reflection** The capacity of a surface to reflect the sun's energy is called its **albedo;** a surface with a high albedo has a high percentage of reflection. The more solar energy that is reflected back into space by Earth's surface, the less that is

**FIGURE 4.15**
The effect of cloud cover on temperatures. (a) By intercepting insolation, clouds produce lower air temperatures during the day. (b) By trapping longwave radiation from Earth, clouds increase air temperatures at night. The overall effect is a great reduction in the diurnal temperature range.
**Desert regions have large diurnal variations in temperature. Why is this so?**

absorbed for heating the atmosphere. Temperatures will be higher at a given location if its surface has a low albedo rather than a high albedo.

As you may know from experience, snow and ice are good reflectors; they have an albedo of 90–95%. This is one reason why glaciers on high mountains do not melt away in the summer or why there may still be snow on the ground on a sunny day in the spring: Solar energy is reflected away. A forest, on the other hand, has an albedo of only 10–12%, which is good for the trees because they need solar energy for photosynthesis. The albedo of cloud

**FIGURE 4.16**
This composite satellite image shows a variety of cloud cover across the globe on this particular day.
**Is there any cloud cover over the Arabian Peninsula? Over the Amazon? Over Australia?**

**FIGURE 4.17**
At low sun angles, water reflects most of the solar radiation that strikes it.
**Why is it so difficult to assign one albedo value to a water surface?**

cover varies, from 40 to 80%, according to the thickness of the clouds. The high albedo of many clouds is why much solar radiation is reflected directly back into space by the atmosphere.

The albedo of water varies greatly, depending on the depth of the water body and the angle of the sun's rays. If the angle of the sun's rays is high, smooth water will reflect little. In fact, if the sun is vertical over a calm ocean, the albedo will be only about 2%. However, a low sun angle, such as just before sunset, causes an albedo of more than 90% from the same ocean surface ( Fig. 4.17). Likewise, a snow surface in winter, when solar angles are lower, can reflect up to 95% of the energy striking it, and skiers must constantly be aware of the danger of severe burns and possible snow blindness from reflected solar radiation. In a similar fashion, the high albedo of sand causes the sides of sunbathers' legs to burn faster when they lie on the beach.

**Horizontal Air Movement** We have already seen that advection is the major mode of horizontal transfer of heat and energy over Earth's surface. Any movement of air due to the wind, whether on a large or small scale, can have a significant short-term effect on the temperatures of a given location. Thus, wind blowing from an ocean to land will generally bring cooler temperatures in summer and warmer temperatures in winter. Large quantities of air moving from polar regions into the middle latitudes can cause sharp drops in temperature, whereas air moving poleward will usually bring warmer temperatures.

**FIGURE 4.18**
Steep, normal, and low atmospheric lapse rates

## Vertical Distribution of Temperature

**Normal Lapse Rates** We have learned that Earth's atmosphere is primarily heated from the ground up as a result of longwave terrestrial radiation, conduction, and convection. Thus, temperatures in the troposphere are usually highest at ground level and decrease with increasing altitude. As noted earlier in the chapter, this decrease in the free air of approximately 6.5°C per 1000 meters (3.6°F/1000 ft) is known as the *normal lapse rate*.

The lapse rate at a particular place can vary for a variety of reasons ( Fig. 4.18). Low lapse rates can exist if denser and colder air is drained into a valley from a higher elevation or if advectional winds bring air in from a cooler region at the same altitude. In each case, the surface is cooled so that its temperature is closer to that at higher elevations directly above it. On the other hand, if the surface is heated strongly by the sun's rays on a hot summer afternoon, the air near Earth will be disproportionately warm, and the lapse rate will be steep. Fluctuations in lapse rates due to abnormal temperature conditions at various altitudes can play an important role in the weather a place may have on a given day.

**Temperature Inversions** Under certain circumstances, the normal observed decrease of temperature with increased altitude might be reversed; temperature may actually *increase* for several hundred meters. This is called a **temperature inversion.**

Some inversions take place 1000 or 2000 meters above the surface of Earth where a layer of warmer air interrupts the normal decrease in temperature with altitude ( Fig. 4.19). Such inversions tend to stabilize the air, causing less turbulence and discouraging both precipitation and the development of storms. Upper air inversions may occur when air settles slowly from the upper atmosphere. Such air is compressed as it sinks and rises in temperature, becoming more stable and less buoyant. Inversions caused by descending air are common at about 30–35° north and south latitudes.

An upper air inversion common to the coastal area of California results when cool marine air blowing in from the Pacific Ocean moves under stable, warmer, and lighter air aloft created by subsidence and compression. Such an inversion layer tends to maintain itself; that is, the cold underlying air is heavier and cannot rise through the warmer air above. Not only does the cold air resist rising or moving, but pollutants, such as smoke, dust particles, and automobile exhaust, created at Earth's surface also fail to disperse. They therefore accumulate in the lower atmosphere. This situation is particularly acute in the Los Angeles area, which is a basin surrounded by higher mountainous areas ( Fig. 4.20). Cooler air blows into the basin from the ocean and then cannot escape either horizontally, because of the landform barriers, or vertically, because of the inversion.

Some of the most noticeable temperature inversions are those that occur near the surface when Earth cools the lowest layer of air through conduction and radiation ( Fig. 4.21). In this situation, the coldest air is nearest the surface and the temperature rises with altitude. Inversions near the surface most often occur on clear nights in the middle latitudes. They may be enhanced by snow cover or the recent advection of cool, dry air into the area. Such conditions produce extremely rapid cooling of Earth's surface at night as it loses the day's insolation through radiation. Then the layers of the atmosphere that are closest to Earth are cooled by radiation and conduction more than those at higher altitudes. Calm air conditions near the surface help produce and partially result from these temperature inversions.

**Surface Inversions: Fog and Frost** Fog and frost often occur as the result of a surface inversion. Especially where Earth's surface is hilly, cold, dense surface air will tend to flow down slope and accumulate in the lower valleys. The

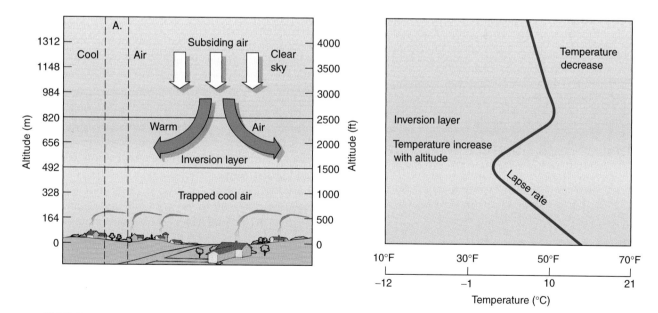

**FIGURE 4.19**
(Left) Temperature inversion caused by subsidence of air. (Right) Lapse rate associated with the column of air (A) in the left-hand drawing.
**Why is the pattern (to the right) called a temperature inversion?**

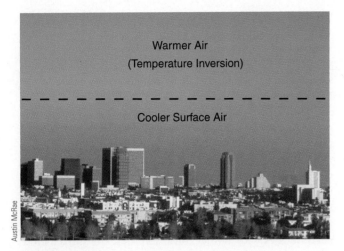

**FIGURE 4.20**
Conditions producing smog-trapping inversion in the Los Angeles area.
**Why is the air clear above the inversion?**

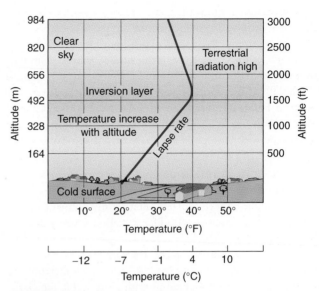

**FIGURE 4.21**
Temperature inversion caused by the rapid cooling of the air above the cold surface of Earth at night.
**What is the significance of an inversion?**

colder air on the valley floors and other low-lying areas sometimes produces fog or, in more extreme cases, a killing frost. Farmers use a variety of methods to prevent such frosts from destroying their crops. For example, fruit trees in California are often planted on the warmer hillsides instead of in the valleys. Farmers may also put blankets of straw, cloth, or some other poor conductor over their plants. This prevents the escape of Earth's heat radiation to outer space and thereby keeps the plants warmer. Large fans and helicopters are sometimes used in an effort to mix surface layers and disturb the inversion ( Fig. 4.22). Huge orchard heaters that warm the air can also be used to disturb the temperature layers. Smudge pots, an older method of preventing frost, pour smoke into the air, which provides a blanket of insulation much like blankets of cloth or straw. However, smudge pots have declined in favor because of their air-pollution potential.

## Controls of Earth's Surface Temperatures

Variations in temperatures over Earth's surface are caused by several **controls.** The major controls are (1) latitude, (2) land and water distribution, (3) ocean currents, (4) altitude, (5) landform barriers, and (6) human activity.

**Latitude**  Latitude is the most important control of temperature variation involved in weather and climate. Recall that, because of the inclination and parallelism of Earth's axis as it revolves around the sun (Chapter 3), there are distinct patterns in the latitudinal distribution of the seasonal and annual receipt of solar energy over Earth's surface. This has a direct effect on temperatures. In general, annual insolation tends to decrease from lower latitudes to higher latitudes (see again Figs. 3.16 and 3.17). Table 4.2 shows the average annual temperatures for several locations in the Northern Hemisphere. We can see that, responding to insolation (with one exception), a poleward decrease in temperature is true for these locations. The exception is near the equator itself. Because of the heavy cloud cover in equatorial regions, annual temperatures there tend to be lower than at places slightly to the north or south, where skies are clearer.

Another very simple way to see this general trend of decreasing temperatures as we move toward the poles is to think about the kinds of clothes we would take along for 1 month—say, January—if we were to visit Ciudad Bolivar, Venezuela; Raleigh, North Carolina; or Point Barrow, Alaska.

**Land and Water Distribution**  Not only do the oceans and seas of Earth serve as storehouses of water for the whole system, but they also store tremendous amounts of heat energy. Their widespread distribution makes them an important atmospheric control that does much to modify the atmospheric elements. All things heat and cool at different rates. This is especially true when comparing land to water, in that land heats and cools faster than water. There are three reasons for this phenomenon. First, the *specific heat* of water is greater than that of land. Specific heat refers to the amount of heat necessary to raise the temperature of 1 gram of any substance 1°C. Water, with a specific heat of 1 calorie/gram degree C, must absorb more heat energy than land with specific heat values of about .2 calories/gram degree C, to be raised the same number of degrees in temperature.

Second, water is *transparent* and solar energy passes through the surface into the layers below, whereas in opaque

**FIGURE 4.22**
(Left) Fans (propellers) are used to protect Washington apple orchards from frost. (Right) Smudge pots were an older method of trying to keep surface temperatures above freezing.

materials like soil and rock the energy is concentrated on the surface. Thus, a given unit of heat energy will spread through a greater volume of water than land. Third, because liquid water circulates and mixes, it can transfer heat to deeper layers within its mass. The result is that as summer changes to winter, the land cools more rapidly than bodies of water, and as winter becomes summer, the land heats more rapidly. Because the air gets much of its heat from the surface with which it is in contact, the differential heating of land and water surfaces produces inequalities in the temperature of the atmosphere above these two surfaces.

The mean temperature in Seattle, Washington, in July is 18°C (64°F), while the mean temperature during the same month in Minneapolis, Minnesota, is 21°C (70°F). Because the two cities are at similar latitudes, their annual pattern and receipt of solar energy are also similar. Therefore, their different temperatures in July must be related to a control other than latitude. Much of this difference in temperature can be attributed to the fact that Seattle is near the Pacific coast, whereas Minneapolis is in the heart of a large continent, far from the moderating influence of an ocean. Seattle stays cooler than Minneapolis in the summer because the surrounding water warms up slowly, keeping the air relatively cool. Minneapolis, on the other hand, is in the center of a large landmass that warms very quickly. In the winter, the opposite is true. Seattle is warmed by the water while Minneapolis is not. The mean temperature in January is 4.5°C (40°F) in Seattle and −15.5°C (4°F) in Minneapolis.

Not only do water and land heat and cool at different rates, but so do various land surface materials. Soil, forest, grass, and rock surfaces all heat and cool differentially and thus vary the temperatures of the overlying air.

**Ocean Currents** Surface ocean currents are large movements of water pushed by the winds. They may flow from a place of warm temperatures to one of cooler temperatures and vice versa. These movements result, as we saw in Chapter 1, from the attempt of Earth systems to reach a balance—in this instance, a balance of temperature and density.

The rotation of Earth affects the movements of the winds, which in turn affect the movement of the ocean currents. In general, the currents move in a clockwise direction in the

## TABLE 4.2
### Temperatures in the Northern Hemisphere

| LOCATION | LATITUDE | Average Annual Temperature (°C) | (°F) |
|---|---|---|---|
| Libreville, Gabon | 0°23′N | 26.5 | 80 |
| Ciudad Bolivar, Venezuela | 8°19′N | 27.5 | 82 |
| Bombay, India | 18°58′N | 26.5 | 80 |
| Amoy, China | 24°26′N | 22.0 | 72 |
| Raleigh, North Carolina | 35°50′N | 18.0 | 66 |
| Bordeaux, France | 44°50′N | 12.5 | 55 |
| Goose Bay, Labrador, Canada | 53°19′N | −1.0 | 31 |
| Markova, Russia | 64°45′N | −9.0 | 15 |
| Point Barrow, Alaska | 71°18′N | −12.0 | 10 |
| Mould Bay, NWT, Canada | 76°17′N | −17.5 | 0 |

**FIGURE 4.23**
A highly simplified map of currents in the Pacific Ocean to show their basic rotary pattern. The major currents move clockwise in the Northern Hemisphere and counterclockwise in the Southern Hemisphere. A similar pattern exists in the Atlantic.
**What path would a hurricane forming off western Africa take as it approached the United States?**

Northern Hemisphere and in a counterclockwise direction in the Southern Hemisphere (Fig. 4.23). Because the temperature of the ocean greatly affects the temperature of the air above it, an ocean current that moves warm equatorial water toward the poles (a warm current) or cold polar water toward the equator (a cold current) can significantly modify the air temperatures of those locations. If the currents pass close to land and are accompanied by onshore winds, they can have a significant impact on the coastal climate.

The Gulf Stream, with its extension, the North Atlantic Drift, is an example of an ocean current that moves warm water poleward. This warm water keeps the coasts of Great Britain, Iceland, and Norway ice free in wintertime and moderates the climates of nearby land areas (Fig. 4.24). We can see the effects of the Gulf Stream if we compare the winter conditions of the British Isles with those of Labrador in northeastern Canada. Though both are at the same latitude, the average temperature in Glasgow, Scotland, in January is 4°C (39°F), while during the same month it is −21.5°C (−7°F) in Nain, Labrador.

The California Current off the west coast of the United States helps moderate the climate of that coast as it brings cold water south. As the current swings southwest away from the coast of central California, cold bottom water is drawn to the surface, causing further chilling of the air

**FIGURE 4.24**
The Gulf Stream (the North Atlantic Drift farther eastward) is a warm current that greatly moderates the climate of northern Europe.
**Use this figure and the information gained in Figure 4.23 to discuss the route sailing ships would follow from the United States to England and back.**

> **Physical Geography Now™** Log on to Physical GeographyNow and select this chapter to work through a Geography Interactive activity on "World Currents" (click Waves, Tides & Current, World Currents).

above. San Francisco's cool summers (July average: 14°C, or 58°F) show the effect of this current.

**Altitude** As we have seen, temperatures within the troposphere decrease with increasing altitude. In Southern California, you can find snow for skiing if you go to an altitude of 2400–3000 meters (8000–10,000 ft). Mount Kenya, 5199 meters (17,058 ft) high and located at the equator, is cold enough to have glaciers. Anyone who has hiked upward 500, 1000, or 1500 meters in midsummer has experienced a decline in temperature with increasing altitude. Even if it is hot on the valley floor, you may need a sweater once you climb a few thousand meters. The city of Quito, Ecuador, only 1° south of the equator, has an average temperature of only 13°C (55°F) because it is located at an altitude of about 2900 meters (9500 ft) ( Fig. 4.25). This concept will be discussed again when dealing with highland climates in Chapter 10.

**Landform Barriers** Landform barriers, especially large mountain ranges, can block movements of air from one place to another and thus affect the temperatures of an area. For example, the Himalayas keep cold, wintertime Asiatic air out of India, giving the Indian subcontinent a year-round tropical climate. Mountain orientation can create some significant differences as well. In North America, for example, southern slopes face the sun and tend to be warmer than the shady north-facing slopes. Snowcaps on the south-facing slopes may have less snow and may exist at a higher elevation. North-facing slopes usually have more snow, and it extends to lower elevations.

**Human Activities** Human beings, too, may be considered "controls" of temperature. The likelihood that they will be attracted to a large city can result in an *urban heat island*. This is a measurable pocket of warm air produced by a large urban area. Cities represent areas where human activity is concentrated, large densities of population live and work in temperature-controlled environments (utilizing heating and air conditioning units), and people develop industries that burn fossil fuels. Rural areas lack the concentration of human population and have no large heat-producing industries. In cities, many thousands of automobiles can add heat to the air, whereas rural areas may experience only limited numbers of farm vehicles. Even the building material of cities (concrete, asphalt, glass, and metal) can heat up quickly (with low specific heat and low albedo values) during the day, as opposed to the grass, trees, and cropland of the surrounding countryside, which react more slowly to insolation.

**FIGURE 4.25**
Snow-capped mountains show the visual evidence that temperatures decrease with altitude.
**How fast do temperatures decrease with height in the troposphere?**

For these reasons and others, cities are generally warmer than the surrounding countryside (Fig. 4.26). In addition, human activities like destroying forests, draining swamps, or creating large reservoirs can significantly affect local climatic patterns and, possibly, world temperature patterns as well.

## Temperature Distribution at Earth's Surface

Displaying the distribution of temperatures over the surface of the Earth requires a mapping device called **isotherms.** Isotherms (from Greek: *isos,* equal; *therm,* heat) are defined as lines that connect points of equal temperature. When constructing isothermal maps showing temperature distribution, we need to account for elevation by adjusting temperature readings to what they would be at sea level. This adjustment means adding 6.5°C for every 1000 meters of elevation (the normal lapse rate). The rate of temperature change on an isothermal map is called the **temperature gradient.** Closely spaced isotherms indicate a steep temperature gradient (a rapid temperature change over a shorter distance), and widely spaced lines indicate a weak one (a slight temperature change over a longer distance).

Figures 4.27 and 4.28 show the horizontal distribution of temperatures for Earth at two critical times, during January and July, when the seasonal extremes of high and low temperatures are most obvious in the Northern and Southern Hemispheres. The easiest feature to recognize on both maps is the general orientation of the isotherms; they run nearly east-west around Earth, as do the parallels of latitude.

A more detailed study of Figures 4.27 and 4.28 and a comparison of the two maps reveal some additional important features. The highest temperatures in January are in the Southern Hemisphere; in July, they are in the Northern Hemisphere. Comparing the latitudes of Portugal and Southern Australia

**FIGURE 4.26**
This image shows heat emitted from Salt Lake City, Utah. The bright orange/red colors display the hotter temperatures. By comparison look at the cooler Wasatch Mountains to the east.
**Where does the heat seem to be concentrated?**

**FIGURE 4.27**
Average sea-level temperatures in January (°C).

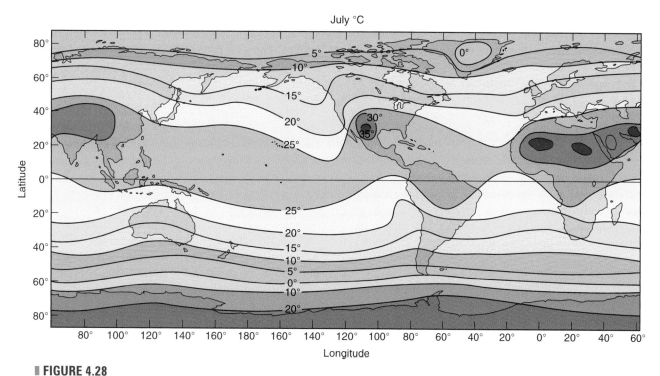

**FIGURE 4.28**
Average sea-level temperatures in July (°C).
**Observe the temperature gradients between the equator and northern Canada in January and July. Which is greater? Why?**

can demonstrate this point. Note on the July map that Portugal in the Northern Hemisphere is nearly on the 20°C isotherm, whereas in Southern Australia in the Southern Hemisphere the average July temperature is around 10°C, even though the two locations are approximately the same distance from the equator. The temperature differences between the two hemispheres are again a product of insolation, this time changing as the sun shifts north and south across the equator between its positions at the two solstices.

Note that the greatest deviation from the east–west trend of temperatures occurs where the isotherms leave large landmasses to cross the oceans. As the isotherms leave the land, they usually bend rather sharply toward the pole in the hemisphere experiencing winter and toward the equator in the summer hemisphere. This behavior of the isotherms is a direct reaction to the differential heating and cooling of land and water. The continents are hotter than the oceans in the summer and colder in the winter. Other interesting features on the January and July maps can be mentioned briefly. Note that the isotherms poleward of 40° latitude are much more regular in their east–west orientation in the Southern than in the Northern Hemisphere. This is because in the Southern Hemisphere (often called the "water hemisphere") there is little land south of 40°S latitude to produce land and water contrasts. Note also that the temperature gradients are much steeper in winter than in summer in both hemispheres. The reason for this

can be understood when you recall that the tropical zones have high temperatures throughout the year, whereas the polar zones have large seasonal differences. Hence, the difference in temperature between tropical and polar zones is much greater in winter than in summer.

As a final point, observe the especially sharp swing of the isotherms off the coasts of eastern North America, southwestern South America, and southwestern Africa in January and off Southern California in July. In these locations, the normal bending of the isotherms due to land–water differences is augmented by the presence of warm or cool ocean currents.

## Annual March of Temperature

Isothermal maps are commonly plotted for January and July because there is a lag of about 30–40 days from the solstices, when the amount of insolation is at a minimum or maximum (depending on the hemisphere), to the time minimum or maximum temperatures are reached. This **annual lag of temperature** behind insolation is similar to the daily lag of temperature explained previously. It is a result of the changing relationship between incoming insolation and outgoing Earth radiation.

Temperatures continue to rise for a month or more after the summer solstice because insolation continues to exceed radiation loss. Temperatures continue to fall after the winter solstice until the increase in insolation finally matches Earth's radiation. In short, the lag exists because it takes time for Earth to heat or cool and for those temperature changes to be transferred to the atmosphere.

The annual changes of temperature for a location can be plotted in a graph ( Fig. 4.29). The mean temperature for each month in a place such as Peoria, Illinois, is recorded and a line drawn connecting the 12 temperatures. The mean monthly temperature is the average of the daily mean temperatures recorded at a weather station during a month. The daily mean temperature is the average of the temperatures for a 24-hour period. Such a temperature graph, depicting the **annual march of temperature,** shows both the decrease in solar radiation, as reflected by a decrease in temperature, from midsummer to midwinter and the increase in temperature from midwinter to midsummer caused by the increase in solar radiation.

## Weather and Climate

We frequently use the words *weather* and *climate* in general conversation, but it is important in a science course to carefully distinguish between the terms. **Weather** refers to the condition of atmospheric elements at a given time and for a specific area. That area could be as large as the New York metropolitan area or as small and specific as a weather observation station. It is the lowest layer, the troposphere, that exhibits Earth's weather and is of the greatest interest to physical geographers and weather forecasters who survey the changing conditions of the atmosphere in the field of study known as **meteorology.**

Many observations of the weather of a place over a period of at least 30 years provide us with a description of its climate. **Climate** describes an area's average weather, but it also includes those common deviations from the normal or average that are likely to occur and the processes that make them so. Climates also include extreme situations, which can be very significant. Thus, we could describe the climate of the southeastern United States in terms of average temperatures and precipitation through a year, but we would also have to include mention of the likelihood of events such as hurricanes and snowstorms during certain periods of the year. **Climatology** is the study of the varieties of climates, both past and present, found on our planet and their distribution over its surface.

Weather and climate are of prime interest to the physical geographer because they affect and are interrelated with all of Earth's environments. The changing conditions of atmospheric elements such as temperature, rainfall, and wind affect soils and vegetation, modify landforms, cause flooding of towns and farms, and in a multitude of other ways, influence the function of systems in each aspect of Earth's physical environments.

**FIGURE 4.29**
The annual march of temperature at Peoria, Illinois, and Sydney, Australia.
**Why do these two locations have opposite temperature curves?**

**FIGURE 4.30**
(a) Weather, such as, a rainy day, is a short-term meteorological event. (b) This tropical rainforest in Peru, however, comes about as the result of a long-term rainy climate.
**Can you make the same weather versus climate comparisons with snow?**

Five basic **elements** of the atmosphere serve as the "ingredients" of weather and climate: (1) solar energy (or insolation), (2) temperature, (3) pressure, (4) wind, and (5) precipitation. We must examine these elements in order to understand and categorize weather and climate. Thus, a weather forecast will generally include the present temperature, the probable temperature range, a description of the cloud cover, the chance of precipitation, the speed and direction of the winds, and air pressure.

In Chapter 3, we noted that the amount of solar energy received at one place on Earth's surface varies during a day and throughout the year. The amount of insolation a place receives is the most important weather element; the other four elements depend in part on the intensity and duration of solar energy.

The temperature of the atmosphere at a given place on or near the surface of Earth is largely a function of the insolation received at that location. It is also influenced by many other factors, such as land and water distribution and altitude. Unless there is some form of precipitation occurring, the temperature of the air may be the first element of weather that we describe when someone asks us what it is like outside.

However, if it is raining or foggy or snowing, we will probably notice and mention that condition first. We are less aware of the amount of water vapor or moisture in the air (except in very arid or humid areas). However, moisture in the air is a vital weather element in the atmosphere, and its variations play an important role in the likelihood of precipitation (Fig. 4.30).

We all know that weather varies. Because it is the momentary state of the atmosphere at a given location, it varies in both time and place. There are even variations in the amount that weather varies. In some places and at some times of the year, the weather changes almost daily. In other places, there may be weeks of uninterrupted sunshine, blue skies, and moderate temperatures followed by weeks of persistent rain. A few places experience only minor differences in the weather throughout the year. The language of the original people of Hawaii is said to have no word for weather because conditions there varied so little.

# Complexity of Earth's Energy Systems

This chapter was designed to show the variations of Earth's energy systems and dynamic balances. These variations are the result of complex interrelationships between the characteristics of Earth and its atmosphere and the energy gained and lost by Earth's environments. The variations are both horizontal across the surface and vertical through our atmosphere. Further, they vary on both daily and seasonal time frames.

Variations of Earth's energy systems impose both diurnal and annual rhythms on our agricultural activities, recreational pursuits, clothing styles, architecture, and energy bills. Human activities are constantly influenced by temperature changes, which reflect the input–output patterns of Earth's energy systems.

# Define & Recall

photosynthesis
greenhouse effect
ozone
troposphere
normal lapse rate
stratosphere
mesosphere
thermosphere
ozonosphere
ionosphere
homosphere
heterosphere
radiation

conduction
convection
advection
latent heat of evaporation
latent heat of condensation
heat
temperature
Fahrenheit scale
Celsius (centigrade) scale
Kelvin scale
daily march of temperature
maritime
continentality

albedo
temperature inversion
control (temperature)
isotherm
temperature gradient
annual lag of temperature
annual march of temperature
weather
meteorology
climate
climatology
element (atmosphere)

# Discuss & Review

1. Why is it useful to think of the atmosphere as a thin film of air? As an ocean of air?
2. Name the gases of which the atmosphere is composed and give the percentage of the total that each supplies.
3. What function does ozone play in the support of life on Earth? Where and how is ozone formed?
4. How is the atmosphere subdivided? In what levels do you live? Have you been in any of the other levels?
5. How does Earth's atmosphere affect incoming solar radiation (insolation)? By what processes is insolation prevented from reaching Earth's surface? What percentages are involved in a generalized situation? What percentages reach the surface, and by what processes?
6. Discuss the role of water in energy exchange. What characteristics of water make it so important?
7. How is the atmosphere heated from Earth's surface? What processes and percentages are involved in a generalized situation?
8. What is meant by Earth's heat energy budget? List and define the important energy exchanges that keep it in balance.
9. What is the temperature in Fahrenheit degrees today in your area? In Celsius degrees?
10. At what time of day does insolation reach its maximum? Its minimum? Compare this to the daily temperature maximum and minimum
11. How is albedo a factor in your selection of outdoor clothes on a hot, sunny day? On a cold, clear winter day?
12. What is a temperature inversion? Give several reasons why temperature inversions occur.
13. Why do citrus growers use wind machines and heaters? Describe any techniques that you are familiar with to prevent frost damage to plants in your area.
14. Would you expect an area like Seattle to have a milder or a harsher winter than Grand Forks, North Dakota? Why?
15. Describe the behavior of the isotherms in Figures 4.27 and 4.28. What factors cause the greatest deviation from an east–west trend? What factors cause the greatest differences between the January and July maps?
16. What is the difference between meteorology and climatology?
17. What are the basic characteristics that we call *atmospheric elements* of weather and climate?
18. What factors cause variation in the elements of weather and climate?

## Consider & Respond

1. Convert the following temperatures to Fahrenheit: 20°C, 30°C, and 15°C.
2. Convert the following temperatures to Celsius: 60°F, 15°F, and 90°F.
3. Convert the temperatures in items 1 and 2 to Kelvin.
4. Refer to Figure 4.10. List the major means by which the atmosphere gains heat and loses heat.
5. What are the major weather and climate controls that operate in your area?
6. The normal lapse rate is 6.5°C/1000 meters. If the surface temperature is 25°C, what is the air temperature at 10,000 meters above Earth's surface? Convert your answer to degrees Fahrenheit.
7. Refer to Figures 4.27 and 4.28. What location on Earth's surface exhibits the greatest annual range of temperature? Why?

CHAPTER 5

The swirling circulation patterns seen on Earth are created by changes in pressure and winds.  NOAA Environmental Visualization Program and M. Manyin NASA

# Atmospheric Pressure, Winds, and Circulation Patterns

PHYSICAL
Geography Now™  This icon, appearing throughout the book, indicates an opportunity to explore interactive tutorials, animations, or practice problems at now.brookscole.com/gabler8

An individual gas molecule weighs almost nothing; however, the atmosphere as a whole has considerable weight and exerts an average pressure of 1034 grams per square centimeter (14.7 lb/sq in.) on Earth's surface. The reason why people are not crushed by this atmospheric pressure is that we have air and water inside us—in our blood, tissues, and cells—exerting an equal outward pressure that balances the inward pressure of the atmosphere. Atmospheric pressure is important because variation in pressure within the Earth–atmosphere system creates our atmospheric circulation and thus plays a major role in determining our weather and climate.

In 1643, Evangelista Torricelli, a student of Galileo, performed an experiment that was the basis for the invention of the *mercury barometer,* an instrument that measures atmospheric (also called barometric) pressure. Torricelli took a tube filled with mercury and inverted it in an open pan of mercury. The mercury inside the tube fell until it was at a height of about 76 centimeters (29.92 in.) above the mercury in the pan, leaving a vacuum

# CHAPTER PREVIEW

**Latitudinal differences in temperature (as a result of differential receipt of insolation) provide a partial explanation for latitudinal differences in pressure.**

- What is the relation between temperature and pressure?
- Why is this only a partial explanation?

**The fact that land heats and cools more rapidly than water is of significance not only to world patterns of temperature but also to world patterns of pressure, winds, and precipitation.**

- How can you explain this fact?
- What effect does this fact have on world patterns?
- In what ways can it affect human beings?

**The horizontal transfer of air (wind) is a direct result of atmospheric movement to adjust for pressure inequalities.**

- How do these inequalities develop?
- How does wind adjust for high and low pressure?

**Convergent and divergent wind systems (cyclones and anticyclones) in the Northern Hemisphere are characterized by air systems spiraling in opposite directions from one another and in opposite directions from their counterparts in the Southern Hemisphere.**

- In what directions (clockwise or counterclockwise) do these systems spiral in the Northern and Southern Hemispheres?

**Planetary (global) wind systems in association with global pressure patterns play a major role in global circulation.**

- What are the primary sources of the planetary (global) winds?
- What are the six major planetary (global) wind belts or zones, and what are their chief characteristics?
- Why do the wind belts migrate with the seasons?

**Upper air winds and atmospheric circulation play a major role in controlling surface weather and climatic conditions.**

- What is upper air circulation like?
- How does it affect surface conditions?
- How are human beings affected?

**Ocean currents constitute a system of global circulation that can exert significant influence on the atmospheric conditions of nearby land areas.**

- What is the driving force behind the ocean current circulation patterns?
- How do ocean currents affect atmospheric conditions of land areas?

**El Niños can have a devastating impact on our global weather.**

- What is an El Niño?
- How does it influence global weather?
- Why have scientists recently become better able to predict the onset of an El Niño?

**FIGURE 5.1**
A simple mercury barometer. Standard sea-level pressure of 1013.2 millibars will cause the mercury to rise 76 centimeters (29.92 in.) in the tube.
**When air pressure increases, what happens to the mercury in the tube?**

bubble at the closed end of the tube (Fig. 5.1). At this point, the pressure exerted by the atmosphere on the open pan of mercury was equal to the pressure from the mercury trying to drain from the tube. Torricelli observed that as the air pressure increased, it pushed the mercury up higher into the tube, increasing the height of the mercury until the pressure exerted by the mercury (under the pull of gravity) would equal the pressure of the air. On the other hand, as the air pressure decreased, the mercury level in the column dropped.

In the strictest sense, a mercury barometer does not actually measure the pressure exerted by the atmosphere on Earth's surface, but instead measures the *response* to that pressure. That is, when the atmosphere exerts a specific pressure, the mercury will respond by rising to a specific height (Fig. 5.2). Meteorologists usually prefer to work with actual pressure units. The unit most often used is the millibar (mb). Standard sea-level pressure of 1013.2 millibars will cause the mercury to rise 76 centimeters (29.92 in.).

Our study of the atmospheric elements that combine to produce weather and climate has to this point focused on the fundamental influence of solar energy on the global distributional patterns of temperature. The unequal receipt of insolation by latitude over Earth's surface produces temperature patterns that vary from the equator to the poles. In this chapter, we learn that these temperature differences are one of the major causes of the development of patterns of higher and lower pressure that also vary with latitude. In addition, we examine patterns of another kind—patterns of movement or,

**FIGURE 5.2**
This mercury barometer is bolted to the wall of the College Heights Weather Station in Bowling Green, Kentucky.
**Why must this instrument be so tall to work properly?**

more properly, circulation, in which both energy and matter travel cyclically through Earth subsystems.

Geographers are particularly interested in circulation patterns because they illustrate spatial interaction, one of geography's major themes introduced in Chapter 1. Patterns of movement between one place and another reveal that the two places have a relationship and prompt geographers to seek both the nature and effect of that relationship. It is also important to understand the causes of the spatial interaction taking place. As we examine the circulation patterns featured later in this chapter, you should make a special effort once again to trace each pattern back to the fundamental influence of solar energy.

# Variations in Atmospheric Pressure

## Vertical Variations in Pressure

Imagine a pileup of football players during a game. The player on the bottom gets squeezed more than a player near the top because he has the weight of all the others on top of him. Similarly, air pressure decreases with elevation, for the higher we go, the more diffused the air molecules become. The increased intermolecular space results in less pressure. In fact, at the top of Mount Everest (elevation 8848 m, or 29,028 ft), the air pressure is only about one third the pressure at sea level.

Humans are usually not sensitive to small, everyday variations in air pressure. However, when we climb or fly to altitudes significantly above sea level, we become aware of the effects of air pressure on our system. When jet aircraft fly at 10,000 meters (33,000 ft), they have to be pressurized and nearly airtight so that a near-sea-level pressure can be maintained. Even then, the pressurization may not work perfectly, so our ears may pop as they adjust to a rapid change in pressure when ascending or descending. Hiking or skiing at heights that are a few thousand meters in elevation will affect us if we are used to the air pressure at sea level. The reduced air pressure means less oxygen is contained in each breath of air. Thus, we sometimes find that we get out of breath far more easily at high elevations until our bodies adjust to the reduced air pressure and corresponding drop in oxygen level.

Changes in air pressure are not solely related to altitude. At Earth's surface, small but important variations in pressure are related to the intensity of insolation, the general movement of global circulation, and local humidity and precipitation. Consequently, a change in air pressure at a given locality often indicates a change in the weather. Weather systems themselves can be classified by the structure and tendency toward change of their pressure.

## Horizontal Variations in Pressure

The causes of horizontal variation in air pressure are grouped into two types: thermal (determined by temperature) and dynamic (related to motion of the atmosphere).

We look at the simpler thermal type first. In Chapter 4, we saw that Earth is heated unevenly because of unequal distribution of insolation, differential heating of land and water surfaces, and different albedos of surfaces. One of the basic laws of gases is that the pressure and density of a given gas vary inversely with temperature. Thus, during the day, as Earth's surface heats the air in contact with it, the air expands in volume and decreases in density. Such air has a tendency to rise as its density decreases. When the warmed air rises, there is less air near the surface, with a consequent decrease in surface pressure. The equator is an area where such low pressure occurs.

In an area with cold air, there is an increase in density and a decrease in volume. This causes the air to sink and pressure to increase. The poles are areas where such high pressures occur. Thus, the constant low pressure in the equatorial zone and the high pressure at the poles are thermally induced.

From this we might expect a gradual increase in pressure from the equator to the poles to accompany the gradual decrease in average annual temperature. However, actual readings taken at Earth's surface indicate that pressure does not increase in a regular fashion poleward from the equator. Instead, there are regions of high pressure in the subtropics and regions of low pressure in the subpolar regions. The dynamic causes of these zones, or *belts,* of high and low pressure are more complex than the thermal causes.

These dynamic causes are related to the rotation of Earth and the broad patterns of circulation. For example, as air rises steadily at the equator, it moves toward the poles. Earth's rotation, however, causes the poleward-flowing air to drift to the east. In fact, by the time it is over the subtropical regions, the air is flowing from west to east. This bending of the flow as it moves poleward impedes movement and causes the air to pile up over the subtropics, which results in increased pressure at Earth's surface there.

With high pressure over the polar and subtropical regions, dynamically induced areas of low pressure are created between them, in the subpolar region. As a result, air flows from the highs to the lows, where it rises. Thus, both the subtropical and subpolar pressure regions are dynamically induced. This example describes horizontal pressure variations on a global scale. We concentrate on this scale later in this chapter.

# Basic Pressure Systems

Before we begin our discussion of circulation patterns leading up to the global scale, we must start by describing the two basic types of pressure systems: the **low,** or **cyclone,** and the **high,** or **anticyclone.** These are represented by the capital letters **L** and **H** that we commonly see on TV, newspaper, and official weather maps.

A low, or cyclone, is an area where air is ascending. As air moves upward away from the surface, it relieves pressure from that surface. In this case, barometer readings will begin to fall. A high, or anticyclone, is just the opposite. In a high, air is descending toward the surface and thus barometer readings will begin to rise, indicating an increased pressure on the surface. Lows and highs are illustrated in Figure 5.3.

## Convergent and Divergent Circulation

As we have just seen, winds blow toward the center of a cyclone and can be said to *converge* toward it. Hence, a cyclone is a closed pressure system whose center serves as the focus for **convergent wind circulation.** The winds of an anticyclone blow away from the center of high pressure and are said to be *diverging.* In the case of an anticyclone, the center

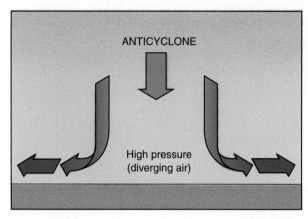

**FIGURE 5.3**
Winds converge and ascend in cyclones (low pressure centers) and descend and diverge from anticyclones (high pressure centers).
**How is temperature related to the density of air?**

of the system serves as the source for **divergent wind circulation.** Figure 5.3 shows converging and diverging winds moving in straight paths. This is not a true picture of reality. In fact, winds moving out of a high and into a low do so in a spiraling motion created by another force, which we cover in the chapter section on wind.

## Mapping Pressure Distribution

Geographers and meteorologists can best study pressure systems when they are mapped. In mapping air pressure, we reduce all pressures to what they would be at sea level, just as we changed temperature to sea level in order to eliminate altitude as a factor. The adjustment to sea level is especially important for atmospheric pressure because the variations due to altitude are far greater than those due to atmospheric dynamics and would tend to mask the more meteorologically important regional differences.

**Isobars** (from Greek: *isos,* equal; *baros,* weight) are lines drawn on maps to connect places of equal pressure. When the isobars appear close together, they portray a significant difference in pressure between places, hence a strong **pressure gradient.** When the isobars are far apart, a weak pressure gradient is indicated. When depicted on a map, high and low pressure cells are outlined by concentric isobars that form a closed system around centers of high or low pressure.

# Wind

**Wind** is the horizontal movement of air in response to differences in pressure. Winds are the means by which the atmosphere attempts to balance the uneven distribution of pressure over Earth's surface. The movements of the wind also play a major role in correcting the imbalances in radiational heating and cooling that occur over Earth's surface.

On average, locations below 38° latitude receive more radiant energy than they lose, whereas locations poleward of 38° lose more than they gain (see again Fig. 4.12). Our global wind system transports energy poleward to help maintain an energy balance. The global wind system also gives rise to the ocean currents, which are another significant factor in equalizing the energy imbalance. Thus, without winds and their associated ocean currents, the equatorial regions would get hotter and the polar regions colder through time.

Besides serving a vital function in the advectional (horizontal) transport of heat energy, winds also transport water vapor from the air above bodies of water, where it has evaporated, to land surfaces, where it condenses and precipitates. This allows greater precipitation over land surfaces than could otherwise occur. In addition, winds exert influence on the rate of evaporation itself. Furthermore, as we become more aware and concerned about the effect that the burning of fossil fuels has on our atmosphere, we look for alternate energy sources. Natural sources such as water, solar energy, and wind become increasingly attractive alternatives to fossil fuels. They are clean, abundant, and renewable.

## Pressure Gradients and Winds

Winds vary widely in velocity, duration, and direction. Much of their strength depends on the size or strength of the pressure gradient to which they are responding. As we noted previously, *pressure gradient* is the term applied to the rate of change of atmospheric pressure between two points (at the same elevation). The greater this change—that is, the steeper the pressure gradient—the greater will be the wind response ( Fig. 5.4). Winds tend to flow down a pressure gradient from high pressure to low pressure, just as water flows down a slope from a high point to a low one. A useful little rhyme,

## Geography's Environmental Science Perspective
### Harnessing the Wind

For centuries, windmills provided the power to pump water and grind grain in rural areas throughout the world. But the widespread availability of inexpensive electricity changed the role of most windmills to that of a nostalgic tourist attraction. Should we then conclude that energy from the wind is only a footnote in the history of power? In no way is that a reasonable assumption if mounting needs for electricity and increasing problems from atmospheric pollution associated with fossil fuels are taken into consideration.

Wind power is an inexhaustible source of clean energy. Although the cost of electrical energy produced by the wind depends on favorable sites for the location of wind turbines, wind power is already cost competitive with power produced from fossil fuels. One expert calls wind generation the fastest-growing electricity-producing technology in the world. During the last decade, power production from the wind increased more than 25%. Much of the growth was in Europe, where most of the world's 17,000 megawatts of wind power is generated. As examples, 13% of Denmark's power and more than 20% of the power in the Netherlands, Spain, and Germany is supplied by the wind.

Two criteria are more important than others in the location of wind turbines. The site must have persistent strong winds, and it must be in an already developed region so that the power from the turbines can be linked directly to an existing electrical grid system. Although individual wind turbines (such as those located on farms scattered throughout the Midwest and Great Plains of the United States) can be found producing electricity, most wind power is generated from wind farms. These are long rows, or more concentrated groups, of as many as 50 or more turbines. Each turbine can economically extract up to 60% of the wind's energy at minimum wind speeds of 20 kilometers (12 mi) per hour, although higher wind speeds are desirable. Because the power generated is proportional to the cube of the wind speed, a doubling of wind velocity increases energy production eight times.

Although North America currently lags far behind Europe in the production of energy from the wind, the continent has great potential. Excellent sites for the location of wind farms exist throughout the open plains of North America's interior and along its coasts from the Maritime Provinces of Canada to Texas and from California to the Pacific Northwest. In addition, the newest wind-power technology places wind farms out of sight and sound in offshore locations that avoid navigation routes and marine-life sanctuaries. And North America has some of the largest coastlines in the world with major adjacent power needs. The sites are available, the technology has been developed, the costs are competitive, and the resolve to shift from fossil fuels is growing. Is it not time for power from the winds to come to North America?

Centuries-old wind power still in use in the Dutch Polders.

Wind farm in coastal Australia. Why is a coastal location an advantage when choosing a site for a wind farm?

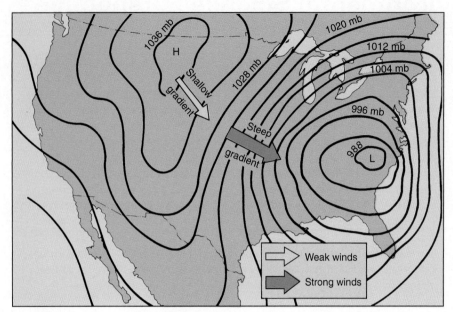

**ACTIVE FIGURE 5.4**

Watch this Active Figure at http://now.brookscole.com/gabler8.
The relationship of wind to the pressure gradient: The steeper the pressure gradient, the stronger will be the resulting wind.
**Where else on this figure (other than the area indicated) would winds be strong?**

object in motion and the latitudinal location of the object. The higher the latitude, the greater will be the Coriolis effect (Fig. 5.5). In fact, not only does the Coriolis effect decrease at lower latitudes, but it does not exist at the equator. Also, the faster the object is moving, the greater will be the apparent deflection, and the greater the distance something must travel, the greater will be the Coriolis effect.

As we have said, anything that moves horizontally over Earth's surface exhibits the Coriolis effect. Thus, both the atmosphere and the oceans are deflected in their movements. Winds in the Northern Hemisphere moving across a gradient from high to low pressure are apparently deflected to the right of their expected path (and to the left in the Southern Hemisphere). In addition, when considering winds at Earth's surface, we must take into account another force. This force, **friction,** interacts with the pressure gradient and the Coriolis effect.

"Winds always blow, from high to low," will always remind you of the direction of surface winds. The steeper the pressure gradients involved, the faster and stronger will be the winds. Yet wind does not flow directly from high to low, as we might expect, because other factors also affect the direction of wind.

## The Coriolis Effect and Wind

Two factors, both related to our Earth's rotation, greatly influence wind direction. First, our fixed-grid system of latitude and longitude is constantly rotating. Thus, our frame of reference for tracking the path of any free-moving object—whether it is an aircraft, a missile, or the wind—is constantly changing its position. Second, the speed of rotation of Earth's surface increases as we move equatorward and decreases as we move toward the poles (see again Fig. 3.12). Thus, to use our previous example, someone in St. Petersburg (60° north latitude), where the distance around a parallel of latitude is about half that at the equator, moves at about 840 kilometers per hour (525 mph) as Earth rotates, while someone in Kampala, Uganda, near the equator, moves at about 1680 kilometers per hour (1050 mph).

Because of these Earth rotation factors, anything moving horizontally appears to be deflected to the right of the direction in which it is traveling in the Northern Hemisphere and to the left in the Southern Hemisphere. This apparent deflection is termed the **Coriolis effect.** The degree of deflection, or curvature, is a function of the speed of the

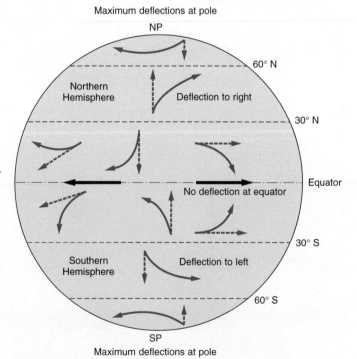

**FIGURE 5.5**

Schematic illustration of the apparent deflection (Coriolis effect) of an object caused by Earth's rotation when an object (or the wind) moves north, south, east, or west in both hemispheres.
**If no Coriolis effect exists at the equator, where would it be at maximum?**

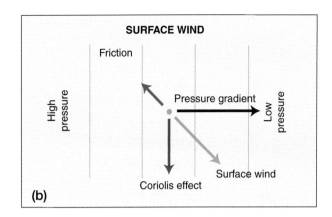

**FIGURE 5.6**
These Northern Hemisphere examples illustrate that (a) in a geostrophic wind, as a parcel of air starts to flow down the pressure gradient, the Coriolis effect causes it to veer to the right until the pressure gradient and Coriolis effect reach an equilibrium and the wind flows between isobars. (b) In a surface wind, this equilibrium is upset by friction, which reduces the wind speed. Because the Coriolis effect is a function of wind speed, it also is reduced. With the Coriolis effect reduced, the pressure gradient dominates, and the wind now flows across isobars in the direction of low pressure.
**If the amount of friction increased, would the surface wind be closer to the pressure gradient, closer to the Coriolis effect, or unchanged?**

Geography Now™ Log on to Physical GeographyNow and select this chapter to work through a Geography Interactive activity on the "Coriolis Force."

## Friction and Wind

Above Earth's surface, frictional drag is of little consequence to wind development. At this level, the wind starts down the pressure gradient and turns 90° in response to the Coriolis effect. At this point, the pressure gradient is balanced by the Coriolis effect, and the wind, termed a **geostrophic wind,** flows parallel to the isobars ( Fig. 5.6a).

However, at or near Earth's surface (up to about 1000 m above the surface), frictional drag is important because it reduces the wind speed. A reduced wind speed in turn reduces the Coriolis effect, but the pressure gradient is not affected. With the pressure gradient and Coriolis effect no longer in balance, the resultant

**FIGURE 5.7**
Illustration of the meaning of windward (facing into the wind) and leeward (facing away from the wind).
**How might vegetation differ on the windward and leeward sides of an island?**

surface wind does not flow between the isobars like its upper-level counterpart. Instead, a surface wind flows obliquely across the isobars toward the low pressure area ( Fig. 5.6b).

## Wind Terminology

Winds are named after their source. Thus, a wind that comes out of the northeast is called a northeast wind. One coming from the south, even though going toward the north, is called a south (or southerly) wind. It is helpful for students to use the phrase "out of" when describing a wind direction. That phrase will help students to keep the correct direction. For example, if the winds are blowing to the south, then by saying, "the winds are out of the north," automatically makes the student think about the direction of the wind's origin.

**Windward** refers to the direction from which the wind blows. The side of something that faces the direction from which the wind is coming is called the *windward* side. Thus, a windward slope is the side of a mountain against which the wind blows ( Fig. 5.7). **Leeward,** on the other hand, means the direction toward which the wind is blowing. Thus, when the winds are coming out of the west, the *leeward* slope of a mountain would be the east slope. We know that winds can blow from any direction, yet in some places winds may tend to blow more from one direction than any other. We speak of these as the **prevailing winds.**

## Cyclones, Anticyclones, and Winds

Imagine a high pressure cell (anticyclone) in the Northern Hemisphere in which the air is moving from the center in all directions down pressure gradients. As it moves, the air will be deflected to the right, no matter which direction it was originally going. Therefore, the wind moving out of an anticyclone in the Northern Hemisphere will move from the center of high pressure in a clockwise spiral ( Fig. 5.8).

Air tends to move down pressure gradients from all directions toward the center of a low pressure area (cyclone). However, because the air is apparently deflected to the right in the Northern Hemisphere, the winds move into the cyclone in a counterclockwise spiral. Because all objects including air and water are apparently deflected to the left in the Southern Hemisphere, spirals there are reversed. Thus, in the Southern Hemisphere, winds moving away from an anticyclone do so in a counterclockwise spiral, and winds moving into a cyclone move in a clockwise spiral.

# Subglobal Surface Wind Systems

As we have seen, winds develop whenever differential heating causes differences in pressure. The global wind system is a response to the constant temperature imbalance between tropical and polar regions. On a smaller, or subglobal, scale, additional wind systems develop. We begin with a discussion of small pressure and wind systems, then move to global systems. Monsoon winds are continental in size and develop in response to the seasonal variations in temperature between large landmasses and adjacent oceans. On the smallest scale are local winds, which develop in response to diurnal (daily) variation in heating.

## Local Winds

Later in this chapter, we discuss the major circulation patterns of Earth's atmosphere. This knowledge is vital to understanding the climatic regions of Earth and the fundamental climatic differences between those regions. Yet we are all aware that there are winds that affect weather on a far smaller scale. These *local winds* are often a response to local landform configurations and add further complexity to the problem of understanding the dynamics of weather.

**Land Breeze–Sea Breeze** The **land breeze–sea breeze cycle** is a diurnal (daily) one in which the differential heating of land and water again plays a role ( Fig. 5.9). During the day, when the land—and consequently the air above it—is heated more quickly and to a

**FIGURE 5.8**
Movement of surface winds associated with low pressure centers (cyclones) and high pressure centers (anticyclones) in the Northern and Southern Hemispheres. Note that the surface winds are to the right of the pressure gradient in the Northern Hemisphere and to the left of the pressure gradient in the Southern Hemisphere.
**What do you think might happen to the diverging air of an anticyclone if there is a cyclone nearby?**

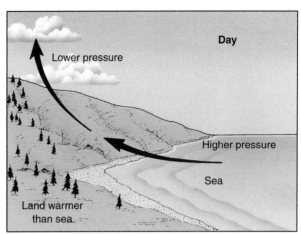

**FIGURE 5.9**
Land and sea breezes. This day-to-night reversal of winds is a consequence of the different rates of heating and cooling of land and water areas. The land becomes warmer than the sea during the day and colder than the sea at night; the air flows from the cooler to the warmer area.
**What is the impact on daytime coastal temperatures of the land and sea breeze?**

**FIGURE 5.10**
Mountain and valley breezes. This daily reversal of winds results from heating of mountain slopes during the day and their cooling at night. Warm air is drawn up slopes during the day, and cold air drains down the slopes at night.
**How might a green, shady valley floor and a bare, rocky mountain slope contribute to these changes?**

higher temperature than the nearby ocean (sea or large lake), the air above the land expands and rises. This process creates a local area of low pressure, and the rising air is replaced by the denser, cooler air from over the ocean. Thus, a sea breeze of cool, moist air blows in over the land during the day. This sea breeze helps explain why seashores are so popular in summer; cooling winds help alleviate the heat. At times, however, sea breezes are responsible for afternoon cloud cover and light rain, spoiling an otherwise sunny day at the shore. These winds can mean a 5°C–9°C (9°F–16°F) reduction in temperature along the coast, as well as a lesser influence on land perhaps as far from the sea as 15–50 kilometers (9–30 mi). During hot summer days, such winds cool cities like Chicago, Milwaukee, and Los Angeles. At night, the land and the air above it cool more quickly and to a lower temperature than the nearby water body and the air above it. Consequently, the pressure builds higher over the land and air flows out toward the lower pressure over the water, creating a land breeze. For thousands of years, sailboats have left their coasts at dawn, when there is still a land breeze, and have returned with the sea breeze of the late afternoon.

**Mountain Breeze–Valley Breeze** Under the calming influence of a high pressure system, there is a daily **mountain breeze–valley breeze cycle** (Fig. 5.10) that is somewhat similar in mechanism to the land breeze–sea breeze cycle just discussed. During the day, when the valleys and slopes of mountains are heated by the sun, the high exposed slopes are heated faster than the lower shadier valley. The air on the slope expands and rises, drawing air from the valley up the sides of the mountains. This warm daytime breeze is the *valley breeze,* named for its place of origin. Clouds, which can often be seen hiding mountain peaks, are actually the visible evidence of condensation in the warm air rising from the valleys. At night, when the valley and slopes are cooled because Earth is giving off more radiation than it is receiving, the air cools and sinks once again into the valley as a cool *mountain breeze.*

**Drainage Winds** Also known as **katabatic winds, drainage winds** are local to mountainous regions and can occur only under calm, clear conditions. Cold, dense air will accumulate in a high valley, plateau, or snowfield within a mountainous area. Because the cold air is very dense, it tends to flow downward, escaping through passes and pouring out onto the land below. Drainage winds can be extremely cold and strong, especially when they result from cold air accumulating over ice sheets such as Greenland and Antarctica. These winds are known by many local names; for example, on the Adriatic coast, they are called the *bora;* in France, the *mistral;* and in Alaska, the *Taku.*

**Chinooks and Other Warming Winds** A fourth type of local wind is also known by several names in different parts of the world—for example, **Chinook** in the Rocky Mountain area and **foehn** (pronounced "fern") in the Alps. Chinook-type winds occur when air originating elsewhere must pass over a mountain range. As these winds flow down the leeward slope after crossing the mountains, the air is compressed and heated at a greater rate than it was cooled when it ascended the windward slope ( Fig. 5.11). Thus, the air enters the valley below as warm, dry winds. The rapid temperature rise brought about by such winds has been known to damage crops, increase forest-fire hazard, and set off avalanches.

An especially hot and dry wind is the **Santa Ana** of Southern California. It forms when high pressure develops over the interior desert regions of Southern California. The clockwise circulation of the high drives the air of the desert southwest over the mountains of Eastern California, accentuating the dry conditions as the air moves down the western slopes. The hot, dry Santa Ana winds are notorious for fanning forest and brush fires, which plague the southwestern United States, especially in California.

There is no question that winds, both local and global, are effective elements of atmospheric dynamics. We all know that a hot, breezy day is not nearly as unpleasant as a hot day without any wind. This difference exists because winds increase the rate of evaporation and thus the rate of removal of heat from our bodies, the air, animals, and plants. For the same reason, the wind on a cold day increases our discomfort.

## Monsoon Winds

The term *monsoon* comes from the Arabic word *mausim,* meaning season. This word has been used by Arab sailors for many centuries to describe seasonal changes in wind direction across the Arabian Sea between Arabia and India. As a meteorological term, **monsoon** refers to the directional shifting of winds from one season to the next. Usually, the monsoon occurs when a humid wind blowing from the ocean toward the land in the summer shifts to a dry, cooler wind blowing seaward off the land in the winter, and it involves a full 180° direction change in the wind.

The monsoon is most characteristic of southern Asia although it occurs on other continents as well. As the large landmass of Asia cools more quickly than the surrounding oceans, the continent develops a strong center of high pressure from which there must be an outflow of air in winter ( Fig. 5.12). This outflow blows across much land toward the tropical low before reaching the oceans. It brings cold, dry air south.

In summer the Asian continent heats quickly and develops a large low

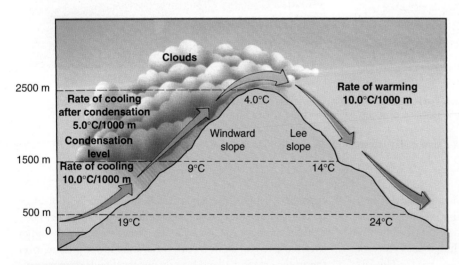

**FIGURE 5.11**
Foehn winds result when air ascends a mountain range and undergoes condensation that dries it out and adds latent heat that slows its rate of cooling. As the dry air descends the leeward side of the range, it is compressed and heated at a greater rate than it was initially cooled. This produces the relatively warm, dry conditions with which foehn winds are associated.
**The term *Chinook,* a type of foehn wind, means "snow eater." Can you offer an explanation for how this name came about?**

# Geography's Spatial Science Perspective
## The Santa Ana Winds and Fire

Wildfires require three factors to occur: *oxygen, fuel,* and an *ignition source.* The conditions for all three factors vary geographically, so their spatial distributions are not equal everywhere. In locations where all three factors have the potential to exist, the danger from wildfires is high. Oxygen in the atmosphere is constant, but winds, which supply more oxygen as a fire consumes it, vary with location, weather, and terrain. High winds cause fires to spread faster and make them difficult to extinguish. Fuel in wildland fires is usually supplied by dry vegetative litter (leaves, branches, and dry annual grasses), and certain environments have more of this fuel than others. Dense vegetation tends to support the spread of fires. Growing vegetation can also become dessicated—dried out by transpiration losses during a drought or an annual dry season. In addition, once a fire becomes large, extreme heat in the areas where it is spreading causes vegetation along the edges of the burning area to lose its moisture through evaporation. Ignition sources are the means by which a fire is started. Lightning and human causes such as campfires provide the main ignition sources for wildfires.

Southern California offers a regional example of how conditions related to these three factors combine with the local physical geography to create an environment that is conducive to wildfire hazard. This is also a region where many people live in forested or scrub-covered locales or along the urban–wildland fringe—areas that are very susceptible to fire. High pressure, warm weather, and low relative humidity dominate the Mediterranean climate of Southern California's coastal region for much of the year. When these conditions occur, the region experiences high fire potential because of the warm dry air and the vegetation that has dried out during the arid summer season.

The most dangerous circumstances for wildfires in Southern California occur when high winds are sweeping the region. When a strong cell of high pressure forms east of Southern California, the clockwise anticyclonic circulation directs winds from the north and east toward the coast. These warm, dry winds (called Santa Ana winds) blow down from nearby high-desert regions, becoming adiabatically warmer and drier as they descend into the coastal lowlands. The Santa Ana winds are most common in fall and winter, and wind speeds can be 50–90 kilometers per hour (30–50 mph) with stronger local wind gusts reaching 160 kmh (100 mph). Just like using a bellows or blowing on a campfire to get it started, the Santa Ana winds produce fire weather that can cause the spread of a wildfire to be extremely rapid after ignition. Most people take great care during these times to avoid or strictly control any activities that could cause a fire to start, but occasionally accidents, acts of arson, or lightning strikes ignite a wildfire. Given the physical geography of the Los Angeles region, when the Santa Ana winds are blowing, the fire danger is especially extreme. Ironically, although the Santa Ana winds create dangerous fire conditions, they also provide some benefits to local residents because the winds tend to blow air pollutants offshore and out of the urban region. In addition, because they are strong winds flowing opposite to the direction of ocean waves, experienced surfers can enjoy higher than normal waves during those periods when Santa Ana winds are present.

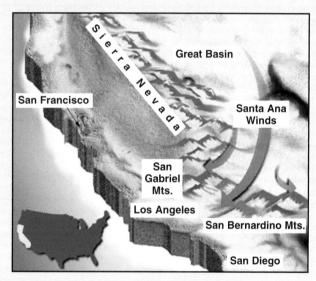

Geographic setting and wind direction for Santa Ana winds.

---

pressure center. This development is reinforced by a poleward shift of the warm, moist tropical air to a position over southern Asia. Warm, moist air from the oceans is attracted into this low. Though full of water vapor, this air does not in itself cause the wet summers with which the monsoon is associated. However, any turbulence or landform barrier that makes this moist air rise and, as a result, cool off will bring about precipitation. This precipitation is particularly noticeable in the foothills of the Himalayas, the western Ghats of India, and the Annamese Highlands of Vietnam. This is the time of year when the rice crop is planted in many parts of Asia.

In the lower latitudes, a monsoonal shift in winds can come about as a reaction to the migration of the direct rays of the sun. For example, the winds of the equatorial

**FIGURE 5.12**
Seasonal changes in surface wind direction that create the Asiatic monsoon system. The "burst" of the "wet monsoon," or the sudden onshore flow of tropical humid air in July, is apparently triggered by changes in the upper air circulation, resulting in heavy precipitation. The offshore flow of dry continental air in winter creates the "dry monsoon" and drought conditions in southern Asia.
**How do the seasonal changes of wind direction in Asia differ from those of the southern United States?**

zone migrate during the summer months northward toward the southern coast of Asia, bringing with them warm, moist, turbulent air. The winds of the Southern Hemisphere also migrate north with the sun, some crossing the equator. They also bring warm, moist air (from their travels over the ocean) to the southern and especially the southeastern coasts of India. In the winter months, the equatorial and tropical winds migrate south, leaving southern Asia under the influence of the dry, calm winds of the tropical Northern Hemisphere. Asia and northern Australia are true monsoon areas, with a full 180° wind shift with changes from summer to winter. Other regions, like the southern United States and West Africa, have "monsoonal tendencies," but are not monsoons in the true meaning of the term.

The phenomenon of monsoon winds and their characteristic seasonal shifting cannot be fully explained by the differential heating of land and water, however, or by the seasonal shifting of tropical and subtropical wind belts. Some aspects of the monsoon system—for example, its "burst" or sudden transition between dry and wet in southern Asia—must have other causes. Meteorologists looking for a more complete explanation of the monsoon are examining the role played by the jet stream (described later in this chapter) and other wind movements of the upper atmosphere.

# Global Pressure Belts

## Idealized Global Pressure Belts

Using what we have learned about pressure on Earth's surface, we can construct a theoretical model of the pressure belts of the world (Fig. 5.13). Later, we see how real conditions depart from our model and examine why these differences occur.

Centered approximately over the equator in our model is a belt of low pressure, or a **trough.** Because this is the region on Earth of greatest annual heating, we can conclude that the low pressure of this area, the **equatorial low (equatorial trough),** is determined primarily by thermal factors, which cause the air to rise.

North and south of the equatorial low and centered on the so-called horse latitudes, about 30°N and 30°S, are cells of relatively high pressure. These are the **subtropical highs,** which are the result of dynamic factors related to the sinking of convectional cells initiated at the equatorial low.

Poleward of the subtropical highs in both the Northern and Southern Hemispheres are large belts of low pressure that extend through the upper-middle latitudes. Pressure decreases through these **subpolar lows** until about 65° latitude. Again, dynamic factors play a role in the existence of subpolar lows.

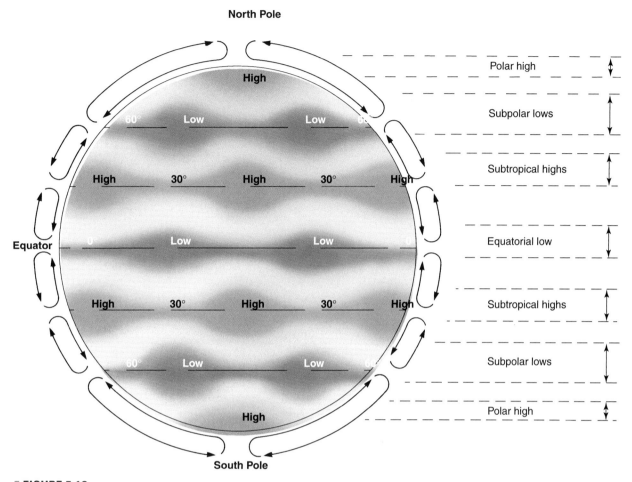

**FIGURE 5.13**
Idealized world pressure belts. Note the arrows on the perimeter of the globe that illustrate the cross-sectional flow associated with the surface pressure belts.

In the polar regions are high pressure systems called the **polar highs.** The extremely cold temperatures and consequent sinking of the dense polar air in those regions create the higher pressures found there.

This system of pressure belts that we have just developed is a generalized picture. Just as temperatures change from month to month, day to day, and hour to hour, so do pressures vary through time at any one place. Our long-term global model disguises these smaller changes, but it does give an idea of broad pressure patterns on the surface of Earth.

## The Global Pattern of Atmospheric Pressure

As our idealized model suggests, the atmosphere tends to form belts of high and low pressure along east–west axes in areas where there are no large bodies of land. These belts are arranged by latitude and generally maintain their bandlike pattern. However, where there are continental landmasses, belts of pressure are broken and tend to form cellular pressure systems.

The landmasses affect the development of belts of atmospheric pressure in several ways. Most influential is the effect of the differential heating of land and water surfaces. In addition, landmasses affect the movement of air and consequently the development of pressure systems through friction with their surfaces. Landform barriers such as mountain ranges also block the movement of air and thereby affect atmospheric pressure.

## Seasonal Variations in the Pattern

In general, the global atmospheric pressure belts shift northward in July and southward in January, following the changing position of the sun's direct rays as they migrate between the Tropics of Cancer and Capricorn. Thus, there are thermally induced seasonal variations in the pressure patterns, as seen in Figures 5.14a and b. These seasonal variations tend to be small at low latitudes, where there is little temperature variation, and large at high latitudes, where there is an increasing contrast in length of daylight and angle of the sun's rays. Furthermore, landmasses tend to alter

**128** CHAPTER 5 ATMOSPHERIC PRESSURE, WINDS, AND CIRCULATION PATTERNS

**FIGURE 5.14**
(a) Average sea-level pressure (in millibars) in January. (b) Average sea-level pressure (in millibars) in July.
**What is the difference between the January and July average sea-level pressures at your location? Why do they vary?**

the general pattern of seasonal variation. This is an especially important factor in the Northern Hemisphere, where land accounts for 40% of the total Earth surface, as opposed to less than 20% in the Southern Hemisphere.

**January** Because continents cool more quickly than the oceans, their temperatures will be lower in winter than those of the surrounding seas. Figure 5.14a shows that in the middle latitudes of the Northern Hemisphere this variation leads to the development of cells of high pressure over the land areas. In contrast, the subpolar lows develop over the oceans because they are comparatively warmer. Over eastern Asia, there is a strongly developed anticyclone during the winter months that is known as the **Siberian High.** Its equivalent in North America, known as the **Canadian High,** is not nearly so well developed because the North American landmass is considerably smaller than the Eurasian continent.

In addition to the Canadian High and the Siberian High, two low pressure centers develop: one in the North Atlantic, called the **Icelandic Low,** and the other in the North Pacific, called the **Aleutian Low.** The air in them has relatively lower pressure than either the subtropical or the polar high systems. Consequently, air moves toward these low pressure areas from both north and south. Such low pressure regions are associated with cloudy, unstable weather and are a major source of winter storms, whereas high pressure areas are associated with clear, blue-sky days; calm, starry nights; and cold, stable weather. Therefore, during the winter months, cloudy and sometimes dangerously stormy weather tends to be associated with the two oceanic lows and clear weather with the continental highs.

We can also see that the polar high in the Northern Hemisphere is well developed. This development is due primarily to thermal factors because January is the coldest time of the year. The subpolar lows have developed into the Aleutian and Icelandic cells described earlier. At the same time, the subtropical highs of the Northern Hemisphere appear slightly south of their average annual position because of the migration of the sun toward the Tropic of Capricorn. The equatorial trough also appears centered south of its average annual position over the geographic equator.

In January in the Southern Hemisphere, the subtropical belt of high pressure appears as three cells centered over the oceans because the belt of high pressure has been interrupted by the continental landmasses where temperatures are much higher and pressure tends to be lower than over the oceans. Because there is virtually no land between 45°S and 70°S latitude, the subpolar low circles Earth as a belt of low pressure and is not divided into cells by any landmasses. There is little seasonal change in this belt of low pressure other than in January (summer in the Southern Hemisphere), when it lies a few degrees north of its July position.

**July** The anticyclone over the North Pole is greatly weakened during the summer months in the Northern Hemisphere, primarily because of the lengthy (24-hour days) heating the oceans and landmasses in that region (Fig. 5.14b). The Aleutian and Icelandic Lows nearly disappear from the oceans, while the landmasses, which developed high pressure cells during the cold winter months, have extensive low pressure cells slightly to the south during the summer. In Asia, a low pressure system develops, but it is divided into two separate cells by the Himalayas (see again Fig. 5.12). The low pressure cell over northwest India is so strong that it combines with the equatorial trough, which has moved north of its position 6 months earlier. The subtropical highs of the Northern Hemisphere are more highly developed over the oceans than over the landmasses. In addition, they migrate northward and are highly influential factors in the climate of landmasses nearby. In the Pacific, this subtropical high is termed the **Pacific High;** this system of pressure plays an important role in moderating the temperatures of the West Coast of the United States. In the Atlantic Ocean, the corresponding cell of high pressure is known as the **Bermuda High** to North Americans and as the **Azores High** to Europeans and West Africans. As we have already mentioned, the equatorial trough of low pressure moves north in July, following the migration of the sun's vertical rays, and the subtropical highs of the Southern Hemisphere lie slightly north of their January locations.

In examining pressure systems at Earth's surface, we have seen that there are essentially seven belts of pressure (two polar highs, two subpolar lows, two subtropical highs, and one equatorial low), which are broken into cells of pressure in some places primarily because of the influence of certain large landmasses. We have also seen that these belts and cells vary in size, intensity, and location with the seasons and with the migration of the sun's vertical rays over Earth's surface. Since these global-scale pressure systems migrate by latitude with the position of the direct sun angle, they are sometimes referred to as *semipermanent pressure systems* because they are never permanently fixed in the same location.

## Global Surface Wind Systems

The planetary, or global, wind system that is a response to the global pressure patterns also plays a role in the maintenance of those same pressures. This wind system, which is the major means of transport for energy and moisture through Earth's atmosphere, can be examined in an idealized state. To do so, however, we must ignore the influences of landmasses and seasonal variations in solar energy. By assuming, for the sake of discussion, that Earth has a homogeneous surface and that there are no seasonal variations in the amount of solar energy received at different latitudes, we can examine a theoretical model of the atmosphere's planetary circulation. Such an understanding will help explain specific features of climate such as the rain and snow of the Sierra Nevada and Cascade Mountains and the existence of arid regions farther to the east. It will also account for the movement of great surface currents in our oceans that are driven by this atmospheric engine.

# Idealized Model of Atmospheric Circulation

Because winds are caused by pressure differences, various types of winds are associated with different kinds of pressure systems. Therefore, a system of global winds can be demonstrated using the model of pressures that we previously developed (see again Fig. 5.13).

The characteristics of convergence and divergence are very important to our understanding of global wind patterns. Surface air diverges from zones of high pressure and converges on areas of low pressure. We also know that, because of the pressure gradient, surface winds always blow from high pressure to low pressure.

Knowing that surface winds originate in areas of high pressure and taking into account the global system of pressure cells, we can develop our model of the wind systems of the world ( Fig. 5.15). This model takes into account differential heating, Earth rotation, and atmospheric dynamics. Note that the winds do not blow in a straight north–south line. The variation is due of course to the Coriolis effect, which causes an apparent deflection to the right in the Northern Hemisphere and to the left in the Southern Hemisphere.

Our idealized model of global atmospheric circulation includes six wind belts, or zones, in addition to the seven pressure zones that we have previously identified. Two wind belts, one in each hemisphere, are located where winds move out of the polar highs and down the pressure gradients toward the subpolar lows. As these winds are deflected to the right in the Northern Hemisphere and to the left in the Southern, they become the **polar easterlies.**

The remaining four wind belts are closely associated with the divergent winds of the subtropical highs. In each hemisphere, winds flow out of the poleward portions of these highs toward the subpolar lows. Because of their general movement from the west, the winds of the upper-middle latitudes are labeled the **westerlies.** The winds blowing from the highs toward the equator have been called the **trade winds.** Because of the Coriolis effect, they are the **northeast trades** in the Northern Hemisphere and the **southeast trades** south of the equator.

Our model does not conform exactly to actual conditions. First, as we know, the vertical rays of the sun do not stay precisely over the equator but migrate as far north as the Tropic of Cancer in June and south to the Tropic of Capricorn in December. Therefore, the pressure systems, and

PHYSICAL
Geography Now™  ■ **ACTIVE FIGURE 5.15**
Watch this Active Figure at http://now.brookscole.com/gabler8.
The general circulation of Earth's atmosphere.

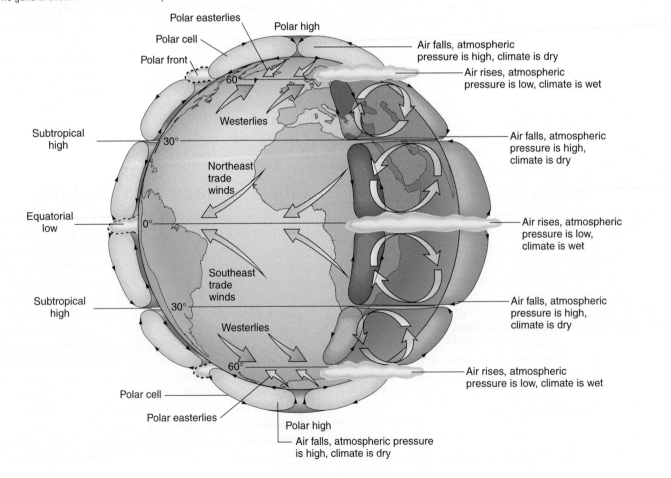

consequently the winds, must move to adjust to the change in the position of the sun. Then, as we have already discovered, the existence of the continents, especially in the Northern Hemisphere, causes longitudinal pressure differentials that affect the zones of high and low pressure.

## Conditions within Latitudinal Zones

**Trade Winds** A good place to begin our examination of winds and associated weather patterns as they actually occur is in the vicinity of the subtropical highs. On Earth's surface, the trade winds, which blow out of the subtropical highs toward the equatorial trough in both the Northern and Southern Hemispheres, can be identified between latitudes 5° and 25°. Because of the Coriolis effect, the northern trades move away from the subtropical high in a clockwise direction out of the northeast. In the Southern Hemisphere, the trades diverge out of the subtropical high toward the tropical low from the southeast, as their movement is counterclockwise. Because the trades tend to blow out of the east, they are also known as the **tropical easterlies.**

The trade winds tend to be constant, steady winds, consistent in their direction. This is most true when they cross the eastern sides of the oceans (near the eastern portion of the subtropical high). The area of the trades varies somewhat during the solar year, moving north and south a few degrees of latitude with the sun. Near their source in the subtropical highs, the weather of the trades is clear and dry, but after crossing large expanses of ocean, the trades have a high potential for stormy weather.

Early Spanish sea captains depended on the northeast trade winds to drive their galleons to destinations in Central and South America in search of gold, spices, and new lands. Going eastward toward home, navigators usually tried to plot a course using the westerlies to the north. The trade winds are one of the reasons that the Hawaiian Islands are so popular with tourists; the steady winds help keep temperatures pleasant, even though Hawaii is located south of the Tropic of Cancer.

**Doldrums** Where the trade winds converge in the equatorial trough (or tropical low) lies a zone of calm and weak winds of no prevailing direction. Here the air, which is very moist and heated by the sun, tends to expand and rise, maintaining the low pressure of the area. These winds, which are roughly between 5°N and 5°S, are generally known as the **doldrums.** This area is called the **intertropical convergence zone (ITCZ),** or the "equatorial belt of variable winds and calms." Because of the converging moist air and high potential for rainfall in the doldrums, this region coincides with the world's latitudinal belt of heaviest precipitation and most persistent cloud cover.

Old sailing ships often remained becalmed in the doldrums for days at a time. A description of a ship becalmed in the doldrums appears in *The Rime of the Ancient Mariner* by Samuel Taylor Coleridge (lines 103–118). The ship is sailing northward from the tropical southeasterlies (trades) when it gets to the doldrums.

> The fair breeze blew,
> the white foam flew,
> The furrow followed free;
> We were the first that ever burst
> Into that silent sea.
>
> Down dropt the breeze,
> the sails dropt down,
> 'Twas sad as sad could be;
> And we did speak only to break
> The silence of the sea!
>
> All in a hot and copper sky,
> The bloody Sun, at noon,
> Right up above the mast did stand,
> No bigger than the Moon.
>
> Day after day,
> day after day,
> We stuck, nor breath nor motion;
> As idle as a painted ship
> Upon a painted ocean.

This is obviously not a happy poem! It is interesting to note that the word *doldrums* in the English language means a bored or depressed state of mind. The sailors were in the doldrums in more ways than one.

**Subtropical Highs** The areas of subtropical high pressure, generally located between latitudes 25° and 35°N and S, and from which winds blow equatorward as the trades, are often called the subtropical belts of variable winds, or the "horse latitudes." This name comes from the occasional need by the Spanish conquistadors to eat their horses or throw them overboard in order to conserve drinking water and lighten the weight when their ships were becalmed in these latitudes. The subtropical highs are areas, like the doldrums, in which there are no strong prevailing winds. However, unlike the doldrums, which are characterized by convergence, rising air, and heavy rainfall, the subtropical highs are areas of sinking and settling air from higher altitudes, which tend to build up the atmospheric pressure. Weather conditions are typically clear, sunny, and rainless, especially over the eastern portions of the oceans where the high pressure cells are strongest.

**Westerlies** The winds that flow poleward out of the subtropical high pressure cells in the Northern Hemisphere are deflected to the right and thus blow from the southwest. Those in the Southern Hemisphere are deflected to the left and blow out of the northwest. Thus, these winds have been

# Career Vision

**Sandra Diaz, Chief Broadcast Meteorologist, KFOX** — **Texas State University, San Marcos**
BS in Physical and Applied Geography

Courtesy Sandra Diaz

I am the chief meteorologist at KFOX, a television station in El Paso, Texas. For 2 years, I was a meteorologist on the Weather Channel, where I was fortunate to work with several of our country's top severe-weather experts. The knowledge and experience that I gained at the national level was invaluable.

I am a full member of the American Meteorological Society (AMS) and a member of the National Weather Association (NWA).

I took my current position in order to return to live in my hometown of El Paso. My job involves understanding and explaining weather patterns for our local area and the world. My position as a meteorologist gives me the opportunity to warn people of incoming inclement weather and a chance to teach about weather and its impact. A strong foundation in physical geography is essential to me because I cover and explain atmospheric topics, topography, wind patterns, seasons, and oceanic issues.

*Why did you decide to major in geography?* I found every class interesting and fun. I loved the combination of science, nature, and math. I have always enjoyed exploring the world around me.

Coursework in physical geography provided a strong foundation for a career in meteorology. I was then able to build on that foundation with specialized meteorological courses. Almost everything I studied comes into play during my daily work now.

Location is very important for forecasting weather patterns. Knowledge of an area's geography can help me better understand how dangerous weather might impact that community—that is how we save lives.

When I was learning the Köppen climate classification system, I was not sure if I would ever really need to know this extensive information in the future. Nonetheless, my instructor wanted us to know this method of understanding world climates in a regional context forward and backward. At the time, I thought this was a tedious exercise, but now I am very happy he did because it has been very useful in understanding the climate and vegetation of places around the world.

Students who wish to prepare for a career like mine should take as many geography courses related to meteorology as possible. Be sure to graduate with a bachelor of science, not a bachelor of arts. Take a public-speaking course to help you get over any fear of speaking to a group of people.

Lastly, I would highly recommend doing an internship with a local television station to get a better understanding about what a broadcast meteorologist does.

*Why do you think that physical geography is an important course for those who are majoring in other fields?* We are all affected by the natural world around us. The more we understand our environment, the better we can face challenges in nature.

In my spare time, I enjoy rock collecting, star gazing, outdoor activities (such as hiking the Franklin Mountains in west Texas), church, and golf.

---

correctly labeled the westerlies. They tend to be less consistent in direction than the trades, but they are usually stronger winds and may be associated with stormy weather. The westerlies occur between about 35° and 65°N and S latitudes. In the Southern Hemisphere, where there is less land than in the Northern Hemisphere to affect the development of winds, the westerlies attain their greatest consistency and strength. Much of Canada and most of the United States—except Florida, Hawaii, and Alaska—are under the influence of the westerlies.

**Polar Winds** Accurate observations of pressure and wind are sparse in the two polar regions; therefore, we must rely on remotely sensed information (mainly by weather satellite imagery). Our best estimate is that pressures are consistently high throughout the year at the poles and that prevailing easterly winds blow from the polar regions to the subpolar low pressure systems.

**Polar Front** Despite our limited knowledge of the wind systems of the polar regions, we do know that the winds can be highly variable, blowing at times with great speed and intensity. When the cold air flowing out of the polar regions and the warmer air moving in the path of the westerlies meet, they do so like two warring armies: One does not absorb the other. Instead, the denser, heavier cold air pushes the warm air upward, forcing it to rise rapidly. The line along which these two great wind systems battle is appropriately known as the **polar front.** The weather that results from the meeting of the cold polar air and the warmer air from the subtropics can be very stormy. In fact, most of the storms that move slowly through the middle latitudes in the path of the prevailing westerlies are born at the polar front.

## The Effects of Seasonal Migration

Just as insolation, temperature, and pressure systems migrate north and south as Earth revolves around the sun, Earth's wind systems also migrate with the seasons. During the summer months in the Northern Hemisphere, maximum insolation is received north of the equator. This condition causes the pressure belts to move north as well, and the wind belts of both hemispheres shift accordingly. Six months later, when maximum heating is taking place south of the equator, the various wind systems have migrated south in response to the migration of the pressure systems. Thus, seasonal variation in wind and pressure conditions is one important way in which actual atmospheric circulation differs from our idealized model.

The seasonal migration will most affect those regions near the boundary zone between two wind or pressure systems. During the winter months, such a region will be subject to the impact of one system. Then, as summer approaches, that system will migrate poleward and the next equatorward system will move in to influence the region. Two such zones in each hemisphere have a major effect on climate. The first lies between latitudes 5° and 15°, where the wet equatorial low of the high-sun season (summer) alternates with the dry subtropical high and trade winds of the low-sun season (winter). The second occurs between 30° and 40°, where the subtropical high dominates in summer but is replaced by the wetter westerlies in winter.

California is an example of a region located within a zone of transition between two wind or pressure systems (Fig. 5.16). During the winter, this region is under the influence of the westerlies blowing out of the Pacific High. These winds, turbulent and full of moisture from the ocean, bring winter rains and storms to "sunny" California. As summer approaches, however, the subtropical high and its associated westerlies move north. As California comes under the influence of the calm and steady high pressure system, it experiences again the climate for which it is famous: day after day of warm, clear, blue, cloudless skies. This alternation of moist winters and dry summers is typical of the western sides of all landmasses between 30° and 40° latitude.

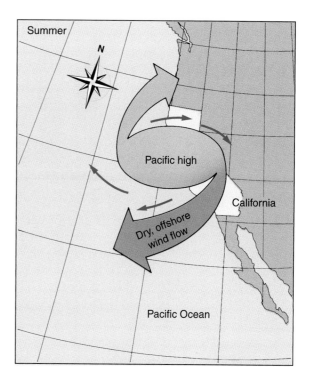

**FIGURE 5.16**
Winter and summer positions of the Pacific anticyclone in relation to California. In the winter, the anticyclone lies well to the south and feeds the westerlies that bring the cyclonic storms and rain from the North Pacific to California. The influence of the anticyclone dominates during the summer. The high pressure blocks cyclonic storms and produces warm, sunny, and dry conditions.
**In what ways would the seasonal migration of the Pacific anticyclone affect agriculture in California?**

## Longitudinal Differences in Winds

We have seen that there are sizable latitudinal differences in pressure and winds. In addition, there are significant longitudinal variations, especially in the zone of the subtropical highs.

As was previously noted, the subtropical high pressure cells, which are generally centered over the oceans, are much stronger on their eastern sides than on their western sides. Thus, over the eastern portions of the oceans (west coasts of the continents) in the subtropics, subsidence and divergence are especially noticeable. The above-surface temperature inversions so typical of anticyclonic circulation are close to the surface, and the air is calm and clear. The air moving equatorward from this portion of the high produces the classic picture of the steady trade winds with clear, dry weather.

Over the western portions of the oceans (eastern sides of the continents), conditions are markedly different. In its passage over the ocean, the diverging air is gradually warmed and moistened; turbulent and stormy weather conditions are likely to develop. As indicated in Figure 5.17, wind movement in the western portions of the anticyclones may actually be poleward and directed toward landmasses. Hence, the trade winds in these areas are especially weak or nonexistent much of the year.

As we have pointed out in discussing Figures 5.14 and 5.15, there are great land–sea contrasts in temperature and pressure throughout the year farther toward the poles, especially in the Northern Hemisphere. In the cold continental winters, the land is associated with pressures that are higher than those over the oceans, and thus there are strong, cold winds from the land to the sea. In the summer, the situation changes, with relatively low pressure existing over the continents because of higher temperatures. Wind directions are thus greatly affected, and the pattern is reversed so that winds flow from the sea toward the land. In North America, this is sometimes referred to as the "North American monsoon."

## Upper Air Winds

Thus far, we have closely examined the wind patterns near Earth's surface. Of equal, or even greater, importance is the flow of air above Earth's surface—in particular, the flow of air at altitudes above 5000 meters (16,500 ft), in the upper troposphere. The formation, movement, and decay of surface cyclones and anticyclones in the middle latitudes depend to a great extent on the flow of air high above Earth's surface.

The circulation of the upper air winds is a far less complex phenomenon than surface wind circulation. In the upper troposphere, an average westerly flow, the *upper air westerlies,* is maintained poleward of about 15°–20° latitude in both hemispheres. Because of the reduced frictional drag, the upper air westerlies move much more rapidly than their surface counterparts. Between 15° and 20°N and S latitudes are the *upper air easterlies,* which can be considered the upper air extension of the trade winds. The flow of the upper air winds became very apparent during World War II when high-altitude bombers moving eastward were found to cover similar distances faster than those flying westward. Pilots had encountered the upper air westerlies, or perhaps even the **jet streams**—very strong air currents embedded within the upper air westerlies.

The upper air westerlies form as a response to the temperature difference between warm tropical air and cold polar air. The air in the equatorial latitudes is warmed, rises convectively to high altitudes, and then flows toward the polar regions. At first this seems to contradict our previous statement, relative to surface winds, that air flows from cold areas (high pressure) toward warm areas (low pressure). This apparent discrepancy disappears, however, if

**FIGURE 5.17**
Circulation pattern in a Northern Hemisphere subtropical anticyclone. Subsidence of air is strongest in the eastern part of the anticyclone, producing calm air and arid conditions over adjacent land areas. The southern margin of the anticyclone feeds the persistent northeast trade winds.
**What wind system is fed by the northern margin?**

you recall that the pressure gradient, down which the flow takes place, must be assessed between two points *at the same elevation*. A column of cold air will exert a higher pressure at Earth's surface than a column of warm air. Consequently, the pressure gradient established at Earth's surface will result in a flow from the cooler air toward the warmer air. However, cold air is denser and more compact than warm air. Thus, pressure decreases with height more rapidly in cold air than in warm air. As a result, at a specific height above Earth's surface, a lower pressure will be encountered above cold surface air than above warm surface air. This will result in a flow (pressure gradient) from the warmer surface air toward the colder surface air at that height. Figure 5.18 illustrates this concept.

Returning to our real-world situation, as the upper air winds flow from the equator toward the poles (down the pressure gradient), they are turned eastward because of the Coriolis effect. The net result is a broad circumpolar flow of westerly winds throughout most of the upper atmosphere ( ▮ Fig. 5.19). Because the upper air westerlies form in response to the thermal gradient between tropical and polar areas, it is not surprising that they are strongest in winter (the low-sun season) when the thermal contrast is greatest. On the other hand, during the summer (the high-sun season) when the contrast in temperature over the hemisphere is much reduced, the upper air westerlies move more slowly.

The temperature gradient between tropical and polar air, especially in winter, is not uniform but rather is concentrated where the warm tropical air meets cold polar air. This boundary, called the polar front, with its stronger pressure gradient, marks the location of the **polar front jet stream**. Ranging from 40 to 160 kilometers (25–100 mi) in width and up to 2 or 3 kilometers (1–2 mi) in depth, the polar front jet stream can be thought of as a faster, internal current of air within the upper air westerlies. While the polar front

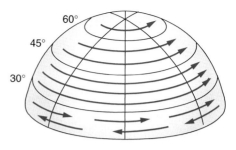

■ **FIGURE 5.19**
The upper air westerlies form a broad circumpolar flow throughout most of the upper atmosphere.

jet stream flows over the middle latitudes, another westerly **subtropical jet stream** flows above the sinking air of the subtropical highs in the lower-middle latitudes. Like the upper air westerlies, both jets are best developed in winter when hemispherical temperatures exhibit their steepest gradient ( ▮ Fig. 5.20). During the summer, both jets weaken in intensity. The subtropical jet stream frequently disappears completely, and the polar front jet tends to migrate northward.

We can now go one step further and combine our knowledge of the circulation of the upper air and surface to yield a more realistic portrayal of the vertical circulation pattern of our atmosphere ( ▮ Fig. 5.21). In general, the upper air westerlies and the associated polar jet stream flow in a fairly smooth pattern ( ▮ Fig. 5.22a). At times, however, the upper air westerlies develop oscillations, termed *long waves,* or **Rossby waves,** after the Swedish meteorologist Carl Rossby who first proposed and then proved their existence ( ▮ Fig. 5.22b). Rossby waves result in cold polar air pushing into the lower latitudes and forming *troughs* of low pressure, while warm tropical air moves into higher latitudes, forming *ridges* of high pressure. It is when the upper air circulation is in this configuration that surface weather is most influenced. We will examine this influence in more detail in Chapter 7.

Eventually, the upper air oscillations become so extreme that the "tongues" of displaced air are cut off, forming upper air cells of warm and cold air ( ▮ Fig. 5.22c). This process helps maintain a net poleward flow of energy from equatorial and tropical areas. The cells eventually dissipate, and the pattern returns to normal (Fig. 5.22a). The complete cycle takes from less than 4–8 weeks. Although it is not completely clear why the upper atmosphere goes into these oscillating patterns, we are

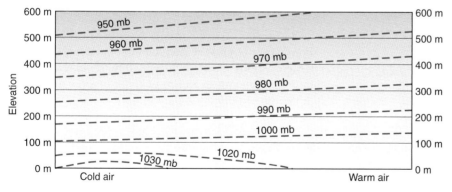

■ **FIGURE 5.18**
Variation of pressure surfaces with height. Note that the horizontal pressure gradient is from cold to warm air at the surface and in the opposite direction at higher elevations (such as 400 m).
**In what direction would the winds flow at 300 meters?**

**ACTIVE FIGURE 5.20**

Watch this Active Figure at http://now.brookscole.com/gabler8.
Approximate location of the subtropical jet stream and area of activity of the polar front jet stream (shaded) in the Northern Hemisphere winter.
**Which jet stream is most likely to affect your home state?**

currently gaining additional insights. One possible cause is variation in ocean-surface temperatures. If the oceans in, say, the northern Pacific or near the equator become unusually warm or cold (for example, El Niño or La Niña, discussed later in this chapter), this apparently triggers oscillations, which continue until the ocean-surface temperature returns to normal. Other causes are also possible.

In addition to this influence on weather, jet streams are important to study for other reasons. They can carry pollutants, such as radioactive wastes or volcanic dust, over great distances and at relatively rapid rates. It was the polar jet stream that carried ash from the Mount St. Helens eruption (in 1980) eastward across the United States and Southern Canada. Nuclear fallout from the Chernobyl incident in the former Soviet Union could be monitored in succeeding days as it crossed the Pacific, and later the United States, in the jet streams. Pilots flying eastward—for example, from North America to Europe—take advantage of the jet stream, so the flying times in this direction may be significantly shorter than those in the reverse direction.

## Ocean Currents

Like the planetary wind system, surface-ocean currents play a significant role in helping equalize the energy imbalance between the tropical and polar regions. In addition, surface-ocean currents greatly influence the climate of coastal locations.

Earth's surface-wind system is the primary control of the major surface currents and drifts. Other controls are the Coriolis effect and the size, shape, and depth of the sea or ocean basin. Other currents may be caused by differences in density due to variations in temperature and salinity, tides, and wave action.

The major surface currents move in broad circulatory patterns, called **gyres,** around the subtropical highs. Because of the Coriolis effect, the gyres flow clockwise in the Northern Hemisphere and counterclockwise in the Southern Hemisphere ( Fig. 5.23). As a general rule, the surface currents do not cross the equator.

Waters near the equator in both hemispheres are driven west by the tropical easterlies or the trade winds. The current thus produced is called the Equatorial Current. At the western margin of the ocean, its warm tropical waters are deflected poleward along the coastline. As these warm waters move into higher latitudes, they move through waters cooler than themselves and are identified as *warm currents* ( Fig. 5.24).

In the Northern Hemisphere, warm currents, such as the Gulf Stream and the Kuroshio Current, are deflected

**FIGURE 5.21**

A more realistic schematic cross section of the average circulation in the atmosphere.

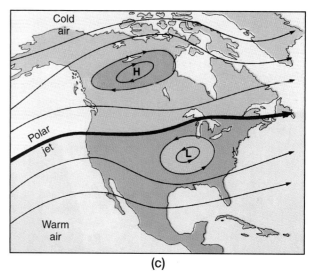

### FIGURE 5.22
Development and dissipation of Rossby waves in the upper air westerlies. (a) A fairly smooth flow prevails. (b) Rossby waves form, with a ridge of warm air extending into Canada and a trough of cold air extending down to Texas. (c) The trough and ridge are cut off and will eventually dissipate. The flow will then return to a pattern similar to (a).

**How are Rossby waves closely associated with the changeable weather of the central and eastern United States?**

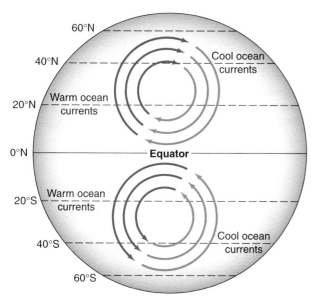

### Geography Now™ ■ ACTIVE FIGURE 5.23
Watch this Active Figure at http://now.brookscole.com/gabler8.
The major ocean currents flow in broad gyres in opposite directions in the Northern and Southern Hemispheres.

**What controls the direction of these gyres?**

more and more to the right (or east) because of the Coriolis effect. At about 40°N, the westerlies begin to drive these warm waters eastward across the ocean, as in the North Atlantic Drift and the North Pacific Drift. Eventually, these currents run into the land at the eastern margin of the ocean, and most of the waters are deflected toward the equator. By this time, these waters have lost much of their warmth, and as they move equatorward into the subtropical latitudes, they are cooler than the adjacent waters. They have become *cool,* or *cold, currents.* These waters complete the circulation pattern when they rejoin the westward-moving Equatorial Current.

On the eastern side of the North Atlantic, the North Atlantic Drift moves into the seas north of the British Isles and around Scandinavia, keeping those areas warmer than their latitudes would suggest. Some Norwegian ports north of the Arctic Circle remain ice free because of this warm water. Cold polar water—the Labrador and Oyashio

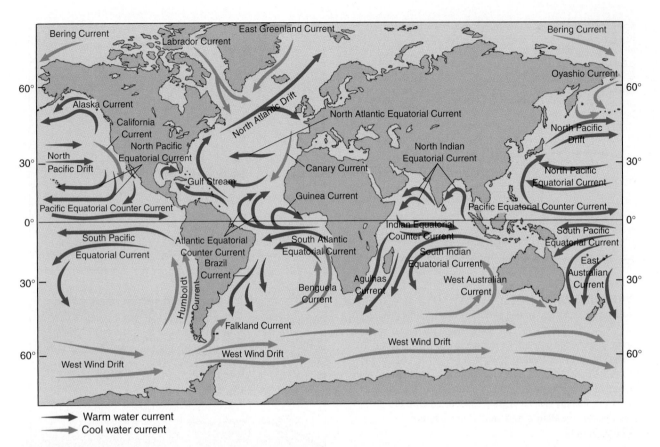

**FIGURE 5.24**
Map of the major world ocean currents, showing warm and cool currents.
**How does this map of ocean currents help explain the mild winters in London, England?**

Currents—flows southward into the Atlantic and Pacific oceans along their western margins.

The circulation in the Southern Hemisphere is comparable to that in the Northern except that it is counterclockwise. Also, because there is little land poleward of 40°S, the West Wind Drift (or Antarctic Circumpolar Drift) circles Earth as a cool current across all three major oceans almost without interruption. It is cooled by the influence of the Antarctic ice sheet (Table 5.1).

In general then, warm currents move poleward as they carry tropical waters into the cooler waters of higher latitudes, as in the case of the Gulf Stream or the Brazil Current. Cool currents deflect water equatorward, as in the California Current and the Humboldt Current. Warm currents tend to have a humidifying and warming effect on the east coasts of continents along which they flow, whereas cool currents tend to have a drying and cooling effect on the west coasts of the landmasses. The contact between the atmosphere and ocean currents is one reason why subtropical highs have a strong side and a weak side. Subtropical highs on the west coast of continents are in contact with cold ocean currents, which cool the air and make the eastern side of a subtropical high more stable and stronger. On the east coasts of continents, contacts with warm ocean currents cause . . . the western sides of subtropical highs to be less stable and weaker.

The general circulation just described is consistent throughout the year, although the position of the currents follows seasonal shifts in atmospheric circulation. In addition, in the North Indian Ocean, the direction of circulation reverses seasonally according to the monsoon winds.

The cold currents along west coasts in subtropical latitudes are frequently reinforced by **upwelling.** As the trade winds in these latitudes drive the surface waters offshore, the wind's frictional drag on the ocean surface displaces the water to the west. As surface waters are dragged away, deeper, colder water rises to the surface to replace them. This upwelling of cold waters adds to the strength and effect of the California, Humboldt (Peru), Canary, and Benguela Currents.

**PHYSICAL Geography Now™** Log on to Physical GeographyNow and select this chapter to work through Geography Interactive activities on "World Currents" (click Waves, Tides & Currents, World Currents) and "Global Ocean."

## TABLE 5.1
### Primary Ocean Currents and Temperature Characteristics

| Pacific Ocean | | Atlantic Ocean | | Indian Ocean | |
|---|---|---|---|---|---|
| Oyashio | Cool | East Greenland | Cool | North Indian monsoon currents (reverse seasonally with the monsoon winds) | |
| Bering | Cool | Labrador | Cool | | |
| North Pacific Drift | Warm | North Atlantic Drift | Warm | North Indian Equatorial | Warm |
| Kuroshio (Japan) | Warm | Gulf Stream | Warm | Indian Equatorial Counter Current | Warm |
| Alaska | Warm | Canary | Cool | South Indian Equatorial | Warm |
| California | Cool | Guinea | Warm | Agulhas (Mozambique) | Warm |
| North Pacific Equatorial | Warm | North Atlantic Equatorial | Warm | West Australian | Cool |
| Pacific Equatorial Counter Current | Warm | North Atlantic Equatorial Counter Current | Warm | | |
| South Pacific Equatorial | Warm | South Atlantic Equatorial | Warm | | |
| Humboldt (Peru) | Cool | Brazil | Warm | | |
| East Australian | Warm | Benguela | Cool | | |
| West Wind Drift (Antarctic Circumpolar Drift; also present in South Atlantic and South Indian Oceans) | Cool | Falkland | Cool | | |

# El Niño

As you can see in Figure 5.24, the cold Humboldt Current flows equatorward along the coasts of Ecuador and Peru. When the current approaches the equator, the westward-flowing trade winds cause upwelling of nutrient-rich cold water along the coast. Fishing, especially for anchovies, is a major local industry.

Every year usually during the months of November and December, a weak warm countercurrent replaces the normally cold coastal waters. Without the upwelling of nutrients from below to feed the fish, fishing comes to a standstill. Fishermen in this region have known of the phenomenon for hundreds of years. In fact, this is the time of year they traditionally set aside to tend to their equipment and await the return of cold water. The residents of the region have given this phenomenon the name **El Niño**, which is Spanish for "The Child," because it occurs about the time of the celebration of the birth of the Christ Child.

The warm-water current usually lasts for 2 months or less, but occasionally the disruption to the normal flow lasts for many months. In these situations, water temperatures are raised not just along the coast but for thousands of kilometers offshore (Fig. 5.25). Over the past decade, the term *El Niño* has come to describe these exceptionally strong episodes and not the annual event. During the past 50 years, as many as 17 years qualify as having El Niño conditions (with sea-surface temperatures 0.5°C, or warmer, than normal for 6 consecutive months). Not only do the El Niños affect the temperature of the equatorial Pacific, but the strongest of them also impact worldwide weather.

> **PHYSICAL Geography Now™** Log on to Physical GeographyNow and select this chapter to work through a Geography Interactivity on "El Niño" (click Weather & Climate, El Niño).

## El Niño and the Southern Oscillation

To completely understand the processes that interact to produce an El Niño requires that we study conditions all across the Pacific, not just in the waters off South America. More than 50 years ago, Sir Gilbert Walker, a British scientist, discovered a connection between surface-pressure readings at weather stations on the eastern and western sides of the

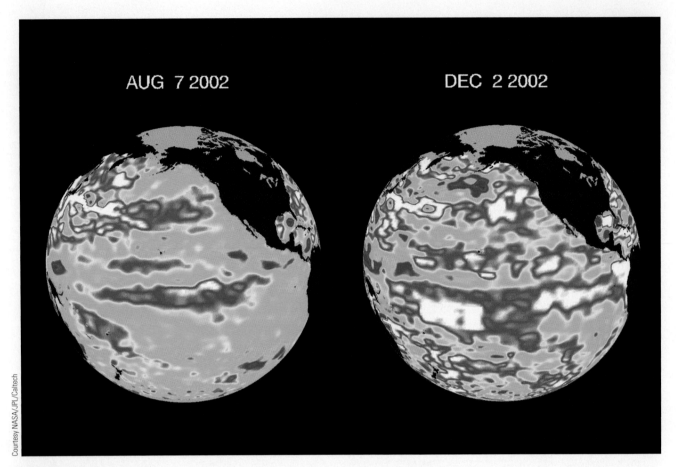

**FIGURE 5.25**
These enhanced satellite images show the development of an El Niño episode in the Tropical Pacific. The red and white shades display the warmer sea surface temperatures, while the blues and greens mark areas of cooler temperatures. From what continent does an El Niño originate?

Pacific. He noted that a rise in pressure in the eastern Pacific is usually accompanied by a fall in pressure in the western Pacific and vice versa. He called this seesaw pattern the **Southern Oscillation.** The link between El Niño and the Southern Oscillation is so great that they are often referred to jointly as ENSO (*E*l Niño/*S*outhern *O*scillation). These days the atmospheric pressure values from Darwin, Australia, are compared to those recorded on the Island of Tahiti, and the relationship between these two values defines the Southern Oscillation.

During a typical year, the eastern Pacific has a higher pressure than the western Pacific. This east-to-west pressure gradient enhances the trade winds over the equatorial Pacific waters. This results in a warm surface current that moves from east to west at the equator. The western Pacific develops a thick, warm layer of water while the eastern Pacific has the cold Humboldt Current enhanced by upwelling.

Then, for unknown reasons, the Southern Oscillation swings in the opposite direction, dramatically changing the usual conditions described above, with pressure increasing in the western Pacific and decreasing in the eastern Pacific. This change in the pressure gradient causes the trade winds to weaken or, in some cases, to reverse. This causes the warm water in the western Pacific to flow eastward, increasing sea-surface temperatures in the central and eastern Pacific. This eastward shift signals the beginning of El Niño.

In contrast, at times and for reasons we do not fully know, the trade winds will intensify. These more powerful trade winds will cause even stronger upwelling than usual to occur. As a result, sea-surface temperatures will be even colder than normal. This condition is known as **La Niña** (in Spanish, "Little Girl," but scientifically simply the opposite of El Niño). La Niña episodes will at times, but not always, bring about the opposite effects of an El Niño episode.

**Physical Geography Now™** Log on to Physical GeographyNow and select this chapter to work through a Geography Interactive activity on the "Southern Oscillation."

## El Niño and Global Weather

Cold ocean waters impede cloud formation. Thus, under normal conditions, clouds tend to develop over the warm waters of the western Pacific but not over the cold waters of the eastern Pacific. However, during an El Niño, when warm water migrates eastward, clouds develop over the entire equatorial region of the Pacific ( ▌ Fig. 5.26). These clouds can build to heights of 18,000 meters (59,000 ft). Clouds of this magnitude can disrupt the high-altitude wind flow above the equator. As we have seen, a change in the upper air wind flow in one portion of the atmosphere will trigger wind flow changes in other portions of the atmosphere. Alterations in the upper air winds result in alterations to surface weather.

Scientists have tried to document as many past El Niño events as possible by piecing together bits of historical evidence, such as sea-surface temperature records, daily observations of atmospheric pressure and rainfall, fisheries' records from South America, and the writings of Spanish colonists living along the coasts of Peru and Ecuador dating back to the 15th century. Additional evidence comes from the growth patterns of coral and trees in the region.

Based on this historical evidence, we know that El Niños have occurred as far back as records go. One disturbing fact is that they are occurring more often. Records indicate that during the 16th century, an El Niño occurred, on average, every 6 years. Evidence gathered over the past few decades indicates that El Niños are now occurring, on average, every 2.2 years. Even more alarming is the fact that they appear to be getting stronger. The record-setting El Niño of 1982–1983 was recently surpassed by the one in 1997–1998.

The 1997–1998 El Niño brought copious and damaging rainfall to the southern United States, from California to Florida. Snowstorms in the northeast portion of the United States were more frequent and stronger than in most years. The warm El Niño winters fueled Hurricane Linda, which devastated the western coast of Mexico. Linda was the strongest hurricane ever recorded in the eastern Pacific.

In recent years, scientists have become better able to monitor and forecast El Niño and La Niña events. An elaborate network of ocean-anchored weather buoys plus satellite observations provide an enormous amount of data that can be analyzed by computer to help predict the formation and strength of El Niño and La Niña events. Our improved observation skills have led to the discovery of the **North Atlantic Oscillation**—a strengthening of the subtropical (in the Azores) high and subpolar (in Iceland) low over the Atlantic Ocean will render a positive index for the NAO. This phase has been associated with mild and wet winters in the United States but cold dry winters in Northern Canada. When the Azores High and Icelandic Low are both weaker, the index becomes negative, associated with colder and snowy winters for the U.S. East Coast. The North Atlantic Oscillation (NAO) is not as well understood as ENSO. Truly, both oscillations require more research in the future if scientists are to better understand how these ocean phenomena affect weather and climate.

Will scientists ever be able to predict the occurrence of such phenomena as ENSO or the NAO? No one can answer that question, but as our technology improves, our forecasting ability will also increase. We have made tremendous progress: In the past few decades, we have come to recognize the close association between the atmosphere and hydrosphere as well as to better understand the complex relationship between these Earth systems.

This chapter began with an examination of the behavior of atmospheric gases as they respond to solar radiation and other dynamic forces. This information enabled a definition and thorough discussion of global pressure systems and their accompanying winds. This discussion in turn permitted a description of atmospheric circulation

▌ **FIGURE 5.26**
During El Niño, the easterly surface winds weaken and retreat to the eastern Pacific, allowing the central Pacific to warm and the rain area to migrate eastward.
**Near what country or countries does El Nino begin?**

patterns that began with the smallest and simplest (diurnal) systems and progressed to seasonal systems and then systems at a global scale. Once again, we can recognize the interactions among Earth's systems. Earth's radiation budget helps create movements in our atmosphere, which in turn help drive ocean circulation, which in turn creates feedback with the atmosphere:

Solar radiation → Atmosphere → Hydrosphere → Back to the atmosphere

In following chapters, we will examine the role of the atmosphere in controlling variations in weather and climate and, later, weather and climate systems as they affect surface landforms.

## Define & Recall

cyclone (low)
anticyclone (high)
convergent wind circulation
divergent wind circulation
isobar
pressure gradient
wind
Coriolis effect
friction
geostrophic wind
windward
leeward
prevailing wind
land breeze–sea breeze cycle
mountain breeze–valley breeze cycle
drainage wind (katabatic wind)
Chinook
foehn
Santa Ana
monsoon
trough
equatorial low (equatorial trough)
subtropical high
subpolar low
polar high
Siberian High
Canadian High
Icelandic Low
Aleutian Low
Pacific High
Bermuda High (Azores High)
polar easterlies
westerlies
trade wind
northeast trades
southeast trades
tropical easterlies
doldrums
intertropical convergence zone (ITCZ)
polar front
jet stream
polar front jet stream
subtropical jet stream
Rossby wave
gyre
upwelling
El Niño
Southern Oscillation
La Niña
North Atlantic Oscillation

## Discuss & Review

1. What is atmospheric pressure at sea level? How do you suppose Earth's gravity is related to atmospheric pressure?
2. Horizontal variations in air pressure are caused by thermal or dynamic factors. How do these two factors differ?
3. How does incoming insolation affect pressure in the atmosphere? Give an example of an area where incoming insolation would create a pressure system. Would high or low pressure occur?
4. What is the difference between a cell and a belt of pressure?
5. What kind of pressure (high or low) would you expect to find in the center of an anticyclone? Describe and diagram the wind patterns of anticyclones in the Northern and Southern Hemispheres.
6. What is the circulation pattern around a center of low pressure (cyclone) in the Northern Hemisphere? In the Southern Hemisphere? Draw diagrams to illustrate these circulation patterns.
7. Explain how surface friction causes the surface winds to flow *across* isobars rather than *parallel to* the isobars, as in the case of a geostrophic wind.
8. Explain how water and land surfaces affect the pressure overhead during summer and winter. How does this relate to the afternoon sea breeze?
9. How are the land breeze–sea breeze and monsoon circulations similar? How are they different?
10. What effect on valley farms could a strong drainage wind have?
11. What effect would foehn-type winds have on farming, forestry, and ski resorts?
12. What are monsoons? Have you ever experienced one? What causes them? Name some nations that are concerned with the arrival of the "wet monsoon."
13. How do landmasses affect the development of belts of atmospheric pressure over Earth's surface?
14. What are the horse latitudes? The doldrums?

15. Why do Earth's wind systems migrate with the seasons?
16. Map the trade winds of the Atlantic Ocean and compare your map with one of trade routes in the 19th century or earlier.
17. Describe the movements of the upper air. How have pilots applied their experience of the upper air to their flying patterns?
18. What is the relationship between the polar front jet stream and upper air westerlies?
19. What is the relationship between ocean currents and global surface wind systems? How does the gyre in the Northern Hemisphere differ from the one in the Southern Hemisphere?
20. Where are the major warm and cool ocean currents located in respect to Earth's continents? Which currents have the greatest effects on North America?
21. What roles do ocean currents and atmospheric pressure differences play in the development of an El Niño?
22. How might an El Niño affect global weather?
23. How can you apply knowledge of pressure and winds in your everyday life?

## Consider & Respond

1. If you wish to sail from London to New York and back, what route will you take? Why?
2. Is the polar front jet stream stronger in the summer or the winter? Why?
3. The amount of power that can be generated by wind is determined by the equation

$$P = \tfrac{1}{2} D \times S^3$$

where $P$ is the power in watts, $D$ is the density, and $S$ is the wind speed in meters per second (m/sec). Because $D = 1.293$ kg/m$^3$, we can rewrite the equation as

$$P = 0.65 \times S^3$$

  a. How much power (in watts) is generated by the following wind speeds: 2 meters per second, 6 meters per second, 10 meters per second, 12 meters per second?
  b. Because wind power increases significantly with increased wind speed, very windy areas are ideal locations for "wind farms." Cities A and B both have average wind speeds of 6 meters per second. However, city A tends to have very consistent winds; in city B, half of its winds tend to be at 2 meters per second and the other half at 10 meters per second. Which site would be the better location for a wind generation plant?

4. Atmospheric pressure decreases at the rate of 0.036 millibar per foot as one ascends through the lower portion of the atmosphere.
  a. The Sears Tower in Chicago, Illinois, is one of the world's tallest buildings at 1450 feet. If the street-level pressure is 1020.4 millibars, what is the pressure at the top of the Sears Tower?
  b. If the difference in atmospheric pressure between the top and ground floor of an office building is 13.5 millibars, how tall is the building?
  c. A single story of a building is 12 feet. You enter an elevator on the top floor of the building and want to descend five floors. The elevator has no floor markings—only a barometer. If the initial reading was 1003.2 millibars, at what pressure reading would you want to get off?
5. Look at the January (Fig. 5.13a) and July (Fig. 5.13b) maps of average sea-level pressure. Answer the following questions:
  a. Why is the subtropical high pressure belt more continuous (linear, not cellular) in the Southern Hemisphere than in the Northern Hemisphere in July?
  b. During July, what area of the United States exhibits the lowest average pressure? Why?

# CHAPTER 6

A New Orleans city bus navigates a street flooded by Hurricane Katrina's torrential rains.  © Mike Theiss/Jim Reed Photography/Corbis

# Moisture, Condensation, and Precipitation

**PHYSICAL Geography Now™** This icon, appearing throughout the book, indicates an opportunity to explore interactive tutorials, animations, or practice problems at now.brookscole.com/gabler8

# CHAPTER PREVIEW

**The hydrologic cycle involves the circulation of water throughout all the major Earth spheres; it is therefore fundamental to the nature and operation of the entire Earth system.**

- How does the hydrologic cycle involve all of Earth's spheres?
- In what ways does the hydrologic cycle affect the Earth system?
- What are the major stages of the hydrologic cycle?

**Although water may evaporate from all Earth surfaces and transpiration may add considerable moisture to the air, the oceans are the most important source of water vapor in the atmosphere.**

- Why is this so?
- What portion (latitudes) of the oceans would have the highest evaporation rates, and why?
- What times of year would evaporation be at its greatest?

**There is only one way for significant condensation, cloud formation, and precipitation to occur: Air must be forced to rise so that sufficient adiabatic cooling will take place.**

- How are condensation, clouds, and precipitation related?
- Why does adiabatic cooling take place in rising air?
- Why might the rate of cooling differ in wet and dry air?

**Stability is associated with low (or weak) environmental lapse rates; instability is associated with high (or steep) environmental lapse rates.**

- What is stability? Instability?
- Why does the lapse rate affect the tendency of air to rise?

**The geographic distribution of precipitation over Earth can be explained by either one or a combination of the following: frontal, cyclonic, orographic, or convectional precipitation.**

- Which is most important in your community?
- Which is most closely related to landforms?
- How is this statement related to the third statement above?

**As a general rule, the less precipitation a region receives, the greater will be the variability of precipitation in that region from year to year.**

- What is the effect of this variability on human beings?
- Where do you think the effect would be greater—in deserts or in semiarid grasslands?
- How might this rule affect a rainforest or a desert?

Water is vital to all life on Earth. Although some living organisms can survive without air, nothing can survive without water. Water is necessary for photosynthesis, cell growth, protein formation, soil formation, and the absorption of nutrients by plants and animals.

Water affects Earth's surface in innumerable ways. The structure of the water molecule is such that water can dissolve an enormous number of substances—so many in fact that it has been called the *universal solvent*. Because water acts as a solvent for so many substances, it is almost never found in a pure state. Even rainwater is filled with impurities picked up in the atmosphere. Indeed, without these impurities to condense around, neither clouds nor precipitation could occur. In addition, rainwater usually contains some dissolved carbon dioxide from the air. Therefore, rain is a very weak form of carbonic acid. We will see later (Chapter 16) that this fact affects how water shapes certain landforms. The weak acidity of rainwater should not be confused with the environmentally damaging *acid rain*, which is at least ten times more acidic.

Not only can water dissolve and transport many minerals, it also can transport solid particles in suspension. These characteristics make water a unique transportation system for Earth. Water supplies nutrients that would not otherwise be available to plants. Water carries minerals and nutrients down streams, through the soil, through the openings in subsurface rocks, and through plants and animals. It deposits solid matter on stream floodplains, in river deltas, and on the ocean floor.

The surface tension of water and the behavior of water molecules to draw together, make possible **capillary action**—the ability of water to pull itself upward through small openings against gravity. Capillary action also permits transport of dissolved material in an upward direction—that is, against the pull of gravity. Capillary action moves water into the stems and leaves of plants—even to the topmost needles of the great California redwoods and the top leaves of the rainforest trees. Capillary action is also important in the movement of blood through our bodies. Without it, many of our cells could not receive the necessary nutrients carried by the blood.

Another important and highly unusual property of water is that it expands when it freezes. Most substances contract when cooled and expand when heated. Water follows these rules until it is cooled below 4°C (39°F); then it begins to expand. Ice is therefore less dense than water and consequently will float on water, as do ice floes and icebergs ( Fig. 6.1).

Finally, compared with solids, water is slow to heat and slow to cool. Therefore, as we learned in Chapter 4, large bodies of water on Earth act as reservoirs of heat during winter and have a cooling effect in summer. This moderating effect on temperature can be seen in the vicinity of lakes as well as on seacoasts.

Earth's water—the *hydrosphere* (from Latin: *hydros,* water)—is found in all three states: as a liquid in rivers, lakes, oceans, and rain; as a solid in the form of snow and ice; and as a vapor in our atmosphere. Even the water temporarily stored in living things can be considered part of the hydrosphere. About 73% of Earth's surface is covered by water, with the largest proportion contained within the world's oceans ( Fig. 6.2). In all, the total water content of the Earth system, whether liquid, solid, or vapor, is about 1.33 billion cubic kilometers (326 million cu mi). Although water cycles in and out of the atmosphere, lithosphere, and biosphere, the total amount of water in the hydrosphere remains constant.

## The Hydrologic Cycle

The circulation of water from one part of the general Earth system to another is known as the **hydrologic cycle.** The air contains water vapor that has entered the atmosphere through evaporation from Earth's surface. When water vapor condenses and falls as precipitation, several things may happen to it. First, it may go directly into a body of water where it is immediately available for evaporation back into the atmosphere. Alternatively, it may fall onto the land where it may run off the surface to form streams, ponds, or lakes. Or it may be absorbed into the ground where it can either be contained by the soil or flow through open spaces, called *interstices,* that exist in loose sand, gravel, silt, and voids in solid rock. Ultimately, much of the water in or on Earth's surface reaches the oceans. Some water that reaches the surface as snow becomes a part of massive ice over Greenland and Antarctica as well as in high mountain glaciers, while some snow remains frozen until spring thaw and then reenters the hydrologic cycle as a liquid. Other water is used by plants and animals and temporarily becomes a part of living things. Thus, there are six storage areas for water in the hydrologic cycle: the atmosphere, the oceans, bodies of fresh water on the surface, plants and animals, open spaces beneath Earth's surface, and glacial ice.

Liquid water is returned to the atmosphere as a gas through evaporation. Water evaporates from all bodies of water on Earth, from plants and animals, and from soils; it can even evaporate from falling precipitation. Once the water is an atmospheric gas again, the cycle can be repeated.

The hydrologic cycle is basically one of evaporation, condensation, and precipitation, and to some extent transportation because these processes are in constant motion. The hydrologic cycle is a *closed system;* that is, water may appear in the system in all three major states—as a liquid, vapor, or solid—and may be transferred from one state to another, but there is no gain or loss of water by the system. This is unlike

**FIGURE 6.1**
Most of Earth's fresh water is locked in glacial ice, as seen in this view of the Northern Hemisphere.
**Can you distinguish between the permanent ice sheets and the seasonal (pack ice) that forms from frozen ocean water?**

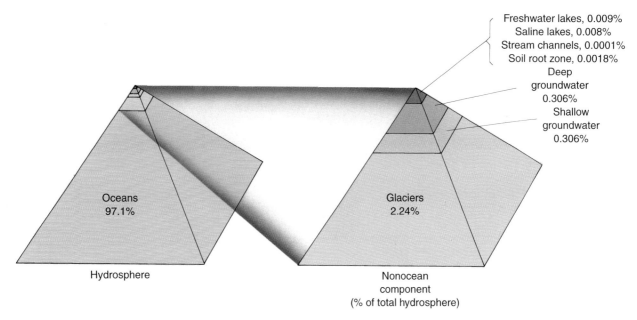

**FIGURE 6.2**
This illustration emphasizes the fact that the vast majority of water in the hydrosphere is salt water, stored in the world's oceans. The bulk of the supply of fresh water is relatively unavailable because it is stored in polar ice sheets.
**How might global warming or cooling alter this figure?**

the radiation budget depicted in Figure 4.10, which is an *open system* where energy can flow both into and out of the system. Although the hydrologic cycle is a closed system, it is not static but exceedingly dynamic. The percentage of water associated with any one component of the system changes constantly over time and place. For example, during the last ice age, evaporation and precipitation were greatly reduced. Also, some changes are human induced; the cutting down of a forest or the damming of a river will cause adjustments among the components.

The hydrologic cycle is one of the most important subsystems of the larger Earth system. It is linked to numerous other subsystems that rely on water as an agent of movement. For example, it plays a major role in the redistribution of energy over Earth's surface. Figure 6.3 provides a schematic illustration of the circulation of water in the hydrologic cycle.

# Water in the Atmosphere
## The Water Budget and Its Relation to the Heat Budget

We are most familiar with and most often take notice of water in its liquid form, as it pours from a tap or as it exists in fine droplets within clouds or fog. Water also exists as a tasteless, odorless, transparent gas known as water vapor, which is mixed with the other gases of the atmosphere in varying proportions. Water vapor is found within approximately the first 5500 meters (18,000 ft) of the troposphere and makes up a small but highly variable percentage of the atmosphere by volume. Atmospheric water is the source of all condensation and precipitation. Through these processes and through evaporation, water plays a significant role as Earth's temperature regulator and modifier. In addition, as we noted in Chapter 4, water vapor in the atmosphere absorbs and reflects a significant portion of both incoming solar energy and outgoing Earth radiation. By preventing great losses of heat from Earth's surface, water vapor helps maintain the moderate range of temperature found on this planet.

As we stated previously, Earth's hydrosphere is a *closed system;* that is, water is neither received from outside the Earth system nor given off from it. Thus, an increase in water within one subsystem must be accounted for by a loss in another. Put another way, we say that the Earth system operates with a *water budget,* in which the total quantity of water remains the same and in which the deficits must balance the gains throughout the entire system.

We know that the atmosphere gives up a great deal of water, most obviously by condensation into clouds, fog, and dew and through several forms of precipitation (rain, snow, hail, sleet). If the quantity of water in the atmosphere remains at the same level through time, the atmosphere must be absorbing from other parts of the system an amount of water equal to that which it is giving up. During 1 minute, over 1 billion tons of water are given up by the atmosphere through some form of precipitation or condensation, while another billion tons are evaporated and absorbed as water vapor by the atmosphere.

If we look again at our discussion of the heat energy budget in Chapter 4, we can see that a part of that budget is

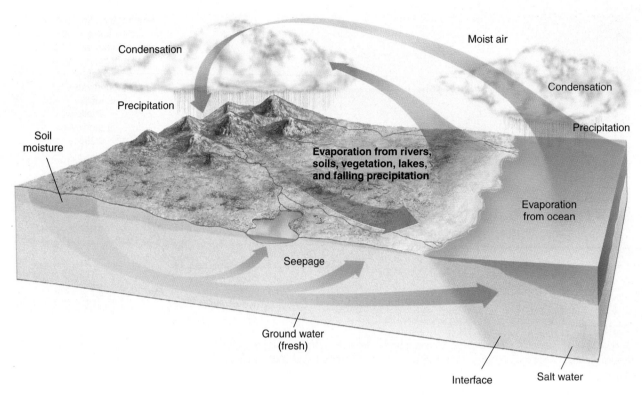

**ACTIVE FIGURE 6.3**
Watch this Active Figure at http://now.brookscole.com/gabler8.
**Environmental Systems: The Hydrologic System—The Water Cycle** The hydrologic system is concerned with the circulation of water from one part of the Earth system to another. The subsystem of the hydrologic system illustrated in this diagram is referred to as the hydrologic cycle. Largely through condensation, precipitation, and evaporation, water is cycled endlessly between the atmosphere, the soil, subsurface storage, lakes and streams, plants and animals, glacial ice, and the principal reservoir—the oceans.

the latent heat of condensation. Of course, this energy is originally derived from the sun. The sun's energy is used in evaporation and is then stored in the molecules of water vapor, to be released only during condensation. Although the heat transfers involved in evaporation and condensation within the total heat energy budget are proportionately small, the actual energy is significant. Imagine the amount of energy released every minute when a billion tons of water condense out of the atmosphere. It is this vast storehouse of energy, the latent heat of condensation, that provides the major source of power for Earth's storms: hurricanes, tornadoes, and thunderstorms.

There are limits to the amount of water vapor that can be held by any parcel of air. A very important determinant of the amount of water vapor that can be held by the air is temperature. The warmer air is, the greater the quantity of water vapor it can hold. Therefore, we can make a generalization that air in the polar regions can hold far less water vapor (approximately 0.2% by volume) than the hot air of the tropics and equatorial regions of Earth, where the air can contain as much as 5% by volume.

## Saturation and Dew Point

When air of a given temperature holds all the water vapor that it possibly can, it is said to be in a state of **saturation** and to have reached its **capacity.** If a constant temperature is maintained in a quantity of air, there will come a point, as more water vapor is added, when the air will be saturated and unable to hold any more water vapor. For example, when you take a shower, the air in the room becomes increasingly humid until a point is reached at which the air cannot contain more water. Then, excess water vapor condenses onto the colder mirrors and walls.

We know that the capacity of air to hold water vapor varies with temperature. In fact, as we can see in Figure 6.4, this capacity of air to contain moisture increases with rising temperatures. Some examples will help illustrate the relationship between temperature and water vapor capacity. If we assume that a parcel of air at 30°C is saturated, then it will contain 30 grams of water vapor in each cubic meter of air (30 g/m$^3$). Now suppose we increase the temperature of the air to 40°C *without* increasing the water vapor content. The parcel is no longer saturated because air at 40°C can

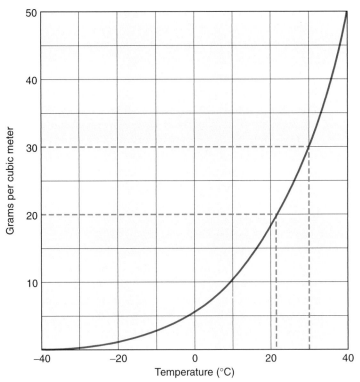

**ACTIVE FIGURE 6.4**

Watch this Active Figure at http://now.brookscole.com/gabler8.
The graph shows the maximum amount of water vapor that can be contained in a cubic meter of air over a wide range of temperatures.

**Compare the change in capacity when the air temperature is raised from 0°C to 10°C with a change from 20°C to 30°C. What does this indicate about the relationship between temperature and capacity?**

hold more than 30 grams per cubic meter of water vapor (actually, 50 g/m³). Conversely, if we decrease the temperature of saturated air from 30°C (which contains 30 g/m³ of water vapor) to 20°C (which has a water vapor capacity of only 17 g/m³), some 13 grams of the water vapor will condense out of the air because of the reduced capacity.

It is also evident that if an unsaturated parcel of air is cooled, it will eventually reach a temperature at which the air will become saturated. This critical temperature is known as the **dew point**—the temperature at which condensation takes place. For example, we know that if a parcel of air at 30°C contains 20 grams per cubic meter of water vapor, it is not saturated because it can hold 30 grams per cubic meter. However, if we cool that parcel of air to 21°C, it would become saturated because the capacity of air at 21°C is 20 grams per cubic meter. Thus, that parcel of air at 30°C has a dew point of 21°C. It is the cooling of air to below its *dew point temperature* that brings about the condensation that must precede precipitation.

Because the capacity of air to hold water vapor increases with rising temperatures, air in the equatorial regions has a higher dew point than air in the polar regions. Thus, because the atmosphere can hold more water in the equatorial regions, there is greater potential for large quantities of precipitation than in the polar regions. Likewise, in the middle latitudes, summer months, because of their higher temperatures, have more potential for large-scale precipitation than do winter months.

## Humidity

The amount of water vapor in the air at any one time and place is called **humidity.** There are three common ways to express the humidity content of the air. Each method provides information that contributes to our discussion of weather and climate.

**Absolute and Specific Humidity** Absolute humidity is the measure of the mass of water vapor that exists within a given volume of air. It is expressed either in the metric system as the number of grams per cubic meter (g/m³) or in the English system as grains per cubic foot (gr/ft³). **Specific humidity** is the mass of water vapor (given in grams) per mass of air (given in kilograms). Obviously, both are measures of the actual amount of water vapor in the air. Because most water vapor gets into the air through the evaporation of water from Earth's surface, it stands to reason that absolute and specific humidity will decrease with height from Earth.

We have also learned that air is compressed as it sinks and expands as it rises. Thus, a given parcel of air changes its volume as it moves vertically, although its weight remains the same and there may be no change in the amount of water vapor in that quantity of air. We can see, then, that absolute humidity, although it measures the amount of water vapor, can vary simply as a result of the vertical movement of a parcel of air. Specific humidity, on the other hand, changes *only* as the quantity of the water vapor changes. For this reason, when assessing the changes of water vapor content in large masses of air, which often have vertical movement, specific humidity is the preferred measurement among geographers and meteorologists.

**Relative Humidity** Probably the best-known means of describing the content of water vapor in the atmosphere—the one commonly given on television and radio weather reports—is **relative humidity.** It is simply the ratio between the amount of water vapor in air of a given temperature and the maximum amount of vapor that the air could hold at that temperature; it is reported as a percentage that expresses how close the air is to saturation. This method of describing humidity has its strengths and weaknesses. Its strength lies in the ease with which it is communicated. Most people can understand the concept of "percent" and are unlikely to understand "grams per cubic meter." However, relative humidity percentages can vary widely through the day, even when the water vapor content of the air remains the same. The unsteady nature of relative humidity is its weakness.

If the temperature and absolute humidity of an air parcel are known, its relative humidity can be determined by using Figure 6.4. For instance, if we know that a parcel of air has a temperature of 30°C and an absolute humidity of 20 grams per cubic meter, we can look at the graph and determine that if it were saturated its absolute humidity would be 30 grams per cubic meter. To determine relative humidity, all we do is divide 20 grams (actual content) by 30 grams (content at capacity) and multiply by 100 (to get an answer in percentage):

$$(20 \text{ grams} \div 30 \text{ grams}) \times 100 = 67\%$$

The relative humidity in this case is 67%. In other words, the air is holding only two thirds of the water vapor it could contain at 30°C; it is only at 67% of its capacity.

Two important factors are involved in the horizontal distribution and variation of relative humidity. One of these is the availability of moisture. For example, air above bodies of water is apt to contain more moisture than similar air over land surfaces because there is simply more water available for evaporation. Conversely, the air overlying a region like the central Sahara Desert is usually very dry because it is far from the oceans and little water is available to be evaporated. The second factor in the horizontal variation of relative humidity is temperature. In regions of higher temperature, relative humidity for the same amount of water vapor will be lower than it would be in a cooler region.

At any one point in the atmosphere, relative humidity varies if the amount of water vapor increases as a result of evaporation *or* if the temperature increases or decreases. Thus, although the quantity of water vapor may not change through a day, the relative humidity will vary with the daily temperature cycle. As air temperature increases from around sunrise to its maximum in midafternoon, the relative humidity decreases as the air becomes capable of holding greater and greater quantities of water vapor. Then, as the air becomes cooler, decreasing toward its minimum temperature just before sunrise, the relative humidity increases ( ▌Fig. 6.5).

Relative humidity affects our comfort through its relationship to the rate of evaporation. Perspiration evaporates into the air, leaving behind a salty residue, which you can taste if you lick your lips after perspiring a great deal. Evaporation is a cooling process because the heat used to change the perspiration to water vapor (which becomes locked in the water vapor as latent heat) is subtracted from your skin. This is the reason that, on a hot August day when the temperature

▌**FIGURE 6.5**
This hygrothermograph illustrates the relationship between air temperature (top half) and relative humidity (bottom half) through a week in May in Bowling Green, Kentucky.
**What do you notice about these lines?**

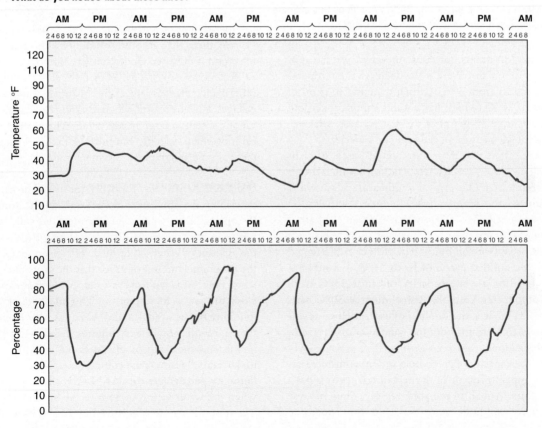

**TABLE 6.1**
**Distribution of Actual Mean Evapotranspiration**

| | Latitude | | | | | |
|---|---|---|---|---|---|---|
| Zone | 60°–50° | 50°–40° | 40°–30° | 30°–20° | 20°–10° | 10°–0° |
| | Northern Hemisphere | | | | | |
| Continents | 36.6 cm (14.2 in.) | 33.0 (13.0) | 38.0 (15.0) | 50.0 (19.7) | 79.0 (31.1) | 115.0 (45.3) |
| Oceans | 40.0 (15.7) | 70.0 (27.6) | 96.0 (37.8) | 115.0 (45.3) | 120.0 (47.2) | 100.0 (39.4) |
| Mean | 38.0 (15.0) | 51.0 (20.1) | 71.0 (28.0) | 91.0 (35.8) | 109.0 (42.9) | 103.0 (40.6) |
| | Southern Hemisphere | | | | | |
| Continents | 20.0 cm (7.9 in.) | NA | 51.0 (20.1) | 41.0 (16.1) | 90.0 (35.4) | 122.0 (48.0) |
| Oceans | 23.0 (9.1) | 58.0 (22.8) | 89.0 (35.0) | 112.0 (44.1) | 119.0 (47.2) | 114.0 (44.9) |
| Mean | 22.5 (8.8) | NA | NA | 99.0 (39.0) | 113.0 (44.5) | 116.0 (45.7) |

approaches 35°C (95°F), you will be far more uncomfortable in Atlanta, Georgia, where the relative humidity is 90%, than in Tucson, Arizona, where it may be only 15% at the same temperature. At 15%, your perspiration will be evaporated at a faster rate at the lower relative humidity, and you will benefit from the resultant cooling effects. When the relative humidity is 90%, the air is nearly saturated, far less evaporation can take place, and less heat is drawn from your skin.

## Sources of Atmospheric Moisture

In our earlier discussion of the hydrologic cycle, we saw that the atmosphere receives water vapor through the process of evaporation. Water evaporates into the atmosphere from many different places, most important of which are the surfaces of Earth's bodies of water. Water also evaporates from wet ground surfaces and soils, from droplets of moisture on vegetation, from city pavements, building roofs, cars, and other surfaces, and even from falling precipitation.

Vegetation provides another source of water vapor. Plants give up water in a complex process known as **transpiration,** which can be a significant source of atmospheric moisture. A mature oak tree, for instance, can give off 400 liters (105 gal) of water per day, and a cornfield may add 11,000–15,000 liters (2900–4000 gal) of water to the atmosphere per day for each acre under cultivation. In some parts of the world—notably tropical rainforests of heavy, lush vegetation—transpiration accounts for a significant amount of atmospheric humidity. Together, evaporation and transpiration, or **evapotranspiration,** account for virtually all the water vapor in the air.

### Rate of Evaporation

The rate of evaporation is affected by several factors. First, it is affected by the amount and temperature of accessible water. Thus, as Table 6.1 shows, the rate of evapotranspiration tends to be greater over the oceans than over the continents. The only place this generalization is not true is in equatorial regions between 0° and 10°N and S, where the vegetation is so lush on the land that transpiration provides a large amount of water for the air.

Second is the degree to which the air is saturated with water vapor. The drier the air and the lower the relative humidity, the greater the rate of evaporation can be. Some of us have had direct experience with this principle. Compare the length of time it takes your bathing suit to dry on a hot, humid day with how long it takes on a day when the air is dry.

Third is the wind, which affects the rate of evaporation. If there is no wind, the air that overlies a water surface will approach saturation as more and more molecules of liquid water change to water vapor. Once saturation is reached, evaporation will cease. However, if there is a wind, it will blow the saturated or nearly saturated air away from the evaporating surface, replacing it with air of lower humidity. This allows evaporation to continue as long as the wind keeps blowing saturated air away and bringing in drier air. Anyone who has gone swimming on a windy day has experienced the chilling effects of rapid evaporation.

Temperature affects the rate of evaporation by affecting the first and second factors above. As air temperature increases, so does the temperature of the water at the evaporation source. Such increases in temperature assure that more energy is available to the water molecules for their escape from a liquid state to a gaseous one. Consequently, more molecules can make the transition. Also as the temperature of the air increases, its capacity to contain moisture also increases. As the air gets warmer, it is energized, the molecules of air separate farther apart, and air density decreases. With more energy and wider spacing between the air molecules, additional water molecules can enter the atmosphere, thus increasing evaporation.

## Potential Evapotranspiration

So far, we have discussed actual evaporation and transpiration (evapotranspiration). However, geographers and meteorologists are also concerned with **potential evapotranspiration** ( Fig. 6.6). This term refers to the idealized conditions in an area under which there would be sufficient moisture for all possible evapotranspiration to occur. Various formulas have been derived for estimating the potential evapotranspiration at a location because it is difficult to measure directly. These formulas commonly use temperature, latitude, vegetation, and soil character (permeability, water-retention ability) as factors that could affect the potential evapotranspiration.

In places where precipitation exceeds potential evapotranspiration, there is a surplus of water for storage in the ground and in bodies of water, and water can even be exported to other places if transportation (like canals) is feasible. When potential evapotranspiration exceeds precipitation, as it does during the dry summer months in California, then there is no water available for storage; in fact, the water stored during previous rainy months evaporates quickly into the warm, dry air ( Fig. 6.7). Soil becomes dry and vegetation turns brown as the available water is absorbed into the atmosphere. For this reason, fires become a potential hazard during the late summer months in California.

Knowledge of potential evapotranspiration is used by irrigation engineers to learn how much water will be lost through evaporation. With that, they can determine whether the water that is left is enough to justify a drainage canal. Farmers, by assessing the daily or weekly relationship between potential evapotranspiration and precipitation, can determine when and how much to irrigate their crops.

## Condensation

Condensation is the process by which a gas is changed to a liquid. In our present discussion of atmospheric moisture and precipitation, condensation refers to the change of water vapor to liquid water.

### FIGURE 6.6
Potential evapotranspiration for the contiguous 48 states.
**Why is potential evapotranspiration so high over the desert southwest?**

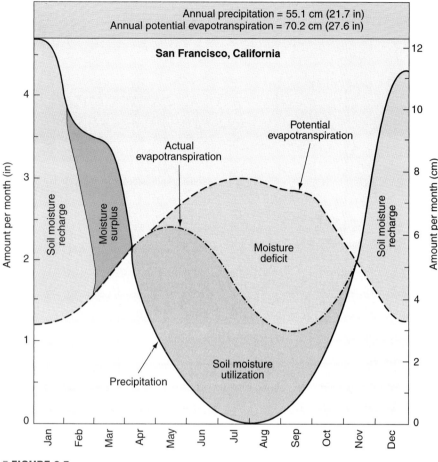

**FIGURE 6.7**
This is an example of the Thornthwaite water budget system, which "keeps score" of the balance between water input by precipitation and water loss to evaporation and transpiration, permitting month-by-month estimates of both runoff and soil moisture.
**When would irrigation be necessary at this site?**

Condensation occurs when air saturated with water vapor is cooled. Viewed in another way, we can say that if we lower the temperature of air until it has a relative humidity of 100% (the air has reached the dew point), condensation will occur with additional cooling. It follows, then, that condensation depends on (1) the relative humidity of the air and (2) the degree of cooling. In the arid air of Death Valley, California, a great amount of cooling must take place before the dew point is reached. In contrast, on a humid summer afternoon in Mississippi, a minimal amount of cooling will bring on condensation.

This is the principle behind the formation of droplets of water on the side of a glass containing a cold drink on a warm afternoon: The temperature of the air is lowered when it comes in contact with the cold glass. Consequently, the air's capacity to hold water vapor is diminished. If air touching the glass is cooled sufficiently, its relative humidity will reach 100%. Any cooling beyond that point will result in condensation in the form of water droplets on the glass.

## Condensation Nuclei

For condensation to occur, one other factor is necessary: the presence of **condensation nuclei.** These are minute particles in the atmosphere that provide a surface upon which condensation can take place. Condensation nuclei are most often sea-salt particles in the air from the evaporation of seawater. They also can be particles of dust, smoke, pollen, or volcanic material. Some particles are said to be more *hygroscopic* (the property to attract moisture) than others. More commonly, they are chemical particles that are the by-products of industrialization. The condensation that takes place on such chemical nuclei is often corrosive and dangerous to human health; when it is, we know it as *smog* (an invented term derived by combining the words *smoke* and *fog*).

Theoretically, if all such particles were removed from a volume of air, we could cool that air below its dew point without condensation occurring. Conversely, if there is a superabundance of such particles, condensation may take place at relative humidity less than 100%. For example, ocean fogs, which are an accumulation of condensation droplets formed on sea-salt particles, can form when the relative humidity is as low as 92%.

In nature, condensation appears in a number of forms. Fog, clouds, and dew are all the results of condensation of water vapor in the atmosphere. The type of condensation produced depends on a number of factors, including the cooling process itself. The cooling that produces condensation in one form or another can occur as a result of radiational cooling, through advection, through convection, or through a combination of these processes.

## Fog

Fogs and clouds appear when water vapor condenses on nuclei and a large number of these droplets form a mass. Not being transparent to light in the way that water vapor is, these masses of condensed water droplets appear to us as fog or clouds, in any of a number of shapes and forms, usually in shades of white or gray.

In the water budget and the hydrologic cycle, fog is a minor form of condensation. Yet, in certain areas of the world, it has important climatic effects. The "drip factor" helps sustain vegetation along desert coastlines where fog occurs. Fog also plays havoc with our modern transportation systems. Navigation on the seas is made more difficult by fog, and air travel can be greatly impeded. In fact, fog sometimes causes major airports to shut down until visibility improves. Highway travel is also greatly hampered by heavy fogs, which can lead to huge, chain-reaction pileups of cars.

**Radiation Fog** Radiational cooling can lead to **radiation fog,** also called **temperature-inversion fog.** This kind of fog is likely to occur on a cold, clear, calm night, usually in the middle latitudes. These conditions allow for maximum outgoing radiation from the ground with no incoming radiation. The ground gets colder during the night as it gives up more of the heat that it has received during the day. As time passes, the air directly above is cooled by conduction through contact with the cold ground. Because the cold surface can cool only the lower few meters of the atmosphere, an inversion is created in which cold air at the surface is overlain by warmer air above. If this cold layer of air at the surface is cooled to a temperature below its dew point, then condensation will occur, usually in the form of a low-lying fog. However, wind strong enough to disturb this inversion layer can prevent the formation of the fog by not allowing the air to stay at the surface long enough to become cooled below its dew point.

The chances of a temperature-inversion fog occurring are increased by certain types of land formations. In valleys and depressions, cold air accumulates through air drainage. During a cold night, this air can be cooled below its dew point, and a fog forms like a pond in the bottom of the valley ( ▮ Fig. 6.8a). It is common in mountainous areas to see an early morning radiation fog on the valley floor while snow-capped mountaintops shine against a clear blue sky. Radiation fog has a diurnal cycle. It forms during the night and is usually the densest at sunrise when temperatures are lowest. It then "burns off" during the day when the heat from the sun slowly penetrates the fog and warms the surface. Earth in turn warms the air directly above it, increasing its temperature and consequently its capacity to hold water vapor. This greater capacity allows the fog to evaporate into the air. As Earth's heat penetrates to higher and higher layers of air, the fog continues to burn off—from the ground up!

Radiation fog often forms even more densely in industrial areas where the high concentration of chemical particles

(a)

**▮ FIGURE 6.8**
(a) Radiation fog forms when cold air drains into a valley. (b) Advection (sea) fog is caused by warm moist air passing over colder water. (c) Upslope fog is caused by moist air adiabatically cooling as it rises up a mountain slope.
**What unique problems might coastal residents face as a result of fog?**

in the air provides abundant condensation nuclei. Such a fog is usually thicker and denser than "natural" radiation fogs and less easily dissipated by wind or sun.

**Advection Fog** Another common type of fog is **advection fog,** which occurs through the movement of warm, moist air over a colder surface, either land or water. When the warm air is cooled below its dew point through heat loss by radiation and conduction from the colder surface below, condensation occurs in the form of fog. Advection fog is usually less localized than radiation fog. It is also less likely to have a diurnal cycle, though if not too thick, it can be burned off early in the day to return again in the afternoon or early evening. More common, however, is the persistent advection fog that spreads itself over a large area for days at a time. Advection fog is a major reason why ski resorts are forced to close. Warm, moist air moving over the cold snow causes the dense fog.

Advection fog forms over land during the winter months in middle latitudes. It forms, for example, in the United States when warm, moist air from the Gulf of Mexico flows northward over the cold, frozen, and sometimes snow-covered upper Mississippi Valley.

During the summer months, advection fog may form over large lakes or over the oceans. Formation over lakes occurs when warm continental air flows over a colder water

(b)

(c)

surface, such as when a warm air mass passes over the cool surface of Lake Michigan. An advection fog also can be formed when a warm air mass moves over a cold ocean current and the air is cooled sufficiently to bring about condensation. This variety of advection fog is known as a sea fog. Such a situation accounts for fogs along the West Coast of the United States. During the summer months, the Pacific subtropical high moves north with the sun, and winds flow out of the high toward the coast where they pass over the cold California Current. When condensation occurs, fogs form that flow in over the shore, pushed from behind by the eastward movement of air and pulled by the low pressure of the warmer land ( Fig. 6.8b). Advection fogs also occur in New England, especially along the coasts of Maine and the Canadian Maritime Provinces, when warm, moist air from above the Gulf Stream flows north over the colder waters of the Labrador Current. Advection fog over the Grand Banks off Newfoundland has long been a hazard for cod fishermen there.

**Upslope Fog** Another type of fog clings to windward sides of mountain slopes and is known as **upslope fog.** Its appearance is sometimes the source of geographic place names—for example, the Great Smoky Mountains where this type of fog is quite common. During early morning hours in middle-latitude locations, a light, moist breeze may ascend a slope and cool to the dew point, leaving a blanket of fog behind ( Fig. 6.8c). In tropical rainforest regions, mountain slopes may be covered in a misty fog any time of day. Because the atmosphere is much more humid, reaching the dew point is easier.

## Other Minor Forms of Condensation

**Dew,** which is made up of tiny droplets of water, is formed by the condensation of water vapor at or near the surface of Earth. Dew collects on surfaces that are good radiators of heat (such as your car or blades of grass). These good radiators give up large amounts of heat during the hours of darkness. When the air comes in contact with these cold surfaces, it cools; if it is cooled to the dew point, droplets of water will form as beads on the surface. When the temperature of the air is below 0°C (32°F), **white frost** forms. It is important to note that frost is not frozen dew but instead represents a sublimation process—water vapor changing directly to the frozen state (see, again, Figure 4.11).

Sometimes, under very still conditions with low air pressure, though air temperatures may be below 0°C (32°F), the liquid droplets that make up clouds or fogs are not frozen into solid particles. When such *supercooled* water droplets come in contact with a surface, such as the edge of an airplane wing or a tree branch or a window, ice crystals are created on that surface in a formation known as **rime.** Icing on the wings, nose, and tail of an airplane is extremely hazardous and has been the cause of aviation disasters through time.

## Clouds

Clouds are the most common form of condensation and are important for several reasons. First, they are the source of all precipitation. **Precipitation** is made up of condensed water particles, either liquid or solid, that fall to Earth. Obviously, not all clouds result in precipitation, but we cannot have precipitation without the formation of a cloud first. Clouds also serve an important function in the heat energy budget. We have already noted that clouds absorb some of the incoming solar energy. They also reflect some of that energy back to space and scatter and diffuse other wavelengths of the incoming energy before it strikes Earth as diffuse radiation. In addition, clouds absorb some of Earth's radiation so that it is not lost to space and then reradiate it back to the surface. Finally, clouds are a beautiful and ever-changing aspect of our environment. The colors of the sky and the variations in the shapes and hues of clouds have provided us all with a beautiful backdrop to the natural scenery here on Earth.

**Cloud Forms** Clouds are composed of billions of tiny water droplets and/or ice crystals so small (some measured in 1000ths of a millimeter) that they can remain suspended in the atmosphere. As you will learn in this and later chapters, clouds are a manifestation of various movements in our atmosphere. Clouds appear white or in shades of gray, even deep gray approaching black. They differ in color depending on how thick or dense they are and if the sun is shining on the surface that we see. The thicker a cloud is, the more sunlight it is able to absorb and thus block from our view, and the darker it will appear. Clouds also seem dark when we are seeing their shaded side instead of their sunlit side.

In 1803 an Englishman by the name of Luke Howard proposed the first cloud classification scheme, which has been modified through time into the current system. Cloud names may (but not always) consist of two parts. The first part of the name refers to the cloud's height: low-level clouds, below 2000 meters (6500 ft), are called **strato**; middle-level clouds, from 2000 to 6000 meters (6500–19,700 ft), are named **alto**; and high-level clouds, above 6000 meters (19,700 ft), are termed **cirro.**

The second part of the name concerns the morphology, or shape, of the clouds. The three basic shapes are termed *cirrus, stratus,* and *cumulus.* Classification systems categorize these cloud formations into many subtypes; however, most subtypes are overlaps of the three basic shapes. Figure 6.9 illustrates the appearance and the general heights of common

**FIGURE 6.9**
Clouds are named based on their height and their form.
**Observe this figure and Figure 6.10; what cloud type is present in your area today?**

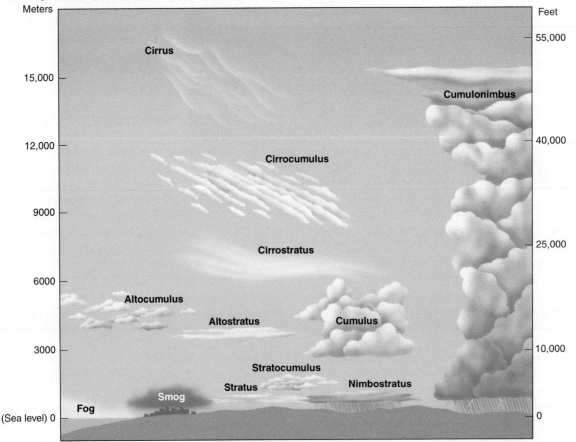

## Geography's Environmental Science Perspective
### Cumulus Congestus: An Aspiring Thunderstorm

According to the World Meteorological Organization's *International Cloud Atlas,* there are ten basic cloud genera, the same cloud types discussed in this chapter. In addition, these genera can break down into about 26 different cloud species and further subdivided into another 32 varieties. In other words, there are a lot of different clouds in our atmosphere. For all practical purposes though, the cloud types detailed in this chapter are all most professionals need to identify. These are the basic clouds and the ones you should learn. Yet, it is also helpful to know one additional cloud species, the *cumulus congestus*.

The stratiform (or layered) clouds that we have discussed are less violent in terms of their life cycles. Their layered appearance indicates that these clouds are rather stable. Stratiform clouds may form by an uplift mechanism, but these flat, layered clouds are not likely to continue their growth. In essence, horizontal layered clouds form in a stable atmosphere, although *nimbostratus* may produce a light rain, or in the wintertime, snow flurries. Unlike stratiform clouds, cumuliform clouds show varying degrees of vertical development, and rising air currents give them their rounded, cauliflower appearance. Updrafts of unstable air cause these clouds to billow upward into what are sometimes called cumulus towers, and continued updrafts can produce cumulus congestus clouds.

The term *cumulus* (Latin: heap or pile) is joined by the term *congestus* (Latin: to heap up or accumulate). So this cloud is building upward and thickening as it grows. As the cumulus congestus continues to develop, it may bring about some rain showers. In the tropics, these clouds may produce heavy rain. Even so, this is still not a fully developed *cumulonimbus,* or thunderstorm cloud. This is why we can informally and affectionately refer to this cloud species as a thunderstorm "wannabe."

If a cumulus congestus cloud continues to grow to great heights and develops the flattened cirrus anvil head, then it becomes a cumulonimbus cloud with its full array of atmospheric violence. But until that stage is reached, it remains a thunderstorm "wannabe." In many parts of the country, especially on hot, humid summer afternoons, it is possible to observe many such clouds looming in the distance. As long as they keep their general size and shape, no severe weather should result. However, if they continue to grow and develop vertically they signal the imminent development of a severe thunderstorm.

A rapidly building cumulus congestus cloud

---

clouds; Figure 6.10 provides a pictorial summary of the major cloud types.

**Cirrus** clouds (from Latin: *cirrus,* a lock or wisp of hair) form at very high altitudes, normally 6000–10,000 meters (19,800–36,300 ft), and are made up of ice crystals rather than droplets of water. They are thin, stringy, white clouds that trail like feathers across the sky. When associated with fair weather, cirrus clouds are scattered white patches in a clear blue sky.

**Stratus** clouds (from Latin: *stratus,* layer) can appear anywhere from near the surface of Earth to almost 6000 meters (19,800 ft). The variations of stratus clouds are based in part on their altitude. The basic characteristic of stratus clouds is their horizontal sheetlike appearance, lying in layers with fairly uniform thickness. The horizontal configuration indicates that they form in stable atmospheric conditions, which inhibit vertical development.

Often stratus clouds cover the entire sky with a gray cloud layer. It is stratus clouds that make up the dull, gray, overcast sky common to winter days in much of the midwestern and eastern United States. The stratus cloud formation may overlie an area for days, and any precipitation will be light but steady and persistent.

**Cumulus** clouds (from Latin: *cumulus,* heap or pile) develop vertically rather than forming the more horizontal structures of the cirrus and stratus forms. Cumulus are massive piles of clouds, rounded or cauliflower in appearance, usually

**FIGURE 6.10**
Cloud types

with a flat base, which can be anywhere from 500 to 12,000 meters (1650–39,600 ft) above sea level. From this base, they pile up into great rounded structures, often with tops like cauliflowers. The cumulus cloud is the visible evidence of an unstable atmosphere; its base is the point where condensation has begun in a column of air as it moves upward.

Examine Figures 6.9 and 6.10 to familiarize yourself with the basic cloud types and their names. Keep in mind that some cloud shapes exist in all three levels—for example, *stratocumulus* (*strato* = low level + *cumulus* = a rounded shape), *altocumulus,* and *cirrocumulus*. These three share the similar rounded or cauliflower appearance of cumulus clouds, which can exist at all three levels. You may notice that *altostratus* (*alto* = middle level + *stratus* = layered shape) and *cirrostratus* have two-part names, but low-level layered clouds are called *stratus* only. Lastly, thin, stringy *cirrus* clouds are found only as high-level clouds, so the term *cirro* (meaning high-level cloud) is not necessary here.

*Cirrus*

*Cumulonimbus*

*Nimbostratus*

*Cumulus*

Other terms used in describing clouds are **nimbo** or **nimbus,** meaning precipitation (rain is falling). Thus, the *nimbostratus* cloud may bring a long-lasting drizzle, and the *cumulonimbus* is the thunderstorm cloud. This latter cloud has a flat top, called an anvil head, as well as a relatively flat base, and it becomes darker as it grows higher and thicker and thus blocks the incoming sunlight. The cumulonimbus is the source of many atmospheric concerns including high-speed winds, torrential rain, thunder, lightning, hail, and possibly tornadoes. This type of cloud can develop in several different ways as we will soon discuss.

> PHYSICAL
> Geography Now™ **Log on to Physical GeographyNow and select this chapter to work through a Geography Interactive activity on "Name That Cloud"**

**Adiabatic Heating and Cooling** The cooling process that leads to cloud formation is quite different from that associated with the other condensation forms that we have already examined. The cooling process that produces fog, frost, and dew is either radiation or advection. On the other hand, clouds usually develop from a cooling process that results when a parcel of air on Earth's surface is lifted into the atmosphere.

The rising parcel of air will expand as it encounters decreasing atmospheric pressure with height. This expansion allows the air molecules to spread out, which causes the parcel's temperature to decrease. This is known as **adiabatic cooling** and occurs at the constant lapse rate of approximately 10°C per 1000 meters (5.6°F/1000 ft). By the same token, air descending through the atmosphere is compressed by the increasing pressure and undergoes **adiabatic heating** of the same magnitude.

However, the rising and cooling parcel of air will eventually reach its dew point—the temperature at which water vapor begins to condense out, forming cloud droplets. From this point on, the adiabatic cooling of the rising parcel will decrease as latent energy released by the condensation process is added to the air. To differentiate between these two adiabatic cooling rates, we refer to the precondensation rate (10°C/1000 m) as the **dry adiabatic lapse rate** and the lower, postcondensation rate as the **wet adiabatic lapse rate.** The latter rate averages 5°C per 1000 meters (3.2°F/1000 ft) but varies according to the amount of water vapor that condenses out of the air.

A rising air parcel will cool at one of these two adiabatic rates. Which rate is in operation depends on whether condensation is (wet adiabatic rate) or is not (dry adiabatic rate) occurring. On the other hand, the warming temperatures of descending air allow it to hold greater quantities of water vapor. In other words, as the air temperature rises farther above the dew point, condensation will not occur, so the heat of condensation will not affect the rate of rise in temperature. Thus, the temperature of air that is descending and being compressed always increases at the dry adiabatic rate.

It is important to note that adiabatic temperature changes are the result of changes in volume and do not involve the addition or subtraction of heat from external sources.

It is also extremely important to differentiate between the *environmental lapse rate* and *adiabatic lapse rates*. In Chapter 4, we found that in general the temperature of our atmosphere decreases with increasing height above Earth's surface; this is known as the environmental lapse rate, or the normal lapse rate. Although it averages 6.5°C per 1000 meters (3.6°F/1000 ft), this rate is quite variable and must be measured through the use of meteorological instruments sent aloft. Whereas the environmental lapse rate reflects nothing more than the vertical temperature structure of the atmosphere, the adiabatic lapse rates are concerned with temperature changes as a parcel of air moves through the atmospheric layers (Fig. 6.11).

**Physical GeographyNow™** Log on to Physical GeographyNow and select this chapter to work through a Geography Interactive activity on "Adiabatic."

## Stability and Instability

Although adiabatic cooling results in the development of clouds, the various forms of clouds are related to differing degrees of vertical air movement. Some clouds are associated with rapidly rising, buoyant air, whereas other forms result when air resists vertical movement.

An air parcel will rise of its own accord as long as it is warmer than the surrounding layer of air. When it reaches a layer of the atmosphere that is the same temperature as itself, it will stop rising. Thus, an air parcel warmer than the surrounding atmospheric air will rise and is said to be *unstable*. On the other hand, an air parcel that is colder than the surrounding atmospheric air will resist any upward movement and will likely sink to lower levels. Then the air is said to be *stable*.

Determining the stability or instability of an air parcel involves nothing more than asking the question, If an air parcel were lifted to a specific elevation (cooling at an adiabatic lapse rate), would it be warmer, colder, or the same temperature as the atmospheric air (determined by the environmental lapse rate at that time) at that same elevation?

If the air parcel is warmer than the atmospheric air at the selected elevation, then the parcel would be unstable and would continue to rise, because warmer air is less dense and therefore buoyant. Thus, under conditions of **instability**, the environmental lapse rate must be *greater than* the adiabatic lapse rate in operation. For example, if the environmental lapse rate is 12°C per 1000 meters and the ground temperature is 30°C, then the atmospheric air temperature at 2000 meters would be 6°C. On the other hand, an air parcel (assuming that no condensation occurs) lifted to 2000 meters would have a temperature of 10°C. Because the air parcel is warmer than the atmospheric air around it, it is unstable and will continue to rise (Fig. 6.12).

Now let's assume that it is another day and all the conditions are the same, except that measurements indicate the environmental lapse rate on this day is 2°C per 1000 meters. Consequently, although our air parcel if lifted to 2000 meters would still have a temperature of 10°C, the temperature of the atmosphere at 2000 meters would now be 26°C. Thus, the air parcel would be colder and would sink back toward Earth as a result of its greater density (see Fig. 6.12). As you can see, under conditions of **stability**, the environmental lapse rate is *less than* the adiabatic lapse rate in operation. If an air parcel, upon being lifted to a specific elevation, has the same temperature as the atmospheric air surrounding it, it is neither stable nor unstable. Instead, it is considered *neutral;* it will neither rise nor sink but will remain at that elevation.

Whether an air parcel will be stable or unstable is related to the amount of cooling and heating of air at Earth's surface. With cooling of the air through radiation and conduction on a cool, clear night, air near the surface will be

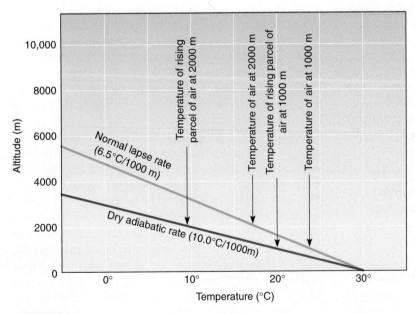

**FIGURE 6.11**
Comparison of the dry adiabatic lapse rate and the environmental lapse rate. The environmental lapse rate is the average vertical change in temperature. Air displaced upward will cool (at the dry adiabatic rate) because of expansion.
**In this example, what is the temperature of the layer of air at 2000 meters?**

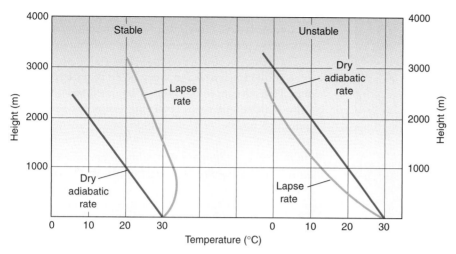

**FIGURE 6.12**
Relationship between lapse rates and air mass stability. When air is forced to rise, it cools adiabatically. Whether it continues to rise or resists vertical motion depends on whether adiabatic cooling is less rapid or more rapid than the prevailing vertical temperature lapse rate. If the adiabatic cooling rate exceeds the lapse rate, the lifted air will be colder than its surroundings and will tend to sink when the lifting force is removed. If the adiabatic cooling rate is less than the lapse rate, the lifted air will be warmer than its surroundings and will therefore be buoyant, continuing to rise even after the original lifting force is removed.
**In these examples, what would be the temperature of the lifted air if it rose to 2000 meters?**

**FIGURE 6.13**
The relative sizes of raindrops, cloud droplets, and condensation nuclei.
**If the diameter of a raindrop is 100 times larger than a cloud droplet, why does it take a million cloud droplets to produce one raindrop?**

relatively close in temperature to that aloft, and the environmental lapse rate will be low, thus enhancing stability. With the rapid heating of the surface on a hot summer day, there will be a very steep environmental lapse rate because the air near the surface is so much warmer than that above, and instability will be enhanced.

Pressure zones can also be related to atmospheric stability. In areas of high pressure, stability is maintained by the slow subsiding air from aloft. In low pressure regions, on the other hand, instability is promoted by the tendency for air to converge and then rise.

# Precipitation Processes

Condensed droplets within cloud formations stay in the air and do not fall to Earth because of their tiny size (0.02 mm, or less than 1000th of an inch), their general buoyancy, and the upward movement of the air within the cloud. These droplets of condensation are so minute that they are kept floating in the cloud formation; their mass and the consequent pull of gravity are insufficient to overcome the buoyant effects of air and the vertical currents, or updrafts, within the clouds. Figure 6.13 shows the relative sizes of a condensation nucleus, a cloud droplet, and a raindrop. It takes about a million cloud droplets to form one raindrop.

Precipitation occurs when the droplets of water, ice, or frozen water vapor grow and develop masses too great to be held aloft. They then fall to Earth as rain, snow, sleet, or hail. The form that precipitation takes depends largely on the method of formation and the temperature during formation. Among the many theories that try to explain the formation of precipitation, the **collision–coalescence process** for warm clouds in low latitudes and the **Bergeron** (or **ice crystal**) **Process** for cold clouds at higher latitudes are the most widely accepted.

Precipitation in the lower latitudes of the tropics and in warm clouds is likely to form by the collision–coalescence process. The collision–coalescence process is one in which the name itself describes the process. By nature, water is quite cohesive (able to stick to itself). When water droplets are colliding in the circulation of the cloud, they tend to coalesce (or grow together). This is especially true as the water droplets begin to fall toward the ground. In falling, the larger droplets overtake the smaller, more buoyant droplets and capture them to form even larger raindrops. The mass of these growing raindrops eventually overcomes the updrafts of the cloud and fall to Earth, under the pull of gravity. This process occurs in the warm section of clouds where all the moisture exists as liquid water ( ■ Fig. 6.14).

**FIGURE 6.14**
Collision and coalescence. (a) In a warm cloud consisting of small cloud droplets of uniform size, the droplets are less likely to collide because they are all falling very slowly and at about the same speed. (b) In a cloud of different-sized droplets, some droplets fall more rapidly and can overtake and capture some of the smaller droplets.
**Why do these tiny droplets fall at different speeds?**

($-40°F$), ice crystals will dominate (▮ Fig. 6.15). It is in relation to these layered clouds that Scandinavian meteorologist Tor Bergeron presented a more complex explanation.

The Bergeron (or ice crystal) Process begins at great heights in the ice crystal and supercooled water layers of the clouds. Here, the supercooled water has a tendency to freeze on any available surface. (It is for this reason that aircraft flying through middle- to high-latitude thunderstorms run the risk of severe icing and invite disaster.) The ice crystals mixed in with the supercooled water in the highest layers of the clouds can become *freezing nuclei* and form the centers of growing ice crystals. (Essentially, this is the process that can also create snow.) As the supercooled water continues to freeze onto these frozen nuclei, their masses grow until gravity begins to pull them toward Earth. As this frozen precipitation enters the lower layer of the clouds, the above-freezing temperatures there melt the ice crystals into liquid rain before they hit the ground. Therefore, according to Bergeron, rain in these clouds begins as frozen precipitation and melts into a liquid before reaching Earth.

As the melted precipitation falls through the lower, warmer section of the cloud, the collision–coalescence process may take over and cause the raindrops to grow even larger as they descend toward the surface.

## Major Forms of Precipitation

**Rain,** consisting of droplets of liquid water, is by far the most common form of precipitation. Raindrops vary in size but are generally about 2–5 millimeters (approximately 0.1–0.25 in.) in diameter (see again Figure 6.13). As we all know, rain can come in many ways: as a brief afternoon shower, a steady rainfall, or the deluge of a tropical rainstorm. When the temperature of an air mass is only slightly below the dew point, the raindrops may be very small (about 0.5 mm or less in diameter) and close together. The result is a fine mist called **drizzle.** Drizzle is so light that it is greatly affected by the direction of air currents and the variability of winds. Consequently, drizzle seldom falls vertically.

**Snow** is the second most common form of precipitation. When water vapor is frozen directly into a solid without first passing through a stage as liquid water (sublimation), it forms minute ice crystals around the freezing nuclei (of the Bergeron Process). These crystals characteristically appear as six-sided, symmetric shapes. Combinations of these ice-crystal shapes make up the intricate patterns of snowflakes. Snow will reach the ground if the entire cloud and the air beneath the cloud maintains below-freezing temperatures.

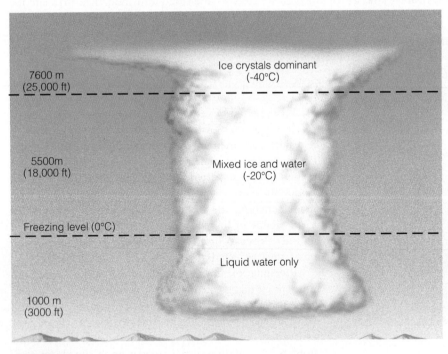

**FIGURE 6.15**
The distribution of water, supercooled water, and ice crystals in a high-latitude storm cloud.
**What is the difference between water and supercooled water?**

At higher latitudes, storm clouds can possess three distinctive layers. The lowermost is a warm layer of liquid water. Here the temperatures are above the freezing point of 0°C (32°F). Above this is the second layer composed of some ice crystals but mainly **supercooled water** (liquid water that exists at a temperature below 0°C). In the uppermost layer of these tall clouds, when temperatures are lower than or equal to $-40°C$

**Sleet** is frozen rain, formed when rain, in falling to Earth, passes through a relatively thick layer of cold air near the surface and freezes. The result is the creation of solid particles of clear ice. In English-speaking countries outside the United States, sleet refers not to this phenomenon of frozen rain but rather to a mixture of rain and snow.

**Hail** is a less common form of precipitation than the three just described. It occurs most often during the spring and summer months and is the result of thunderstorm activity. Hail appears as rounded lumps of ice, called hailstones, which can vary in size from 5 millimeters (0.2 in.) in diameter and up to the sizes larger than a baseball. The world record is a hailstone 30 centimeters (12 in.) in diameter that fell in Australia ( Fig. 6.16). Dropping from the sky, hailstones can be highly destructive to crops and other vegetation, as well as to cars and buildings. Though primarily a property destroyer, hailstones have been known to kill animals and humans. Children think it is strange that they must leave a pool or lake where they are swimming simply because hailstones are falling, but this is a sensible precaution because the atmospheric conditions that produce hailstones also produce thunder and lightning. In fact, these phenomena often occur in conjunction with one another.

Hail forms when ice crystals are lifted by strong updrafts in a cumulonimbus (thunderstorm) cloud. Then, as these ice crystals fall through the cloud, supercooled water droplets attach themselves and are frozen as a layer. Sometimes these pellets are lifted up into the cold layer of air and then dropped again and again. The resulting hailstone, made up of concentric layers of ice, has a frosty, opaque appearance when it finally breaks out of the strong updrafts of the cloud formation and falls to Earth. The larger the hailstone, the more times it is cycled through the freezing process and accumulated additional frozen layers.

On occasion, a raindrop can form and have a temperature below 0°C (32°F). This will occur when there is a shallow layer of below-freezing temperatures all the way to the ground so that the liquid rain can reach a supercooled state. These supercooled droplets will freeze the instant they fall onto a surface that is also at a below-freezing temperature. The resulting icy covering on trees, plants, and telephone and power lines is known as **freezing rain** (or **glaze**). People usually call the rain and its blanket of ice an "ice storm" ( Fig. 6.17). Because of the weight of ice, glazing can break off large branches of trees, bringing down telephone and power lines. It can also make roads practically impassable. A small counterbalance against the negative effects of glazing is the beauty of the natural landscape after an ice storm. Against the background of a clear blue sky, sunlight catches on the ice, reflecting and making a diamond-like surface covering the most ordinary weeds and tree branches.

## Factors Necessary for Precipitation

Three factors are necessary for the formation of any type of precipitation on Earth. The first is the presence of *moist air* on the surface. This air obviously represents the source of moisture (for the precipitation) and energy (in the form of latent heat of condensation). Second are the *condensation nuclei* around which the water vapor can condense, discussed previously in this chapter. Third is a *mechanism of uplift*. These **uplift mechanisms** are responsible for forcing the air higher into the atmosphere so that it can cool down (by the dry adiabatic rate) to the dew point. These uplift mechanisms are vital to the process of precipitation.

A parcel of air can be forced to rise in four major ways. All the precipitation that falls anywhere on Earth can be traced back to one of these four uplift mechanisms ( Fig. 6.18):

- **Convectional precipitation** results from the displacement of warm air upward in a convectional system.
- **Frontal precipitation** takes place when a warm air mass rises after encountering a colder and denser air mass.

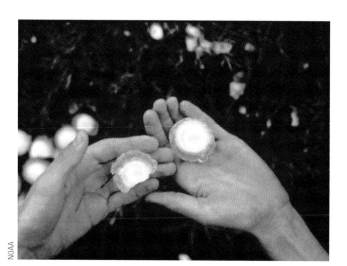

**FIGURE 6.16**
Hailstones can be the size of golf balls, or even larger.
**What gives them their spherical appearance?**

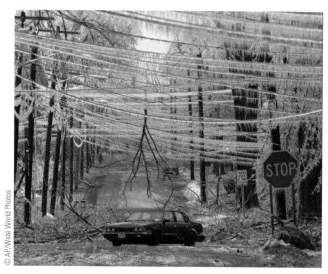

**FIGURE 6.17**
An ice storm can cover a city with a dangerous glazing of ice.
**Why are power failures a common occurrence with ice storms?**

## Geography's Physical Science Perspective

### The Lifting Condensation Level (LCL)

When you look at clouds in our atmosphere, it is often quite easy to see their relatively flat bases. Cloud tops may appear quite irregular, but cloud bases are often flat. If the cloud bases do not seem flat, it will be obvious that the clouds you see all seem to be formed at the same level above the surface. This level represents the altitude to which the air must be lifted (and cooled at the dry adiabatic rate) before saturation is reached. Any additional lifting and clouds will form and build upward. Therefore, the height at which clouds form from lifting is called the lifting condensation level (LCL) and can be estimated by the equation:

$$\text{LCL (in meters)} = 125 \text{ meters} \times (\text{Celsius temperature} - \text{Celsius dew point})$$

For example, if the surface temperature is 7.2°C (45°F) and the dew point temperature is 4.4°C (40°F), then the LCL is estimated at 350 meters (1148 ft) above the surface.

*Caution:* Keep in mind that different layers of clouds may exist at the same time. Low, middle, and high clouds as defined in this chapter may all appear on the same afternoon. These clouds may have formed in other regions and be only passing overhead. The formula presented here is best used with the lowest level of cloud cover that appears overhead.

The stratocumulus clouds (bottom layer) show the lifting condensation level (LCL).

M. Trapasso

Convectional

Cyclonic

Frontal

Orographic

**FIGURE 6.18**
The principal cause of precipitation is upward movement of moist air resulting from convectional, frontal, cyclonic, or orographic lifting.
**What do all four diagrams have in common?**

- **Cyclonic precipitation** occurs when air is lifted up into a low pressure system.
- **Orographic precipitation** results when a moving air mass encounters a land barrier, usually a mountain, and must rise above it in order to pass.

**Convectional Precipitation** The simple explanation of convection is that when air is heated near the surface it expands, becomes lighter, and rises. It is then displaced by the cooler, denser air around it to complete the convection cycle. The important factor in convection for our discussion of precipitation is that the heated air rises and thus fulfills the one essential criterion for significant condensation and, ultimately, precipitation.

To enlarge our understanding of convectional precipitation, let's apply what we have learned about instability and stability. Figure 6.19 illustrates two different cases in which air rises due to convection. In both, the lapse rate in the free atmosphere is the same; it is especially high during the first few thousand meters but slows after that (as on a hot summer day).

In the first case (Fig. 6.19a), the air parcel is not very humid, and thus the dry adiabatic rate applies throughout its ascent. By the time the air reaches 3000 meters (9900 ft), its temperature and density are the same as those

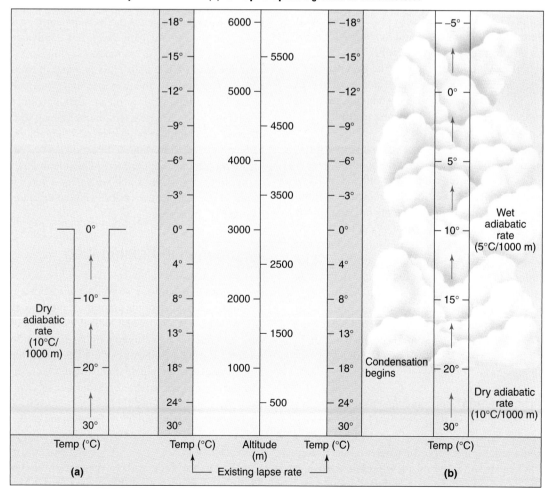

**PHYSICAL Geography Now™ ACTIVE FIGURE 6.19**
Watch this Active Figure at http://now.brookscole.com/gabler8.
Effect of humidity on air mass stability. (a) Warm, dry air rises and cools at the dry adiabatic rate, soon becoming the same temperature as the surrounding air, at which point convectional uplift terminates. Because the rising dry air did not cool to its dew point temperature by the time that convectional lifting ended, no cloud formed. (b) Rising warm, moist air soon cools to its dew point temperature. The upward-moving air subsequently cools at the wet adiabatic rate, which keeps the air warmer than the surrounding atmosphere so that the uplift continues. Only when all moisture is removed by condensation will the air cool rapidly enough at the dry adiabatic rate to become stable.
**What would be necessary for the cloud in (b) to stop its upward growth at 4500 meters?**

of the surrounding atmospheric air. At this point, convectional lifting stops.

In the second case (Fig. 6.19b), we have introduced the latent heat of condensation. Here again, the unsaturated rising column of air cools at the dry adiabatic rate of 10°C per 1000 meters (5.6°F/1000 ft) for the first 1000 meters (3300 ft). However, because the air parcel is humid, the rising air column soon reaches the dew point, condensation takes place, and cumulus clouds begin to form. As condensation occurs, the heat locked up in the water vapor is released and heats the moving parcel of air, retarding the adiabatic rate of cooling so that the rising air is now cooling at the wet adiabatic rate (5°C/1000 meters). Hence, the temperature of the rising air parcel remains warmer than that of the atmospheric layer it is passing through, and the air parcel will continue to rise on its own. In this case, which incorporates the latent heat of condensation, we have massive condensation, towering cumulus clouds, and a thunderstorm potential.

Convectional precipitation is most common in the humid equatorial and tropical areas that receive much of the sun's energy and in summer in the middle latitudes. Though differential heating of land surfaces plays an important role in convectional precipitation, it is not the sole factor. Other factors, such as surface topography and atmospheric dynamics associated with the upper air winds, may provide the initial upward lift for air that is potentially unstable. Once condensation begins in a convectional column, additional energy is available from the latent heat of condensation for further lifting.

This convectional lifting can result in the heavy precipitation, thunder, lightning, and tornadoes of summer afternoon thunderstorms. When the convectional currents are strong in the characteristic cumulonimbus clouds, hail can result.

### Frontal Precipitation

The zones of contact between relatively warm and relatively cold bodies of air are known as **fronts**. When two large bodies of air that differ in density, humidity, and temperature meet, the warmer one is lifted above the colder. When this happens, the major criterion for large-scale condensation and precipitation is once again met. Frontal precipitation thus occurs as the moisture-laden warm air rises above the front caused by contact with the cold air. Continuous frontal precipitation has caused some devastating floods through time.

To fully understand fronts, we must examine what causes unlike bodies of air to come together and what happens when they do. This will be discussed in Chapter 7, where we will take a more detailed look at frontal disturbances and precipitation.

### Cyclonic Precipitation

The third mechanism, the cyclonic, was first introduced in Chapter 5 (see Fig. 5.3). When air enters a low pressure system, or *cyclone*, it does so (in a counterclockwise fashion in the Northern Hemisphere) from all directions. When air converges on a low pressure system, it has little option but to rise. Therefore, clouds and possible precipitation are common around the center of a cyclone.

### Orographic Precipitation

As was the case with convectional rainfall, orographic rainfall has a simple definition and a somewhat more complex explanation. When land barriers—such as mountain ranges, hilly regions, or even the escarpments (steep edges) of plateaus or tablelands—lie in the path of prevailing winds, large portions of the atmosphere are forced to rise above these barriers. This fills the one main criterion for significant precipitation—that large masses of air are cooled by ascent and expansion until large-scale condensation takes place. The resultant precipitation is termed *orographic* (from Greek: *oros*, mountains). As long as the air parcel rising up the mountainside remains stable (cooling at a greater rate than the environmental lapse rate), any resulting cloud cover will be a type of stratus cloud. However, the situation can be complicated by the same circumstances illustrated in Figure 6.19b. A potentially unstable air parcel may need only the initial lift provided by the orographic barrier to set it in motion. In this case, it will continue to rise of its own accord (no longer forced) as it seeks air of its own temperature and density. Once the land barrier provides the initial thrust, it has performed its function as a lifting mechanism.

Because the air deposits most of its moisture on the windward side of a mountain, there will normally be a great deal less precipitation on the leeward side; on this side, the air will be much drier and the dew point consequently much lower. Also, as air descends the leeward slope, its temperature warms (at the dry adiabatic rate), and condensation ceases. The leeward side of the mountain is thus said to be in the **rain shadow** (Fig. 6.20). Just as being in the shade, or in shadow, means that you are not receiving any direct sun, so being in the rain shadow means that you do not receive much rain. If you live near a mountain range, you can see the effects of orographic precipitation and the rain shadow in the pattern of vegetation (Figs. 6.20b and c). The windward side of the mountains (say, the Sierra Nevada in California) will be heavily forested and thick with vegetation. The opposite slopes in the rain shadow will usually be drier and the cover of vegetation sparser.

## Distribution of Precipitation

The precipitation a region receives can be described in different ways. We can look at average annual precipitation to get an overall picture of the amount of moisture that a region gets during a year. We can also look at its number of raindays—days on which 1.0 millimeter (0.01 in.) or more of rain is received during a 24-hour period. Less than this amount is known as a **trace** of rain. If we divide the number of raindays in a month or year by the total number of days in that period, the resulting figure represents the probability of rain. Such a measure is important to farmers and to ski or summer resort owners whose incomes may depend on precipitation or the lack of it.

We can also look at the average monthly precipitation. This provides a picture of the seasonal variations in precipitation (Fig. 6.21). For instance, in describing the climate of the west coast of California, average annual precipitation would not give the full story because this figure would not show the distinct wet and dry seasons that characterize this region.

**FIGURE 6.20**
(a) Orographic uplift over the windward (western) slope of the Sierras produces condensation, cloud formation, precipitation, resulting in (b) dense stands of forest. (c) Semiarid or rain-shadow conditions occur on the leeward (eastern) slope of the Sierras.
**Can you identify a mountain range in Eurasia in which the leeward side of that range is in the rain shadow?**

## Horizontal Distribution of Precipitation

Figure 6.22 shows average annual precipitation for the world's continents. We can see that there is great variability in the distribution of precipitation over Earth's surface. Although there is a zonal distribution of precipitation related to latitude, this distribution is obviously not the only factor involved in the amount of precipitation an area receives.

The likelihood and amount of precipitation are based on two factors. First, precipitation depends on the degree of lifting that occurs in air of a particular region. This lifting, as we have already seen, may be due to the convergence of different air masses, to the entrance into a low pressure system, to differential heating of Earth's surface, to the lifting that results when an air mass encounters a rise in Earth's surface, or to a combination of these factors. Also, precipitation depends on the internal characteristics of the air itself, including its degree of instability, its temperature, and its humidity.

**FIGURE 6.21**
Average monthly precipitation in San Francisco, California, is represented by colored bars along the bottom of the graph. Such a graph of monthly precipitation figures gives a much more accurate picture than the annual precipitation total, which does not tell us that nearly all the precipitation occurs in only half of the year.
**How would this rainfall pattern affect agriculture?**

**168** CHAPTER 6 MOISTURE, CONDENSATION, AND PRECIPITATION

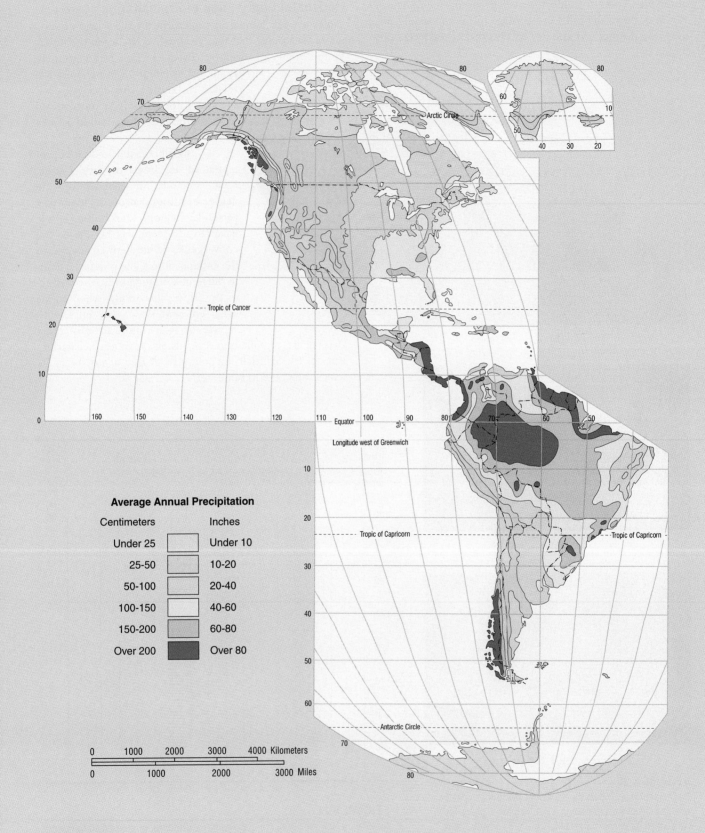

**FIGURE 6.22**
World map of average annual precipitation.
**In general, where on Earth's surface does the heaviest rainfall occur? Why?**

# PRECIPITATION PROCESSES 169

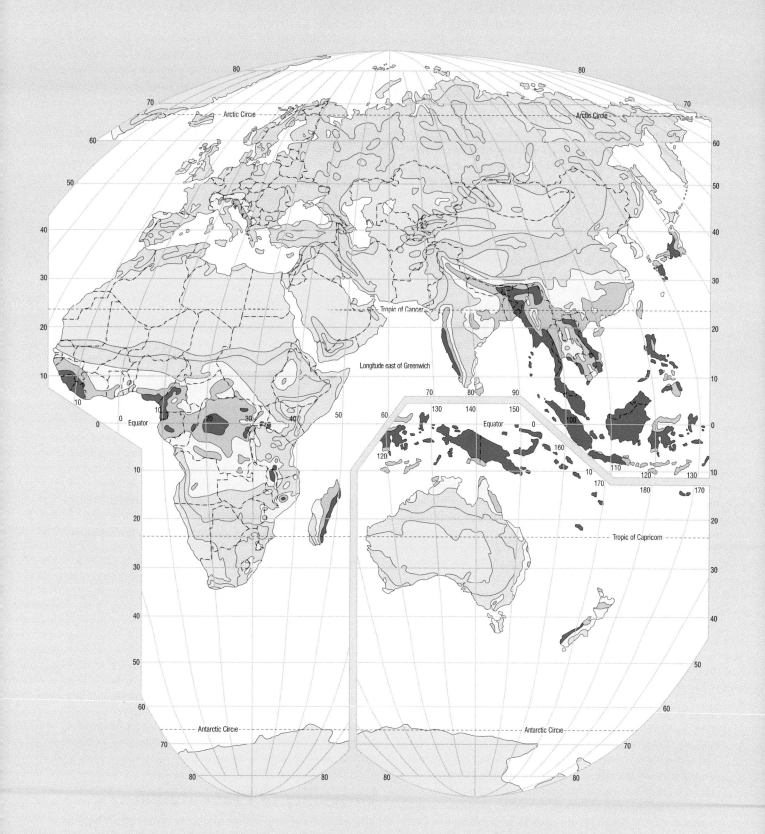

A Western Paragraphic Projection
developed at Western Illinois University

Because higher temperatures, as we have seen, allow air masses to hold greater amounts of water vapor and because, conversely, cold air masses can hold less water vapor, we can expect a general decrease of precipitation from the equator to the poles that is related to the unequal zonal distribution of incoming solar energy discussed in Chapter 3.

However, if we look again at Figure 6.22, we see a great deal of variability in average annual precipitation beyond the general pattern of a decrease with increased latitude. In the following discussion, we examine some of these variations and give the reasons for them. We also apply what we have already learned about temperature, pressure systems, wind belts, and precipitation.

**Distribution within Latitudinal Zones** The equatorial zone is generally an area of high precipitation—more than 200 centimeters (79 in.) annually—largely due to the zone's high temperatures, high humidity, and the instability of its air. High temperatures and instability lead to a general pattern of rising air, which in turn allows for precipitation. This tendency is strongly reinforced by the convergence of the trades as they move toward the equator from opposite hemispheres. In fact, the intertropical convergence zone is one of the two great zones where air masses converge. (The other is along the polar front within the westerlies.)

In general, the air of the trade wind zones is stable compared with the instability of the equatorial zone. Under the control of these steady winds, there is little in the way of atmospheric disturbances to lead to convergent or convectional lifting. However, because the trade winds are basically easterly, when they move onshore along east coasts or islands with high elevations, they bring moisture from the oceans with them. Thus, within the trade wind belt, continental east coasts tend to be wetter than continental west coasts.

In fact, where the air of the equatorial and trade wind regions—with its high temperatures and vast amounts of moisture—moves onshore from the ocean and meets a landform barrier, record rainfalls can be measured. The windward slope of Mount Waialeale on Kauai, Hawaii, at approximately 22°N latitude, holds the world's record for greatest average annual rainfall—1146 centimeters (451 in.).

Moving poleward from the trade wind belts, we enter the zones of subtropical high pressure where the air is subsiding. As it sinks lower, it is warmed adiabatically, increasing its moisture-holding capacity and consequently reducing the amount of precipitation in this area. In fact, if we look at Figure 6.22, which shows average annual precipitation on a latitudinal basis, we can see a dip in precipitation level corresponding to the latitude of the subtropical high pressure cells. These areas of subtropical high pressure are in fact where we find most of the great deserts of the world: in northern and southern Africa, Arabia, North America, and Australia. The exceptions to this subtropical aridity occur along the eastern sides of the landmasses where, as we have already noted, the subtropical high pressure cells are weak and wind direction is often onshore. This exception is especially true of regions affected by the monsoons.

In the zones of the westerlies, from about 35° to 65°N and S latitude, precipitation occurs largely as a result of the meeting of cold, dry polar air masses and warm, humid subtropical air masses along the polar front. Thus, there is much cyclonic and frontal precipitation in this zone.

Naturally, the continental interiors of the middle latitudes are drier than the coasts because they are farther away from the oceans. Furthermore, where air in the prevailing westerlies is forced to rise, as it is when it crosses the Cascades and Sierra Nevada of the Pacific Northwest and California, especially during the winter months, there is heavy orographic precipitation. Thus, in the middle latitudes, continental west coasts tend to be wet, and precipitation decreases with movement eastward toward continental interiors. Along eastern coasts within the westerlies, precipitation usually increases once again because of proximity to humid air from the oceans.

In the United States, the interior lowlands are not as dry as we might expect within the prevailing westerlies. This is because of the great amount of frontal activity resulting from the conflicting northward and southward movements of polar and subtropical air. If there were a high east–west mountain range extending from central Texas to northern Florida, the lowlands of the continental United States north of that range would be much drier than they actually are because they would be cut off from moist air originating in the Gulf of Mexico.

Also characteristic of the belt of the westerlies are desert areas that occur in the rain shadows of prevailing winds that are forced to rise over mountain ranges. This effect is in part responsible for the development and maintenance of California's Death Valley, as well as the desert zone of eastern California and Nevada in the United States, the mountain-ringed deserts of eastern Asia, and Argentina's Patagonian Desert, which is in the rain shadow of the Andes. Note in Figure 6.23 that there is greater precipitation in the middle latitudes of the Southern Hemisphere than there is in the Northern middle latitudes. This occurs largely because there is a lot more ocean and less landmass in the Southern Hemisphere westerlies than in the corresponding zone of the Northern Hemisphere.

Moving poleward, we find that temperatures decrease, along with the moisture-holding capacity of the air. The low temperatures also lead to low evaporation rates. In addition, the air in the polar regions shows a general pattern of subsidence that yields areas of high pressure. This settling of the air in the polar regions is the opposite of the lifting needed for precipitation. All these factors combine to cause low precipitation values in the polar zones.

# Variability of Precipitation

The rainfall depicted in Figure 6.22 is an annual average. It should be remembered, however, that in many parts of the world there are significant variations in precipitation, both within any one year and between years. For example, areas like the Mediterranean region, California, Chile, South Africa, and Western Australia, which are on the west sides of the continents and roughly between 30° and 40° latitude, get much more rain in the winter than in the summer. There are also areas between 10° and 20° latitude that get much more of their precipitation in the summer (high-sun season) than in the winter (low-sun season).

Rainfall totals can change markedly from one year to the next, and tragically for many of the world's people, the drier a place is on the average, the greater will be the statistical variability in its precipitation (compare Fig. 6.24 with Fig. 6.22). To make matters worse for people in dry areas, a year with a particularly high amount of rainfall may be balanced with several years of below-average precipitation. This situation has occurred recently in West Africa's Sahel, the Russian steppe, and the American Great Plains.

Thus, there are years of drought and years of flood, each bringing its own kind of disaster upon the land. Farmers, resort owners, construction workers, and others whose economic well-being depends in one way or another on weather can determine only a probability of rainfall on an annual, monthly, or even a seasonal basis.

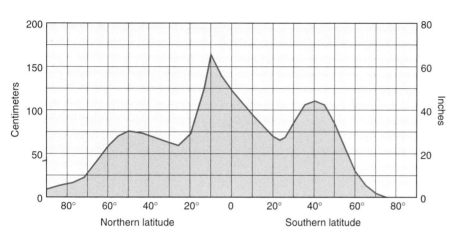

■ **FIGURE 6.23**
Latitudinal distribution of average annual precipitation. This figure illustrates the four distinctive precipitation zones: high precipitation caused by convergence of air in the tropics and in the middle latitudes along the polar front and low precipitation caused by subsidence and divergence of air in the subtropical and polar regions.
**Compare this figure with Figure 5.13. What is the relationship between world rainfall patterns and world pressure distribution?**

■ **FIGURE 6.24**
Precipitation variability. The greatest variability in year-to-year precipitation totals is in the dry regions, accentuating the critical problem of moisture supply in those parts of the world.
**Compare this figure with Figure 6.22. What are some of the similarities and differences?**

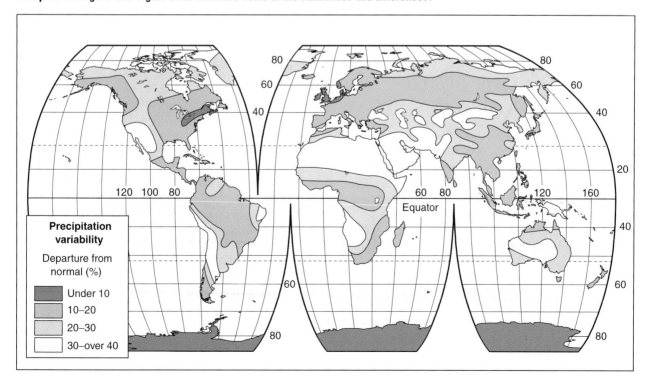

Meteorologists cannot predict rainfall with 100% accuracy. This inability is due to the many factors involved in causing precipitation—temperature, available moisture, atmospheric disturbances, landform barriers, frontal activity, air mass movement, upper air winds, and differential surface heating, among others. In addition, the interaction of these factors in the development of precipitation is very complex and not completely understood.

Study of the hydrologic system helps scientists understand how changes in one subsystem can greatly impact other subsystems. For example, it is theorized that changes in the radiation budget of the Earth system can produce changes in the hydrologic cycle that in turn can explain changes in glacial subsystems and bring about continental glaciation (an ice age). If the energy available to heat Earth's atmosphere were reduced, the colder temperatures would affect the hydrologic cycle. A colder atmosphere would store less water vapor. Thus, there would be less precipitation, but the colder temperatures would result in more snowfall. With time, glacial ice would increase, and sea level would drop. The high albedo of the ice would further reduce the energy available to heat our atmosphere, which would allow glacial ice to increase and eventually result in continental ice sheets covering large portions of Earth.

Even without a dramatic example like the one described above, our planet's hydrologic system relates to all other Earth systems. Later chapters will demonstrate other aspects of this theme, which runs throughout this textbook.

## Define & Recall

capillary action
hydrologic cycle
saturation
capacity
dew point
humidity
absolute humidity
specific humidity
relative humidity
transpiration
evapotranspiration
potential evapotranspiration
condensation nuclei
radiation fog (temperature-inversion fog)
advection fog
upslope fog
dew

white frost
rime
precipitation
strato
alto
cirro
cirrus
stratus
cumulus
nimbus (nimbo)
adiabatic cooling
adiabatic heating
dry adiabatic lapse rate
wet adiabatic lapse rate
instability
stability
collision–coalescence process

Bergeron (ice crystal) Process
supercooled water
rain
drizzle
snow
sleet
hail
freezing rain (glaze)
uplift mechanism
convectional precipitation
frontal precipitation
cyclonic precipitation
orographic precipitation
front
rain shadow
trace

## Discuss & Review

1. How is the hydrologic cycle related to Earth's water budget?
2. Is the saturation temperature, or dew point, in your area generally higher or lower than that in Siberia? Honduras?
3. What is the difference between absolute and specific humidity? What is relative humidity?
4. Why is the inside of a greenhouse generally more humid than an ordinary room?
5. Why is wearing wet clothes sometimes bad for your health? Under what conditions might wearing wet clothes be good for your health?
6. Imagine that you are deciding when, in your daily schedule, to water the garden. What time of day would be best for conserving water? Why?
7. What factors must be taken into consideration when calculating potential evapotranspiration? How is evapotranspiration related to the water budget of a region?
8. What factors affect the formation of temperature-inversion fogs?
9. What kinds of fogs occur in the area where you live?
10. Describe the most recent cloud formations in your area. With what kinds of weather (temperature, humidity, precipitation) were the cloud types associated?
11. What causes adiabatic cooling? Differentiate between the environmental lapse rate and adiabatic lapse rates.
12. How and why does the wet adiabatic lapse rate differ from the dry adiabatic lapse rate?
13. How is atmospheric stability related to the adiabatic lapse rates?

14. What atmospheric conditions are necessary for precipitation to occur?
15. Find out how many inches of precipitation have fallen in your area this year. Is that average or unusually high or low?
16. Compare and contrast convectional, orographic, cyclonic, and frontal precipitation.
17. How is rainfall variability related to total annual rainfall? How might this relationship be considered a double problem for people?

## Consider & Respond

1. Refer to Figure 6.4.
   a. What is the water vapor capacity of air at 0°C? 20°C? 30°C?
   b. If a parcel of air at 30°C has an absolute humidity (actual water vapor content) of 20.5 grams per cubic meter, what is the parcel's relative humidity?
   c. If the relative humidity of a parcel of air is 33% and the air temperature is 15°C, what is the absolute humidity of the air in grams per cubic meter?
   d. A major concern of northern climate residents is the low relative humidity within their homes during the winter. Low relative humidity is not healthy, and it has an adverse effect on the homes' furnishings. The problem results when cold air, which can hold little water vapor, is brought indoors and heated up. The following example will illustrate the problem: Assume that the air outside is 5°C and has a relative humidity of 60%. What is the actual water vapor content of this air? If it is brought indoors (through the doors, windows, and cracks in the home) and heated to 20°C, with no increase in water vapor content, what is the new relative humidity?
2. Recall that as a parcel of air rises, it expands and cools. The rate of cooling, termed the dry adiabatic lapse rate, is 10°C per 1000 meters. (A descending parcel of air will always warm at this rate.) In addition, the dew point temperature decreases 2°C per 1000 meters within a rising parcel of air. At the height at which the dew point temperature is reached, condensation begins and, as discussed in the text, the wet adiabatic lapse rate of 5°C per 1000 meters becomes operational. When the wet adiabatic lapse rate is in operation, the dew point temperature will be the same as the air temperature. When a parcel descends through the atmosphere, its dew point temperature increases 2°C per 1000 meters. The height at which condensation begins, termed the lifting condensation level (LCL), can be determined by using the formula found in Geography's Physical Science Perspective: The Lifting Condensation Level (LCL), page 164.
   a. A parcel of air has a temperature of 25°C and a dew point temperature of 14°C. What is the height of the LCL? If that parcel were to rise to 4000 meters, what would be its temperature?
   b. A parcel of air at 6000 meters has a temperature of −5°C and a dew point of −10°C. If it descended to 2000 meters, what would be its temperature and dew point temperature?

# CHAPTER 7

An enhanced satellite view of Hurricane Katrina as it swirls toward the New Orleans area. NASA

# Air Masses and Weather Systems

**Physical Geography Now™** This icon, appearing throughout the book, indicates an opportunity to explore interactive tutorials, animations, or practice problems at now.brookscole.com/gabler8

If we are to understand the types of weather that rule the middle latitudes, we must first come to grips with the vital parts of our basic weather systems. In the previous four chapters, we have looked at the elements of the atmosphere and investigated some of the controls that act upon those elements, causing them to vary from place to place and through time. However, even more is involved in the examination of weather. We have not yet looked at storms (atmospheric disturbances)—their types and characteristics, their origin, and their development. Storms are an important part of the weather story. They help illustrate the interactions among the weather elements. Further, they represent a major means of energy exchange within the atmosphere.

# CHAPTER PREVIEW

**The movement of relatively large bodies of the air (air masses) is responsible for the transportation of distinguishable characteristics of temperature and humidity to regions far from their original sources.**

- How is this important to the operation of Earth systems?
- What is the significance of air mass movement to human beings?
- How might air masses be modified?

**The meeting of the leading edges of two unlike air masses occurs along a sloping surface of discontinuity called a front.**

- How do air masses differ?
- What kinds of air masses meet along the polar front?
- Why are fronts important in explaining middle-latitude weather?

**The major explanation for the variable and nearly unpredictable weather of the middle latitudes may be found in the irregular migration of relatively short-lived low pressure systems (cyclones) in the path of the prevailing westerlies.**

- Why do cyclones play such a significant role?
- In which direction would you normally look to obtain some forewarning of future weather?
- What are the human consequences of variable and unpredictable weather?

**Meteorology is an inexact science, and there is much yet to be learned about the behavior of air masses, fronts, and pressure systems. We should therefore anticipate that weather forecasting will remain a complicated art.**

- What questions about the weather remain unanswered?
- How accurate is weather prediction?
- How successful are humans at altering the weather?

**The latent heat of condensation is the major source of energy behind the violent atmospheric disturbances of the middle latitudes, such as thunderstorms, tornadoes, and hurricanes.**

- What is latent heat?
- How might this information be useful?
- What have we done in the United States to protect against violent storms?

**Tropical cyclones and extratropical cyclones are among the largest weather systems in the world. These two weather systems are known by other names.**

- What are they?
- How do they differ?

# Air Masses

Before we begin to study weather systems, we should understand the nature and significance of air masses. In themselves, air masses provide a straightforward way of looking at the weather. An **air mass** is a large body of air, at times subcontinental in size, that moves over Earth's surface with distinguishable characteristics. An air mass is relatively homogeneous in temperature and humidity; that is, at approximately the same altitude within the air mass, the temperature and humidity will be similar. As a result of this temperature and moisture uniformity, the density of air will be much the same throughout any one level within an air mass. Of course, because an air mass may extend over 20 or 30 degrees of latitude, we can expect some slight variations due to changes in sun angle and its corresponding insolation, which are significant over that distance. Changes caused by contact with differing land and ocean surfaces also affect the characteristics of air masses.

The similar characteristics of temperature and humidity within an air mass are determined by the nature of its **source region**—the place where the air mass originates. Only a few areas on Earth make good source regions. For the air mass to have similar characteristics throughout, the source region must have a nearly homogeneous surface. For example, it can be a desert, an ice sheet, or an ocean body, but not a combination of surfaces. In addition, the air mass must have sufficient time to acquire the characteristics of the source region. Hence, gently settling, slowly diverging air will mimic a source region, whereas converging, rising air will not.

Air masses are identified by a simple letter code. The first is always a lowercase letter. There are two choices: The letter *m*, for maritime, means the air mass originates over the sea and is therefore relatively moist. The letter *c*, for continental, means the air mass originates over land and is therefore dry. The second letter is always a capital. These help to locate the latitude of the source region. *E* stands for Equatorial; this air is very warm. The letter *T* identifies a Tropical origin and is therefore warm air. A *P* represents Polar; this air can be quite cold. Lastly, an *A* identifies Arctic air, which is very cold. These six letters can be combined to give us the classification of air masses first described in 1928 and still used today: **Maritime Equatorial (*mE*), Maritime Tropical (*mT*), Continental Tropical (*cT*), Continental Polar (*cP*), Maritime Polar (*mP*),** and **Continental Arctic (*cA*)**. These six types are described more fully in Table 7.1. From now on, we will use the symbols rather than the full names as we discuss each type of air mass.

## Modification and Stability of Air Masses

As a result of the general circulation patterns within the atmosphere, air masses do not remain stationary over their source regions indefinitely. When an air mass begins to move over Earth's surface along a path known as a trajectory, for the most part it retains its distinct and homogeneous characteristics. However, modification does occur as the air mass gains or loses some of its thermal energy and moisture content to the surface below. Although this modification is generally slight, the gain or loss of thermal energy can make an air mass more stable or unstable.

An air mass is further classified by whether it is warmer or colder than the surface over which it travels because this has a bearing on its stability. If an air mass is colder than the surface over which it passes, then the surface will heat the air mass from below. This will in turn increase the environmental lapse rate, enhancing the prospect of instability. To describe such a situation, the letter *k* (from German: *kalt*, cold) is added to the other letters that symbolize the air mass. For example, an *mT* air mass originating over the Gulf of Mexico in summer that moves onshore over warmer land would be denoted *mTk*. Such an air mass is often unstable and can produce copious convective precipitation. On the other hand, this same *mT* air mass moving onshore during the winter would be warmer than the land surface. Consequently, the air mass would be cooled from below, decreasing its environmental lapse rate, which enhances the prospect of stability. We describe this situation with the letter *w* (from German: *warm*, warm), and the air mass would be denoted *mTw*. In this case, stratiform, not convective, precipitation is most likely.

The modification of air masses can also involve moisture content. For example, during the early-winter to midwinter seasons, cold, dry *cP* or *cA* air from Canada can move southeastward across the Great Lakes region. While passing over the lakes, this air mass can pick up moisture, thus increasing its humidity level. This modified *cP* or *cA* air reaches the frigid land on the leeward shores of the Great Lakes and precipitates, at times, large amounts of *lake-effect snows*. These snowfall areas may appear as *snow belts* or bands of snow, extending downwind from the lakes. The chances for lake-effect snow events diminish in late winter as the surfaces of the lakes freeze, thus cutting off the moisture supply to the air masses flowing across them.

## North American Air Masses

Most of us are familiar with the weather in at least one region of the United States or Canada; therefore, in this chapter we will concentrate on the air masses of North America and their effects on weather. What we learn will be applicable to the rest of the world, and as we examine climate regions in some of the following chapters, we will be able to understand that weather everywhere is most often affected by the movements of air masses. Especially in middle-latitude regions, the majority of atmospheric disturbances result from the confrontations of different air masses.

Five types of air masses (*cA, cP, mP, mT,* and *cT*) influence the weather of North America, some more than others. Air masses assume characteristics of their source regions (▌Fig. 7.1). Consequently, as the source regions change with the seasons, primarily because of changing insolation, the air masses also will vary.

## TABLE 7.1
### Types of Air Masses

| | Source Region | Usual Characteristics at Source | Accompanying Weather |
|---|---|---|---|
| Maritime Equatorial (*mE*) | Equatorial oceans | Ascending air, very high moisture content | High temperature and humidity, heavy rainfall; never reaches the United States |
| Maritime Tropical (*mT*) | Tropical and subtropical oceans | Subsiding air; fairly stable but some instability on western side of oceans; warm and humid | High temperatures and humidity, cumulus clouds, convectional rain in summer; mild temperatures, overcast skies, fog, drizzle, and occasional snowfall in winter; heavy precipitation along *mT/cP* fronts in all seasons |
| Continental Tropical (*cT*) | Deserts and dry plateaus of subtropical latitudes | Subsiding air aloft; generally stable but some local instability at surface; hot and very dry | High temperatures, low humidity, clear skies, rare precipitation |
| Maritime Polar (*mP*) | Oceans between 40° and 60° latitude | Ascending air and general instability, especially in winter; mild and moist | Mild temperatures, high humidity; overcast skies and frequent fogs and precipitation, especially during winter; clear skies and fair weather common in summer; heavy orographic precipitation, including snow, in mountainous areas |
| Continental Polar (*cP*) | Plains and plateaus of subpolar and polar latitudes | Subsiding and stable air, especially in winter; cold and dry | Cool (summer) to very cold (winter) temperatures, low humidity; clear skies except along fronts; heavy precipitation, including winter snow, along *cP/mT* fronts |
| Continental Arctic (*cA*) | Arctic Ocean, Greenland, and Antarctica | Subsiding very stable air; very cold and very dry | Seldom reaches United States, but when it does, bitter cold, subzero temperatures, clear skies, often calm conditions |

**Continental Arctic Air Masses** The frigid, frozen surface of the Arctic Ocean and the land surface of northern Canada and Alaska serve as source regions for this air mass. It is extremely cold, very dry, and very stable. Though it will affect parts of Canada, even during the winter when this air mass is best developed, it seldom travels far enough south to affect the United States. However, on those few occasions when it does extend down into the midwestern and southeastern United States, its impact is awesome. Record-setting cold temperatures often result. If the *cA* air mass remains in the Midwest for an extended period, vegetation—not accustomed to the extreme cold—can be severely damaged or killed.

**Continental Polar Air Masses** At its source in north-central North America, a *cP* air mass is cold, dry, and stable because it is warmer than the surface beneath it; the weather of a *cP* air mass is cold, crisp, and clear. Because there are no east–west landform barriers in North America, *cP* air can migrate south across Canada and the United States. A tongue of *cP* air can sometimes reach as far south as the Gulf of Mexico or Florida. When winter *cP* air extends into the United States, its temperature and humidity are raised only slightly. The movement of such an air mass into the Midwest and South brings with it a cold wave characterized by colder-than-average temperatures and clear, dry air and can cause freezing temperatures as far south as Florida and Texas.

The general westerly direction of atmospheric circulation in the middle latitudes rarely allows a *cP* air mass to break through the great western mountain ranges to the West Coast of the United States. When such an air mass does reach the Washington, Oregon, and California coasts, it brings with it unusual freezing temperatures that do great damage to agriculture.

**Maritime Polar Air Masses** During winter months, the oceans tend to be warmer than the land, so an *mP* air mass tends to be warmer than its counterpart on land (the *cP* air mass). Much *mP* air is originally cold, dry *cP* air that has moved to a position over the ocean. There, it is modified by the warmer water and collects heat and moisture. Thus, *mP* air is cold (although not as cold as *cP* air) and damp, with a tendency toward instability. The northern Pacific Ocean serves as the source region for *mP* air masses, which, because of the general westerly circulation of the atmosphere in the

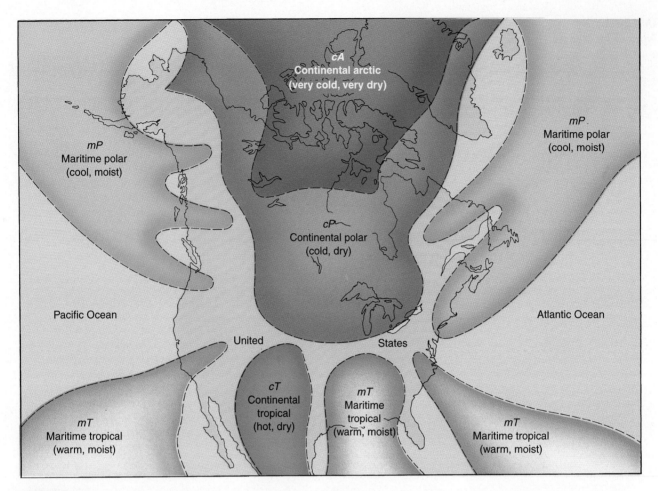

**FIGURE 7.1**
Source regions of North American air masses. Air mass movements import the temperature and moisture characteristics of these source regions into far distant areas.
**Use Table 7.1 and this figure to determine which air masses affect your location. Are there seasonal variations?**

middle latitudes, affect the weather of the northwestern United States and southwestern Canada. When this $mP$ air meets an uplift mechanism (such as a mass of colder, denser air or coastal mountain ranges), the result is usually very cloudy weather with a great deal of precipitation. An $mP$ air mass may still be the source of many midwestern snowstorms even after crossing the western mountain ranges.

Generally, an $mP$ air mass that develops over the northern Atlantic Ocean does not affect the weather of the United States because such an air mass tends to flow eastward toward Europe. However, on some occasions, there may be a reversal of the dominant wind direction accompanying a low pressure system, and New England can be made miserable by the cool, damp winds, rain, and snow of a weather system called a *nor'easter*.

**Maritime Tropical Air Masses** The Gulf of Mexico and subtropical Atlantic and Pacific Oceans serve as source regions for $mT$ air masses that have a great influence on the weather of the United States and at times southeastern Canada. During winter, the waters are warm, and the air above is warm, and moist. As the warm, moist air moves northward up the Mississippi lowlands, it travels over increasingly cooler land surfaces. The lower layers of air are chilled, and dense advection fog often results. When it reaches the $cP$ air migrating southward from Canada, the warm $mT$ air is forced to rise over the colder, drier $cP$ air, and significant precipitation can occur.

The longer days and more intense insolation of summer months modify an $mT$ air mass at the source region by increasing its temperature and moisture content. However, during summer, the land is warmer than the nearby waters, and as the $mT$ air mass moves onto the land, the instability of the air mass is increased. This air mass is a factor in the formation of great thunderstorms and convective precipitation on hot, humid days, and it is also responsible for much of the hot, humid weather of the southeastern and eastern United States.

Other $mT$ air masses form over the Pacific Ocean in the subtropical latitudes. These air masses tend to be slightly cooler than those that form over the Gulf of Mexico and the Atlantic, partly because of their passage over the cooler California Current. A Pacific $mT$ air mass is also more stable because of the strong subsidence associated with the eastern portion of the Pacific subtropical high. This air mass contributes to the dry summers of southern California and occasionally brings moisture in winter as it rises over the mountains of the Pacific Coast.

**Continental Tropical Air Masses** A fifth type of air mass may affect North America, but it is the least important to the weather of the United States and Canada. This is the $cT$ air mass that develops over large, homogeneous land surfaces in the subtropics. The Sahara Desert of North Africa is a prime example of a source region for this type of air mass. The weather typical of the $cT$ air mass is usually very hot and dry, with clear skies and major heating from the sun during daytime.

In North America, there is little land in the correct latitudes to serve as a source region for a $cT$ air mass of any significant proportion. A small $cT$ air mass can form over the deserts of the southwestern United States and central Mexico in the summer. In its source region, a $cT$ air mass provides hot, dry, clear weather. When it moves eastward, however, it is usually greatly modified as it comes in contact with larger and stronger air masses of different temperature, humidity, and density values. At times, $cT$ air from Mexico and Texas meets with $mT$ air from the Gulf of Mexico. This boundary is known as a *dry line*. Here, the drier air is denser and will lift the moister air over it. This mechanism of uplift may act as a trigger for precipitation episodes and perhaps thunderstorm activity.

# Fronts

We have seen that air masses migrate with the general circulation of the atmosphere. Over the United States, which is influenced primarily by the westerlies, there is a general eastward flow of the air masses. In addition, air masses tend to diverge from areas of high pressure and converge toward areas of low pressure. This tendency means that the tropical and polar air masses, formed within systems of divergence, tend to flow toward areas of convergence within the United States. As previously noted, an important feature of an air mass is that it maintains the primary characteristics first imparted to it by its source region although some slight modification may occur during its migration.

When air masses differ, they do so primarily in their temperature and in their moisture content, which in turn affect the air masses' density and atmospheric pressure. As we saw in Chapter 6, when different air masses come together, they do not mix easily but instead come in contact along sloping boundaries called *fronts*. Although usually depicted on maps as a one-dimensional boundary line separating two different air masses, a front is actually a three-dimensional surface with length, width, and height. To emphasize this concept, a front is sometimes referred to as a **surface of discontinuity.** This surface of discontinuity is a zone that can cover an area from 2 to 3 kilometers (1–2 mi) wide to as wide as 150 kilometers (90 mi). Hence, it is more accurate to speak of a frontal zone rather than a frontal line.

The sloping surface of a front is created as the warmer and lighter of the two contrasting air masses is lifted or rises above the cooler and denser air mass. Such rising, known as *frontal uplift,* is a major source of precipitation in middle-latitude countries like the United States and Canada (as well as middle-latitude European and Asian countries) where contrasting air masses are most likely to converge.

The steepness of the frontal surface is governed primarily by the degree of difference between the two converging air masses. When there is a sharp difference between the two air masses, as when an $mT$ air mass of high temperature and moisture content meets a $cP$ air mass with its cold, dry characteristics, the slope of the frontal surface will be steep. With a steep slope, there will be greater frontal uplift. Provided other conditions (for example, temperature and moisture content) are equal, a steep slope with its greater frontal uplift will produce heavier precipitation than will a gentler slope.

Fronts are differentiated by determining whether the colder air mass is moving in on the warmer one, or vice versa. The weather that occurs along a front also depends on which air mass is the "aggressor."

## Cold Front

A **cold front** occurs when a cold air mass actively moves in on a warmer air mass and pushes it upward. The colder air, denser and heavier than the warm air it is displacing, stays at the surface while forcing the warmer air to rise. As we can see in Figure 7.2, a cold front usually results in a relatively steep slope in which the warm air may rise 1 meter vertically for every 40–80 meters of horizontal distance. If the warm air mass is unstable and has a high moisture content, heavy precipitation can result, sometimes in the form of violent thunderstorms. A **squall line** may result when several storms align themselves on (or in advance of) a cold front. In any case, cold fronts are usually associated with strong weather disturbances or sharp changes in temperature, air pressure, and wind.

## Warm Front

When a warmer air mass is the aggressor and invades a region occupied by a colder air mass, a **warm front** results. At a warm front, the warmer air, as it slowly pushes against the cold air, also rises over the colder, denser air mass, which again stays in contact with Earth's surface. The slope of the surface of discontinuity that results is usually far gentler than that occurring in a cold front. In fact, the warm air may rise only 1 meter vertically for every 100 or even 200 meters of horizontal distance. Thus, the frontal uplift that develops will not be as great as that occurring along a cold front. The result is that the warmer weather associated with the passage of a warm front tends to be less violent and the changes less abrupt than those associated with cold fronts.

If we look at Figure 7.3, we can see why the advancing warm front affects the weather of areas ahead of the actual

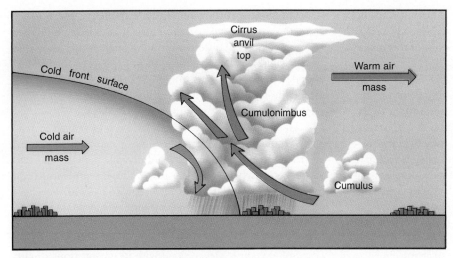

**ACTIVE FIGURE 7.2**
Watch this Active Figure at http://now.brookscole.com/gabler8.
Cross section of a cold front. Cold fronts generally move rapidly, with a blunt forward edge that drives adjacent warmer air upward. This can produce violent precipitation from the warmer air.

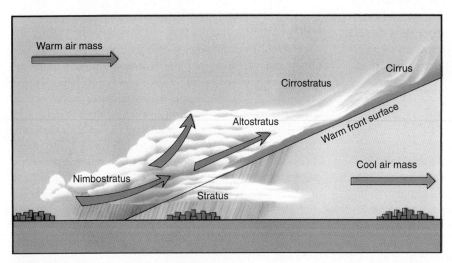

**ACTIVE FIGURE 7.3**
Watch this Active Figure at http://now.brookscole.com/gabler8.
Cross section of a warm front. Warm fronts advance more slowly than cold fronts and replace rather than displace cold air by sliding upward over it. The gentle rise of the warm air produces stratus clouds and gentle drizzles.
**Compare Figures 7.2 and 7.3. How are they different? How are they similar?**

surface location of the frontal zone. Changes in the weather from approaching fronts can sometimes be indicated by the series of cloud types that precede them.

## Stationary and Occluded Fronts

When two air masses have converged and formed a frontal boundary but then neither moves, we have a situation known as a *quasi-stationary* or, as it is more commonly called, a **stationary front.** Locations under the influence of a stationary front are apt to experience clouds, drizzle, and rain (or possible thunderstorms) for several days. In fact, a stationary front and its accompanying weather will remain until either the contrasts between the two air masses are reduced or the circulation of the atmosphere finally causes one of the air masses to move. If a stationary front holds a position for a length of time, then regional flooding is likely to occur.

An **occluded front** occurs when a faster-moving cold front overtakes a warm front, pushing all of the warm air aloft. This frontal situation usually occurs in the latter stages of a middle-latitude cyclone, which will be discussed next. Map symbols for the four frontal types are shown in Figure 7.4.

## Atmospheric Disturbances

Embedded within the wind belts of the general atmospheric circulation (see Chapter 5) are secondary circulations. These are made up of storms and other atmospheric disturbances. We use the term **atmospheric disturbance** because it is more general than a storm and includes variations in the secondary circulation of the atmosphere that cannot be correctly classified as storms.

Partly because our primary interest is in the weather of North America, we concentrate on an examination of **middle-latitude disturbances,** sometimes known as **extratropical disturbances.**

Shortly after World War I, Norwegian meteorologists Jacob Bjerknes and Halvor Solberg put forth the *polar front theory,* which provided insight into the development, movement, and dissipation of middle-latitude storms. They recognized the middle latitudes as an area of convergence where unlike air masses, such as cold polar air and warm subtropical air, commonly meet at a boundary called the *polar front.* Though the polar front may be a continuous boundary circling the entire globe, it is most often fragmented into several individual line segments. Furthermore, the polar front tends to move north and south with the seasons and is apt to

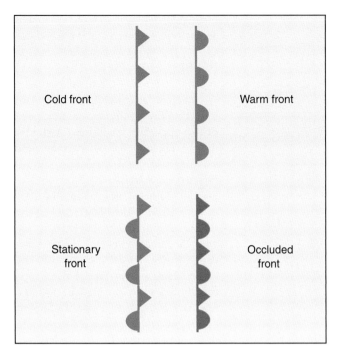

**FIGURE 7.4**
The four major frontal symbols used on weather maps

be stronger in winter than in summer. It is along this wavy polar front that the upper air westerlies, also known as the *Polar Front Jet Stream,* develop and flow.

Middle-latitude storms develop at the front and then travel along it. These migrating storms, with their opposing cold, dry polar air and warm, humid tropical air, can cause significant variation in the day-to-day weather of the locations over which they pass. It is not unusual in some parts of the United States and Canada for people to go to bed at the end of a beautiful warm day in early spring and wake up to falling snow the next morning. Such variability is common for middle-latitude weather, especially during certain times of the year when the weather changes from a period of cold, clear, dry days to a period of snow, only to be followed by one or two more moderate but humid days.

## Cyclones and Anticyclones

### Nature, Size, and Appearance on Maps
We have previously distinguished cyclones and anticyclones according to differences in pressure and wind direction. Also, when studying maps of world pressure distribution, we identified large areas of semipermanent cyclonic and anticyclonic circulation in Earth's atmosphere (the subtropical high, for example). Now, when examining middle-latitude atmospheric disturbances, we use the terms *cyclone* and *anticyclone* to describe the moving cells of low and high pressure, respectively, that drift with varying regularity in the path of the prevailing westerly winds. As systems of higher pressure, anticyclones are usually characterized by clear skies, gentle winds, and a general lack of precipitation. As centers for converging, rising air, cyclones create the storms of the middle latitudes, with associated fronts of various types.

As we know from experience, no middle-latitude cyclonic storm is ever exactly like any other. The storms vary in their intensity, their longevity, their speed, the strength of their winds, their amount and type of cloud cover, the quantity and kind of their precipitation, and the surface area they affect.

Because there are an endless variety of cyclones, we describe "model cyclones" in the following discussions. Not every storm will act in the way we describe, but certain generalizations are helpful in understanding middle-latitude cyclones.

A cyclone has a low pressure center; thus, winds tend to converge toward that center in an attempt to equalize pressure. If we visualize air moving in toward the center of the low pressure system, we can see that the air that is already at the center must be displaced upward. Incoming *mT* air spirals upward, and the lifting that occurs in a cyclone results in clouds and precipitation.

Anticyclones are high pressure systems in which atmospheric pressure decreases toward the outer limits of the system. Visualizing an anticyclone, or high, we can see that air in the center of the system must be subsiding, in turn displacing surface air outward, away from the center of the system. Hence, an anticyclone has diverging winds. In addition, an anticyclone tends to be a fair-weather system; the subsiding air in its center increases in temperature and stability, reducing the opportunity for condensation.

We should note here that the pressures we are referring to in these two systems are relative. What is important is that in a cyclone, pressure decreases toward the center, and in an anticyclone, pressure increases toward the center. Furthermore, the intensities of the winds involved in these systems depend on the steepness of the *pressure gradients* (the change in pressure over a horizontal distance) involved. Thus, if there is a steep pressure gradient in a cyclone, with the pressure much lower at the center than at the outer portions of the system, the winds will converge toward the center with considerable velocity.

The situation is easier to visualize if we imagine these pressure systems as landforms. A cyclone is shaped like a basin ( Fig. 7.5). If we are filling the basin with water, we know that the water will flow in faster the steeper the sides and the deeper the depression. If we visualize an anticyclone as a hill or mountain, then we can also see that just as water flowing down the sides of such landforms will flow faster with increased height and steepness, so will the air blowing out of an area of very high pressure move rapidly.

On a surface weather map, cyclones and anticyclones are depicted by concentric isobars of increasing pressure toward the center of a high and of decreasing pressure toward the center of a low. Usually a high will cover a larger area than a low,

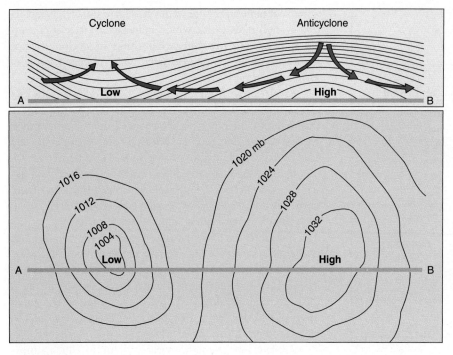

**FIGURE 7.5**
Close spacing of isobars around a cyclone or anticyclone indicates a steep pressure gradient that will produce strong winds. Wide spacing of isobars indicates a weaker system.
**Where would be the strongest winds in this figure? Where would be the weakest winds?**

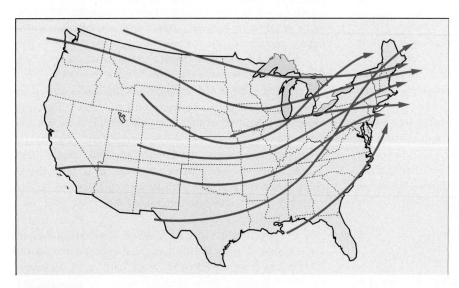

**FIGURE 7.6**
Common storm tracks for the United States. Virtually all cyclonic storms move from west to east in the prevailing westerlies and swing northeastward across the Atlantic coast. Storm tracks originating in the Gulf of Mexico represent tropical hurricanes.
**What storm tracks influence your location?**

but both pressure systems are capable of covering and affecting extensive areas. There are times when nearly the entire midwestern United States is under the influence of the same system. The average diameter of an anticyclone is about 1500 kilometers (900 mi); that of a cyclone is about 1000 kilometers (600 mi).

**General Movement** The cyclones and anticyclones of the middle latitudes are steered, or guided, along a path reflecting the configuration and speed of the upper air westerlies (or the jet stream). The upper air flow can be quite variable with wild oscillations. However, a general west-to-east pattern does prevail. Consequently, people in most of the eastern United States look at the weather occurring to the west to see what they might expect in the next few days. Most storms that develop in the Great Plains or Far West move across the United States during a period of a few days at an average speed of about 36 kilometers per hour (23 mph) and then travel on into the North Atlantic before occluding.

Although neither cyclones nor anticyclones develop in exactly the same places at the same times each year, they do tend to develop in certain areas or regions more frequently than in others. They also follow the same general paths, known as **storm tracks** ( Fig. 7.6). These storm tracks vary with the seasons. In addition, because the temperature variations between the air masses are stronger during the winter months, the atmospheric disturbances that develop in the middle latitudes during those months are greater in number and intensity.

**Cyclones** Now let's look more closely at cyclones—their origin, development, and characteristics. Warm and cold air masses meet at the polar front where most cyclones develop. These two contrasting air masses do not merge but may move in opposite directions along the frontal zone. Although there may be some slight uplift of the warmer air along the edge of the denser, colder air, the uplift will not be significant. There may be some cloudiness and precipitation along such a frontal zone, though not of storm caliber.

For reasons not completely understood but certainly related to the wind flow in the upper troposphere, a wavelike kink may develop along the polar front. This is the initial step in the formation of a fully grown middle-latitude cyclone (Fig. 7.7). At this bend in the polar front, we now have warm air pushing poleward (a warm front) and cold air pushing equatorward (a cold front), with a center of low pressure at the location where the two fronts are joined.

As the contrasting air masses jockey for position, the clouds and precipitation that exist along the fronts are greatly intensified, and the area affected by the storm is much greater. Along the warm front, precipitation will be more widespread but less intense than along the cold front. One factor that can vary the kind of precipitation occurring at the warm front is the stability of the warm air mass. If it is stable, then its uplifting over the cold air mass may cause only a fine drizzle or a light, powdery snow if the temperatures are cold enough. On the other hand, if the warm air mass is moist and unstable, the uplifting may set off heavier precipitation. As you can see by referring again to Figure 7.3, the precipitation that falls at the warm front may *appear* to be coming from the colder air. Though weather may feel cold and damp, the pre-cipitation is actually coming from the overriding warmer air mass above, then falling through the colder air mass to reach Earth's surface.

Because a cold front usually moves faster, it will eventually overtake the warm front. This produces the situation we previously identified as an *occluded front*. Because additional warm, moist air will not be lifted after occlusion, condensation and the release of latent heat energy will diminish, and the system will soon die. Occlusions are usually accompanied by rain and are the major process by which middle-latitude cyclones dissipate.

**Cyclones and Local Weather** Different parts of a middle-latitude cyclone exhibit different weather. Therefore, the weather that a location experiences at a particular time depends on which portion of the middle-latitude cyclone is over the location. Also, because the entire cyclonic system tends to travel from west to east, a specific sequence of

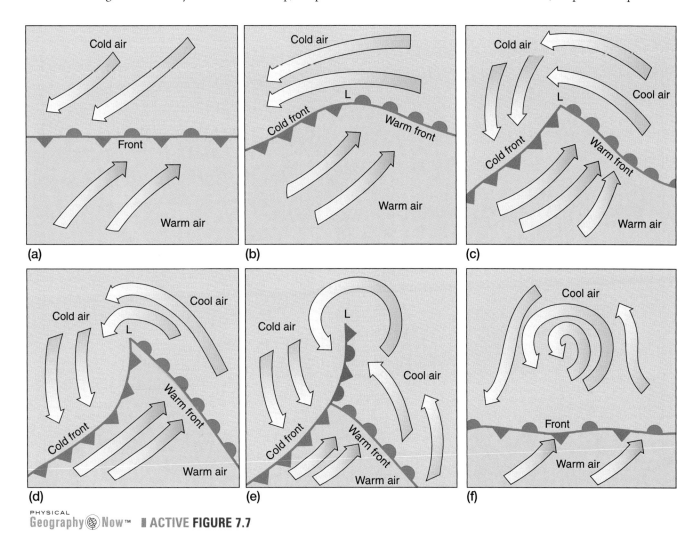

**PHYSICAL Geography⊛Now™ ■ ACTIVE FIGURE 7.7**
Watch this Active Figure at http://now.brookscole.com/gabler8.
Stages in the development of a middle-latitude cyclone. Each view represents the development somewhat eastward of the preceding view as the cyclone travels along its storm track. Note the occlusion in (e).
**In (c), where would you expect rain to develop? Why?**

weather can be expected at a given location as the cyclone passes over that location.

Let's assume that it is late spring. A cyclonic storm has originated in the southeast corner of Nebraska and is following a track (refer again to Figure 7.6) across northern Illinois, northern Indiana, northern Ohio, through Pennsylvania, and finally out over the Atlantic Ocean. A view of this storm on a weather map, at a specific time in its journey, is presented in Figure 7.8a. Figure 7.8b shows a cross-sectional view north of the center of the cyclone, and Figure 7.8c shows a cross-sectional view south of the center of the cyclone. As the storm continues eastward, at 9–13 meters per second (20–30 mph), the sequence of weather will be different for Detroit, where the warm and cold fronts will pass just to the south, than for Pittsburgh, where both fronts will pass overhead. To illustrate this point, let's examine, element by element, the variation in weather that will occur in Pittsburgh, with reference, where appropriate, to the differences that occur in Detroit as the cyclonic system moves east.

As we have previously stated, atmospheric temperature and pressure are closely related. As temperature increases, air expands and pressure decreases. Therefore, these two elements are discussed together. Because a cyclonic storm is composed of two dissimilar air masses, there are usually significant temperature contrasts. The sector of warm, humid $mT$ air between the two fronts of the cyclone is usually considerably warmer than the cold $cP$ air surrounding it. The temperature contrast is accentuated in the winter when the source region for $cP$ air is the cold cell of high pressure normally found in Canada at that time of year. During the summer, the contrast between these air masses is greatly reduced.

As a consequence of the temperature difference, the atmospheric pressure in the warm sector is considerably lower than the atmospheric pressure in the cold air behind the cold front. Far in advance of the warm front, the pressure is also high, but as the warm front (see again Figure 7.3) approaches, increasingly more cold air is replaced by overriding warm air, thus steadily reducing the surface pressure.

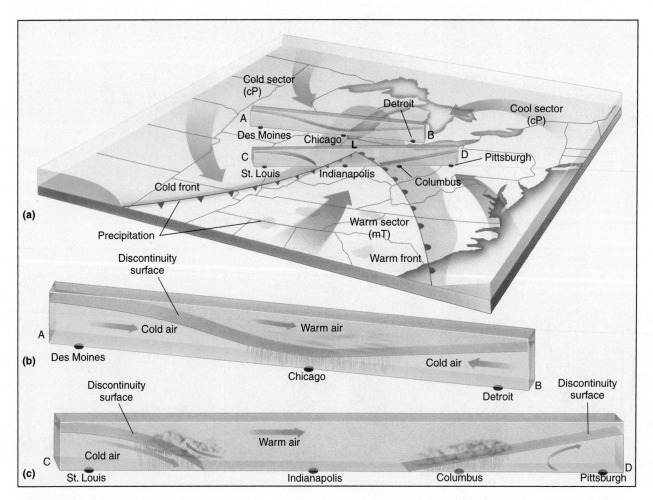

**FIGURE 7.8**

Environmental Systems: Middle-Latitude Cyclonic Systems. This diagram models a middle-latitude cyclone positioned over the Midwest as the system moves eastward: (a) a map view of the weather system; (b) a cross section along line AB north of the center of low pressure; (c) a cross section along line CD South of the center of low pressure.

## Career Vision

**Thomas G. Novak, Meteorological Computer Training Consultant, Meteorlogix, University of Minnesota, BA in Geography.**

At the University of Minnesota, where I received my bachelor of arts degree in geography, no schools in the Twin Cities offered a degree in meteorology. Weather is my first love; my second love is maps and making maps. I decided to go with geography. Geography gives you the whole gamut—meteorology, geology, climatology, and cartography.

My work experience includes an Internship with the National Weather Service, president of the Upper Midwest Weather Consultants, and weather broadcaster for KAAL-TV. We provide our weather data to a variety of industries including broadcast companies, TV stations, cable companies, farmers, and aviation. My job is to go to each customer who needs training on our equipment and provide expert training. Training involves computer data management and how to create an aesthetically pleasing weather map. My job actually focuses more on geography than meteorology.

We've recently started implementing our weather data with GIS software. Being able to incorporate our weather data within a GIS system is going to be key to our future. In geography, I learned about meteorology, climatology, and cartography. What was stressed to me, as I was leaving the University of Minnesota, was that GIS would play a significant technological role in the future. In general, people like to visualize data through maps, instead of just having statistics and numbers in front of them. GIS gives people the opportunity to visually see what is happening in their industry and in the market around them.

The one aspect of my geography degree that has helped me the most at Meteorlogix is knowing how to make an aesthetically pleasing map. It's a talent that most people don't possess. Creating an informative and aesthetically pleasing weather map usually incorporates knowledge of cartography, meteorology, and climatology. A geography degree puts all these backgrounds together; it's like having a full basket rather than just a partial basket of knowledge. I often hear a lot of professional individuals say, "How did you make such a nice-looking map?" Well, it comes down to years upon years of studying this kind of stuff, seeing what actually looks good. I wouldn't have that knowledge if it weren't for geography.

Number one, I would tell people to get some GIS knowledge and then wrap all the other geography knowledge base (cartography, meteorology, climatology, hydrology) in with it. Tell people to turn on the television and watch their local news. The map that the weather broadcasters are pointing to, I likely had a hand in creating it and the animation.

A degree in geography offers many different courses to choose from, and it is your responsibility to correctly navigate through them. I headed more for a physical geography background, but others may choose the human side of geography by getting into statistics, demographics, and the like. I've always believed in being well rounded rather than having knowledge in only one thing. A degree in geography offers a well-rounded education and can certainly open doors for you in several different career paths.

---

Therefore, as the warm front of this late-spring cyclonic storm approaches Pittsburgh, the pressure will decrease. After the warm front passes through Pittsburgh where the temperature may have been 8°C (46°F) or more, the pressure will stop falling, and the temperature may rise up to 18°–20°C (64°F–69°F) as $mT$ air invades the area. At this point, Indianapolis has already experienced the passage of the warm front. After the cold front passes, the pressure will rise rapidly and the temperature will drop. In this late-spring storm, the $cP$ air temperature behind the cold front might be 2°C–5°C (35°F–40°F). Detroit, which is to the north of the center of the cyclone, will miss the warm air sector entirely and therefore will experience a slight increase in pressure and a temperature change from cool to cold as the cyclone moves to the east.

Changes in wind direction are one signal of the approach and passing of a cyclonic storm. Because a cyclone is a center of low pressure, winds flow counterclockwise toward its center. Also, winds are caused by differences in pressure. Therefore, the winds associated with a cyclonic storm are stronger in winter when the pressure (and temperature) differences between air masses are greatest.

In our example, Pittsburgh is located to the south and east of the center of low pressure and ahead of the warm front, and it is experiencing winds from the southeast. As the entire cyclonic system moves east, the winds in Pittsburgh will shift to the south-southwest after the warm front passes. Indianapolis is currently in this position. After the cold front passes, the winds in Pittsburgh will be out of the north-northwest. St. Louis has already experienced the passage of the cold front and currently has winds from the northwest. The changing direction of wind, clockwise around the compass from east to southeast to south to southwest to west and northwest, is called a **veering wind shift** and indicates that your position is south of the center of a low. On the other hand, Detroit, which is also experiencing winds from the southeast, will undergo a completely different sequence of directional wind changes as the cyclonic storm moves eastward. Detroit's winds will shift to the northeast as the center of the storm passes to the south. Chicago has just undergone this shift. Finally, after the storm has passed, the winds will blow from the northwest. Des Moines, to the west of the storm, currently has northwest winds. Such a change of wind direction, from east to northeast to north to northwest, is called a **backing wind shift,** as the wind "backs" counterclockwise around the compass. A backing wind shift indicates that you are north of the cyclone's center.

The type and intensity of precipitation and cloud cover also vary as a cyclonic disturbance moves through a location. In Pittsburgh, the first sign of the approaching warm front will be high cirrus clouds. As the warm front continues to approach, the clouds will thicken and lower. When the warm front is within 150–300 kilometers (90–180 mi) of Pittsburgh, light rain and drizzle may begin, and stratus clouds will blanket the sky.

After the warm front has passed, precipitation will stop and the skies will clear. However, if the warm, moist *mT* air is unstable, convective showers may result, especially during the spring and summer months with their high afternoon sun angles.

As the cold front passes, warm air in its path will be forced to move aloft rapidly. This may mean that there will be a cold, hard rain, but the band of precipitation normally will not be very wide because of the steep angle of the surface of discontinuity along a cold front. In our example, the cold front and the band of precipitation have just passed St. Louis.

Thus, Pittsburgh can expect three zones of precipitation as the cyclonic system passes over its location: (1) a broad area of cold showers and drizzle in advance of the warm front; (2) a zone within the moist, subtropical air from the south where scattered convectional showers can occur; and (3) a narrow band of hard rainfall associated with the cold front ( Fig. 7.8c). However, locations to the north of the center of the cyclonic storm, such as Detroit, will usually experience a single, broad band of light rains resulting from the lifting of warm air above cold air from the north ( Fig. 7.8b). In winter, the precipitation is likely to be snow, especially in locations just to the northwest of the center of the storm, where the humid *mT* air overlies extremely cold *cP* air.

As you can see, different portions of a middle-latitude cyclone are accompanied by different weather. If we know where the cyclone will pass relative to our location, we can make a fairly accurate forecast of what our weather will be like as the storm moves east along its track (see Map Interpretation: Weather Maps).

**Cyclones and the Upper Air Flow** The upper air wind flow greatly influences our surface weather. We have already discussed the role of these upper air winds in the steering of surface storm systems. Another less obvious influence of the upper air flow is related to the undulating, wavelike flow so often exhibited by the upper air. As the air moves its way through these waves, it undergoes divergence or convergence because of the atmospheric dynamics associated with curved flow. This upper air convergence and divergence greatly influence the surface storms below.

The region between a ridge and the next downwind trough (A–B in  Fig. 7.9) is an area of upper-level convergence. In our atmosphere, an action taken in one part of the atmosphere is compensated for by an opposite reaction somewhere else. In this case, the upper air convergence is compensated for by divergence at the surface. In this area, anticyclonic circulation is promoted as the air is pushed downward. This pattern will inhibit the formation of a middle-latitude storm or cause an existing storm to weaken or even dissipate. On the other hand, the region between a trough and the next downwind ridge (B–C in  Fig. 7.9) is an area of upper-level divergence, which in turn is compensated for by surface convergence. This is an area where air is drawn upward and cyclonic circulation is encouraged. Convergence at the surface will certainly enhance the prospects of storm development or strengthen an already existing storm.

In addition to storm development or dissipation, upper air flow will have an impact on temperatures as well. If we assume that our "average" upper air flow is from west to east, then any deviation from that pattern will cause either colder air from the north or warmer air from the south to be advected into an area. Thus, after the atmosphere has been in a wavelike pattern for a few days, the areas in the vicinity of a trough (area B in  Fig. 7.9) will be colder than normal as polar air from higher latitudes is brought into that area. Just the opposite occurs at locations near a ridge (area C in  Fig. 7.9).

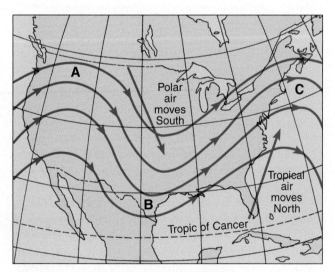

PHYSICAL
Geography Now™  ACTIVE FIGURE 7.9
Watch this Active Figure at http://now.brookscole.com/gabler8.
The upper air wind pattern, such as that depicted here, can have a significant influence on temperatures and precipitation on Earth's surface.
**Where would you expect storms to develop?**

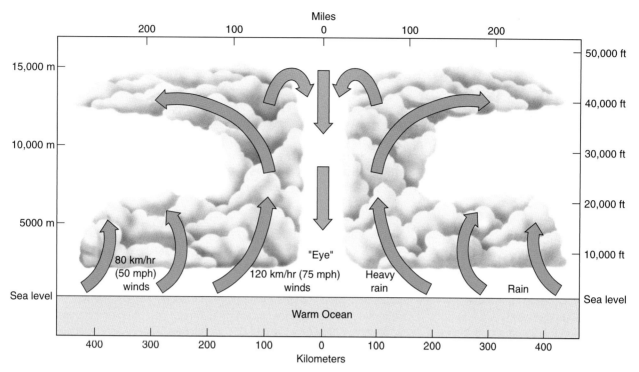

**ACTIVE FIGURE 7.22**
Watch this Active Figure at http://now.brookscole.com/gabler8.
Circulation patterns within a hurricane, showing inflow of air in the spiraling arms of the cyclonic system, rising air in the towering circular wall cloud, and outflow in the upper atmosphere. Subsidence of air in the storm's center produces the distinctive calm, cloudless "eye" of the hurricane.
**Why is this so?**

**TABLE 7.3**
**Saffir–Simpson Hurricane Scale**

| Scale Number (CATEGORY) | Central Pressure (MILLIBARS) | Wind Speed (KPH) | Wind Speed (MPH) | Storm Surge (METERS) | Storm Surge (FEET) | Damage |
|---|---|---|---|---|---|---|
| 1 | ≥ 980 | 119–153 | 74–95 | 1.2–1.5 | 4–5 | Minimal |
| 2 | 965–979 | 154–177 | 96–110 | 1.6–2.4 | 6–8 | Moderate |
| 3 | 945–964 | 178–209 | 111–130 | 2.5–3.6 | 9–12 | Extensive |
| 4 | 920–944 | 210–250 | 131–155 | 3.7–5.4 | 13–18 | Extreme |
| 5 | <920 | >250 | >155 | >5.4 | >18 | Catastrophic |

which we might have been led to expect from its circular winds.

The Saffir–Simpson Hurricane Scale provides a means of classifying hurricane intensity and potential damage by assigning a number from 1 to 5 based on a combination of central pressure, wind speed, and the height of the storm surge (Table 7.3).

Although a great deal of time, effort, and money has been spent on studying the development, growth, and paths of hurricanes, much is still not known. For example, it is not

# Geography's Spatial Science Perspective
## Hurricane Paths and Landfall Probability Maps

Hurricanes (called typhoons in Asia) are generated over tropical or subtropical oceans and build strength as they move over regions of warm ocean water. Ships and aircraft regularly avoid hurricane paths by navigating away from these huge violent storms. People living in the path of an oncoming hurricane try to prepare their belongings, homes, and other structures and may have to evacuate if the hazard potential of the impending storm is great enough. Landfall refers to the location where the eye of the storm encounters the coastline. Storm surges present the most dangerous hazard associated with hurricanes, where the ocean violently washes over and floods low-lying coastal areas. In 1900, 6000 residents of Galveston Island in Texas were killed when a hurricane pushed a 7-meter-high wall of water over the island. Much of the city was destroyed by this storm, the worst natural disaster ever to occur in the United States in terms of lives lost.

Today we have sophisticated technology for tracking and evaluating tropical storms. Computer models, developed from maps of the behavior of past storms, are used to indicate a hurricane's most likely path and landfall location, as well as the chance that it may strike the coast at other locations. The nearer a storm is to the coast, the more accurate the predicted landfall site should be, but in some cases a hurricane may begin to move in a completely different direction. In general, hurricanes that originate in the North Atlantic Ocean tend to move eastward toward North America and then turn northward along the Atlantic or Gulf Coasts.

Nature still remains unpredictable, so potential landfall sites are shown on probability maps, which show the degree of likelihood for the hurricane path. These maps help local authorities and residents decide what course of action is best to take in preparing for the approach of a hurricane. A 90% probability means that nine times out of ten storms under similar regional weather conditions have moved onshore in the direction indicated by that level on the map. Regions where the hurricane is considered likely to move next are represented on the map by color shadings that correspond to varying degrees of probability for the storm path. A 60% probability means that six of ten hurricanes moved as indicated, and so forth. In recent years, the National Weather Service has worked hard to develop computer models that will yield better predictions of hurricane paths and landfalls. If you live in a coastal area affected by hurricanes and tropical disturbances, understanding these maps of landfall probability may be very important to your safety and your ability to prepare for a coming storm.

Landfall probability map for Hurricane Charley on Wednesday, August 11, 2004, showing the most likely place for landfall at the time the map was produced. This was 2 days before the hurricane struck Florida's west coast.

This map, produced after the storm had dissipated, shows the actual path of Hurricane Charley and its severity. In this case, the landfall probability map proved fairly accurate. From landfall along the Gulf Coast, Hurricane Charley crossed the Florida peninsula, was downgraded to tropical storm status, but then struck Atlantic coastal areas of the Carolinas with strong winds, heavy rain, and coastal flooding.

**FIGURE 7.23**
A world map showing major "Hurricane Alleys".
**Which coastlines seem unaffected by these tracks?**

yet possible to predict the path of a hurricane, even though it can be tracked with radar and studied by planes and weather satellites. In addition, meteorologists can list factors favorable for the development of a hurricane but cannot say that in a certain situation a hurricane will definitely develop and travel along a particular path. As with tornadoes, there are also "Hurricane Alleys," or areas where their development is more likely to occur ( Fig. 7.23).

Among the factors leading to hurricane development are a warm ocean surface of about 25°C (77°F) and warm, moist overlying air. These factors are probably the reasons why hurricanes occur most often in the late summer and early fall when air masses have maximum humidity and ocean surface temperatures are highest. Also, the Coriolis effect must be sufficient to support the rapid spiraling of the hurricane. Therefore, hurricanes neither develop nor survive in the equatorial zone from about 8°S to 8°N, for there the Coriolis effect is far too weak. Hurricanes begin as weak tropical disturbances such as the easterly wave and in fact will not develop without the impetus of such a disturbance. It is further speculated that some sort of turbulence in the upper air may play a part in a hurricane's initial development.

Names are assigned to storms once they reach *tropical storm* status, with wind speeds between 17 and 33 meters per second (39–74 mph). Each year the names are selected from a different alphabetical list of alternating female and male names—one list for the North Atlantic and one for the North Pacific. If a hurricane is especially destructive and becomes a part of recorded history, its name is retired and never used again.

Hurricanes do not last long over land because their source of moisture (and consequently their source of energy) is cut off. Also, friction with the land surface produces a drag on the whole system. North Atlantic hurricanes that move first toward the west with the trade winds and then north and northeast as they intrude into the westerlies become polar cyclones if they remain over the colder region of the ocean and eventually die out. Over land, they will also become simple cyclonic storms, but even when they have lost some of their power, hurricanes can still do great damage.

Hurricanes can occur over most subtropical and tropical oceans and seas (the South Atlantic was the exception, though it was not known why). On March, 26, 2004, tropical Cyclone Catarina was the first to attack the southern coast of Brazil, much to the amazement of atmospheric scientists all over the world! In Australia and the South Pacific, they are called tropical cyclones (or willy-willies). Near the Philippines, they are known as *bagyos,* but in most of East Asia they are called **typhoons.** In the Bay of Bengal, they are referred to as *cyclones*.

**FIGURE 7.24**
This montage of satellite images shows the remarkable similarity of the hurricane tracks as they approached Florida in 2004.
**Do you know any people who were affected by these hurricanes?**

The year 2004 was certainly one for the record books. Typhoon Tokage struck the Japanese Coast near Tokyo with significant loss of life. A total of ten tropical cyclones pounded Japan in 2004 alone. In our own Caribbean and Gulf of Mexico, three hurricanes, Charley, Frances, and Jeanne, directly hit Florida. A fourth, Ivan, whose center struck Gulf Shores, Mississippi, caused devastation in Florida's western panhandle (Fig. 7.24). The damages from these storms is estimated to reach $23 billion. This exceeds the cost of Hurricane Andrew in 1992, that at the time was called the costliest natural disaster in U.S. history and caused more than $20 billion in damage.

However, for the United States the worst was yet to come. In 2005, Hurricane Katrina, with 140 mph winds and a storm surge rising over 16 feet in height, struck the Gulf of Mexico coasts of Louisiana, Mississippi, and Alabama. Katrina caused the deaths of hundreds of people and the areas destroyed by the winds and floodwaters will have long-term recovery costs estimated to be as much as $200 billion (see the section titled Catastrophic Waves in Chapter 20). Some people have suggested that we seek ways to control these destructive storms. On the other hand, hurricanes are a major source of rainfall and an important means of transferring energy within Earth's system away from the tropics. Eliminating them might cause unwanted and unforeseen climate changes.

In this chapter, we combined some of the elements of the atmosphere learned in previous chapters to explain our major weather systems. We have noted how temperature and moisture differences characterize air masses and their leading edges, called fronts. With their additional pressure and wind components, air masses and fronts form the recognizable weather systems that all of us deal with from day to day. These systems may be relatively small, like tornadoes, or cover large areas, like middle-latitude cyclones. But most important, these weather systems affect our lives, whether in small ways, like an inconvenient forecast, or in devastating ways, as in life-threatening storms.

# Define & Recall

air mass
source region
Maritime Equatorial (*mE*)
Maritime Tropical (*mT*)
Continental Tropical (*cT*)
Continental Polar (*cP*)
Maritime Polar (*mP*)
Continental Arctic (*cA*)
surface of discontinuity
cold front
squall line
warm front
stationary front
occluded front
atmospheric disturbance
middle-latitude disturbance
   (extratropical disturbance)
storm track
veering wind shift
backing wind shift
convective thunderstorm
orographic thunderstorm
frontal thunderstorm
snowstorm
blizzard
tornado
Doppler radar
easterly wave
storm surge
hurricane
typhoon

# Discuss & Review

1. What is an air mass?
2. Do all areas on Earth produce air masses? Why or why not?
3. What letter symbols are used to identify air masses? How are these combined? What air masses influence the weather of North America? Where and at what time of the year are they most effective?
4. Use Table 7.1 and Figure 7.1 to find out what kinds of air masses are most likely to affect your local area. How do they affect weather in your area?
5. What forces modify the behavior of air masses? What kinds of weather may be produced when an air mass begins to move?
6. Why do you suppose air masses can be classified by whether they develop over water or over land?
7. Why does *mP* air affect the United States? Are there any deviations from this tendency?
8. What kind of air mass forms over the southwestern United States in summer? Have you ever experienced weather in such an air mass? What was it like? What kind of weather might you expect to experience if such an air mass met an *mP* air mass?
9. What is a front? How does it occur?
10. In a meeting of two contrasting air masses, how can the aggressor be determined?
11. Compare warm and cold fronts. How do they differ in duration and precipitation characteristics?
12. What kind of weather often results from a stationary front? What kinds of forces tend to break up stationary fronts?
13. How does the westerly circulation of winds affect air masses in your area? What kinds of weather result?
14. What are the major differences between middle-latitude cyclones and anticyclones?
15. Can you draw a diagram of a mature (fully developed) middle-latitude cyclone that includes the center of the low with several isobars, the warm front, the cold front, wind direction arrows, appropriate labeling of warm and cold air masses, and zones of precipitation?
16. If a wind changes to a clockwise direction, what is the shift called? Where does it locate you in relation to the center of a low pressure system? Explain why this happens.
17. How does the configuration of the upper air wind patterns play a role in the surface weather conditions?
18. How can an understanding of cyclones be used to help forecast weather changes? For how long in advance? What changes might occur to spoil your forecast?
19. Describe the sequence of weather events over a 48-hour period in St. Louis, Missouri, if a typical low pressure system (cyclone) passes 300 kilometers (180 mi) north of that location in the spring.
20. List three major causes of thunderstorms. How might the storms that develop from each of these causes differ?
21. Have you ever experienced a tornado or hurricane? Describe your feelings during it and the events surrounding it. How do the news media prepare us for such natural disasters?

# Consider & Respond

1. Collect a 3-day series of weather maps from your local newspapers. Based on the migration of high and low pressure systems during that period, discuss the likely pattern of the upper air winds.
2. List the ideal conditions for the development of a hurricane.
3. Look at Figure 7.8a. Assume you are driving from point A to point B. Describe the changes in weather (temperature, wind speed and direction, barometric pressure, precipitation, and cloud cover) you would encounter on your trip. Do the same analysis for a trip from point C to point D.
4. The location of the polar front changes with seasons. Why? In what way is Figure 7.16b related to the seasonal migration of the polar front?
5. Redraw Figure 7.7 so that it depicts a Southern Hemisphere example.

# Map Interpretation
## Weather Maps

One of the most widely used weather maps is a *surface weather map*. These maps, which portray meteorological conditions over a large area at a given moment in time, are important for current weather descriptions and forecasting. Simultaneous observations of meteorological data are recorded at weather stations across the United States (and worldwide). This information is electronically relayed to the National Centers for Environmental Prediction near Washington, D.C., where the surface data are analyzed and mapped.

Meteorologists at the Centers then use the individual pieces of information to depict the general weather picture over a larger area. For example, isobars (lines of equal air pressure) are drawn to reveal the locations of cyclones (L) and anticyclones (H), and to indicate frontal boundaries. Areas that were receiving precipitation at the time the map depicts are shaded in green so that these areas are highlighted. The end result is a map of surface weather conditions that can be used to forecast changes in weather patterns. This map is accompanied by a satellite image that was taken on the same date.

1. Isobars are lines of equal atmospheric pressure expressed in millibars. What is the interval (in millibars) between adjacent isobars on this weather map?
2. What kind of front is passing through central Florida at this time?

A satellite image of the atmospheric conditions shown on the accompanying weather map. ©NOAA

3. Which Canadian high pressure system is stronger—the one located over British Columbia or the one near Newfoundland?
4. Which state is free of precipitation at this time: Nebraska, Connecticut, Mississippi, or Kentucky?
5. What kind of front is located over Nevada and Utah?
6. Does the surface map accurately depict the cloud cover indicated on the accompanying satellite image?
7. Can you find the low pressure systems on the satellite image?
8. Is there any cloud cover over West Virginia? How can you tell?
9. Do the locations of the fronts and areas of precipitation depicted on the map agree with the idealized relationship represented in Figure 7.8 (Middle-Latitude Cyclonic Systems)?
10. On the map, what kind of frontal symbol lies off the U.S. coast between New Jersey and Connecticut?
11. Looking at the satellite image, comment about the cold frontal symbols extending out to the northeast from Florida, and extending out to the southeast from the New England coast.
13. Are both of the low pressure systems depicted on this map occluding?

# CHAPTER 8

In recent years, huge blocks of ice have calved off from Antarctic coastal ice sheets. Many scientists regard this as a sure sign of global warming. Photo courtesy of Pedro Skvarca, Instituto Antártico Argentino, 13 March 2002.

# Global Climates and Climate Change

**PHYSICAL Geography Now™** This icon, appearing throughout the book, indicates an opportunity to explore interactive tutorials, animations, or practice problems at now.brookscole.com/gabler8

If someone asked you "What's the weather like where you live?" how might you respond? Would you talk about the storm that occurred last week or say that summers are very warm where you live? You may find that a question dealing with local atmospheric conditions is difficult to answer. Is the question referring to *weather* or *climate*? It is essential that you can distinguish between the two. Weather and climate were defined briefly in Chapter 4. In Chapter 7, we discussed the fundamentals of weather. In this chapter, we begin the study of climate in much greater detail.

Unlike weather, which describes the state of the atmosphere over short periods of time, climatic analysis relies heavily on averages and statistical probabilities involving data accumulated for the atmospheric elements over periods of many years. Climatic descriptions include such things as averages, extremes, and patterns of change for temperature, precipitation, pressure, sunshine, wind velocity and direction, and other weather elements throughout the year.

# CHAPTER PREVIEW

**Atmospheric elements vary so greatly from place to place over Earth's landmasses that scientists have classified climates by combining those elements with similar statistics to identify a manageable number of groups or types.**

- What two atmospheric elements are most often used when classifying climates?
- Why were these elements selected?
- Why is there a need for more than one system of climate classification?

**Knowledge of the controls of weather and climate is sufficient preparation to permit a reasonably accurate prediction of the broad distribution of climatic types over Earth's land surfaces.**

- What control has the greatest influence over climate distribution?
- What other controls produce significant deviation from the pattern established by the major control?

**On a global, or macro, scale, the Köppen system of climate classification is one of the most widely used.**

- What are some of the advantages and disadvantages of the Köppen system?
- What are some advantages of the Thornthwaite system, and how does it differ from the Köppen system?
- At what other scales can climate be studied?

**Geographers use regions for much the same reasons that scholars in other disciplines use arbitrary systems for the organization of information—to create an orderly presentation of diverse phenomena.**

- How does a geographer identify and define a region?
- What type of phenomena can be organized into regions?
- Where is a region best defined?

**There are a number of plausible causes of climate change, and it is therefore difficult for scientists to determine which cause is responsible for a specific climatic event.**

- What are some of the major causes?
- Which causes seem to best explain some of the most important climatic events of the past?
- In which of the causes is human activity most likely involved?

**Despite the best scientific methods and the most modern technology, accurately predicting future climate remains exceedingly difficult.**

- Why is this so?
- What methods and technology are most commonly used?
- What climate changes are most likely in the immediate and more distant future?

In the first part of this chapter, we introduce the characteristics and classification of modern climates. Because climate can be defined at different scales, from a single hillside to a region as large as the Sahara stretching across much of northern Africa, two systems of describing and classifying climates are discussed. The first is introduced because it is one of the most effective systems available to various scientists for the classification of climates on a more local scale.

The second system has been widely adopted by physical geographers and other scientists; in a modified version, it will be the basis for the worldwide regional study of present-day climates in Chapters 9 and 10. The remainder of this chapter focuses on climate change. Climates in the past were not the same as they are today, and there is every reason to believe that future climates will be different as well. It is now also widely recognized that humans can also alter Earth's climate.

For decades scientists have realized that Earth has experienced major climate shifts during its history. It was believed that these shifts were gradual and could not be detected by humans during a lifetime or two. However, recent research reveals that climate has shifted repeatedly between extremes over some exceedingly short intervals. Moreover, the research has revealed that climate during the most recent 10,000 years has been extraordinarily stable compared to similar intervals in the past.

To predict future climates, it is critical that we examine the details of past climate changes, including both the magnitude and rates of prehistoric climate change. Earth has experienced both ice ages and lengthy periods that were warmer than today. These fluctuations serve as indicators of the natural variability of climate in the absence of significant human impact. Using knowledge of present and past climates, as well as models of how and why climate changes, we conclude this chapter with some predictions of future climate trends.

# Classifying Climates

Knowledge that climate varies from region to region dates to ancient times. The ancient Greeks (such as Aristotle, circa 350 BC) classified the known world into Torrid, Temperate, and Frigid zones based on their relative warmth. It was also recognized that these zones varied systematically with latitude and that the flora and fauna reflected these changes as well. With the further exploration of the world, naturalists noticed that the distribution of climates could be explained using factors such as sun angles, prevailing winds, elevation, and proximity to large water bodies.

The two weather variables used most often as indicators of climate are temperature and precipitation. To classify climates accurately, climatologists require a minimum of 30 years to describe the climate of an area. The invention of an instrument to reliably measure temperature—the thermometer—dates only to Galileo in the early 17th century. European settlement of and sporadic collection of temperature and precipitation data from distant colonies began in the 1700s but was not routine until the mid-19th century. This was soon followed in the early 20th century by some of the first attempts to classify global climates using actual temperature and precipitation data.

As we have seen in earlier chapters, temperature and precipitation vary greatly over Earth's surface. Climatologists have worked to reduce the infinite number of worldwide variations in atmospheric elements to a comprehensible number of groups or varieties by combining elements with similar statistics ( ▌Fig. 8.1). That is, they can classify climates strictly on the basis of atmospheric elements, ignoring the causes of those variations (such as the frequency of air mass movements). This type of classification, based on statistical parameters or physical characteristics, is called an **empirical classification.** A classification based on the causes, or *genesis,* of climate variation is known as a **genetic classification.**

Ordering the vast wealth of available climatic data into descriptions of major climatic groups, on either an empirical or a genetic basis, enables geographers to concentrate on the larger-scale causes of climatic differentiation. In addition, they can examine exceptions to the general relationships, the causes of which are often one or more of the other atmospheric controls. Finally, differentiating climates helps explain the distribution of other climate-related phenomena of importance to humans.

Despite its value, climate classification is not without its problems. Climate is a generalization about observed facts based on the averages and probabilities of weather. It does not describe a real weather situation; instead, it presents a composite weather picture. Within such a generalization, it is impossible to include the many variations that actually exist. On a global scale, generalizations, simplifications, and compromises are made to distinguish among climate types and regions.

## The Thornthwaite System

The first system for classifying climates concentrates on a more local scale. This system is very useful for soil scientists, water resources specialists, and agriculturalists. For example, for a farmer interested in growing a specific crop in a particular area, a system classifying large regions of Earth is inadequate. Identifying the major vegetation type of the region and the annual range of both temperature and precipitation does not provide a farmer with information concerning the amounts and timing of annual soil moisture surpluses or deficits. From an agricultural perspective, it is much more important to know that moisture will be available in the growing season, whether it comes directly in the form of precipitation or from the soil.

**FIGURE 8.1**
(a) This map shows the diversity of climates possible in a relatively small area, including portions of Chile, Argentina, Uruguay, and Brazil. The climates range from dry to wet and from hot to cold, with many possible combinations of temperature and moisture characteristics. (b) The Argentine Patagonian Steppe. (c) A meadow in the Argentine, Tierra del Fuego.

**What can you suggest as the causes for the major climate changes as you follow the 40° south latitude line from west to east across South America?**

Developed by an American climatologist, C. Warren Thornthwaite, the **Thornthwaite system** establishes moisture availability at the subregional scale (Fig. 8.2). It is the system preferred by those examining climates on a local scale. Development of detailed climate classification systems such as the Thornthwaite system became possible only after temperature and precipitation data were widely collected at numerous locations beginning in the latter half of the 19th century.

The Thornthwaite system is based on the concept of **potential evapotranspiration (potential ET),** which

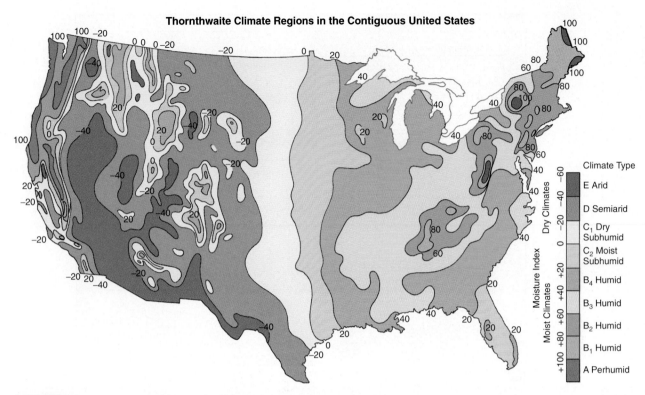

**FIGURE 8.2**
Thornthwaite climate regions in the contiguous United States are based on the relationship between precipitation (P) and potential ET. The moisture index (MI) for a region is determined by this simple equation:

$$MI = 100 \times \frac{P - \text{Potential ET}}{\text{Potential ET}}$$

Where precipitation exceeds potential ET, the index is positive; where potential ET exceeds precipitation, the index is negative.
**What are the moisture index and Thornthwaite climate type for coastal California?**

approximates the water use of plants with an unlimited water supply. (Evapotranspiration, discussed in Chapter 6, is a combination of evaporation and transpiration, or water loss through vegetation.) Potential ET is a theoretical value that increases with increasing temperature, winds, and length of daylight and decreases with increasing humidity. **Actual evapotranspiration (actual ET),** which reflects actual water use by plants, can be supplied during the dry season by soil moisture if the soil is saturated, the climate is relatively cool, and/or the day lengths are short. Thus, measurements of actual ET relative to potential ET and available soil moisture are the determining factors for most vegetation and crop growth.

The Thornthwaite system recognizes three climate zones based on potential ET values: low-latitude climates, with potential ET greater than 130 centimeters (51 in.); middle-latitude climates, with potential ET less than 130 but greater than 52.5 centimeters (20.5 in.); and high-latitude climates, with potential ET less than 52.5 centimeters. Climate zones may be subdivided based on how long and by how much actual ET is below potential ET. Moist climates have either a surplus or a minor deficit of less than 15 centimeters (6 in.). Dry climates have an annual deficit greater than 15 centimeters.

Thornthwaite's original equations for potential ET were based on analyses of data collected in the midwestern and eastern United States. The method was subsequently used with less success in other parts of the world. Over the past few decades, many attempts have been made to improve the accuracy of the Thornthwaite system for regions outside the United States.

## The Köppen System

The most widely used climate classification is based on temperature and precipitation patterns. It is referred to as the **Köppen system** after the German botanist and climatologist who developed it. Wladimir Köppen recognized that major vegetation associations reflect the area's climate. Hence, his climate regions were formulated to coincide with well-defined vegetation regions, and each climate region was described by the natural vegetation most often found there. Evidence of the strong influence of Köppen's system is seen in the wide usage of his climatic terminology, even in nonscientific literature (for example, steppe climate, tundra climate, rainforest climate).

### Advantages and Limitations of the Köppen System

Not only are temperature and precipitation two of the easiest weather elements to measure, but they are also measured more often and in more parts of the world than any other variables. By using temperature and precipitation statistics to define his boundaries, Köppen was able to develop precise definitions for each climate region, eliminating the imprecision that can develop in verbal and sometimes in genetic classifications.

Moreover, temperature and precipitation are the most important and effective weather elements. Variations caused by the atmospheric controls will show up most obviously in temperature and precipitation statistics. At the same time, temperature and precipitation are the weather elements that most directly affect humans, other animals, vegetation, soils, and the form of the landscape.

Köppen's climate boundaries were designed to define the vegetation regions. Thus, Köppen's climate boundaries reflect "vegetation lines." For example, the Köppen classification uses the 10°C (50°F) monthly isotherm because of its relevance to the timberline—the line beyond which it is too cold for trees to thrive. For this reason, Köppen defined the treeless polar climates as those areas where the mean temperature of the warmest month is below 10°C. Clearly, if climates are divided according to associated vegetation types and if the division is based on the atmospheric elements of temperature and precipitation, then the result will be a visible association of vegetation with climate types. The relationship with the visible world in Köppen's climate classification system is one of its most appealing features to geographers.

There are of course limitations to Köppen's system. For example, Köppen considered only average monthly temperature and precipitation in making his climate classifications. These two elements permit estimates of precipitation effectiveness but do not measure it with enough precision to permit comparison from one specific locality to another. In addition, for the purposes of generalization and simplification, Köppen ignored winds, cloud cover, intensity of precipitation, humidity, and daily temperature extremes—much, in fact, of what makes local weather and climate distinctive.

### Simplified Köppen Classification

The Köppen system, as modified by later climatologists, divides the world into six major climate categories. The first four are based on the annual range of temperatures: humid tropical climates (*A*), humid mesothermal (mild winter) climates (*C*), humid microthermal (severe winter) climates (*D*), and polar climates (*E*). Another category, the arid and semiarid climates (*BW* and *BS*), identifies regions that are characteristically dry based on both temperature and precipitation values. Because plants need more moisture to survive as the temperature increases, the arid and semiarid climates include regions where the temperatures range from cold to very hot. The final category, highland climates (*H*), identifies mountainous regions where vegetation and climate vary rapidly as a result of changes in elevation and exposure.

Within each of the first five major categories, individual climate types and subtypes are differentiated from one another by specific parameters of temperature and precipitation. (Table 1 in the "Graph Interpretation" exercise, pages 230–231, outlines the letter designations and procedures for determining the types and subtypes of the Köppen classification system.) This table can be used with any Köppen climate type presented in this chapter as well as Chapters 9 and 10.

### The Distribution of Climate Types

Five of the six major climate categories of the Köppen classification include enough differences in the ranges, total amounts, and seasonality of temperature and precipitation to produce the 13 distinctive climate types listed in Table 8.1. The tropical and arid climate types are discussed in some detail in the next chapter; the mesothermal, microthermal, and polar climates are presented in Chapter 10, along with a brief coverage of undifferentiated highland climates.

### Tropical (*A*) Climates

Near the equator we find high temperatures year-round because the noon sun is never far from 90° (directly overhead). Humid climates of this type with no winter season are Köppen's **tropical climates**. As his boundary for tropical climates, Köppen chose 18°C (64.4°F) for the average temperature of the coldest month because it closely coincides with the geographic limit of certain tropical palms.

Table 8.1 shows that there are three humid tropical climates, reflecting major differences in the amount and distribution of rainfall within the tropical regions. Tropical climates extend poleward to 30° latitude or higher in the continent's interior but to lower latitudes near the coasts because of the moderating influence of the oceans on coastal temperatures.

Regions near the equator are influenced by the intertropical convergence zone (ITCZ). However, the convergent and rising air of the ITCZ, which brings rain to the tropics, is not anchored in one place; instead, it follows the 90° sun angle (see again "The Analemma," Chapter 3), migrating with the seasons. Within 5°–10° latitude of the equator, rainfall occurs year-round because the ITCZ moves through

**TABLE 8.1**
**Simplified Köppen Climate Classes**

| Climates | Climograph Abbreviation |
|---|---|
| Humid Tropical Climates (*A*) | |
|    Tropical Rainforest Climate | Tropical Rf. |
|    Tropical Monsoon Climate | Tropical Mon. |
|    Tropical Savanna Climate | Tropical Sav. |
| Arid Climates (*B*) | |
|    Steppe Climate | Low-lat./Mid-lat. Steppe |
|    Desert Climate | Low-lat./Mid-lat. Desert |
| Humid Mesothermal (Mild Winter) | |
|    Climates (*C*) | |
|    Mediterranean Climate | Medit. |
|    Humid Subtropical Climate | Humid Subt. |
|    Marine West Coast Climate | Marine W.C. |
| Humid Microthermal (Severe Winter) | |
|    Climates (*D*) | |
|    Humid Continental, Hot Summer Climate | Humid Cont. H.S. |
|    Humid Continental, Mild Summer Climate | Humid Cont. M.S. |
|    Subarctic Climate | Subarctic |
| Polar Climates (*E*) | |
|    Tundra Climate | Tundra |
|    Ice-sheet Climate | Ice-sheet |
| Highland Climates (*H*) | |

twice a year and is never far away (Fig. 8.3). Poleward of this zone, precipitation becomes seasonal. When the ITCZ is over the region during the high-sun period (summer), there is adequate rainfall. However, during the low-sun period (winter), the subtropical highs and trade winds invade the area, bringing clear, dry weather.

We find the tropical rainforest climate in the equatorial region flanked both north and south by the dry-winter tropical savanna climate (Fig. 8.4). Finally, along coasts facing the strong, moisture-laden inflow of air associated with the summer monsoon, we find the tropical monsoon climate (Fig. 8.5). The atmospheric processes that produce the various tropical (*A*) climates are discussed in Chapter 9. Note the equatorial regions of Africa and South America shown in Figure 8.6.

**Polar (*E*) Climates** Just as the tropical climates lack winters (cold periods), the polar climates—at least statistically—lack summers. **Polar climates,** as defined by Köppen, are areas in which no month has an average temperature exceeding 10°C (50°F). Poleward of this temperature boundary, trees cannot survive. The 10°C isotherm for the warmest month more or less coincides with the Arctic Circle, poleward of which the sun does not rise above the horizon in midwinter and in summer strikes at a low angle.

The polar climates are subdivided into tundra and ice-sheet climates. Tundra climate occurs where at least one month averages above 0°C (32°F); in the ice-sheet climate, no

■ **FIGURE 8.3**
Tropical rainforest climate: island of Jamaica

**FIGURE 8.4**
Tropical savanna climate: East African high plains

**FIGURE 8.5**
Tropical monsoon climate: Himalayan foothills, West Bengal, India

month has an average temperature above 0°C (Fig. 8.7). Look at the far northern regions of Eurasia, North America, and Antarctica in Figure 8.6. The processes creating the polar (*E*) climates are explained in Chapter 10.

## Mesothermal (*C*) and Microthermal (*D*) Climates

Except where arid climates intervene, the lands between the tropical and polar climates are occupied by the transitional middle-latitude mesothermal and microthermal climates. As they are neither tropical nor polar, the mild and severe winter climates must have at least 1 month averaging below 18°C (64.4°F) and 1 month averaging above 10°C (50°F). Although both middle-latitude climate categories have distinct temperature seasons, the **microthermal climates** have severe winters with at least 1 month averaging below freezing. Once again, vegetation reflects the climatic differences. In the severe-winter climates, all broadleaf and even a species of needle-leaf trees defoliate naturally during the winter (generally, needle-leaf trees do not defoliate in winter) because soil water is temporarily frozen and unavailable. Much of the natural vegetation of the mild-winter **mesothermal climates** retains its foliage throughout the year because liquid water is always present in the soil. The line separating mild from severe winters usually lies in the vicinity of the 40th parallel.

A number of important internal differences within the mesothermal and microthermal climate groups produce individual climate types based on precipitation patterns or seasonal temperature contrasts. The Mediterranean, or dry summer, mesothermal (*C*) climate (like southern California or southern Spain in Figure 8.6) appears along west coasts between 30° and 40° latitude (Fig. 8.8). On the east coasts, in generally the same latitudes, the humid subtropical climate is found (Fig. 8.9). This type of climate is found in regions like the southeastern U.S. and southeastern China in Figure 8.6.

The distinction between the humid subtropical and marine west coast climates illustrates a second important criterion for the internal subdivisions of middle-latitude climates: seasonal contrasts. Both mesothermal climates have year-round precipitation, but humid subtropical summer temperatures are much higher than those in the marine west coast climate. Therefore, summers are hot. In contrast, the mild summers of the marine west coast climate, located poleward of the Mediterranean climate along continental west coasts, often extend beyond 60° latitude (Fig. 8.10). Some examples of marine west coast climates, shown in Figure 8.6, are along the northwest coast of the United States and extending into Canada, the west coast of Europe, and the British Isles.

Another example of internal differences is found among microthermal (*D*) climates, which usually receive year-round precipitation associated with storms traveling along the polar front. Internal subdivision into climate types is based on summers that become shorter and cooler and winters that become longer and more severe with increasing latitude and continentality. Microthermal climates are found exclusively in the Northern Hemisphere (Fig. 8.11) because there is no land in the Southern Hemisphere latitudes that would normally be occupied by these climate types. In the Northern Hemisphere, these climates progress poleward through the humid continental, hot-summer climate, to the humid continental, mild-summer climate, and finally, to the subarctic climate (note the eastern United States and Canada or northward through eastern Europe in Figure 8.6).

## Arid (*B*) Climates
Climates that are dominated by year-round moisture deficiency are called **arid climates.** These climates will penetrate deep into the continent, interrupting the latitudinal zonation of climates that would otherwise exist. The definition of climatic aridity is that precipi-

■ **FIGURE 8.7**
Polar ice climate: glacier in the Alaska Mountain range

■ **FIGURE 8.8**
Mediterranean mesothermal climate: village in southern Spain

■ **FIGURE 8.9**
Humid subtropical climate: grapefruit grove in central Florida

■ **FIGURE 8.10**
Marine west coast climate: North Sea coast of Scotland

■ **FIGURE 8.11**
Microthermal, severe winter climate: winter in Illinois

tation received is less than potential ET. Aridity does not depend solely on the amount of precipitation received; potential ET rates and temperature must also be taken into account. In a low-latitude climate with relatively high temperatures, the potential ET rate is greater than in a colder, higher-latitude climate. As a result, more rain must fall in the lower latitudes to produce the same effects (on vegetation) that smaller amounts of precipitation produce in areas with lower temperatures and, consequently, lower potential ET rates.

Arid climates are concentrated in a zone from about 15°N and S to about 30°N and S latitude along the western coasts, expanding much farther poleward over the heart of each landmass. The correspondence between the arid climates and the belt of subtropical high pressure systems is quite unmistakable (like in the southwestern United States, central Australia, and north Africa in Figure 8.6), and the poleward expansion is a consequence of remoteness from the oceanic moisture supply.

In desert (*BW*) climates, the annual amount of precipitation is less than half the annual potential ET (■ Fig. 8.12). Bordering the deserts are steppe (*BS*) climates—semiarid climates that are transitional between the extreme aridity of the deserts and the moisture surplus of the humid climates (■ Fig. 8.13). The definition of the

**FIGURE 8.12**
Desert climate: Sonoran Desert of Arizona

**FIGURE 8.14**
Highland climate: Teton Mountain Range, Wyoming

**FIGURE 8.13**
Steppe climate: Sand Hills of Nebraska

steppe climate is an area where annual precipitation is less than potential ET but more than half the potential ET. *B* climates and the processes that create them are discussed in more detail in Chapter 9.

**Highland (*H*) Climates** The pattern of climates and extent of aridity are affected by irregularities in Earth's surface, such as the presence of deep gulfs, interior seas, or significant highlands. The climatic patterns of Europe and North America are quite different because of such variations.

Highlands can channel air mass movements and create abrupt climatic divides. Their own microclimates form an intricate pattern related to elevation, cloud cover, and exposure (❙ Fig. 8.14). One significant effect of highlands aligned at right angles to the prevailing wind direction is the creation of arid regions extending tens to hundreds of kilometers leeward. Look at the mountain ranges in Figure 8.6: the Rockies, the Andes, the Alps, and the Himalaya show *H* climates. These undifferentiated **highland climates** are discussed in more detail in Chapter 10.

## Climate Regions

Each of our modified Köppen climate types is defined by specific parameters for monthly averages of temperature and precipitation; thus, it is possible to draw boundaries between these types on a world map. The areas within these boundaries are examples of one type of world region. The term **region,** as used by geographers, refers to an area that has recognizably similar internal characteristics that are distinct from those of other areas. A region may be described on any basis that unifies it and differentiates it from others.

As we examine the climate regions of the world in the chapters that follow, you should make frequent reference to the map of world climate regions (Fig. 8.6). It shows the patterns of Earth's climates as they are distributed over each continent. However, a word of caution is in order. On a map of climate regions, distinct lines separate one region from another. Obviously, the lines do not mark points where there are abrupt changes in temperature or precipitation conditions. Rather, the lines signify **zones of transition** between different climate regions. Furthermore, these zones or boundaries between regions are based on monthly and annual averages and may shift as temperature and moisture statistics change over the years.

The actual transition from one climate region to another is gradual, except in cases in which the change is brought about by an unusual climate control such as a mountain barrier. It would be more accurate to depict climate regions and their zones of transition on a map by showing one color fading into another. Always keep in mind, as we describe Earth's climates, that it is the core areas of the regions that best exhibit the characteristics that distinguish one climate from another.

Now, let's look more closely at Figure 8.6. One thing that is immediately noticeable is the change in climate with latitude. This is especially apparent in North America when we examine the East Coast of the United States moving north into Canada. We can also see that similar climates usually appear in similar latitudes and/or in similar locations with respect to landmasses, ocean currents, or topography. These climate patterns emphasize the close relationship among climate, the weather elements, and the climate controls. There is an order to Earth's atmospheric conditions and so also to its climate regions.

A striking variation in these global climate patterns becomes apparent when we compare the Northern and Southern Hemispheres. The Southern Hemisphere lacks the large landmasses of the Northern Hemisphere; thus, no climates in the higher latitudes (in land regions) can be classified as humid microthermal, and only one small peninsula of Antarctica can be said to have a tundra climate.

## Climographs

It is possible to summarize the nature of the climate at any point on Earth in graph form, as shown in Figure 8.15. Given information on mean monthly temperature and rainfall, we can express the nature of the changes in these two elements throughout the year simply by plotting their values as points above or below (in the case of temperature) a zero line. To make the pattern of the monthly temperature changes clearer, we can connect the monthly values with a continuous line, producing an annual temperature curve. To avoid confusion, monthly precipitation amounts are usually shown as bars reaching to various heights above the line of zero precipitation. Such a display of a location's climate is called a **climograph**. To read the graph, one must relate the temperature curve to the values given along the left side and the precipitation amounts to the scale on the right. Other information may also be displayed, depending on the type of climograph used. Figure 8.15 represents the type that we use in Chapters 9 and 10. This climograph can be used to determine the Köppen classification of the station as well as to show its specific temperature and rainfall regimes. The climate-type abbreviations relating to all climographs are found in Table 8.1.

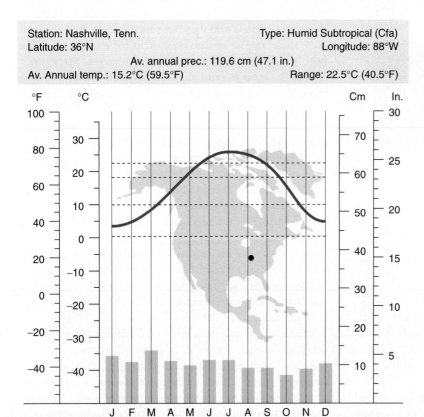

■ **FIGURE 8.15**
A standard climograph showing average monthly temperature (curve) and rainfall (bars). The horizontal index lines at 0°C (32°F), 10°C (50°F), 18°C (64.4°F), and 22°C (71.6°F) are the Köppen temperature parameters by which the station is classified.
**What specific information can you read from the graph that identifies Nashville as a specific climate type (humid subtropical) in the Köppen classification?**

## Scale and Climate

Climate can be measured at different scales (macro, meso, or micro). The climate of a large (macro) region, such as the Sahara, may be described correctly as hot and dry. Climate can also be described at mesoscale levels; for example, the climate of coastal southern California is sunny and warm, with dry summers and wet winters. Finally, climate can be described at local scales, such as on the slopes of a single hill. This is termed a **microclimate**.

At the microclimate level, many factors will cause the climate to differ from nearby areas. For example, in the United States and other regions north of the Tropic of Cancer, south-facing slopes tend to be warmer and drier than north-facing slopes because they receive more sunlight ( ■ Fig. 8.16). This variable is referred to as *slope aspect*—the direction a mountain slope faces in respect to the sun's rays. Microclimatic differences such as slope aspect can cause significant differences in vegetation and soil moisture. In what is sometimes called *topoclimates,* tall mountains often possess vertical zones of vegetation that reflect changes in the microclimates as one ascends from the base of the mountain (which may be surrounded by a tropical-type vegetation) to higher slopes with middle latitude–type vegetation to the summit covered with ice and snow.

Human activities can influence microclimates as well. Recent research indicates that

**FIGURE 8.16**
This aerial photograph, facing eastward over valleys in the Coast Ranges of California, illustrates the significance of slope aspect. South-facing slopes on the left sides of valleys receive direct rays of the sun, and are hotter and drier than the more shaded north-facing slopes. The south-facing slopes support grasses and only a few trees, while the shaded, north-facing slopes are tree covered.
**Why do the differing angles that the sun's rays strike the two opposite slopes affect temperatures?**

the construction of a large reservoir leads to greater annual precipitation immediately downwind of this impounded water. This occurs because the lake supplies additional water vapor to passing storms, which intensifies the rainfall or snows immediately downwind of the lake. These microclimatic effects are similar to the lake-effect snows that occur downwind of the Great Lakes in the early winter when the lakes are not frozen (discussed in Chapter 7). Another example of human impact on microclimates is the urban heat-island effect (discussed in Chapter 4), which leads to changes in temperature (urban centers tend to be warmer than their outlying rural areas), rainfall, wind speeds, and many other phenomena.

# Climates of the Past

To try to predict future climates, it is critical to understand the magnitude and frequency of previous climate changes. Knowledge that Earth experienced major climate changes in the past is not new. In 1837, Louis Agassiz, a European naturalist, proposed that Earth had experienced major periods of *glaciation,* periods known as ice ages, when large areas of the continents were covered by huge sheets of ice. He presented evidence that glaciers (flowing ice) had once covered most of England, northern Europe, and Asia, as well as the foothill regions of the Alps. Agassiz arrived in the United States in 1846 and found similar evidence of widespread glaciation throughout North America.

## The Ice Ages

Until the 1960s, it was widely believed that Earth had experienced four major glacial advances followed by warmer interglacial periods. These glacial cycles occurred during the geologic epoch known as the *Pleistocene* (from 1.6 million years up to 10,000 years ago). In Europe, these glacial epochs were termed the Günz (oldest), Mindel, Riss, and Würm. Likewise in North America, evidence of four glacial periods was recognized; these were termed the Nebraskan (oldest), Kansan, Illinoian, and Wisconsinan glaciations ( Fig 8.17).

A major problem with studying the advance and retreat of glaciers on land is that each subsequent advance of the glaciers tends to destroy, bury, or greatly disrupt the sedimentary evidence of the previous glacial period. The evidence of the fourfold record of glacial advances was largely recognized on the basis of minor glacial deposits lying beyond the limit of the more recent glaciations. Evidence of "average" glacial advances that were subsequently overridden by more recent glaciers was rarely recognized.

Before the advent of radiometric-dating techniques (mineral and organic material can be dated by measuring the extent to which radioactive elements in the material have decayed through time), the timing of the glacial advances in both Europe and the United States was only crudely known. For example, estimates of the age of the last interglacial period were based on the rates at which Niagara Falls had eroded headward after the areas were first exposed when the glaciers retreated. Calculations ranging from 8000 to 30,000 years ago were produced.

## Modern Research

Two major advances in scientific knowledge about climate change occurred in the 1950s. First, radiometric techniques, such as radiocarbon dating, that measured the absolute ages of landforms produced by the glaciers began to be widely used. Radiocarbon dating conclusively showed that the last ice advance peaked a mere 18,000 years ago. This ice sheet covered essentially all of Canada and the United States, extending down to the Ohio and Missouri Rivers and covering modern-day city sites such as Boston, New York, Indianapolis, and Des Moines.

**FIGURE 8.17**
This map identifies the extensive areas of Canada and the northern United States that were covered by moving sheets of ice as recently as 18,000 years ago.
**Why does the ice move in various directions in different regions of the continent?**

The second major discovery was that evidence of detailed climate changes has been recorded in the sediments on the ocean floors. Unlike the continental record, the deep-sea sedimentary record had not been disrupted by subsequent glacial advances. Rather, the slow, continuous sediment record provides a complete history of climate changes during the past several million years. The most important discovery of the deep-sea record is that Earth has experienced numerous major glacial advances during the Pleistocene, not just the four that had been identified previously. Today, the names of only two of the North American glacial periods, the Illinoian and Wisconsinan, have been retained.

Because the deep-sea sedimentary record is so important to climate-change studies, it is important to understand how the record is deciphered. The deep-sea mud contains the microscopic record of innumerable surface-dwelling marine animals that built tiny shells for protection. When they died, these tiny shells sank to the seafloor, forming the layers of mud. Different species thrive in different surface-water temperatures; therefore, the stratigraphic record of the tiny fossils produces a detailed history of water-temperature fluctuations.

These tiny seashells are composed of calcium carbonate ($CaCO_3$); therefore, the analyses also record the oxygen composition of the seawater in which they were formed. One common measurement technique for determining oxygen composition is known as **oxygen-isotope analysis.** Two of the most common isotopes of oxygen are $O_{16}$ and the heavier $O_{18}$. Modern seawater has a fixed ratio of the two oxygen isotopes. The $O_{18}/O_{16}$ ratios will indicate changes in ocean temperatures relating to glacial cycles. A review of the oxygen-isotope record indicates that the last glacial advance about 18,000 years ago was only one of many major glacial advances during the past 2.4 million years.

Today, climatologists are aware that the present climate is but a short interval of relative stability in a time of major climate shifts. Moreover, the modern climate epoch, known as the *Holocene* (10,000 years ago to the present), is a time of extraordinarily stable, warm temperatures compared to most of the last 2.4 million years ( Fig. 8.18). Based on the deep-sea record, it appears that global climates tend to rest at one of two extremes: a very cold interval characterized by major glaciers and lower sea levels and shorter intervals between the glacial advances marked by unusually warm temperatures and high sea levels.

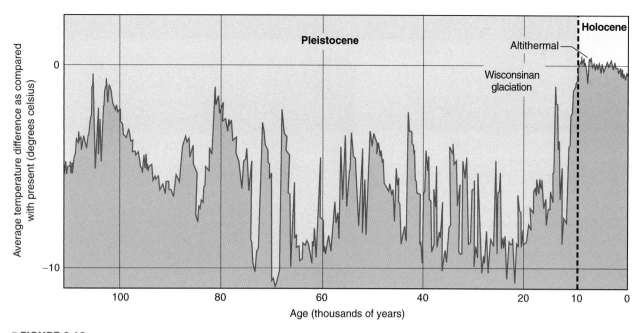

**FIGURE 8.18**
Analyses of oxygen-isotope ratios in ice cores taken from the glacial ice of Antarctica and Greenland provide evidence of surprising shifts of climate over short periods of time.
**Has the general trend of temperatures on Earth been warmer or colder during the Holocene?**

With the realization that global climates have changed dramatically numerous times, two obvious questions arise: How quickly does global climate change from one extreme to the other, and what causes global climate to change so often?

## Rates of Climate Change

Through most of Canada and into the United States, glaciers covered most areas north of the Missouri and Ohio Rivers 18,000 years ago. In the west, freshwater lakes more than 500 feet deep covered much of Utah and Nevada. The United States was mostly glacier free, and the western lake basins were dry, by about 9000 years ago. Abundant evidence has even been found that the climate about 7000 years ago (a time known as the **Altithermal**) was hotter than today (see again Fig. 8.18).

For glaciers several thousand feet thick to melt completely and for deep lakes to evaporate, a substantial increase in insolation is required over a few thousand years. Where did so much extra energy come from?

To answer questions about such rapid rates of climate change requires a more detailed record of climate than the deep-sea sediments can provide. This is because the deep-sea sedimentary record is extraordinarily slow—a few centimeters of sea mud accumulates in a thousand years. Rapid shifts in climate during periods of a few hundred years are not recorded clearly in the seafloor sediments. This problem has been solved by coring the thick glaciers covering Antarctica and Greenland. Glacial ice records yearly amounts of snowfall and is much more likely to provide short-term evidence of climate changes. Analyses of oxygen isotopes in the glacial ice of Antarctica and, most recently, Greenland have revealed a detailed record of climate changes during the past 250,000 years (see again Figure 8.18).

A surprising discovery of the ice-sheet analyses is the speed at which climate changes. Rather than changing gradually from glacial to interglacial conditions over thousands of years, the ice record indicates that the shifts can occur in a few years or decades. Thus, whatever is most responsible for major climate changes can develop rapidly. This probably requires a *positive feedback system,* which means, as explained in Chapter 1, that a change in one variable will cause changes in other variables that magnify the amount of original change. For example, most glaciers have high albedos, reflecting significant amounts of sunlight back to space. However, if the ice sheets retreat for whatever reason, low-albedo land begins to absorb more insolation, increasing the amount of energy available to melt the ice. Thus, the more ice that melts, the more energy is available to melt the ice further, magnifying the initial glacial retreat.

In contrast, a *negative feedback system,* where changes in one of the variables induce the system to remain stable, also affects the likelihood or rate of climate change. For example, increasing global temperatures cause evaporation rates to increase.

## Geography's Physical Science Perspective
### Determining Past Climates

In trying to understand oxygen-isotope analysis as a means of reconstructing paleoclimates (ancient climates), it is helpful to review some basic definitions in physics. *Isotopes* are defined as atoms with the same atomic number but different atomic mass. The atomic number is equal to the positive charge of the nucleus—essentially, the number of *protons* in the nucleus. The atomic mass (or atomic weight) is equal to the number of *protons* and *neutrons* that comprise the nucleus of the atom. *Electrons,* which orbit the atom, are negatively charged particles that possess no appreciable mass (or weight). In a neutral atom, the number of electrons should equal the number of protons. When an electron is gained or lost to the atom, then a net (−) or (+) charge, respectively, will result, and the particle is then classified as an *ion.*

When dealing with isotopes, the atomic number (proton number) must remain the same (giving the atom its identity in the Periodic Table of Elements), but the number of neutrons can vary. In the example of oxygen isotopes, the atomic number is always 8. This is necessary to identify the atom as oxygen. However, the atom may contain 8 neutrons ($O^{16}$, the lighter isotope) or 10 neutrons ($O^{18}$, the heavier isotope). In oxygen-isotope analysis, the ratio of $O^{18}$ to $O^{16}$ is measured and compared to normal values. We have already discussed the $O^{18}/O^{16}$ ratios of seafloor ($CaCO_3$) sediment; however, when dealing with yearly layers within Greenland and Antarctic ice cores, we must use them in a different way. When water evaporates from the ocean, slightly more of the $O^{16}$ than $O^{18}$ evaporates because water containing the lighter-weight oxygen evaporates more readily. During an ice age, the evaporated water is stored in the form of glacial ice rather than returned to the oceans and the $O^{18}/O^{16}$ ratio in the ocean changes slightly to reflect the $O^{16}$-enriched water being stored in the glaciers. In this way $O^{18}/O^{16}$ ratios can help reconstruct climates of the past from glacial ice layers.

Through time, paleoclimatologists have discovered new and different ways to

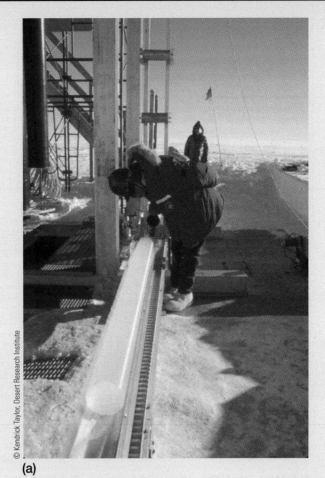

(a) Paleoclimates can be reconstructed using $O^{18}/O^{16}$ ratios from layers found in ice cores. Other methods include the identification and characterization of (b) pollen samples and (c) tree rings.

The more water that evaporates from the ocean surface, the more clouds will form. The more clouds that exist, the more insolation is reflected back to space, cooling Earth's surface. (A counterargument to this effect is that clouds also operate as a greenhouse blanket, trapping heat in the lower atmosphere.) Thus, for climate changes to occur rapidly, negative feedback cycles such as this one must be overwhelmed by positive feedback cycles.

## Causes of Climate Change

Although theories about the causes of climate change are numerous, they can be organized into four broad categories: (1) astronomical variations in Earth's orbit; (2) changes in atmospheric composition; (3) changes in oceanic circulation; and (4) changes in landmasses that affect albedo and oceanic circulation.

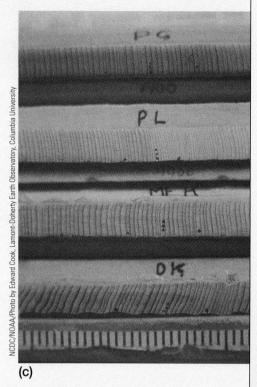

determine climates of the past. The oxygen-isotope analysis is one of the most widely accepted methods, but there are other well-established methods as well. Two are worth a brief discussion; they are dendrochronology and palynology.

*Dendrochronology* (or tree-ring dating) has been used for decades. This analysis calls for the examination of tree rings exposed by cores taken through the middle of certain species of trees. The core (small enough so as not to harm the tree) will reveal each yearly tree ring. The rings are counted back through time to supply a time scale for the analysis. Each ring, by its thickness, color, and texture, can reveal the climate conditions (temperature and precipitation characteristics) during that particular year of the tree's growth. Thus, a short-term climate record can be determined through careful examination of these wooden rings.

*Palynology* (or pollen-analysis dating) is also a well-established way of reconstructing past climates. Though pollen samples can be recovered from a variety of environments, organic bogs are the best place to conduct this analysis. A core is removed from a bog showing the layers of organic material all the way to the bottom layer. Then each organic bog layer is radiocarbon dated to identify its age. Then, all the pollen is removed from each layer of organic material and analyzed to identify its tree or plant type. Thus, pollen is used to identify the numbers, types, and relative distribution of the trees and plants from which the pollen came. It is then left to the paleoclimatologist to determine what type of climate would be required to sustain a forest at the time of the one described by the analysis of each bog layer.

## Orbital Variations

Astronomers have detected slow changes in Earth's orbit that affect the distance between the sun and Earth as well as the deviation of Earth's axis on the plane of the ecliptic. These orbital cycles produce regular changes in the amount of solar energy that reaches Earth. The longest is known as the **eccentricity cycle,** which is a 100,000-year variation in the shape of Earth's orbit around the sun. In simple terms, Earth's orbit changes from an ellipse (oval), to a more circular orbit, and then back, affecting Earth–sun distance. A second cycle, termed the **obliquity cycle,** represents a 41,000-year variation in the tilt of Earth's axis from a maximum 24.5° to a minimum of 22.0° and then back. The more Earth is tilted, the greater is the seasonality at middle and high latitudes. Finally, a **precession cycle** has been recognized with a periodicity of 21,000 years. The precession cycle determines the time of year that perihelion occurs. Today, Earth is closest to the sun on January 3 and, as a result, receives about 3.5% greater insolation than the average in January. When aphelion occurs on January 3 in about 10,500 years, the Northern Hemisphere winters should be somewhat colder ( Fig. 8.19).

These cycles operate collectively, and the combined effect of the three cycles can be calculated. The first person to examine all three of these cycles in detail was the mathematician Milutin Milankovitch, who completed the complex mathematical calculations and showed how these changes in Earth's orbit would affect insolation. Milankovitch's calculations indicated that numerous glacial cycles should occur during 1 million-year intervals.

By the 1980s, most paleoclimate (ancient climate) scientists were convinced that an unusually good correlation existed between the deep-sea record and Milankovitch's predictions. This suggests that the primary driving force behind glacial cycles is regular orbital variations, and it indicates that long-term climate cycles are entirely predictable! Unfortunately for humans, the Milankovitch theory indicates that the warm Holocene interglacial will soon end and that Earth is destined to experience full glacial conditions (glacial ice possibly as far south as the Ohio and Missouri Rivers) in about 20,000 years.

## Changes in Earth's Atmosphere

Many theories attribute climate changes to variations in atmospheric dust levels. The primary villain is volcanic activity, which pumps enormous quantities of ash into the stratosphere, where strong winds spread it around the world. The dust reduces the amount of insolation reaching Earth's surface for periods of 1–3 years ( Figs. 8.20 and 8.21).

**Volcanic Activity** The climatic cooling effect of volcanic activity is unquestioned: All of the coldest years on record over the past two centuries have occurred in the year following a major eruption. Following the massive eruption of Tambora (in Indonesia) in 1815, 1816 was known as "the year without a summer." Killing frosts in July ruined crops in New England and Europe, resulting in famines. Several decades later, following the massive eruption of Krakatoa (also in Indonesia) in 1883, temperatures decreased significantly during 1884. Although no 20th-century eruptions have approached the magnitude of these two, the 1991 eruption of Mt. Pinatubo (in the Philippines) produced a substantial respite of cool conditions in an otherwise continuous series of record warm years ( Fig. 8.22).

**Atmospheric Gases** Another phenomenon closely correlated with average global temperatures is the composition of atmospheric gases. Scientists have known for many years

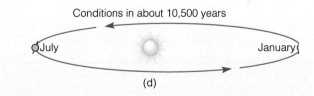

**PHYSICAL Geography Now™** ■ **ACTIVE FIGURE 8.19**
Watch this Active Figure at http://now.brookscole.com/gabler8.
Milankovitch calculated the periodicities for (a) eccentricity, (b) obliquity, and (c) precession.
**What effect should these changes in receipt of insolation have on global climates?**

that carbon dioxide ($CO_2$) acts as a "greenhouse gas." There is no question that $CO_2$ is transparent to incoming shortwave radiation and blocks outgoing longwave radiation, similar to the effect of the glass panes in a greenhouse or in your automobile on a sunny day (refer again to the greenhouse

**FIGURE 8.20**
Volcanic activity at Mt. St. Helens pumps various gases and particulates into the atmosphere.
**Besides affecting the climate, what other hazards result from volcanic explosions?**

discussion in Chapter 4). Thus, as the atmospheric content of greenhouse gases rises, so will the amount of heat trapped in the lower atmosphere.

Captured in the glacial ice of Antarctica and Greenland are air bubbles containing minor samples of the atmosphere that existed at the time that the ice formed. One of the important discoveries of the ice-core projects is that prehistoric atmospheric $CO_2$ levels increased during interglacial periods and decreased during major glacial advances.

The fact that average global temperatures and $CO_2$ levels are so closely correlated suggests that Earth will experience record warmth as the atmospheric level of $CO_2$ increases. The present level of approximately 370 parts per million of $CO_2$ is already higher than at any time in the past million years.

Carbon dioxide is not the only greenhouse gas. Molecule for molecule, methane ($CH_4$) is more than 20 times more effective than $CO_2$ as a greenhouse gas but is considered less important because the atmospheric concentrations and the length of time the molecules of gas remain in the atmosphere (residence time) is much smaller. Garbage dump emissions and termite mounds both produce substantial quantities of $CH_4$. But a much more important source of atmospheric methane may come from the tundra regions or the deep sea. If warming the tundra or ocean water indeed releases large amounts of methane as is theorized, the resulting positive feedback cycle of warming could be enormous.

Other greenhouse gases include CFCs (chlorofluorocarbons) and $N_2O$ (nitrous oxide). The relative greenhouse contribution of common greenhouse gases and their average residence times in the atmosphere are presented in Figure 8.23.

## Changes in the Ocean

Oceans cover over 70% of Earth's surface. Their enormous volume and high heat capacity make the oceans the single largest buffer against changes in Earth's climate. Whenever changes occur in oceanic temperatures, chemistry, or circulation, significant changes in global climate are certain to follow.

Surface oceanic currents are driven mostly by winds. However, a much slower circulation deep below the surface moves large volumes of water between the oceans. A major driving force of the deep circulation appears to be differences in water buoyancy caused by differences in salinity (salt content). Where surface evaporation is rapid, the rising salinity content causes the seawater density to increase, inducing subsidence. On the other hand, when major influxes of freshwater flow from adjacent continents or concentrations of melting icebergs flood into the oceans, the salinity is reduced, thereby increasing the buoyancy of the water. When the surface water is buoyant, deep-water circulation slows. In many cases, the freshwater influx is immediately followed by a major flow of warm surface waters into the North Atlantic, causing an abrupt warming of the Northern Hemisphere. Subsurface ocean currents are also affected by water temperature. Extremely cold Arctic and Antarctic waters are quite dense and tend to subside, whereas tropical water is warmer and may tend to rise. Therefore, salinity and temperature taken together bring about rather complex subsurface flows deep within our ocean basins.

In modern times, short-term changes in Pacific circulation are primarily responsible for El Niño and La Niña events (discussed in Chapter 5). The onset of El Niño/Southern Oscillation (ENSO) climatic events is both rapid and global in extent, and it is widely believed that changes in oceanic circulation may be responsible for similar rapid climate changes during the last 2.4 million years.

## Changes in Landmasses

The final category of climate change theories involves changes in Earth's surface to explain lengthy periods of cold climates. Ice ages, some with more than 125 glacial advances,

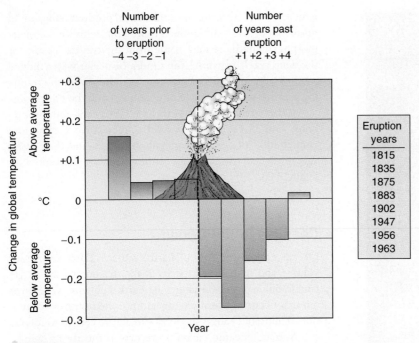

**FIGURE 8.21**
Examination of global temperatures within 4 years before and after major volcanic eruptions provides compelling evidence that volcanic activity can have a direct effect upon the amounts of insolation reaching Earth's surface.
**At what period after an eruption year does the effect seem the greatest?**

occurred during Earth's history. To explain some of the previous glacial periods, scientists have proposed several factors that might be responsible. For example, one characteristic that all of these glacial periods have in common with the Pleistocene is the presence of a continent in polar latitudes. Polar continents permit glaciers to accumulate on land, which results in lowered sea levels and consequent global effects.

Another geologic factor sometimes invoked as a cause of climate change is the formation, disappearance, or movement of a landmass that restricts oceanic or atmospheric circulation. For example, eruptions of volcanoes and the formation of the Isthmus of Panama severed the connection between the Atlantic and Pacific, thereby closing a pathway of significant ocean circulation. This redirection of ocean water created the Gulf Stream Current/North Atlantic Drift (see again Figure 5.23). Another example is the uplift of the Himalayas, altering atmospheric flows, and monsoonal effects in Asia. Both of these events, and several other significant

**FIGURE 8.22**
This graph shows the gradual warming trend in global temperatures since 1880. It also documents the sharp reversal of the trend and the cooling of temperatures after the eruptions of Krakatoa in 1883 and Mt. Pinatubo in 1991.
**Is volcanic activity responsible for all of these temperature reversals?**

## Geography's Spatial Science Perspective
### Climate Change and Its Impact on Coastlines

When we look at a map or a globe, one of the pieces of geographic information that we see is so obvious and basic that we often take it for granted, perhaps failing to recognize that it is spatial information. This is the location of coastlines—the boundaries between land and ocean regions. One of the most basic aspects of our planet that a world map shows us is where landmasses exist and where the oceans are located, as well as the generally familiar shape of these major Earth features. But maps of our planet today only show where the coastline is currently located. We know that sea level has changed over time and that it rose 20–30 centimeters (8–12 in.) during the 20th century. The hydrologic system on Earth is a closed system because the total amount of water (as a gas, liquid, and solid) on our planet is fixed. When the climate supports more glacial ice, sea level falls. When world climates experience a warming tendency, sea level rises. More ice in glaciers means less water in the oceans, and vice versa. If global warming trends continue at their present rate, the U.S. Environmental Protection Agency predicts that sea level will rise 31 centimeters (1 ft) in the next 25–50 years. That amount of sea-level rise will cause problems for low-lying coastal areas; the populations of some coral islands in the Pacific are already concerned, as their homelands are barely above the high-tide level.

Looking at a map of world population distribution and comparing it to a world physical map shows a strong link between settlement density and coastal areas. For low-lying coastal regions, sea-level rise is a major concern, and the more gentle the slope of the coast is, the farther inland the inundation would be with every increment of sea-level rise. Scientists at the United States Geological Survey have determined that if all the glaciers on Earth were to melt, sea level would rise 80 meters (263 ft) and that a 10-meter (33-ft) rise would displace 25% of the United States population.

Before the ice ages of the Pleistocene, worldwide climates were generally warmer; through much of Earth's history, no glaciers have existed on the planet. During times like those, when Earth was ice free, sea level would have been at a maximum, and that might occur again in the distant future under similar climatic conditions. At the time when glaciers were most extensive, during the maximum advance of Pleistocene glaciers, sea level fell to about 100 meters (330 ft) below today's level. Maps that create the positions of coastlines and the shape of continents during times of major environmental change show how temporary and vulnerable coastal areas can be.

The accompanying maps below show the pre-Pleistocene maximum rise of sea level (light green), the Pleistocene drop in sea level (light blue), and the impact on North American coastlines. It seems odd to think of the world maps as temporary, but they only show us where the coast is located today. Like other natural boundaries, coasts can shift over time.

These maps show how changes in sea level would affect the coastline. Light green shows the coastline if all glaciers on Earth were to melt. Light blue shows the coastline if glaciers expanded to the level of maximum extent during the Pleistocene.

---

changes, immediately predate the onset of the modern series of glaciations. Which ones caused climate changes and which changes are simply coincidences is yet to be determined.

A final group of theories involve changes in albedo, caused either by major snow accumulations on high-latitude landmasses or by large oceanic ice sheets drifting into lower latitudes. The increased reflection of sunlight starts a positive feedback cycle of cooling that may end when the polar oceans freeze, shutting off the primary moisture source for the polar ice sheets.

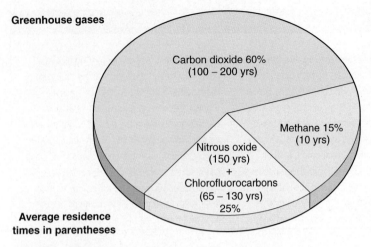

**FIGURE 8.23**
Gases, other than carbon dioxide, released to the atmosphere by human activity contribute approximately 40% to the greenhouse effect. The figures in parentheses indicate the average number of years that the different gases remain in the atmosphere and contribute to temperature change.
**Which gas has the longest residence time?**

## Future Climates

With so many variables potentially responsible for climate change, reliably predicting future climate is a very difficult proposition at best. The primary difficulty in climate prediction is posed by natural variability. Figure 8.24 displays the frequency and magnitudes of climate changes that have occurred naturally over the past 150,000 years. Although the Holocene has been the most stable interval of the whole period, a detailed examination of the Holocene record reveals a wide range of climates. For example, a long interval of climates, hotter than today's, occurred during the Altithermal. This interval was characterized by the dominance of grasslands in the Sahara and severe droughts on the Great Plains. Other warm intervals occurred during the Bronze Age, during the second half of the Roman Empire, and in medieval times. An unusually cold interval began with the eruption of Santorini (the site of a civilization that some believe was the basis for the Atlantis myth) in the Aegean Sea. Other cold periods occurred during the Dark Ages and again beginning about 1197 in the North Atlantic and 1309 in continental Europe. This last episode has been termed the **Little Ice Age.** The Little Ice Age had major impacts on civilizations—from the isolation of the Greenland settlements established during the medieval warm period to the abandonment of the Colorado Plateau region by the Anasazi cultures. An important point to remember is that, with the exception of the cold interval that began with the eruption of Santorini, climatologists do not know what variables changed to cause each of these major climate fluctuations.

Attempts to predict future climates are complicated further by the operation of many feedback cycles. Simply increasing the amount of heat that is trapped by atmospheric gases in the lower atmosphere may or may not result in long-term warming. Negative feedback processes such as increased cloud formation and increased plant uptake of $CO_2$ may operate to counteract the warming. On the other hand, warming of the oceans and tundra may release additional greenhouse gases, setting into motion some significant positive feedback cycles. Which feedback mechanisms will dominate is not certain; therefore, all predictions must be tentative.

There have been numerous attempts to simulate the variables affecting climate. *General circulation models* (GCMs) are complex computer simulations based on the relationships among variables discussed throughout this book: sun angles, temperature, evaporation rates, land versus water effects, energy transfers, and so on. Although the complexity and usefulness of GCMs is increasing rapidly, so far none of the models has accurately predicted most previous climate changes, and their ability to predict future climate changes is therefore still suspect. Nonetheless, GCMs appear to do a good job in predicting how conditions will change in specific regions as the globe warms or cools, and they have added new insights into how some climatic variables interact.

Based on the record of climate changes during the past, only one thing can be concluded about future climates: They will change. Looking into the distant future, the Milankovitch cycles indicate that another glacial cycle is probably on the way. The most rapid cooling should occur between 3000 and 7000 years from now. In the near term, however, global warming is most likely. The rise of atmospheric $CO_2$, the widespread destruction of vegetation, and the feedback cycles that will most likely result are bound to increase the average global temperature for the foreseeable future. An average increase of 1°C (nearly 2°F) would be equivalent to the change that has occurred since the end of the Little Ice Age in about 1850. A 2°C warming would be greater than anything that has happened in the Holocene, including the Altithermal. A 3°C warming would exceed anything that has happened in the past million years. Current estimates and the most reliable GCMs predict a 1°C to 3.5°C (2° to 6°F) warming in the 21st century.

It is clear that not all areas will be affected equally. One of the most important effects is expected to be a more vigorous hydrologic cycle, fueled largely by increases in evaporation from the ocean. Intense rainfalls will be more likely in many regions, as will droughts in other regions such as the Great Plains. Temperatures will rise most in the polar regions, mainly during the winter months. As a result of the warming, sea levels will rise, mostly because of the thermal expansion of ocean water and melting ice sheets. By 2100, sea levels should be between 15 and 95 centimeters (0.5–3.1 ft) higher than today. In addition, the ranges of

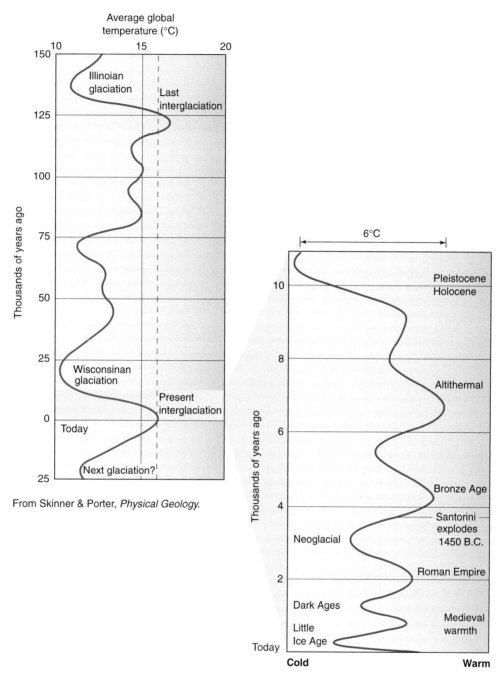

**FIGURE 8.24**

This figure shows broad climate trends of the past 150,000 years, with significant details for the Holocene. Climatologists have been remarkably successful in dating recent climate change, but predicting future climates remains extremely difficult.
**Why is this so?**

tropical diseases will expand toward higher latitudes, tree lines will rise, and many alpine glaciers will continue to disappear (Fig. 8.25).

However, some greenhouse effects will be beneficial to humans. Growing seasons in the high latitudes will increase in length. The increase in atmospheric $CO_2$ will help some crops such as wheat, rice, and soybeans grow larger faster. In the United States, a 1°C increase in average temperature should decrease heating bills by about 11%.

**FIGURE 8.25**
With few exceptions, mountain glaciers worldwide are retreating. This valley is leading to the terminus of the Franz-Josef Glacier in New Zealand. In 1865 this entire valley was filled with glacial ice.
**Is this a sign of global warming?**

Many scientists believe that the warming is already occurring. The five hottest years of the 20th century occurred during the 1990s, and each subsequent year usually sets a new record. Average annual global temperatures have already risen between 0.3°C and 0.6°C (0.5°F–1.1°F), and sea level has risen between 10 and 25 centimeters (4–10 in.) during the past 100 years. Given the long residence times of many greenhouse gases (see Fig. 8.23) and the heat capacity of the oceans, some warming is inevitable and will likely continue. On the other hand, the lesson of this chapter is that short-term climate trends and some longer-term climate trends will never be reliably predictable. Major volcanic eruptions, changing oceanic circulation, or human impacts (such as massive deforestation and urbanization) can significantly disrupt climate trends at any time. Moreover, because sudden, major shifts in climate systems that were not caused by human activity are abundant in the record of the last 2.4 million years, predicting climate will always be a risky endeavor.

## Define & Recall

empirical classification
genetic classification
Thornthwaite system
potential evapotranspiration (potential ET)
actual evapotranspiration (actual ET)
Köppen system
tropical climate
polar climate
microthermal climate
mesothermal climate
arid climate
highland climate
region
zone of transition
climograph
microclimate
oxygen-isotope analysis
Altithermal
eccentricity cycle
obliquity cycle
precession cycle
Little Ice Age

**Physical Geography Now™** I. Log on to Physical GeographyNow and select this chapter to work through a Geography Interactive activity on "Temperature Trends."

## Discuss & Review

1. Why is it important to study the nature and possible causes of past climates when attempting to predict future climate change?
2. What is the difference between genetic and empirical classification? Is a classification based on statistics empirical or genetic?
3. Why are temperature and precipitation the two atmospheric elements most widely used as the sources of statistics for climate classification? How are these two elements used in the Köppen system to identify six major climate categories?
4. What are the advantages and disadvantages of the Köppen system for geographers? Why are the Köppen climate boundaries often referred to as "vegetation lines"?
5. What is a climograph? What is its function?
6. How does the Thornthwaite system of climate classification differ from the Köppen system? What are the advantages of the Thornthwaite system?
7. What examples of microclimates can be found in your local area? How did you decide that these are examples of microclimates?

8. Why is the occurrence, frequency, and dating of glacial advances and retreats so important to the study of past climates? How has modern research changed earlier theories of glacial coverage and associated climate change during the Pleistocene?
9. In what way has the Holocene been unusual compared to climatic conditions during most of the rest of the Pleistocene?
10. How have scientists been able to document the rapid shifts of climates that have occurred during the latter part of the Pleistocene?
11. What are the major possible causes of global climate change? What contribution did the mathematician Milankovitch make to theories regarding glaciation?
12. What is the evidence that volcanic activity can affect global temperatures? How does this occur?
13. What effects can changes in the amounts of $CO_2$ and other greenhouse gases in the atmosphere have on global temperatures? How can past changes in amounts of $CO_2$ be determined?
14. How might changes in Earth's oceans and landmasses affect global climates?
15. What is the primary difficulty for any climatologist who attempts to predict future climates? How effective has the use of computer simulations been in the prediction of future climate change?
16. What changes are likely to occur in Earth's major subsystems if global warming continues for the near term, as many scientists believe?

## Consider & Respond

1. Study the definitions for the individual Köppen climate types described in the "Graph Interpretation" exercise. Why do you think Köppen and later climatologists who modified the system selected the particular temperature and precipitation parameters that separate the individual types from one another?
2. Examine the climograph in Figure 8.15. During what month does Nashville experience the greatest precipitation? What major change would immediately identify this graph as representing a Southern Hemisphere location? What do you think the four horizontal dashed lines represent?
3. Review Figures 8.18 and 8.24. State in your own words the general conclusions you would draw from a study of these two figures.
4. After studying Chapters 7 and 8 in your textbook, which do you believe is more important to you now and in the future—the subject of weather or of climate? Defend your answer.

# Graph Interpretation

## The Köppen Climate Classification System

The key to understanding any system of classification is found by personally practicing use of the system. This is one reason why the Consider & Respond review sections of both Chapters 9 and 10 are based on the classification of data from sites selected throughout the world. Correctly classifying these sites in the modified Köppen system may seem complicated at first, but you will find that, after applying the system to a few of the sample locations, the determination of a correct letter symbol and associated climate name for any other site data should be routine.

Before you begin, take the time to familiarize yourself with Table 1. You will note that there are precise definitions in regard to temperature or precipitation that identify a site as one of the five major climate categories in the Köppen system (*A*, tropical; *B*, arid; *C*, mesothermal; *D*, microthermal; *E*, polar). Furthermore, you will note that the additional letters required to identify the actual climate type also have precise definitions or are determined by the use of the graphs. In other words, Table 1 is all you need to classify a site if monthly and annual means of precipitation and temperature are available. Table 1 should be used in a systematic fashion to determine first the major climate category and, once that is determined, the second and third letter symbols (if needed) that complete the classification. As you begin to classify, it is strongly recommended that you use the following procedure. (After a few examples you may find that you can omit some steps with a glance at the statistics.)

### TABLE 1
### Simplified Köppen Classification of Climates

| First Letter | Second Letter | Third Letter |
|---|---|---|
| *E* Warmest month less than 10°C (50°F) POLAR CLIMATES *ET*—Tundra *EF*—Ice Sheet | *T* Warmest month between 10°C (50°F) and 0°C (32°F) *F* Warmest month below 0°C (32°F) | NO THIRD LETTER (with polar climates) SUMMERLESS |
| *B* Arid or semiarid climates ARID CLIMATES *BS*—Steppe *BW*—Desert | *S* Semiarid climate (see Graph 1) *W* Arid climate (see Graph 1) | *h* Mean annual temperature greater than 18°C (64.4°F) *k* Mean annual temperature less than 18°C (64.4°F) |

Even distribution

Summer concentration

Winter concentration

Graph 1 Humid/Dry Climate Boundaries

# TABLE 1
## Simplified Köppen Classification of Climates (Continued)

| First Letter | Second Letter | Third Letter |
|---|---|---|
| A  Coolest month greater than 18°C (64.4°F)<br>TROPICAL CLIMATES<br>Am—Tropical monsoon<br>Aw—Tropical savanna<br>Af—Tropical rainforest | f  Driest month has at least 6 cm (2.4 in.) of precipitation<br><br>m  Seasonally, excessively moist (see Graph 2)<br><br>w  Dry winter, wet summer (see Graph 2) | NO THIRD LETTER<br>(with tropical climates)<br>WINTERLESS |
| C  Coldest month between 18°C (64.4°F) and 0°C (32°F); at least one month over 10°C (50°F)<br>MESOTHERMAL CLIMATES<br>Csa, Csb—Mediterranean<br>Cfa, Cwa—Humid subtropical<br>Cfb, Cfc—Marine west coast | s  (DRY SUMMER) Driest month in the summer half of the year, with less than 3 cm (1.2 in.) of precipitation and less than one third of the wettest winter month | a  Warmest month above 22°C (71.6°F)<br><br>b  Warmest month below 22°C (71.6°F), with at least four months above 10°C (50°F) |
| D  Coldest month less than 0°C (32°F); at least one month over 10°C (50°F)<br>MICROTHERMAL CLIMATES<br>Dfa, Dwa—Humid continental, hot summer<br>Dfb, Dwb—Humid continental, mild Summer<br>Dfc, Dwc, Dfd, Dwd—Subarctic | w  (DRY WINTER) Driest month in the winter half of the year, with less than one tenth the precipitation of the wettest summer month<br><br>f  (ALWAYS MOIST) Does not meet conditions for s or w above | c  Warmest month below 22°C (71.6°F), with one to three months above 10°C (50°F)<br><br>d  Same as c, but coldest month is below {neg}38°C ({neg}36.4°F) |

Graph 2   Rainfall of driest month (cm)

Step 1. Ask: Is this a polar climate (*E*)? Is the warmest month less than 10°C (50°F)? If so, is the warmest month between 10°C (50°F) and 0°C (32°F) (*ET*) or below 0°C (32°F) (*EF*)? If not, move on to:

Step 2. Ask: Is there a seasonal concentration of precipitation? Examine the monthly precipitation data for the driest and wettest summer and winter months for the site. Take careful note of temperature data as well because you must determine whether the site is located in the Northern or Southern Hemisphere. (April to September are summer months in the Northern Hemisphere but winter months in the Southern Hemisphere. Similarly, the Northern Hemisphere winter months of October to March are summer south of the equator.) As the table indicates, a site has a dry summer (*s*) if the driest month in summer has less than 3 cm (1.2 in.) of precipitation and less than one third of the precipitation of the wettest winter month. It has a dry winter (*w*) if the driest month in winter has less than one tenth the precipitation of the wettest summer month. If the site has neither a dry summer nor a dry winter, it is classified as having an even distribution of precipitation (*f*). Move on to:

Step 3. Ask: Is this an arid climate (*B*)? Use one of the small graphs (included in Graph 1) to decide. Based on your answer in Step 2, select one of the small graphs and compare mean annual temperature with mean annual precipitation. The graph will indicate whether the site is an arid (*B*) climate or not. If it is, the graph will indicate which one (*BW* or *BS*). You should further classify the site by adding *h* if the mean annual temperature is above 18°C (64.4°F) and *k* if it is below. If the site is neither *BW* nor *BS*, it is a humid climate (*A*, *C*, or *D*). Move on to:

Step 4a. Ask: Is this a tropical climate (*A*)? The site has a tropical climate if the temperature of the coolest month is higher than 18°C (64.4°F). If so, use Graph 2 in Table1 to determine which tropical climate the site represents. (Note that there are no additional lowercase letters required.) If not, move on to:

Step 4b. Ask: Which major middle-latitude climate group does that site represent, mesothermal (*C*) or microthermal (*D*)? If the temperature of the coldest month is between 18°C (64.4°F) and 0°C (32°F), the site has a mesothermal climate. If is below 0°C (32°F), it has a microthermal climate. Once you have answered the question, move on to:

Step 5. Ask: What was the distribution of precipitation? This was determined back in Step 2. Add *s*, *w*, or *f* for a *C* climate or *w* or *f* for a *D* climate to the letter symbol for the climate. Then, move on to:

Step 6. Ask: What is needed to express the details of seasonal temperature for the site? Refer again to Table 1 and the definitions for the letter symbols. Add *a*, *b*, or *c* for the mesothermal (*C*) climates or *a*, *b*, *c*, or *d* for the microthermal (*D*) climates, and you have completed the classification of your climate. However, note that you may not have come this far because you might have completed your classification at Steps 1, 3, or 4a.

We should now be ready to try out the use of Table 1, following the steps we have recommended. Data for Madison, Wisconsin, is presented below for our example.

|  | J | F | M | A | M | J | J | A | S | O | N | D | Year |
|---|---|---|---|---|---|---|---|---|---|---|---|---|---|
| T(°C) | −8 | −7 | −1 | 7 | 13 | 19 | 21 | 21 | 16 | 10 | 2 | −6 | 7 |
| P (cm) | 3.3 | 2.5 | 4.8 | 6.9 | 8.6 | 11.0 | 9.6 | 7.9 | 8.6 | 5.6 | 4.8 | 3.8 | 77.0 |

The correct answer is derived below:
Step 1. We must determine whether or not our site has an *E* climate. Because Madison has several months averaging above 10°C, it does not have an *E* climate.

Step 2. We must determine if there is a seasonal concentration of precipitation. Because Madison is driest in winter, we compare the 2.5 centimeters of February precipitation with the precipitation of June (1/10 of 11.0 cm, or 1.1 cm) and conclude that Madison has neither a dry summer nor a dry winter but instead has an even distribution of precipitation (*f*). (*Note:* The 2.5 cm of February precipitation is not less than 1/10 [1.1 cm] of June precipitation.)

Step 3. Next we assess, through the use of Graphs 1(*f*), 1(*w*), or 1(*s*), whether our site is an arid climate (*BW*, *BS*) or a humid climate (*A*, *C*, or *D*). Because we have previously determined that Madison has an even distribution of precipitation, we will use Graph 1(*f*). Based on Madison's mean annual precipitation (77.0 cm) and mean annual temperature (7°C), we conclude that Madison is a humid climate (*A, C,* or *D*).

Step 4. Now we must assess which humid climate type Madison falls under. Because the coldest month (−8°C) is below 18°C, Madison does *not* have an *A* climate. Although the warmest month (21°C) is above 10°C, the coldest month (−8°C) is not between 0°C and 18°C, so Madison does *not* have a *C* climate. Because the warmest month (21°C) is above 10°C and the coldest month is below 0°C, Madison *does* have a *D* climate.

Step 5. Because Madison has a *D* climate, the second letter will be *w* or *f*. Because precipitation in the driest month of winter (2.5 cm) is not less that one tenth of the amount of the wettest summer month (1/10 × 11.0 cm = 1.1 cm), Madison does *not* have a *Dw* climate. Madison therefore has a *Df* climate.

Step 6. Because Madison is a *Df* climate, the third letter will be *a*, *b*, *c*, or *d*. Because the average temperature of the warmest month (21°C) is not above 22°C, Madison does *not* have a *Dfa* climate. Because the average temperature of the warmest month is below 22°C, with at least 4 months above 10°C, Madison is a *Dfb* climate.

# CHAPTER 9

This scene was depicted in the film *Star Wars* as an exotic planet in a galaxy far, far away. It is actually the rainforest canopy surrounding the ancient Mayan City of Tikal, in Guatemala.  ©Simeone Huber/Getty Images

# Low-Latitude and Arid Climate Regions

**PHYSICAL Geography Now™**  This icon, appearing throughout the book, indicates an opportunity to explore interactive tutorials, animations, or practice problems at now.brookscole.com/gabler8

# CHAPTER PREVIEW

**Although the humid tropical climates are all characterized by high temperatures throughout the year, they exhibit significant differences based on either the amounts or the distribution of the precipitation they receive.**

- What temperature parameter do these climates have in common?
- How do they differ from one another on the basis of precipitation?
- What controlling factors explain these differences?

**Except for a few unusual circumstances, the tropical rainforest climate regions are among the least populated areas of the world, despite being coincident with the belt of heaviest rainfall, insolation, and vegetative growth.**

- Why are there so few people in these regions?
- What are the exceptions?
- In what ways are these regions valuable to humankind?

**Although rainfall is seasonal in both tropical savanna and tropical monsoon climate regions, the differences in total rainfall between the two climates cause major dissimilarities in their appearance, resource characteristics, and human use.**

- What are the chief dissimilarities?
- How is rainfall responsible?
- How are humans affected?

**Earth's arid regions are created by various processes that are found in a wide range of latitudes across the globe.**

- What are some of these latitudes?
- Where are the most extensive arid regions found?
- What processes create these different regions?

**Knowledge of the location of the world's deserts is similar to an understanding of the distribution of the world's steppe regions.**

- What is the association between deserts and steppes?
- How do the regions differ?
- In what ways are they similar?

**Living in harsh environments like rainforests and deserts produces a series of hazards and risks to human inhabitants.**

- What are some of these hazards?
- Which hazards and risks are found in these two different environments?
- How have humans adapted to these risks?

The odds are overwhelming that within your lifetime, if you have not already done so, you will travel. You may travel extensively either within or beyond the borders of North America to destinations far from where you are living today. You may be moving to a new home or place of employment. You may be traveling on business or simply for pleasure. Whatever the reason, it is likely that you will ask the question almost every other traveler asks: "I wonder what the weather will be like?"

Realistically, as you have learned from reading previous chapters, the question should probably be, "I wonder what the climate is like?" The constant variability associated with weather in many areas of the world makes it difficult to predict. However, the long-term ranges and averages upon which climate is based allow geographers to provide the traveler with a general idea of the atmospheric conditions likely to be experienced at specific locations throughout the world during different times of the year.

Of course, there are many reasons other than travel why knowledge of climate and its variation over Earth's surface is a valuable asset. An understanding of climate in other areas of the world helps us understand the adaptations to atmospheric conditions that have been made by the people who live there. We can better appreciate some of their economic activities and certain aspects of their cultures. In addition, the climate of any place on Earth has a dominant effect on native vegetation and animal life. It influences the rate and manner by which rock material is destroyed and soil is formed. It is a contributing factor in the way landforms are reduced and physical landscapes are sculptured. In short, knowledge of climates provides endless clues to not only atmospheric conditions but also numerous other aspects of the physical environment.

In this chapter and the next, you will be provided with a broad descriptive survey of world climates: their locations, distributions, general characteristics, formation processes, associated features, and related human activities. The information contained in these two chapters can serve as a valuable knowledge base as you prepare for the future. Throughout these chapters, we use the modified version of the Köppen climate classification that was introduced in Chapter 8. It is interesting to note that each of the climates discussed can be found within North America, so even if you never travel beyond this continent's borders, you will still find the discussions of climates valuable preparation as you move about your own country.

## Humid Tropical Climate Regions

We have already learned a good deal about the climate regions of the humid tropics through our preliminary discussions of these climate types in Chapter 8. In addition, we can review the location of these climates in relation to other climate regions through regular examination of the world map of climates (Figure 8.6). It now remains for us to identify the major characteristics of each humid tropical climate type in turn, along with its associated world regions.

Study of Figure 9.1 and a careful reading of Table 9.1 will provide the locations of the humid tropical climates and a preview of the significant facts associated with them. The table also reminds us that, although each of the three humid tropical climates has high average temperatures throughout the year, they differ greatly in the amount and distribution of precipitation.

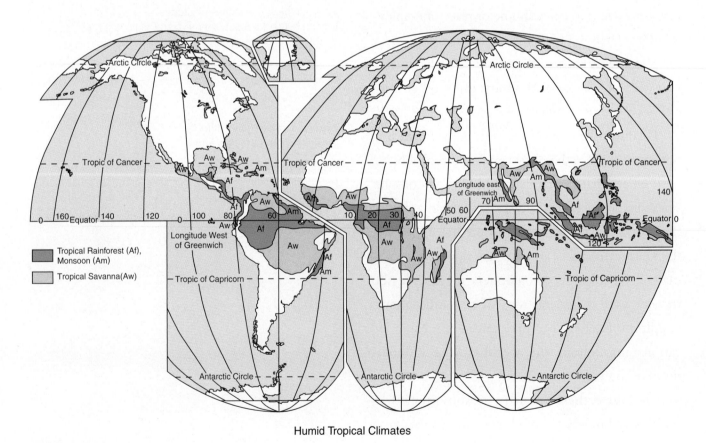

**FIGURE 9.1**
Index map of humid tropical climates

**TABLE 9.1**
**The Humid Tropical Climates**

| Name and Description | Controlling Factors | Geographic Distribution | Distinguishing Characteristics | Related Features |
|---|---|---|---|---|
| *Tropical Rainforest* | | | | |
| Coolest month above 18°C (64.4°F); driest month with at least 6 cm (2.4 in.) of precipitation | High year-round insolation and precipitation of doldrums (ITCZ); rising air along trade wind coasts | Amazon R. Basin, Congo R. Basin, east coast of Central America, east coast of Brazil, east coast of Madagascar, Malaysia, Indonesia, Philippines | Constant high temperatures; equal length of day and night; lowest (2°C–3°C/3°F–5°F) annual temperature ranges; evenly distributed heavy precipitation; high amount of cloud cover and humidity | Tropical rainforest vegetation (selva); jungle where light penetrates; tropical iron-rich soils; climbing and flying animals, reptiles, and insects; slash-and-burn agriculture |
| *Tropical Monsoon* | | | | |
| Coolest month above 18°C (64.4°F); one or more months with less than 6 cm (2.4 in.) of precipitation; excessively wet during rainy season | Summer onshore and winter offshore air movement related to shifting ITCZ and changing pressure conditions over large landmasses; also transitional between rainforest and savanna | Coastal areas of southwest India, Sri Lanka, Bangladesh, Myanmar, southwestern Africa, Guyana, Surinam, French Guiana, northeast and southeast Brazil | Heavy high-sun rainfall (especially with orographic lifting), short low-sun drought; 2°C–6°C (3°F–10°F) annual temperature range, highest temperature just prior to rainy season | Forest vegetation with fewer species than tropical rainforest; grading to jungle and thorn forest in drier margins; iron-rich soils; rainforest animals with larger leaf-eaters and carnivores near savannas; paddy rice agriculture |
| *Tropical Savanna* | | | | |
| Coolest month above 18°C (64.4°F); wet during high-sun season, dry during lower-sun season | Alternation between high-sun doldrums (ITCZ) and low-sun subtropical highs and trades caused by shifting winds and pressure belts | Northern and eastern India, interior Myanmar and Indo-Chinese Peninsula; northern Australia; borderlands of Congo R., south central Africa; llanos of Venezuela, campos of Brazil; western Central America, south Florida, and Caribbean Islands | Distinct high-sun wet and low-sun dry seasons; rainfall averaging 75–150 cm (30–60 in.); highest temperature ranges for humid tropical climates | Grasslands with scattered, drought-resistant trees, scrub, and thorn bushes; poor soils for farming, grazing more common; large herbivores, carnivores, and scavengers |

## Tropical Rainforest Climate

The **tropical rainforest climate** probably comes most readily to mind when someone says the word *tropical*. Hot and wet throughout the year, the tropical rainforest climate has been the stage for many stories of both fact and fiction. One cannot easily forget the life-and-death struggle with the elements portrayed by Humphrey Bogart and Katharine Hepburn in the classic film *The African Queen*. More recent films about the Vietnam War also depict the difficulties of moving and fighting in such formidable environments. Upon visiting this type of climate, one would easily feel the high temperatures, oppressive humidity, and the frequent heavy rains, which sustain the massive vegetative growth for which it is known ( Fig. 9.2).

### Constant Heat and Humidity
Most weather stations in the tropical rainforest climate regions record average monthly temperatures of 25°C (77°F) or more ( Fig. 9.3). Because these regions are usually located within 5° or 10° of the equator, the sun's noon rays are always close to being directly overhead. Days and nights are of almost equal length, and the amount of insolation received remains nearly constant throughout the year. Consequently, no appreciable temperature variations can be linked to the sun angle and therefore be considered seasonal. In other words, the concept of summer and winter as being hot and cold seasons, respectively, does not exist here.

The **annual temperature range**—the difference between the average temperatures of the warmest and coolest months of the year—reflects the consistently high angle of the sun's rays. As indicated in Figure 9.3, the annual range is seldom more than 2°C or 3°C (4°F or 5°F). In fact, at Ocean Island in the central Pacific, the annual range is 0°C because of the additional moderating influence of the ocean on the nearly uniform pattern of insolation.

One of the most interesting features of the tropical rainforest climate is that the **daily (diurnal) temperature ranges**—the differences between the highest and lowest temperatures during the day—is usually far greater than the annual range. Highs of 30°C–35°C (86°F–95°F) and lows of 20°C–24°C (68°F–75°F) produce daily ranges of 10°C–15°C (18°F–27°F). However, the drop in temperature at night is small comfort. The high humidity causes even the cooler evenings to seem oppressive. (Recall, water vapor is a greenhouse gas and helps retain the heat energy.)

The climographs of Figure 9.3 illustrate that significant variations in precipitation can occur even within rainforest regions. Although most rainforest locations receive more than 200 centimeters (80 in.) a year of precipitation and the average is in the neighborhood of 250 centimeters (100 in.), some locations record an annual precipitation of more than 500 centimeters (200 in.). Ocean locations, near the greatest source of moisture, tend to receive the most rain. Mt. Waialeale, in the Hawaiian Islands, receives a yearly average of 1146 centimeters (451 in.), making it the wettest spot on Earth. As a group, climate stations in the humid tropics experience much higher annual totals than typical humid middle-latitude stations. Compare, for example, the 365 centimeters in Akassa, Nigeria, with the average 112 centimeters received annually in Portland, Oregon, or the 61 centimeters received in London, England.

We should recall that the heavy precipitation of the tropical rainforest climate is associated with the warm, humid air of the doldrums and the unstable conditions along the ITCZ (intertropical

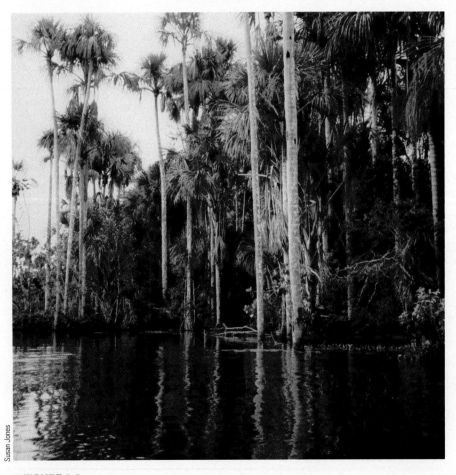

**FIGURE 9.2**
Typical vegetation of a tropical rainforest climate as seen along the Madre de Dios River in Peru.
**How many different levels of treetops can you see in this photo?**

**FIGURE 9.3**
Climographs for tropical rainforest climate stations.
**Why is it difficult, without looking at the climograph key, to determine whether each station is located in the Northern or Southern Hemisphere?**

convergence zone). Both convection and convergence serve as uplift mechanisms, causing the moist air to rise and condense and resulting in the heavy rains that are characteristic of this climate. These processes are enhanced on the east coasts of continents where warm ocean currents allow humid tropical climates to extend farther poleward. On west coasts where cold ocean currents flow, these mechanisms of uplift are somewhat inhibited. There is heavy cloud cover during the warmer, daylight hours when convection is at its peak, although the nights and early mornings can be quite clear. Variations in rainfall can usually be traced to the ITCZ and its low pressure cells of varying strength. Many tropical rainforest locations (Akassa, for example) exhibit two maximum precipitation periods during the year, one during each appearance of the ITCZ as it follows the migration of the sun's direct rays (recall, the sun crosses the equator on March 21 and September 22). In addition, although no season can be called dry, during some months it may rain on only 15 or 20 days.

**Cloud Forests** Highland areas near a sea coast in both the tropical regions like Costa Rica (Fig. 9.4) and in the middle latitudes (the coastal Washington state) may contain **cloud**

**FIGURE 9.4**
A cloud forest in the highlands of Costa Rica.
**Can a constant, misty rain drop as much water as less frequent rainstorms?**

**forests.** Here, moisture-laden maritime air is lifted up the windward slopes of mountains. This orographic precipitation may not fall as heavy rain showers or thunderstorms but rather as an almost constant misty fog. Through time, this cloudy environment can precipitate enough moisture to qualify as a rainforest, but without the oppressive heat found in the rainforests

**FIGURE 9.5**
At ground level, a true rainforest offers considerable open space.
**Why is this so?**

of the low-altitude tropics. One advantage of the cloud forests is a scarcity of flying insects that cannot survive in the colder temperatures.

**A Delicate Balance** The most common vegetation of tropical rainforest climate regions is multistoried, broadleaf evergreen forest made up of many species whose tops form a thick, almost continuous canopy cover that blocks out much of the sun's light (see the chapter-opening photo). This type of rainforest is sometimes called a **selva.** Within the selva, there is usually little undergrowth on the forest floor because sunlight cannot penetrate enough to support much low-growing vegetation ( Fig. 9.5).

The relationship between the soils beneath the selva and the vegetation that the soils support is so close that there exists a nearly perfect ecological balance between the two, threatened only by people's efforts to earn a living from the soil. The trees of the selva supply the tropical soils with the nutrients that the trees themselves need for growth. As leaves, flowers, and branches fall to the ground or as roots die, the numerous soil-dwelling animals and bacteria act on them, transforming the forest litter into organic matter with vital nutrients. However, if the trees are removed, there is no replenishment of these nutrients and no natural barrier (forest litter and root systems) to prevent large amounts of rain from percolating through the soil. This percolating water can dissolve and remove nutrients and minerals from the topsoil. The intense activities of microorganisms, worms, termites, ants, and other insects cause rapid deterioration of the remaining organic debris, and soon all that remains is an infertile mixture of insoluble manganese, aluminum, and iron compounds. In essence, without the vegetation to protect and feed the soils, rainforest soils are quite barren.

In recent years, there has been large-scale harvesting of the tropical rainforests by the lumber industry, and land has been cleared for agriculture and livestock production, especially in the Amazon River basin. Such deforestation can have a significant and, unfortunately, permanent impact on the delicate balance that exists among Earth's systems.

Environmental conditions vary from place to place within climate regions; therefore, the typical rainforest situation that we have just described does not apply everywhere in the tropical rainforest climate. Some regions are covered by true jungle, a term often misused when describing the rainforest. **Jungle** is a dense tangle of vines and smaller trees that develops where direct sunlight does reach the ground, as in clearings and along streams ( Fig. 9.6). Other regions have soils that remain fertile or have bedrock that is chemically basic and provides the soils above with a constant supply of soluble nutrients through the natural weathering processes. Examples of the former region are found along major river floodplains; examples of the latter are the volcanic regions of Indonesia and the limestone areas of Malaysia and Vietnam. Only in such regions of continuous soil fertility can agriculture be intensive and continuous enough to support population centers in the tropical rainforest climate.

**Human Activities** Throughout much of the tropical rainforest climate, humans are far outnumbered by other forms of animal life. Though there are few large animals of any kind, a great variety of smaller tree-dwelling and aquatic species live in the rainforest. Small predatory cats, birds, monkeys, bats, alligators, crocodiles, snakes, and amphibians such as frogs of many varieties abound. Animals that can fly or climb into the food-rich leaf canopy have become the dominant animals in this world of trees.

Most common of all, though, are the insects. Mosquitoes, ants, termites, flies, beetles, grasshoppers, butterflies, and bees live everywhere in the rainforest. Insects can breed continuously in this climate without danger from cold or drought.

Besides the insects, there are genuine health hazards for human inhabitants of the tropical rainforest. Not only does the oppressive, sultry weather impose uncomfortable living conditions, but also any open wound would heal more slowly in the steamy environment. This climate also allows a variety of parasites and disease-carrying insects to threaten human survival. Malaria, yellow fever, dengue fever, and

## Geography's Environmental Science Perspective
### The Amazon Rainforest

Currently, an area of Earth's virgin tropical forest somewhat larger than a football field is being destroyed every second. During a recent decade, the rate of deforestation doubled, and in the Amazon Basin, where more than half of the rainforest resources are located, the rate nearly tripled. Simple mathematics indicates that even at the present rate of deforestation, the Amazon rainforest will virtually disappear in 150 years.

From the point of view of a developing country, there are basic economic reasons for clearing the Amazon rainforest, and Peru, Ecuador, Colombia, and Brazil (where most of the forest is located) are developing countries. Tropical timber sales provide short-term income to finance national growth and repay staggering debts owed to foreign banks. In addition, Brazil, in particular, has viewed the Amazon rainforest as a frontier land available for agricultural development and for resettlement of the poor from overcrowded urban areas.

No one questions that decisions concerning the future of the Amazon rainforest rest with the governments of nations that control these resources. So why has the rainforest become a serious international issue? The answer is found in the concerns of physical geographers and other environmental scientists throughout the world.

When the tropical rainforests are removed, the hydrologic cycle and energy budgets in the previously forested areas are dramatically altered, often irreversibly. In tropical rainforests, the canopy shades the forest floor, thus helping to keep it cooler. In addition, the huge mass of vegetation provides a tremendous amount of water vapor to the atmosphere through transpiration. The water vapor condenses to form clouds, which in turn provide rainfall to nourish the forest. With the forests removed, transpiration is diminished, which leads to less cloud cover and less rainfall. With fewer clouds and no forest canopy, more solar energy reaches Earth's surface.

The unfortunate outcome is that areas that are deforested soon become hotter and drier, and any ecosystem in place is seriously damaged or destroyed. Once the rainforests are removed, the soils lose their source of plant nutrients, and this precludes the growth of any significant crops or plants. In addition, the rainforest, which was in harmony with the soil, cannot reestablish itself. Thus, the multitude of flora and fauna species indigenous to the rainforest is lost forever. It is impossible to calculate the true cost of this reduction in biodiversity (the total number of different plant and animal species in the Earth system).

The lost species may have held secrets to increased food production; a cure for AIDS, cancer, or other health problems; or a base for better insecticides that do not harm the environment. Similar services to humanity already have been provided by tropical forest species.

Tropical deforestation is also threatening the natural chemistry of the atmosphere. Rainforests are a major source of the atmospheric oxygen so essential to all animal life. And deforestation encourages global warming by enhancing the greenhouse effect because forests act as a major reservoir of carbon dioxide. It has been estimated that forest clearing since the mid-1800s has contributed more than 130 billion tons of carbon to the atmosphere; more than two thirds as much as has been added by the burning of coal, oil, and natural gases combined.

What can be done? The reasons for tropical deforestation and the solutions to the problem may be economic, but the issues are extremely complex. It is not sufficient for the rest of the world to point out to governments of tropical nations that their forests are a major key to human survival. It is unacceptable for scientists and politicians from nations where barely one fourth of the original forests remain to insist that the citizens of the tropics cease cutting trees and establish forest plantations on deforested land. These are desired outcomes, but it is first the responsibility of all the world's people to help resolve the serious economic and social problems that have prevented most tropical nations from considering their forests as a sustainable resource.

A section of Amazon rainforest cleared by slash-and-burn techniques for potential farming or grazing

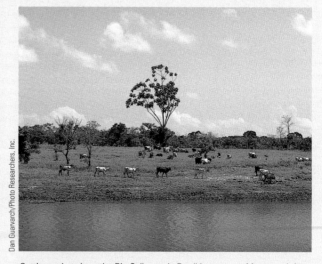

Cattle grazing along the Rio Salimoes in Brazil in an area of former rainforest

**FIGURE 9.6**
A jungle along the Usumacinta River on the Mexico–Guatemala border.
**Why might this photograph give an incorrect impression about vegetation at ground level inside the forest?**

**FIGURE 9.7**
An example of subsistence slash-and-burn agriculture: preparing an area for planting in Ecuador.
**Would you expect shifting cultivation to be on the increase or decrease in tropical rainforests?**

sleeping sickness are all insect-borne (sometimes fatal) diseases of the tropics and uncommon in the middle latitudes.

Whenever native populations have existed in the rainforest, subsistence hunting and gathering of fruits, berries, small animals, and fish have been important. Since the introduction of agriculture, land has been cleared, and crops such as manioc, yams, beans, maize (corn), bananas, and sugarcane have been grown. It has been the practice to cut down the smaller trees, burn the resulting debris, and plant the desired crops. With the forest gone, this kind of farming is possible for only 2 or 3 years before the soil is completely exhausted of its small supply of nutrients and the surrounding area is depleted of game. At this point, the native population moves to another area of forest to begin the practice over again. This kind of subsistence agriculture is known as **slash-and-burn** or simply **shifting cultivation.** Its impact on the close ecological balance between soil and forest is obvious in many rainforest regions. Sometimes the damage done to the system is irreparable, and only jungle, thornbushes, or scrub vegetation will return to the cleared areas (Fig. 9.7).

In terms of numbers of people supported, the most important agricultural use of the tropical rainforest climate is the wet-field (paddy) rice agriculture on the river floodplains of southeastern Asia. However, this type of agriculture is best developed in the monsoon variant of this climate. Commercial plantation agriculture is also significant. The principal plantation crops are rubber, sugarcane, and cacao, all of which originally grew with abundance in the forests of the Amazon Basin but are now of greatest importance elsewhere—rubber in Malaysia and Indonesia, sugarcane and cacao in West Africa and the Caribbean area.

## Tropical Monsoon Climate

We associate the **tropical monsoon climate** most closely with the peninsula lands of Southeast Asia. Here the alternating circulation of air (from sea to land in summer and from land to sea in winter) is strongly related to the shifting of the ITCZ. During the summer, the ITCZ moves north into the Indian subcontinent and adjoining lands to latitudes of 20° or 25°. This is due in part to the attracting force of the deep low pressure system of the Asian

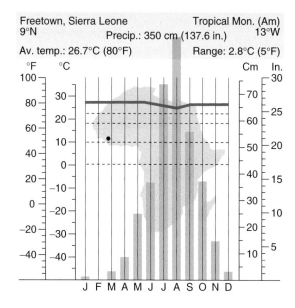

**FIGURE 9.8**
Climographs for tropical monsoon climate stations.
**What is the estimated range of precipitation between the highest and lowest months?**

continent. However, as we have previously noted, the mechanism is complex and involves changes in the upper air flow as well as in surface currents. Several months later, the moisture-laden summer monsoon is replaced by an outflow of dry air from the massive Siberian high pressure system that develops in the winter season over central Asia. By this time, the ITCZ has shifted to its southernmost position (see again Figure 5.12).

Figures 9.1 and 9.8 confirm that climate regions outside of Asia fit the simplified Köppen classification of tropical monsoon as well. A modified version of the monsoonal wind shift occurs at Freetown, Sierra Leone, in Africa, but the climate there might also be described as transitional between the constantly wet rainforest climate and the sharply seasonal wet and dry conditions of the tropical savanna.

### Distinctions between Rainforest and Monsoon

Whatever the factors are that produce tropical monsoon climate regions, these regions have strong similarities to those classified as tropical rainforest. In fact, although their core regions are distinctly different, the two climates are often intermixed over *zones of transition*. A major reason for the similarity between monsoon and rainforest climates is that a monsoon area has enough precipitation to allow continuous vegetative growth with no dormant period during the year. Rains are so abundant and intense and the dry season so short that the soils usually do not dry out completely. As a result, this climate and its soils support a plant cover much like that of the tropical rainforests.

However, there are clear distinctions between rainforest and monsoon climate regions. The most important distinction of course concerns precipitation, including both distribution and amount. The monsoon climate has a short dry season, whereas the rainforest does not. Perhaps even more interesting, the average rainfall in monsoon regions varies more widely from place to place. It usually totals between 150 and 400 centimeters (60 to 150 in.) and may be massive where the onshore monsoon winds are forced to rise over mountain barriers. Mahabaleshwar, altitude 1362 meters (4467 ft), on the windward side of India's Western Ghats, averages more than 630 centimeters (250 in.) of rain during the 5 months of the summer monsoon.

The annual march of temperature of the monsoon climate differs appreciably from the monotony of the rainforest climate. The heavy cloud cover of the rainy monsoon reduces insolation and temperatures during that time of year. During the period of clear skies just prior to the onslaught of the rains, higher temperatures are recorded. As a result, the annual temperature range in a monsoon climate is 2°C–6°C (compared with 2°C–3°C in the tropical rainforest).

Some additional distinctions between monsoon and rainforest regions can be found in vegetation and animal life. Toward the wetter margins, the tropical monsoon forest resembles the tropical rainforest, but fewer species are present and certain ones become dominant. The seasonality of rainfall in the monsoon narrows the range of species that will prosper. Toward the drier margins of the climate, the trees grow farther apart, and the monsoon forest often gives way to jungle or a dwarfed thorn forest. The composition of the animal kingdom here also changes. The climbing and flying species that dominate the forest are joined by larger, hooved leaf eaters and by larger carnivores such as the famous tigers of Bengal.

## Tropical Savanna Climate

Located well within the tropics (usually between latitudes 5° and 20° on either side of the equator), the **tropical savanna climate** has much in common with the tropical rainforest and monsoon. The sun's vertical rays at noon are never far from overhead, the receipt of solar energy is nearly at a maximum, and temperatures remain constantly high. Days and nights are of nearly equal length throughout the year, as they are in other tropical regions.

However, as previously noted, its distinct seasonal precipitation pattern identifies the tropical savanna. As the latitudinal wind and pressure belts shift with the direct angle of the sun, savanna regions are under the influence of the rain-producing ITCZ (doldrums) for part of the year and the rain-suppressing subtropical highs for the other part. In fact, the poleward limits of the savanna climate are approximately the poleward limits of migration of the ITCZ, and the equatorward limits of this climate are the equatorward limits of movement by the subtropical high pressure systems.

As you can see in Figure 9.1 and Table 9.1, the greatest areas of savanna climate are found peripheral to the rainforest climates of Central and South America and Africa. Lesser but still important savanna regions occur in India, peninsular Southeast Asia, and Australia. In some instances, the climate extends poleward of the tropics, as it does in the southernmost portion of Florida.

**Transitional Features of the Savanna** Of particular interest to the geographer is the transitional nature of the tropical savanna. Often situated between the humid rainforest climate on one side and the rain-deficient steppe climate on the other, the savanna experiences some of the characteristics of both. During the rainy, high-sun season, atmospheric conditions resemble those of the rainforest, whereas the low-sun season can be as dry as nearby arid lands are all year. The gradational nature of the climate causes precipitation patterns to vary considerably (Fig. 9.10). Savanna locations close to the rainforest may have rain during every month, and total annual precipitation may exceed 180 centimeters (70 in.). In contrast, the drier margins of the savanna have longer and more intensive periods of drought and lower annual rainfalls, less than 100 centimeters (40 in.).

Other characteristics of the savanna help demonstrate its transitional nature. The higher temperatures just prior to the arrival of the ITCZ produce annual temperature ranges 3°C–6°C (5°F–11°F) wider than those of the rainforest, but still not as wide as those of the steppe and desert. Although the typical savanna vegetation (known as *llanos* in Venezuela, *campos* in Brazil, and *pampas* in Argentina) is a mixture of grassland

**FIGURE 9.9**
Crop selection and agricultural production adjust to the (a) wet and (b) dry seasons throughout Southeast Asia.
**Would it be beneficial to the people of Southeast Asia if the traditional rice-farming methods were replaced by mechanized rice agriculture as practiced in the United States?**

**Effects of Seasonal Change** The seasonal precipitation of the tropical monsoon climate is of major importance for economic reasons, especially to the people of Southeast Asia and India. Most of the people living in those areas are farmers, and their major crop is rice, which is the staple food for millions of Asians. Rice is most often an irrigated crop, so the monsoon rains are very important to its growth. Harvesting, on the other hand, must be done during the dry season (Fig. 9.9).

Each year, an adequate food supply for much of South and Southeast Asia depends on the arrival and departure of the monsoon rains. The difference between famine and survival for many people in these regions is very much associated with the climate.

**FIGURE 9.10**
Climographs for tropical savanna climate stations.
**Consider the differences in climate and human use of the environment between Key West and Kano. Which are more important in the geography of the two places, the physical or the human factors?**

and trees, scrub, and thorn bushes, there is considerable variation. Near the equatorward margins of these climates, grasses are taller, and trees, where they exist, grow fairly close together ( Fig. 9.11). Toward the drier, poleward margins, trees are more widely scattered and smaller, and the grasses are shorter. Soils, too, are affected by the climatic gradation as the iron-rich reddish soils of the wetter sections are replaced by darker-colored more organic-rich soils in the drier regions.

Both vegetation and soils have made special adaptations to the alternating wet–dry seasons of the savanna. During the wet (high-sun) period, the grasslands are green, and the trees are covered with foliage. During the dry (low-sun) period, the grass turns brown, dry, and lifeless, and most of the trees lose their leaves as an aid in reducing moisture loss through transpiration. The trees develop deep roots that can reach down to water in the soil during the dry season. They are also fire resistant, an advantage for survival in the savanna where the grasses may burn during the winter drought.

**Savanna Potential** Conditions within tropical savanna regions are not well suited to agriculture although many of our domesticated grasses (grains) are presumed to have grown wild there. Rainfall is far less predictable than in the rainforest

**FIGURE 9.11**
The pampas of Argentina after some rainfall displays lush tropical savanna vegetation.
**Can you guess why Argentina is such a major exporter of beef cattle?**

or even the monsoon climate. For example, Nairobi, Kenya, has an average rainfall of 86 centimeters (34 in.). Yet from year to year, the amount of rain received may vary from 50 to 150 centimeters (20–60 in.). As a rule, the drier the savanna station, the more unreliable the rainfall becomes. However, the rains are essential for human and animal survival in savanna regions. When they are late or deficient, as they have

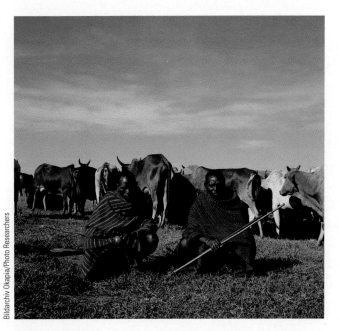

**FIGURE 9.12**
The grasslands of the East African savanna are well suited to the support of large numbers of grazing animals. The Masai cattle herders shown here in Kenya count their wealth by the numbers of animals they own.
**What problems are created for the Kenyan government as cattle herds grow?**

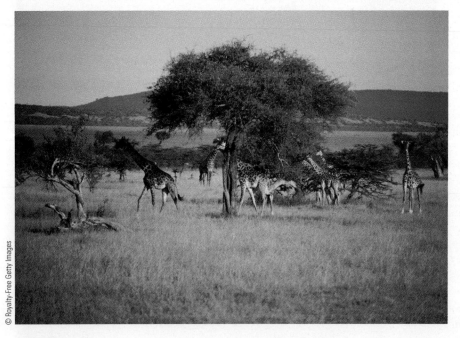

**FIGURE 9.13**
Giraffes have always been a majestic sight in the savanna climate of East Africa.
**Looking in the background, can you see the faint outline of Mount Kilima Njaro, the tallest peak in Africa?**

Savanna soils (except in areas of recent stream deposits) also limit productivity. During the rains of the wet season, they may become gummy; during the dry season, they are hard and almost impenetrable. Consequently, people in the savannas have often found the soils better suited to grazing than to farming. The Masai, a tribe of cattle herders and fierce warriors of East Africa, are world-famous examples ( Fig. 9.12). However, even animal husbandry has its problems. Many savanna regions make poor pasturelands, at least during the dry part of the year.

The savannas of Africa have exhibited the greatest potential of the world's savanna regions. They have been veritable zoological gardens for the larger tropical animals, to such an extent that the popularity of classic photo safaris has made the African savannas a major center for tourism. The grasslands support many different herbivores (plant eaters), such as the elephant, rhinoceros, giraffe, zebra, and wildebeest ( Fig. 9.13). The herbivores in turn are eaten by the carnivores (flesh eaters), such as the lion, leopard, and cheetah. Lastly, scavengers, such as hyenas, jackals, and vultures, devour what remains of the carnivore's kill. During the dry season, the herbivores find grasses and water along stream banks and forest margins and at isolated water holes. The carnivores follow the herbivores to the water, and a few human hunters and scavengers still follow them both.

## Arid Climate Regions

Arid climate regions in the simplified Köppen system are widely distributed over Earth's surface. A brief study of Figure 9.14 confirms that they are found from the vicinity of the equator to more than 50°N and S latitude. There are two major concentrations of desert lands, and each illustrates one of the important causes of climatic aridity. The first is centered on the Tropics of Cancer and Capricorn (23½°N and S latitudes) and extends 10°–15° poleward and equatorward from there. This region contains the most extensive areas of arid climates in the world. The second is located poleward of the first and occupies continental interiors, particularly in the Northern Hemisphere.

The concentration of deserts in the vicinity of the two tropic lines is directly related to the subtropical high pressure systems. Although the boundaries of the subtropical highs may migrate north and south with the direct rays of the sun, their influence remains constant in these latitudes. We have already learned that the subsidence and divergence of air associated with these systems is strongest along the eastern

been in West Africa in recent years, severe drought and famine result. On the other hand, when the rains last longer than usual or are excessive, they can cause major floods, often followed by outbreaks of disease.

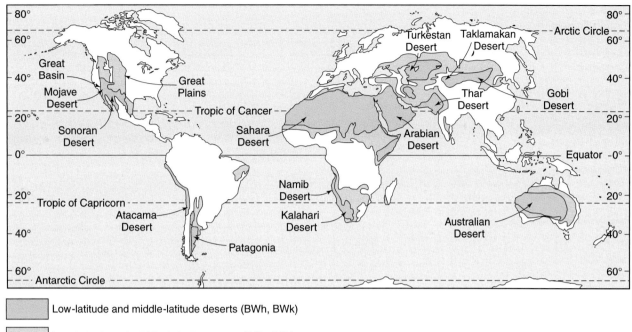

**FIGURE 9.14**
A map of the world's arid lands.
**What does a comparison of this map with the Map of World Population Density (inside back cover) suggest?**

portions of the oceans (recall, cold ocean currents off the western coasts of continents help stabilize the atmosphere). Hence, the clear weather and dry conditions of the subtropical high pressure extend inland from the western coasts of each landmass in the subtropics. The Atacama, Namib, and Kalahari Deserts and the desert of Baja California are restricted in their development by the small size of the landmass or by landform barriers to the interior. However, the western portion of North Africa and the Middle East comprises the greatest stretch of desert in the world and includes the Sahara, Arabian, and Thar Deserts. Similarly, the Australian Desert occupies most of the interior of the Australian continent.

The second concentration of deserts is located within continental interiors remote from moisture-carrying winds. Such arid lands include the vast cold-winter deserts of inner Asia and the Great Basin of the western United States. The dry conditions of the latter region extend northward into the Columbia Plateau and southward into the Colorado Plateau and are increased by the mountain barriers that restrict the movement of rain-bearing air masses from the Pacific. Similar rain-shadow conditions help explain the Patagonia Desert of Argentina and the arid lands of western China.

Both wind direction and ocean currents can accentuate aridity in coastal regions. When prevailing winds blow parallel to a coastline instead of onshore, desert conditions are likely to occur because little moisture is brought inland. This seems to be the case in eastern Africa and in northeastern Brazil. Where a cold current flows next to a coastal desert, foggy conditions may develop. Warm, moist air from the ocean may be cooled to its dew point as it passes over the cooler current. A temperature inversion is created, increasing stability and preventing the upward movement of air required for precipitation. The unique, fog-shrouded coastal deserts in Chile (the Atacama), southwest Africa (the Namib), and Baja California have the lowest precipitation of any regions on Earth.

Figure 9.14 shows deserts of the world to be core areas of aridity, usually surrounded by the slightly moister steppe regions. Hence, our explanations for the location of deserts hold true for the steppes as well. The steppe climates either are subhumid borderlands of the humid tropical, mesothermal, and microthermal climates or are transitional between these climates and the deserts. As previously noted, we classify both steppe and desert on the basis of the relation between precipitation and potential ET. In the **desert climate,** the amount of precipitation received is less than half the potential ET. In the **steppe climate,** the precipitation is more than half but less than the total potential ET.

The criterion for determining whether a climate is desert, steppe, or humid is *precipitation effectiveness.* The amount of precipitation actually available for use by plants and animals is the *effective* precipitation. Precipitation effectiveness is related to temperature. At higher temperatures, it takes more precipitation to have the same effect on vegetation and soils than at lower

**TABLE 9.2**
**The Arid Climates**

| Name and Description | Controlling Factors | Geographic Distribution | Distinguishing Characteristics | Related Features |
|---|---|---|---|---|
| *Desert* | | | | |
| Precipitation less than half of potential evapotranspiration mean annual temperature above 18°C (64.4°F) (low-lat.), below (mid-lat.) | Descending, diverging circulation of subtropical highs; continentality often linked with rain-shadow location | Coastal Chile and Peru, southern Argentina, southwest Africa, central Australia, Baja California and interior Mexico, North Africa, Arabia, Iran, Pakistan and western India (low-lat.); inner Asia and western United States (mid-lat.) | Aridity; low relative humidity; irregular and unreliable rainfall; highest percentage of sunshine; highest diurnal temperature range; highest daytime temperatures; windy conditions | Xerophytic vegetation; often barren, rocky, or sandy surface; desert soils; excessive salinity; usually small, nocturnal, or burrowing animals; nomadic herding |
| *Steppe* | | | | |
| Precipitation more than half but less than potential evapotranspiration mean annual temperature above 18°C (64.4°F) (low-lat.), below (mid-lat.) | Same as deserts; usually transitional between deserts and humid climates | Peripheral to deserts, especially in Argentina, northern and southern Africa, Australia, central and southwest Asia, and western United States | Semiarid conditions, annual rainfall distribution similar to nearest humid climate; temperatures vary with latitude, elevation, and continentality | Dry savanna (tropics) or short grass vegetation; highly fertile black and brown soils; grazing animals in vast herds, predators and smaller animals; ranching, dry farming |

temperatures. The result is that areas with higher temperatures that promote greater ET can receive more precipitation than cooler regions and yet have a more arid climate.

Because of the temperature influence, precipitation effectiveness depends on the season in which an arid region's meager precipitation is concentrated. Obviously, precipitation received during the low-sun period will be more effective than that received during the high-sun period when temperatures are higher because less will be lost through ET. The simplified Köppen graphs based on the concept of precipitation effectiveness are included in the "Graph Interpretation" exercise (at the end of Chapter 8) and may be used to determine whether a particular location has a desert, steppe, or humid climate.

## Desert Climates

The deserts of the world extend through such a wide range of latitudes that the simplified Köppen system recognizes two major subdivisions. The first are low-latitude deserts where temperatures are relatively high year-round and frost is absent or infrequent even along poleward margins; the second are middle-latitude deserts, which have distinct seasons, including below-freezing temperatures during winter (Table 9.2). However, the significant characteristic of all deserts is their aridity. The relative unimportance of temperature is emphasized by the small number of occasions on which we will distinguish between low-latitude and middle-latitude deserts in the discussion that follows.

**Land of Extremes** By definition, deserts are associated with a minimum of precipitation, but they represent the extremes in other atmospheric conditions as well. With few clouds and little water vapor in the air, as much as 90% of insolation reaches Earth in desert regions. This is why the highest insolation and highest temperatures are recorded in low-latitude desert areas and not in the more humid tropical climates that are closer to the equator. Again because of light cloud cover, there is little atmospheric effect, and much of the energy received by Earth during the day is radiated back to the atmosphere at night. Consequently, night temperatures in the desert drop far below

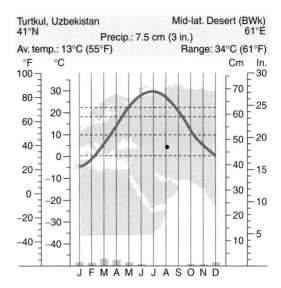

**FIGURE 9.15**
Climographs for desert climate stations.
**If you consider the serious limitations of desert climates, how do you explain why some people choose to live in desert regions?**

their daytime highs. This excessive heating and cooling give low-latitude deserts the greatest diurnal temperature ranges in the world, and middle-latitude deserts are not far behind. In the spring and fall, these ranges may be as great as 40°C (72°F) in a day. More common diurnal temperature ranges in deserts are 22°C–28°C (40°F–50°F).

The sun's rays are so intense in the clear, dry desert air that temperatures in shade are much lower than those a few steps away in direct sunlight. (Keep in mind that all temperatures for meteorological statistics are recorded in the shade.) Khartoum, Sudan (in the Sahara), has an *average* annual temperature of 29.5°C (85°F), which is a *shade* temperature. Temperatures in the bright desert sun under cloudless skies at Khartoum are often 43°C (110°F) or more. Soil temperatures rise to close to 95°C (200°F) in midsummer in the Mojave Desert of southern California.

During low-sun or winter months, deserts experience colder temperatures than more humid areas at the same latitude, and in summer they experience hotter temperatures. Just as with the high diurnal ranges in deserts, these high annual temperature ranges can be attributed to the lack of moisture in the air.

Annual temperature ranges are usually greater in middle-latitude deserts, such as the Gobi in Asia, than in low-latitude deserts because of the colder winters experienced at higher latitudes. Compare, for example, the climograph for

**FIGURE 9.16**
A rainstorm in the Mojave Desert of California produces a double rainbow.
**What environmental clues suggest that rainfall is an infrequent event?**

Aswan in south central Egypt—at 24°N, a low-latitude desert location—with the climograph for Turtkul, Uzbekistan—at 41°N, a middle-latitude desert location ( Fig. 9.15). The annual range for Aswan is 17°C (31°F); in Turtkul, it is 34°C (61°F).

Precipitation in the desert climate is irregular and unreliable, but when it comes, it may arrive in an enormous cloudburst, bringing more precipitation in a single rainfall than has been recorded in years ( Fig. 9.16). This happened in the extreme at the port of Walvis Bay, on the coast of the

# Geography's Spatial Science Perspective
## Desertification

Although desertification is sometimes confused with drought, the two terms describe distinctly different processes. Drought—a longer than normal period of little or no rainfall—is a naturally recurring climatic event. It is especially common in arid and semiarid regions but can occur in subhumid and sometimes even humid climates. As a general rule, the drier a climate is, the greater are the variability of rainfall and the risk of drought.

Desertification, by contrast, is the natural process of desert expansion caused by climatic change but accelerated by human activities. It is a more serious situation than drought because it involves long-term environmental and human consequences. Desertification expands the margins of the desert when rare rains cause gully erosion, sheet erosion, and loss of soil. It also increases wind erosion, causing dust storms and sand dune movement into grassland and farmland areas. Desertification is pronounced in regions of the world where humans have accelerated the expansion of desert climate and landform features into former grassland and woodland regions. Although climate change may be the trigger, the process is accelerated by deforestation, overcultivation, soil salinization due to irrigation, and overgrazing by cattle, sheep, and goats.

The Sahel region of Africa, shown here in a light-tan color, is the transition zone between the extreme aridity of the Sahara and the tropical humid areas of Africa (in green tones). In recent years, the Sahel has experienced desertification through climate change and overuse of this marginal land by human activities.

Namib Desert, a cold-current coastal desert of southwest Africa. The equivalent of 10 years' rain was received in one night when a freak storm dumped 3.2 centimeters (1.3 in.) of rain.

The actual amount of water vapor in the air may be high in desert areas. However, the hot daytime temperatures increase the capacity of the air to hold moisture so that the relative humidity during daylight hours is quite low (10–30%). Desert nights are a different story. Radiation of energy is rapid in the clear air. As temperature drops, relative humidity increases and the formation of dew in the cool hours of early morning is common. Where measurements have been made, the amount of dew formed has sometimes considerably exceeded the annual rainfall for that location. It has been suggested, in fact, that dew may be of great importance to plant and animal life in the desert. Studies are now being carried out on the use of dew as a moisture source for certain crops, thereby minimizing the need for large-scale irrigation.

The convection currents set up by the intense heating of the land during the day help make the desert a windy place. In addition, the sparseness of vegetation and the absence of topographic interruptions in some deserts allow winds to sweep across these arid lands unimpeded. Sand and dust are carried by the desert winds, lowering visibility and irritating eyes and throats ( Fig. 9.17).

Desertification is not new. Archeological evidence from Israel and Jordan indicates that as far back as 4000 BC early farming communities may have destroyed the soil and deforested the hills, causing desertification. Recent research into ancient environmental catastrophes has shown a similar pattern of denudation of the hilly landscape of Greece as early as 3000 BC. Today, evidence of desertification is visible in areas of Spain that exhibit deep gully erosion, in northwestern India as the Thar Desert expands into Rajasthan's farming areas, and throughout much of the Middle East, northern China, and Africa. Along with the threat to the human population, desertification endangers habitats for wildlife.

It was not until the 1970s, however, that desertification became well known, as television revealed starving and suffering citizens of the nations of the African Sahel. It showed bone-thin cattle trying to find a blade of grass in a barren landscape. The TV also revealed villages being invaded by sand dunes. The Sahel is the semiarid zone bordering the southern margin of the Sahara. It extends across northern Africa from Mauritania on the Atlantic coast to Somalia on the Indian Ocean. The term *desertification* was popularized at a U.N. conference dealing with problems like those of the Sahel, and most people associate the term with the continuing plight of the people in the region.

The United Nations Environment Program (UNEP) includes a cost estimate of up to $20 billion annually for 20 years in order to successfully fight worldwide desertification. In 1994, 87 nations signed the Desertification Convention in Paris. When ratified by 50 nations, this treaty will budget funds to help protect the fertility of lands that are at the greatest risk of desertification. It will take the support of all the world's nations for antidesertification programs to be successful. Only a major international effort can deal with a natural hazard that causes such large-scale environmental deterioration and human suffering.

Dune sand encroaches on grazing land in Mali, in the west African part of the Sahel region. Desertification reduces the amount of land that is directly usable for agriculture and grazing.

**Adaptations by Plants and Animals** Deserts tend to have sparse vegetation, and large tracts may be barren bedrock, sand, or gravel. The plants that do exist are **xerophytic,** or adapted to extreme drought. They may have thick bark, thorns, little foliage, and waxy leaves, all of which reduce loss of water by transpiration. Another characteristic adaptation is the storage of moisture in stem or leaf cells, as in the cactus (saguaro and barrel cactus, prickly pear) ( ❚ Fig. 9.18). Some plants, such as creosote bush, mesquite, and acacias, have deep root systems to reach water; others, such as the Joshua tree, spread their roots widely near the surface for their moisture supply.

Even humans, the most adaptable of animals, find the desert environment a lasting challenge. For the most part, people have been hunters and gatherers, nomadic herders, and subsistence farmers wherever there was a water supply from wells, oases, or exotic streams (streams bearing water from outside the region), such as the Nile, Tigris, Euphrates, Indus, and Colorado Rivers ( ❚ Fig. 9.19). Desert people have learned to adjust their habits to the environment. For example, they wear loose clothing to protect themselves from the burning rays of the sun and to prevent moisture loss by evaporation from the skin. At night, when the temperatures drop, the clothing keeps them warm by insulating and minimizing the loss of body heat.

**FIGURE 9.17**
The dust storm invades a marketplace in Sudan.
**How might storms like these effect human health? What about machinery and vehicles?**

**FIGURE 9.18**
Vegetation adapted to the arid conditions in the Sonoran Desert, Arizona.
**What physical characteristics of cacti help them to survive the heat, drought, and evaporation rates of the desert?**

Permanent agriculture has been established in desert regions all around the world wherever river or well water is available. Some produce mainly subsistence crops, but others have become significant producers of commercial crops for export.

## Steppe Climates

Further study of Figure 9.14 and Table 9.2 provides a reminder that the distribution of the world's steppe lands is closely related to the location of deserts. Both moisture-deficient climate types share the controlling factors of continentality, rain-shadow location, the subtropical high pressure systems, or some combination of the three. The transitional nature of the steppes may make them seem like better-watered deserts at one time and like slightly subhumid versions of their humid-climate neighbors at another. Herein lies the major problem of steppe regions: How and to what extent should these variable and unpredictable climate regimes be used by humans?

### Similarities to Deserts
We are already aware that steppe regions are differentiated from deserts by their greater precipitation. Whereas most low-latitude desert locations receive fewer than 25 centimeters (10 in.) of rain annually, low-latitude steppe regions usually receive between 25 and 50 centimeters (10–20 in.). However, the similarities between deserts and steppes are often greater than the differences. In both climates, the potential ET exceeds the precipitation. As in the deserts, precipitation in steppe regions is unpredictable and varies widely in total amount from year to year. Annual rainfall differs significantly from place to place within both desert and steppe regions, and vegetation varies accordingly.

To be more specific, both the general precipitation pattern and the nature of the vegetation of a steppe region are usually closely related to the more humid climate immediately adjacent to it. Thus, when the steppe is located between desert and tropical savanna, the steppe's rains come with the high-sun season. Next to a Mediterranean climate, the steppe receives primarily winter precipitation. Similarly, the short, shallow-rooted grasses most commonly associated with the steppe climate occur in the areas of transition from mesothermal and microthermal climates to desert. However, in the areas of transition from tropical savanna to desert, the vegetation is the dry savanna type, including scrub tree and bush growth, which becomes more stunted and sparse toward the drier margins until the typical desert shrub–vegetation type is dominant.

Both low-latitude and middle-latitude steppe varieties are identified by mean annual temperature (Table 9.2). As in the desert, steppe temperatures vary throughout the climate type with latitude, distance from the sea, and elevation. The climographs of Figure 9.20 demonstrate that although summer temperatures are high in all steppe regions, the differences in winter temperatures can produce annual ranges in middle-latitude steppes that are two or three times as great as those in low-latitude steppes.

**A Dangerous Appeal** Although the surface cover is often incomplete, in more humid regions of the middle-latitude steppes the grasses have been excellent for pasture. In North America, this was the realm of the bison and antelope; in Africa, it was the domain of wildebeests and zebra. Steppe soils are usually high in organic matter and soluble minerals. Attributes such as these have attracted farmers and herders alike to the rich grasslands—but not without penalty to both humans and land.

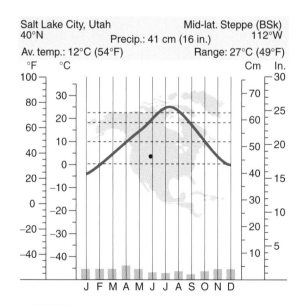

■ **FIGURE 9.19**
This oasis supplies enough water to support a small suburb of Ica, in southern Peru.
**What can you see in the background that proves this is a desert environment?**

■ **FIGURE 9.20**
Climographs for steppe climate stations.
**What are the chief differences between the climographs for Salt Lake City and Daly Waters? What are the causes of these differences?**

The climate is dangerous for agriculture, even in the middle-latitude steppe, and people take a sizable risk when they attempt to farm. Although dry-farmed wheat and drought-resistant barley and sorghum can be successfully raised because both farming methods and crops are adapted to the environment, the use of techniques employed in more humid regions can lead to serious problems. During dry cycles, crops fail year after year, and with the land stripped of its natural sod, the soil is exposed to wind erosion. Even using the grasses for grazing domesticated animals is not always the answer, for overgrazing can just as quickly create "Dust Bowl" conditions (as occurred during the extreme droughts of the 1930s in the southwestern United States).

The difficulties in making steppe regions more productive point out again the sensitive ecological balance of Earth's systems. The natural rains in the steppe are usually sufficient to support a vegetation cover of short grasses that can feed the roaming herbivores that graze on them. The herbivore population in turn is kept in check by the carnivores who prey on them (Fig 9.21). When people enter the scene, however, sending out more animals to graze, plowing the land, or merely killing off the predators, the ecological balance is tipped, and sometimes the results are disastrous.

**FIGURE 9.21**
This wildebeest has more to worry about than an attack by a predator.
**What other types of dangers might threaten this environment?**

# Define & Recall

tropical rainforest climate
annual temperature range
daily (diurnal) temperature range
cloud forest
selva
jungle
slash-and-burn (shifting cultivation)
tropical monsoon climate
tropical savanna climate
desert climate
steppe climate
xerophytic

# Discuss & Review

1. What can you say about the amount of insolation as it relates to tropical climates?
2. Describe the three tropical (A) climates. Why does each exist?
3. How do ocean currents affect tropical and arid climates?
4. Describe the delicate balance between vegetative growth and soil fertility in a tropical rainforest environment.
5. What is the difference between a rainforest and a jungle? What factor differentiates these two forest types?
6. How does the ITCZ play a role in differentiating tropical rainforest from the tropical monsoon? How does the tropical savanna fit into this role?
7. Where do you find the greatest difference in temperature between direct sunlight and shade, the rainforest or the savanna? Why are these two so different?
8. What is the major factor producing the dryness of arid climates?
9. What exceptions are there to the general rule that the pattern of climate regions is produced by Earth–sun relationships? How do the Northern and Southern Hemispheres differ in climate patterns, and why?
10. What are the controlling factors that explain the tropical rainforest climate? Give a brief verbal description of this climate type.

11. What aspects of tropical rainforests are favorable to human use? Unfavorable? Describe the delicate balance between forest and soil. How might humans affect that balance?
12. Explain the seasonal precipitation pattern of the tropical savanna climate. State some of the transitional features of this climate. How have vegetation and soils adapted to the wet–dry seasons?
13. What conditions give rise to desert climates?
14. How do steppes differ from deserts? Why might human use of steppe regions in some ways be more hazardous than use of deserts?
15. Do each of the following for the listed climates: tropical rainforest, tropical monsoon, tropical savanna, desert, and steppe.

a. Identify the climate from a set of data or a climograph provided in the chapter indicating average monthly temperature and precipitation for a representative station within a region typical of that climate.
b. Match the climate type with a written statement that includes one or more of the following: the statistical parameters of the climate in the modified Köppen classification; the particular climate controls (controlling factors) that produce the climate; and the geographic distributions of the climate, stated in terms of physical or political location.
c. Distinguish between the important subtypes (if any) of each climate by identifying the characteristics that separate them from one another.

## Consider & Respond

1. Based on the classification scheme presented in the "Graph Interpretation" exercise (end of Chapter 8), classify the following climate stations from the data provided.

|   |   | J | F | M | A | M | J | J | A | S | O | N | D | Yr. |
|---|---|---|---|---|---|---|---|---|---|---|---|---|---|---|
| a. | Temp. (°C) | 20 | 23 | 27 | 31 | 34 | 34 | 34 | 33 | 33 | 31 | 26 | 21 | 29 |
|   | Precip. (cm) | 0.0 | 0.0 | 0.0 | 0.0 | 0.2 | 0.1 | 1.1 | 0.2 | 0.0 | 0.0 | 0.0 | 0.0 | 1.6 |
| b. | Temp. (°C) | 21 | 24 | 28 | 31 | 30 | 28 | 26 | 25 | 26 | 27 | 25 | 22 | 26 |
|   | Precip. (cm) | 0.0 | 0.0 | 0.2 | 0.8 | 7.1 | 11.9 | 20.9 | 31.1 | 13.7 | 1.4 | 0.0 | 0.0 | 87.2 |
| c. | Temp. (°C) | 19 | 20 | 21 | 23 | 26 | 27 | 28 | 28 | 27 | 26 | 22 | 20 | 24 |
|   | Precip. (cm) | 5.1 | 4.8 | 5.8 | 9.9 | 16.3 | 18.0 | 17.0 | 17.0 | 24.0 | 20.0 | 7.1 | 3.0 | 149.0 |
| d. | Temp. (°C) | 2 | 4 | 8 | 13 | 18 | 24 | 26 | 24 | 21 | 14 | 7 | 3 | 14 |
|   | Precip. (cm) | 1.0 | 1.0 | 1.0 | 1.3 | 2.0 | 1.5 | 3.0 | 3.3 | 2.3 | 2.0 | 1.0 | 1.3 | 20.6 |
| e. | Temp. (°C) | 27 | 26 | 27 | 27 | 27 | 27 | 27 | 27 | 27 | 27 | 27 | 27 | 27 |
|   | Precip. (cm) | 31.8 | 35.8 | 35.8 | 32.0 | 25.9 | 17.0 | 15.0 | 11.2 | 8.9 | 8.4 | 6.6 | 15.5 | 243.8 |
| f. | Temp. (°C) | 13 | 17 | 17 | 19 | 22 | 24 | 26 | 26 | 26 | 24 | 19 | 15 | 20 |
|   | Precip. (cm) | 6.6 | 2.0 | 2.0 | 0.5 | 0.3 | 0.0 | 0.0 | 0.0 | 0.3 | 1.8 | 4.6 | 6.6 | 26.7 |

2. The following six locations are represented by the data in the previous table, although not in the order listed: Albuquerque, New Mexico; Belém, Brazil; Benghazi, Libya; Faya, Chad; Kano, Nigeria; Miami, Florida. Use an atlas and your knowledge of climates to match the climatic data with the locations.
3. Benghazi, Libya, and Albuquerque, New Mexico, both exhibit steppe climates, but the cause (or control) of their dry conditions is quite different. Describe the primary factor responsible for the dry conditions at each location.
4. Tropical B climates are located more poleward than A climates, yet their daytime high temperatures are often higher. Why?
5. Kano, Nigeria, has copious amounts of rainfall during the summer season. There are two sources, or causes, of this rainfall. What are they?

CHAPTER 10

The forested North Shore of Lake Superior in Minnesota reflects the influence of a humid continental, middle-latitude climate.   David Hansen, University of Minnesota

# Middle-Latitude, Polar, and Highland Climate Regions

PHYSICAL
Geography Now™   This icon, appearing throughout the book, indicates an opportunity to explore interactive tutorials, animations, or practice problems at now.brookscole.com/gabler8

In this chapter, we further explain the world climates as they dominate higher latitudes and higher altitudes. As we have noted, the tropical climates exhibit the constant characteristic of heat, whereas polar climates exhibit the constant characteristic of cold. The arid climates have inadequate precipitation in common. But what is the constant characteristic of the mesothermal and microthermal middle-latitude climates, which we are about to examine? If there is one, then it is the oxymoron "constant inconsistency." Each of the middle-latitude climates is dominated by the changing of the seasons and the variability of atmospheric conditions associated with migrating air masses or cyclonic activity along the polar front.

# CHAPTER PREVIEW

**Perhaps the most distinctive feature of Mediterranean climate regions is that they receive most of their precipitation during the season opposite that during which greatest plant growth normally occurs.**

- Why is this so?
- What plant adaptations have developed as a result?
- What human use patterns have developed in response to these climatic conditions?

**The locations of the Mediterranean and the humid subtropical climates on opposite sides of continents in the lower-middle latitudes provide convincing evidence of the dominance of the subtropical highs as the climatic control of these latitudes.**

- In what ways are the subtropical highs responsible for these contrasting climates?
- How do the highs influence climate both poleward and equatorward of these latitudes?
- How are the oceans involved?

**Some geographers refer to the marine west coast climate as the "temperate oceanic," or simply the "marine," climate.**

- Why do we call this climate the marine west coast climate?
- Why do all these names emphasize maritime terms?
- What factors produce the climate?

**Although humid continental hot summer and mild summer regions have more in common than not, there are distinct differences in natural and cultivated vegetation between the two climates, particularly in North America.**

- What are these differences, and why do they exist?
- How do people deal with these differences?
- What does this statement suggest about scientific classification?

**The climatic and related physical characteristics that distinguish the subarctic climate and the two polar climates are accompanied by human utilization patterns significantly different from those found in the humid continental climates.**

- In what ways does the physical environment in each of these high-latitude climates affect human use patterns?
- In what ways are cultural factors more influential?
- How delicate are these climates?

**Highlands are occupied by a complex pattern of widely varying microclimates far too intricate to be shown on anything but large-scale maps.**

- What factors are most important in explaining the existence of a particular highland microclimate?
- What other phenomena are most directly affected by the different microclimates?
- What are the major peculiarities of mountain climates?

## Middle-Latitude Climates

If change is constant in the mesothermal and microthermal climate groups, then degree of change is what distinguishes one climate type from another. In one or another of the middle-latitude climates, summers vary from hot to cool and winters from mild to extremely cold. Some climates receive adequate monthly precipitation year-round; others experience winter drought or, even more challenging to humans who live there, dry months during the normal summer growing season. Despite the changing atmospheric conditions of middle-latitude climates, their regions are home to the majority of Earth's people, and they are major factors in some of Earth's most attractive, interesting, and productive physical environments.

## Humid Mesothermal Climate Regions

When we use the term *mesothermal* (from Greek: *mesos,* middle) in describing climates, we are usually referring to the moderate temperatures that characterize such regions. However, we could also be referring to their middle position between those climates that have high temperature throughout the year and those that experience severe cold. By definition, the mesothermal climates experience seasonality, with distinct summers and winters that distinguish them from the humid tropics. Their summers are long and their winters mild, and this separates them climatically from the microthermal climates, which lie poleward.

Table 10.1 presents the three distinct mesothermal climates introduced in Chapter 8. In all three, the annual precipitation exceeds the annual potential evapotranspiration, but the Mediterranean climate has a lengthy period of precipitation deficit in the summer season that distinguishes it from the humid subtropical and the marine west coast climates. The latter two are further differentiated by the fact that the humid subtropical regions have hot summers, whereas the marine west coast regions experience mild summers.

### Mediterranean Climate

The **Mediterranean climate** is one of the best arguments for organizing a study of the environment or developing an understanding of world regions, based on climate classification. Such a climate appears with remarkable regularity in the vicinity of 30° to 40° latitude along the west coasts of each landmass ( Fig. 10.1). The alternating controls of subtropical high pressure in summer and westerly wind movement in winter are so predictable that all Mediterranean lands have notably similar and easily recognized temperature and precipitation characteristics ( Fig. 10.2). The special appearance, combinations, and climatic adaptations of Mediterranean vegetation not only are unusual but also are clearly distinguishable from those of other

**FIGURE 10.1**
Index map of humid mesothermal climates

Humid Mesothermal Climates

**TABLE 10.1**
**The Mesothermal Climates**

| Name and Description | Controlling Factors | Geographic Distribution | Distinguishing Characteristics | Related Features |
|---|---|---|---|---|
| *Mediterranean* | | | | |
| Warmest month above 10°C (50°F); coldest month between 18°C (64.4°F) and 0°C (32°F); summer drought; hot summers (inland), mild summers (coastal) | West coast location between 30° and 40°N and S latitudes; alternation between subtropical highs in summer and westerlies in winter | Central California; central Chile; Mediterranean Sea borderlands, Iranian highlands; Capetown area of South Africa; southern and southwestern Australia | Mild, moist winters and hot, dry summers inland with cooler, often foggy coasts; high percentage of sunshine; high summer diurnal temperature range; frost danger | Sclerophyllous vegetation; low, tough brush (chaparral); scrub woodlands; varied soils, erosion in Old World regions; winter-sown grains, olives, grapes, vegetables, citrus, irrigation |
| *Humid Subtropical* | | | | |
| Warmest month above 10°C (50°F); coldest month between 18°C (64.4°F) and 0°C (32°F); hot summers; generally year-round precipitation, winter drought (Asia) | East coast location between 20° and 40°N and S latitudes; humid onshore (monsoonal) air movement in summer, cyclonic storms in winter | Southeastern United States; southeastern South America; coastal southeast South Africa and eastern Australia; eastern Asia from northern India through south China to southern Japan | High humidity; summers like humid tropics; frost with polar air masses in winter; precipitation 62–250 cm (25–100 in.), decreasing inland; monsoon influence in Asia | Mixed forests, some grasslands, pines in sandy areas; soils productive with regular fertilization; rice, wheat, corn, cotton, tobacco, sugarcane, citrus |
| *Marine West Coast* | | | | |
| Warmest month above 10°C (50°F); coldest month between 18°C (64.4°F) and 0°C (32°F); year-round precipitation; mild to cool summers | West coast location under the year-round influence of the westerlies; warm ocean currents along some coasts | Coastal Oregon, Washington, British Columbia, and southern Alaska; southern Chile; interior South Africa; southeast Australia and New Zealand; northwest Europe | Mild winters, mild summers, low annual temperature range; heavy cloud cover, high humidity; frequent cyclonic storms, with prolonged rain, drizzle, or fog; 3- to 4-month frost period | Naturally forested, green year-round; soils require fertilization; root crops, deciduous fruits, winter wheat, rye, pasture and grazing animals; coastal fisheries |

**FIGURE 10.2**
Climographs for Mediterranean climate stations.
**In what way do these climographs differ? What causes the differences?**

climates. Agricultural practices, crops, recreational activities, and architectural styles all exhibit strong similarities within Mediterranean lands.

**Warm, Dry Summers; Mild, Moist Winters** The major characteristics of the Mediterranean climate are a dry summer; a mild, moist winter; and abundant sunshine (90% of possible sunshine in summer and as much as 50–60% even during the rainy winter season). Summers are warm throughout the climate, but there are enough differences between the monthly temperatures in coastal and interior locations to recognize two distinct subtypes. The moderate-summer subtype has the lower summer temperatures associated with a strong maritime influence. The hot-summer subtype is located further inland and reflects an increased influence of *continentality*. The inland version has higher summer and daytime temperatures and slightly cooler winter and nighttime temperatures than its coastal counterpart, and it has greater annual and diurnal temperature ranges as well. Compare, for example, the annual range for Red Bluff, California, an inland station, with that of San Francisco, a coastal station about 240 kilometers (150 mi) farther south (see Fig. 10.2).

Whichever the subtype, Mediterranean summers clearly show the influence of the subtropical highs. Weeks go by without a sign of rain, and evapotranspiration rates are high. Effective precipitation is lower than actual precipitation, and the summer drought is as intense as that of the desert. Days are warm to hot, skies are blue and clear, and sunshine is abundant. The high percentage of insolation coupled with nearly vertical rays of the noon sun may drive daytime temperatures as high as 30°C–38°C (86°F–100°F), except where moderated by a strong ocean breeze or coastal fog.

Fog is common throughout the year in coastal locations and is especially noticeable during the summer. As moist maritime air moves onshore, it passes over the cold ocean currents that typically parallel west coasts in Mediterranean latitudes. The air is cooled, condensation takes place, and fog regularly creeps in during the late afternoon, remains through the night, and "burns off" during the morning hours. As in the desert, radiation loss is rapid at night, and even in summer, nighttime temperatures are commonly only 10°C–15°C (50°F–60°F).

Winter is the rainy season in the Mediterranean climate. The average annual rainfall in these regions is usually between 35 and 75 centimeters (15–30 in.), with 75% or more of the total rain falling during the winter months. The precipitation results primarily from the cyclonic storms and frontal systems common with the westerlies. Annual amounts increase with elevation and decrease with increased distance from the ocean. Only because the rain comes during the cooler months, when evapotranspiration rates are lower, is there sufficient precipitation to make this a humid climate.

Despite the rain during the winter season, there are often many days of fine, mild weather. Insolation is still usually above 50%, and the average temperature of the coldest month rarely falls below 4°C–10°C (40°F–50°F). Frost is uncommon and, because of its rarity, many less hardy tropical varieties of fruits and vegetables are grown in these regions. However, when frost does occur, it can do great damage.

**Special Adaptations** The summer drought, not frost, is the great challenge to vegetation in Mediterranean regions. The natural vegetation reflects the wet–dry seasonal pattern of the climate. During the rainy season, the land is covered with lush, green grasses that turn golden and then brown under the summer drought. Only with winter and the return of the rains does the landscape become green again. Much of the natural vegetation is **sclerophyllous** (hard-leafed) and drought resistant. Like xerophytes, these plants have tough surfaces, shiny, thick leaves that resist moisture loss, and deep roots to help combat aridity.

One of the most familiar plant communities is made up of many low, scrubby bushes that grow together in a thick tangle. In the western United States, this is called **chaparral** ( Fig. 10.3). Most chaparral is less than 25 years old because of the frequent fires that occur in this dry brush. The fires help perpetuate the chaparral because the associated heat is required to open seedpods and allow many chaparral species to reproduce. People often remove chaparral as a preventive measure against fires, but the removal can have disastrous results because the chaparral acts as a check against erosion of soils during the rainy season. With the chaparral removed, soils wash or slide down hillsides during the heavy rains of winter, frequently taking homes with them.

Brush similar to the chaparral of California is called *mallee* in Australia, *mattoral* in Chile, and *maquis* in France. In fact, the "maquis" French Underground that fought against Nazi occupation in World War II literally meant "underbrush."

Trees that appear in the Mediterranean climate also respond to moisture conditions. Because of their drought-resistant qualities, the needle-leaf pines are among the more common species. Groves of deciduous and evergreen oaks appear in depressions where moisture collects and on the shady north sides of hills where evapotranspiration rates are lower. Where the summer drought is more distinct, the scrub and woodlands open up to parklands of grasses and scattered oak trees. Even the great redwood forests of northern California probably could not survive without the heavy fogs that regularly invade the coastal lands in summer ( Fig. 10.4).

In response to the dry summers, most soils in Mediterranean regions are high in soluble nutrients, but the potential for agriculture differs widely from one region to another. The lime-rich soils around the Mediterranean Sea originally were highly productive, but destructive agricultural practices and overgrazing over thousands of years of human use have caused serious erosion problems. Today, bare white limestone is widely visible on hillsides in Spain, Italy, Greece, Crete, Syria, Lebanon, Israel, and Jordan. In the Mediterranean regions of California, the soils are dense clays, gluelike when wet and hard as concrete when dry. During the Spanish period in California, the clay was formed into adobe bricks

### FIGURE 10.4
California redwoods (*Sequoia sempervirens*) may reach heights of 100 meters (330 ft) and live for thousands of years.
**Why is it considered unusual to find redwoods growing in a Mediterranean climate?**

### FIGURE 10.3
Typical remnant chaparral vegetation as found in southern Spain.
**Why would you expect to find little "native" vegetation left in Old World Mediterranean lands?**

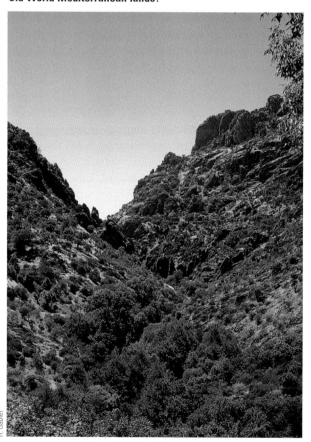

and used for building material. These clay soils are hazardous on slopes because they absorb so much water during the wet season that they can become mudflows destructive to roads and homes in areas where vegetation has been removed by fire or human interference.

In all Mediterranean regions, the most productive areas are the lowlands covered by stream deposits. Here farmers have made special adaptations to climatic conditions. There is sufficient rainfall in the cool season to permit fall planting and spring harvesting of winter wheat and barley. These grasses originally grew wild in the eastern Mediterranean region. Grapevines, fig and olive trees, and the cork oak, which undoubtedly were also native to Mediterranean lands, are especially well adapted to the dry summers because of their deep roots and thick, well-insulated stems or bark (Fig. 10.5). Where water for irrigation is available, an incredible diversity of crops may be seen. These include, in addition to those already mentioned, oranges, lemons, limes, melons, dates, rice, cotton, deciduous fruits, various types of nuts, and countless vegetables. California, blessed both with fertile valleys for growing fruits, vegetables, and flowers and with snow meltwater for irrigation, is probably the most agriculturally productive of the Mediterranean regions.

Even the houses of these regions show people's adaptation to the climate. Usually white or pastel in color, they gleam in the brilliant sunshine against clear blue skies. Many have shuttered windows to reduce incoming sunlight and to keep the houses cool in summer. On the other hand, much less attention is paid to keeping homes warm during the cooler winter months. This is true even in the United States, where many midwesterners and easterners are surprised at the lack of insulation in California homes and the small number and size of heating devices.

## Humid Subtropical Climate

The **humid subtropical climate** extends inland from continental east coasts between 15° and 20° and 40°N and S latitude (refer again to Table 10.1 and Fig. 10.1). Thus, it is located within approximately the same latitudes and in a similar transitional position as the Mediterranean climate but on the eastern instead of the western continental margins. There is ample evidence of this climatic transition. Summers in the humid subtropics are similar to the humid tropical climates further equatorward. When the noon sun is nearly overhead, these regions are subject to the importation of moist tropical air masses. High temperatures, high relative humidity, and frequent convectional showers are all characteristics that they share with the tropical climates. In contrast, during the winter months, when the pressure and wind belts have shifted equatorward, the humid subtropical regions are more commonly under the influence of the cyclonic systems of the continental middle latitudes. Polar air masses can bring colder temperatures and occasional frost.

### Comparison with the Mediterranean Climate
Like the inland version of the Mediterranean climate, the humid subtropical climate has mild winters and hot summers. But it has no dry season. Whereas the Mediterranean lands are under the drought-producing eastern flank of the subtropical high pressure sytems, the humid subtropical regions are located on the weak western sides of the subtropical highs. Subsidence and stability are greatly reduced or absent, even during the summer months. Here again, ocean temperatures play a significant role. The warm ocean currents that are commonly found along continental east coasts in these latitudes also moderate the winter temperatures and warm the lower atmosphere, thus increasing lapse rates, which enhances instability. Furthermore, a modified monsoon effect (especially in Asia and to some extent in the southern United States) increases summer precipitation as the moist, unstable tropical air is drawn in over the land.

As might be expected from the year-round rainfall, average annual precipitation for humid subtropical locations usually exceeds that for Mediterranean stations and may vary more widely as well. Humid subtropical regions receive anywhere from 60 to 250 centimeters (25–100 in.) a year. Precipitation generally decreases inland toward continental interiors and away from the oceanic sources of moisture. It is not surprising that these regions are noticeably

**FIGURE 10.5**
Grapes adapt well to the Mediterranean climate. This vineyard is located in California.
**These are not tall trees. What factors of this climate might limit their growth?**

drier the closer they are to steppe regions inland toward their western margins.

Both the Mediterranean and humid subtropical climates receive winter moisture from cyclonic storms, which travel along the polar front. As we have noted, the great contrast occurs in the summer when the humid subtropics receive substantial precipitation from convectional showers, supplemented in certain regions by a modified monsoon effect. In addition, because of the shift in the sun and wind belts during the summer months, the humid subtropical climates are subject to tropical storms, some of which develop into hurricanes (or typhoons), especially in late summer. These three factors—the modified monsoon effect, convectional activity, and tropical storms—combine in most of these regions to produce a precipitation maximum in late summer. The climographs for New Orleans, Louisiana, and Brisbane, Australia, illustrate these effects ( Fig. 10.6).

A subtype of the humid subtropical climate is found most often on the Asian continent, where the monsoon effect is most pronounced because of the magnitude of the seasonal pressure changes over this immense landmass. There, the low-sun period, or winter season, is noticeably drier than the high-sun period. High pressure over the continent blocks the importation of moist air so that some months receive less than 3 centimeters (1.2 in.) of precipitation.

Temperatures in the humid subtropics are much like those of the Mediterranean regions. Annual ranges are similar, despite a greater variation among climatic stations in the humid subtropical climate, primarily because the climate covers a far larger land area. Mediterranean stations record higher summer daytime temperatures, but summer months in both climates average around 25°C (77°F), increasing to as much as 32°C (90°F) as maritime influence decreases inland. Winter months in both climates average around 7°C–14°C (45°F–57°F). Frost is a similar problem. The long growing season in the warmest humid subtropical regions enables farmers to grow such delicate crops as oranges, grapefruit, and lemons, but, as in the Mediterranean climates, farmers must be prepared with various means to protect their more sensitive crops from the danger of freezing. The growers of citrus crops in Florida are concentrated in the Central Lake District to take advantage of the moderating influence of nearby bodies of water.

Variation in humidity greatly affects the effective temperatures—the temperatures we feel—in the humid subtropical and Mediterranean climates. The summer temperatures in humid subtropical regions feel far warmer than they are because of the high humidity there. In fact, summers in this climate are oppressively hot, sultry, and uncomfortable. Nor is there the relief of lower night temperatures, as in the Mediterranean regions. The high humidity of the humid subtropical climate prevents much radiative loss of heat at night. Consequently, the air remains hot and sticky. Diurnal temperature ranges, in winter as well as in summer, are far smaller in the humid subtropical than in the drier Mediterranean regions. Despite the relatively mild temperatures, humid subtropical winters seem cold and damp, again because of the high humidity.

**A Productive Climate** Vegetation generally thrives in humid subtropical regions, with their abundant rainfall, high temperatures, and long growing season. The wetter portions support forests of broad-leaf deciduous trees, pine forests on sandy soils, and mixed forests ( Fig. 10.7). In the drier interiors near the steppe regions, forests give way to grasslands,

**FIGURE 10.6**
Climographs for humid subtropical climate stations.
**What hemispheric characteristics are shown in these graphs?**

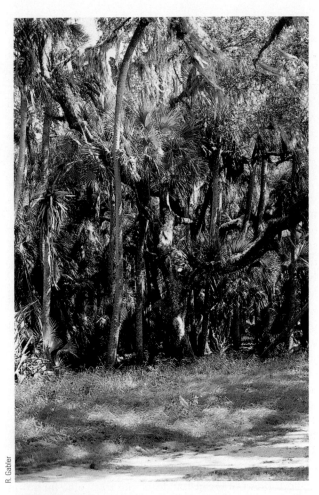

**FIGURE 10.7**
Forest vegetation similar to this in Myakka State Park originally covered much of humid subtropical central Florida.
**How has the physical landscape of central Florida been changed by human occupancy?**

which require less moisture. There is an abundant and varied fauna. A few of the common species are deer, bears, foxes, rabbits, squirrels, opossums, raccoons, skunks, and birds of many sizes and species. Bird life in lake and marsh areas is incredibly rich. Alligators inhabit the American swamps in North Carolina, Georgia, and Florida.

As in the tropics, soils tend to have limited fertility because of rapid removal of soluble nutrients. However, there are exceptions in drier grassland areas, such as the pampas of Argentina and Uruguay, which constitute South America's "bread basket." Whatever the soil resource, the humid subtropical regions are of enormous agricultural value because of their favorable temperature and moisture characteristics. They have been used intensively for both subsistence crops, such as rice and wheat in Asia, and commercial crops, such as cotton and tobacco in the United States. When we consider that this climate (with its monsoon phase) is characteristic of south China as well as of the most densely populated portions of both India and Japan, we realize that this climate regime contains and feeds far more human beings than any other type (Fig. 10.8). The care with which agriculture has been practiced and the soil resource conserved over thousands of years of intensive use in eastern Asia is in sharp contrast to the agricultural exploitation of the past 200 years in the corresponding area of the United States. The traditional system of cotton and tobacco farming, in particular, devastated the land by exhausting the soil and triggering massive sheet and gully erosion. Over much of the old cotton and tobacco belt, extending from the Carolinas through Georgia and Alabama to the Mississippi delta, all topsoil or a significant part of it has been lost, and we see only the red clay subsoil and occasionally bare rock where crops formerly flourished. In the remaining areas, practices have had to change to conserve the soil that is left. Heavy applications of fertilizer, scientific crop rotation, and careful tilling of the land are now the rule.

Where forests still form the major natural vegetation, forest products, such as lumber, pulpwood, and turpentine, are important commercially (Fig. 10.9). The long-leaf and slash pines of the southeastern United States are a source of lumber, as well as the resinous products of the pine tree (pitch, tar, resin, and turpentine). The absence of temperature and moisture limitations strongly favors forest growth. In Georgia, for example, trees may grow two to four times faster than in colder regions such as New England. This means that trees can be planted and harvested in much less time than in cooler forested regions, offering distinct commercial advantages.

The fact that living things thrive in the humid subtropical climate presents certain problems, however, for parasites and disease-carrying insects thrive along with other forms of life. African "killer bees," for example, have migrated through Mexico and now pose a serious new threat to humans and livestock in the southern United States.

Agriculture and lumbering are not the only important industries in the humid subtropical climates. In the southeastern United States, livestock raising has been increasing greatly in importance. Despite the commercial advantages of this climate, people often find it an uncomfortable one in which to live. The development and spread of air conditioning helps mitigate this problem. Where the ocean offers relief from the summer heat, as in Florida, the humid subtropical climate is an attractive recreation and retirement region. The beauty of its more unusual features, such as its cypress swamps and forests draped with Spanish moss, has to be experienced to be fully appreciated.

## Marine West Coast Climate

Proximity to the sea and prevailing onshore winds make the **marine west coast climate** one of the most temperate in the world. Thus, it is sometimes known as the *temperate oceanic climate*. Found in those middle-latitude regions (between 40° and 65°) that are continuously influenced by the westerlies, the marine west coast climate receives ample precipitation throughout the year. However, unlike the humid subtropical climate

**FIGURE 10.8**
The terraced fields in this humid subtropical climate region near Wakayama, Japan, are ideally suited for rice production.
**Why is rice a preferred crop in Japan?**

**FIGURE 10.9**
A commercial tree farm in northern Georgia. Note that the pine trees are planted in orderly rows to expedite cultivation and harvest.
**Why are tree farms common in the U.S. Southeast and Pacific Northwest but not in New England or the upper Midwest?**

just discussed, it has mild to cool summers. The climograph for Bordeaux, France, in Figure 10.10 is representative of the mild-summer marine west coast climate, and the climograph for Reykjavik, Iceland, represents the cool-summer variety.

**Oceanic Influences** As they travel onshore, the westerlies carry with them the moderating marine influence on temperature, as well as much moisture. In addition, warm ocean currents, such as the North Atlantic Drift, bathe some of the coastal lands in the latitudes of the marine west coast climate, further moderating climatic conditions and accentuating humidity. This latter influence is particularly noticeable in Europe where the marine west coast climate extends along the coast of Norway to beyond the Arctic Circle (see Fig. 10.1).

In this climate zone, the marine influence is so strong that temperatures decrease little with poleward movement. Thus, the influence of the oceans is even stronger than latitude in determining these temperatures. This is obvious when we examine isotherms in the areas of marine west coast climates on a map of world temperatures (see again Figures 4.27 and 4.28). Wherever the marine west coast climate prevails, the isotherms swing poleward, parallel to the coast, clearly demonstrating the dominant *marine influence*.

Another result of the ocean's moderating effect is that the annual temperature ranges in the marine west coast climates are relatively small, considering the latitude. For an illustration of this, compare the monthly temperature graphs for Portland, Oregon, and Eau Claire, Wisconsin (Fig. 10.11). Though these two cities are at the same latitude, the annual range at Portland is 15.5°C (28°F), while at Eau Claire it is 31.5°C (57°F). The moderating effect of the ocean on Portland's temperatures is clearly contrasted to the effect of *continentality* on the temperatures in Eau Claire.

Diurnal temperature ranges are also smaller than they are in other climate regions at similar latitudes and in more arid climates. Heavy cloud cover and high humidity in both summer and winter diminish daytime heating and prevent much radiational cooling at night. Consequently, the difference between the daily maximum and minimum temperatures is small.

Of course, these climographs and climate statistics represent averages, which can be misleading. Marine west coast climate regions

**FIGURE 10.10**
Climographs for marine west coast climate stations.
**How do you explain why Reykjavik has a lower temperature range than Bordeaux?**

**FIGURE 10.11**
Effect of maritime influence on climates of two stations at the same latitude. Portland, Oregon, exemplifies the maritime influences dominating marine west coast climates. Eau Claire, Wisconsin, shows the effect of location in the continental interior. The difference in temperature ranges for the two stations are significant, but note also the interesting differences in precipitation distribution.
**How do you explain these differences?**

experience the unpredictable weather conditions associated with the polar front. Occasional invasions of a tropical air mass in summer or a polar air mass in winter can move against the general westerly flow of air in these latitudes and produce surprising results. For example, under just such weather conditions, temperatures in Seattle, Washington, have reached a high of 38°C (100°F) and a low of −19°C (−3°F).

Despite the insulating effect of cloudy skies and high-moisture content of the air, slowing heat loss at night, frost is a significant factor in the marine west coast climate. It occurs more often, may last longer, and is more intense than in other mesothermal regions. The growing season is limited to 8 months or less, but even during the months when freezing temperatures may occur, only half the nights or fewer may experience them. The possibility of frost and the frequency of its occurrence increase inland far more rapidly than they do poleward, once more illustrating the importance of the marine influence.

As final evidence of oceanic influences, study the distribution of the marine west coast climate in Figure 10.1. Where mountain barriers prevent the movement of maritime air inland, this climate is restricted to a narrow coastal strip, as in the Pacific Northwest of North America and in Chile. Where the land is surrounded by water, as in New Zealand, or where the air masses move across broad plains, as in much of northwestern Europe, the climate extends well into the interior of the landmass.

**Clouds and Precipitation** The marine west coast has a justly deserved reputation as one of the cloudiest, foggiest, rainiest, and stormiest climates in the world. This is particularly true during the winter season. Rain or drizzle may last off and on for days, though the amount of rain received is small for the number of rainy days recorded. Even when rain is not falling, the weather is apt to be cloudy or foggy. Advection fog may be especially common and long lasting in the winter months when air masses pass over warm ocean currents and pick up considerable moisture, which is then condensed as a fog when the air masses move over colder land. The cyclonic storms and frontal systems are also strongest in the winter when the subtropical highs have shifted equatorward. Conspicuous winter maximums in rainfall occur near the coasts and near boundaries with the Mediterranean climate. However, farther inland a summer maximum may occur.

Though all parts of this climate type receive ample precipitation, there is much greater site-to-site variation in precipitation averages than in temperature statistics. Precipitation tends to decrease very gradually as one moves inland, away from the oceanic source of moisture. It also decreases equatorward, especially during summer months, as the influence of the subtropical highs increases and the influence of the westerlies decreases. This can bring about periods of beautiful, clear weather, something rarely associated with this climate but not uncommon in our Pacific Northwest.

The most important factor in the amount of precipitation received is local topography. When a mountain barrier such as the Cascades in the Pacific Northwest or the Andes in Chile parallels the coast, abundant precipitation, both cyclonic and orographic, falls on the windward side of the mountains. Valdivia, Chile, located windward of the Andes, receives an average of 267 centimeters (105 in.) of precipitation a year. A similar location in Canada, Henderson Lake, British Columbia, averages 666 centimeters (262 in.) of rain a year, the highest figure for the entire North American continent. During the colder Pleistocene Epoch, these high precipitation amounts, falling largely as winter snow, produced large mountain glaciers. In many cases, these came down to the sea, excavating deep troughs that now appear as elongated inlets or *fjords*. Fjord coasts are present in Norway, British Columbia, Chile, and New Zealand—all areas of marine west coast climates today (❙ Fig. 10.12). In contrast, where there are lowlands and no major landforms of high elevation, precipitation is spread more evenly over a wide area, and the amount received at individual stations is more moderate, around 50–75 centimeters (20–30 in.) annually. This is the situation in much of the Northern European Plain, extending from western France to eastern Poland.

Two aspects of precipitation are directly related to the moderate temperatures of this climate regime. Snow falls infrequently and, when it does, it melts or turns to slush as soon as it hits the ground. Snow is especially rare in lowland regions of this climate zone. Paris averages only 14 snow days a year; London, 13; and Seattle, 10. In addition, thunderstorms and convectional showers are uncommon although they occur occasionally. Even in summer, surface heating is rarely sufficient to produce the towering cumulonimbus clouds.

❙ **FIGURE 10.12**
The scenic fjords of coastal Norway, shown here, were produced by glacial erosion during the Pleistocene ice advance.
**In what other areas of the world are fjords common?**

**FIGURE 10.13**
Reliable precipitation makes a diversified type of agriculture possible in the marine west coast climate areas, with emphasis on grain, orchard crops, vineyards, vegetables, and dairying. The village in the photograph is Iphofen, Germany.
**Although the climate is similar, why would a photo taken in a marine west coast agricultural region of the United States depict a scene that is significantly different from this one?**

**Resource Potential** There is little doubt that this climate offers certain advantages for agriculture. The small annual temperature ranges, mild winters, long growing seasons, and abundant precipitation all favor plant growth. Many crops, such as wheat, barley, and rye, can be grown farther poleward than in more continental regions. Although the soils common to these regions are not naturally rich in soluble nutrients, highly successful agriculture is possible with the application of natural or commercial fertilizers ( Fig. 10.13). Root crops (such as potatoes, beets, and turnips), deciduous fruits (such as apples and pears), berries, and grapes join the grains previously mentioned as important farm products. Grass in particular requires little sunshine, and pastures are always lush. The greenness of Ireland—the Emerald Isle—is evidence of these favorable conditions, as is the abundance of herds of beef and dairy cattle.

The magnificent forests that form the natural vegetation of the marine west coast regions have been a readily available resource. Some of the finest stands of commercial timber in the world are found along the Pacific coast of North America where pines, firs, and spruces abound, commonly exceeding 30 meters (100 ft) in height. Europe and the British Isles were once heavily forested, but most of those forests (even the famous Sherwood Forest of Robin Hood fame) have been cut down for building material and have been replaced by agricultural lands and urbanization.

# Humid Microthermal Climate Regions

Our definition of humid microthermal includes temperatures high enough during part of the year to have a recognizable summer and cold enough 6 months later to have a distinct winter. In between are two periods, called spring and fall, when all life, and especially vegetation, makes preparations for the temperature extremes. Thus, in this section, we talk about climate regions that clearly display four readily identifiable seasons.

Seasonality is not the only reason we often use the word *variable* when describing the humid microthermal climates. As Figure 10.14 indicates, these climates are generally located between 35°N and 75°N on the North American and Eurasian landmasses. (Since there are no large landmasses at similar latitudes in the Southern Hemisphere, these climate types exist only in the Northern Hemisphere.)

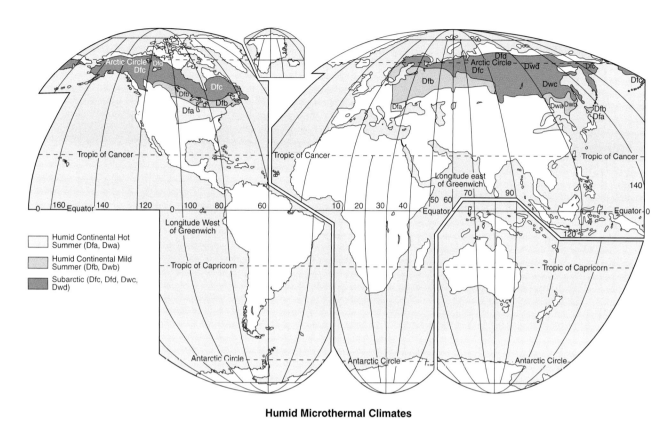

**FIGURE 10.14**
Index map of humid microthermal climates

Thus, they share the westerlies and the storms of the polar front with the marine west coast climate. However, their position in the continental interiors and in high latitudes prevents them from experiencing the moderating influence of the oceans. In fact, the dominance of continentality in these climates is best demonstrated by the fact that they do not exist in the Southern Hemisphere where there are no large landmasses in the appropriate latitudes.

The recognition of three separate microthermal climates is based mainly on latitude and the resulting differences in the length and severity of the seasons (Table 10.2). Winters tend to be longer and colder toward the poleward margins because of latitude and toward interiors throughout the microthermal climates because of the continental influence. Summers inland are also inclined to be hotter, but they become progressively shorter as the winter season lengthens poleward. Thus, the three microthermal climates can be defined as humid continental, hot summer; humid continental, mild summer; and subarctic, with a cool summer and, in extreme cases, a long, bitterly cold winter.

All microthermal climates have several features in common. By definition, they all experience a surplus of precipitation over potential evapotranspiration, and they have year-round precipitation. An exception to this rule lies in an area of Asia where the Siberian High causes winter droughts to occur. The greater frequency of maritime tropical air masses in summer and continental polar air masses in winter, combined with the monsoonal effect and strong summer convection, produce a precipitation maximum in the summer. Although the length of time that snow remains on the ground increases poleward and toward the continental interior (❙ Fig. 10.15), all three microthermal climates experience significant snow cover. This decreases the effectiveness of insolation and helps explain their cold winter temperatures. Finally, the unpredictable and variable nature of the weather is especially apparent in the humid microthermal climates and is present throughout all microthermal regions.

With these generalizations in mind, let's compare the microthermal climates with the mesothermal climates that we have just examined. Regions with microthermal climates have more severe winters, a lasting snow cover, shorter summers, shorter growing seasons, shorter frost-free seasons, a truer four-season development, lower nighttime temperatures, greater average annual temperature ranges, lower relative humidity, and much more variable weather than do the mesothermal climatic regions.

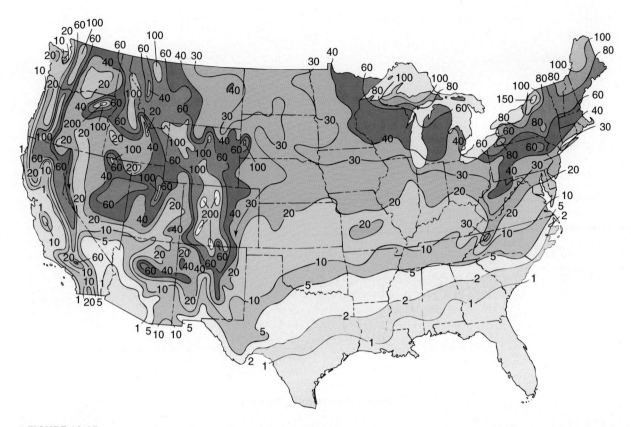

**FIGURE 10.15**
Map of the contiguous United States, showing average annual number of days of snow cover.
**What areas of the United States average the greatest number of days of snow cover?**

## Humid Continental, Hot Summer Climate

Unlike the other two microthermal climates, the **humid continental, hot summer climate** is relatively limited in its distribution on the Eurasian landmass (see Table 10.2). This is unfortunate for the people of Europe and Asia because it has by far the greatest agricultural potential and is the most productive of the microthermal climates. In the United States, this climate is distributed over a wide area that begins with the eastern seaboard of New York, New Jersey, and southern New England and stretches continuously across the heartland of the eastern United States to encompass much of the American Midwest. It is one of the most densely populated, highly developed, and agriculturally productive regions in the world.

In terms of environmental conditions, the hot summer variety of microthermal climate has some obvious advantages over its poleward counterparts. Its higher summer temperatures and longer growing season permit farmers to produce a wide variety of crops. Those lands within the hot-summer region that were covered by ice sheets are far enough equatorward that there has been sufficient time for most negative effects of continental glaciation to be removed, and primarily positive effects remain. Soils are inclined to be more fertile, especially under forest cover, where the typical soil-forming processes are not as extreme and where deciduous trees are more common than the acid-associated pine. Of course, some advantages are matched by liabilities. The lower fuel bills of winter in the humid continental, hot summer climate are often more than offset by the cost of air conditioning during the long, hot summers not found in other microthermal climates.

**Internal Variations** From place to place within the humid continental, hot summer climate, there are significant differences in temperature characteristics. The length of the growing season is directly related to latitude, varying from 200 days equatorward to as little as 140 days along poleward margins of the climate. In addition, the degree of continentality can have an effect on both summer and winter temperatures and, as a result, on temperature range. Ranges are consistently large, but they become progressively larger toward continental interiors. Especially near the coasts in this climatic region, temperatures may be modified by a slight marine influence so that temperatures are milder, in both summer and winter, than those at inland locations at comparable latitudes. Large lakes may cause a similar effect. Even the size of the continent exerts an influence. Galesburg, Illinois, a typical station in the United States, has a significantly lower temperature range than Shenyang,

**TABLE 10.2**
**The Microthermal Climates**

| Name and Description | Controlling Factors | Geographic Distribution | Distinguishing Characteristics | Related Features |
|---|---|---|---|---|
| *Humid Continental, Hot Summer* | | | | |
| Warmest month above 10°C (50°F); coldest month below 0°C (32°F); hot summers; usually year-round precipitation, winter drought (Asia) | Location in the lower middle latitudes (35°–45°); cyclonic storms along the polar front; prevailing westerlies; continentality; polar anticyclone in winter (Asia) | Eastern and midwestern U.S. from Atlantic coast to 100°W longitude; east central Europe; northern and northeast (Manchuria) China, northern Korea, and Honshu (Japan) | Hot, often humid summers; occasional winter cold waves; rather large annual temperature ranges; weather variability; precipitation 50–115 cm (20–45 in.), decreasing inland and poleward; 140- to 200-day growing season | Broad-leaf deciduous and mixed forest; moderately fertile soils with fertilization in wetter areas; highly fertile grassland and prairie soils in drier areas; "corn belt," soybeans, hay, oats, winter wheat |
| *Humid Continental, Mild Summer* | | | | |
| Warmest month above 10°C (50°F); coldest month below 0°C (32°F); mild summers; usually year-round precipitation, winter drought (Asia) | Location in the middle latitudes (45°–55°); cyclonic storms along the polar front; prevailing westerlies; continentality; polar anticyclone in winter (Asia) | New England, the Great Lakes region, and south central Canada; southeastern Scandinavia; eastern Europe, west central Asia; eastern Manchuria (China) and Hokkaido (Japan) | Moderate summers; long winters with frequent spells of clear, cold weather; large annual temperature ranges; variable weather with less total precipitation than further south; 90- to 130-day growing season | Mixed or coniferous forest; moderately fertile soils with fertilization in wetter areas; highly fertile grassland and prairie soils in drier areas; spring wheat, corn for fodder, root crops, hay, dairying |
| *Subarctic* | | | | |
| Warmest month above 10°C (50°F); coldest month below 0°C (32°F); cool summers, cold winters poleward; usually year-round precipitation, winter drought (Asia) | Location in the higher middle latitudes (50°–70°); westerlies in summer, strong polar anticyclone in winter (Asia); occasional cyclonic storms; extreme continentality | Northern North America from Newfoundland to Alaska; northern Eurasia from Scandinavia through most of Siberia to the Bering Sea and the Sea of Okhotsk | Brief, cool summers; long, bitterly cold winters; largest annual temperature ranges; lowest temperatures outside Antarctica; low precipitation, 20–50 cm (10–20 in.); unreliable 50- to 80-day growing season; permafrost common | Northern coniferous forest (taiga); strongly acidic soils; poor drainage and swampy conditions in warm season; experimental vegetables and root crops |

northeast China (Manchuria), which is located at almost the same latitude but which experiences the greater seasonal contrasts of the Eurasian landmass ( Fig. 10.16).

The amount and distribution of precipitation are also variable from station to station. The total precipitation received decreases both poleward and inland ( Fig. 10.17). A move in either direction is a move away from the source regions of warm maritime air masses that provide much of the moisture for cyclonic storms and convectional showers. This decrease can be seen in the average annual precipitation

**FIGURE 10.16**
Climographs for humid continental, hot summer climate stations.
**What are the reasons for the differences in temperature and precipitation between the two stations?**

figures for the following cities, all at a latitude of about 40°N: New York (longitude 74°W), 115 centimeters (45 in.); Indianapolis, Indiana (86°W), 100 centimeters (40 in.); Hannibal, Missouri (92°W), 90 centimeters (35 in.); and Grand Island, Nebraska (98°W), 60 centimeters (24 in.). Most stations have a precipitation maximum in summer when the warm, moist air masses dominate. In certain regions of Asia, not only does a summer maximum of precipitation exist, but also the monsoon circulation inhibits winter precipitation to such a degree that they experience winter drought (see Shenyang, Fig. 10.16).

As might be expected, vegetation and soils vary with the climatic elements, especially precipitation. In the wetter regions, forests and forest soils predominate. At one time, in certain sections of the American Midwest, tall prairie grasses grew where precipitation was sufficient to support forests; in all the drier portions of the climate, grasslands are the natural vegetation. The soils that developed under these grasslands are among the richest in the world.

## Seasonal Changes

The four seasons are highly developed in the humid microthermal, hot summer climate. Each is distinct from the other three and has a character all its own. The winter is cold and often snowy; the spring is warmer, with frequent showers that produce flowers, budding leaves, and green grasses; the hot, humid summer brings occasional violent thunderstorms; and the fall has periods of both clear and rainy weather, with mild days and frosty nights in which the green leaves of summer turn to colorful reds, oranges, yellows, and browns before falling to the ground.

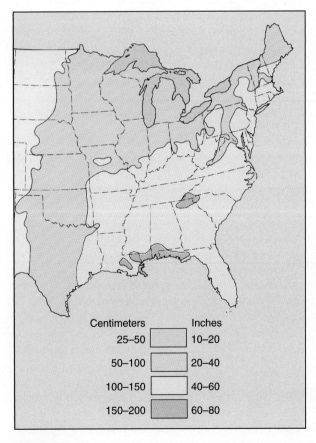

**FIGURE 10.17**
Decreases of precipitation inland from coastal regions are clearly evident in this map of the eastern United States.
**Why does precipitation decrease inland and poleward?**

Of course, the most significant differences are between summer and winter. In all regions, the summers are long, humid, and hot. The centers of the migrating low pressure systems usually pass poleward of these regions, and they are dominated by tropical maritime air. So-called hot spells can go on day and night for a week or more, with only temporary relief available from convectional thunderstorms or an occasional cold (cool) front. Asia, in particular, experiences the heavy summer precipitation associated with the monsoonal effect. Conditions are usually ideal for vigorous vegetative growth. The summer heat and humidity are also the ideal formula for insects; mosquitoes, flies, gnats, and bugs of all kinds abound.

Winters are not as severe as those further poleward, but average January temperatures are usually between −5°C and 0°C (23°F–32°F) or below. Once again, the averages tell only part of the story. There is invariably a prolonged invasion of cold, dry, arctic air once or twice during the winter. This often occurs just after a storm has passed and the ground is covered with snow. The sky remains clear and blue for days at a time; the temperatures will stay near −18°C (0°F) and may occasionally dip to 30°C or 35°C (20°F or 30°F) *below zero* at night. The ground remains frozen for long periods, and snow cover may be present for several days, even weeks, at a time. However, these characteristics do not last continuously because the greatest frequency of cyclones occurs during winter and sudden weather contrasts are common. Cold air precedes warmer air, and thaw follows freeze. Vegetation remains dormant throughout the winter season but bursts into life again with the return of consistently warmer temperatures. Throughout its early growth, vegetation is in constant danger of late-spring frost.

As should be apparent from this description, the atmospheric changes within seasons are just as significant as those between seasons. The humid continental, hot summer climate is the classic example of variable middle-latitude weather. This is the domain of the polar front. Cyclonic storms are born as tropical air masses move northward and confront polar air masses migrating to the south. The daily weather in these regions is dominated by days of stormy frontal activity followed by the clear conditions of a following anticyclone. Above the land is a battlefield in which storms mark the struggles of air masses for dominance, and as in battle, the conflict occurs at the *front*. The general circulation of the atmosphere in these latitudes continuously carries the cyclones and anticyclones toward the east along the polar front. When the polar front is most directly over these regions, as it is in winter and spring, one storm and its associated fronts seem to follow directly behind another with such speed and regularity that the only safe weather prediction is that the weather will change (see again Chapter 7).

## Humid Continental, Mild Summer Climate

If you review the relative distributions of the humid continental, hot summer and mild summer climates in Figure 10.14, the close relationship between the two is unmistakable. Where one is found, the other is found as well; in each situation, the mild summer climate invariably lies adjacent to and poleward from the hot summer climate.

In most instances, the **humid continental, mild summer climate** is a more continental or severe-winter version of its equatorward counterpart. It is characterized by distinct seasonality. There is significant climatic variation, particularly with respect to precipitation, from place to place within the climate. Variable weather is the rule, and storms along the polar front provide most of the precipitation within this climate type. Of course, there are differences between the neighboring climates. These are especially apparent when we examine certain aspects of temperature, growing season, vegetation, and human activity.

**Mild Summer-Hot Summer Comparison** In the microthermal climates, precipitation tends to decrease poleward; therefore, the humid continental, mild summer climate tends to have less precipitation than the hot summer regions closer to the equator. Precipitation continues to decrease throughout this climate type toward the poleward margins and from the coasts toward the arid continental interiors. As in its hot summer counterpart, the monsoon effect in the mild summer climate is strong enough in Asia to produce a dry-winter season (see Vladivostok, Fig. 10.18).

Winters in the mild summer climate are more severe and longer than in its neighbor to the south. Summers, on the other hand, are not as long or hot. The combination of more severe winter and shorter summers makes for a growing season of between 90 and 130 days, which is 1–3 months shorter than in the hot summer climate. In addition, although overall precipitation totals—50 to 100 centimeters (20–40 in.)—are generally lower, snowfall is greater, and snow cover is both thicker and longer lasting (Fig. 10.19).

The humid continental, mild summer regions exhibit seasonal changes just as clearly as the hot summer regions. Annual temperature ranges are generally larger. Vigorous polar and tropical air mass interaction makes weather change a common occurrence. However, the more poleward position of the mild summer climate brings about a greater dominance of the colder air masses and explains why temperature variability is not as abrupt or as great as it is farther south. Under normal conditions, tropical air is strongly modified by the time it reaches mild summer regions: Even in the high-sun season, intrusions of warm, humid air rarely last more than a few days at a time. By contrast, winter invasions of cold arctic air periodically bring several successive days or weeks of clear skies and frigid temperatures.

As in the hot summer climate, the wetter regions of the mild summer climate are associated with natural forest vegetation. However, many trees common in the hot summer climate, such as oaks, hickories, and maples, find it difficult to compete with firs, pines, and spruces toward the colder, polar

**FIGURE 10.18**
Climographs for humid continental, mild summer climate stations.
**What characteristics of these climographs distinguish them from the climographs of Figure 10.16?**

**FIGURE 10.19**
People, animals, and plants living in the humid continental, mild summer regions have learned to cope with abundant snow, which may be present continuously for long periods of time. Here is a typical winter scene near Buffalo, New York.
**The snow presents a variety of problems for some people living in humid continental, mild summer regions. Can you name a few?**

margins of these regions. Fortunately for agriculture, northward extensions of the grasslands and the rich soils that accompany them in the hot summer regions are found in the drier portions of the mild-summer climate.

## Human Activity in the Humid Continental Climates

Perhaps the greatest contrast between the hot summer and mild summer humid continental regions is exhibited in agriculture. Despite the unpredictability of the weather, the humid continental hot summer agricultural regions are among the finest in the world. The favorable combination of long, hot summers, ample rainfall, and highly fertile soils has made the American Midwest a leading producer of corn, beef cattle, and hogs. Soybeans, which are native to similar climate regions in northern China, are now second to corn throughout the Midwest as feed for animals and as a raw material for the food-processing, plastics, and vegetable oil industries. Wheat, barley, and other grains are especially important in European and Asian regions, and winters are sufficiently mild that fall-sown varieties can be raised in the United States. In the mild summer climate, on the other hand, a shorter growing season imposes certain limitations on agriculture and restricts the crops that can be grown. Farmers rely more on quick-ripening varieties, grazing animals, orchard products, and root crops. Dairy products—milk, cheese, butter, cream—are mainstays in the economies of Wisconsin, New York, and northern New England. The moderating effect of the Great Lakes or other water bodies permits the growth of deciduous fruits, such as apples, plums, and cherries.

The length of the growing season is the most obvious reason for the differences in agriculture between the two humid continental climates, but there is another climate-related reason as well. The great ice sheets of the Pleistocene Epoch have had significant but different effects on mild summer and hot summer regions, especially in North America. In the hot summer regions, the ice sheets thinned and receded, releasing the enormous load of soil and solid rock debris they had

**FIGURE 10.20**
Although the effect of glacial action farther south was to deposit material, here in New Hampshire, we see an area of glacial erosion—Newfound Lake, a former glacial trough.
**How might this area be considered an economic resource?**

(50°F), and its poleward limit roughly coincides with the 10°C isotherm for the warmest month of the year. As you may recall from our earlier discussion of the simplified Köppen system, forests cannot survive where at least 1 month does not have an average temperature over 10°C. Thus, the poleward boundary of the subarctic climate is the poleward limit of forest growth as well.

As Figure 10.14 indicates, the subarctic climate, like the other microthermal climates, is found exclusively in the Northern Hemisphere. It covers vast areas of subpolar Eurasia and North America. Conditions vary widely over such great distances. Extremely severe winter regions are located along the polar margins or deep in the interior of the Asian landmass, and climate subtypes with winter drought

stripped off the lands nearer to their centers of origin. The material was laid down in a blanket hundreds of feet thick in the areas of maximum glacial advance. As the ice retreated northward, less and less debris was deposited, much of it flushed away by meltwater streams. The more southerly, hot summer region consequently has an undulating topography underlain by thick masses of glacial debris. The soils formed on this debris are well developed and fertile, and plant nutrients are more likely to be evenly distributed because steep slopes are lacking. The more northerly, mild summer region, on the other hand, mainly shows the effects of glacial erosion ( Fig. 10.20). Rockbound lakes and marshy lowlands alternate with ice-scoured rock hills. Soils are either thin and stony or waterlogged. Because of its lower agricultural potential, large sections of this area remain in forest.

However, because of its wilderness character and the abundance of lakes in basins produced by glacial erosion, recreational possibilities in a mild summer region far exceed those of a more subdued hot summer region. Minnesota calls itself the "Land of 10,000 Lakes," and in New York and New England, lakes, rough mountains, and forest combine to produce some of the most spectacular scenery east of the Rocky Mountains ( Fig. 10.21).

## Subarctic Climate

The **subarctic climate** is the farthest poleward and most extreme of the microthermal climates. By definition, it has at least 1 month with an average temperature above 10°C

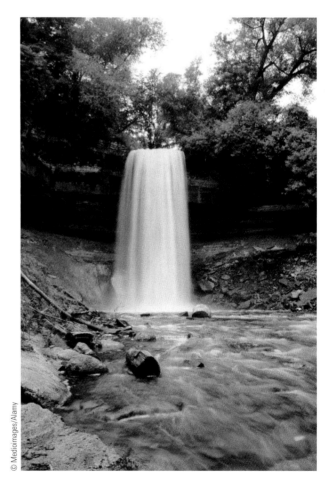

**FIGURE 10.21**
Minnehaha Falls near St. Paul, Minnesota, helps feed one of the area's many lakes.
**What can you say about the water resources of states like New York and Minnesota?**

## Geography's Physical Science Perspective
### Effective Temperatures

*Effective temperatures* (formerly known as *sensible temperatures*) are temperatures as they might be experienced by a person at rest, in ordinary clothing, in a motionless atmosphere. In other words, at any given time, how comfortable does a temperature feel to the individual experiencing it? This temperature value cannot be obtained by simply reading a thermometer.

Several factors come into play when considering effective temperatures. These factors can be divided into two categories: atmospheric factors and human factors. Of the atmospheric factors, four are most important. First is the actual temperature; obviously, thermometers will help distinguish between cold and warm days. Second is humidity; because the evaporation of sweat is a cooling process for the human body, humid days feel warmer than dry days. Third is wind speed; winds not only carry heat from the body but can also accelerate the evaporation of sweat. Fourth is the percentage of clear sky; shady areas are cooler than sunny areas.

Of the many human factors, the following stand out as important. First is respiration; breathing in a lungful of cold air will make one

| Air temperature (°F) \ Relative humidity (%) | 0 | 5 | 10 | 15 | 20 | 25 | 30 | 35 | 40 | 45 | 50 | 55 | 60 | 65 | 70 | 75 | 80 | 85 | 90 | 95 | 100 |
|---|---|---|---|---|---|---|---|---|---|---|---|---|---|---|---|---|---|---|---|---|---|
| 140 | 125 | | | | | | | | | | | | | | | | | | | | |
| 135 | 120 | 128 | | | | | | | | | | | | | | | | | | | |
| 130 | 117 | 122 | 131 | | | | | | | | | | | | | | | | | | |
| 125 | 111 | 116 | 123 | 131 | 141 | | | | | | | | | | | | | | | | |
| 120 | 107 | 111 | 116 | 123 | 130 | 139 | 148 | | | | | | | | | | | | | | |
| 115 | 103 | 107 | 111 | 115 | 120 | 127 | 135 | 143 | 151 | | | | | | | | | | | | |
| 110 | 99 | 102 | 105 | 108 | 112 | 117 | 123 | 130 | 137 | 143 | 150 | | | | | | | | | | |
| 105 | 95 | 97 | 100 | 102 | 105 | 109 | 113 | 118 | 123 | 129 | 135 | 142 | 149 | | | | | | | | |
| 100 | 91 | 93 | 95 | 97 | 99 | 101 | 104 | 107 | 110 | 115 | 120 | 126 | 132 | 138 | 144 | | | | | | |
| 95 | 87 | 88 | 90 | 91 | 93 | 94 | 96 | 98 | 101 | 104 | 107 | 110 | 114 | 119 | 124 | 130 | 136 | | | | |
| 90 | 83 | 84 | 85 | 86 | 87 | 88 | 90 | 91 | 93 | 95 | 96 | 98 | 100 | 102 | 106 | 109 | 113 | 117 | 122 | | |
| 85 | 78 | 79 | 80 | 81 | 82 | 83 | 84 | 85 | 86 | 87 | 88 | 89 | 90 | 91 | 93 | 95 | 97 | 99 | 102 | 105 | 108 |
| 80 | 73 | 74 | 75 | 76 | 77 | 77 | 78 | 79 | 79 | 80 | 81 | 81 | 82 | 83 | 85 | 86 | 86 | 87 | 88 | 89 | 91 |
| 75 | 69 | 69 | 70 | 71 | 72 | 72 | 73 | 73 | 74 | 74 | 75 | 75 | 76 | 76 | 77 | 77 | 78 | 78 | 79 | 79 | 80 |
| 70 | 64 | 64 | 65 | 65 | 66 | 66 | 67 | 67 | 68 | 68 | 69 | 69 | 70 | 70 | 70 | 70 | 71 | 71 | 71 | 71 | 72 |

Heat index (or apparent temperature)

A chart of effective temperatures can be used to determine the heat index, which combines temperature and humidity conditions to find an apparent temperature.

---

are found in association with the Siberian high and its clear skies, bitter cold, and strong subsidence of air over interior Asia during winter. Other subarctic regions experience less severe winters or year-round precipitation.

Further study of Figure 10.14 suggests two additional observations. First, ocean currents tend to influence the distribution of the subarctic climate. Along the west coasts of the continents, especially in North America, the warm ocean currents modify temperatures sufficiently to permit the marine west coast climate to extend into latitudes normally occupied by the subarctic and to cause the subarctic to be found well beyond the Arctic Circle. Along east coasts, where cold ocean currents help reduce winter temperatures, the subarctic is situated farther south. Second, the development of the subarctic climate is not as extensive in North America as it is in Eurasia. This is because (1) the Eurasian continent is a larger landmass, which increases the effect of continentality, and (2) the large water surface of Hudson Bay in Canada provides a modifying marine influence inland, which tends to counter the effect of continentality there.

### The Effects of High Latitude and Continentality

Subarctic regions experience short, cool summers and long, bitterly cold winters ( ■ Fig. 10.22). The rapid heating and cooling associated with continental interiors in the higher latitudes allow little time for the in-between seasons of spring and fall. At Eagle, Alaska, a station in the Klondike region of the Yukon River Valley, the temperature climbs 8°C–10°C (15°F–20°F) per month as summer approaches and drops just as rapidly prior to the next winter season. At Verkhoyansk in Siberia, the change between the seasonal extremes is even more rapid, averaging 15°C–20°C (30°F–40°F) per month.

feel colder. Second is perspiration; the evaporative-cooling process is quite efficient for the human body, but it differs from one individual to another. Third is the amount of activity involved; physical work or playing a physical sport can heat the body rapidly. Fourth is the amount of exposed skin; tank tops versus sweatshirts can make a world of difference.

Effective temperatures are established by considering the interplay between these two sets of factors. For example, the well-known heat index (also called apparent temperature) takes into account the temperature and humidity of a summer day and calculates how it might feel. The equally well-known wind chill index considers both temperature and wind speed to establish how cold it might feel on a winter day. Keep in mind that these are only theoretical values. No one can predict exactly how comfortable a particular person will feel on a given day. However, these temperature indices can help guide us when dealing with seasonally extreme days.

**Wind Chill Chart**
Temperature (°F)

| Wind (mph) \ Calm | 40 | 35 | 30 | 25 | 20 | 15 | 10 | 5 | 0 | −5 | −10 | −15 | −20 | −25 | −30 | −35 | −40 | −45 |
|---|---|---|---|---|---|---|---|---|---|---|---|---|---|---|---|---|---|---|
| 5 | 36 | 31 | 25 | 19 | 13 | 7 | 1 | −5 | −11 | −16 | −22 | −28 | −34 | −40 | −46 | −52 | −57 | −63 |
| 10 | 34 | 27 | 21 | 15 | 9 | 3 | −4 | −10 | −16 | −22 | −28 | −35 | −41 | −47 | −53 | −59 | −66 | −72 |
| 15 | 32 | 25 | 19 | 13 | 6 | 0 | −7 | −13 | −19 | −26 | −32 | −39 | −45 | −51 | −58 | −64 | −71 | −77 |
| 20 | 30 | 24 | 17 | 11 | 4 | −2 | −9 | −15 | −22 | −29 | −35 | −42 | −48 | −55 | −61 | −68 | −74 | −81 |
| 25 | 29 | 23 | 16 | 9 | 3 | −4 | −11 | −17 | −24 | −31 | −37 | −44 | −51 | −58 | −64 | −71 | −78 | −84 |
| 30 | 28 | 22 | 15 | 8 | 1 | −5 | −12 | −19 | −26 | −33 | −39 | −46 | −53 | −60 | −67 | −73 | −80 | −87 |
| 35 | 28 | 21 | 14 | 7 | 0 | −7 | −14 | −21 | −27 | −34 | −41 | −48 | −55 | −62 | −69 | −76 | −82 | −89 |
| 40 | 27 | 20 | 13 | 6 | −1 | −8 | −15 | −22 | −29 | −36 | −43 | −50 | −57 | −64 | −71 | −78 | −84 | −91 |
| 45 | 26 | 19 | 12 | 5 | −2 | −9 | −16 | −23 | −30 | −37 | −44 | −51 | −58 | −65 | −72 | −79 | −86 | −93 |
| 50 | 26 | 19 | 12 | 4 | −3 | −10 | −17 | −24 | −31 | −38 | −45 | −52 | −60 | −67 | −74 | −81 | −88 | −95 |
| 55 | 25 | 18 | 11 | 4 | −3 | −11 | −18 | −25 | −32 | −39 | −46 | −54 | −61 | −68 | −75 | −82 | −89 | −97 |
| 60 | 25 | 17 | 10 | 3 | −4 | −11 | −19 | −26 | −33 | −40 | −48 | −55 | −62 | −69 | −76 | −84 | −91 | −98 |

Frostbite occurs in 15 minutes or less

A chart of effective temperatures can be used to determine wind chill, which shows what the combination of cold temperatures and wind speed will make the temperature outside feel like.

Because of the high latitudes of these regions, summer days are quite long, and nights are short. The noon sun is as high in the sky during a subarctic summer as during a subtropical winter. The combination of a moderately high angle of the sun's rays and many hours of daylight means that some subarctic locations receive as much insolation at the time of the summer solstice as the equator does. As a result, temperatures during the 1–3 months of the subarctic summer usually average 10°C–15°C (50°F–60°F), and on some days they may even approach 30°C (86°F). Thus, the brief summer in the subarctic climate is often pleasantly warm, even hot, on some days.

The winter season in the subarctic is bitter, intense, and lasts for as long as 8 months. Eagle, Alaska, has 8 months with average temperatures below freezing. In the Siberian subarctic, the January temperatures regularly *average* −40°C to −50°C (−40°F to −60°F). The coldest temperatures in the Northern Hemisphere—officially, −68°C (−90°F) at both Verkhoyansk and Oymyakon; unofficially, −78°C (−108°F) at Oymyakon—have been recorded there. In addition, the winter nights, with an average 18–20 hours of darkness that extend well into one's working hours, can be mentally depressing and can increase the impression of climatic severity.

As a direct result of the intense heating and cooling of the land, the subarctic has the largest annual temperature ranges of any climate. Average annual ranges in equatorward margins of the climate vary from near 40°C (72°F) to more than 45°C (80°F). The exceptions are near western coasts, where warm ocean currents and the marine influence may significantly modify winter temperatures. Average annual ranges for poleward stations are even greater. The climograph

**FIGURE 10.22**
Climographs of the subarctic climate stations.
**Why would people settle in such severe-winter climate regions?**

for Verkhoyansk, which indicates a range of 64°C (115°F), is an extreme example.

As with subarctic temperature, latitude and continentality influence subarctic precipitation. These climate controls combine to limit annual precipitation amounts to less than 50 centimeters (20 in.) for most regions and to 25 centimeters (10 in.) or less in northern and interior locations. The low temperatures in the subarctic reduce the moisture-holding capacity of the air, thus minimizing precipitation during the occasional passage of cyclonic storms. Location toward the center of large landmasses or near lee coasts increases distance from oceanic sources of moisture. Finally, the higher latitudes occupied by the subarctic climate are dominated by the polar anticyclone, especially in the winter season. The subsidence and divergence of air in the polar anticyclone limit the opportunity for lifting and hence for precipitation in the subarctic regions. This high pressure system also blocks the entry of moist air from warmer areas to the south.

Subarctic precipitation is cyclonic or frontal, and because the polar anticyclone is weaker and farther north during the warmer summer months, more precipitation comes during that season. The meager winter precipitation falls as fine, dry snow. Though there is not as much snowfall as in less severe climates, the temperatures remain so cold for so long that the snow cover lasts for as long as 7 or 8 months. During this time, there is almost no melting of snow, especially in the dark shadows of the forest.

**A Limiting Environment** The climatic restrictions of subarctic regions place distinct limitations on plant and animal life and on human activity. The characteristic vegetation is coniferous forest, adapted to the severe temperatures; the physiologic drought associated with frozen soil water; and the infertile soils. Seemingly endless tracts of spruce, fir, and pine thrive over enormous areas, untouched by humans (❙ Fig. 10.23). In Russia, the forest is called the *taiga* (or *boreal forests* in other regions), and this name is sometimes given to the subarctic climate type itself.

The brief summers and long, cold winters severely limit the growth of vegetation in subarctic regions. Even the trees are shorter and more slender than comparable species in less severe climate regimes. There is little hope for agriculture. The growing season averages 50–75 days, and frost may occur even during June, July, or August. Thus, in some years, a subarctic location may have no truly frost-free season. Although scientists are working to develop plant species that can take advantage of the long hours of daylight in summer, only minimal success has been achieved, in southern parts of the climate, with certain vegetables such as cabbage and root crops such as potatoes.

A particularly vexing problem to people in subarctic (as well as tundra) regions is **permafrost,** a permanently frozen layer of subsoil and underlying rock that may extend to a depth of 300 meters (1000 ft) or more in the northernmost sections of the climate. Permafrost is present over much of the subarctic climate, but it varies greatly in thickness and is often discontinuous. Where it occurs, the land is frozen completely from the surface down in winter. The warm temperatures of spring and summer cause the top few feet to thaw out, but because the land beneath this thawed top layer remains frozen, water cannot percolate downward, and the thawed soil becomes sodden with moisture, especially in spring, when there is an abundant supply of water from the melting snow. Permafrost poses a

**FIGURE 10.23**
Seemingly endless tracts of taiga (boreal forest) are typical of the vegetation throughout much of the Siberian subarctic. This photo was taken between the Russian towns of Vanavara and Tura.
**Why are these virgin forests currently of little economic value?**

**FIGURE 10.24**
Frost polygons, or patterned ground, as seen in the subarctic region of Alaska. Repeated freezing and thawing causes the soil to produce polygonal shapes.
**In this photo, what part of the polygon is heaved upward, the middle or the edges?**

problem to agriculture by preventing proper soil drainage. The seasonal freeze and thaw of the surface layer above the permafrost also poses special problems for construction engineers. The cycle causes repeated expansion and contraction, heaving the surface up and then letting it sag down. The effects of this cycle break up roads, force buried pipelines out of the ground, cause walls and bridge piers to collapse, and even swivel buildings off their foundations. Landscapes called **patterned ground,** or **frost polygons,** are commonly found in regions of subarctic yearly freezing and thawing (Fig. 10.24).

With agriculture a questionable occupation at best, there is little economic incentive to draw humans to subarctic regions. Logging is unimportant because of small tree size. Even use of the vast forests for paper, pulp, and wood products is restricted by their interior location far from world markets. Miners occasionally exploit rich ore deposits; others, many of them native to the subarctic regions, pursue hunting, trapping, and fishing of the relatively limited wildlife. Fur-bearing animals such as mink, fox, wolf, ermine, otter, and muskrat are of greatest value.

## Polar Climate Regions

The polar climates are the last of Köppen's humid climate subdivisions to be differentiated on the basis of temperature. These climate regions are situated at the greatest distance from the equator, and they owe their existence primarily to the low annual amounts of insolation they receive. No polar station experiences a month with average temperatures as high as 10°C (50°F), and hence all are without a warm summer (Table 10.3). Trees cannot survive in such a regimen. In the regions where at least 1 month averages above 0°C (32°F), they are replaced by tundra vegetation. Elsewhere, the surface is covered by great expanses of frozen ice. Thus, there are two polar climate types, tundra and ice sheet.

There are two important points to keep in mind in the discussion of polar climate regions. First, these regions have a large net annual radiation loss; that is, they give up much more radiation or energy than they receive from the sun during a year, resulting in a major radiation deficiency. The transfer of heat from lower to higher latitudes to make up this deficiency is the driving force of the general atmospheric circulation. Without this compensating poleward transfer of heat from the lower latitudes, the polar regions would become too cold to permit any form of life, and the equatorial regions would heat to temperatures no organism could survive.

A second and equally important characteristic of polar climates is the unique pattern of day and night. At the poles, 6 months of relative darkness, caused when the sun is positioned below the horizon, alternate with 6 months of daylight during which the sun is above the horizon. Even when the sun is above the horizon, however, the sun's rays are at a sharply oblique angle, and little insolation is received for the number of hours of daylight. Moving outward from the poles, the lengths of periods of continuous winter night and continuous summer day decrease rapidly from 6 months at the poles to 24 hours at the Arctic and Antarctic Circles (66°N and S). Here the 24-hour night or day occurs only at the winter and summer solstices, respectively.

**TABLE 10.3**
**The Polar Climates**

| Name and Description | Controlling Factors | Geographic Distribution | Distinguishing Characteristics | Related Features |
|---|---|---|---|---|
| *Tundra* | | | | |
| Warmest month between 0°C (32°F) and 10°C (50°F); precipitation exceeds potential evapotranspiration | Location in the high latitudes; subsidence and divergence of the polar anticyclone; proximity to coasts | Arctic Ocean borderlands of North America, Greenland, and Eurasia; Antarctic Peninsula; some polar islands | At least 9 months average below freezing; low evaporation; precipitation usually below 25.5 cm (10 in.); coastal fog; strong winds | Tundra vegetation; tundra soils; permafrost; swamps and bogs during melting period; life most common in nearby seas; Inuit; mineral and oil resources; defense industry |
| *Ice-sheet* | | | | |
| Warmest month below 0°C (32°F); precipitation exceeds potential evaporation | Location in the high latitudes and interior of landmasses; year-round influence of the polar anticyclone; ice cover; elevation | Antarctica; interior Greenland; permanently frozen portions of the Arctic Ocean and associated islands | Summerless; all months average below freezing; world's coldest temperature; extremely meager precipitation in the form of snow, evaporation even less; gale-force winds | Ice- and snow-covered surface; no vegetation; no exposed soils; only sea life or aquatic birds; scientific exploration |

## Tundra Climate

Compare the location of the **tundra climate** with that of the subarctic climate in Figure 8.6. You can see that although the tundra climate is situated closer to the poles, it is also along the periphery of landmasses, and, with the exception of the Antarctic Peninsula, it is everywhere adjacent to the Arctic Ocean. Even though temperature ranges in the tundra are large, they are not as large as in the subarctic because of the maritime influence. Winter temperatures in particular are not as severe in the tundra as they are inland ( ▌ Fig. 10.25).

It almost seems inappropriate to call the unpleasantly chilly and damp conditions of the tundra's warmer season "summer." Temperatures average around 4°C (40°F) to 10°C (50°F) for the warmest month, and frosts occur regularly. The air does warm sufficiently to melt the thin snow cover and the ice on small bodies of water, but this only causes marshes, swamps, and bogs to form across the land because drainage is blocked by permafrost ( ▌ Fig. 10.26). Out of this soggy landscape, known as **muskeg** in Canada and Alaska, swarm clouds of black flies, mosquitoes, and gnats. The one bright note in the landscape is provided by the enormous number of migratory birds that nest in the arctic regions at this time of year and feed on the insects. However, as soon as the shrinking days of autumn approach, these birds depart for warmer climates.

Winters are cold and seem to last forever, especially in tundra locations where the sun is below the horizon for days at a time. The climograph for Barrow, Alaska, illustrates the low temperatures of this climate. Note that average monthly temperatures are *below freezing* 9 months of the year. The average annual temperature is −12°C (10°F). The low-growing tundra vegetation survives despite the forbidding environment. It consists of lichens, mosses, sedges, flowering herbaceous plants, small shrubs, and grasses. In particular, the plants have adjusted to the conditions associated with nearly universal permafrost ( ▌ Fig. 10.27).

The tundra regions exhibit several other significant climatic characteristics. Diurnal temperature ranges are small because insolation is uniformly high during the long summer days and uniformly low during the long winter nights. Precipitation is generally low, except in eastern Canada and Greenland, because of exceedingly low absolute humidity and the influence of the polar anticyclone. Icy winds sweep

**FIGURE 10.25**
Climographs for tundra climate stations.
**Why is it not surprising that both stations are located in the Northern Hemisphere?**

**FIGURE 10.26**
Permafrost regions, such as this area at the base of the Alaska Range, become almost impenetrable swampland during the brief Alaskan summer. Travel over land is feasible only in the winter season.
**What is the preferred means of travel in the summer?**

across the open land surface and are an added factor in eliminating the trees that might impede their progress. Coastal fog is characteristic in marine locations, where cool polar maritime air drifts onshore and is chilled below the dew point by contact with the even colder land.

## Ice-Sheet Climate

The **ice-sheet climate** is the most severe and restrictive climate on Earth. As Table 10.3 indicates, it covers large areas in both the Northern and Southern Hemispheres, a total of

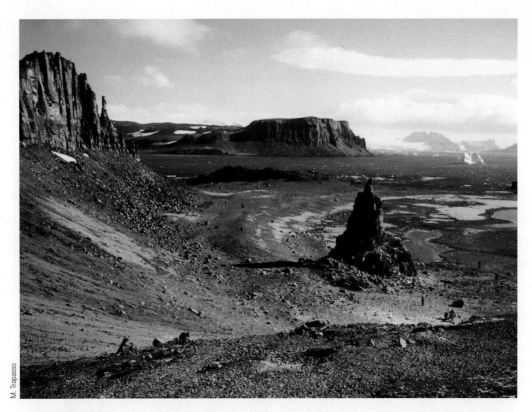

**FIGURE 10.27**
One of the Hecho Islands off the Antarctic Peninsula displays thick, bright patches of moss as its most complex vegetation.
**What weather elements might help to form this stark landscape?**

about 16 million square kilometers (6 million sq mi)—nearly the same area as the United States and Canada combined. All average monthly temperatures are below freezing, and because most surfaces are covered with glacial ice, no vegetation can survive in this climate. It is a virtually lifeless region of perpetual frost.

Antarctica is the coldest place on Earth (although Siberia sometimes has longer and more severe periods of cold in winter). The world's coldest temperature, −88°C (−127°F), was recorded at Vostok, Antarctica. Consider the climographs for Little America, Antarctica, and Eismitte, Greenland, for a fuller picture of the cold ice-sheet temperatures ( Fig. 10.28).

The primary reason for the low temperatures of ice-sheet climates is the minimal insolation received in these regions. Not only is little or no insolation received during half the year, but also the sun's radiant energy that is received arrives at sharply oblique angles. In addition, the perpetual snow and ice cover of this climate reflects nearly all incoming radiation. A further factor, in both Greenland and Antarctica, is elevation. The ice sheets covering both regions rise more than 3000 meters (10,000 ft) above sea level ( Fig. 10.29). Naturally, this elevation contributes to the cold temperatures.

The polar anticyclone severely limits precipitation in the ice-sheet climate to the fine, dry snow associated with occasional cyclonic storms. Precipitation is so meager in this climate that regions within this regime are sometimes incorrectly referred to as "polar deserts." However, because of the exceedingly low evaporation rates (evapotranspiration rates are not considered in this climate because there is no plant life to account for any transpiration) associated with the severely cold temperatures, precipitation still exceeds potential evaporation, and the climate can be classified as humid. The annual precipitation surplus produces glaciers, which export snowfall similar to the way rivers export rainfall.

The strong and persistent polar winds are another staple of the harsh ice-sheet climate. Mawson Base, Antarctica, for example, has approximately 340 days a year with gale-force winds of 15 meters per second (33 mph) or more. *Katabatic winds,* which are caused by the downslope drainage of heavy cold air accumulated over ice sheets, are common along the edges of the polar ice. The winds of these regions can result in whiteouts—periods of zero visibility due to blowing fine snow and ice crystals.

**FIGURE 10.28**
Climographs for ice-sheet climate stations.
**If you were to accept an offer for an all-expense-paid trip to visit either Greenland or Antarctica, which would you choose, and why would you go?**

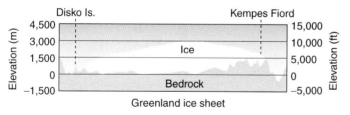

**FIGURE 10.29**
The Antarctic ice sheet resulting from the ice-sheet climate of the south polar regions is as much as 4000 meters (13,200 ft) thick locally. Where it is thinnest, it is floated by seawater to produce an ice shelf. The smaller Greenland ice sheet is about 3000 meters (10,000 ft) thick.
**What reasons might be given for the fact that more land in Greenland than in Antarctica is free of glacial ice?**

## Human Activity in Polar Regions

The climatic severity that limits animal life in polar regions to a few scattered species in the tundra is just as restrictive on human settlement. The celebrated Lapps of northern Europe migrate with their reindeer to the tundra from the adjacent forest during warmer months. They join the musk ox, arctic hare, fox, wolf, and polar bear that manage to make a home there despite the prohibitive environment. Only the Inuit (Eskimos) of Alaska, northern Canada, and Greenland have in the past succeeded in developing a year-round lifestyle adapted to the tundra regime. Yet even this group relies less on the resources of the tundra than on the large variety of fish and sea mammals, such as cod, salmon, halibut, seal, walrus, and whale, that occupy the adjacent seas.

As their communication with the rest of the world has increased and they have become acquainted with alternative lifestyles, the permanent population of Inuit living in the tundra has greatly diminished, and life for those remaining has changed drastically. Some have gained new economic security through employment at defense installations or at sites where they join other skilled workers from outside the region to exploit mineral or energy resources. However, the new population centers based on the construction and maintenance of radar and missile defense stations or, as in the case of Alaska's North Slope, on the production and transportation of oil, cannot be considered

**FIGURE 10.30**
Mile-marker 0 on the Alaska pipeline, near Prudhoe Bay. This is one very profitable venture for humans in the North Slope oil fields. However in 1989, the *Exxon Valdez* spilled 11 million gallons of crude oil into Prince William Sound, Alaska.
**Considering the vulnerability of Alaska's physical environment, should development of the North Slope oil fields have been permitted?**

**FIGURE 10.31**
Greenland's ice sheet covers about 85% of its surface. Here we see ice swallowing up the landscape.
**What kind of activities might bring individuals from other regions to an ice-sheet climate?**

permanent (Fig. 10.30). Workers depend on other regions for support and often inhabit this region only temporarily.

The ice-sheet climate cannot serve as a home for humans or other animals. Even the penguins, gulls, leopard seals, and polar bears are coastal inhabitants. It is without question the harshest, most restrictive, most nearly lifeless climate zone on Earth (Fig. 10.31). Yet, especially in Antarctica, it is of strategic importance and of great scientific interest. Scientists study the oxygen-isotope ratios of the Antarctic ice cores and the gas bubbles trapped within them to help them reconstruct past climates. They have also noted that ozone concentrations have decreased over Antarctica every fall for several decades. The result is a hole in the ozone layer above Antarctica that is as large as the North American continent. This decrease in the ozone layer is a major environmental concern. Antarctica's strategic value is so widely recognized that the world's nations have voluntarily given up claims to territorial rights on the continent in exchange for cooperative scientific exploration on behalf of all humankind.

# Highland Climate Regions

As we saw in Chapter 4, temperature decreases with increasing altitude at the rate of about 6.5°C per 1000 meters (3.6°F per 1000 ft). Thus, you might suspect that highland regions exhibit broad zones of climate based on changes in temperature with elevation that roughly correspond to Köppen's climate zones based on change of temperature with latitude (Fig. 10.32). This is indeed the case, with one important exception: Seasons only exist in highlands if they also exist in the nearby lowland regions. For example, although zones of increasingly cooler temperature occur at progressively higher elevations in the tropical climate regions, the seasonal changes of Köppen's middle-latitude climates are not present.

**FIGURE 10.32**
Note the similarities between (a) Slovar, Norway, 68°N latitude (an Arctic location), at sea level, and (b) Machu Picchu, 13°S latitude (a tropical location), at 2590 meters (8500 feet) above sea level.
**In what ways might high latitudes be similar to high altitudes?**

slopes; others are lee slopes or are sheltered behind higher topography. The nature of the wind, its temperature, and its moisture content depend on whether the mountain is (1) in a coastal location or deep in a continental interior and (2) at a high or low latitude within or beyond the reaches of cyclonic storms and monsoon circulation. In the middle and high latitudes, mountain slopes and valley walls that face the equator receive the direct rays of the sun and are warm; poleward-facing slopes are shadowed and cool. West-facing slopes feel the hot afternoon sun, whereas east-facing slopes are sunlit only in the cool of the morning. This factor, known as **slope aspect,** affects where people live in the mountains and where particular crops will do best. The higher one rises in the mountains, the more important direct sunlight is as a source of warmth and energy for plant and animal life processes.

Complexity is the hallmark of highland climates. Every mountain range of significance is composed of a mosaic of climates far too intricate to differentiate on a world map or even on a map of a single continent. Highland climates are therefore undifferentiated, signifying climate complexity. Highland climates are indicated on Figure 8.6 wherever there is marked local variation in climate as a consequence of elevation, exposure, and slope aspect. We can see that these regions are distributed widely over Earth but are particularly concentrated in Asia, central Europe, and western North and South America.

The areas of highland climate on the world map are cool, moist islands in the midst of the zonal climates that dominate the areas around them. Consequently, highland areas are also biotic islands, supporting a flora and fauna adapted to cooler and wetter conditions than those of the surrounding lowlands. This coolness is part of the highland charm, particularly where mountains rise cloaked with forests above arid plains, as do the Canadian Rocky Mountains and California's Sierra Nevada.

Elevation is only one of several controls of highland climates; **exposure** is another. Just as some continental coasts face the prevailing wind, so do some mountain

# Geography's Environmental Science Perspective
## The Effects of Altitude on the Human Body

The changes in climate, soils, and vegetation as an individual climbs into higher altitudes are discussed elsewhere in the text. However, it is also interesting to note some of the ways in which high altitudes affect the human body. Besides food and water, the metabolism of a human being also depends on the consumption of oxygen. A lessening of the normal amount of oxygen intake can have some profound physiological effects. One way to look at the amount of oxygen available for human respiration is to consider something called the *partial pressure of oxygen gas* ($pO_2$). Partial pressure in this case refers to the portion of total atmospheric pressure attributed to oxygen alone. At sea level, atmospheric pressure is 1013.2 millibars, and the $pO_2$ is about 212 millibars. In other words, 212 of the 1013.2 millibars are attributed to oxygen gas. At an altitude of 10 kilometers (6.2 mi) the $pO_2$ drops to only 55 millibars! Even at moderate altitudes, the effects of *hypoxia* (oxygen starvation) can cause headaches and nausea. Above 6 kilometers (3.7 mi), this lack of oxygen can seriously affect the brain. Since the body's need for oxygen does not alter with major changes in altitude, any significant drop in $pO_2$ with height can cause severe body stress. Pressure-controlled cabins of high-flying aircraft and the use of oxygen by mountain climbers provide dramatic examples of ways to meet the vital need for oxygen.

Fortunately, the human body can acclimatize to moderate changes in altitude, but it takes time (days to sometimes weeks, depending on the altitude). During this time, however, one may experience some of the following symptoms: sleeplessness, headaches, loss of weight, thyroid deficiency, increased excitability, muscle pain, gastrointestinal disturbances, swelling of the lungs, severe infections, psychological and mental disturbances, and others. At the very least, when visiting higher altitude locations, most people can expect headaches (perhaps leading to nausea), a drop in physical endurance (climbing stairs will become more difficult), and hampered mental function (solutions to simple problems may be difficult to grasp). There is little need to worry, though; humans do acclimatize and these symptoms will pass.

It is also important to remember that, in addition to hypoxia, higher altitudes may cause increased susceptibility to severe sunburn. If you climb to a higher altitude, this means that more of Earth's atmosphere is below you. Correspondingly, there is less of the protective atmosphere above you to filter harmful ultraviolet radiation. Keeping latitude constant, sunburn of exposed skin is more likely on a high-altitude mountainside than on a sea-level beach.

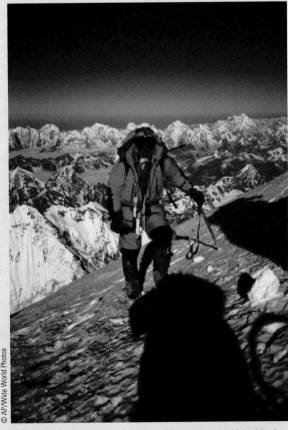

Mountain climbers carry an oxygen supply to help with high altitudes.

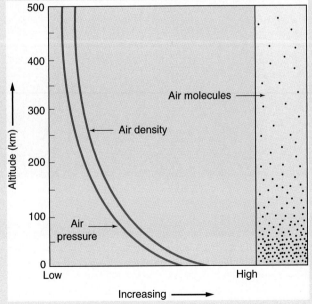

Both air pressure and air density decrease rapidly with increasing altitude.

Highlands stimulate moisture condensation and precipitation by forcing moving air masses to rise over them (Fig. 10.33). Where mountain slopes are rocky and forest free, their surfaces grow warm during the day, causing upward convection, which often produces afternoon thundershowers. Mountains receive abundant precipitation and are the source area for multitudes of streams that join to form the great rivers of all of the continents.

There are few streams of significance whose headwaters do not lie in rugged highlands. Much of the stream flow on all continents is produced by the summer melting of mountain snowfields. Thus, the mountains not only wring water from the atmosphere but also store much of it in a form that gradually releases it throughout summer droughts when water is most needed for irrigation and for municipal and domestic uses.

## Peculiarities of Mountain Climates

A general characteristic of mountain weather is its variability from hour to hour as well as from place to place. Strong orographic flow over mountains often causes clouds to form very quickly, leading to thunderstorms and longer rains that do not affect surrounding cloud-free lowlands. Where the cloud cover is diminished, diurnal temperature ranges over mountains are far greater than those over lowlands. Because mountains penetrate upward beyond the densest part of the atmosphere, the greenhouse effect is less developed there than anywhere else on Earth. The thinner layer of low-density air above a mountain site does not greatly impede insolation, thus allowing surfaces to warm dramatically during the daytime. By the same token, the atmosphere in these areas does little to impede longwave radiation loss at night. Consequently, air temperatures overnight are cooler than the elevation alone would indicate. Because the atmospheric shield is thinnest at high elevations, plants, animals, and humans receive proportionately more of the sun's shortwave radiation at high altitudes. Violet and ultraviolet radiation are particularly noticeable; severe sunburn is one of the real hazards of a day in the high country.

In the middle and high latitudes, mountains rise from mesothermal and microthermal climates into tundra and snow-covered zones. The lower slopes of mountains are commonly forested with conifers, which become more stunted as one moves upward, until the last dwarfed tree is passed at the **tree line**. This is the line beyond which low winter temperatures and severe wind stress eliminate all forms of vegetation except those that grow low to the ground, where they can be protected by a blanket of snow (Fig. 10.34). Where mountains are high enough, snow or ice permanently covers the land surface. The line above which summer melting is insufficient to remove all of the preceding winter's snowfall is called the **snow line**.

In tropical mountain regions, the vertical zonation of climate is even more pronounced. Both tree line and snow line occur at higher elevations than in middle latitudes. Any

**FIGURE 10.33**

Variation in precipitation caused by uplift of air crossing the Sierra Nevada range of California from west to east. The maximum precipitation occurs on the windward slope because air in the summit region is too cool to retain a large supply of moisture. Note the strong rain shadow to the lee that gives Reno a desert climate.

**Taking into consideration the locations of the recording stations, during what season of the year does the maximum precipitation on the windward slope occur?**

seasonal change is mainly restricted to rainfall; temperatures are stable year-round, regardless of elevation. Each climate zone has its own particular association of natural vegetation and has given rise to a distinctive crop combination where agriculture is practiced ( Fig. 10.35). In South America, four vertical climate zones are recognized: *tierra caliente* (hot lands), *tierra templada* (temperate lands), *tierra fría* (cool lands), and *tierra helada* (frozen lands).

## Highland Climates and Human Activity

In middle-latitude highlands, soils are poor, the growing season is short, and the winter snow cover is heavy in the conifer zone, which dominates the lower and middle mountain slopes. Therefore, little agriculture is practiced, and permanent settlements in the mountains are few. However, as the winter snow melts off the high ground just below the bare rocky peaks, grass springs into life, and humans drive herds of cattle and flocks of sheep and goats up from the warmer valleys. The high pastures are lush throughout the summer, but in early fall they are once again vacated by the animals and their keepers, who return to the valleys. This seasonal movement of herds and herders between alpine pastures and villages in the valleys, termed *transhumance*, was once common in the European highlands (the Alps, Pyrenees, Carpathian Mountains, and mountains throughout Scandinavia) and is still practiced there on a reduced scale.

Otherwise, the middle-latitude highlands serve mainly as sources of timber and of minerals formed by the same geologic forces that elevated the mountains and as arenas for recreation—both summer and winter. Recreational use of the highlands is a relatively recent phenomenon, resulting both from new interest in mountain areas and from new access routes by road, rail, and air that did not exist a century ago.

In contrast to poleward mountain regions, tropical highlands may actually experience more favorable climatic conditions and are often a greater attraction to human settlement (from ancient to modern times) than adjacent lowlands. In fact, large permanent populations are supported throughout the tropics where topography and soil favor agriculture in the vertical climate zones. Highland climates are at such a premium in many areas that steep mountain slopes have been extensively terraced to produce level land for agricultural use. Spectacular agricultural terraces can be seen in Peru, Yemen, the Philippines, and many other tropical highlands. Where the climate is appropriate and population pressure is high, people have created a topography to suit their needs, although they have had to hack it out of mountainsides.

### FIGURE 10.34
As in this example in the Colorado Rocky Mountains, the last tree species found at the tree line are stunted, prostrate forms, which often produce an elfin forest. Where the trees are especially gnarled and misshapen by wind stress, the vegetation is called *krummholz* (crooked wood).
**What do you see in the photograph that indicates prevailing wind direction?**

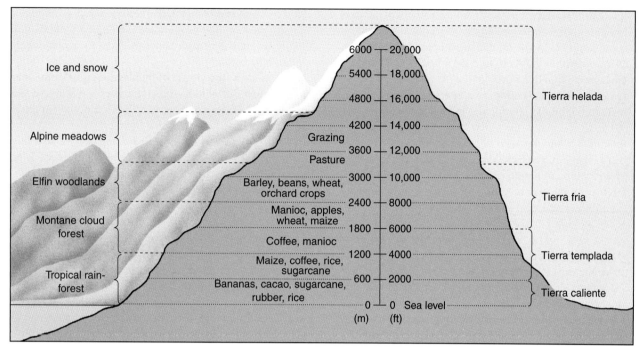

**FIGURE 10.35**
Natural vegetation, vertical climate zones, and agricultural products in tropical mountains. Note that this example extends from tropical life zones to the zone of permanent snow and ice. There is little seasonal temperature change in tropical mountains, which allows life-forms sensitive to low temperatures to survive at relatively high elevations.
**When Europeans first settled in the highlands of tropical South America, in which vertical climate zone did they prefer to live?**

## Define & Recall

Mediterranean climate
sclerophyllous
chaparral
humid subtropical climate
marine west coast climate
humid continental, hot summer climate
humid continental, mild summer climate
subarctic climate
permafrost
patterned ground (frost polygons)
tundra climate
muskeg
ice-sheet climate
exposure
slope aspect
tree line
snow line

## Discuss & Review

1. What characteristics make the Mediterranean climate readily distinguishable from all others? What controls are responsible for producing these characteristics?
2. Summarize the special adaptations of vegetation and soils in Mediterranean regions. How have humans also adapted to these regions?
3. Compare the humid subtropical and Mediterranean climates. What are their most obvious similarities and differences?
4. What factors combine to cause a precipitation maximum in late summer in most of the humid subtropical regions?
5. What factors serve to make the humid subtropical climate one of the world's most productive? What are some of its handicaps?
6. How are temperature, precipitation, and geographic distribution of marine west coast regions linked to the controlling factors for this climate?
7. The marine west coast climate has long had its supporters and critics. Give reasons why you would or would not wish to live in such a climate region.
8. Explain why the microthermal climates are limited to the Northern Hemisphere.

9. List several features that all humid microthermal climates have in common. How do these features differ from those displayed by the humid mesothermal climates?
10. Why is weather in the humid continental, hot summer climate so variable?
11. Describe the relationship between vegetation and climate in the humid continental, mild summer regions.
12. What are the major differences between the humid continental, hot summer and mild summer climates? Contrast the human use of regions occupied by the two climates.
13. How have past climate changes helped bring about differences in the configuration of the land between some hot summer and mild summer regions?
14. Refer to Figure 10.22. Using the climographs for Eagle, Alaska, and Verkhoyansk, Russia, describe the temperature patterns of the subarctic regions.
15. What factors limit precipitation in the subarctic regions?
16. How does permafrost affect human activity in both the subarctic and tundra regions?
17. Identify and compare the controlling factors of the tundra and ice-sheet regions. How do these controlling factors affect the distribution of these climates?
18. What kind of plant and animal life can survive in the polar climates? What special adaptations must this life make to the harsh conditions of these regions?
19. How do elevation, exposure, and slope aspect affect the microclimates of highland regions? What are the major climatic differences between highland regions and nearby lowlands?
20. Describe and compare the vertical zonation of highland climates in the tropics and in the middle latitudes. How are the climate zones in tropical highlands related to agriculture?
21. How does human use of highland regions in the middle latitudes differ from that in the tropics? What special human adaptations have aided utilization of these regions?
22. Do each of the following for these climates: Mediterranean; humid subtropical; marine west coast; humid continental, hot summer; humid continental, mild summer; subarctic; tundra; ice-sheet.
   a. Identify the climate from a set of data or a climograph indicating average monthly temperature and precipitation for a representative station within a region of that climate.
   b. Match the climate type with a written statement that includes one or more of the following: the statistical parameters of the climate in the modified Köppen classification; the particular climate controls (controlling factors) that produce the climate; the geographic distribution of the climate as stated in terms of physical or political location; the unique climate characteristics or combination of characteristics that distinguishes the climate from others; types of plants, animals, and soils associated with the climate; and the human utilization typical of the climate.
   c. Distinguish between the important subtypes (if any) of each climate by identifying the characteristics that separate them from one another.

# Consider & Respond

1. Based on the classification scheme presented in the "Graph Interpretation" exercise at the end of Chapter 8, classify the following climate stations from the data provided.

|  |  | J | F | M | A | M | J | J | A | S | O | N | D | Yr |
|---|---|---|---|---|---|---|---|---|---|---|---|---|---|---|
| a. | Temp. (°C) | −42 | −27 | −40 | −31 | −20 | 15 | −11 | −18 | −22 | −36 | −43 | −39 | −30 |
|  | Precip. (cm) | 0.3 | 0.3 | 0.5 | 0.3 | 0.5 | 0.8 | 2.0 | 1.8 | 0.8 | 0.3 | 0.8 | 0.5 | 8.6 |
| b. | Temp. (°C) | 3 | 3 | 5 | 8 | 19 | 13 | 15 | 14 | 13 | 10 | 7 | 5 | 9 |
|  | Precip. (cm) | 4.8 | 3.6 | 3.3 | 3.3 | 4.8 | 4.6 | 8.9 | 9.1 | 4.8 | 5.1 | 6.1 | 7.4 | 65.8 |
| c. | Temp. (°C) | 23 | 23 | 22 | 19 | 16 | 14 | 13 | 13 | 14 | 17 | 19 | 22 | 18 |
|  | Precip. (cm) | 0.8 | 1.0 | 2.0 | 4.3 | 13.0 | 18.0 | 17.0 | 14.5 | 8.6 | 5.6 | 2.0 | 1.3 | 88.1 |
| d. | Temp. (°C) | −27 | −28 | −26 | −18 | −8 | 1 | 4 | 3 | −1 | −8 | −18 | −24 | 12 |
|  | Precip. (cm) | 0.5 | 0.5 | 0.3 | 0.3 | 0.3 | 1.0 | 2.0 | 2.3 | 1.5 | 1.3 | 0.5 | 0.5 | 10.9 |
| e. | Temp. (°C) | −4 | −2 | 5 | 14 | 20 | 24 | 26 | 25 | 20 | 13 | 3 | −2 | 12 |
|  | Precip. (cm) | 0.5 | 0.5 | 0.8 | 1.8 | 3.6 | 7.9 | 24.4 | 14.2 | 5.8 | 1.5 | 1.0 | 0.3 | 62.2 |
| f. | Temp. (°C) | 9 | 9 | 9 | 10 | 12 | 13 | 14 | 14 | 14 | 12 | 11 | 9 | 11 |
|  | Precip. (cm) | 17.0 | 14.8 | 13.3 | 6.8 | 5.5 | 1.9 | 0.3 | 0.3 | 1.6 | 8.1 | 11.7 | 17.0 | 97.6 |
| g. | Temp. (°C) | −3 | −2 | 2 | 9 | 16 | 21 | 24 | 23 | 19 | 13 | 4 | −2 | 11 |
|  | Precip. (cm) | 4.8 | 4.1 | 6.9 | 7.6 | 9.4 | 10.4 | 8.6 | 8.1 | 6.9 | 7.1 | 5.6 | 4.8 | 84.8 |
| h. | Temp. (°C) | 0 | 0 | 4 | 9 | 16 | 21 | 24 | 23 | 20 | 14 | 8 | 2 | 12 |
|  | Precip. (cm) | 8.1 | 7.4 | 10.7 | 8.9 | 9.4 | 8.6 | 10.2 | 12.7 | 10.7 | 8.1 | 8.9 | 8.1 | 111.5 |

2. The data in the previous table represent the following eight locations, although not in this order: Beijing, China; Point Barrow, Alaska; Chicago, Illinois; Eismitte, Greenland; Eureka, California; Edinburgh, Scotland; New York, New York; Perth, Australia. Use an atlas and your knowledge of climates to match the climatic data with the locations.
3. Eureka, Chicago, and New York are located within a few degrees latitude of one another, yet they represent three distinctly different climate types. Discuss these differences and identify the primary cause, or source, of the differences.
4. The precipitation recorded at Albuquerque, New Mexico (see Consider and Respond, Chapter 9), is almost twice that recorded at Point Barrow, Alaska, yet Albuquerque is considered a dry climate and Point Barrow a humid climate. Why?
5. What differentiates the $Dw$ climate from the other $D$ climates?
6. Why is the $Dw$ climate type found only in Asia?
7. $Csa$ climate regions and $Cfa$ climate regions are both under the influence of subtropical high pressure cells during the summer, yet $Csa$ climates are dry during the summer and $Cfa$ climates are wet. Why?

# CHAPTER 11

Spatial patterns of vegetation reflect different environmental conditions, as seen in this autumn scene in Uncompahgre National Park, Colorado. © Willard Clay/Getty Images

# Biogeography

**PHYSICAL Geography Now™** This icon, appearing throughout the book, indicates an opportunity to explore interactive tutorials, animations, or practice problems at now.brookscole.com/gabler8

# CHAPTER PREVIEW

**Plants, animals, and the environments in which they live are interdependent, each affecting the others.**

- In what ways are plants as a group and animals as a group mutually dependent on one another?
- In what ways do humans have a much greater impact on ecosystems than all other life forms?
- In what specific ways do humans impact plants and animals?

**As the basic producers, autotrophs are generally considered to be the most important component of an ecosystem.**

- What are the other components?
- What are the differences between autotrophs and heterotrophs?
- In what ways are the autotrophs affected by the other components?

**Plant and animal distributions present a mosaic on the landscape, with the most broadly distributed plant associations typically comprising the matrix of the landscape.**

- Why isn't a matrix based on animal distributions?
- What are patches and corridors within the matrix?
- How do successional processes relate to the concept of patches within a mosaic?

**Although other environmental controls may be more important on a local scale, climate has the greatest influence over ecosystems on a worldwide basis.**

- What is the evidence that this is the case?
- Which climatic factors have the greatest effect on plants and animals?
- How does climate influence the shape and size of animals and their appendages?
- What environmental controls other than climate are important?

**Earth's major terrestrial ecosystems (biomes) are classified on the basis of the dominant vegetation types that occupy the ecosystems.**

- What are these vegetation types?
- Why not base the classification on animals?
- How does this statement relate to the second statement above?

**Freely floating plants called phytoplankton are the most important link in the ocean food chain.**

- What role do they play?
- What function in relation to the atmosphere do phytoplankton perform?
- Where in the ocean are phytoplankton concentrated?

As was pointed out in Chapter 1, biogeographers are those physical geographers who specialize in the study of natural and human-modified environments and the ecological processes that influence each environment's nature and distribution. Along with ecologists from a host of other science disciplines, biogeographers often focus their research on *ecosystems*—communities of organisms that function together in an interdependent relationship with the environments that they occupy.

In almost all respects, the study of ecosystems provides the ideal opportunity to demonstrate the multiple perspectives of physical geography among the sciences: the spatial science perspective, the physical science perspective, and the environmental science perspective. Biogeographers examine the locations, distribution, spatial patterns, and spatial interactions of all plant and animal life. They delineate the boundaries and study the characteristics of the ecosystems that they identify and also monitor the flow of energy and material through each system. In addition, they pay particular attention to the impact of humans on each ecosystem's living and nonliving physical environment.

Animal life would not exist without plants as basic food, and most plants could not survive without some animals. Together, plants and animals must adapt to their physical environment. Humans alone have the intelligence and the capacity to alter, either carelessly or deliberately, the plant–animal–physical environment relationship.

What lasting effects will increased industrialization have on the water that animals in the 21st century will drink or on the atmosphere in which plants will grow? What will be the consequences for life-forms if concrete and steel replace additional square kilometers of forest? What could happen to marine life if toxic wastes accumulate in the world's oceans? The understandings that come from the study of ecosystems by biogeographers can provide answers to these questions and can help humans learn to work with, and not against, nature to sustain and improve life as we know it on planet Earth today.

# Organization within Ecosystems

It can be said that ecology is an old science. The great voyages of exploration that began in the 15th century carried colonists and adventurers to uncharted lands with exotic environments. The more scholarly observers within each group made careful note of the flora and fauna found in each new part of the world. It soon became apparent that certain plants and animals were found together and that they bore a direct relationship to the climate in which they lived. As information about various world environments became more reliable and readily available, early biologists began to study plant communities and classify vegetation types. As the relationships of animals to these plant communities were recognized, naturalists in the early 20th century began dividing Earth's life-forms into *biotic associations*. Recently, the functional relationships of plants, animals, and their physical environment have been the primary focus of attention, and the concept of the ecosystem has become widely used.

Our definition of an ecosystem is both broad and flexible. The term can be used in reference to the Earth system in its entirety (the ecosphere) or to any group of organisms occupying a given area and functioning together with their nonliving environment. An ecosystem may be large or small, marine or terrestrial (on land), short lived or long lasting ( ▮ Fig. 11.1). It may even be an artificial ecosystem, such as a farmer's field. When a farmer plants crops, spreads fertilizer, practices weed control, and sprays insecticides, a new ecosystem is created, but this does not alter the fact that plants and animals are living together in an interdependent relationship with the soil, rainfall, temperatures, sunshine, and other factors that constitute the physical environment ( ▮ Fig. 11.2).

As noted in Chapter 1, ecosystems are *open systems*. There is free movement of both energy and materials into and out of these systems. They are usually so closely related to nearby ecosystems or so integrated with the larger ecosystems of which they are a part that they are not isolated in nature or readily delimited. Nevertheless, the concept of the ecosystem is a valuable model for examining the structure and function of life on Earth.

## Major Components

Ecosystems are many and varied, but the typical ecosystem has four basic components. The first of these is the nonliving, or **abiotic,** part of the system. This is the physical environment in which the plants and animals of the system live. In an aquatic ecosystem (a pond, for example), the abiotic component would include such inorganic substances as calcium, mineral salts, oxygen, carbon dioxide, and water. Some of these would be dissolved in the water, but the majority would lie at the bottom as sediments—a natural reservoir of nutrients for both plants and animals. In a terrestrial ecosystem, the abiotic component provides life-supporting elements and compounds in the soil, groundwater, and atmosphere.

▮ **FIGURE 11.1**
This woodland ecosystem in New Hampshire on the slopes of Mount Cardigan demonstrates the close relationship between living organisms and their nonliving environment.
**Why might it be difficult for a biogeographer to determine boundaries for this ecosystem?**

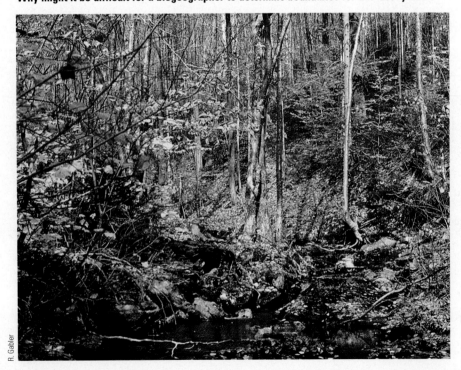

The second and perhaps most important component of an ecosystem consists of the basic producers, or **autotrophs** (meaning "self-nourished"). Plants, the most important autotrophs, are essential to virtually all life on Earth because they are capable of using energy from sunlight to convert water and carbon dioxide into organic molecules through the process known as photosynthesis (see again Chapter 4). The sugars, fats, and proteins produced by plants through photosynthesis are the foundation for the food supply that supports other forms of life. It should be noted that some bacteria are also capable of photosynthesis and hence are classed as autotrophs along with plants. Sulfur-dependent organisms that dwell at ocean-bottom thermal vents are also classified as autotrophs.

A third component of most ecosystems consists of consumers, or **heterotrophs** (meaning "other-nourished"). These are animals that survive by eating plants or other animals. Heterotrophs are classified on the basis of their feeding habits. **Herbivores** eat only living plant material; **carnivores** eat other animals; **omnivores** feed on both plants and animals. Animals make an essential contribution to the Earth ecosystem of which they are a part. They use oxygen in their respiration and return as an end product to the atmosphere the carbon dioxide that is required for photosynthesis by plants. They can influence soil development through their digging and trampling activities, and those activities in turn may affect local plant distributions.

We might assume that plants, animals, and a supporting environment are all that are required for a functioning ecosystem, but such is not the case. Without the fourth component of ecosystems, the decomposers, plant growth would soon come to a halt. The **decomposers,** or **detritivores,** feed on dead plant and animal material and waste products. They promote decay and return mineral nutrients to the soil and sea in a form that plants can use.

**■ FIGURE 11.2**
Hybrid seed, fertilizers, insecticides, and, on occasion, irrigation systems may be used by the farmer to ensure the success of this artificial ecosystem in the Corn Belt.
**How does the role of humans in an artificial ecosystem differ from that in a natural one?**

## Trophic Structure

From the discussion of the autotrophs and heterotrophs, it becomes apparent that there is a definite arrangement of the major components of an ecosystem. The components form a sequence in their levels of eating: Herbivores eat plants, carnivores may eat herbivores or other carnivores, and decomposers feed on dead plants and animals and their waste products. The pattern of feeding in an ecosystem is called the **trophic structure,** and the sequence of levels in the feeding pattern is referred to as a **food chain.** The simplest food chain would include only plants and decomposers. However, the chain usually includes at least four steps—for example, grass–field mouse–owl–fungi (plants–herbivore–carnivore–decomposer). More complex food chains may include six or more levels as carnivores feed on other carnivores—for example, zooplankton eat plants, small fish eat zooplankton, larger fish eat small fish, bears eat larger fish, and decomposers consume the bear after it dies.

Organisms within a food chain are often identified by their **trophic level,** or the number of steps they are removed from the autotrophs or plants in a food chain (Table 11.1). Plants occupy the first trophic level, herbivores the second, carnivores feeding on herbivores the third, and so forth until the last level, the decomposers, is reached. Omnivores may belong to several trophic levels because they eat both plants and animals.

In reality, linear food chains do not operate in isolation; they overlap and interact to form a feeding mosaic within an ecosystem called a **food web** (■ Fig. 11.3). Both food chains and food webs merit careful study because they can be used to trace the movement of food and energy from one level to another in the ecosystem.

## Nutrient Cycles

Some biologists and ecologists find it helpful to separate the trophic structure into specific nutrient cycles. There are several such cycles, which at times intertwine and help explain the routing for most of the nutrients through our ecosystems. Cycles have been developed for: water, carbon, nitrogen, oxygen, sulfur, and phosphorous. Some of these cycles may be

**TABLE 11.1**
**Trophic Structure of Ecosystems**

| Ecosystem Component | Trophic Level | Examples |
|---|---|---|
| Autotroph | First | Trees, shrubs, grass |
| Heterotroph | Second | Locust, rabbit, field mouse, deer, cow, bear |
| | Third | Praying mantis, owl, hawk, coyote, wolf, bear |
| | Fourth, etc. | Bobcat, wolf, hawk, bear |
| Decomposer | Last | Fungi, bacteria |

familiar to you. Parts of the oxygen and carbon cycles were discussed in various sections of Chapter 4. The water (hydrologic) cycle was highlighted in Chapter 6. Though each of these cycles can be singled out individually, Figure 11.4 shows a summary diagram, which incorporates the major processes involved in these cycles. Knowledge of chemical nutrient cycles is essential to an understanding of energy flow in ecosystems.

## Energy Flow

When physical geographers study ecosystems, they trace the flow of energy through the system just as they do when they study energy flow in other systems, such as streams or glaciers. Just as in other systems, the laws of thermodynamics apply to ecosystems. For example, as the first law of thermodynamics states, energy cannot be created or destroyed; it can only be changed from one form to another. Energy comes to the ecosystem in the form of sunlight, which is used by plants in photosynthesis. This energy is stored in the system in the form of organic material in plants and animals. It flows through the system along food chains and webs from one trophic level to the next. It is finally released from the system when oxygen is combined with the chemical compounds of the organic material through the process of oxidation. Respiration, which involves the combination of oxygen with chemical compounds

**FIGURE 11.3**
Ecosystems are worthwhile subjects of study by physical geographers. They clearly illustrate the interdependence of the variables in systems, especially the close relationships between the living components of systems (the biosphere) and the nonliving or abiotic components in systems (the atmosphere, hydrosphere, and lithosphere).
**Can you trace a trophic structure through this diagram?**

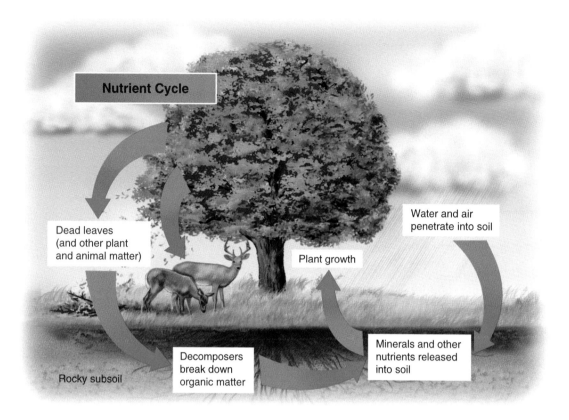

### FIGURE 11.4
This simplified diagram displays the processes used by the nutrient cycles to travel through the ecosystem.
**What kinds of processes are taking place beneath the soil surface?**

in living cells and can occur at any trophic level, is the major form of oxidation. Fire is yet another form.

The total amount of living material in an ecosystem is referred to as the **biomass.** Because the energy of a system is stored in the biomass, scientists measure the biomass at each trophic level to trace the energy flow through the system. They usually find that the biomass decreases with each successive trophic level ( Fig. 11.5). There are a number of explanations for this, each involving a loss of energy. The first instance occurs between trophic levels. The second law of thermodynamics states that whenever energy is transformed from one state to another there will be a loss of energy through heat. Hence, when an organism at one trophic level feeds on an organism at another, not all of the food energy is used. Some is lost to the system. Additional energy is lost through respiration and movement. At each successive trophic level, the amount of energy required is greater. A deer may graze in a limited area, but the wolf that preys on the deer must hunt over a much larger territory. Whatever the reason for energy loss, it follows that as the flow of energy decreases with each successive trophic level, the biomass also decreases. This principle also applies to agriculture. A great deal more biomass (and food energy) is available in a field of corn than there is in the cattle that eat the corn.

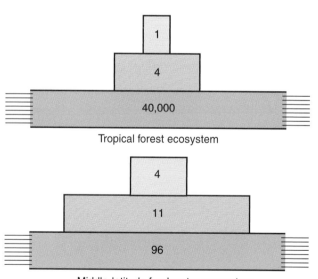

### FIGURE 11.5
Trophic pyramids showing biomass of organisms at various trophic levels in two contrasting ecosystems. Dry weight is used to measure biomass because the proportion of water to total mass differs from one organism to another.
**How can you explain the exceptionally large loss of biomass between the first and second trophic levels of the tropical forest ecosystem?**

**TABLE 11.2**
**Net Primary Productivity of Selected Ecosystems**

| TYPE OF ECOSYSTEM | NORMAL RANGE | Net Primary Productivity, gm² per year MEAN |
|---|---|---|
| Tropical rainforest | 1000–3500 | 2200 |
| Tropical seasonal forest | 1000–2500 | 1600 |
| Middle-latitude evergreen forest | 600–2500 | 1300 |
| Middle-latitude deciduous forest | 600–2500 | 1200 |
| Boreal forest (taiga) | 400–2000 | 800 |
| Woodland and shrubland | 250–1200 | 700 |
| Savanna | 200–2000 | 900 |
| Middle-latitude grassland | 200–1500 | 600 |
| Tundra and alpine | 10–400 | 140 |
| Desert and semidesert scrub | 10–250 | 90 |
| Extreme desert, rock, sand, and ice | 0–10 | 3 |
| Cultivated land | 100–3500 | 650 |
| Swamp and marsh | 800–3500 | 2000 |
| Lake and stream | 100–1500 | 250 |
| Open ocean | 2–400 | 125 |
| Upwelling zones | 400–1000 | 500 |
| Continental shelf | 200–600 | 360 |
| Algal beds and reefs | 500–4000 | 2500 |
| Estuaries | 200–3500 | 1500 |

Source: R. H. Whittaker, *Communities and Ecosystems* (2nd ed.). New York: Macmillan, 1975.

## Productivity

Productivity in an ecosystem is defined as the rate at which new organic material is created at a particular trophic level. **Primary productivity** refers to the formation of new organic matter through photosynthesis by autotrophs; **secondary productivity** refers to the rate of formation of new organic material at the heterotroph level.

**Primary Productivity** Just how efficient are plants at producing new organic matter through photosynthesis? The answer to this question depends on a number of variables.

Photosynthesis requires sunlight, the amount of which depends on the length of day and the angle of the sun's rays, which in turn differ widely with latitude. Photosynthesis is also affected by factors such as soil moisture, temperature, the availability of mineral nutrients, the carbon dioxide content of the atmosphere, and the age and species of the individual plants.

Most studies of productivity in ecosystems have been concerned with measuring the net biomass at the autotroph level (Table 11.2). Wherever figures have been compiled on the efficiency of photosynthesis, the efficiency has

**FIGURE 11.6**
Worldwide vegetation patterns revealed through a color index derived from environmental satellite observations. Compare this image with the world map of natural vegetation in Figure 11.21.
**What color on this map represents desert vegetation?**

been surprisingly low. Most studies indicate that less than 5% of the available sunlight is used to produce new biomass in ecosystems. For Earth as a whole, the figure is probably less than 1%. Nonetheless, the net primary productivity of the ecosphere is enormous. It is estimated to be in the range of 170 billion metric tonnes (a metric tonne is about 10% greater than a U.S. ton) of organic matter annually. Even though oceans cover approximately 70% of Earth's surface, slightly more than two thirds of net annual productivity is from terrestrial ecosystems and less than one third is from marine ecosystems. Perhaps even more surprising is the fact that humans consume less than 1% of Earth's primary productivity as plant food. However, humans also use biomass, in a variety of other ways—for example, using lumber for construction and paper production and using biomass energy for feedstock and as fodder for range animals.

Table 11.2 illustrates the wide range of net primary productivity displayed by various ecosystems. The latitudinal control of insolation and the subsequent effect on photosynthesis can be easily recognized when comparing figures for terrestrial ecosystems. There is a noticeable decrease in terrestrial productivity from tropical ecosystems to those in middle and higher latitudes. Even the tropical savannas, which are dominated by grasses, produce more biomass in a year than the boreal forests, which are found in the colder climates. Today, satellites monitor Earth's biological productivity and give us a global perspective on our biosphere (Fig. 11.6).

The reasons for differences among aquatic, or water-controlled, ecosystems are not quite as apparent. Swamps and marshes are especially well supplied with plant nutrients and therefore have a relatively large biomass at the first trophic level. On the other hand, depth of water has the greatest impact on ocean ecosystems. Most nutrients in the open ocean sink to the bottom, beyond the depth where sunlight can penetrate and make photosynthesis possible. Hence, the most productive marine ecosystems are found in the sunlit, shallow waters of estuaries, continental shelves, or coral reefs or in areas where ocean upwelling carries nutrients nearer to the surface (Fig. 11.7).

Some artificial ecosystems associated with agriculture can be fairly productive when compared with the natural ecosystems they have replaced. This is especially true in the warmer latitudes where farmers may raise two or more crops in a year or in arid lands where irrigation supplies the water essential to growth. However, Table 11.2 indicates that mean productivity for cultivated land does not approach that of forested land and is just about the same as that of middle-latitude grasslands. Most quantitative studies have

**FIGURE 11.7**
The nutrient-rich waters weave through the Great Barrier Reef off Mackay, Australia.
**Why are the most productive marine ecosystems found in the shallow waters bordering the world's continents?**

shown that agricultural ecosystems are significantly less productive than natural systems in the same environment.

**Secondary Productivity** As we have seen, secondary productivity results from the conversion of plant materials to animal substances. We have also noted that the ecological efficiency, or the rate of energy transfer from one trophic level to the next, is low ( Fig. 11.8). The efficiency of transfer from autotrophs to heterotrophs varies widely from one ecosystem to another. The amount of net primary productivity actually eaten by herbivores may range from as high as 15% in some grassland areas to as low as 1 or 2% in certain forested regions. In ocean ecosystems, the figure may be much higher, but there is a greater loss during the digestion process. Once the food is eaten, energy loss through respiration or body movement reduces secondary productivity to a small fraction of the biomass available as net primary productivity.

Most authorities consider 10% to be a reasonable estimate of ecological efficiency for both herbivores and carnivores. If both herbivores and carnivores have ecological efficiencies of only 10%, the ratio of biomass at the first trophic level to biomass of carnivores at the third trophic level is several thousand to one. It obviously requires a huge biomass at the autotroph level to support one animal that eats only meat. As human populations grow at increasing rates and agricultural production lags behind, it is indeed fortunate that human beings are omnivores and can adopt a more vegetarian diet ( Fig. 11.9).

## Ecological Niche

There are a surprising number of species in each ecosystem, except for those ecosystems severely restricted by adverse environmental conditions. Yet each organism performs a specific role in the system and lives in a certain location, described as its **habitat.** The combination of role and habitat for a particular species is referred to as its **ecological niche.** A number of factors influence the ecological niche of an organism. Some species are **generalists** and can survive on a wide variety of food. The North American brown, or grizzly, bear, as an omnivore, will eat berries, honey, and fish. On the other hand, the koala of Australia is a specialist and eats only the leaves of certain eucalyptus trees. Specialists do well when their particular food supply is abundant, but they cannot adapt to changing environmental conditions. The generalists are in the majority in most ecosystems because their broader ecological niche allows survival on alternative food supplies.

Some generalists among species occupy an ecological niche in one ecosystem that is quite different from the niche they occupy in another. As food supply varies with habitat, so varies the ecological niche. Humans are the extreme example of the generalist: In some parts of Earth, they are carnivores; in some parts, herbivores; and in some parts, omnivores. It is also true that different species may occupy the same ecological niche in habitats that are similar but located in separate ecosystems.

## Succession and Climax Communities

Up to this point, we have been discussing ecosystems in general terms. In the remainder of this chapter, we note that it is the species that occupy the ecosystem that give the ecosystem its character. At least for terrestrial ecosystems, it is the autotrophs—the plant species at the first trophic level—that most easily distinguish one ecosystem from another. All other species in an ecosystem depend on the autotrophs for food, and the association of all living organisms determines the energy flow and the trophic structure of the ecosystem. It should also be noted that the species that

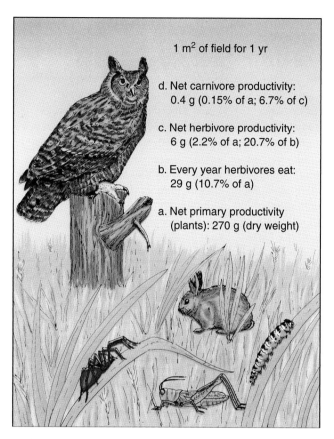

**FIGURE 11.8**
Productivity at the autotroph, herbivore, and carnivore trophic levels as measured in a Tennessee field. The figures represent productivity for 1 square meter of field in 1 year. Note the extremely small proportion of primary productivity that reaches the carnivore level of the food chain.
**In this example, which group—carnivores or herbivores—is more efficient (produces the greater percentage of energy available at the trophic level immediately below it in the food chain)?**

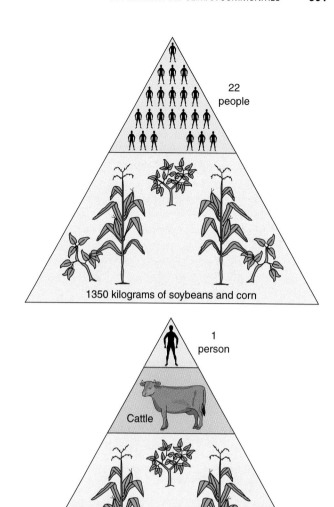

**FIGURE 11.9**
The triangles illustrate the advantages of a vegetarian diet as we experience another century of rapid population growth. It is fortunate that humans are omnivores and can choose to eat grain products. The same 1350 kilograms of grain that will support, if converted to meat, only 1 person will support 22 people if cattle or other animals are omitted from the food chain.
**In what areas of the world today do grain products constitute nearly all of the total food supply?**

occupy the first trophic level are greatly influenced by climate; again we see the interconnections between the major Earth subsystems.

If the plants that comprise the biomass of the first trophic level are allowed to develop naturally without obvious interference from or modification by humans, the resulting association of plants is called **natural vegetation.** These plant associations, or **plant communities,** are compatible because each species within the community has different requirements in relation to major environmental factors such as light, moisture, and mineral nutrients. If two species within a community were to compete, one would eventually eliminate the other. The species forming a community at any specific place and time will be an aggregation of those that together can adapt to the prevailing environmental conditions.

## Succession

Natural vegetation of a particular location develops in a sequence of stages involving different plant communities. This developmental process, known as **succession,** usually begins with a relatively simple plant community. Two types of succession, *primary* and *secondary,* are recognized. In primary

# Geography's Environmental Science Perspective
## The Theory of Island Biogeography

Biogeographers are intrigued by the life forms and the diversity of species that exist on islands—areas that are isolated from larger landmasses by ocean environments that are not inhabited by these life forms. How can there be land plants and animals living on an island surrounded by a wide expanse of sea? How did this assortment of flora and fauna become established and flourish on these often distant and geologically recent terrains (volcanic islands, for example, which were barren after they formed)? The farther an island is from the nearest landmasses, the more difficult it is for species to migrate to the island and to establish a viable population there.

The seeds of some plants are carried by the wind, by birds, or by currents to islands, and germinate to develop the vegetative environments on these isolated landmasses. Many other plant and animal species living on islands were introduced to these remote locations by humans as they migrated to these islands. But why did the species adapt and survive?

The theory of island biogeography offers an explanation for how natural factors interact to affect successful colonization or extinction of species that initially come to live on an island. The theory considers the degree of isolation of the island (the distance from a mainland source of migrating species), the size of the island, and the number of species living on an island. Generally, the diversity of life forms on islands is low compared to mainland locations with a similar climate and other environmental characteristics. Low diversity of species typically also means that the floral and faunal populations of that place exist in an environmentally sensitive location. Many extinctions have occurred on islands because of the

Diverse species exist on this island despite its extreme isolation. Moorea Island in French Polynesia

introduction of some factor that made the habitat nonviable for that species to survive.

Several major factors affect the species diversity on islands (as long as other environmental conditions such as climate are comparable):

1. The farther an island is from the area from which species must migrate, the lower the species diversity. Islands nearer to large landmasses tend to have higher diversity than those that are more distant.
2. The larger the island, the greater the species diversity. This is partly because larger islands tend to offer a wider variety of environments to colonizing organisms than smaller islands do. Larger islands also offer more space for species to occupy.
3. The species diversity of an island results from an equilibrium between the rates of extinction of species on the island and the colonization of species. If the islands' extinction rate is higher, only a few hardy species will live there; if the extinction rate is lower compared to the colonization rate, then more species will thrive, and the diversity will be higher.

The theory of island biogeography has also been useful in understanding the ecology and biota of many other kinds of isolated environments, such as high mountain areas that stand, much like islands, above surrounding deserts. In those regions, plants and animals adapted to cool wet environments live in isolation from similar populations on nearby mountains, separated by inhospitable arid environments.

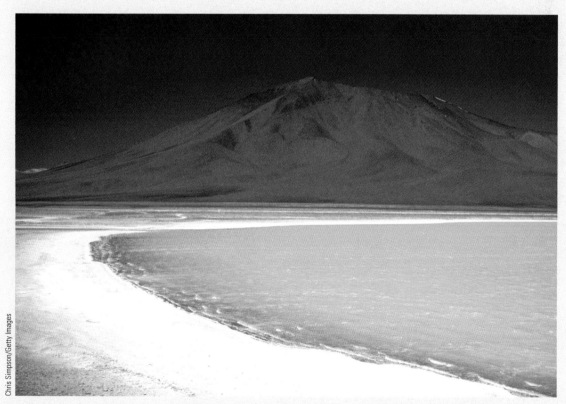

The volcanic peak known as Tunupa rises from the vast Uyuni Salt Flats, in Bolivia. Tunupa has a variety of ecosystems unique to its slopes.

succession, a bare substrate is the beginning point. No soil or seedbed exists at this point. A *pioneer* community invades the bare substrate (whether it be volcanic lava, glacially deposited sediment, or a bare beach, among others) and begins to alter the environment. As a result, the species structure of the ecosystem does not remain constant. In time, the alterations of the environment become sufficient to allow a new plant community (a community that could not have survived under the original conditions) to appear and eventually to dominate the original community. The process continues with each succeeding community rendering further changes to the environment. Because of the initial absence of soil, primary succession can take hundreds or even a few thousand years. Secondary succession begins when some natural process, such as a forest fire, tornado, or landslide, has destroyed or damaged a great deal of the existing vegetation. Ecologists refer to this process as **gap** creation. Even with such damage, however, soil still typically exists, and seeds may be lying dormant in that soil ready to invade the newly opened gap. Secondary succession, therefore, can occur much more quickly than primary succession.

A common form of secondary succession associated with agriculture in the southeastern United States is depicted in Figure 11.10. After agriculture has ceased, weeds and grasses are the first vegetative types to adapt to the somewhat adverse conditions associated with bare fields. These low-growing plants will stabilize the topsoil, add organic matter, and in general pave the way for the development of hardwood brush such as sassafras, persimmon, and sweet gum. During the brush stage, the soil will become richer in nutrients and organic matter, and its ability to retain water will increase. These conditions encourage the development of pine forests, the next stage in this vegetative evolutionary process. Pine forests thrive in the newly created environment and will eventually dwarf and dominate the weeds, grasses, and brush that preceded them.

Ironically, the dominance of the pine forest leads to its demise. Pine trees require much sunlight if their seeds are to germinate. When competing with low-lying brush, grasses, and weedy annuals, there is no problem in getting enough sunlight, but once the pines dominate the landscape, their seeds will not germinate in the shade and litter that their dense foliage creates. Thus, pines eventually will give way to hardwoods, such as oak and hickory, whose seeds can germinate under those conditions. These seeds may have been present and dormant in the soil, or blown into the forest, or carried into it by animals. In this example, then, the end result is an oak–hickory forest. If the changes continue uninterrupted, it is estimated that the complete succession will take 100–200 years. In other ecosystems, such as a tropical rainforest, this complete process may take many hundreds of years.

## The Climax Community

The theory of plant succession was introduced early in the 20th century. However, some of the original ideas have undergone considerable modification. Succession was considered to be an orderly process that included various *predictable* steps or phases and ended with a dominant vegetative cover that would remain

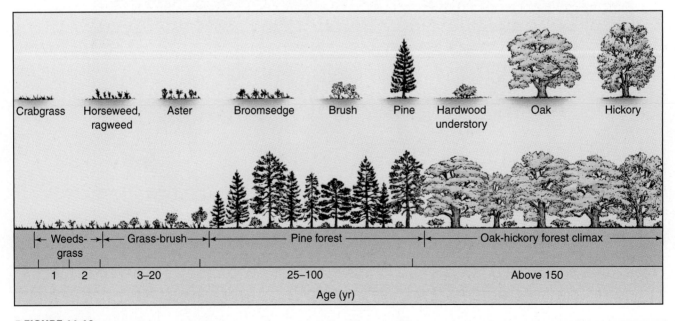

**FIGURE 11.10**
A common plant succession in the southeastern United States. Each succeeding vegetation type alters the environment in such a way that species having more stringent environmental requirements can develop.
**Why would plant succession be quite different in another region of the United States?**

in balance with the environment until disturbed by human activity—or until there were major changes in the environment. The final step in the process of succession has been referred to as the **climax community.** It was generally agreed that such a community was self-perpetuating and had reached a state of equilibrium or stability with the environment. In our illustration of plant succession in the southeastern United States, the oak–hickory forest would be considered the climax community.

Succession is still a useful model in the study of ecosystems. Why, then, have some of the original ideas been challenged, and what is the most recent thought on the subject? For one thing, early proponents of succession emphasized the predictable nature of the theory. One plant community would follow another in regular order as the species structure of the ecosystem evolved. But it has been demonstrated on many occasions that changes in ecosystems do not occur in such a rigid fashion. More often than not, the movement of species into an area to form a new community occurs in a random fashion and may be largely a function of chance events.

Many scientists today no longer believe that only one type of climax vegetation is possible for each major climate region of the world. Some suggest that one of several different climax communities might develop within a given area, influenced not only by climate but also by drainage conditions, nutrients, soil, or topography. The dynamic nature of climate is now also much more fully understood than it was when the theories of succession and climax were developed, so it is now seen that by the time the species structure of a plant community has adjusted to new climatic conditions, the climate may change again. Because of the dynamic nature of each habitat, no one climax community can exist in equilibrium with the environment for an indefinite period of time. Many biogeographers and ecologists today view plant communities and their ecosystems as a *landscape* that is an expression of all the various environmental factors functioning together. They view the landscape of an area as a **mosaic** of interlocking parts, much like the tiles in a mosaic artwork. In a pine forest, for example, other plants also exist, and some areas do not support pine trees. The dominant area of the mosaic—that is, the pine forest—is referred to as the **matrix.** Gaps within the matrix, resulting from areas of different soil conditions or gaps created by natural processes, are referred to as **patches** within the matrix ( Fig. 11.11). Relatively linear features cutting across the mosaic, including natural features such as rivers and human-created structures such as roads, fence lines, power lines, and hedgerows, are termed **corridors** ( Fig. 11.12). Each particular habitat is unique and constantly

**FIGURE 11.11**
This patch in the tropical rainforest matrix of Jamaica is the result of land cleared for shifting (slash-and-burn) cultivation.
**What types of human activity might be responsible for patches in the matrix of a middle-latitude forest?**

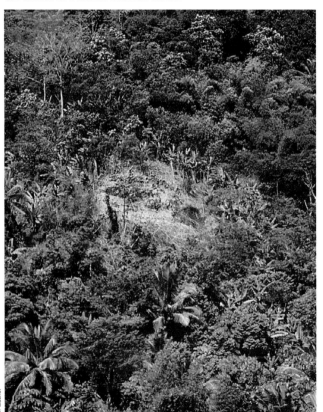

**FIGURE 11.12**
This aerial view of Taskinas Creek, Virginia, shows riparian corridors passing through a forest matrix on the Atlantic coastal plain.
**How does a corridor differ from a patch?**

changing, and resultant plant and animal communities must constantly adjust to these changes. The dominant environmental influence is climate.

As discussed in Chapter 8, climates are also changing. Over both relatively short time periods, on the scales of decades and centuries, and much longer times measured in millennia, climate changes occur. They may be subtle, or they may be sufficiently drastic to create ice ages or warm periods between ice ages. Plant and animal communities must be able to respond to these ongoing changes, or they will not survive. Many modern biogeographers are deeply involved in the reconstruction of the vegetation communities of past climate periods, through the examination of evidence such as tree rings, pollen, insect fossils, and plant fossils. By determining how past climate changes affected and induced change in Earth's ecosystems, biogeographers hope to be able to suggest how future changes may develop as climate continues to change.

# Environmental Controls

As was discussed, the plant and animal species occupying a particular ecosystem at a given time are those that are most successful in adjusting to the unique environment that constitutes their habitat. Each species of living organism has a range within which it can adapt to environmental factors. For example, some plants can survive under a wide range of temperature conditions, whereas others have narrow temperature requirements. Biogeographers and ecologists refer to this characteristic as an organism's range of **tolerance** for a particular environmental condition. The ranges of tolerance for a species will determine where on Earth that species may be found, and species with wide ranges of tolerance will be the most widely distributed. The *ecological optimum* refers to the environmental conditions under which a species will flourish ( Fig. 11.13). As a species moves away from its ecological optimum or as one moves away from the geographic core of a plant or animal community, the environmental conditions become increasingly difficult for that species or community to survive. At the same time, those conditions may be more amenable for another species or community. The **ecotone** is the overlap, or zone of transition, between two plant or animal communities (see Fig. 11.13).

Climate has the greatest influence over natural vegetation when we observe plant communities on a worldwide basis. The major types of terrestrial ecosystems, or **biomes**, are each associated with specific ranges of temperature and critical precipitation characteristics such as annual amounts and seasonal distribution. Climate influences leaf shape and size in trees and determines if trees can even exist in a region. At the local scale, however, other environmental factors can be as important as climate. A plant's range of tolerance for the acidity of the soil, the drainage of the land, or the salinity of the water may be the critical environmental factor in determining whether that plant is a part of the ecosystem. The discussion that follows serves to illustrate how the major environmental factors influence the organization and structure of ecosystems.

## Climatic Factors

Of all the various climatic factors that influence the ecosystem, sunlight conditions are often the most critical. Sunlight is the vital source of energy for photosynthesis in plants and a control of life patterns for animals. The competition for light makes forest trees grow taller and limits growth on the forest floor to plants, such as ferns, that can tolerate shade conditions. Leaf sizes, shapes, and even colors may reflect this variation in light reception, with large leaves in areas of limited light reception. The *quality* of light is important, especially in mountain areas, where plant growth may be severely retarded by excess ultraviolet radiation. This radiation does significant damage in the thin air at higher elevations but is effectively screened out by the denser atmosphere at lower elevations. Light *intensity* affects the rate of photosynthesis and hence the rate of primary productivity in an ecosystem. The more intense light of the low latitudes produces a higher energy input and greater biomass in the tropical forest than does the less intense light of the higher latitudes in Arctic regions. The *duration* of daylight, in association with the changing seasons, has a profound effect on the flowering of plants, the activity patterns of insects, and the migration and mating habits of animals.

A second important climatic control of ecosystems is the availability of water. Virtually all organisms require water to survive. Plants require water for germination, growth, and reproduction, and most plant nutrients are dissolved in soil water before they can be absorbed by plants. Marine plants are adapted to living completely in water (seaweed); some plants, such as mangrove ( Fig. 11.14) and bald cypress ( Fig. 11.15), rise from coastal marshes and inland swamps; others thrive in the constantly wet rainforest. Certain tropical plants drop their leaves and become dormant during dry seasons; others store water received during periods of rain in order to survive seasons of drought. Desert plants, such as cacti, are especially adapted to obtaining and storing water when it is available while minimizing their water loss from transpiration.

Animals, too, are severely restricted when water is in short supply. In arid regions, animals must make special adaptations to environmental conditions. Many become inactive during the hottest and driest seasons, and most leave their burrows or the shade of plants and rocks only at night. Others, like the camel, can travel for great distances and live for extended periods without a water supply.

Organisms are affected less by temperature variations than by sunlight and water availability. Many plants can tolerate a wide range of temperatures although each species has optimum conditions for germination, growth, and reproduction. These functions, however, can be impeded by temperature extremes. Temperatures may also have indirect effects on vegetation. For example, high temperatures will lower the relative humidity, thus increasing transpiration. If a

**FIGURE 11.13**
The concepts of *ecotone, ecological optimum, range of tolerance, mosaic, matrix, patch,* and *corridor* are illustrated in the diagram.
**What effect would a change to drier climatic conditions throughout the area have on the relative sizes of the two ecosystems as well as the position of the ecotone?**

plant's root system cannot extract enough moisture from the soil to meet this increase in transpiration, the plant will wilt and eventually die.

Because of their mobility, animals are not as dependent on the vagaries of climate as are plants. Despite the great advantage afforded by mobility, however, animals are nevertheless subject to climatic stress. The geographic distribution of some groups of animals reflects this sensitivity to climate. Cold-blooded animals are, for example, more widespread in warmer climates and more restricted in colder climates. Some warm-blooded animals develop a layer of fat or fur and are able to shiver to protect themselves against the cold. In hot periods, they may sweat, shed fur, or lick their fur in an attempt to stay cool. In extremely cold or arid regions, animals may hibernate. During hibernation, the body temperature of the animal changes roughly in response to outside and ground temperatures. Cold-blooded animals such as the desert rattlesnake move in and out of shade in response to temperature change. Warm-blooded animals may migrate great distances out of environmentally harsh areas.

Some warm-blooded animals exhibit an interesting linkage between body shape and size and variations in average environmental temperature. These adaptations have been described by biologists as *Bergmann's Rule* and *Allen's Rule.* Bergmann's Rule states that, within a warm-blooded species, the body size of the subspecies usually increases with the decreasing mean temperature of its habitat; Allen's Rule notes that, in warm-blooded species, the relative size of exposed portions of the body decreases with the decrease of mean temperature. These rules essentially boil down to the fact that

**FIGURE 11.14**
Mangrove thicket along the Gulf of Mexico coast of southern Florida.
**How might this vegetation type have influenced the routes that were followed by the Spanish adventurers who first explored Florida?**

**FIGURE 11.15**
This stand of bald cypress trees is located in extreme southern Illinois, at the poleward limit of growth for this type of vegetation.
**What Köppen climatic type does this site represent?**

members of the same species living in colder climates eventually evolve shorter or smaller appendages (ears, noses, arms, legs, and so on) than their relatives in warmer climates (Allen's Rule) and that in cold climates body size will be larger with more mass to provide the body heat needed for survival and for protection of the main trunk of the body where vital organs are located (Bergmann's Rule). In cold climates, small appendages are advantageous because they reduce the amount of exposed area subject to temperature loss, frostbite, and cellular disruption ( Fig. 11.16). In warm climates, large body sizes are not necessary for protection of internal organs, but long limbs, noses, and ears allow for heat dissipation in addition to that provided by panting or fur licking.

Although most significant in areas such as deserts, polar regions, coastal zones, and highlands, wind can also serve as a climatic control. Wind may cause direct injury to vegetation or may have an indirect effect by increasing the rate of evapotranspiration. To prevent water loss in the areas of severe wind stress, plants will twist and grow close to the ground to minimize the degree of their exposure ( Fig. 11.17). During severe winters, they are better off buried by snow than exposed to icy gales. In some coastal regions, the shoreline may be devoid of trees or other tall plants; where trees do grow, they are often misshapen or swept bare of leaves and branches on their windward sides.

## Soil and Topography

In terrestrial ecosystems, the soils in which plants grow supply much of the moisture and minerals that are transformed into plant tissues. Soil variations are among the most conspicuous influences on plant distribution and often produce sharp boundaries in vegetation type. This is partly a consequence of the varying chemical requirements of different plant species and partly a reflection of other factors such as soil texture. In a particular area, clay soil may retain too much moisture for certain plants, whereas sandy soil retains too little. It is well known that pines thrive in sandy soils, grasses in clays, cranberries in acid soils, and wheat and chili peppers in alkaline soils. The subject of soils and their influence on vegetation will be explored more thoroughly in Chapter 12.

In the discussion of highland climate regions in Chapter 10, we learned that topography influences ecosystems

## Natural Catastrophes

The distributions of plants and animals are also affected by a diversity of natural processes frequently termed *catastrophes*. It should be noted, however, that this term is applied from a strictly human perspective. What may be catastrophic to a human, such as a hurricane, forest fire, landslide, or tsunami ( ❙ Fig. 11.18), is simply a natural process operating to produce openings (gaps) in the prevailing vegetative mosaic of a region. The resulting successional processes, whether primary or secondary in nature, produce a variety of patch habitats within the broader regional matrix of vegetation. The study of natural catastrophes and the resulting patch dynamics among the plant and animal residents of an area is a topic of strong interest and ongoing research in modern landscape biogeography.

## Biotic Factors

Although they might tend to be overlooked as environmental controls, other plants and animals may be the critical factors in determining whether a given organism is a part of an ecosystem. Some interactions between organisms may be beneficial to both species involved, whereas others may have an adverse effect on one or both. Because most ecosystems are suitable to a wide variety of plants and animals, there is always competition between species and among members of a given species to determine which organisms will survive. The greatest competition occurs between species that occupy the same ecological niche, especially during the earliest stages of life cycles when organisms are most vulnerable. Among plants, the greatest competition is for light. Those trees that become dominant in the forest are those that grow the tallest and partially shade the plants growing beneath them. Other competition occurs underground, where the roots compete for soil water and plant nutrients.

Interactions between plants and animals and competition both within and among animal species also have significant effects on the nature of an ecosystem. Animals are often helpful to plants during pollination or the dispersal of seeds, and plants are the basic food supply for many animals. The simple act of grazing may help determine the species that make up a plant community. During dry periods, herbivores may be forced to graze an area more closely than usual, with the result that the taller plants are quickly grazed out. Plants that grow close to the ground, that are unpalatable, or that have the strongest root development are the ones that survive. Hence, grazing is a part of the natural selection process, but serious overgrazing rarely results under natural conditions because wild animal populations increase or decrease with the available food supply. To be more precise, the number of animals of a given species will fluctuate between the maximum number that can be supported when its food supply is greatest and the minimum number required for reproduction of the species. For most

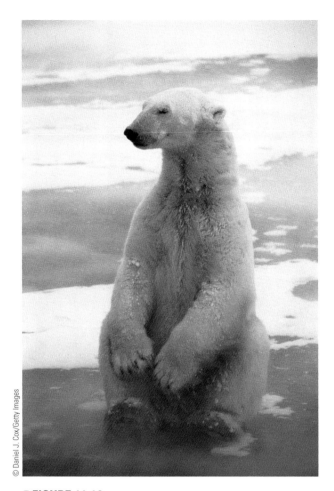

### ❙ FIGURE 11.16
The Arctic polar bear provides an excellent example of both Allen's and Bergmann's Rules.

**What warm-climate animal might you suggest as a prime example of Allen's Rule?**

indirectly by providing many microclimates within a relatively small area. Plant communities vary significantly from place to place in highland areas in response to the differing nature of the climatic conditions. Some plants thrive on the sunny south-facing slopes of highland areas in the Northern Hemisphere; others survive on the colder, shaded, north-facing slopes. The steepness and shape of a slope also affect the amount of time water is present there before draining downslope.

Luxuriant forests tower above the well-watered windward sides of mountain ranges such as the Sierra Nevada and Cascades, but semiarid grasslands and sparse forests cover the leeward sides. Spatial variations in precipitation drainage and resulting vegetation differences would not exist were it not for the presence of the topographic barrier inducing orographic uplift. Each major increase in elevation also produces a different mixture of plant species that can tolerate the lower ranges of temperature found at the higher elevation.

**FIGURE 11.17**
Krummholz vegetation at the upper reaches of the subalpine zone on Pennsylvania Mountain in the Mosquito Range of the Colorado Rockies. The healthy green vegetation has been covered by snow much of the year and has been protected from the bitterly cold temperatures associated with gale-force winter winds. Note the flag trees, which give a clear indication of wind direction.
**What type of vegetation would be found at elevations higher than the one depicted in this photograph?**

## Human Impact on Ecosystems

Throughout human history, we have modified the natural development of ecosystems. Except in regions too remote to be altered significantly by civilization, humans have eliminated much of Earth's natural vegetation. Farming, fire, grazing of domesticated animals, deforestation and afforestation, road building, urban development, dam building and irrigation, raising and lowering of water tables, mining, and the filling in or draining of wetlands are just a few ways in which humans have modified the plant communities around them. Overgrazing by domesticated animals can seriously harm marginal environments in semiarid climates. Trampling and compaction of the soil by grazing herbivores may reduce the soil's ability to absorb moisture, leading to increased surface runoff of precipitation. In turn, the decreased absorption and increased runoff may respectively lead to *land degradation* and gully erosion.

It should be noted that ecosystems are not the only victims as humans alter natural environments; the changes can often produce long-term negative effects on humans themselves. The desertification of large tracts of semiarid portions of East Africa has resulted periodically in widespread famine in countries such as Ethiopia and Somalia (Fig. 11.19). Elsewhere, the continuing destruction of wetlands not only eliminates valuable plant and animal communities but also often seriously threatens the quality and reliability of the water supplies for the people who drained the land.

**FIGURE 11.18**
The vegetation mosaic of this area is coniferous forest, but frequent snow avalanches passing down the gully keep rigid-stemmed conifers from invading the patch of open, low shrubs and grasses. The effects of attempted colonization are seen at right, where a swath of conifers has been destroyed by a recent snow avalanche.
**What other environmental factors might account for the limited amount of coniferous vegetation on the left, south-facing slope?**

animals, predators are also an important control of numbers. Fortunately, when the predator's favorite species is scarce, it will seek an alternative species for its food supply.

# Geography's Spatial Science Perspective
## Introduction and Diffusion of an Exotic Sepcies: Fire Ants

Exotic species are plants or animals that have been introduced to a new environment from where they originated, usually by human activities. Some exotics are regarded as beneficial; many landscaping plants used in North America were brought in from other continents or regions and planted as decorative foliage. Problems have occurred, however, with many other exotics, introduced either purposefully or inadvertently, that have adapted to their new environment to the detriment of the native plants or animals. If the exotic is unable to adapt to its new home, then it dies off and will not be a problem. Many exotics, however, thrive at the expense of native populations. One example of a destructive and harmful exotic is the imported red fire ant, which causes a painful bite and kills off many native insect and other populations (often beneficial insects) in the areas they invade. Fire ants came to the United States in the 1930s from Brazil, as "hitchhikers" on a ship that docked at Mobile, Alabama. Since that time, the geographic distribution of fire ant colonies has been closely documented as they have spread throughout the southern states from their point of introduction.

A map of the expansion of fire ants outward from this location illustrates the rapid impact that the introduction of a species to a new environment can have. This spread of fire ants is an example of spatial diffusion—the expansion of a distribution over an area. Climatic factors will influence or limit the eventual distribution of fire ant populations in the United States, but as of now they are still expanding their geographic range. Cold to the north and aridity to the west are the limiting environmental factors, but these ants also have been discovered in Southern California and in settled areas of the arid West, where lawn or agricultural irrigation provides adequate moisture for their survival. The fire ants may have been spread to these locations by potted nursery plants or turf grasses shipped in from areas already infested with this pest.

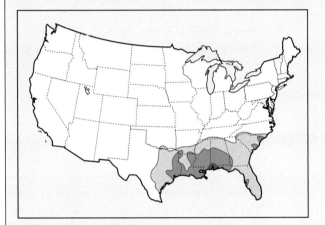

Invasion of Fire Ants 1931–1998

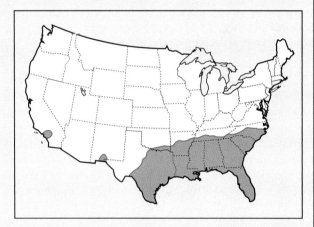

Invasion of Fire Ants 2001. Note the small incursions in the arid southern regions of California and New Mexico.

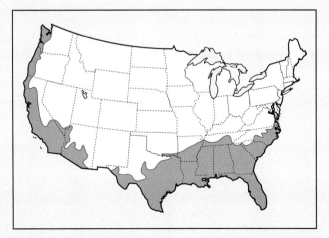

Possible Future Fire Ant Infestation Area. Suburban and agricultural irrigation provides adequate moisture for the potential spread of this invasive exotic species. However, winter temperature colder than 10°F is a limiting factor.

**FIGURE 11.19**
Overgrazing by cattle has helped promote this example of desertification in the Lake Chad basin near the Sahara.
**What are some of the other causes of desertification?**

We have in fact so changed the vegetation in some parts of the world that we can characterize classes of cultivated vegetation cared for by humans—for example, flowers, shrubs, and grasses to decorate our living areas and grains, vegetables, and fruits that we raise for our own food and to feed the animals that we eat. Our focus in the remainder of this chapter, however, is on the major ecosystems of Earth as we assume they would appear without human modification.

# Classification of Terrestrial Ecosystems

The geographic classification of natural vegetation is as difficult as the classification of any other complex phenomena influenced by a variety of factors. However, plant communities are among the most highly visible of natural phenomena, so they can be categorized on the basis of form and structure or gross physical characteristics. Of course, the composition of the natural vegetation changes from place to place in a transitional manner, just as temperature and rainfall do, and although distinctly different types are apparent, there may be broad transition zones (ecotones) between them. Nevertheless, over the world there are distinctive recurring plant communities, indicating a consistent botanical response to systematic controls that are essentially climatic. It is the dominant vegetation of these plant communities that we recognize when we classify Earth's major terrestrial ecosystems (biomes).

All of Earth's biomes can be categorized into one of four easily recognized types: forest, grassland, desert, and tundra ( Fig. 11.20). Because vegetation adapted to cold climates may occur not only in high-latitude regions but also at high elevations at any latitude, biogeographers often refer to the last of the major types as arctic and alpine tundra. However, the forests of the equatorial lowlands are an entirely different world from those of Siberia or of New England, and the original grasslands of Kansas bore little resemblance to those of the Sudan or Kenya. Hence, the four major types of ecosystems can be subdivided into distinctive biomes, each of which is an association of plants and animals of many different species. Pure stands of particular trees, shrubs, or even grasses are extremely rare and are limited to small areas having peculiar soil or drainage conditions.

Earth's major biomes are mapped in Figure 11.21 on the basis of the dominant associations of natural vegetation that give each its distinctive character and appearance. The direct influence of climate on the distribution of these biomes is immediately apparent. Temperature (or latitudinal effect on temperature and insolation) and the availability of moisture are the key factors in determining the location of major biomes on the world scale ( Fig. 11.22).

# Forest Biomes

Forests are easily recognized as associations of large, woody, perennial tree species, generally several times the height of a human, and with a more or less closed canopy of leaves overhead. They vary enormously in density and physical appearance. Some are evergreen and either needle- or broadleaf; others are deciduous, dropping their leaves to reduce moisture losses during dry seasons or when soil water is frozen. Forests are found only where the annual moisture balance is positive—where moisture availability considerably exceeds potential evapotranspiration in the growing season. Thus, they occur in the tropics, where either the ITCZ (intertropical convergence zone) or the monsoonal circulation brings plentiful rainfall, and in the middle latitudes, where precipitation is associated with cyclones along the polar front, with summer convectional rainfall, or with orographic uplift.

Tropical and middle-latitude forests have evolved different characteristics in response to the nature of the physical limitations in each area. In general, tropical forests have developed in less restrictive forest environments. Temperatures are always high, though not extreme, in the humid tropics, encouraging rapid and luxuriant growth. Middle-latitude forests, on the other hand, must adapt to combat either seasonal cold (ranging from occasional frosts to subzero temperatures) or seasonal drought (which may occur at the worst possible time for vegetative processes).

**FIGURE 11.20**
The four major types of Earth biomes: (a) forest biome near Dornbirn, Austria; (b) grassland biome in Kenya; (c) desert biome in Big Bend National Park, Texas; (d) tundra biome in Coastal Greenland.

**314** CHAPTER 11 BIOGEOGRAPHY

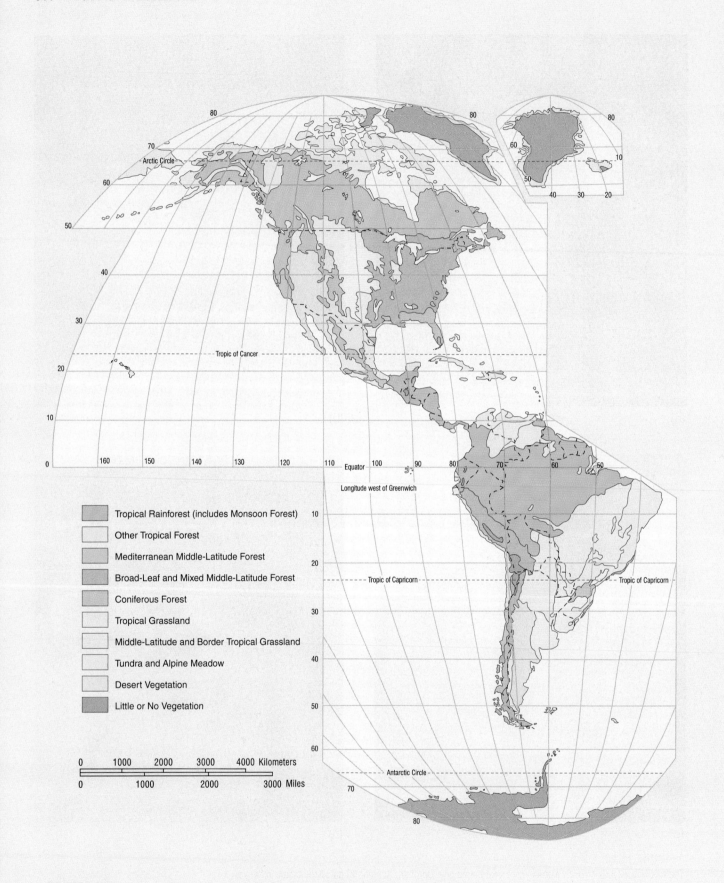

**FIGURE 11.21**
World map of natural vegetation

# FOREST BIOMES 315

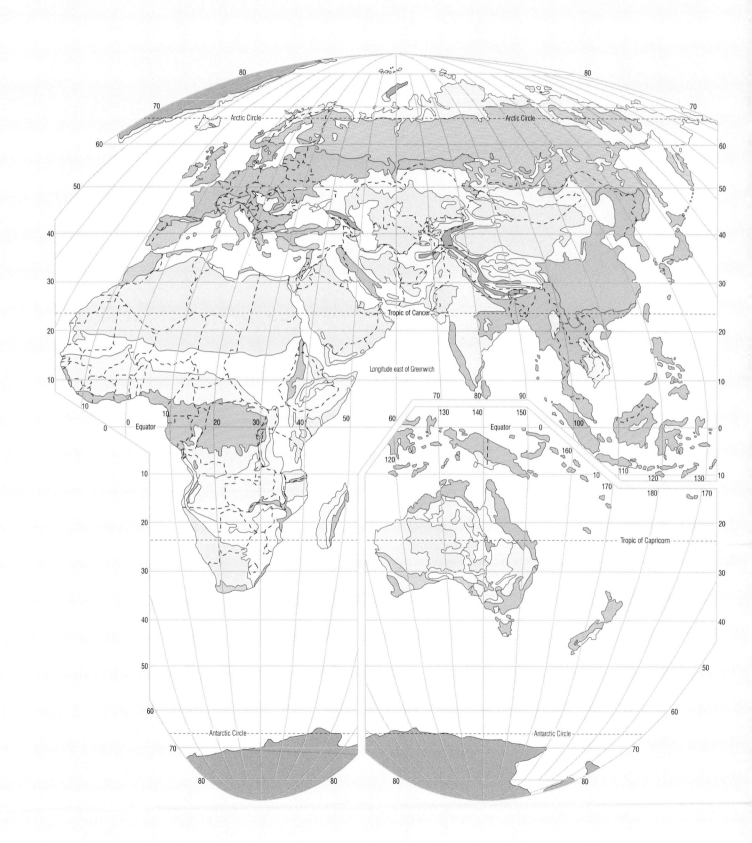

A Western Paragraphic Projection
developed at Western Illinois University

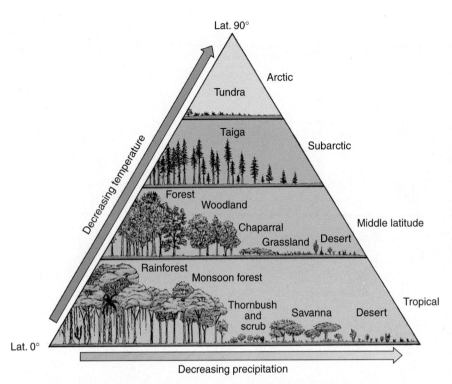

**FIGURE 11.22**
This schematic diagram shows distribution of Earth's major biomes as they are related to temperature (latitude) and the availability of moisture. Within the tropics and middle latitudes, there are distinctly different biomes as total biomass decreases with decreasing precipitation.
**What major biome dominates the wetter margins of all latitudes but the Arctic?**

## Tropical Forests

The forests of the tropics are far from uniform in appearance and composition. They grade poleward from the equatorial rainforests, which support Earth's greatest biomass, to the last scattering of low trees that overlook seemingly endless expanses of tall grass or desert shrubs on the tropical margins. We have subdivided the tropical forests into three distinct biomes: the tropical rainforest, the monsoon rainforest, and other tropical forest types, primarily thornbush and scrub. Of course, there are gradations (ecotones) between the different types, as well as distinctive variations that are found in individual localities only.

**Tropical Rainforest** In the equatorial lowlands dominated by Köppen's tropical rainforest climate, the only physical limitation for vegetation growth is competition between adjacent species. The competition is for light. Temperatures are high enough to promote constant growth, and water is always sufficient. Thus, we find forests consisting of an amazing number of broad-leaf evergreen tree species of rather similar appearance because special adaptations are not required. A cross section of the forest often reveals concentrations of leaf canopies at several different levels. The trees composing the distinctive individual tiers have similar light requirements—lower than those of the higher tiers but higher than those of the lower tiers (❙ Fig. 11.23). Little or no sunlight reaches the forest floor, which may support ferns but is often rather sparsely vegetated. The forest is literally bound together by vines, **lianas,** which climb the trunks of the forest trees and intertwine in the canopy in their own search for light. Aerial plants may cover the limbs of the forest giants, deriving nutrients from the water and the plant debris that falls from higher levels. Light and variable wind conditions in the rainforest (associated with the latitudinal belt of the doldrums) preclude wind from being an effective agent of seed and pollen dispersal, so large colorful fruits and flowers, designed to attract animals that will unwittingly carry out seed and pollen dispersal, prevail.

The forest trees commonly depend on widely flared or buttressed bases for support because their root systems are shallow. This is a consequence of the richness of the surface soil and the poverty of its lower levels. The rainforest vegetation and soil are intimately associated. The forest litter is quickly decomposed, its nutrients released and almost immediately reabsorbed by the forest root systems, which consequently remain near the surface. In this way, a rainforest biomass and the available soil nutrients maintain an almost *closed system*. Tropical soils that maintain the amazing biomass of the rainforest are fertile only as long as the forest remains undisturbed. Clearing the forest interrupts the crucial cycling of nutrients between the vegetation and the soil; the copious amounts of water percolating through the soil leach away its soluble constituents, leaving behind only inert iron and aluminum oxides that cannot support forest growth. The present rate of clearing threatens to wipe out the worldwide tropical rainforests within the foreseeable future. The largest remaining areas of unmodified rainforests are in the upper Amazon Basin where they cover hundreds of thousands of square kilometers.

Because of the darkness and extensive root systems present on the forest floor, animals of the tropical rainforest are primarily arboreal. A wide variety of species of tree-dwelling monkeys and lemurs, snakes, tree frogs, birds, and insects characterize the rainforest. Even the large herbivorous and carnivorous mammals—such as sloths, ocelots, and jaguars, respectively—are primarily arboreal.

Within the areas of rainforest and extending outward beyond its limits along streams are patches or strips of jungle.

# Career Vision

**Roxie Anderson, Physical Geographer/GIS Specialist, Environmental Wetlands Mapping, Ducks Unlimited,
University of Wisconsin, Eau Claire Bachelor's degree.**

My early interests were camping, cross-country skiing, biking, and anything outdoors. I only recall having one geography class in high school. But I've always loved maps and cartography, and I used topographic maps for hiking, biking, and camping. I've been with Ducks Unlimited since graduating. I also work on campus as an adventure trip leader with the outdoor recreation department.

I was one of those people who didn't know what I wanted to do for a long time because I wanted to do everything! When I finally found geography, I knew that this field would allow me to expand my horizons and become a jack-of-all-trades, so to speak.

*Why did you decide to major in geography?*
Geography allowed me to choose from a diverse scope of opportunities. Geography can be used to tie together and integrate the environmental sciences and can offer a larger view of the effects of human/environment interaction.

*How do you, as a physical geographer, fit in with your company's mission?*
Ducks Unlimited works to conserve, restore, and manage wetlands and associated habitats for North America's waterfowl. These habitats also benefit other wildlife and people.

As a physical geographer, I use tools such as remote sensing and GIS to do mapping using satellite imagery. These maps help identify habitats that need to be conserved, restored, or managed. To date, Ducks Unlimited has mapped about 162 million acres (almost 50% of Alaska).

I provide GIS support for the biologists and engineers. This includes creating project area maps, managing and collecting GIS data, and using GIS suitability analysis models.

*What did you learn in your physical geography education that is important in your career?*
Learning the geomorphology of landscapes has been the most important lesson for me. My cover mapping depends heavily on what I know about a landscape and how I interpret it. Geographers have a broad sense of the environment, and as a geographer I am able to work with different disciplines and professionals to get a project done.

*What academic preparation would you recommend to students who are interested in a career similar to yours?*
Take as much physical geography as possible! I also took a remote-sensing course, GIS courses, soils, surveying, and GPS/cartography courses. If could go back, I would have taken an additional soils class and switched my minor to biology or ecology.

Get outside, and observe the landscape around you. Try to interpret the geologic/geomorphic history of the area to understand the way the landscape is today. Field trips are excellent for getting outside and seeing real-life examples of geomorphology with an experienced physical geographer there to explain the landscape.

*Why do you think physical geography is an important course for those who are majoring in other fields?*
Physical geography, especially through applications of GIS, is becoming essential in many disciplines. In my experience, I've worked jointly with biologists, archaeologists, agriculture specialists, water specialists, geologists, and engineers.

---

Jungle consists of an almost impenetrable tangle of vegetation contrasting strongly with the relatively open nature (at ground level) of the true rainforest. It is often composed of secondary growth that quickly invades the rainforest where a clearing has allowed light to penetrate to the forest floor. Jungle commonly extends into the drier areas beyond the forest margins along the courses of streams. There it forms a gallery of vegetative growth closing over the watercourse and hence has been called **galleria forest.**

**Monsoon Rainforest** In areas of monsoonal circulation, there is an alternation between the dry monsoon season, when the dominant flow of air is from the land to the sea, and the wet monsoon season, when the atmospheric circulation reverses, bringing moist air onshore along tropical coasts. The wet monsoon season rainfall may be very high, even hundreds of centimeters where air is forced upward by topographic barriers. In any case, it is sufficient to produce a forest that, once established, remains despite

**FIGURE 11.23**
A rainforest on the Rio Negro, Brazil. The dense nature of the forest canopy effectively conceals the vast number of different evergreen tree species and relatively open forest floor.
**How might this rainforest differ from the rainforests of the Pacific Northwest of the United States?**

the dry monsoon season. Monsoon forests may have discernible tiers of vegetation related to the varying light demands of different species, and they are included with the tropical rainforest in Figure 11.21. However, the number of species is less than in the true rainforest, and the overall height and density of vegetation are also somewhat less. Some of the species are evergreen, but many are deciduous.

### Other Tropical Forests, Thornbush, and Scrub

Where seasonal drought has precluded the development of true rainforest or where soil characteristics prevent the growth of such vegetation, variant types of tropical forests have developed. These tend to be found on the subtropical margins of the rainforests and on old plateau surfaces where soils are especially poor in nutrients. The vegetation included in this category varies enormously but is generally low growing in comparison to rainforest, without any semblance of a tiered structure, and is denser at ground level. It is commonly thorny, indicating defensive adaptation against browsing animals, and it shows resistance to drought in that it is generally deciduous, dropping its leaves to conserve moisture during the dry winter season. Ordinarily, grass is present beneath the trees and shrubs. As we move away from the equatorial zone, we find the trees more widely spaced and the grassy areas be-

coming dominant. Along tropical coastlines, a specially adapted plant community, known as mangrove, thrives (see again Fig. 11.14). Here trees are able to grow in salt water.

## Middle-Latitude Forests

The forest biomes of the middle latitudes differ from those of the tropics because the dominant trees have evolved mechanisms to withstand periods of water deprivation due to low temperatures and annual variations in precipitation. Evergreen and deciduous plants are present, equipped to cope with seasonal extremes not encountered in tropical latitudes.

**Mediterranean Sclerophyllous Woodland** Surrounding the Mediterranean Sea and on the southwest coasts of the continents between approximately 30° and 40°N and S latitude, we have seen that a distinctive climate exists—Köppen's mesothermal hot- and dry-summer type (Mediterranean). Here annual temperature variations are moderate, and freezing temperatures are rare. However, little or no rainfall occurs during the warmest months, and plants must be drought resistant. This requirement has resulted in the evolution of distinctive vegetation that is relatively low growing, with small, hard-surfaced leaves and roots that probe deeply for water. The leaves must be capable of photosynthesis with minimum transpiration of moisture. The general look of the vegetation is a thick scrub plant community, called chaparral in the western United States and *maquis* in the Mediterranean region (see again Fig. 10.3). Wherever moisture is concentrated in depressions or on the cooler north-facing hill slopes, deciduous and evergreen oaks occur in groves ( ▌Fig. 11.24). Drought-resistant needle-leaf trees, especially pines, are also part of the overall vegetation association. Thus, the vegetation is a mosaic related to site characteristics and microclimate. Nevertheless, the similarity of the natural vegetative cover in such widely separated areas as Spain, Turkey, and California is astonishing. (Note the location of Mediterranean middle-latitude forest, Fig. 11.21.)

**Broad-Leaf Deciduous Forest** The humid regions of the middle latitudes experience a seasonal rhythm dominated by warm tropical air in the summer and invasions of cold polar air in the winter. To avoid frost damage during the colder

**FIGURE 11.24**
The distinctive sclerophyllous evergreen vegetation type encountered wherever hot, dry summers alternate with rainy winters. Oaks commonly occupy relatively damp sites such as the gullies and windward slopes seen here in southern California.
**What are the general characteristics of sclerophyllous vegetation?**

winters and to survive periods of total moisture deprivation when the ground is frozen, trees whose leaves have large transpiring surfaces drop these leaves and become dormant, coming to life and producing new leaves only when the danger period is past. A large variety of trees have evolved this mechanism; certain oaks, hickory, chestnut, beech, and maples are common examples. The seasonal rhythms produce some beautiful scenes, particularly during the periods of transition between dormancy and activity, with the sprouting of new leaves in the spring and the brilliant coloration of the fall as chemical substances draw back into the plant for winter storage (❙ Fig. 11.25).

The trees of the deciduous forest may be almost as tall as those of the tropical rainforests and, like them, produce a closed canopy of leaves overhead or, in the cold season, an interlaced network of bare branches. However, lacking a multistoried structure and having lower density as a whole, the middle-latitude deciduous forests allow much more light to reach ground level. Forests of this type are the natural vegetation in much of western Europe, eastern Asia, and eastern North America. To the north and south, they merge with mixed forests composed of broad-leaf deciduous trees and conifers. (Broad-leaf deciduous forests and mixed forests are combined in Figure 11.21.) Both the broad-leaf deciduous and mixed forests have been largely logged off or cleared for agricultural land, and the original vegetation of these regions is rarely seen.

**Broad-Leaf Evergreen Forest**
Beyond the tropics, broad-leaf trees remain evergreen (active throughout the year) in significant areas only in certain Southern Hemisphere locations. Here the mild maritime influence is strong enough to prevent either dangerous seasonal droughts or severely low winter temperatures. Southeastern Australia and portions of New Zealand, South Africa, and southern Chile are the principal areas of this type. In the Northern Hemisphere, broad-leaf evergreen forest may once have been significant in eastern Asia, but it has long since been cleared for cultivation. Limited areas occur in the United States in Florida and along the Gulf Coast as a belt of evergreen oak and magnolia.

**Mixed Forest** Poleward and equatorward, the broad-leaf deciduous forests in North America, Europe, and Asia gradually merge into mixed forests, including needle-leaf coniferous trees, normally pines. In general, where conditions permit the growth of broad-leaf deciduous trees, coniferous trees cannot compete successfully with them. Thus, in mixed forests, the conifers, which are actually more adaptable to soil and moisture deficiencies, are found in the less hospitable sites: in sandy areas, on acid soils, or where the soil itself is thin. The northern mixed forests reflect the transition to colder climates with increasing latitude; eventually, conifers become dominant in this direction. The southern mixed forests are more problematic in origin. In the United States, they are transitional to pine forests situated on sandy soils of the coastal plain. In Eurasia, they coincide with highlands dominated by conifers during a stage in the plant succession that began with the change of climate environments at the end of the ice ages, some 10,000 years ago.

**Coniferous Forest** The coniferous forests occupy the frontiers of tree growth. They survive where most of the broad-leaf species cannot endure the climatic severity and impoverished soils. The hard, narrow needles of coniferous species transpire much less moisture than do broad leaves so that needle-leaf species can tolerate conditions of physiologic drought (unavailability of moisture because of excessive soil permeability, a dry season, or frozen soil water) without defoliation. Pines, in particular, also demand little from the soil in the form of soluble plant nutrients, especially basic elements

(a)

(b)

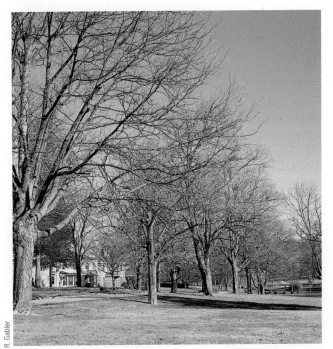
(c)

### FIGURE 11.25
The appearance of hardwood forests in middle-latitude regions with cold winters changes dramatically with the seasons. The green leaves of summer (a) change to reds, golds, and browns in fall (b) and drop to the forest floor in winter (c). Leaf dropping in areas of cold winters, such as this example in western Illinois, is a means of minimizing transpiration and moisture loss when the soil water is frozen.

**What length of growing season (frost-free period) is associated with the climate of western Illinois?**

such as calcium, magnesium, sodium, and potassium. Thus, they grow in sandy places and where the soil is acid in character. As a whole, conifers are particularly well adapted to regions having long, severe winters combined with summers warm enough for vigorous plant growth. Because all but a few exceptions retain their leaves (needles) throughout the year, they are ready to begin photosynthesis as soon as temperatures permit without having to produce a new set of leaves to do the work.

Thus, we find a great band of coniferous forests (the **boreal forests,** or **taiga**) dominated by spruce and fir species, with pines on sandy soils, sweeping the full breadth of North America and Eurasia northward of the 50th parallel of latitude, approximately occupying the region of Köppen's subarctic climate (see again Fig. 8.6). Conifers differ from other trees in that their seeds are not enclosed in a case or fruit but are carried naked on cones. All are needle leaf and drought resistant, but a few are not evergreen. Thus, a large portion of eastern Siberia is dominated by larch which produces a mix of deciduous, coniferous forest. In this area, January mean temperatures may be −35°C to −51°C (−30°F to −60°F). This is the most severe winter climate in which trees can maintain themselves, and even needle-leaf foliage must be shed for the vegetation to survive. Hardy broad-leaf deciduous birch trees share this extreme climate with the indomitable larches.

Extensive coniferous forests are not confined to high-latitude areas of short summers. Higher elevations in middle-latitude mountains of the Northern Hemisphere have forests of pine, hemlock, and fir ( Fig. 11.26), with subalpine larch and specially adapted pine species characterizing the harshest and highest forest sites. The forests along the sandy coastal plain of the eastern United States are there in part because of the sandy soils, but they may also reflect a stage in plant succession that in time will lead to domination by broad-leaf types. Similarly, a temporary stage in the postglacial vegetation succession of the Great Lakes area included magnificent forests of white pine and hemlock that were completely logged off during the late 19th century.

A more maritime coniferous forest occupies the west coast of North America extending from southern Alaska to central California. It is made up of sequoias, Douglas fir, cedar, hemlock, and, farther north, Sitka spruce. Many of the California sequoias are thousands of years old and more than 100 meters (330 ft) high. The southern regions experience summer drought, and farther north, sandy, acidic, or coarse-textured soils dominate.

# Grassland Biomes

Grasses, like conifers, appear in a variety of settings and are part of many diverse plant communities. They are in fact an initial form in most plant successions. However, there are enormous, continuous expanses of grasslands on Earth. In general, it is thought that grasses are dominant only where trees and shrubs cannot maintain themselves because of either excessive or deficient moisture in the soil. On the global scale, grassland biomes are located in continental interiors where most, if not all, of the precipitation falls in the summer. Two great geographic realms of grasslands are generally recognized: the tropical and the middle-latitude grasslands. However, it is difficult to define grasslands of either type using any specific climatic parameter, and geographers suspect that human interference with the natural vegetation has caused expansion of grasslands into forests in both the tropical and middle latitudes.

## Tropical Savanna Grasslands

The tropical grassland biome differs from grassland biomes of the middle latitudes in that it ordinarily includes a scattering of trees; this is implied in the term **savanna** ( Fig. 11.27). In fact, the demarcation between tropical scrub forest and savanna is seldom a clear one. The savanna grasses tend to be tall and coarse with bare ground visible between the individual tufts. The related tree species generally are low growing and wide-crowned forms, having both drought- and fire-resisting qualities, indicating that fires frequently sweep the savannas during the drought season. Savannas occur under a variety of temperature and rainfall conditions but generally fall within the limits of Köppen's tropical savanna type. They commonly occur on red-colored soils, leached of all but iron and aluminum oxides, which become bricklike when dried. They likewise coincide with areas in which the level of the water table (the zone below which all soil and rock pore space is saturated by water) fluctuates dramatically. The up-and-down movement of the water table may in itself inhibit forest development, and it is no doubt a factor in the peculiar chemical nature of savanna soils. Large migratory herds of grazing animals and associated predators, responding to the periodically abundant grasses followed

**FIGURE 11.26**
As this photograph in the Gallatin National Forest of Montana indicates, evergreen coniferous forest is the characteristic vegetation of higher-elevation regions in the middle latitudes. The needle-leaf trees are an adaptation to the physiologic drought of the winter season, which is longer and more severe at higher elevations.
**Why are needle-leaf trees better adapted to physiologic drought than broad-leaf trees?**

**FIGURE 11.27**
The savanna biome of eastern Africa is a classic landscape of grasses and scattered trees (background). In the foreground, vultures (detritivores) eat the carcass of a zebra (herbivore) left behind by a lion (carnivore).
**What organisms will take over when the vultures are satisfied?**

by seasonal drought, characterized the savanna prior to widespread human disruption.

## Middle-Latitude Grasslands

The middle-latitude grasslands occupy the zone of transition between the middle-latitude deserts and forests. On their dry margins, they pass gradually into deserts in Eurasia and are cut off westward by mountains in North America. However, on their humid side, they terminate rather abruptly against the forest margin, again raising questions as to whether their limits are natural or have been created by human activities, particularly the intentional use of fire to drive game animals. The middle-latitude grasslands of North America, like the African savannas, formerly supported enormous herds of grazing animals—in this case, antelope and bison, which were the principal means of support of the American Plains Indians.

Like the savannas, the middle-latitude grasslands were diverse in appearance. They, too, consisted of varying associations of plant species that were never uniform in composition. In North America, the grasses were as much as 3 meters (10 ft) tall in the more humid sections, as in Iowa, Indiana, and Illinois, but only 15 centimeters (6 in.) high on the dry margins from New Mexico to western Canada. Thus, the middle-latitude grassland biomes are usually divided into tall-grass and short-grass prairie, often with a zone of mixture recognized between them. Unlike growth in the tropical savannas, the germination and growth of middle-latitude grasses are attuned to the melting of winter snows, followed by summer rainfall. Whether the grasses are annuals that complete their life cycle in one growing season or perennials that grow from year to year, they are dormant in the winter season. Also, unlike in the savannas, the soils beneath these grasslands are extremely rich in organic matter and soluble nutrients. As a consequence, most of the middle-latitude grasslands have been completely transformed by agricultural activity. Their wild grasses have been replaced by domesticated varieties—wheat, corn, and barley—and they have become the "breadbaskets" of the world.

**Tall-Grass Prairie** The **tall-grass prairies** were an impressive sight; in some better-watered areas, they made up endless seas of grass moving in the breeze, reaching higher than a horse's back. Flowering plants were conspicuous, adding to the effect. Unfortunately, this tall-grass prairie scene, which inspired much vivid description by those first encountering it, is no longer visible anywhere. The tall-grass prairies, which once reached continuously from Alberta to Texas, have been destroyed. Their tough sod, formed by the dense grass root network, defeated the first wooden plows, which had served well enough in breaking up the forest soils. But the steel plow, invented in the 1830s, subdued the sod and was aided by the introduction of subsurface tile for draining the nearly flat uplands and by the simultaneous appearance of well-digging machinery and barbed wire. These four innovations transformed the tall-grass prairie from grazing land to cropland.

In North America, the tall-grass prairie pushed as far eastward as Lake Michigan. Why trees did not invade the prairie in this relatively humid area remains an unanswered question. Farther west, shallow-rooted grass cover is fully understandable because the lower soil levels, to which tree roots must penetrate for adequate support and sustenance, are bone dry. In such areas, trees can survive only along streams or where depressions collect water.

In Eurasia, tall-grass prairies were found on a large scale in a discontinuous belt from Hungary, through Ukraine, Russia, and central Asia, to northern China. The grasslands are known in South America as the pampas of Uruguay and Argentina and in South Africa as the Veldt. Today, all these areas have been changed by agriculture. The factor that seems to account best for the tall-grass prairie—precipitation that is both moderate and variable in amount from year to year—is the principal hazard in the use of these regions as farmland. However, this hazard becomes much greater in the areas of short-grass prairie.

**Short-Grass Prairie** West of the 100th meridian in the United States and extending across Eurasia from the Black Sea to northern China, roughly coinciding with the areas of Köppen's middle-latitude steppe climate, are vast, nearly level grasslands composed of a mixture of tall and short grass, with short grass becoming dominant in the direction of lower annual precipitation totals ( Fig.11.28). On the Great Plains between the Rocky Mountains and the tall-grass prairies, the **short-grass prairie** zone more or less coincides with the zone in which moisture rarely penetrates more than 60 centimeters (2 ft) into the soil, so the subsoil is permanently dry. Moving toward the drier areas, the grassland vegetation association dwindles in diversity and, more conspicuously, in height to less than 30 centimeters (1 ft). This is a consequence of reduction in numbers of tall-growing species and greater abundance of shallow-rooted and lower-growing types. The total amount of ground cover also declines toward the drier margins as the deeply rooted, sod-forming grasses of the prairie grassland give way to bunchgrass species (so called because, instead of forming a continuous grass cover, they occur in isolated clumps or bunches). In their natural conditions, the short-grass prairies of North America and Eurasia supported higher densities of grazing animals—bison and antelope in the former, wild horses in the latter—than did the tall-grass prairies. Indeed, it is suspected that the specific plant association of the short-grass regions may have been a consequence of overgrazing under natural conditions. The short-grass prairies cannot be cultivated without the use of irrigation or dry farming methods, so they remain primarily the domain of wide-ranging grazing animals; however, today's animals are domesticated cattle, not the thundering self-sufficient herds of wild species that formerly made these plains one of Earth's marvels.

## Desert

Eventually, lack of precipitation can become too severe even for the hardy grasses. Where evapotranspiration demands greatly exceed available moisture throughout the year, as in Köppen's desert climates, either special forms of plant life have evolved or the surface is bare. Plants that actively combat low precipitation are equipped to probe deeply or widely for moisture, to reduce moisture losses to the minimum, or to store moisture when it is available. Other plants evade drought by merely lying dormant, perhaps for years, until enough moisture is available to ensure successful growth and reproduction. The desert biome is recognized by the presence of plants that are either drought resisting or drought evading ( Fig. 11.29). In extremely dry deserts, only a few plants can survive, and ground cover is much less common ( Fig. 11.30).

Plants that have evolved mechanisms to combat drought are known as *xerophytes*. They are perennial shrubs whose root systems below ground are much more extensive than their visible parts or that have evolved tiny leaves with a waxy covering to combat transpiration. They may have leaves that are needlelike or trunks and limbs that

**FIGURE 11.28**
Short-grass prairie vegetation of the Nebraska Sand Hills. Although the tall-grass prairies have been completely transformed by humans, vast areas of short-grass prairie remain because of their low and unpredictable precipitation, which makes them hazardous for agriculture.
**What would have been obviously different if a photograph had been taken here 200 years ago?**

excessively saline water or shrubs that shed their leaves until sufficient moisture is available for new leaf growth. The nonxerophytic vegetation consists mainly of short-lived annuals that germinate and hurry through their complete life cycle of growth— leaf production, flowering, and seed dispersal—in a matter of weeks when triggered by moisture availability. Like other species, these ephemeral plants also require days of a certain length, so they appear only in particular months; therefore, the month-to-month and year-to-year variation in form and appearance of desert vegetation is enormous. Animals of the deserts are primarily nocturnal to avoid the searing heat of the daytime, and many have evolved long ears, noses, legs, and tails that allow for greater blood circulation and cooling. The similar lifeforms and habits of the different plant and animal species found in the deserts of widely separated continents are a remarkable display of repeated evolution to ensure survival in similar climatic settings.

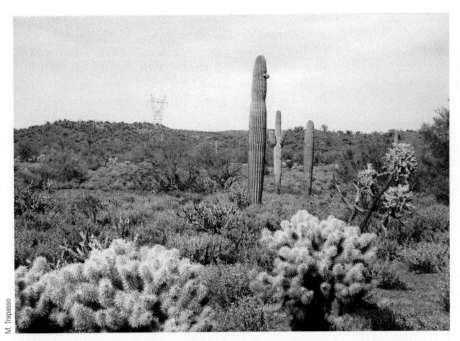

■ **FIGURE 11.29**
A host of drought-resistant plants can be seen across parts of Arizona.
**What drought-resistant adaptations can be easily observed?**

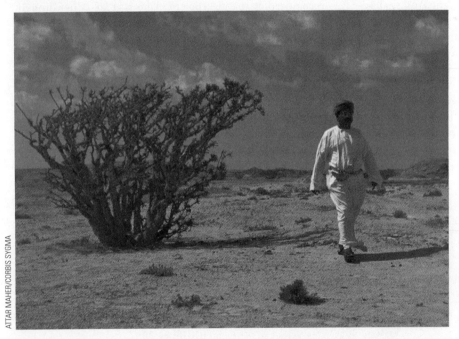

■ **FIGURE 11.30**
The scarcity of ground cover in this photograph of the desert in Oman is a clear indication of the extremely low annual rainfall experienced in this region.
**Are there any reasons why humans might be found in such desolate regions?**

## Arctic and Alpine Tundra

Proceeding upward in elevation and poleward in latitude, we finally come to regions in which the growing season is too brief to permit tree growth. Nearing the poles, we enter a vast realm dominated by subfreezing temperatures and thin snow cover much of the year, so the ground is frozen to depths of hundreds of meters. Only the top 36–60 centimeters (15–25 in.) thaw during the short summer interval. Still, vegetation survives here and in fact forms a nearly complete cover over the surface. Such vegetation must be equipped to tolerate frozen subsoil (permafrost), icy winds, a low sun angle, summer frosts, and soil that is waterlogged during the short growing season. The result is **tundra**—a mixture of grasses, flowering herbs, sedges, mosses, lichens, and occasional low-growing shrubs. Most of the plants are perennials that produce buds

photosynthesize like leaves or that have expandable tissues or accordion-like stems to store water when it is plentiful (the succulent cacti). They may be plants that can tolerate close to or beneath the soil surface, protected from the wind. Many of the plants show xerophytic adaptation—such as small, hard leaves—in response to extreme physiologic drought resulting from wind stress. This is particularly true in areas of alpine tundra where extended periods of waterlogging of soils are uncommon. The effect of wind is evident from the fact that the less exposed valleys within the tundra region are often occupied by coniferous woodlands.

In a band of varying width reaching across northern Alaska, Canada, Scandinavia, and northern Russia, several types of tundra are recognized: *bush tundra,* consisting of dwarf willow, birch, and alder, which grow along the edge of the coniferous forest; *grass tundra,* which is hummocky and water soaked during the summer ( ▍Fig. 11.31); and *desert tundra,* in which expanses of bare rocks may be covered by colorful lichens. In a few ice-free valleys of Antarctica, only desert tundra occurs.

Alpine conditions are not exactly like those in the Arctic latitudes. The deeper snow cover of the high mountains prevents the development of permafrost, and the summer sun results in considerably more evaporation. However, many high areas are swept clean of snow by wind and are thus extremely exposed, supporting only desert tundra. Microclimate becomes an important control of vegetation because of the varying exposures to sun and wind. Nevertheless, the short growing season and severe wind stress produce an overall plant community similar to that in the Arctic regions ( ▍Fig. 11.32).

Animals of the tundra obviously must cope with long periods of extreme cold as well as darkness. Many animals hibernate through the long Arctic winters (or the cold and windy alpine winters); others, such as the caribou of Alaska and Canada, migrate into the boreal forest to escape the extreme cold exacerbated by Arctic winds. Year-round residents, such as the polar bear and musk ox, must have extremely large fat reserves around their large chests, in addition to extremely efficient fur. Many insects, such as the ubiquitous mosquito, cope with the extreme climate by emerging from eggs in the spring, maturing and laying eggs, and dying within one short Arctic summer. Enough eggs survive, buried under insulating snow, to continue the cycle in the following year.

## Marine Ecosystems

The living organisms of the ocean can be divided into three groups according to where or how they live in the ocean. The first group is called **plankton.** Plankton is made up mostly of the ocean's smallest—usually microscopic—plants (**phytoplankton**) and animals (**zooplankton**). These tiny plants and animals float freely with the movements of ocean water, are true "drifters," and form the basis of the oceanic food chain. The second group is composed of the animals that swim in the water. This group, called **nekton,** includes fish, squid, marine reptiles, and marine mammals such as whales and seals ( ▍Fig. 11.33). The third group is composed of the plants and animals that live on the ocean floor. This group is called the **benthos.** It includes corals, sponges, and many algae; such burrowing or crawling animals as the barnacle, crab, lobster, and oyster; and attached plants such as turtle grass and kelp.

Life in the ocean depends on the sun's energy and on the nutrients available in the water. Phytoplankton are the most important link in the ocean food chain. At the base of the marine food web, phytoplankton take the dissolved nutrients in the water and, through the process of photosynthesis, produce

▍**FIGURE 11.31**
A variety of tundra grasses grow during the Arctic summer in Alaska. The photograph was taken near the Alaskan Oil Pipeline at Prudhoe Bay.
**Why might alpine tundra be of a different use to humans?**

**FIGURE 11.32**
Close examination of the tundra biome of the mountainous western United States reveals a mixture of grasses, sedges, rushes and wildflowers.
**Why does all vegetation, including the occasional shrub or stunted tree, grow so close to the ground?**

**FIGURE 11.33**
Elephant seals bask in the nutrient-rich and cool upwelled waters off the San Benitos Islands, Mexico.
**To which group of ocean organisms do these elephant seals belong?**

oxygen and foods needed by zooplankton and by the smallest nekton. Phytoplankton are the only food source for these animals, which in turn form the food source for larger carnivorous fish and marine mammals. These in turn are prey for still larger animals. For example, tiny shrimplike creatures known as *krill* are nicknamed the "power food of the Antarctic." These crustaceans feed on plankton and then become the main food source for birds, penguins, seals, and whales.

The marine food chain just described is part of a full food cycle in the ocean because, through the excretions of animals and the decomposition of both plants and animals in the ocean, chemical nutrients are returned to the water and are again made available for transformation by phytoplankton into usable foods. Phytoplankton also play an essential role in the production of oxygen for Earth's atmosphere. In fact, the greatest concern of scientists who study the annual Antarctic "ozone hole" (see Chapter 4) is that excessive ultraviolet radiation will destroy or diminish the phytoplankton in the Antarctic and other oceans.

The uneven distribution of nutrients in the oceans and the fact that sunlight can penetrate only to a depth of about 120 meters (400 ft), depending on the clarity of the water, means that the distribution of marine organisms is also variable. Most organisms are concentrated in the upper layers of the ocean where the most solar energy is available. In deep waters, where the ocean floor lies below the level to which sunlight penetrates, the benthos organisms depend on whatever nutrients and plant and animal detritus filter down to them. For this reason, benthos animals are scarce in deep and dark ocean waters. They are most common in shallow waters near coasts, such as coral reefs and tide pools, where there is sunlight and a rich supply of nutrients and where phytoplankton and zooplankton are abundant as well.

The waters of the continental shelf have the highest concentration of marine life. The supply of chemical nutrients is greater in waters near the continents where nutrients are washed into the sea from rivers. Marine organisms are also concentrated where deeper waters rise (upwelling) to the surface layers where sunlight is available. Such vertical exchanges are sometimes the result of variations in salinity or density. A similar situation occurs where convection causes bottom layers of water to rise and mix with top layers, as is the case in cold polar waters and in middle-latitude waters during the colder winter months. Marine life is also abundant in areas where there is a mixing of cold and warm ocean currents, as there is off the northeastern coast of the United States (■ Fig. 11.34).

During the 1977 dives of the manned submersible vessel *Alvin* off the East Pacific Rise, scientists for the first time observed abundant sea life on the floor of the ocean at depths of more than 2500 meters (8100 ft). It was previously presumed that these cold (2°C/36°F), dark waters were a virtual biological desert. However, the undersea volcanic mountain range produces vents of warm, mineral-rich waters, which nourish bacteria and large colonies of crabs, clams, mussels, and giant 3-meter (10-ft) red tube worms. The discovery of this deep-ocean ecosystem, which exists without the benefit of sunlight, has caused scientists to rethink old theories about the ocean and its chemistry. This unusual "chemosynthetic" vent community has been observed more recently at several other oceanic ridge sites in the Pacific and Atlantic Oceans.

■ **FIGURE 11.34**
Distribution of chlorophyll-producing marine plankton mapped by space observations of ocean color. Purple to yellow to red indicates increasing chlorophyll concentration.
**Where do these plankton seem to be located?**

# Define & Recall

abiotic
autotroph
heterotroph
herbivore
carnivore
omnivore
decomposer (detritivore)
trophic structure
food chain
trophic level
food web
biomass
primary productivity
secondary productivity

habitat
ecological niche
generalist
natural vegetation
plant community
succession
gap
climax community
mosaic
matrix
patch
corridor
tolerance
ecotone

biome
liana
galleria forest
boreal forest (taiga)
savanna
tall-grass prairie
short-grass prairie
tundra
plankton
phytoplankton
zooplankton
nekton
benthos

# Discuss & Review

1. Give reasons why the study of ecosystems is important in the world today. Give several examples of natural ecosystems within easy driving distance of your own residence.
2. What are the four basic components of an ecosystem?
3. How can an organism belong to more than one trophic level? Give an example.
4. How does the biomass at one trophic level usually compare in weight with that at the next level?
5. What is the difference between primary and secondary productivity? Which is most important to the ecosystem?
6. What factors are most critical in affecting the net primary productivity of a terrestrial ecosystem?
7. Why are generalists in the majority in ecosystems?
8. In what ways has the original theory of succession been modified?
9. How do the terms *mosaic, matrix, patch, corridor,* and *ecotone* relate to each other and to a vegetation landscape?
10. What are the important environmental controls of ecosystems? Which are the most important on a worldwide basis?
11. What are the four major types of Earth biomes?
12. What important climate characteristics are related to each of the major biomes?
13. Describe a true tropical rainforest. How does such a forest differ from jungle?
14. What are the distinctive features of chaparral vegetation? What climate conditions are associated with chaparral?
15. What are the major differences between tropical and middle-latitude forests?
16. What conditions of climate or soil might be anticipated for each of the following in the middle latitudes: broad-leaf deciduous forest; broad-leaf evergreen forest; needle-leaf coniferous forest?
17. What factors contribute to the development and maintenance of savannas?
18. How have xerophytes adapted to desert climate conditions?
19. What are the chief characteristics of tundra vegetation? What adaptations to climate does tundra vegetation make?
20. Which link in ocean food chains is most important?
21. The highest concentrations of marine life (see Fig. 11.34) are found in which parts of the oceans? Why?

# Consider & Respond

1. Arrange the following organisms into six logical trophic levels in order from the first to the last.
   a. Owl, fungi, vulture, grass, bobcat, rabbit
   b. Herbivorous zooplankton, small fish, phytoplankton, shark, bottom bacteria, large fish
2. Refer to Table 11.2. Based on the means, which is more important for terrestrial ecosystem productivity between the equator and the polar regions: annual temperatures or the availability of water? Use examples from the table to explain your choice.
3. What broad groups might you use if you were to classify Earth's vegetation on the basis of latitudinal zones? Why, do you suppose, did your textbook authors choose not to do this, electing instead to identify broad groups based on major terrestrial ecosystems (biomes)?
4. Refer to Figure 11.21, which shows considerable variation in the natural vegetation of the United States between 30° and 40° north latitude. Describe the broad changes in vegetation that occur within that latitudinal band as you move from the East Coast to the West Coast of the United States.
5. Examine the climate variation (see Fig. 8.6) within the same latitudinal band described in Question 4. Does there appear to be a relationship between natural vegetation and climate? If so, describe this relationship.

# CHAPTER 12

Soil is a vital natural resource, essential to life as we know it. © Paul Hardy/ CORBIS

# Soils and Soil Development

**Physical Geography Now™** This icon, appearing throughout the book, indicates an opportunity to explore interactive tutorials, animations, or practice problems at now.brookscole.com/gabler8

## CHAPTER PREVIEW

**Soil is an outstanding example of the interrelationships among Earth's subsystems.**

- Which of the four major subsystems are involved?
- Why is soil such a good example of subsystem interaction and integration?
- Why should soil be considered an open system?

**Soil water is the means by which plants receive dissolved nutrients that are essential for growth.**

- Why is gravitational water such an effective agent of solution?
- How is capillary water important during periods of drought?

**Soil fertility depends on many factors, and a soil that is fertile for one vegetation type may not be for another.**

- What factors determine a soil's fertility?
- How are acidity and alkalinity related to soil fertility?
- How is vegetation involved in soil fertility?

**On a global scale, climate exerts a major influence on the formation and characteristics of soils.**

- How do temperature, precipitation, and moisture regimes affect the development of soils?
- How does climate affect soils indirectly through vegetation?
- Why is the regional distribution of soils similar to regions of climate and vegetation?

**Soils are most productive when no single soil-forming factor is dominant.**

- Why is this so?
- What often results when a single soil-forming process proceeds unchecked?
- What may happen to a soil if its vegetation cover is stripped away?

**Soils are among the world's most critical and widely abused yet least understood natural resources.**

- Why is there so much ignorance concerning soils?
- How are soils abused?
- What can be done about it?

Most people do not realize that a critically important natural resource that is essential for sustaining Earth's environments, and to their life on Earth, lies right below their feet. This is a serious oversight, particularly in an age of environmental awareness and concern. That vital resource, one that we could not live without, is *soil*. Composed primarily of decomposed rock materials and varying amounts of water, gases, and organic materials, soil covers most land surfaces with a fragile and indispensable mantle. Soil has been called the "skin of the Earth," and its condition and nature reflect the environment under which it formed and exists today. A soil also functions as a system. As with most environmental systems, soils reflect and respond to many natural processes. The life-forms that live in or on a soil play a significant role in its development and characteristics, and human impacts on soils have increased dramatically with population growth and the spread of expanding civilizations.

**Soil** is a dynamic natural body capable of supporting a vegetative cover. It contains chemical solutions, gases, organic refuse, and an active fauna. The complex physical, chemical, and biological processes that take place among the components of a soil are integral parts of its dynamic character. Soil responds to climatic conditions (especially to temperature and moisture), to the land surface configuration, to vegetative cover and composition, and to animal activity. Because a soil develops in response to its environment and its characteristics reflect the conditions under which it has formed, soil is an exceptional example of the integration and overlap among Earth's systems (Fig. 12.1). In fact, because soils integrate the major subsystems of Earth so well, they are sometimes considered a separate system called the *pedosphere* (from Greek: *pedon,* ground). Soil development and characteristics depend on a great number of factors. But when soils are viewed on a world regional scale, the dominant influence is climate. The relationships between climates and soil types, as well as the association of soils with climate-controlled vegetation, were considered in Chapters 9 and 10.

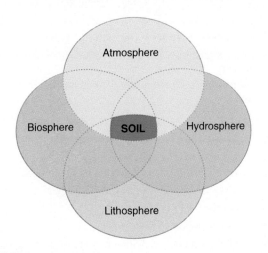

**FIGURE 12.1**
The intertwined links between soil and the major Earth subsystems.
**Why is soil considered to be such an integrator of Earth systems?**

## Major Soil Components

What is soil actually made of? What does the scoop of a bulldozer contain when it shovels up a load of soil? What soil characteristics support the variation in Earth's vegetational environments? Soils function as *open systems*. Matter and energy flow in and out of a soil, and they are also held in storage (Fig. 12.2). Understanding these flows—inputs and outputs, the components and processes involved, and how

**FIGURE 12.2**
Soil as an open system: the relation of soil to environmental factors. Soil is an example of an open system because it receives inputs of matter and energy, stores part of these inputs, and outputs matter and energy. Note the inputs and outputs on this diagram.
**What are some examples of energy and matter that flow into and out of the soil system?**

they vary from soil to soil—is a key to appreciating the complexities of this critical natural resource. Soils contain four major components, and there are many processes that act on these components. The four major components of soil are inorganic materials, soil water, soil air, and organic matter.

## Inorganic Materials

Soil consists of insoluble materials, minerals that will not dissolve in water, as well as soluble minerals—chemicals in solution. Most minerals found in soils are combinations of the most common elements of Earth's surface rocks: silicon, aluminum, oxygen, and iron. Some of these constituents occur as solid chemical compounds, and others are found in the air and water that are essential components of a soil. A large number of chemical elements and compounds are necessary to sustain the Earth's ecosystems. In addition to the four elements listed previously, carbon, hydrogen, nitrogen, sodium, potassium, zinc, copper, iodine, and compounds of these elements are important in soils.

A soil's chemical constituents typically come from many sources. Some are derived from the breakdown of underlying rocks or from accumulations of loose sediments, and others enter as solutions in water. Still others are in the air found in soils or are derived from organic activities, which help to disintegrate rocks, release gases, and create new chemical compounds.

Plants need many chemical substances for growth, and having a knowledge of a soil's mineral and chemical content is necessary to determine its productive potential. Frequently, soil deficiencies in a specific substance can be rectified by fertilization to increase the soil's productivity ( Fig.12.3). **Soil fertilization** is the process of adding certain nutrients or other constituents to the soil in order to meet the conditions that certain plants require.

## Soil Water

Soil water supplies both an ingredient and a catalyst for chemical reactions that sustain life, and it provides nutrients in a form that can be extracted by vegetation. Plants need air, water, and minerals to function, live, and grow, and they depend on soil for much of these necessities. Soil water is not pure but is a water solution that contains soluble nutrients.

The original source of soil moisture is precipitation. When precipitation falls on the land, most of it is either absorbed into the ground or runs downslope. Water seeping through a soil washes over and through various soil materials, dissolving some of these materials and carrying them through the soil.

The water in a soil is found in several different circumstances (see again Fig. 12.2). Water that percolates down through a soil, under the force of gravity, is called **gravitational water.** Gravitational water moves downward through voids between soil particles toward the *water table*—the level below which all available spaces in the soil and rock are filled with water. As a consequence, water cannot percolate any deeper. The quantity of gravitational water in a soil is related to several conditions, including the amount of precipitation and the time since it fell, evaporation rates, how easily the water can move through the soil, and the space available for water storage.

Gravitational water functions in the soil in several ways ( Fig. 12.4). First, gravitational water moving down through a soil takes with it the finer particles (clay and silt) from the upper soil layers. This downward removal of soil components from the topsoil by water is called **eluviation.** Eventually, as gravitational water percolates downward, it deposits fine materials, which were removed from the topsoil, at a lower level in the soil. This process of deposition by water in the subsoil is called **illuviation.** Gravitational water also mixes the soil particles as it moves them downward. One result of eluviation is that the topsoil's texture becomes coarser as the finer particles are removed. Consequently, the topsoil's ability to retain water is reduced, while illuviation enhances the subsoil's ability to retain water. Illuviation may eventually cause the subsoil to become dense and compact, forming a clay **hardpan.**

As gravitational water percolates downward, it dissolves soluble inorganic minerals and carries them into deeper levels of the soil, perhaps to the zone where all open spaces are saturated. Depleting nutrients in the upper soil by the through flow of water is a process called **leaching.** In regions of heavy

**FIGURE 12.3**
Fertilizers increase the productivity of soils.
**Why can soil fertilizer be either useful or detrimental when it is introduced into the soil system?**

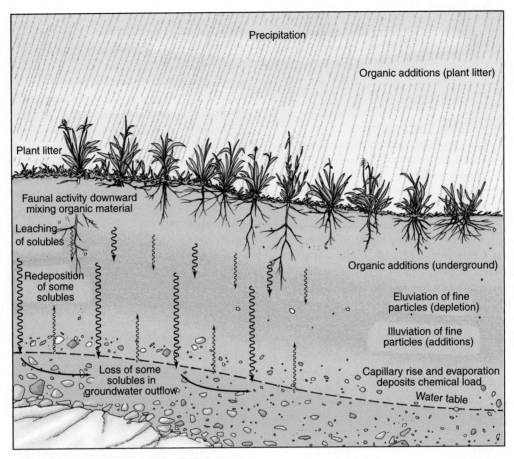

**FIGURE 12.4**
The role of water in soil development processes. Water is important in moving nutrients and particles vertically, both up and down, in the soil layer.
**How does deposition by capillary water differ from deposition (illuviation) by gravitational water?**

rainfall, leaching is common and can be intense, robbing a topsoil of all but the insoluble substances.

Leaching and eluviation both strongly influence the characteristic **stratification,** or layered changes with depth, found in soils. Fine particles and certain substances dissolved from the upper soil are deposited in lower levels, which become dense and may be strongly colored by accumulated iron compounds.

Soil water adheres to both soil particles and soil clumps by surface tension (the property that causes small water droplets to form rounded beads instead of spreading out in a thin film). This soil water, called **capillary water,** serves as a stored water supply for plants. Capillary water can move in all directions through soil because it migrates from areas with more water to areas with less. Thus, during dry periods, when there is no gravitational water flowing through the soil, capillary water can move upward or horizontally to supply plant roots with moisture and dissolved nutrients.

Capillary water migrating upward moves minerals from the subsoil toward the surface. If this water evaporates, the formerly dissolved minerals are left behind, generally as alkaline or saline deposits in the topsoil. Such mineral deposits can be detrimental to plants and animals existing in the soil. Lime (calcium carbonate) deposited by evaporating soil water can build up to produce a cementlike layer, called *caliche,* which like a clay hardpan prevents the downward percolation of water.

Soil water is also found as a very thin film, invisible to the naked eye, that is bound to the surfaces of soil particles by strong electrical forces. Because this water, called **hygroscopic water,** does not move through the soil, it does not supply plants with the moisture that they need.

## Soil Air

Much of a soil—in some cases, approaching 50%—consists of voids between soil particles and between clumps (aggregates of soil particles). When these voids are not filled with water, they contain air. Compared to the composition of the lower atmosphere, the air in a soil is likely to have less oxygen, more carbon dioxide, and a fairly high relative humidity because of the presence of capillary and hygroscopic water.

For most microorganisms and plants that live in the ground, soil air supplies oxygen and carbon dioxide necessary for life processes. The problem with a water-saturated soil is not necessarily excess water but that, if all pore spaces are

**FIGURE 12.10**
The pH scale of acidity, neutrality, and alkalinity. The degree of acidity or of alkalinity, called pH, can be better understood when numbers on the scale are linked to common substances. Low pH means acidic, and high pH means alkaline; a reading of 7 is neutral.

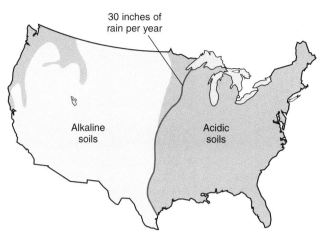

**FIGURE 12.11**
Alkaline and acidic soils in the United States, based on general climate. In general, soils in the East tend to be acidic and those in the West, alkaline. The dividing line corresponds fairly well with the 30-inch annual precipitation isohyet.
**What environmental factors might cause this east–west variation? How might you explain the places west of the 30-inch line that are acidic?**

**FIGURE 12.12**
A soil profile is examined by digging a pit with vertical walls to clearly show the variations in color, structure, composition, and other characteristics that occur with depth. This soil is in a grassland region of northern Minnesota.

## Development of Soil Horizons

Soil development begins when plants and animals colonize rocks or deposits of rock fragments, the **parent material** on which soil will form. Once organic processes begin among mineral particles or disintegrated rock fragments, differences start to develop from the surface down through the parent material.

Initially, this vertical differentiation results from a surface accumulation of organic litter and the removal of fine particles and dissolved minerals from upper layers by percolating water that deposits these materials at a lower level. As climate, vegetation, animal life, and characteristics of the land surface affect soil formation over time, this vertical differentiation becomes more and more apparent. The vertical cross section of a soil from the surface down to the parent material is called a **soil profile** ( Fig. 12.12). The differences among

**FIGURE 12.13**
Soils are categorized by the degree of development and the physical characteristics of their horizons. *Regolith* is a generic term for broken bedrock fragments at or very near the surface.
**Which soil profiles shown in Figures 12.25–12.28 display horizons that are easy to recognize?**

the infinite variety of soils that exist are apparent in an examination of their profiles and the vertical differences that soil profiles contain.

## Soil Horizons

Well-developed soils typically exhibit zones or distinct layers, called **soil horizons,** that are distinguished by their physical and chemical properties. Most well-developed soils have several recognizable horizons. Soil scientists group and classify soils largely on the basis of physical and chemical differences in their horizons and in the processes involved with those differences (❚ Fig. 12.13). Soil horizons are designated by letters that refer to their composition, dominant process, and/or position in the soil profile.

At the surface, in regions with sufficient vegetation and moderate or modest rates of organic decomposition, is the O horizon. This is a layer of organic debris and humus ("O" refers to its high organic content). Immediately below is the A horizon, commonly referred to as "topsoil." In general, the A horizon is dark because of a concentration of decomposed organic matter. Beneath the A horizon, some soils have a lighter-colored E horizon, named for the eluvial processes that dominate. Below this is a zone of accumulation, the B horizon, where much of the materials removed from the A and E horizons are deposited. Except in soils that have a high organic content accompanied by vertical mixing, the B horizon generally has little humus. The C horizon is the weathered parent material from which the soil has developed—either fragments of the bedrock directly beneath or rock materials that have been transported and deposited. The C horizon does not reflect the movements of matter or organic activities in the higher zones.

The lowest layer, sometimes called the R horizon, is unchanged parent material, either bedrock or rock fragments transported to the site by water, wind, glacial, or other surface processes. Horizons in some soils may not be as well developed as others, and some horizons may be missing altogether. Because soils and the processes that form them vary and can be transitional between horizons, the boundaries between horizons may be either sharp or gradational. Variations in color and texture within a horizon are not unusual.

## Factors Affecting Soil Formation

Because of the great variety among soil components and processes of formation, no two soils are identical in all of their characteristics. One important process (actually a set of processes that will be discussed in more detail in a later chapter) is weathering. Rock *weathering* refers to the many natural processes that act to break down rocks into smaller fragments. Two major categories of weathering occur: chemical reactions that cause rocks and minerals to decompose and physical processes that cause the breakup of rocks. Just as statues, monuments, and buildings become "weather-beaten" over time, rocks exposed to the elements eventually break up and decompose. The factors controlling the formation and distribution of different soil types are parent material, organic processes, climate, land surface configuration, and time.

Hans Jenny, a distinguished soil scientist, observed that soil development was a function of climate, organic matter, relief, parent material, and time—factors that are easy to remember by their initials arranged in the following order: **Cl, O, R, P, T.** Of these factors, parent material is distinctive because it is the raw material. The other factors influence the type of soil that forms from the parent material.

## Parent Material

All soil contains weathered fragments of rock material. If these weathered rock particles have accumulated in place—through the physical and chemical breakdown of bedrock directly beneath the soil—we refer to the fragments as **residual parent material.**

If the rock fragments that form a soil have been carried to the site and deposited by streams, waves, winds, gravity, or glaciers, this mass of deposits is called **transported parent material.** The development and action of organic matter through the life cycles of organisms and the climatic environment are primarily responsible for differentiating soil from the fragmentary rocks or other parent material that is present beneath it.

Parent material influences the characteristics of a soil derived from it in varying degrees. Some parent materials, such as sandstone that contains extremely hard and resistant sand-sized fragments, are far less subject to weathering than others. Soils that develop from weathering-resistant rocks tend to have a high level of similarity to their parent materials. Some kinds of bedrock are easily weathered, however, and the soils that develop from them tend to bear a greater similarity to other soils in regions of similar climate than to those of comparable parent materials that formed in a different climate.

It is common to find that certain soil-forming factors may have a greater influence on the soil than does the parent material. In fact, on a worldwide basis, climate and the associated plant communities produce greater variations in soil characteristics than do parent materials. Soil differences that are related to variations in parent material are most visible on a local level and are often studied by soil scientists and agriculturists.

As a soil develops, becoming more mature, the influence of parent material on its characteristics declines. Given the same soil-forming conditions, a young or less mature soil will show more similarity to its parent material, compared to a soil that has become well developed by forming over a long time.

Both residual and transported parent materials affect the soils that develop from them in very specific ways. First, the chemicals and nutrients available to plants and animals living in a soil are derived from that soil's parent material ( Fig. 12.14). For example, calcium-deficient parent materials will produce soils that are low in calcium, and its natural fauna and plant cover will be of types that require little calcium. Likewise, a parent material with a high aluminum content will produce a soil that is rich in aluminum. In fact, the main source of metallic aluminum is bauxite ore found in tropical soils where it has been concentrated by intense leaching that has removed the other bases.

The particle sizes that result from weathering of parent material are a prime determinant of soil texture and structure. A rock material such as sandstone, which contains little clay and weathers into relatively coarse fragments, will produce a soil of coarse texture. Parent materials are also an important influence on the availability of air and water to a soil's living population.

**FIGURE 12.14**
Despite strong leaching under a wet tropical climate, Hawaiian soils remain high in nutrients because their parent material is of recent volcanic origin.
**What other parent materials provide the basis for continuously fertile soils in wet tropical climates?**

## Organic Activity

Plants and animals affect soil formation in many ways. The life processes of the dominant plants are as important to a soil as its microorganisms—the microscopic plants and animals that live in most soils.

Generally, a dense vegetative cover protects a soil from being removed through erosion by running water or wind. Forests form a protective canopy and produce a mulch of surface litter, which keeps rain from beating directly on the soil and increases the proportion of rainwater entering the soil rather than running off its surface. Variations in vegetation species and density of cover can also affect the evapotranspiration rate. A sparse vegetative cover will allow greater evaporation of soil moisture than will thick protective vegetation. This evaporation in turn increases the movement of capillary water toward the surface.

The nature of a plant community determines the nutrient cycles that are involved in soil formation. Certain nutrients, absorbed by plants, are returned to the soil after the plants die

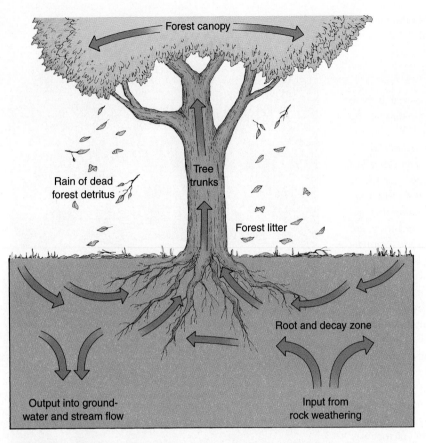

**FIGURE 12.15**
The nutrient cycle in a forest. In a wet climate, if trees do not take up soluble nutrients in the soil, they are flushed away in groundwater and lost permanently. The more demanding the forest vegetation, the richer will be the resulting soil. Pines are notoriously undemanding of soluble bases; consequently, in pine forests, bases are lost by leaching and are not replaced by the vegetation, resulting in soil deterioration.

**In what way does the destruction of tropical rainforests offer a notorious illustration of the nutrient cycle?**

The process of decay, aided by bacterial and fungal activity, transforms organic matter into the jellylike mass called humus. As noted earlier, humus is a very important soil component for several reasons. For one, humus is the primary food supply for soil microorganisms. Humus also affects soil structure by enhancing its water retention and workability. Microorganisms acting on humus return to the soil organic and inorganic materials that are necessary for plant life. Consequently, in most soils there is a direct correlation between humus content and fertility.

In terms of their contribution to soil formation, bacteria are perhaps the most important microorganisms that live in soils. Bacteria break down organic matter, humus, and the debris of living things into organic and inorganic components, allowing the formation of new organic compounds that promote plant growth. It is difficult to estimate the number of bacteria, fungi, and other microscopic plants and animals that live in a soil, though it has been suggested that there may be 1 billion per gram (a fifth of a teaspoon) of soil. Whatever the number, it is enormous. The activities and remains of these microorganisms, minute though they are individually, add considerably to the organic content of a soil.

Earthworms, nematodes, ants, termites, wood lice, centipedes, burrowing rodents, snails, and slugs also stir up the soil, mixing mineral components from lower levels with organic components from the upper portion. Earthworms contribute greatly to soil development because they take soil in, pass it through their digestive tracts, and excrete it in casts. The process not only helps mix the soil but also changes the texture, structure, and chemical qualities of the soil. In the late 1800s, Charles Darwin estimated that earthworm casts produced in a year would equal as much as 10–15 tons per acre. As for the number of earthworms, a study suggested that the total weight of earthworms beneath a pasture in New Zealand equaled the weight of the sheep grazing above them.

## Climate

The discussions of climate regions and processes in Chapters 9 and 10 clearly demonstrated that, on a world scale, climate is a major factor in soil formation. Of course, many soil variations that are related to nonclimatic factors are apparent on a local basis. Obviously, if the climate is the same in a local region but the soils vary, other factors must be responsible for the local variation in soils. The differences that are

and decompose. Soluble nutrients that are not used by plants can be lost through leaching, which can impoverish the soil (Fig. 12.15). The roots of plants break up the soil structure, making it more porous, and roots also absorb water and nutrients from the soil.

Vegetation (leaves, bark, branches, flowers, and root networks) contributes to the organic composition of a soil, through litter and through the remains of dead plants. The organic content of soil depends on the nature of its associated plant life. For example, a grass-covered prairie supplies much more organic matter than the sparse vegetative cover found in desert regions. There is some question, however, as to whether forests or grasslands (with their thick root network and annual life cycle) furnish the soil with greater organic content. But there is no question that many of the world's grassland regions, like the North American prairies, provide some of the world's most fertile soils for cultivation in part because of the high amount of organic matter that a grass cover generates.

## Career Vision

**Amy Jo Steffen, Cartographer/soils, Environmental Mapping, Environmental Services, Inc.**
**University of Wisconsin, Eau Claire**
BS in Geography (emphasis in Natural Resource Management, minor in Geology)

I have always had an appreciation for the landscape because I still remember first seeing the Rocky Mountains when I was 7 years old. I always loved the outdoors but didn't realize my passion for the environment and the landscape until I took a geography course my sophomore year—that's when I declared Geography as my major.

I am a cartographer for Environmental Services, Inc., in Raleigh, North Carolina. I manage and process GPS data, print field maps, and create maps and graphics using GIS, CAD, and design programs. My company mainly deals with wetland and stream delineation and mitigation for transportation routes. We also deal with other environmental concerns such as hazardous materials sites, borrow pits, endangered species, and cultural resources.

As a geographer, I am an important link between my company and our clients. The maps I create display spatial data in pictures that are understood by our clients. The concepts that I use the most include coordinate systems and map scales. We receive data from clients in different coordinate systems and have to convert our data to fit theirs so that it can be represented properly. I also have to choose the right scale for each map, so that the right amount of detail is portrayed.

For what I'm doing, courses in GPS, GIS, cartography, remote sensing, and physical and cultural landscapes are most important. However, working with biologists, archaeologists, and geologists has made me realize that I can more easily understand what it is they want to portray because I've also had some courses in those disciplines. Even an art course in drawing, color, or graphic design is helpful. You can create a map with all the right information, but if its presentation isn't clear, it doesn't have the same value.

My company was hired by the North Carolina Department of Transportation to do a mitigation site search for the Lumber River Basin. We gathered the soils, vegetation, and topographic information and used GIS to overlay them to find possible wetland mitigation areas. We chose areas that fit the criteria and were of a certain size and printed them on aerial photos to be field checked. We are now in the process of preserving, enhancing, and reclaiming many streams and wetlands in the basin.

Almost every subject of study can be related back to geography. In biology, it could be the location and migration of certain animals. In geology, it's the spatial relationship between different rock bodies that help tell Earth's history. In economics, when the locations of high- and low-income areas are looked at, it's geography. In English, the spread of a specific phrase or language uses geography. Every time you get into your car to go somewhere, you use geography to guide you—direction, distance, physical landmarks, and so on. Everyone uses geography!

---

apparent at a local level tend to reflect the influence of other factors such as parent material, land surface configuration, vegetation type, and time.

Temperature directly affects the activity of soil microorganisms, which in turn affects the rates of decomposition of organic matter. In hot equatorial regions, intense activities by soil microorganisms preclude thick accumulations of organic debris or humus. Figure 12.16 shows that the amounts of organic matter and humus in a soil increase toward the middle latitudes and away from the tropics. In the Köppen mesothermal $C$ and microthermal $D$ climates, soil microorganism activity is slow enough to allow decaying organic matter and humus to accumulate in rich layers. Moving poleward into

### FIGURE 12.16
Relationship of temperature to production and destruction of organic matter in the soil.
**What range of mean annual temperatures is most favorable for the accumulation of humus?**

colder regions, the combination of retarded microorganism activity and limited plant growth results in thin accumulations of organic matter that may be either partially decomposed or unaffected by decomposition.

Temperature also influences the rates of chemical reactions in soil, many of which make nutrients available for plant growth. Chemical activity tends to increase and decrease directly with temperature, given equal availability of moisture. As a result, parent materials of soils in hot, humid equatorial regions are altered to a far greater degree by chemical means than are parent materials in colder zones ( Fig. 12.17).

Temperature affects soil indirectly through its influence on vegetation. We know that vegetation associations are adapted to certain temperature regimes. Soils generally reflect the character of plant cover because of nutrient cycles that tend to keep both vegetation and soil in chemical equilibrium.

Moisture conditions affect the development and character of soils more clearly than any other climatic factor. Without precipitation—and, consequently, soil water and the chemicals dissolved therein—terrestrial plant life is impossible. An absence of plants greatly diminishes the organic content and thereby the fertility of a soil.

We have already discussed the effects of both gravitational and capillary water on soil structure, texture, color, and development. Because precipitation is the original source of soil water (disregarding the minor contribution of dew), the amount of precipitation received by a soil affects degrees of leaching, eluviation, and illuviation and thereby the rates of soil formation and horizon development.

When considering the effects of precipitation on the quantity and movement of soil water, we should note that the evaporation rate is a very important factor as well. Salt and gypsum deposits from the upward migration of capillary water are more extensive in hot, dry regions—such as the southwestern United States where evaporation rates are high—than in colder, dry regions (see again Fig. 12.17).

Just as temperature affects soil development indirectly through its influence on vegetation, so too does the moisture regime of a region. In regions where warm seasons are dry, soils tend to be alkaline. Where the summers are warm and extremely wet and the winter season warm and dry, intense chemical weathering and fluctuating water tables produce a peculiar type of soil crust known as **laterite**. This crust is common in the tropics where it is quarried for building material ( Fig. 12.18).

## Land Surface Configuration

The slope of the land, its relief, and its aspect (the direction it faces) all influence soil development both directly and indirectly. Steep slopes are generally better drained than gentler ones, and they are also subject to rapid runoff of surface water. As a consequence, there is less infiltration of water on steeper slopes, which retards the soil-forming processes. This retardation inhibits the development of soils, sometimes to the extent that no soil will develop on the parent material. In addition, rapid runoff on steep slopes can erode surfaces as fast or faster than soil can develop on them. On gentler slopes, because there is less runoff and more infiltration, more water is available for soil-forming processes and to support vegetative growth, and erosion is not as intense. In fact, erosion rates in areas of gently rolling hills may be just enough to offset the production of soil from parent material. In general, well-developed, mature soils of the most ideal characteristics typically form on land that has a gentle slope.

Valley floors and flatlands are often poorly drained. When this is the case and water levels in the soil are near the surface, gravitational water cannot percolate downward, and capillary water may move harmful concentrations of salt and alkaline substances to the surface. This condition presents a constant problem in the irrigation of flatlands because irrigation tends to raise the water table. Artificial drainage ditches must be provided to lower high water tables in such instances.

*Slope aspect* has a direct effect on microclimate in areas other than the equatorial tropics. North-facing slopes in the middle and high latitudes of the Northern Hemisphere have microclimates that are cooler and wetter than those on south-facing exposures, which receive the sun's rays more directly and are therefore warmer and drier. Local variations in soil depth, texture, and profile development result directly from microclimate differences.

Land surface configuration also influences soil development, indirectly, through its effects on vegetation. Steep slopes prevent the development of a mature soil that would support abundant vegetation, and a modest plant cover yields less organic debris for the soil.

## Time

Soils have a tendency to develop toward a state of equilibrium with their environment. A soil is mature when it has reached such a condition of equilibrium. Young soils are still in the process of alteration to achieve equilibrium with their environmental conditions. Mature soils have well-developed and stable horizons that indicate the conditions under which they developed. Young or immature soils typically have poorly developed horizons or none at all. Very old soils may have well-developed horizons that present problems. Such soils frequently contain dense pans or crusts in their *B* horizons. These horizons may consist of eluviated clays, concentrated calcium carbonate (lime), silica, or oxides of iron and aluminum. Soils on ancient low-relief surfaces in the tropics, where soil formation is rapid, frequently present such problems. To use these soils for agriculture, farmers may need to break up the crusts with dynamite or, less dramatically, by deep plowing.

Another effect of time is that, as soils mature, their influence by parent material decreases and they increasingly

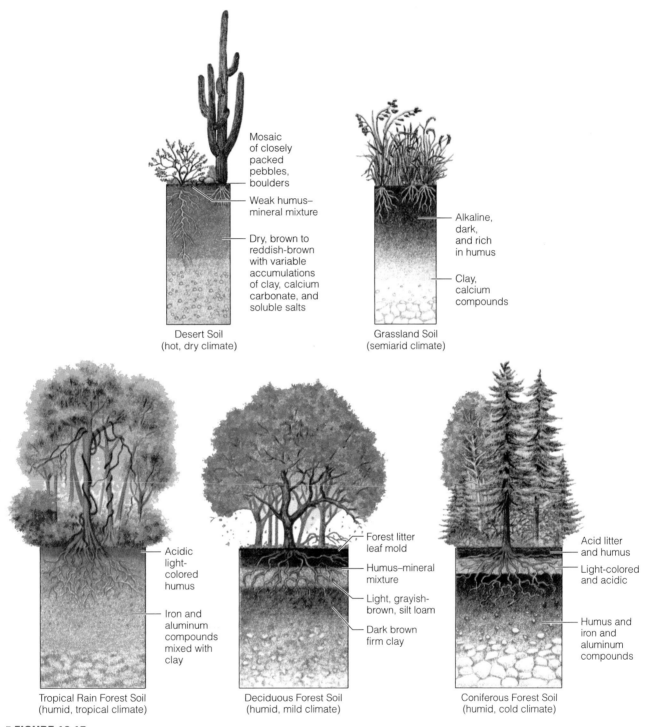

**FIGURE 12.17**
Idealized diagrams of five different soil profiles illustrate the effects of climate and vegetation on the development of soils and their horizons.
**Which two environments produce the most humus and which two produce the least?**

reflect their climate and vegetative environments. On a global scale, climate is typically the most apparent of all influences on soils, provided sufficient time has passed for the soils to reach maturity ( Fig. 12.19).

The importance of time in soil formation is especially clear in soils developed on transported parent materials. Depositional surfaces are in many cases quite recent in geologic terms and have not been exposed to weathering long

**346** CHAPTER 12 SOILS AND SOIL DEVELOPMENT

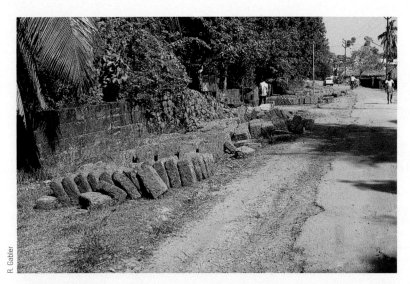

■ **FIGURE 12.18**
Laterite cut for building stone and stacked along a village road in the state of Orissa, India.
**Why is building with brick or stone rather than wood so important in heavily populated, less developed nations such as India?**

■ **FIGURE 12.19**
The time that a soil has been developing is important to its composition and physical character. Given enough time and the proper environmental conditions, soils will become more maturely developed with a deeper profile and stronger horizon development.
**What major changes occur as the soil illustrated here becomes better developed over time?**

enough for a mature soil to develop. Typically, recent deposits of transported materials have not yet been leached of their soluble nutrients, nor has their soil developed undesirable characteristics. Deposition occurs in a variety of settings: on river floodplains where the accumulating sediment is known as *alluvium;* downwind from dry areas where dust settles out of the atmosphere to form blankets of wind-deposited silts, called *loess;* and in volcanic regions showered by ash and covered by lava. Ten thousand years ago, glaciers withdrew from vast areas, leaving behind jumbled and often complex deposits of rocks, sand, silt, and clay. In terms of agricultural productivity, the world's best soils are found on alluvium, loess, volcanic ash and lava, and certain types of glacial deposits.

Because of the great number and variability of materials and processes involved in the formation of soils, there is no fixed amount of time that it takes for a soil to become mature. The Natural Resources Conservation Service, however, estimates that it takes about 500 years on the average to develop 1 inch of soil in the agricultural regions of the United

States. Generally, though, it takes thousands of years for a soil to reach maturity.

## Soil-Forming Regimes

Although a nearly infinite number of factors are involved in soil development, certain processes can be classified into a few general **soil-forming regimes.** The characteristics that make major soil types distinctive and different from one another result from their soil-forming regimes. These soil-forming regimes vary mainly because of differences in climate and vegetation. At the broadest scale of generalization, climate differences produce three primary soil-forming regimes: laterization, podzolization, and calcification ( Fig. 12.20).

### Laterization

**Laterization** is a soil-forming regime that occurs in humid tropical and subtropical climates as a result of high temperatures and abundant precipitation. These climatic environments encourage rapid breakdown of rocks and decomposition of nearly all minerals. Despite the dense vegetation that is typical of these climates, little humus is incorporated into the soil because of rapid decomposition of plant litter and enormous numbers of soil microorganisms. Because of abundant moisture, eluviation and leaching are dominant soil-forming processes in the humid tropics.

Lateritic soils are generally reddish in color from iron oxides; the term *laterite* means "bricklike." Laterites do not have an O horizon, and the A horizon loses fine soil particles as well as most minerals and bases except for iron and aluminum compounds, which are insoluble primarily because of the absence of organic acids ( Fig. 12.20a). As a result, the topsoil is reddish, coarse textured, and tends to be porous. In contrast to the A horizon, the B horizon in a lateritic soil has a heavy concentration of illuviated materials.

In the tropical forests, soluble nutrients released by weathering are quickly absorbed by vegetation, which eventually returns them to the soil where they are reabsorbed by plants. This rapid cycling of nutrients prevents the total leaching away of bases, leaving the soil only moderately acidic. Removal of vegetation permits total leaching of bases, resulting in the formation of crusts of iron and aluminum compounds (laterites), as well as accelerated erosion of the A horizon.

Laterization is a year-round process because of the small seasonal variations in temperature or soil moisture in the humid tropics. This continuous activity and strong weathering of parent material cause some tropical soils to develop to depths of as much as 8 meters (25 ft) or more.

### Podzolization

**Podzolization** occurs mainly in the high middle latitudes where the climate is moist with short, cool summers and long, severe winters. The coniferous forests of these climate regions are an integral part of the podzolization process.

Where temperatures are low much of the year, microorganism activity is reduced enough that humus does accumulate; however, because of the small number of animals living in the soil, there is little mixing of humus below the surface. Leaching and eluviation by acidic solutions remove the soluble bases and aluminum and iron compounds from the A horizon ( Fig. 12.20b). The remaining silica gives a distinctive ash-gray color to the E horizon (*podzol* is derived from a Russian word meaning "ashy"). The needles that coniferous trees drop are chemically acidic and contribute to the soil acidity. It is difficult to determine whether the soil is acidic because of the vegetative cover or whether the vegetative cover is adapted to the acidic soil.

Podzolization can take place outside the typical cold, moist climate regions if the parent material is highly acidic—for example, on the sandy areas common along the East Coast of the United States. The pine forests that grow in such acidic conditions return acids to the soil, promoting the process of podzolization.

### Calcification

The third distinctive soil-forming regime is called **calcification.** In contrast to both laterization and podzolization, which require humid climates, calcification occurs in regions where evapotranspiration significantly exceeds precipitation.

In many areas of low precipitation, the air is often loaded with alkali dusts such as calcium carbonate ($CaCO_3$). When calm conditions prevail or when it rains, the dust settles across the landscape and accumulates in the soil. The rainfall produces an amount of soil water that is just sufficient to translocate these materials to the B horizon ( Fig. 12.20c). Over hundreds to thousands of years, the $CaCO_3$-enriched dust concentrates in the B horizon, forming hard layers of *caliche*. Much thicker accumulations called *calcretes* ( Fig. 12.21) form by the upward (capillary) movement of dissolved calcium in groundwater when the water table is near the surface.

Calcification is important in the climate regions where moisture penetration is shallow. The subsoil is too dry to support tree growth, and shallow-rooted grass or shrubs are the primary forms of vegetation. Calcification is enhanced as grasses use calcium, drawing it up from lower soil layers and returning it to the soil when the grasses die. Grasses and their dense root networks provide large amounts of organic matter, which typically is mixed deep into the soil by burrowing animals. Middle-latitude grassland soils are rich both in bases and in humus and are the world's most productive agricultural soils. The deserts of the American West generally have no humus, and the rise of capillary water can leave not only calcium carbonate but also sodium chloride (salt) at the surface.

# Geography's Environmental Science Perspective
## Slash-and-Burn Agriculture Adds Nutrients to Tropical Soils

The lush vegetation and forest cover of the tropical rainforests would suggest that the soils of the tropics are very fertile and rich with nutrients. Actually the opposite is true—tropical soils are relatively infertile. The heavy rainfall of these regions leaches the nutrients out of the soils. Given this circumstance, agricultural peoples living in these areas have devised methods for being able to grow crops and provide nutrients to the soil.

A technique called slash-and-burn (previously mentioned in Chapter 9) is used for farming in tropical rainforest regions. This seemingly destructive method is used for small-scale subsistence farming for two reasons. First, it is necessary to clear a plot of land in the dense forest area for cultivation. Second, the method is a way of treating the topsoil to improve its crop productivity.

This agricultural method is typically used in rural tropical regions, where other fertilizers may not be available, as a way of adding nutrients to the soils. The technique is quite simple. First, a small area is slashed and cleared of standing native vegetation, typically a dense cover to tropical forest. The cut vegetation is left on the ground and allowed to dry (the drying process is sometimes delayed by humidity and rainfall). The next step is completed when the vegetation is ignited and burned totally to ashes. Finally, the ashes are mixed in with the topsoil.

Standard fertilizers usually (but not always) contain $NO_3$ (nitrates), $K_2O$ (potassium oxide), and $P_2O_5$ (diphosphorus pentoxide). When natural rainforest vegetation is burned, its ashes contain compounds of N (nitrogen), K (potassium), and P (phosphorus), among other chemical constituents. When the ashes are worked into the upper soil layer, they act as a low-grade fertilizer. The soil becomes moderately fertile soil, but the use of the nutrients by crops and continued leaching by heavy precipitation typically means that the soil will support crops for only 2–3 years. The slash-and-burn process must then be shifted to another patch of the forest.

If slash-and-burn fields are small and if they are kept separated, they will recover from the burning and agricultural use. The surrounding native vegetation will encroach and once again take over. However, when massive burning is used to clear vast acreage of rainforest for mining or for pasture for grazing, then recovery of the natural vegetation is unlikely, and wide areas of the rainforest may be lost forever.

To increase the fertility of the soil, it is essential that the ashes from burned vegetation be mixed into the topsoil.

**FIGURE 12.26**
Soil profile examples: (a) Andisols, (b) Gelisols, and (c) Aridisols

are high in bases, and contain considerable organic material derived from the grasslands or savanna vegetation with which they are normally associated. Although they harden when dry and become sticky and difficult to cultivate when swollen with moisture, Vertisols can be agriculturally productive.

**Mollisols** are most closely associated with grassland regions and are among the best soils for sustained agriculture ( Fig. 12.27b). Because they are located in semiarid climates, Mollisols are not heavily leached, and they have a generous supply of bases, especially calcium. The characteristic horizon of a Mollisol is a mollic epipedon—a thick, dark-colored surface layer rich in organic matter from the decay of abundant root material. Grasslands and associated Mollisols served as the grazing lands for countless herds of antelope, bison, and horses. Before the invention of the steel plow, the thick root material made this soil nearly uncultivable in the United States and thus led to the widespread public image of the Great Plains as a "Great American Desert." Today, Mollisols support most of the grain production from domesticated grasses (wheat, rye, oats, corn, barley, and sorghum).

In regions of adequate precipitation, such as the tallgrass prairies of the American Midwest, the combination of soils and climate is unexcelled for agriculture. In areas of lesser precipitation, periodic drought is a constant threat, and the temptation of fertile soils was the downfall of many farmers prior to the advent of center pivot irrigation.

**Alfisols** occur in a wide variety of climate settings. They are characterized by a subsurface clay horizon (argillic *B* horizon), a medium to high base supply, and a light-colored ochric epipedon ( Fig. 12.27c). The five suborders of Alfisols reflect climate types and exemplify the hierarchical nature of the classification system: *Aqualfs* are seasonally wet and can be found in mesothermal areas such as Louisiana, Mississippi, and Florida; *Boralfs* are found in moist, microthermal climates such as Montana, Wyoming, and Minnesota; *Udalfs* are common in both microthermal and mesothermal climates that are moist enough to support agriculture without irrigation, such as Wisconsin, Ohio, and Tennessee; *Ustalfs* are found in mesothermal climates that are intermittently dry, such as Texas and New Mexico; and *Xeralfs* are found in California's Mediterranean climate, which is characterized by wet winters and long, dry summers.

Because of their abundant bases, Alfisols can be very productive agriculturally if local deficiencies are corrected: irrigation for the dry suborders, properly drained fields for the wet suborders.

**Spodosols** are most closely associated with the podzolization soil-forming process. They are readily identified by their strong horizon development ( Fig. 12.28a). There is often a white or light-gray *E* horizon (albic horizon) covered with a thin, black layer of partially decomposed humus and underlain by a colorful *B* horizon enriched in relocated iron and aluminum compounds (spodic horizon).

Spodosols are generally low in bases and form in porous substrates such as glacial drift or beach sands. In New England and Michigan, Spodosols are also acidic. In these regions, as well as in similar regions in northern Russia, Scandinavia, and Poland, only a few types of agricultural

(a)            (b)            (c)

**■ FIGURE 12.27**
Soil profile examples: (a) Vertisols, (b) Mollisols, and (c) Alfisols

plants, such as cucumbers and potatoes, can tolerate the microthermal climates and sandy, acidic soils. Consequently, the cuisine of these regions directly reflects the Spodosols that dominate the areas.

**Ultisols,** like Spodosols, are also low in bases because they develop in moist or wet regions. Ultisols are characterized by a subsurface clay horizon (argillic horizon) and are often yellow or red because of residual iron and aluminum oxides in the *A* horizon (■ Fig. 12.28b). In North America, the Ultisols are most closely associated with the southeastern United States. When first cleared of forests, these soils can be agriculturally productive for several decades. But a combination of high rainfall with the associated runoff and erosion from the fields decreases the natural fertility of the soils. Ultisols remain productive only with the continuous application of fertilizers. Today, forests cover many former cotton and tobacco fields of the southeastern United States because of a reduction in soil fertility and extensive soil erosion (see again Fig. 10.9).

**Oxisols** have developed over long periods of time in tropical regions with high temperatures and heavy annual rainfall. They are almost entirely leached of soluble bases and are characterized by a thick development of iron and aluminum oxides (■ Fig. 12.28c). The soil consists mainly of minerals that resist weathering (for example, quartz, clays, hydrated oxides). Oxisols are most closely associated with the humid tropics, but they also extend into savanna and tropical thorn forest regions. In the United States, Oxisols are present only in Hawaii. Oxisols are dominated by laterization and retain their natural fertility only as long as the soils and forest cover maintain their delicate equilibrium. The bases in the tropical rainforests are stored mainly in the vegetation. When a tree dies, epiphytes and insects must recycle the bases rapidly before the heavy rainfall leaches them from the system.

The burning of vegetation associated with slash-and-burn agriculture in rainforests releases the nutrients necessary for crop growth but quickly results in their loss from the ecosystem. Many tropical Oxisols that once supported lush forests are now heavily dissected by erosion and only support a combination of weeds, shrubs, and grasses.

## Soils as a Critical Natural Resource

Regardless of their composition, origin, or state of development, Earth's soils remain one of our most important and vulnerable resources. The word *fertility,* so often associated with soils, has a meaning that takes into consideration the usefulness of a soil to humans. Soils are fertile in respect to their effectiveness in producing specific vegetation types or associations. Some soils may be fertile for corn and others for potatoes. Other soils retain their fertility only as long as they remain in delicate equilibrium with their vegetative cover.

It is clearly the responsibility of all of us who enjoy the agricultural end products of farm, ranch, and orchard, as well as appreciate the natural beauty of Earth's diverse biomes, to help protect our valuable soils. Although space does not permit a thorough examination of problems associated with soil erosion, soil depletion, and the mismanagement of land, we should be conscious of their existence in the world today (■ Fig. 12.29). We should also be aware that these problems have reasonable solutions (■ Fig. 12.30). Maintaining soil fertility and usefulness is a critical challenge to humanity that is essential in our continuing struggle to protect the resources of our natural environment.

SOILS AS A CRITICAL NATURAL RESOURCE   357

(a)                  (b)                  (c)

**FIGURE 12.28**
Soil profile examples: (a) Spodosols, (b) Ultisols, and (c) Oxisols

**FIGURE 12.29**
Gully erosion is one of the more spectacular examples of poor agricultural practices. It produces permanent alteration of the landscape and guarantees that the original productivity of the land cannot be regained.
**What could have been done to prevent this example of land mismanagement?**

**FIGURE 12.30**
The contour farming techniques used on this farm are excellent examples of conservation methods designed to preserve the soil resource.
**What other soil conservation practices are often used to preserve the soil resource?**

# Define & Recall

| | | |
|---|---|---|
| soil | loam | soil taxonomy |
| soil fertilization | infiltrate | soil survey |
| gravitational water | soil ped | soil order |
| eluviation | porosity | subsurface horizon |
| illuviation | permeability | epipedon |
| hardpan | pH scale | Entisol |
| leaching | parent material | Inceptisol |
| stratification | soil profile | Histosol |
| capillary water | soil horizon | Andisol |
| hygroscopic water | Cl, O, R, P, T | Gelisol |
| humus | residual parent material | Aridisol |
| soil texture | transported parent material | Vertisol |
| clayey | laterite | Mollisol |
| clay | soil-forming regime | Alfisol |
| silty | laterization | Spodosol |
| silt | podzolization | Ultisol |
| sandy | calcification | Oxisol |
| sand | salinization | |
| soil grade | gleization | |

# Discuss & Review

1. Why is soil an outstanding example of integration and interaction among Earth's subsystems?
2. Describe the different circumstances in which water is found in soil.
3. What is the effect of eluviation if carried to an extreme? What is the effect of illuviation if carried to an extreme?
4. Under what conditions does leaching take place? What is the effect of leaching on the soil and, consequently, on the vegetation that it supports?
5. How can capillary water contribute to the formation of caliche? What is the effect of caliche on drainage?
6. How might soil air differ from air in the atmosphere? What is the effect on life when air is excluded from water-saturated soils?
7. How is humus formed? What relation does humus have to soil fertility?
8. What conclusions can you draw from the color of the soil in your area? How might color relate to fertility?
9. How is texture used to classify soils? Describe the ways scientists have classified soil structure.
10. What pH range indicates soil suitable for most complex plants?
11. What are the general characteristics of each horizon in a soil profile? How are soil profiles important to scientists?
12. What factors are involved in the formation of soils? Which is most important on a global scale?
13. How does transported parent material differ from residual parent material? List those factors that help determine how much effect the parent material will have on the soil.
14. What are the most important effects of parent material on soil?
15. List a number of ways in which humus is important to soils.
16. How does the presence of earthworms alter soil?
17. Describe the various ways in which temperature and precipitation are related to soil formation.
18. The Bonneville Salt Flats in Utah are well known as a natural soil formation that provides a perfect surface for auto racing. How do you suppose these salt flats were formed?
19. Describe the three major soil-forming regimes.
20. Which soil orders of the NRCS soil classification system have the most agricultural potential? Why?

# Consider & Respond

1. Refer to Figure 12.8. Using the texture triangle, determine the textures of the following soil samples.

   |    | Sand | Silt | Clay |
   |----|------|------|------|
   | a. | 35%  | 45%  | 20%  |
   | b. | 75%  | 15%  | 10%  |
   | c. | 10%  | 60%  | 30%  |
   | d. | 5%   | 45%  | 50%  |

   What are the percentages of sand, silt, and clay of the following soil textures? (*Note:* Answers may vary, but they should total 100%.)
   e. Sandy clay
   f. Silty loam

2. Refer to Figure 12.13 and associated pages in the text.
   a. What horizons make up the zone of eluviation?
   b. What are two processes that occur in the zone of eluviation?
   c. The various *B* horizons are in what zone?
   d. Weathered parent material is the major constituent of what horizon?
   e. Partly decomposed organic debris makes up which horizon?

3. Refer to Table 12.1.
   a. What materials accumulate in an argillic horizon?
   b. Which would generally be better suited for agriculture—a soil with an ochric epipedon or a mollic epipedon? Why?
   c. What name would be given to a 7-inch-thick horizon that contained at least 2% salt?

4. Where would you rank soils in terms of importance among a nation's environmental resources?

5. Give your opinion of the overall value of soils in the United States and the extent to which these soils are preserved and protected.

# CHAPTER 13

Ninety million years ago, Earth had a very different distribution of land and sea. No ice existed at the poles, so sea level was higher than it is today. In North America, most of the Great Plains, much of the Midwest, and the present Pacific Coast region were beneath shallow seas. R. Blakey, Northern Arizona University

# Earth Materials, the Lithosphere, and Plate Tectonics

**PHYSICAL Geography Now™** This icon, appearing throughout the book, indicates an opportunity to explore interactive tutorials, animations, or practice problems at now.brookscole.com/gabler8

# CHAPTER PREVIEW

**Knowledge of the origin, nature, and structure of rocks is important to physical geographers because these three factors influence the character of Earth's landscapes and the processes that operate on our planet's surface.**

- What surface characteristics does bedrock influence?
- How are humans affected by bedrock properties?

**The rock cycle is a good conceptual model for understanding the internal and external processes that create, alter, and recycle the Earth materials that make up our planet's surface.**

- What are these processes and materials, and how do they interact to produce the major types of rocks?
- Could the rock cycle be considered an Earth subsystem? Why or why not?
- How could human activities affect parts of the rock cycle, or vice versa?

**Only eight chemical elements account for almost 99% of Earth's crust by weight, and the most common minerals are combinations of these same elements.**

- What does this suggest about mineral classification?
- What does it suggest concerning the most common rocks in Earth's crust?

**Earth's surface configuration and the locations of continents, oceans, and major landforms have changed through geologic time and continue to change today, "with no clear beginning and no definite end in sight."**

- What does this quotation mean?
- What is its significance?
- How has continental movement affected global climates and environments?

**For a scientific theory to be accepted, it is not sufficient merely to describe what happens; the theory must also explain *how* it happens and be supported by adequate evidence.**

- How is this statement related to the concept of continental drift?
- What processes explain plate tectonics?
- What evidence on Earth's surface supports plate tectonics?

**Plate tectonics theory has an all-encompassing relationship to the Earth sciences, similar to that which evolution has to the life sciences.**

- What is the importance of such a theory?
- Why is science so concerned with theory?
- How is this theory all encompassing to an understanding of the Earth system?

If we could go back in time to view Earth 100 million years ago, we would see a spatial distribution of oceans and landmasses that varied considerably from what exists on our planet at the present time. The shapes of oceans and continents, their locations, and their orientations relative to the poles and equator—were very different from those of today. Where extensive plains or major mountain ranges exist today, vast inland seas existed. Fossil fish that are known to have lived in ocean waters have been found in midcontinental locations, sometimes thousands of feet above sea level. Fossilized remains of tropical plants have been discovered in the polar regions, and ancient tropical soils exist where it is not hot or humid today. These dramatic environmental changes, along with the current shapes, sizes, and distribution of our planet's major surface features, require scientific explanations. They also demonstrate the interrelatedness of the four major subsystems of Earth: atmosphere, hydrosphere, biosphere, and lithosphere.

The remainder of this textbook focuses on the lithosphere and hydrosphere but does not neglect the importance of the atmosphere and biosphere as they interact with and affect Earth's surface environments and topography. Understanding the geography of the past and the processes that shaped ancient environmental conditions helps us think about and consider in scientific terms the geography of the future as well as the interactions between the processes and Earth materials that affect our planet's surface today. As a prelude to studying Earth's surface, its features, and the processes that modify them, we need to examine the materials—rocks and minerals—that make up our planet's solid exterior. This outer skin, Earth's **crust,** is the rocky surface on which we live.

## Minerals and Rocks

**Minerals** are naturally occurring inorganic substances that are the building blocks of rocks. They are well-defined combinations of atomic elements with distinct chemical formulas and recognizable physical characteristics. A **rock** is an aggregate (a whole made up of parts) of mineral particles. The most common elements of Earth's crust (and therefore the minerals and rocks that make up the crust) are oxygen and silicon, followed by aluminum and iron, and the bases: calcium, sodium, potassium, and magnesium. As you can see in Table 13.1, these 8 chemical elements—out of the more than 100 known—account for almost 99% of the Earth's crust by weight. The most common minerals are combinations of these 8 elements.

### Minerals

Every mineral has distinctive and recognizable physical characteristics that aid in its identification—for example, a particular color, luster, hardness, tendency to fracture, and specific gravity. Minerals are crystalline in nature although the crystals may only be evident when viewed through a microscope. Mineral crystals also have consistent geometric shapes that express their molecular structure ( Fig. 13.1). Chemical bonds hold together the atoms and molecules that compose a mineral. The strength and nature of these chemical bonds affect the resistance and hardness of minerals and of the rocks that they form. Minerals with weak internal bonds undergo chemical alteration most easily. Ions may leave or be traded within their structure, producing physical changes that are the chemical basis of *rock weathering*.

There are discrete families of minerals that are based on their chemical composition. Certain elements combine readily with a variety of others, and the most active of these elements are silicon, oxygen, and carbon. Consequently, the most common mineral groups are silicates, oxides, and carbonates. **Silicate** minerals are compounds of oxygen and silica with one or more metals and/or bases. The silicates are by far the largest and most common mineral family and constitute 92% of Earth's crust. They are generally created by the cooling of *lava* or **magma** (a molten subsurface mass containing the elements

**TABLE 13.1**
**Most Common Elements in Earth's Crust**

| Element | Percentage of the Earth's Crust by Weight |
|---|---|
| Oxygen (O) | 46.60 |
| Silicon (Si) | 27.72 |
| Aluminum (Al) | 8.13 |
| Iron (Fe) | 5.00 |
| Calcium (Ca) | 3.63 |
| Sodium (Na) | 2.83 |
| Potassium (K) | 2.70 |
| Magnesium (Mg) | 2.09 |
| Total | 98.70 |

*Source:* J. Green, "Geochemical Table of the Elements for 1953," *Bulletin of the Geological Society of America 64* (1953).

**FIGURE 13.1**
The geometric arrangement of atoms determines the crystal form of a mineral. These are crystals of the mineral halite (common salt) that formed in rocks from an ancient evaporating sea.
**What is a mineral?**

from which minerals can form), which causes the crystallization of different minerals at successively lower temperatures and pressures ( Fig. 13.2). Dark, heavy, iron-rich silicate minerals crystallize first (at high temperatures), and light-colored,

MINERALS AND ROCKS **363**

**FIGURE 13.2**
The crystallization of silicate minerals from magma. In general, dark, iron-rich minerals form solid crystals first and tend to sink in the magma. Lighter minerals, richer in silica, tend to crystallize later at lower temperatures and pressures as the magma cools into rock.
**How is this crystallization process related to the freezing of ice in water?**

**FIGURE 13.3**
Most rocks consist of several minerals. This piece of cut and polished granite displays intergrown crystals of differing composition, color, and size that give the rock a distinctive appearance.

lighter-weight, silica-rich minerals crystallize later at cooler temperatures. The order of mineral crystallization parallels their relative stability in a rock because silicate minerals that crystallize later tend to be more resistant. In the weathering of granite (a rock that includes a variety of silicates), the dark, heavy minerals are the first to decompose, and the lighter, silica-rich minerals decompose later. Quartz is one of the last silicate minerals to form from magma, and it is relatively hard and resistant.

## Rocks

Although a few rock types are composed of many particles of a single mineral, most rocks consist of several minerals (▮ Fig. 13.3). Each mineral remains separate and retains its own distinctive characteristics, and the properties of a rock reflect its mineral composition. The number of rock-forming minerals that are common is limited, but they combine through a multitude of processes to produce an enormous variety of rock types (refer to Appendix C for information and pictures of common rocks mentioned in the text). Rocks are the fundamental building materials of the lithosphere. They are lifted, pushed down, and deformed by tectonic processes, and they are weathered and eroded, to be deposited as sediment elsewhere.

Solid rock that underlies loose surface material is called **bedrock.** Where bedrock is not exposed at the surface, a cover of broken and decomposed rock fragments called **regolith** overlies it. Soil may or may not have formed on the regolith (see again Fig. 12.12). On steep slopes, regolith may be absent because running water, gravity, or some other surface process may have exposed the bedrock by removing the weathered material. A mass of exposed bedrock is called an **outcrop** (▮ Fig. 13.4).

Three major categories of rocks can be defined, based on their origin. These rock types are igneous, sedimentary, and metamorphic.

**FIGURE 13.4**
Outcrops are masses of solid rock exposed at the surface.
**What physical characteristics of these rock outcrops caused them to protrude above the general land surface?**

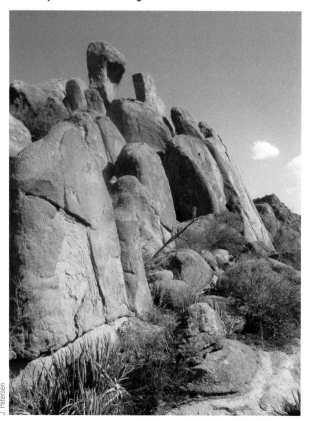

**The Rock Cycle** Like landforms, many rocks do not remain in their original form indefinitely but instead, over a long term, tend to undergo processes of transformation. The **rock cycle** is a conceptual model for understanding processes that generate, alter, transport, and deposit mineral materials to form different kinds of rocks (▌Fig. 13.5). The term *cycle* means that existing rocks supply the materials to make new and sometimes very different rocks. Existing rocks can be "recycled" to form new rocks. The geologic age of a rock is based on the time when it assumed its current state; metamorphism, or melting, and other rock-forming processes reset the age of origin.

Although a complete cycle is shown in Figure 13.5, many rocks do not go through all steps of the rock cycle, as shown by the arrows that cut across the diagram. **Igneous rocks** form by the cooling and crystallizing of molten lava or magma. Igneous rocks can be remelted and recrystallized to form new rocks; can be changed into metamorphic rocks by heat, pressure, and/or chemical action; or weathered into fragments that are eroded, transported, and deposited to form sedimentary rocks. **Sedimentary rocks** consist of particles and deposits derived from any of the three basic rock types. **Metamorphic rocks** are rocks that have been changed through heat and pressure; they can be created out of igneous or sedimentary rock or metamorphosed further into a new rock type. In addition, metamorphic rocks can be heated sufficiently to melt into magma and cool as igneous rocks.

> PHYSICAL
> Geography◉Now™ Log on to Physical GeographyNow and select this chapter to work through a Geography Interactive activity on the "Rock Cycle" (click Rocks and the Rock Cycle → Rock Cycle).

**Igneous Rocks** When molten rock material cools and solidifies, igneous rocks are formed. Below Earth's surface, this melt is called *magma*. The igneous material with which we are most familiar is **lava,** the surface form of magma (▌Fig. 13.6).

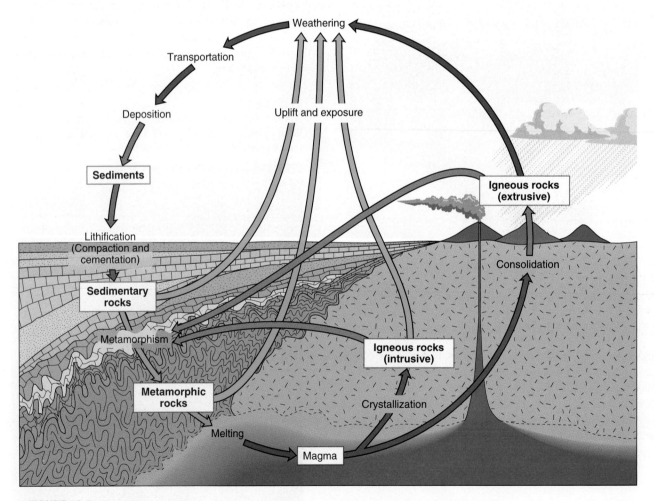

▌**FIGURE 13.5**
The rock cycle helps explain the formation of igneous, sedimentary, and metamorphic rocks. Note the links that bypass some of the parts of the cycle.
**What conditions are necessary to change igneous rock to metamorphic rock, and metamorphic rock to igneous rock?**

Lava is spewed forth from volcanoes or fissures in the crust at temperatures as high as 1090°C (2000°F). There are three major categories of igneous rocks: extrusive, intrusive, and pyroclastic.

Molten material that emerges and solidifies at Earth's surface forms **extrusive igneous rock,** also called volcanic rock, such as solidified lavas. If rising magma does not break through to the surface but solidifies within Earth, the resulting internally cooled and crystallized rock is known as **intrusive igneous rock.** When intrusive masses solidify deep beneath the surface, they are sometimes referred to as **plutonic rocks** (after Pluto, the Roman god of the underworld). Igneous rocks vary in chemical composition, tendency to fracture, texture, crystalline structure, and presence or absence of layering. They may be grouped or classified in terms of their crystal size or texture as well as their mineral composition.

The chemical composition of igneous rocks varies from *felsic,* also called *sialic* (rich in light-colored, lighter-weight minerals, especially silica *si* = silica, and *al* = aluminium), to *mafic* (lower in silica and rich in heavy minerals, such as compounds of magnesium and iron—*ma* = magnesium and *f* = iron). Granite, a sialic, coarse-grained, intrusive rock, has the same chemical and mineral composition as rhyolite, a fine-grained, extrusive lava ( Fig. 13.7). Likewise, the intrusive mafic counterpart of fine-grained basalt is gabbro, a dark, heavy, coarse-grained rock. Both of these igneous rock types formed from mafic magma, but basalt cools at the surface and gabbro solidifies at depth. Intrusive rocks cool very slowly because surrounding masses of rock retard the loss of heat from molten magma. Slow cooling allows more time for larger

**FIGURE 13.6**
Basalt, a fine-grained extrusive igneous rock, forms these fresh lava flows on the island of Hawaii.
**What visible characteristics in this picture suggest that the lava flow is recent?**

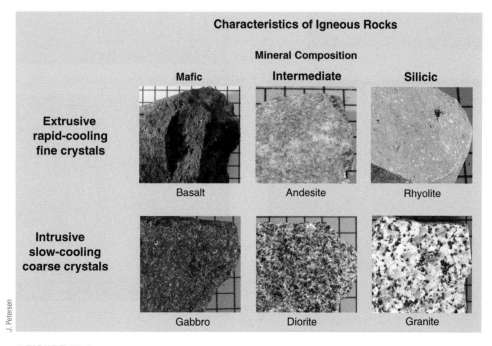

**FIGURE 13.7**
The types of igneous rocks are based on origin and mineral composition. Igneous rocks that cooled rapidly, at or near the surface, have fine crystals. Rocks that cooled slowly have a coarser crystalline structure. Mineral content can be mafic, silicic, or intermediate.
**What is the difference between granite and basalt?**

crystal formation prior to solidification. Rocks formed in this manner are coarse-grained and have crystal grains that are easily visible without magnification. *Granite* and *gabbro* illustrate this coarse texture.

Igneous rocks also form with an intermediate composition, a rough balance between sialic and mafic minerals. The intrusive rock is called *diorite,* and the extrusive rock is *andesite,* named after the Andes where many volcanoes erupt lava of this composition.

Extrusive rocks, as well as some intrusive rocks that have pushed close to the surface undergo rapid cooling and solidification. Rapid cooling forms fine-grained igneous rocks, typical of lavas, such as *basalt* or *rhyolite* (see again Fig. 13.7). Small crystals develop under conditions of rapid cooling because there is little time for crystal growth prior to solidification. In some instances, cooling is so rapid that the resulting rock has a glassy texture, as in *obsidian,* basically volcanic glass ( ▌ Fig. 13.8a). **Pyroclastics** (fire fragments) are another category of extrusive igneous rocks that formed by the accumulation of volcanic fragments, dust-sized or larger, that settled out of the air after being ejected during a volcanic eruption ( ▌ Fig. 13.8b).

Many igneous rocks are broken along fractures that may be spaced or arranged in regular geometric patterns. In the Earth sciences, simple fractures or cracks in bedrock are called **joints.** Although joints are common in all rock types, one way they develop in igneous rocks is by a molten mass shrinking in volume and fracturing as it cools and solidifies. Lavas in particular tend to have many fractures. Basalt is famous for its distinctive jointing, which commonly forms hexagonal columns of solidified lava. Devil's Postpile in California and Devil's Tower in Wyoming are well-known landforms that consist of hexagonal columns of lava formed by **columnar joints** ( ▌ Fig. 13.9). Jointing in rocks of all types may also be caused by regional stresses in the crust and can be a major influence in the development of landforms.

▌ **FIGURE 13.8**
(a) Obsidian is volcanic glass—lava that cooled too rapidly to form crystals. (b) Pyroclastic rocks are made of fragments ejected during a volcanic eruption.
**In terms of composition, how are these two volcanic rocks different from other igneous rocks?**

▌ **FIGURE 13.9**
Hexagonal columnar jointing forms basalt columns: (a) Devil's Postpile National Monument, California. (b) Devil's Tower National Monument, Wyoming.
**Why are the cliffs shown in the photographs so steep?**

> **PHYSICAL**
> **Geography Now™** Log on to Physical GeographyNow and select this chapter to work through a Geography Interactive activity on "Igneous Rocks" (click Rocks and the Rock Cycle → Rock Laboratory and select the Igneous tab).

**Sedimentary Rocks** As their name implies, sedimentary rocks are derived from accumulated sediments—unconsolidated mineral materials that have been eroded, transported, and deposited as a result of gradational processes. After the materials have accumulated in horizontal layers, pressure from the material above compacts the sediment, expelling water and reducing pore space. Cementation occurs when silica, calcium carbonate, or iron oxide accumulates and solidifies between particles of sediment. The processes of compaction and cementation transform (lithify) sediments into solid, coherent layers of rock. There are three major categories of sedimentary rocks: clastic, organic, and chemical precipitates.

Most sedimentary materials are fragments of previously existing rocks (cobbles, pebbles, sand, silt, or clay; refer to Appendix C and see again Fig. 12.7) that were deposited in a riverbed, on a beach or sand dune, on a lake bottom, on the ocean floor, or in other environments where deposition occurs. Sedimentary rocks that form from fragments of preexisting rocks are called **clastic** (from Latin: *clastus*, broken).

Clastic sedimentary rocks include *conglomerate, sandstone, siltstone,* and *shale* (■ Fig. 13.10). Conglomerate is a solid mass of cemented, roughly rounded pebbles, cobbles, or boulders and may have sand, silt, or clay filling in spaces between large particles. A somewhat similar rock that has cemented fragments that are angular rather than rounded is called *breccia*. Conglomerate and breccia tend to be hard rocks, commonly much like concrete, and relatively resistant to weathering. Sandstone is formed of cemented sand-sized particles, most commonly grains of quartz. Sandstone is usually granular, porous, and resistant to weathering, but the cementing material influences its strength and hardness. If cemented by substances other than silica (such as calcium carbonate or iron

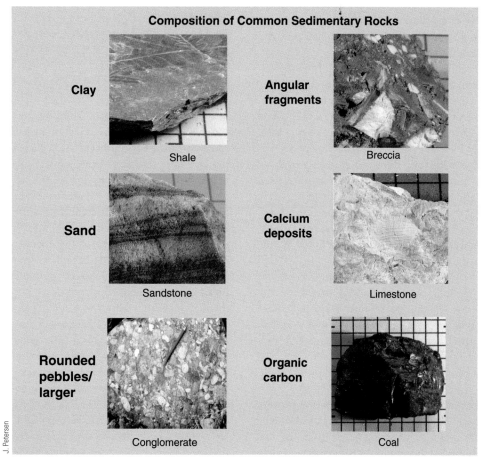

■ **FIGURE 13.10**
Clastic sedimentary rocks are classified by the size and/or shape of the sediment particles they contain.
**Why do the shapes and sizes of sediments in sedimentary rocks differ?**

oxide), sandstone tends to be more easily weathered. Siltstone is similar to sandstone, though much finer textured, being formed of smaller silt-sized particles. Shale is produced from the compaction of very fine-grained sediments, especially clays. Shale is finely bedded, smooth-textured, and impermeable. It is also brittle and easily cracked, broken, or flaked apart.

Sedimentary rocks may be further classified by their origin as either marine or terrestrial (continental). Marine sandstones have typically formed originally in near-shore coastal zones; terrestrial sandstones generally originate in desert or floodplain environments on land. The nature and arrangement of sediments in a sedimentary rock provide a great deal of evidence for the kind of environment under which it was deposited and the processes of deposition, whether in a streambed, on a beach, or on the deep-ocean floor.

**Organic sedimentary rocks** form from the remains of organisms, both plants and animals. *Coal,* for example, was formed by the accumulation and compaction of decayed vegetation in acidic, swampy environments where water-saturated ground prevented oxidation. The initial transformation of such organic material produces peat, which, when subjected to deeper burial and further compaction, is lithified to produce coal. Most of the world's greatest coal deposits originated about 300 million years ago during a geologic time known as the Mississippian and Pennsylvanian Periods, called the Carboniferous Period in Europe.

Other organic sedimentary rocks developed from the remains of organisms in lakes and seas. The remains of shellfish, corals, and microscopic drifting organisms called plankton sank to the bottom of such water bodies where they were cemented and compacted together to form *limestones,* which typically contain fossils such as shells and coral fragments.

At times in Earth history, dissolved minerals such as calcium carbonate (lime) accumulated in the seas until they became saturated and the lime (or other dissolved mineral) was precipitated out to form a deposit on the bottom. Many fine-grained limestones formed as **chemical sedimentary rocks** in this manner, including the chalk seen in the White Cliffs of Dover along the English Channel ( ▌Fig. 13.11). Limestone therefore may vary from a jagged and cemented complex of visible shells or fossil skeletal material to a smooth-textured rock. Where magnesium is an important constituent along with calcium carbonate, the rock is called *dolomite.* Because the calcium carbonate in limestone can slowly dissolve in water, limestone in arid or semiarid climates tends to be resistant, but in humid environments it tends to be weak.

Mineral salts that have reached saturation in evaporating seas or lakes will precipitate out to form a variety of sedimentary deposits that are useful to humans. These include *gypsum* (used in wallboard), *halite* (common salt), and *borates,* which are important in hundreds of products from fertilizer, fiberglass, and pharmaceuticals to detergents and photographic chemicals.

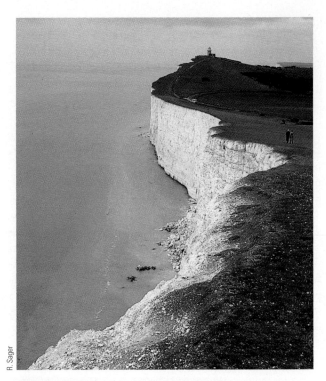

▌**FIGURE 13.11**
The White Cliffs of Dover. These striking, steep white cliffs along the English Channel are made of chalky limestone from the skeletal remains of microscopic marine organisms.
**What special type of limestone is this very white variety?**

Sediments accumulate in layers, and most sedimentary rocks display distinctive layering, or **stratification.** Each layer represents the depositional record of a past environment and the layers, or **strata,** at a site provide evidence of the local geologic history. For example, a layer of sandstone representing an ancient beach may lie directly beneath shale layers that represent an offshore environment, suggesting that first this was a beach and later the sea covered the site. The varieties of sedimentary deposits produce differentiated strata, or beds. The **bedding planes,** or boundaries between sedimentary layers, indicate changes in the depositional environment but no real break in the sequence of deposition ( ▌Fig. 13.12). Where a marked mismatch and an irregular, eroded surface occur between beds, the contact between the rocks is called an **unconformity.** This indicates a time when erosion occurred rather than deposition, and the erosion removed some of the older deposits before sedimentation was renewed sometime later. Within some sedimentary rocks, especially sandstones, thin "microbedding" may occur. A type of microbedding, called **cross-bedding,** is characterized by a pattern of thin layers that accumulated at an angle to the main strata, often reflecting shifts of direction by waves along a coast, currents in streams, or winds over sand dunes ( ▌Fig. 13.13).

Many sedimentary rocks become jointed, or fractured, as they lithify or when they are subjected to crustal stresses.

MINERALS AND ROCKS 369

**FIGURE 13.12**
Bedding planes and an unconformity. Different types of sedimentary rocks may be deposited at various time periods. Each sequence of rocks consists of strata (layers) separated by bedding planes, such as these at Grand Canyon National Park, Arizona.
**Where would the youngest strata be located in this photo?**

**FIGURE 13.14**
Vertical jointing of sandstone strata at Arches National Park, Utah.
**What would cause joints such as the one in the upper strata in this photo?**

of physical landscapes because these structures are weak points in the rock where weathering and erosion can attack. Joints allow water to penetrate deeply into some rock masses, causing them to be removed at a faster rate than the surrounding rock.

> **PHYSICAL Geography Now™** Log on to Physical GeographyNow and select this chapter to work through a Geography Interactive activity on "Sedimentary Rocks" (click Rocks and the Rock Cycle → Rock Laboratory and select the Sedimentary tab).

**Metamorphic Rocks** *Metamorphic* means "changed form." Enormous heat and pressure deep in Earth's crust can form a rock type that is completely different from the original by reconstituting the composition and arrangement of its minerals. Usually, the resulting rock is harder and more compact, has a reoriented crystalline structure, and is more resistant to weathering than it was before being metamorphosed. There are two major types of metamorphic rocks, based on the presence (foliated) or absence (nonfoliated) of platy surfaces or wavy alignments of light and dark minerals that formed during metamorphism.

Metamorphism occurs most commonly where crustal rocks are subjected to great pressures by tectonic processes, or deep burial, or where rising magma generates heat, pressure, solutions, and gases that modify the nearby rock. Metamorphism either wholly or partially fuses the affected rock so that it can be deformed, or flow slightly, without becoming molten. This process causes minerals to recrystallize with an orientation at right angles to the direction of pressure or stress

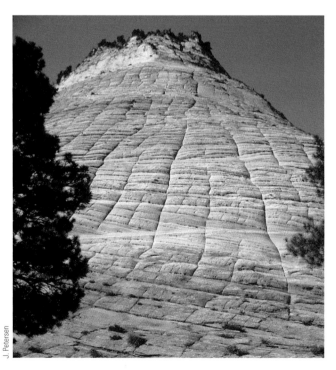

**FIGURE 13.13**
Cross-bedding in sandstone at Zion National Park, Utah.
**This zigzag pattern usually indicates what type of condition during deposition?**

The impressive "fins" of rock at Arches National Park, Utah, owe their vertical, tabular shape to joints in great beds of sandstone (Fig. 13.14). Structures such as bedding planes, cross-bedding, and joints are important in the development

**FIGURE 13.15**
Metamorphic rocks and foliations. Metamorphic foliations develop at right angles to the stress directions. (a) Layered rocks under moderate pressure. (b) Greater pressure can cause metamorphism and the development of platy foliations in rocks. (c) Stronger metamorphism can cause the foliations to widen into wavy bands of light and dark minerals.
**How are foliations different from bedding planes?**

( Fig. 13.15). Such metamorphism produces rocks whose minerals are segregated in wavy bands, or platy surfaces known as **foliations.** Some shales produce a hard metamorphic rock known as *slate,* which exhibits a tendency to break apart or cleave along smooth, flat surfaces that are actually extremely thin foliations ( Fig. 13.16a). Where the foliations are moderately thin, individual minerals have a flattened but wavy, "platy" structure, and the rocks tend to flake apart along these bands. A common metamorphic rock with thin foliations is called *schist*.

Where the foliations develop into broad mineral bands, the rock is extremely hard and is known as *gneiss* (pronounced "nice"). Coarse-grained rocks such as granite generally metamorphose into gneiss, whereas finer-grained rocks may produce schists.

Rocks that originally were composed of one dominant mineral are not foliated by metamorphism ( Fig. 13.16b). Limestone is metamorphosed into much denser *marble,* and impurities in the rock can produce a beautiful variety of colors. Silica-rich sandstones fuse into solid layers of quartz, known as *quartzite*. Quartzite is brittle, harder than steel, and almost inert chemically. Thus, it is virtually immune to chemical weathering and commonly forms cliffs or rugged mountain peaks after the surrounding, less resistant rocks have been removed by erosion. The physical and chemical characteristics of rocks are important factors in the development of landforms.

**PHYSICAL Geography Now™** Log on to Physical GeographyNow and select this chapter to work through a Geography Interactive activity on "Metamorphic Rocks" (click Rocks and the Rock Cycle → Rock Laboratory and select the Metamorphic tab).

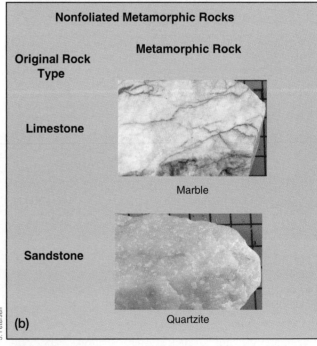

**FIGURE 13.16**
Metamorphic rocks. (a) Slate, schist, and gneiss illustrate differences in metamorphism and the size of foliations. (b) Marble and quartzite are nonfoliated metamorphic rocks with a harder, recrystallized composition of the rock from which they were formed.

# Earth's Planetary Structure

Physical geography has a dominant focus on Earth's external characteristics and systems, yet a broad knowledge of the planet's internal structures and processes is required to understand Earth's surface features and landscapes. Earth has a radius of about 6400 kilometers (4000 mi), but scientists have only been able to penetrate and examine directly the crust. Through direct means such as mining and drilling, we have gained a modest knowledge of Earth's interior. The lure of gold has taken miners to a depth of 3.5 kilometers (2.2 mi) in South Africa, and drilling for oil and gas has penetrated to several times that distance. These explorations have been helpful in providing information about Earth's uppermost layers, but they are really only scratches in the planet's surface. Scientists are continually working to better understand the deep interior of Earth. A more detailed knowledge of the structure, composition, and processes going on within Earth will help answer more questions about crustal motion, earthquakes, volcanic eruptions, the formation of mineral deposits, and the origins of continents and of our planet itself.

Most of what we have learned about Earth's internal structure and composition has been deduced through indirect means. The most important evidence that scientists have used to gain such indirect knowledge is the behavior of earthquake waves and other shock waves (generated by controlled explosions) as they pass through Earth. A sensitive instrument called a **seismograph** ( Fig. 13.17) can record these seismic waves, even when the earthquake is centered thousands of kilometers away or even on the opposite side of the planet.

There are two major types of seismic waves, which travel at different speeds in materials of varying densities and states. These are P (primary) waves, which travel fastest and arrive first at the seismograph recording an earthquake, and S (secondary) waves, which travel more slowly. Repeated seismograph records of P and S waves suggest that these waves are refracted, or bent, as they pass through boundaries of major density change between zones within Earth's interior. The waves speed up in denser material and slow down in material that is less dense. P waves also pass through molten rock, but S waves do not ( Fig. 13.18a). By analyzing worldwide patterns of many earthquake waves for decades, scientists have developed a general model of Earth's interior. Such information, supplemented by studies of Earth's magnetism and gravitational pull, suggests a series of layers, or zones, in Earth's internal structure. These zones, from the innermost to the surface, are known as the core, the mantle, and the crust ( Fig. 13.18b).

## Earth's Core

The **core** forms one third of Earth's mass and has a radius of about 3360 kilometers (2100 mi). Earth's core is under enormous pressure—several million times atmospheric pressure at sea level (more than 100 million pounds per square inch). The core is believed to be composed primarily of iron and nickel in two distinct sections. The **outer core** is 2400 kilometers (1500 mi) thick and is made of iron and nickel. Because the outer core blocks the passage of seismic S waves that will not travel through fluids, Earth scientists assume that the outer core is molten.

The **inner core** of Earth, however, appears to be solid iron. Scientists have an explanation for why it is solid. The melting point of mineral material depends not only on temperature but also on pressure. The pressure on this innermost part of Earth is so great that the inner core remains solid; that is, its melting point has been raised to a temperature above even the high temperatures found there. The outer core, though its temperatures are lower, is under less pressure and can exist in a molten state. Estimates of internal temperatures are 4800°C (8643°F) at the core–mantle boundary with an increase to 6900°C (12,423°F) at the very center of Earth.

Density at the boundary between the core and the mantle is estimated at 10 grams per cubic centimeter, increasing to 13 grams per cubic centimeter at Earth's center. These high core densities contrast with the lower densities of Earth's mantle (3.3–5.5 g/cm$^3$) and crust (2.7–3.0 g/cm$^3$) and give our planet an average density of 5.5 grams per cubic centimeter. The high density of Earth's core is one reason that scientists believe that iron is its primary component; the density and composition of meteorites also support this theory.

**FIGURE 13.17**
Seismographs record earthquake waves for scientific study. This seismograph trace shows multiple earthquakes associated with volcanic eruptions.

**PHYSICAL Geography Now™** Log on to Physical GeographyNow and select this chapter to work through a Geography Interactive activity on "Core Studies" (click Earth's Layers → Core Studies).

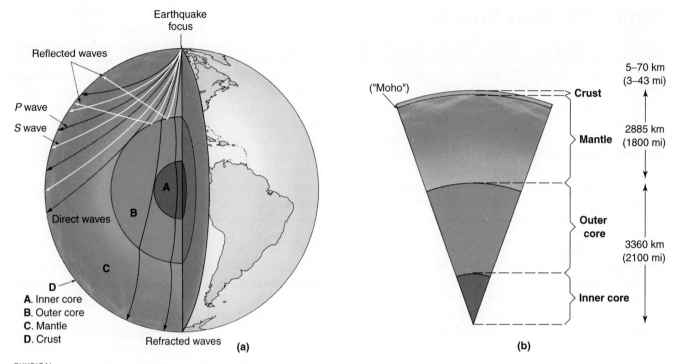

**ACTIVE FIGURE 13.18**
Watch this Active Figure at http://now.brookscole.com/gabler8.
The internal structure of Earth as revealed by seismic waves. (a) The presence of internal changes is revealed by the refraction of P (primary) seismic waves and the inability of S (secondary) waves to pass through Earth's liquid outer core. (b) Cross section through Earth's internal structural zones.
**How does the thickness of the crust compare with the thicknesses of the core and mantle?**

## Earth's Mantle

The **mantle** is about 2885 kilometers (1800 mi) thick and constitutes about two thirds of Earth's mass. Earthquake waves that pass through the mantle indicate that this zone of Earth's interior is composed of dense rocky material that is solid, in contrast to the molten outer core that lies beneath it. Although most of the mantle is solid, part of the upper mantle material has the characteristics of a **plastic solid,** meaning that it can easily deform and can "flow" a few centimeters per year (an inch or two per year). Scientists agree that the mantle consists of dark, heavy silicate rocks that are high in iron and magnesium.

The mantle contains several layers, or zones, of differing strength and rigidity. The uppermost layer of the mantle is solid and together with the crust forms the **lithosphere.** The term *lithosphere* has traditionally been used to describe the entire solid Earth (as discussed in Chapter 1). In recent decades, the term *lithosphere* has been used in an important and more precise way to refer to the solid outer shell of Earth, including the crust and the rigid upper mantle down to the plastic layer ( Fig. 13.19). Beneath the lithosphere at a depth of about 100–700 kilometers (62–435 mi), in the upper mantle, is the **asthenosphere** (from Greek: *asthenias,* without strength), a thick layer of plastic mantle material.

The material in the asthenosphere flows both vertically and horizontally, dragging segments of the overlying, rigid lithosphere along with it. Many Earth scientists now believe that the energy for tectonic forces that break and deform the crust comes from movement within the asthenosphere produced by thermal convection currents originating within the mantle and heated by decaying radioactive materials in the planet's interior.

The interface between the crust and upper mantle is marked by a significant change of density, or a discontinuity, as indicated by an abrupt increase in the velocity of earthquake waves descending through this internal boundary. Scientists have labeled this zone the **Mohorovicic discontinuity,** or **Moho** for short, after the Croatian geophysicist who first detected it in 1909. The Moho does not lie at a constant depth but generally mirrors the surface topography, being deepest under mountain ranges where the crust is thick and rising to within 8 kilometers (5 mi) of the ocean floor (see again Fig. 13.19). As of 2005, no geologic drilling has penetrated to the Moho and into the mantle. However, an international scientific partnership called the Integrated Ocean Drilling Program has come close and plans to drill to that depth in the near future. Rock samples retrieved from the drill cores should add

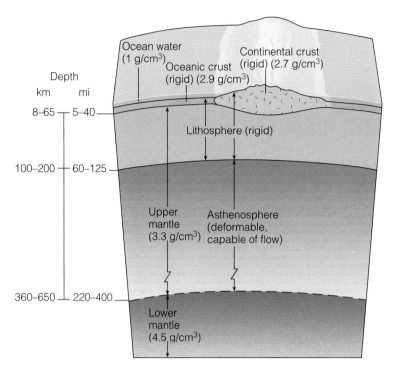

**FIGURE 13.19**
A cross section of the lithosphere, asthenosphere, and lower mantle. The lithosphere is the solid outer part of Earth including the crust, and solid upper mantle, down to the asthenosphere.

greatly to our understanding of Earth's lithospheric composition and structure.

## Earth's Crust

Earth's solid and rocky exterior is the *crust*, which is composed of a great variety of rocks and minerals that respond in diverse ways and at varying rates to Earth-shaping processes. The crust is the only portion of the lithosphere of which Earth scientists have direct knowledge, yet its related surface materials form only about 1% of Earth's planetary mass. Earth's crust forms the exterior of the lithosphere and is of primary importance in understanding surface processes and landforms. Earth's deep interior components, the core and mantle, are of concern to physical geography because they are responsible for and can help explain changes in the lithosphere, particularly the crust that forms the ocean floors and continents.

Earth's crust is less dense than either the core or the mantle. It is also thin in comparison to the size of the planet. Two kinds of Earth crust, oceanic and continental, can be distinguished by their thickness, location, and composition ( Fig. 13.20). Crustal thickness varies from 3 to 5 kilometers (1.9–3 mi) in the ocean basins to as much as 70 kilometers (43 mi) under some continental mountain systems. The average thickness of continental crust is about 32–40 kilometers (20–25 mi). The crust is relatively cold, rigid, and brittle compared to the mantle. It responds to stress by fracturing, wrinkling, and raising or lowering rocks into upwarps and downwarps.

**Oceanic crust** is composed of basaltic rocks, which are solidified, dark-colored, fine-grained, and iron-rich lavas. In comparison to the crust that forms the continents, oceanic crust is thinner, and because it is composed of darker, heavier, mafic rocks, it has a higher density (3.0 g/cm$^3$). Forming the deep ocean floors, its most common minerals are compounds of silica (Si) that are relatively high in iron (Fe) and magnesium (Mg). Basalt is the most common rock on Earth and appears in great lava outflows on all of the continents as well as on the seafloor.

**Continental crust** forms the major landmasses; it is much thicker, less dense (2.7 g/cm$^3$), more sialic, and lighter in color than oceanic crust. Although rocks of every type and of all geologic ages exist on the continents, the average density of continental crust is similar to that of granite, a common coarse-grained intrusive igneous rock. Thus, continental crust is regarded as granitic, in contrast to the basaltic oceanic crust. Granite forms from a molten state deep underground, as opposed to basalt, which cools and solidifies at the surface. Where the surface elevation is high, such as where there are mountains or plateaus, the crust is thicker; at lower elevations, continental crust is thinner, but not as thin as the crust of the seafloor.

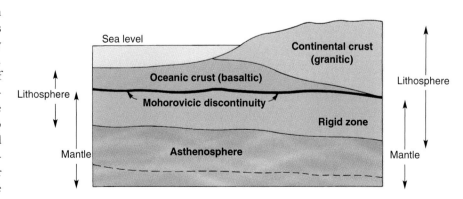

**FIGURE 13.20**
Earth has two distinct types of crust: oceanic and continental. Together with the rigid upper zone of the mantle, these form the lithosphere. Below the lithosphere is the plastic zone in the mantle called the asthenosphere.
**What two zones are separated by the Mohorovicic discontinuity (Moho)?**

# Continents in Motion: The Search for a Unifying Theory

Scientists in all disciplines constantly search for broad explanations that shed light on the detailed facts, recurring patterns, and interrelated processes that they observe and analyze. Is there one broad theory that can help explain how and why Earth's lithospheric processes work? Can it explain such diverse phenomena as the growth of continents, the movement of solid rock beneath Los Angeles, the location of great mountain ranges, differing patterns of temperature in the rocks of the seafloor, and the violent volcanic eruptions on the island of Montserrat in the West Indies? The answer is yes, the concept of the continual movement of landmasses on Earth's surface over millions of years of time.

Sometimes the refinement of scientific understanding may take years of study to develop and test a theory until it is broadly acceptable and more fully understood. As more data and information are gathered and analyzed, new methods and technologies contribute to the process of testing hypotheses via the scientific method, and bit-by-bit an acceptable explanatory framework emerges. Over the past few decades, the theoretical framework of *continental drift* has been refined into a well-established theory called *plate tectonics,* which has been tested by a great deal of evidence in the lithosphere. Plate tectonics has revolutionized the Earth sciences and our understanding of Earth's history.

At one time scientists believed that Earth's landscapes were created in great cataclysms—that the Grand Canyon, for example, split open one violent day and has remained that way ever since or that the Rocky Mountains appeared overnight. This theory, called **catastrophism,** is now rejected. For almost two centuries, geographers, geologists, and other Earth scientists have accepted the theory of **uniformitarianism**—the idea that processes, internal and external to Earth's surface, are operating today in the same manner as they have for millions of years.

Uniformitarianism, however, does not mean that these processes operate at the same rates, either all the time or equally everywhere on Earth. In fact, our planet's surface features are the result of variations in the intensity of internal and external processes influenced by their geographic location. These processes have also varied in intensity and location throughout Earth's history. The accumulation of regular or episodic changes in the Earth system that seem relatively small to us can dramatically change a landscape or our planet's surface after progressing, even on an irregular basis, for millions of years.

## Continental Drift

Most of us have noted on a world map that the Atlantic coasts of South America and Africa look as if they fit together. In fact, if a globe were made into a spherical jigsaw puzzle, several widely separated landmasses could fit alongside each other without large gaps or overlaps ( Fig. 13.21). Is there a scientific explanation?

In the early 1900s, Alfred Wegener, a German climatologist, proposed the theory of **continental drift,** the idea that continents and other landmasses have shifted their positions during Earth history. Wegener's evidence for continental drift included the close fit of continental coastlines on opposite sides of oceans and the trends of mountain ranges on land areas that also match across oceans. He cited similar geographic patterns of fossils and rock types found on distant continents that could not result from chance and did not reflect current climatic conditions. He reasoned that the continents must once have been joined so as to explain the spatial distributions of these ancient features. Wegener also cited evidence of great climate change (for instance, ancient evidence of glaciation where the Sahara Desert is today and tropical fossils found in Antarctica) that could be explained best by large landmasses moving from one climate zone to another.

Wegener hypothesized that all the continents had once been part of a single supercontinent, which he called **Pangaea,** that later divided into two large landmasses, one

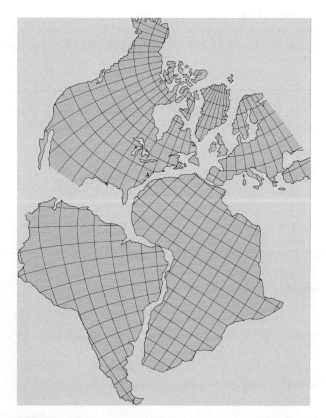

**FIGURE 13.21**
The geographic basis for Wegener's continental drift hypothesis. Note the close correlation of the edges of the continents that face one another across the width of the Atlantic Ocean. The actual fit is even closer if the continental slopes are matched.

**About how long ago is it now assumed that the continents fit together as one landmass?**

in the Southern Hemisphere (Gondwana), and one in the Northern Hemisphere (Laurasia). Later, these two supercontinents also broke apart into sections (the present continents) and drifted to their current positions. Laurasia in the Northern Hemisphere consisted of North America, Europe, and Asia. Gondwana in the Southern Hemisphere was made up of South America, Africa, Australia, Antarctica, and India ( Fig. 13.22). Continued continental movement formed the geographic configuration of the landmasses that exist on Earth today.

The reaction of the scientific community to Wegener's proposal ranged from skepticism to ridicule. A major objection to his hypothesis was that neither he nor anyone else could provide an acceptable explanation for the energy needed to break up huge landmasses and slide them over the rigid crust and across vast oceans.

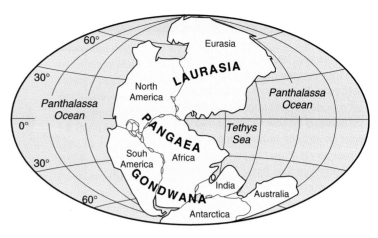

**FIGURE 13.22**
The supercontinent of Pangaea included all of today's major landmasses joined together. Pangaea later split to form Laurasia in the Northern Hemisphere and Gondwana in the Southern Hemisphere. Further plate motion has produced the continents as they are today.

**How has continental movement affected the climates of landmasses?**

## Supporting Evidence for Continental Drift

About a half-century later, in the late 1950s and 1960s, Earth scientists began giving serious consideration to Wegener's notion of moving continents. New information appeared from research in oceanography, geophysics, and other Earth sciences, aided by sonar, radioactive dating of rocks, and improvements in equipment for measuring Earth's magnetism. These scientific efforts discovered much new evidence that indicated the movement of portions of the lithosphere (including the continents).

As one example, scientists were originally unable to explain the unusual orientations of magnetic fields found in basaltic rocks that had cooled millions of years before. Minute iron minerals within these rocks displayed magnetic fields that were oriented in directions that commonly pointed away from today's magnetic north pole and sometimes in a completely opposite direction, toward the south magnetic pole. Known as **paleomagnetism**, the orientation of this magnetism is locked into rocks as they solidify, recording the direction to magnetic north at the time of cooling.

Rocks of different ages and on different continents show magnetic orientations at a variety of angles to Earth's magnetic field as it is today. At first it was assumed that Earth's magnetic poles had wandered, which they do but not enough to explain this much variation, or the varying paleomagnetism in rocks of the same age but at different locations. It was also discovered that many times in the geologic past the polarity of Earth's magnetic field had reversed.

A detailed model of the continents was projected backward in time so that the magnetic orientations of the rocks coincided with Earth's magnetic field during past periods of geologic history ( Fig. 13.23). Paleomagnetism indicated an almost perfect fit of the continental jigsaw puzzle

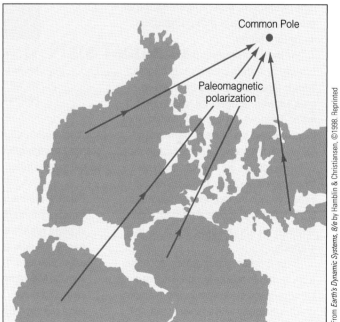

**FIGURE 13.23**
Paleomagnetic properties of rocks that formed when the Northern Hemisphere continents were joined point to the location of magnetic north at that time. It requires rejoining the continents to their original positions, as shown on this map, in order for the magnetic orientations to point to a common magnetic pole. From Garrison, *Oceanography*, 5th ed.

**How do these orientations point today with the continents in their present positions?**

about 200 million years ago when the two supercontinents began to split apart to form the beginnings of the modern Atlantic Ocean.

**FIGURE 13.25**
Earth's solid exterior is broken into giant segments called plates. This map shows Earth's major tectonic plates, large sections of the lithosphere, and their general directions of movement. Most tectonic and volcanic activity occurs along plate boundaries where plates separate, collide, or slide past each other.
**Do the edges of continents and plate boundaries correlate closely? Do all the plates have a continent located on them?**

## Seafloor Spreading and Convection Currents

In the 1960s, several keys to plate tectonics theory were found while studying and mapping the ocean floors. First, detailed undersea mapping was conducted on a system of midocean ridges (also called oceanic ridges or rises) with configurations that are remarkably similar to the continental coastlines ( Fig. 13.26). Second, it was discovered that parallel bands of basaltic seafloor displayed matching patterns of magnetic properties in rocks of the same age but on opposite sides of midocean ridges in the Atlantic and Pacific Oceans. Third, scientists made the surprising discovery that although some continental rocks are 3.6 billion years old, rocks on the ocean floor are all geologically young—they have been in existence less than 250 million years. Fourth, the oldest rocks of the seafloor are under the deepest ocean waters or nearest to the continents, and rock ages become progressively younger toward the midocean ridges, where the most recent basaltic rocks are found. Finally, temperatures of rocks on the ocean floor vary significantly, being hottest near the ridges and becoming progressively cooler farther away.

Only one logical explanation emerged to fit all of the new evidence. It became apparent that new oceanic crust is being formed at the midocean ridges while older oceanic crust is being destroyed along certain margins of ocean basins. The emergence of this new oceanic crust is associated with the movement of great sections or plates of lithosphere away from the midocean ridges. This phenomenon, a major advance in our understanding of how continents move, is called **seafloor spreading** ( Fig. 13.27). The lithospheric plates move laterally and separate at an average rate of 2–5 centimeters (1–2 in.) per year above the flowing plastic asthenosphere in the mantle. The young age of oceanic crust results from the creation of new basaltic rock at undersea ridges and the movement of the seafloor with lithospheric plates toward remelting and destruction of older rock at the margins of ocean basins. As molten basalt cooled and crystallized in the seafloor, the iron minerals that they contained became magnetized in a manner that replicated the orientation of Earth's magnetic field at that time. Like a tape recorder, the iron-rich basalts of the seafloor preserved a historical record of the Earth's magnetic field, including **polarity reversals** (times when the positive pole became negative, and vice versa).

**FIGURE 13.26**
The oceanic ridge system and the age of the seafloor. The red areas are the youngest, and blue represents the oldest seafloor. Detailed mapping and study of the ocean floors yielded much evidence to support plate tectonics, through the process of seafloor spreading.

**FIGURE 13.27**
Seafloor spreading at an oceanic ridge creates new seafloor.

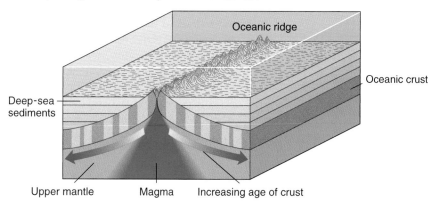

Plate tectonics includes a plausible explanation of the mechanism for the movement of continents that had eluded Wegener. The mechanism is **convection** —the transfer of hot mantle material upward toward Earth's surface and cooler material downward as part of huge subcrustal convection cells (Fig. 13.28). Mantle material flowing in the asthenosphere rises to the solid base of the lithosphere and spreads laterally when it hits that boundary. Mantle material flowing in opposite directions drags the lithospheric plates with it, breaking the brittle midocean ridge open. Molten basalt wells up into the fractures, cooling and sealing them to form new seafloor. In this process, the ocean becomes wider by the width of the now sealed fracture. The convective motion continues as solidified crustal material moves away from the ridges. Apparently, in a time frame of about 250 million years, older oceanic crust is consumed in the deep trenches near plate boundaries where some sections of the lithosphere meet and are recycled into Earth's interior.

## Tectonic Plate Movement

The shifting of tectonic plates relative to one another provides an explanation for many of Earth's surface features. Plate tectonics theory enables physical geographers to better understand not only our planet's ancient geography but also quakes, volcanic activity, zones of crustal movement, and major landform features (Fig. 13.29). Let's briefly examine the three ways in which lithospheric plates relate to one another along their boundaries as a result of tectonic movement, by pulling apart, pushing together, or sliding alongside each other.

**Plate Divergence** The pulling apart of plates, tectonic **plate divergence,** is directly related to seafloor spreading (see again Fig. 13.27). Tension-producing forces cause the crust to thin and weaken. Shallow earthquakes are often associated with crustal stretching, and asthenospheric magma wells up between crustal fractures, forming new crustal ridges and ocean floor as plates move away from each other. The formation of new crust in these spreading centers gives the label *constructive plate margins* to these zones. Occasional "oceanic" volcanoes like those of Iceland, the Azores, and Tristan da Cunha mark such boundaries (Fig. 13.30).

Though most plate divergence occurs along oceanic ridges, this process can also break continents apart, forming smaller landmasses (Fig. 13.31a). The Atlantic Ocean floor formed as the continent that included South America and Africa broke up and moved apart 2–4 centimeters (1–2 in.) per year over millions of years. The Atlantic Ocean continues to grow today at about the same rate. The best modern example of divergence on a continent is the rift valley system of East Africa, stretching from the Red Sea to Lake Malawi. Crustal blocks that have moved downward with respect to the land on either side, with lakes occupying many of the depressions, characterize the entire system, including the Sinai Peninsula and the Dead Sea. Measurable widening of the Red Sea suggests that it may be the beginning of a future ocean that is forming between Africa and the Arabian Peninsula, similar to the young Atlantic between Africa and South America about 200 million years ago (Fig. 13.31b).

**Plate Convergence** A wide variety of crustal activity occurs at areas of tectonic **plate convergence.** Despite the relatively slow rates of plate movement (in terms of human perception), the incredible energy involved in convergence causes the crust to crumple as one plate overrides another. The denser plate is forced deep below the surface in a process called **subduction.** Subduction is most common where dense oceanic crust collides with and descends beneath lighter continental crust

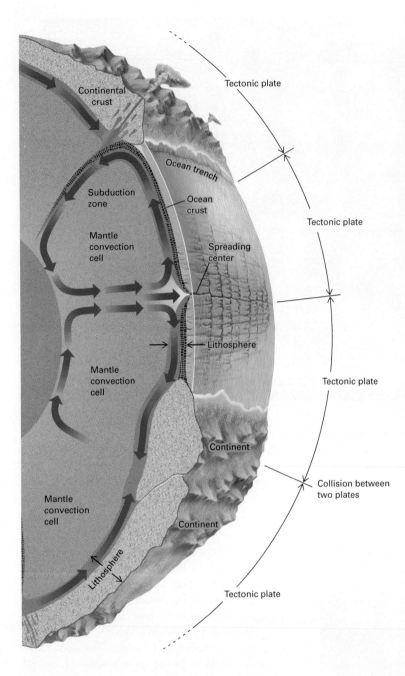

**PHYSICAL Geography Now™ ACTIVE FIGURE 13.28**
Watch this Active Figure at http://now.brookscole.com/gabler8.
Convection: the mechanism for plate motion. Convection currents in the mantle rise to the base of the solid lithosphere and begin to flow laterally, causing the plates to move because of friction with the flowing material in the asthenosphere. The continents ride like passengers on the shifting plates.

the modern global distributions and spatial relationships among such diverse but often related phenomena as earth-

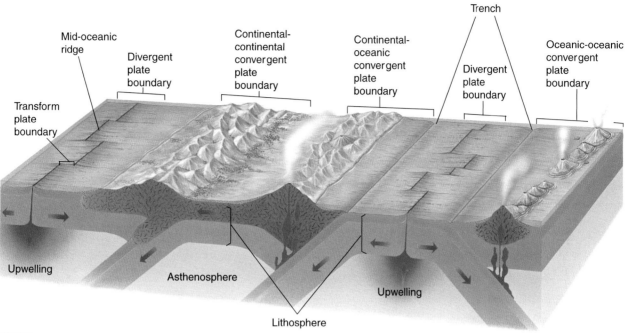

### PHYSICAL Geography Now™ ▮ ACTIVE FIGURE 13.29

Watch this Active Figure at http://now.brookscole.com/gabler8.

**Environmental Systems: Plate Tectonic Movement** Unlike the other major Earth systems, the plate tectonics system does not derive its energy from the sun. Instead, movements of the lithosphere are a result of heat energy derived from Earth's interior. As lithospheric plates move, they interact with adjoining plates, forming different boundary types, each having distinct landform features. This diagram shows three major plate boundary types: spreading centers, subduction zones, and collision zones. Spreading centers, at divergent plate boundaries, are shown on the left, and to the right of the center of the diagram are constructive boundaries as they form new crustal material at the surface along active rift zones. Over time, newer material pushes older rock progressively away from the active rift zone in both directions. Earth's oceanic divergent plate boundaries form the midocean ridge system, which runs through all the major oceans.

Subduction zones (right side of the diagram) occur where two plates are converging. This is a destructive boundary where crustal material returns to Earth's interior. The denser oceanic plate is forced by gravity and plate movement to subduct beneath the thicker but less dense continental plate. Surface features common to subduction zones are deep ocean trenches and volcanic mountain ranges or island arcs. The best examples of subduction are found around the Pacific Ring of Fire, such as those in Japan, Chile, New Zealand, and the northwest coast of the United States.

Continental collision zones (on the left), are found where two continental plates collide. Massive mountain building occurs as the crust thickens because of compression, and volcanoes tend to be absent in these regions. The world's highest mountains, the Himalayas, were formed when the Indian plate collided with Eurasia. The Alps were formed in a similar manner in a collision between the African and Eurasian plates.

( ▮ Fig. 13.32). This is the situation along South America's Pacific coast, where the Nazca plate subducts beneath the South American plate, and in Japan, where the Pacific plate dips under the Eurasian plate. As oceanic crust, and the lithospheric plate of which it forms a part, is subducted, it descends into the asthenosphere to be melted and recycled into Earth's interior.

Deep ocean trenches, such as the Peru–Chile trench and the Japanese trench, form where the crust is dragged downward into the mantle. Frequently, hundreds of meters of sediments that are deposited on the seafloor or continental margins are carried into these trenches to later form sedimentary rocks. At convergent boundaries such as subduction zones, rocks can be squeezed and contorted between colliding plates, becoming uplifted and greatly deformed or metamorphosed. These processes at convergent plate margins have produced many great mountain ranges, such as the Andes. A subducting plate is heated as it plunges downward into the mantle. Its rocks are

**FIGURE 13.30**
Iceland is part of the Mid-Atlantic Ridge that stands above sea level to form a volcanic island.
**Why would Iceland experience frequent earthquakes?**

**ACTIVE FIGURE 13.31**
Watch this Active Figure at http://now.brookscole.com/gabler8.
(a) A continental divergent plate boundary breaks continents into smaller landmasses. (b) The Sinai Peninsula, looking along the Mediterranean coast of Egypt, illustrates the breakup of a continental landmass. North is to the right corner in this image. To the east of the Sinai is the spreading center that forms the Gulf of Aqaba. To the west is the Red Sea rift, also a zone of spreading. The irrigated valley of the Nile River is visible as it flows northward across the desert into the Mediterranean Sea.
**What two major landmasses are separated by the Red Sea?**

melted, and the resulting hot magmas migrate upward into the overriding plate. Where molten rock reaches the surface, it forms a series of volcanic peaks in mountain regions, as in the Cascades Range of the northwestern United States. Where two oceanic plates meet, the older, denser one will subduct below the younger, less dense oceanic plate, and volcanoes may develop, forming major **island arcs** on the overriding plate between the continents and the ocean trenches. The Aleutians, the Kuriles, and the Marianas are all examples of island arcs near oceanic trenches that border the Pacific plate.

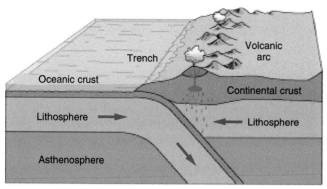

Oceanic-continental convergence

**PHYSICAL Geography Now™** ■ ACTIVE **FIGURE 13.32**

Watch this Active Figure at http://now.brookscole.com/gabler8.
An oceanic–continental convergent plate boundary where continent and seafloor collide. The west coast of South America is an excellent example of this kind of plate margin. Collision has contributed to the development of the Andes and a deep ocean trench offshore.

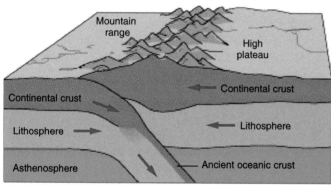

Continental-continental convergence

**PHYSICAL Geography Now™** ■ ACTIVE **FIGURE 13.33**

Watch this Active Figure at http://now.brookscole.com/gabler8.
Continental collision along a convergent plate boundary fuses two landmasses together. The Himalayas, the world's highest mountains, were formed when India drifted northward to collide with Asia.

As the subducting plate grinds downward, enormous friction is produced, which explains the occurrence of major earthquakes in these regions. Subduction zones are sometimes referred to as Benioff zones, after the seismologist Hugo Benioff, who first plotted the existence of earthquakes extending downward at a steep angle on the leading edge of a subducting plate (see again Fig. 13.29).

**Continental collision** causes two continents or landmasses to fuse or join together, creating a new larger landmass ( ■ Fig. 13.33). This process, which closes an ocean basin that once separated the colliding landmasses, has been called *continental suturing* (suturing is a medical term for closing a wound). Where two continental masses collide, the result is massive folding and crustal block movement rather than volcanic activity. This crustal thickening generally produces major mountain ranges at sites of continental collision. The Himalayas, the Tibetan Plateau, and other high Eurasian ranges formed in this way as the plate containing the Indian subcontinent collided with Eurasia some 40 million years ago. India is still pushing into Asia today to produce the highest mountains in the world. The Alps were formed as the African plate was thrust against the Eurasian plate.

Zones where plates are converging mark locations of major and more tectonically active landforms on our planet: huge mountain ranges, chains of volcanoes, and deep ocean trenches. The distinctive spatial arrangement of these features worldwide can best be understood within the framework of plate tectonics.

**Transform Movement** Lateral sliding along plate boundaries, called **transform movement,** occurs where plates neither pull apart nor converge but instead slide alongside each other as they move in opposite directions. Such a boundary exists along the San Andreas Fault zone in California ( ■ Fig. 13.34). Mexico's Baja peninsula and southern California are west of the fault on the Pacific plate. San Francisco and other parts of California east of the fault zone are on the North American plate. In the fault zone, the Pacific plate is moving laterally northwestward in relation to the North American plate at a rate of about 8 centimeters (3 in.) a year (80 km or about 50 mi per million years). If this movement continues at this rate, Los Angeles will move alongside San Francisco (450 mi northwest) in about 10 million years and eventually pass that city on its way to finally colliding with the Aleutian Islands at a subduction zone.

Another type of lateral plate movement occurs on ocean floors in areas of plate divergence. As plates pull apart, they usually do so along a series of fracture zones that tend to form at right angles to the major zone of plate contact. These crosshatched plate boundaries along which lateral movement takes place are called *transform faults.* Transform faults, or fracture zones, are common along midocean ridges, but examples can also be seen elsewhere, as on the seafloor offshore from the Pacific Northwest coast between the Pacific and Juan de Fuca plates (see again Fig. 13.34). Transform faults are caused as adjacent plates move at variable rates, causing lateral movement of one plate relative to the other. The most rapid plate motion is on the East Pacific rise where the rate of movement is more than 17 centimeters (5 in.) per year.

**PHYSICAL Geography Now™** **Log on to Physical GeographyNow and select this chapter to work through a Geography Interactive activity on "Plate Boundaries" (click Plate Tectonics → Plate Boundaries).**

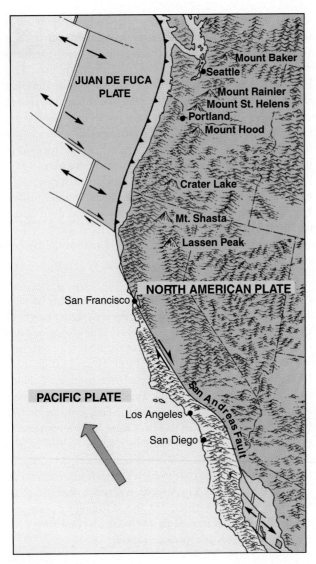

**FIGURE 13.34**
A lateral plate boundary. The Pacific plate moves laterally along the North American plate in western North America forming the San Andreas Fault system. The area west of the fault is moving northwestward in relation to the plate on the opposite side.
**Note that north of San Francisco the boundary type changes. What type of boundary is located there, and what type of surface features indicates this change?**

## Hot Spots in the Mantle

The Hawaiian Islands, like many major landform features, owe their existence to processes associated with plate tectonics. As the Pacific plate moves toward the northwest, it passes over a mass of molten rock in the mantle that does not move with the lithospheric plate. Called **hot spots,** these stationary molten masses also occur in a few other places in both continental and oceanic locations. Melting in the upper mantle and the oceanic crust causes undersea eruptions and the outpouring of basaltic lava on the seafloor, eventually building a volcanic island. This process is responsible for building the Hawaiian Islands, as well as the chain of islands and undersea volcanoes that extend for thousands of miles northwest of Hawaii. Today the hot spot causes active volcanic eruptions on the "big island" of Hawaii. The other islands in the Hawaiian chain also have a similar origin, having formed over the hot spot as well, but these volcanoes have now drifted along with the Pacific plate away from their magmatic source. Evidence of the plate motion is indicated by the fact that the youngest islands of the Hawaiian chain, Hawaii and Maui, are to the southeast, and the older islands, such as Kauai and Oahu, are located to the northwest ( ❙ Fig. 13.35). A newly forming undersea volcano, named Loihi, is now developing and will someday be the next member of the Hawaiian chain, located to the southeast of the island of Hawaii.

## Growth of Continents

The origin of continents remains a mystery. It is clear that the continents tend to have a core area of very old igneous and metamorphic rocks that may represent the deeply eroded roots of ancient mountains. These core regions have been worn down by hundreds of millions of years of erosion to form areas of relatively low relief that are located far from active plate boundaries and have a history of tectonic stability over an immense period of time. These ancient crystalline rock areas are called **continental shields** ( ❙ Fig. 13.36). The Canadian, Scandinavian, and Siberian shields are outstanding examples. Around the peripheries of the exposed shields, flat-lying, younger sedimentary rocks indicate the presence of a stable and rigid mass beneath, as in the American Midwest, western Siberia, and much of Africa.

Most continents appear to grow outward by mountain building and uplift around the sediment-covered margins of ancient continental shields. This process is clearly related to plate tectonics because, as the edge of a lithospheric plate descends in an ocean trench along a continent, new molten material is generated to form deep intrusive masses and volcanoes on the continental edge. Simultaneously, the continental edges tend to be buckled and uplifted by plate convergence and associated subduction processes. Oceanic rocks and marine sediments peel off and pile against continental margins to form ranges such as the Coast Ranges of California and Oregon and the Olympic Mountains of Washington. Except where two continents have collided to become one, mountain building is generally restricted to plate boundaries where igneous intrusions, surface volcanism, and the squeezing upward of marine sediments add new crustal material to the continental mass.

Earth scientists have also suggested that continents grow by **accretion**—that is, by adding numerous chunks of crust to the main continent by collision. North America may have grown in this manner over the past 200 million years by adding segments of crust, known as **microplate terranes**

# Geography's Physical Science Perspective
## Isostasy: Balancing the Earth's Lithosphere

Plate tectonics explains that the continents are passengers on lithospheric plates that act like rafts, gliding on the asthenosphere. The solid upper mantle, oceanic crust, and continental crust are suspended and buoyant above the flowing asthenosphere. The mantle material in the asthenosphere flows at about 2–5 centimeters (1–2 in.) per year, like a very thick fluid. Gravity does not cause the lithosphere to sink because Earth's solid exterior is floating on the denser asthenosphere. The principle of buoyancy tells us that an object will sink if its total density is higher than the fluid. If it floats, the proportion floating above the surface equals the percentage of density difference between it and the fluid. The volume of fluid displaced by a floating object will weigh the same as the volume of the floater that is submerged. No matter how much cargo a ship has on board, this balance (equilibrium) will be maintained, unless it becomes heavier than the water it displaces and sinks. It will float higher when empty and lower when full of cargo.

*Isostasy* is the term for the equalization of hydrostatic pressure (fluid balance) that affects Earth's lithosphere and in turn its topography. One concept of isostasy suggests that the lithosphere exists in equilibrium in terms of the density of materials that exist above the asthenosphere. What this means is that a column of lithosphere (and the overlying hydrosphere) anywhere on Earth weighs about the same as a column of equal diameter from anywhere else regardless of thickness. Thicker lithosphere will contain a higher percentage of lighter materials, compared to areas where the lithosphere is thinner. Continental crust is lighter than oceanic crust, which is why oceanic crust is subducted along ocean trenches.

If an additional load is placed in an area by a massive accumulation of sediments, lake waters, or glacial ice, the crust will subside to a new equilibrium level. If these materials are later removed, the region will tend to rise in a process called *isostatic rebound*. But neither uplift or subsidence of the crust will be instantaneous because flow in the asthenosphere is only a few centimeters per year. Imagine a waterbed filled with molasses. If you lie on it, you will sink slowly, because molasses is thicker than water, until you reach a floating equilibrium. When you get out of this bed, the depression that you made will slowly rise back up as the molasses fills in the space from below.

Isostasy also suggests that mountains are made of relatively light crustal materials but exist in areas of very thick crust and low elevation regions have thin crust. Here the analogy is like that of an iceberg: A tall iceberg requires a massive amount of ice below the surface in order to expose ice so high above sea level, and as ice above the surface melts, ice from below will rise above sea level to replace it until the iceberg has completely melted.

Isostatic balance helps to explain many aspects of Earth's surface, including the following:

- Why most of the continents' crust exists above sea level
- Why wide areas of the seafloor are at the same depth
- Why certain mountains continue to rise while erosion removes their tops
- Why some regions where rivers are depositing great amounts of sediments are subsiding
- Why the crust in areas that were covered by 2–3000 meters of glacial ice during the Pleistocene subsided and now continues to rebound after deglaciation

The density of ice is 90% of water; thus, icebergs (and ice cubes) float with nine tenths of their volume below the surface and 10% above.

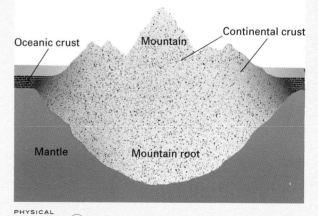

Watch this Active Figure at http://now.brookscole.com/gabler8.

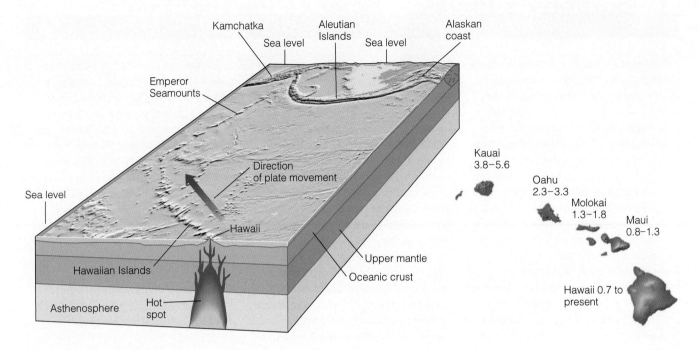

**ACTIVE FIGURE 13.35**

Watch this Active Figure at http://now.brookscole.com/gabler8.
Mantle hot spots and the Hawaiian Islands. A stationary zone of molten material in the mantle has caused volcanoes to form at the same location in the Pacific Ocean for millions of years. As the Pacific plate has drifted, each of the Hawaiian Islands has moved with it to the northwest, away from the active volcanic zone. Away from the hot spot (today on the island of Hawaii), the islands are progressively older to the northwest.

**It is about 300 kilometers from the big island of Hawaii to Honolulu (Oahu). How long did it take the Pacific plate to move Oahu to its current position?**

(a term that should not be confused with the term *terrain*), as it moved westward over the Pacific and former oceanic plates. Though their exact origin is still speculative, parts of western North America from Alaska to California may have originated south of the equator. Terranes, which have their own distinct geology from that of the continent to which they are now joined, may have originally been offshore island arcs, undersea volcanoes, or islands made of continental fragments, such as New Zealand or Madagascar.

## Paleogeography

The study of past geographic environments is known as **paleogeography**. The goal of paleogeography is to try to reconstruct the past environment of a geographic region based on geologic and climatic evidence. For students of physical geography, it generally seems that the present is complex enough without trying to know what the geography of ancient times was like. However, peering into the past helps us forecast and prepare for changes in the future.

The immensity of geologic time over which major events or processes (such as plate tectonics, ice ages, or the formation and erosion of mountain ranges) have taken place is difficult to picture in a human time frame of days, months, and years. The geologic timescale is a calendar of Earth history (Table 13.2). It is divided into *Eras*, the most important and typically longest units of time, such as the Mesozoic Era (which means "middle life"), and Eras are divided into *Periods*, such as the Cretaceous Period. *Epochs*, such as the Pleistocene Epoch (recent ice ages), are shorter time units and are used to subdivide the Periods of the Cenozoic Era ("recent life"), for which geologic evidence is more abundant. Today we are in the Holocene Epoch (last 10,000 years), of the Quaternary Period (last 1.6 million years), of the Cenozoic Era (last 64 million years). In a sense, these divisions are used like we would use days, months, and years to record time.

If we took a 24-hour day to represent the approximately 4.6 billion-year history of Earth, the Precambrian would consume the first 21 hours, an era of which we know very little. The current Period, the Quaternary, which has lasted about 1.6 million years, would take less than 35 seconds, and human beginnings, over about the last 4 million years, about 1 minute.

### FIGURE 13.36
Map of North America showing the continental shield and the general ages of rocks.
**Going outward from the shield toward the coast, what generally happens to the ages of rocks? What does this suggest about the size of the continent during the time span of the rocks along the continental margins?**

If we look at evidence for the paleogeography of the Mesozoic Era (245 million to 65 million years ago), for instance, we would find a much different physical geography than exists today. This was a time when the supercontinents, Gondwana and Laurasia, each gradually split apart as new ocean floors widened, creating the continents that are familiar to us today.

Global and local Mesozoic climates were very different from those of today but were changing as North America drifted northwest. During the Cretaceous Period, much of the present United States experienced warmer climates than today. Fern and conifer forests were common. The Mesozoic was the "age of the dinosaurs," large reptiles that ruled the land and the sea. Other life also thrived, including marine

**TABLE 13.2**
Geologic Timescale

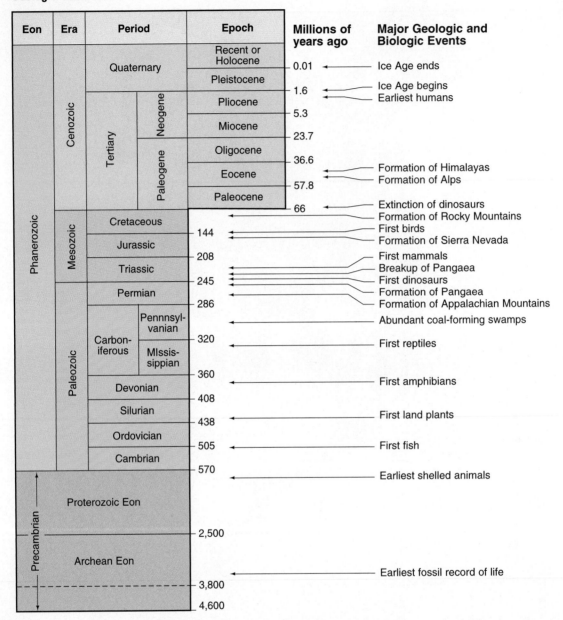

plants and invertebrates, insects, and the earliest birds and mammals.

The Mesozoic Era ended with an episode of great extinctions, including the end of the dinosaurs. Geologists, paleontologists, and paleogeographers are not in agreement as to what caused these great extinctions. Some of the strongest evidence points to a large meteorite striking Earth 65 million years ago, disrupting global climate and causing global environmental change.

Other evidence points to plate tectonic changes in the distribution of oceans and continents or increased volcanic activity, either of which could cause rapid climate changes that might possibly trigger mass extinctions. Each Era and Period in Earth's geologic history had a unique paleogeography with its own distribution of land and sea, climate regions, plants, and animal life. Early humans lived under different environmental conditions than we do today.

Our maps of Earth in early geologic times show only approximate and generalized patterns of mountains, plains, coasts, and oceans, with the addition of some environmental characteristics. These maps portray a general picture of how global geography has changed through geologic time

(Fig. 13.37). Much of the evidence and the rocks that bear this information have been lost through metamorphism or erosion, buried under younger sediments or lava flows, or recycled into the Earth's interior. The further back in time and the older the time period, the sketchier is the paleoenvironmental information presented on the map. Paleomaps, like other maps, are simplified models of the regions and times they represent.

As time passes and additional evidence is collected, paleogeographers may be able to fill in more of the empty spaces on those maps of the past that are so unfamiliar to us.

These paleogeographic studies aim not only at understanding the past but also at understanding today's environments and physical landscapes, how they have developed, and how processes act to change them today. We can also study the far more abundant evidence of Earth processes that are ongoing today or that operated in the recent geologic past. By applying the concept of uniformitarianism and the theory of plate tectonics to our knowledge of how the Earth system and its subsystems function, we can gain a better understanding of our planet's geologic past, as well as its present, and this will facilitate better forecasts of its potential future.

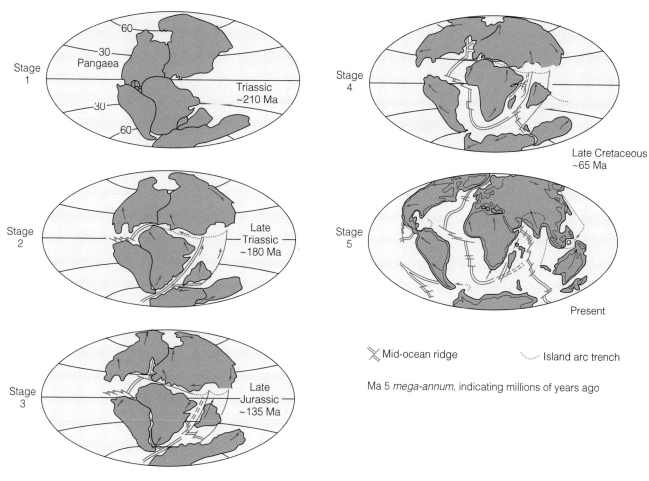

**PHYSICAL Geography Now™ ■ ACTIVE FIGURE 13.37**

Watch this Active Figure at http://now.brookscole.com/gabler8.

Paleomaps showing Earth's tectonic history over the last 250 million years of geologic time. A preponderance of evidence from paleomagnetism, ages and distributions of rocks and fossils, patterns of earthquakes and volcanoes, configurations of landmasses and mountain ranges, and studies of the ocean floors supports plate tectonics. These lines of evidence make it possible to produce a generalized historic sequence of how Earth's global geography has changed over that time frame.

**How has the environment at the location where you live changed through geologic time?**

## Define & Recall

| | | |
|---|---|---|
| crust | organic sedimentary rock | uniformitarianism |
| mineral | chemical sedimentary rock | continental drift |
| rock | stratification | Pangaea |
| silicate | strata | paleomagnetism |
| magma | bedding plane | plate tectonics |
| bedrock | unconformity | lithospheric plate |
| regolith | cross-bedding | seafloor spreading |
| outcrop | foliation | polarity reversal |
| rock cycle | seismograph | convection |
| igneous rock | core | plate divergence |
| sedimentary rock | outer core | plate convergence |
| metamorphic rock | inner core | subduction |
| lava | mantle | island arc |
| extrusive igneous rock | plastic solid | continental collision |
| intrusive igneous rock | lithosphere (crust and upper mantle) | transform movement |
| plutonic rock | asthenosphere | hot spot |
| pyroclastic | Mohorovicic discontinuity (Moho) | continental shield |
| joint | oceanic crust | accretion |
| columnar joint | continental crust | microplate terrane |
| clastic | catastrophism | paleogeography |

## Discuss & Review

1. List the most common elements in Earth's crust. What is a mineral? What is a rock?
2. Describe the three major classifications of rock and the means by which they are formed. Give an example of each.
3. Describe the rock cycle. What is magma and how is it related to the rock cycle?
4. Identify the major zones of Earth from the center to the surface. How do these zones differ from one another? What is the special significance of the asthenosphere?
5. Define the differences between continental crust, oceanic crust, lithosphere, and asthenosphere.
6. What evidence did Wegener rely on in the formulation of his theory of continental drift? What evidence did he lack? What evidence has since been found to support his theory?
7. What type of plate boundary is found near the Andes, along the San Andreas Fault, in Iceland, and near the Himalayas?
8. How does the formation of the Hawaiian Islands support plate tectonics theory?
9. Define paleogeography. Why are geographers interested in this topic?

## Consider & Respond

1. Explain why the eastern United States has relatively little tectonic activity in comparison to the western United States.
2. How is the concept of uniformitarianism related to plate tectonics theory and the arrangement of the continents today? How is it related to the geographic distribution of tectonically active regions?
3. List four of the major plate boundary types. Name a country that is located along each one of these plate boundary types.

# CHAPTER 14

This landscape in Iceland was created by a combination of faulting and volcanism. © Yann Arthus-Bertrand/CORBIS

# Volcanoes, Earthquakes, and Tectonic Landforms

**Physical Geography Now™** This icon, appearing throughout the book, indicates an opportunity to explore interactive tutorials, animations, or practice problems at now.brookscole.com/gabler8

## CHAPTER PREVIEW

**Earth's topography results from the interaction of internal and external processes that act to either increase or decrease elevation differences on the land surface.**

- Why are certain areas high in elevation and others low?
- What two major internal processes act to uplift the land surface?
- How are external processes involved in this leveling or building up of the land?

**Many natural processes that build and shape Earth's surface operate in a manner that is not predictable, steady, and continuous but rather unpredictable, variable, and episodic.**

- Can humans expect to control these natural forces?
- How should humans deal with these forces and the landforms that they create?

**Intense natural processes such as volcanic eruptions and earthquakes can lead to great human suffering and economic loss, but each occurrence has a direct cause and a rational explanation well understood by Earth scientists.**

- How is this linked to disaster prediction?
- Why do humans often avoid clear warning signs?

**Tectonic uplift and volcanism can produce mountains; in doing so, these processes also affect other parts of the Earth system.**

- What other environmental changes in the Earth system may occur because of mountain building?
- How is this important to human beings?

**The locations of tectonic plate boundaries have a direct correlation with the spatial distribution of some of Earth's great "hazard zones."**

- In what ways are volcanism and faulting involved in this relationship?
- Why do humans persist in living in or near areas subject to volcanic and earthquake activity?

Our planet's surface is both intriguing and complex. Its landforms and topography—mountains, valleys, plateaus, plains, canyons, coastlines, and ocean basins—strongly influence the beauty and diversity of Earth's environments. National, state, and local parks attract millions of visitors annually who seek to observe and experience firsthand spectacular examples of terrain and associated environmental features. Landforms are the surface expression of the lithosphere and owe their development to processes and materials that originate within Earth's interior, at its surface, or both.

Understanding landforms and landscapes—where they exist, how they formed, and their significance in a global, regional, or local context—are some of the scientific goals in **geomorphology,** a major subfield of physical geography. Geomorphologists seek explanations for the shape, origin, spatial distribution, and development of terrain features of all kinds, as well as the processes that create, modify, or destroy landforms. Variations in landforms result from interactions among the processes that elevate or disrupt Earth's surface, creating topographic inequalities, and the processes that wear down, fill in, and tend to level the landscape. The mechanisms that uplift portions of Earth's surface gain their energy from Earth's interior and are mainly either **igneous processes** (from Latin: *ignis,* fire) or **tectonic processes** (from Greek: *tekton,* carpenter, builder). **Gradation processes** are the many surficial processes that act to wear away high places and fill in low areas, mainly through erosion and deposition.

The geographic distributions of landform features are not random. Volcanoes and associated igneous landforms are most common along plate boundaries, as are major earthquake zones and the largest mountain ranges. Although tectonic and volcanic forces originate from Earth's interior, they are major controls of topography—affecting the appearance, shape, height, and size of features on the surface. These forces can build extensive mountain systems, but they also produce a great variety of other landforms. Areas of the crust can be uplifted by internal processes or built up by materials that are ejected from Earth's interior onto the surface.

Volcanic eruptions and earthquakes have produced many scenic landscapes, but they can also present serious natural hazards to humans and their property. This chapter and those that follow focus on understanding how certain landforms and landscapes develop and how potential hazards or other environmental concerns are related to the natural processes that cause them. Having an awareness that we exist on a dynamic planet and understanding how different processes interact to shape landforms and physical landscapes are important to human welfare and to the preservation of our planet's environments.

## Landforms and Geomorphology

Landscapes and landforms are often described by their **relief**—a term for the difference in elevation between the highest and lowest points within a specified area or on a particular surface feature. Earth's **landforms**—its geomorphology or terrain features—result from processes that act to increase relief by raising or lowering a land surface and processes that work to reduce relief through erosion of high places and the filling in of lower areas ( ▌Fig. 14.1). Without variations in relief, our planet would be a smooth, featureless sphere and certainly much less interesting. It is hard to imagine Earth without dramatic terrain as seen in the mountainous regions of the Himalayas, Alps, Andes, Rockies, and Appalachians or in the Grand Canyon.

Mountains and other high-relief landforms are typically made of resistant rocks that undergo slow rates of **degradation** (weathering and erosional removal), or they have formed so recently that degradation has not yet had time to wear them away. In some cases, recent uplift and resistant rocks are both involved. Typically, where uplift operates (or has operated)

▌**FIGURE 14.1**
The Teton range of Wyoming stands high above the valley floor because rates of uplift exceed the forces of weathering and erosion, which operate to reduce their height.

faster than degradation, positive relief features such as mountains and hills are formed, and where degradation operates (or has operated) faster than uplift, valleys and low-relief features develop. Relief is also reduced by **aggradation,** the filling in of low places with sediments that have been eroded from higher landforms.

Natural processes operate at a wide variety of rates, usually expressed by an average number of centimeters or millimeters of change per year, averaged over years of record keeping on that process. However, it is important to remember that these are average rates, and most processes tend to operate in an episodic manner. Periods of relative calm continue until a certain threshold condition is reached, resulting in short "bursts" of rapid activity that cause significant environmental modification. Earthquakes and volcanoes provide good examples of this type of Earth change. Instead of functioning in a continuous, steady, and easily predictable manner, activity is usually quiescent but occasionally interrupted by bursts of intense activity. The timing of these periods of increased activity is difficult or impossible to predict with certainty. Many physical and life processes in the Earth system operate in this manner, termed **punctuated equilibrium** ( Fig. 14.2). After an event of major change, the landscape adjusts to the newly created conditions and stays in relative equilibrium, experiencing only modest change, until the next major event again disrupts the system. For example, major earthquake faulting or volcanic eruption can dramatically change a landscape that had previously been undergoing slow or moderate change by gradational processes. Many gradational processes (river floods, landslides) also tend to operate episodically, with extended periods of modest activity and episodic, short periods of intense process ( Fig. 14.3).

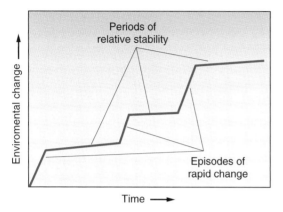

**▌FIGURE 14.2**
Punctuated Equilibrium. Many Earth processes operate slowly most of the time but are punctuated by events that cause relatively major change in a landscape or natural system. For example, many small earthquakes (or volcanic eruptions) occur over time, resulting in little or modest landscape change. Occasionally, large events (earthquakes or eruptions) occur that cause major changes in a short period of time. A great variety of processes that affect the Earth system operate in this episodic manner.
**What Earth processes can you think of that operate this way?**

**▌FIGURE 14.3**
(a) Mt. Vesuvius overlooks the ancient city center of Pompeii, near Naples, Italy. The eruption of Vesuvius in A.D. 79, which destroyed Pompeii, was an example of an episodic process. It is often difficult for humans to fully comprehend the potential danger from Earth processes that operate with bursts of intense activity, separated by years, decades, centuries, or millennia of relative quiescence. (b) Plaster cast of a victim who attempted to cover his face from hot gases and the volcanic ash that buried Pompeii.

(a)  (b)

In this chapter, we study the internal and external processes related to the uplift or buildup of land surfaces through volcanic and tectonic activity and examine the landforms and rock structures associated with these activities. Later chapters focus on landforms, landscapes, and processes associated with various types of gradation.

# Igneous Processes and Landforms

**Volcanism** refers to the rise of magma and its cooling at Earth's surface. It also includes the extrusive rocks and landforms created by this surface activity, such as **volcanoes**—mountains or hills formed by the eruptive expulsion of molten or solid rock material from beneath the surface. **Plutonism** refers to igneous processes that occur below Earth's surface. It also includes the intrusive rocks and rock masses formed from magma that exists or cools in the subsurface.

## Volcanoes and Eruptions

There are few spectacles in nature as awesome as an explosive volcanic eruption ( Fig. 14.4). Although large violent eruptions tend to be infrequent events, they can also devastate the surrounding environment and completely change the nearby terrain. Yet volcanic eruptions are natural processes and should not be unexpected by people who live in the vicinity of active volcanoes.

Eruptions can vary greatly in their size and character, and the volcanic landforms that result are extremely diverse. Variations in eruptive styles and in the landforms produced by volcanism result mainly from temperature and chemical differences in the magma that feeds the eruption. The mineral composition (mafic to sialic) that exists in a magma source is an important factor. Silica-rich magmas tend to be cooler in temperature while molten and have a viscous (thick—resistant to flowing) consistency. Mafic magmas produce iron-rich minerals with a high melting temperature and tend to be extremely hot, less viscous, and flow readily in comparison to silica-rich magmas. Magmas contain large amounts of gases that remain dissolved when under high pressure at great depths. As molten rock rises closer to the surface, the pressure decreases, which tends to release expanding gases. If the gases trapped beneath the surface cannot be readily vented to the atmosphere or do not remain dissolved in the magma, explosive expansion of gases can produce a violent eruption. Viscous silica-rich lavas and magmas (with a granitic composition) have the potential to produce violent eruptions. Basaltic lavas and mafic magmas are hotter and more fluid (less viscous) and tend to vent these gases more readily. When basaltic magma is forced to the surface, the resulting eruptions are not highly explosive although enormous amounts of fluid lava may be produced.

**▌ FIGURE 14.4**
Few natural events are as spectacular as a volcanic eruption. This lava fountain in Hawaii was 300 meters (1000 ft) high.

In addition to emitting lava flows and gases, volcanic eruptions frequently hurl molten material into the air that may solidify in flight or at the surface, and they also expel solid lava fragments of various sizes. These **pyroclastic materials** (from Greek: *pyros,* fire; *clastus,* broken), also referred to as **tephra,** vary from huge volcanic "bombs" to cinders to **volcanic ash,** which ranges in size from sand-sized particles to extremely fine volcanic dust. In the most explosive eruptions, dust-sized volcanic ash may be hurled into the atmosphere to an altitude of 10,000 meters (32,800 ft) or more. The 1991 eruptions of Mount Pinatubo in the Philippines ejected a volcanic aerosol cloud that circled the globe ( Fig. 14.5). The suspended material caused spectacular reddish orange sunsets and may have lowered global temperatures slightly by reflecting solar energy back to space.

PHYSICAL
Geography Now™ Log on to Physical GeographyNow and select this chapter to work through a Geography Interactive activity on "Magma Chemistry and Explosivity" (click Volcanism → Magma Chemistry and Explosivity).

# IGNEOUS PROCESSES AND LANDFORMS 397

**FIGURE 14.5**
The explosive eruption of Mount Pinatubo in the Philippines was a clear demonstration of the hazards associated with living near an active volcano.

formed topographically high areas of low relief called **basalt plateaus.** In the geologic past, at some world regions, huge amounts of basalt have poured out of fissures, eventually burying existing landscapes under thousands of meters of lava flows. The Columbia Plateau in Washington, Oregon, and Idaho, covering 520,000 square kilometers (200,000 sq mi), is a major example of such a basalt plateau ( Fig. 14.7), as is the Deccan Plateau in India.

## Types of Volcanoes

There are four basic types of volcanoes: shield volcanoes, cinder cones, composite cones, and plug domes ( Fig. 14.8a–d). Volcanoes differ according to the dominant type of volcanic material that they erupt (lava or pyroclastics), their internal composition and structure, their eruptive style, their shape, and their size.

**Shield volcanoes** are built by the accumulation of many basaltic lava flows but relatively minor eruptions of ash or cinders (Fig. 14.8a). The gently sloping, dome-shaped cones of Hawaii best illustrate this type of volcano ( Fig. 14.9). They erupt extremely hot, mafic lava with temperatures of more than 1090°C (2000°F). Escape of gases and steam may hurl molten lava a few hundred meters into the air, with some buildup of cinders (fragments or lava clots that congeal in the air), but the major feature is the outpouring of fluid basaltic lava flows. Compared to other volcano types, these eruptions are mild although still potentially damaging and dangerous. The extremely hot and fluid basalt can flow long distances before solidifying, and the accumulation of

## Lava Flows

**Lava flows** are layers of molten rock that flow over the land surface, cooling and solidifying. Lava flows can form from any lava type (see Appendix C), but basalt is by far the most common. The surface characteristics of lava flows vary. Fluid lavas can flow rapidly and for long distances before cooling with a ropy surface called **pahoehoe.** More viscous lavas that flow along more slowly with the appearance of piled up, sharp-edged, jagged blocks is a flow type called **aa**. Both terms are of Hawaiian origin ( Fig. 14.6).

Not all lava flows emanate from a volcano. **Fissure flows** originate from deep cracks in the crust (like the lavas of the seafloor). Some continental areas are covered with enormous accumulations of lava, called **flood basalts,** that consist of hundreds of overlapping lava flows. Accumulations of flood basalts have

**FIGURE 14.6**
The names of the two major types of lava flow surfaces use Hawaiian terminology. Hotter, more fluid lava forms pahoehoe, or ropy flow surfaces (lower portion of the photo). Thicker, viscous lava forms block flows, called aa (upper portion of the photo).

flow layers develops broad, dome-shaped volcanoes with very gentle slopes. On the "big island" of Hawaii, active shield volcanoes also erupt lava from fissures on their flanks so that living on the island's edges, away from the summit craters, does not guarantee safety from volcanic hazards. Neighborhoods in Hawaii have been destroyed or threatened by lava flows in recent years. The Hawaiian volcanic shields form the largest volcanoes on Earth in terms of both their height (beginning at the ocean floor) and diameter.

The smallest type of volcano, typically only a few hundred to more than a thousand feet high, is known as a **cinder cone** and typically consists largely of pebble- to cobble-sized tephra (see Fig. 14.8b). Gas-charged eruptions "spray" molten lava and solid pyroclastic fragments into the air. Falling under the influence of gravity, these particles accumulate in a large pile of cinders. Each eruptive burst ejects more cinders that fall and cascade down the sides to build an internally layered volcanic cone. Cinder cone volcanoes typically have a basaltic composition. The form of a cinder cone is very distinctive, with steep straight sides and a large crater in the center, given the size of the

**FIGURE 14.7**
Columbia Plateau flood basalts in southwestern Idaho. River erosion has cut a deep canyon to expose the uppermost layers of basalt underlying the Columbia Plateau.

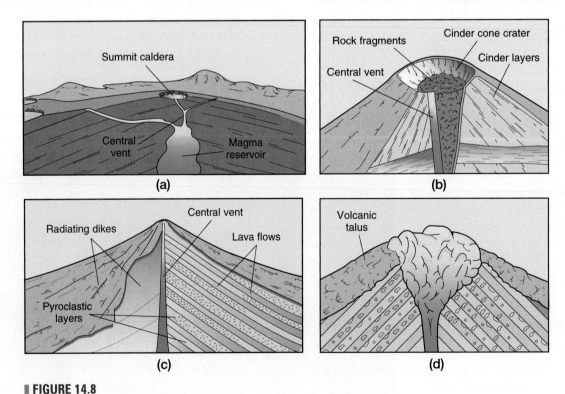

**FIGURE 14.8**
The four basic types of volcanoes: (a) shield volcano, (b) cinder cone, (c) composite cone or stratovolcano, and (d) plug dome.
**What are the key differences in their shapes?**

**FIGURE 14.9**
Mauna Kea, at 4204 meters (13,792 ft) above sea level, is the tallest volcano on the "big island" of Hawaii and clearly displays the dome shape of a classic shield volcano. A 9-kilometer-thick (5.6 mi) accumulation of basaltic lava flows built this volcano up from the bottom of the Pacific Ocean.
**Why are Hawaiian volcanoes less explosive than those of the Cascades or Andes?**

**FIGURE 14.10**
Cinder cones form by the gas-charged expulsion of volcanic fragments. A cinder cone stands among lava flows in Lassen Volcanic National Park.
**Why is the crater so prominent on this volcano?**

volcano ( Fig. 14.10). Cinder cone examples include the Craters of the Moon area in Idaho, Mount Capulin in New Mexico, and Sunset Crater, Arizona. In 1943 a remarkable cinder cone called Paricutín grew from a fissure in a Mexican cornfield to a height of 92 meters (300 ft) in 5 days and to more than 360 meters (1200 ft) in a year. Eventually, the volcano began erupting basaltic lava flows, which buried a nearby village except for the top of a church steeple.

**Composite cones** are formed by eruptions that vary between a mix of lava flows and pyroclastic materials (Fig. 14.8c). The topographic profile of a composite cone presents a classic volcano shape, with curved slopes that go from gentle downslope to steep near the top. Most of the world's best-known volcanoes are composite cones: Fujiyama in Japan, Cotopaxi in Ecuador, Vesuvius in Italy, Mount Rainier in Washington, and Mount Shasta in California ( Fig.14.11). Composed of great volumes of intermediate composition lava—particularly andesite—as well as ash and larger tephra, composite cones, also known as *stratovolcanoes* (consisting of stratified pyroclastics and lava), are potentially dangerous. The viscous andesitic lava of composite cones can trap gases until the pressure from expanding gases produces an explosive eruption. Andesite is named for the Andes of South America, which include the highest volcano on Earth, Nevados Ojos del Salado, which rises to 6887 meters (22,595 ft) on the border between Chile and Argentina.

On May 18, 1980, residents of the Pacific Northwest—and soon after, people throughout the United States—were shocked by the realization that every age in history is a volcanic age. Mount St. Helens, a composite cone in southwestern Washington that had been venting steam and ash for several weeks, exploded with incredible force. A menacing bulge had been growing on the side of Mount St. Helens, and Earth scientists warned of a possible major eruption, but no one could forecast the magnitude or the exact timing of the blast. Within minutes, nearly 400 meters (1300 ft) of the mountain's north summit had disappeared by being blasted into the sky and down the mountainside ( Fig. 14.12). Unlike most volcanic eruptions, in which the eruptive force is directed vertically, much of the explosion blew pyroclastic debris laterally outward from where the bulge had been. An intensely hot cloud of steam,

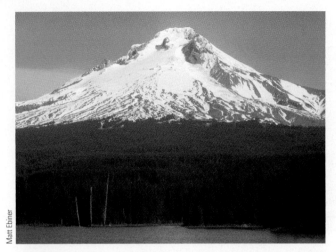

**FIGURE 14.11**
Composite cones are formed from layers of lava and pyroclastic material. Oregon's Mount Hood is a composite cone in the Cascade Range.
**Along what type of tectonic plate boundary is this volcano located?**

(a)

(b)

(c)

**FIGURE 14.12**
Mount St. Helens, Washington, in the Cascade Range of the Pacific Northwest. (a) A view of Mount St. Helens and Spirit Lake prior to the eruption. (b) On May 18, 1980, at 8:32 a.m., Mount St. Helens erupted violently. The massive landslide and blast removed more than 4.2 cubic kilometers (1 cu mi) of material from the mountain's north slope, leaving a crater more than 400 meters (1300 ft) deep. The blast cloud and monstrous mudflows destroyed the surrounding forests and lakes and took 60 human lives. (c) A view of Mount St. Helens 2 years after the eruption.
**What type of volcano is Mount St. Helens?**

noxious gases, and volcanic ash blasted outward at more than 320 kilometers (200 mi) per hour, obliterating forests, lakes, streams, and campsites for nearly 32 kilometers (20 mi). Volcanic ash and water from melted ice formed huge mudflows that choked streams, buried valleys, and engulfed everything in their paths. More than 500 square kilometers (193 sq mi) of magnificent forests and recreational lands were destroyed. Hundreds of homes were buried or badly damaged. Choking ash several centimeters thick covered nearby cities, untold numbers of wildlife were killed, and more than 60 people lost their lives in the eruption. It was a minor event in Earth's history but a sharp reminder of the awesome power of natural forces.

Some of the worst natural disasters in history have occurred in the shadows of composite cones. Mount Vesuvius, in Italy, killed more than 20,000 people in the cities of Pompeii and Herculaneum in A.D. 79. Mount Etna, on the Italian island of Sicily, destroyed 14 cities in 1669, killing more than 20,000 people. Today, Mount Etna is active much of the time. The greatest volcanic eruption in recent history was the explosion of Krakatoa in the Dutch East Indies (now Indonesia) in 1883. The explosive eruption killed more than 36,000 persons, many as a result of the subsequent *tsunamis* (seismic sea waves, discussed in Chapter 20) that swept the coasts of Java and Sumatra. In 1985 the Andean composite cone Nevado del Ruiz, in the center of Colombia's coffee-growing region, erupted and melted its snowcap, sending torrents of mud and debris down its slopes to bury cities and villages, resulting in a death toll in excess of 23,000. The 1991 eruption of Mount Pinatubo in the Philippines killed more than 300 people and caused suspected climatic effects for years following the eruption. In 1997 a series of violent eruptions from the Soufriere Hills volcano destroyed more than half of the Caribbean

IGNEOUS PROCESSES AND LANDFORMS 401

**FIGURE 14.13**
Volcanic destruction on Montserrat. The Caribbean island of Montserrat was struck by a series of volcanic eruptions beginning in 1995 that devastated much of the island. The town of Plymouth, shown here, has been completely abandoned because of the destruction and threat of future eruptions. Prior to the 1995 disaster, the volcano had not erupted for 400 years.

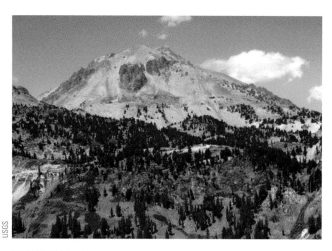

**FIGURE 14.14**
Plug dome volcanoes extrude stiff silica-rich lava and form steep slopes. Lassen Peak, located in northern California, is a plug dome and the southernmost volcano in the Cascade Range. The lava plugs are the darker areas protruding from the volcanic peak. Lassen was last active between 1914 and 1921.
**Why are plug dome volcanoes considered dangerous?**

island of Montserrat with volcanic ash and pyroclastic flows (Fig. 14.13). In recent years, Mexico City, perhaps the world's most populous urban area, has been threatened by Popocatepetl, a large, active stratovolcano that is 80 kilometers (50 miles) away. At this distance, ash falls from a major eruption would be the most severe hazard to be expected. Volcanic ash is much like tiny slivers of glass and can cause breathing problems as well as stall vehicles by choking the air intakes of combustion engines. The heavy weight of significant ash accumulations on building roofs can cause them to collapse.

**Plug dome** volcanoes extrude extremely stiff, taffylike lava (rhyolite) that fills an initial pyroclastic cone, generally without flowing beyond it (see Fig. 14.8d). Plug domes are volcanoes with summits (plugs) composed of solid lava and steep sloping sides formed by accumulations of jagged blocks that fractured away as the lava cooled. These volcanoes become plugged by thick lava, which allows great pressures to build, so plug domes, like composite cones, have the potential for violent explosive eruptions. In 1903 Mount Pelée, a plug dome on the French West Indies island of Martinique, destroyed in a single blast all but one person from a town of 30,000. Lassen Peak in California is a large plug dome that has been active in the last 100 years (Fig. 14.14). Other plug domes exist in Japan, Guatemala, the Caribbean, and the Aleutian Islands.

Occasionally, a volcano may expel so much eruptive material that its summit collapses into its emptied magma chamber, forming a large crater known as a **caldera.** The best-known caldera in North America is Crater Lake in Oregon, a circular body of water 10 kilometers (6 mi) across and almost 610 meters (2000 ft) deep, surrounded by near-vertical cliffs as much as 610 meters (2000 ft) high. The caldera that contains Crater Lake formed by the prehistoric collapse of a composite cone volcano. A cinder cone, Wizard Island, has built up from the floor of the caldera and rises above the lake's surface (Fig. 14.15). The area of Yellowstone National Park is the site of three ancient calderas, and the Valles Caldera in New Mexico is another excellent example. Krakatoa in Indonesia and Santorini (Thera) in Greece have left island remnants of their calderas. Calderas are also found in the Philippines, the Azores, Japan, Nicaragua, Tanzania, and Italy, many of them occupied by deep lakes.

> PHYSICAL
> Geography Now™ **Log on to Physical Geography Now and select this chapter to work through a Geography Interactive activity on "Volcanic Landforms" (click Volcanism → Volcanic Landforms).**

## Intrusive Activity and Intrusions

Bodies of magma that exist beneath the surface or masses of intrusive igneous rock that cooled and solidified beneath the surface are called igneous **intrusions.** A great variety of shapes and sizes of magma bodies can result from igneous activity. Smaller intrusions when first formed have little or no effect on the surface terrain. Massive intrusions, however, may be associated with uplift of the land surface under which they form. **Plutons** are intrusive igneous bodies that eventually cool and solidify at great depth.

The many different kinds of intrusions are classified by their size, shape, and relationship to the surrounding rocks (Fig. 14.16). After millions of years of uplift and erosion, intrusions may become exposed at the surface to become

**FIGURE 14.15**
Crater Lake National Park, Oregon. (a) Crater Lake is the best-known caldera in North America. It formed when a violent eruption of Mount Mazama about 6000 years ago caused the summit to collapse, forming a deep crater. (b) Later the crater filled with water, forming the 610-meter-deep (2000 ft) Crater Lake. Wizard Island is a secondary volcanic peak formed in the caldera.
**Do you think other Cascade volcanoes could repeat this kind of event?**

part of the landscape. Uplifted intrusions composed of granite or other intrusive igneous rocks are eventually exposed by erosion and tend to stand higher than the landscape around them because of their resistance compared to the surrounding rocks.

A relatively small, irregularly shaped intrusion is called a **stock.** A stock is usually limited in area to less than 100 square kilometers (40 sq mi). The largest intrusions, called **batholiths,** are larger than 100 square kilometers and are complex masses of solidified magma, usually granite. Batholiths melt, metamorphose, or push aside other rocks as they develop within Earth's crust. Batholiths vary in size; some are as much as several hundred kilometers across and

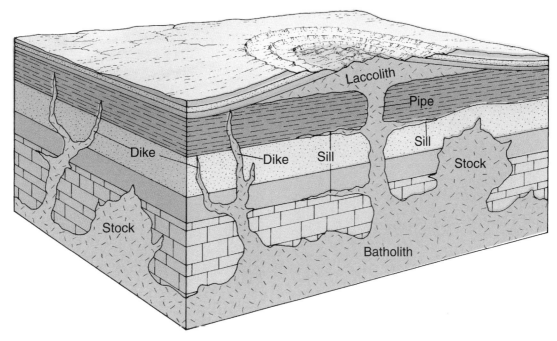

**FIGURE 14.16**
The major intrusive features. These igneous features crystallize below the surface. Later these forms may be exposed by erosion.
**What three criteria are used to differentiate among those features illustrated here?**

thousands of meters thick, and they form kilometers beneath Earth's surface. Batholiths form the core of many major mountain ranges, primarily because uplift has caused older covering rocks to be eroded away and granite tends to be resistant to degradation processes. The Sierra Nevada, Idaho, Rocky Mountain, Coast, and Baja California batholiths cover areas of hundreds of thousands of square kilometers of granite landscapes in western North America.

Magma can also force its way into fractures and between rock layers to form other kinds of igneous intrusions, without melting the surrounding rock. A **laccolith** develops when molten magma flows horizontally between rock layers, bulging the layers nearer to the surface upward to form a solidified mushroom-shaped structure. Laccoliths have a mushroomlike shape because they are usually connected to a magma source by a pipe or stem. The resulting uplift on Earth's surface is like a giant blister, with magma beneath the overlying layers comparable to the fluid beneath the skin of a blister. Laccoliths are typically much smaller than batholiths, but both of these types of intrusions typically form the core of mountains or hills after erosion has worn away the overlying sedimentary rocks. The LaSal, Abajo, and Henry Mountains in southern Utah are formed from exposed laccoliths, as are many other mountains in the American West (Fig. 14.17).

Smaller but no less interesting landforms created by intrusive activity may be exposed at the surface through erosion. Magma can intrude between rock layers without bulging them upward, solidifying into a horizontal sheet of

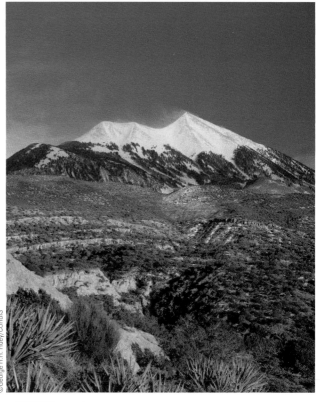

**FIGURE 14.17**
The LaSal Mountains in Southern Utah, near Moab, were formed by a laccolith that was exposed at the surface by uplift and subsequent erosion of the overlying sedimentary rocks.

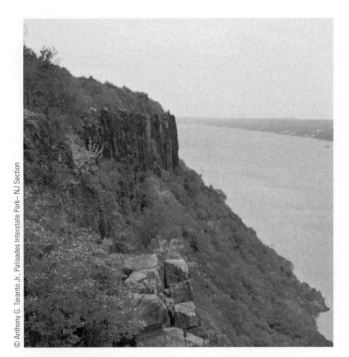

### FIGURE 14.18
Sills form where magma intrudes between parallel layers of surrounding rocks. This photo shows one of the most famous sills in the United States, the Palisades of the Hudson River.
**What is the major difference between a dike and a sill?**

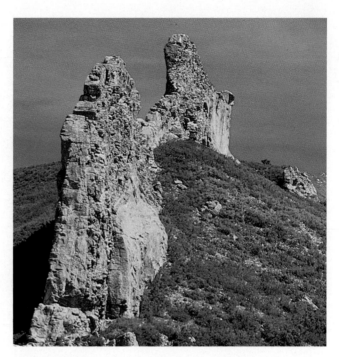

### FIGURE 14.19
An exposed dike in Colorado. This igneous rock was intruded into softer sandstone. Later the sandstone was eroded away, leaving the resistant dike exposed.
**How do you know that the dike is made of resistant rock?**

igneous rock called a **sill.** The Palisades, along New York's Hudson River, provide an example of a sill made of basaltic rock (Fig. 14.18). More commonly, molten rock under pressure intrudes into a vertical fracture that runs across the general trend of the surrounding rocks. As it solidifies, it forms a wall-like sheet of igneous rock known as a **dike.** When exposed by erosion, dikes appear as vertical walls of resistant rock rising above the surrounding topography (Fig. 14.19). At Shiprock, in New Mexico, resistant dikes many kilometers long rise vertically to more than 90 meters (300 ft) above the surrounding plateau (Fig. 14.20). Shiprock is a **volcanic neck,** a spire or mountain that formed from the exposed (formerly subsurface) pipe that fed a long-extinct volcano situated above it about 30 million years ago. Erosion has removed the volcanic cone, exposing the resistant dikes and neck that were once internal features of the volcano at Shiprock.

## Tectonic Processes, Rock Structure, and Landforms

Tectonic processes include the bending, folding, warping, and fracturing of Earth's crust that are largely related to the conflicting motions of Earth's lithospheric plates. Most of our information about such deformation comes from observing

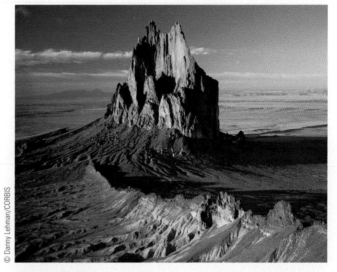

### FIGURE 14.20
Shiprock, New Mexico, is a volcanic neck. Volcanic necks form when the resistant intrusive pipe of the volcano is exposed by erosion.
**Why do you think this feature is called Shiprock?**

aspects of bedrock **structure**—the nature, configuration, and arrangement of rocks as they extend into the subsurface. Sedimentary rocks offer useful information because we know that they generally formed as nearly horizontal rock layers and the youngest layers are deposited on top of older ones. If layers of sedimentary rocks are tilted, bent, or displaced, then we can assume that some kind of deformation has taken place.

Earth scientists describe the inclination of rock layers or of other rock surfaces as **dip.** The dip of a rock surface is measured in degrees of angle from the horizontal (the degree of tilt). The **strike** of such beds is their compass direction (in map view), taken at right angles to the dip. Thus, we might say that certain layers of outcropping rocks have a dip of 40° to the southeast and a northeast–southwest strike (Fig. 14.21). Earth's crust has been subjected to tectonic pressures and tensions throughout its history, although crustal stresses have been greater during some geologic periods and have varied widely from one region compared to others. Crustal rocks have responded to these stresses by warping, folding, and fracturing, as well as by being pushed up or by sinking down. Most of these changes have occurred over hundreds of thousands or millions of years (again episodically), but others have been rapid and cataclysmic. The response of crustal rocks to tectonic stress can yield a variety of configurations in the rock structure, depending on the nature of deformation. Gradational processes act on these rock structures as they appear at Earth's surface, working to wear them down, fill them in, and level the landscape.

Tectonic stresses may involve uplift or subsidence of large crustal sections and may result from either **compression,** which tends to shorten and thicken the crust, or **tension,** which tends to stretch and fracture crustal rocks (Fig. 14.22). Compression that operates slowly or at great depth typically results in folding. Eventually, however, if the stress is great enough, the rocks may break and shear or slide over themselves horizontally. Tensional stress tends to cause rock fracturing, and crustal blocks may move downward in order to accommodate the expansion or pulling apart of the crust.

**FIGURE 14.21**
Dip and strike in layers of sedimentary rocks. Dip is the angle and direction that rock layers tilt, measured from the horizontal. Strike is the compass direction of the line marking the intersection of the rock layers with Earth's surface. The strike is always at right angles to the dip.
**What are the dip and strike of the upper strata of sandstone in this diagram? (North is toward the top of the diagram.)**

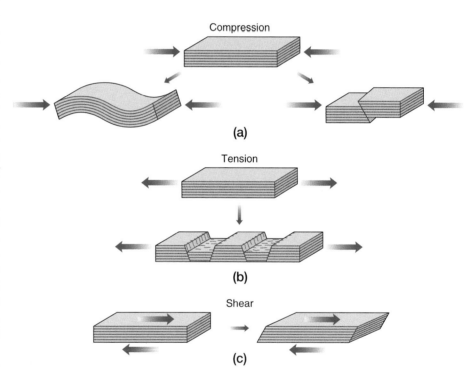

**FIGURE 14.22**
Three types of stress and the deformation of rock layers. (a) Compression can cause rocks to bend and fold or break and shift along a fault. (b) Tension, or pulling apart, generally causes rocks to break and shift along a fault. (c) Shear stress pulls or pushes rocks in opposite directions and may cause blocks of rock to break and slide alongside each other.

## Warping of Earth's Crust

Broad, gentle deformation over large, regional areas is called **crustal warping.** Florida and Mexico's Yucatán Peninsula were raised to their present levels from below the ocean by crustal upwarping. The Florida Peninsula is composed of

## Geography's Spatial Science Perspective
### Spatial Relationships between Plate Boundaries, Volcanoes, and Earthquakes

The geographic distributions of volcanism and earthquake activity are highly similar and tend to be concentrated in linear patterns along plate boundaries. Although there are exceptions, the locations of volcanoes and earthquakes match fairly well, yet their nature and severity differ from place to place. In general, the frequency and severity of volcanic eruptions or earthquakes vary according to their proximity to a certain type of plate boundary or by their central location on a tectonic plate.

Plate divergence, whether it breaks continents or seafloor, creates fractures that provide avenues for molten rock to reach the surface. The divergent midocean ridges are volcanic and tend to have rather mild volcanic eruptions and small to moderate earthquakes that originate at a shallow depth. These tectonic activities affect ocean islands associated with midocean ridges, such as the Azores and Iceland.

Volcanoes also tend to occur where continental crust is breaking and diverging. In these regions, earthquakes tend to be small to moderate, but continental crust mixed with mafic magma produces a wider variety of volcano types, some of which are potentially quite violent. Examples in the East African rift valleys include Mount Kilimanjaro and Mount Kenya.

The potential severity of earthquakes and eruptions is much greater where plates are converging. Along the ocean trenches where crustal material is subducted, volcanoes typically develop along the edge of the overriding plate. The largest region where this occurs is the "Pacific Ring of Fire," a volcanically active and earthquake-prone rim around the Pacific Ocean. When oceanic crust subducts beneath continental crust along an ocean trench, it melts into magma that moves upward under the continental borders. Subduction along the Pacific is associated with volcanoes in the Andes, the Cascades, and the Aleutians; the Kuril Islands and the Kamchatka Peninsula in Russia; and Japan, the Philippines, New Guinea, Tonga, and New Zealand. Many of these volcanoes erupt andesite and can be dangerously explosive. Earthquakes are common events along the Pacific Rim, mostly small to moderate, but the largest earthquakes ever recorded have been related to subduction in this region. The depth of earthquakes becomes deeper under the overriding plates, tracking the subducting plates downward to where they are recycled into the mantle.

Another volcanic and seismic belt marks the collision zone between northward-moving Southern Hemisphere plates and the Eurasian plate. The volcanoes of the Mediterranean region, Turkey, Iran, and Indonesia are located along this collision zone, seismic activity is common, and occasionally major earthquakes occur.

Transform plate boundaries, where lateral sliding occurs, also mark the locations of many earthquakes. The potential for major earthquakes mainly exists in places such as along the San Andreas Fault zone in California where thick continental crust is resistant to sliding easily. Volcanic activity along transform boundaries is moderate on the seafloor to modest in continental locations.

Certain active earthquake and volcanic areas do not occur near active plate boundaries, rather they exist within the central parts of plates. The Hawaiian Islands, the Galapagos Islands, and the Yellowstone National Park area are examples of intraplate "hot spots." Some may be associated with a plume of magma rising from the mantle. Oceanic areas that are over hot spots tend to have strong volcanic activity and modest earthquake activity. In midcontinental areas where large earthquakes have occurred, these regions are suture zones where continents are pushing together, such as in the Himalayas, or where broken edges of ancient landmasses shift, today located in midcontinent but deeply buried by more recent rocks.

Volcanic and earthquake activities that are not located along active plate margins are intriguing and show that we still have much to learn about Earth's internal processes and their impact on the surface. Yet, plate tectonics has greatly influenced our understanding of the variations in volcanism, earthquake activity, and the landforms associated with these processes that are discussed in this chapter.

---

sedimentary rocks that represent undersea environments and contain many marine fossils as evidence of its origin. The extensive Colorado Plateau, centered on the "four corners" region of the Southwest, has been uplifted thousands of feet in similar fashion ( Fig.14.23). Slow downwarping of the crust and consequent inundation of low plains by the ocean have created the basin occupied by the North Sea. Crustal warping is an important process of landform change that affects broad regions of Earth's surface. Although warping is a tectonic process, it is often not directly related to widespread tension or compression in Earth's crust; rather, it proceeds from vertical movements probably related to lateral flow of material in the asthenosphere.

## Folding

The wrinkling of Earth's crust, known as **folding,** usually occurs in response to compression. Rock layers that are deeply buried in the crust and under great pressure are particularly susceptible to behaving plastically (deforming without breaking) and become folded, rather than breaking, under compressive stress. Folds in rock layers may be very small, covering a few centimeters, or they may be enormous

The spatial correspondence among plate margins, active volcanoes, earthquake activity, and hot spots is not coincidental but is strongly related to tectonic plate boundaries. (a) Plate boundaries and the global distribution of active volcanoes (1960–1994). (b) Plate boundaries and the global distribution of earthquake activity (magnitude 4.5+, 1990–1995).

with vertical distances between the upfolds and downfolds measured in kilometers. Folds can be tight or broad, symmetrical or asymmetrical ( Fig. 14.24). Much of the Appalachian Mountain system is an example of folding on a large scale (see also Fig. 15.20). Spectacular folds exist in the Rocky Mountains of Colorado, Wyoming, Montana and in the Canadian Rockies. Highly complex folding formed the Alps where folds are overturned, sheared off, and piled on top of one another. Almost all mountain systems exhibit some degree of folding, though the folding is usually found in connection with some faulting (fracturing of the crust).

As elements of rock structure, upfolds are called **anticlines,** and downfolds are called **synclines** ( Fig. 14.25).

The rock layers that form the flanks of anticlinal crests and synclinal troughs are the fold limbs. Folds are typically symmetrical (each limb has about the same dip angle) if they formed by compressional stresses that are relatively equal from both sides. If these compressive forces are stronger from one side, the folds may become asymmetrical, with the dip of one limb being much steeper than the other. Eventually, asymmetrically folded rocks may become overturned and perhaps so compressed that the fold lies horizontally; these are known as **recumbent folds** (see again Fig 14.24). In some cases, crustal rocks under compression may no longer yield to folding, and the upper part of the fold shears or breaks, sliding over the lower rock layers. This process is called

**408** CHAPTER 14 VOLCANOES, EARTHQUAKES, AND TECTONIC LANDFORMS

overthrusting, and the rock structure is referred to as an **overthrust** (thrust fault). Major overthrusts occur along the northern Rocky Mountains and in the southern Appalachians. Recumbent folds and overthrusts are important rock structures that have formed in complex mountain ranges such as the Andes, Alps, and Himalayas.

## Faulting

Rock layers that are near the surface tend to be brittle and too rigid to form folds, so they rupture when under crustal stress. Both tension and compression can cause rocks to fracture and move differentially with respect to one another. **Faulting** is slippage or displacement of rocks along a fracture surface, and the fracture along which movement has occurred is called a **fault.** The instantaneous movement along a fault during an earthquake can vary from fractions of a centimeter to several meters. Faulting can move rocks vertically (up and down), laterally, or both. The maximum horizontal displacement along the

**FIGURE 14.23**
This satellite image of the Grand Canyon in the Colorado Plateau shows that the deepest part of the canyon cuts though a gentle upwarp in the surrounding rock layers. The green areas are forests, growing on the upwarp, surrounded by lower elevations that are warmer, more arid, and cannot support forest vegetation. The heavily visited south rim is on the left side of the canyon as viewed here (left of photo center).

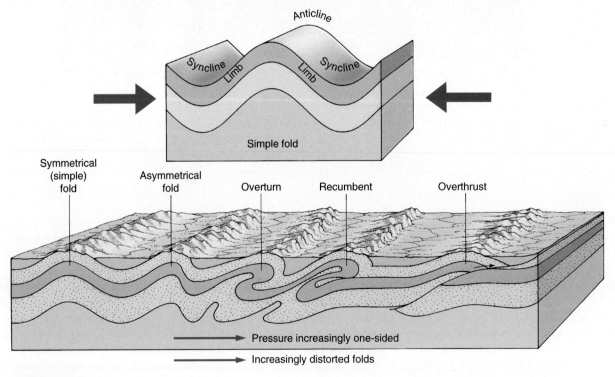

**FIGURE 14.24**
Diagram of the major types of folds

TECTONIC PROCESSES, ROCK STRUCTURE, AND LANDFORMS   409

■ FIGURE 14.25
Folded sedimentary strata at Ecola Beach, Oregon.
**Identify the anticlines and synclines in this photo.**

San Andreas Fault in California during the 1906 San Francisco earthquake was more than 6 meters (21 ft). A vertical displacement of more than 10 meters (33 ft) occurred during the Alaskan earthquake of 1964. Over millions of years, the cumulative displacement along a major fault may be tens of kilometers vertically or hundreds of kilometers horizontally, though the majority of faults have offsets that are much smaller.

Vertical displacement along a fault occurs when the rocks on one side move up or drop down in relation to rocks on the other side. This kind of movement, up or down along the *dip* of the fault plane extending into Earth, is known as **dip slip.** Faults that have vertical displacement resulting from tension are called **normal faults** (■ Fig. 14.26a). Faults resulting

■ FIGURE 14.26
Diagrams of the major types of faults.

from compression are called **reverse faults** (Fig. 14.26b). When compression pushes one side up along a narrow angle to override rocks on the other side, the displacement is called a **thrust fault**, or overthrust (Fig. 14.26c).

The displacement along some faults is horizontal rather than vertical. All such faults descend at a near-vertical angle into Earth's crust, but the slippage is parallel to the surface trace, or *strike*, of the fault. This lateral movement creates a **strike-slip fault** (Fig. 14.26d). Active strike-slip faults can cause horizontal displacement of roads, railroad tracks, fences, streambeds, and other features that are astride the fault. The San Andreas Fault, which runs through much of California, has strike-slip movement (Fig. 14.27). A long valley, formed by erosion of rocks that have been crushed and weakened by faulting, runs along the strike of the San Andreas Fault zone.

The topographic escarpment of an uplifted block where vertical displacement of bedrock offsets the land surface is called a **fault scarp**, or **fault escarpment**. Fault scarps can account for spectacular mountain walls, especially in regions like much of the western United States with a history of recent tectonic activity. The east face of the 645-kilometer-long (405 mi) Sierra Nevada Range in California is a classic example of a fault scarp that rises steeply 3350 meters (11,000 ft) above the desert (Fig. 14.28). In contrast, the west side of the Sierra (the "back slope") descends very gently over a distance of 100 kilometers (60 mi) through rolling foothills. The Sierra Nevada Range is a great **tilted fault block** where the east side was faulted upward and the west side tilted down (Fig. 14.29). The equally dramatic Grand Tetons of Wyoming also rise along a fault

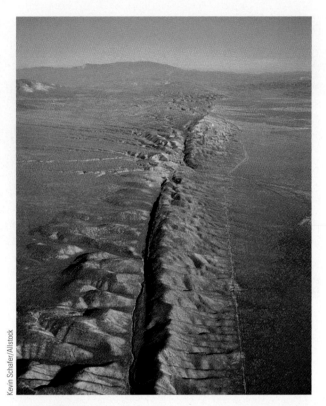

**FIGURE 14.27**
The San Andreas Fault along the Carrizo Plain in central California. The area west of the fault is moving northwestward, in relation to the area on the eastern side. Note that erosion has formed a long, narrow valley in the faulting-weakened rock along the fault zone.
**What type of fault is the San Andreas?**

**FIGURE 14.28**
The east front of the Sierra Nevada in California is a steep fault scarp.

**FIGURE 14.29**
A diagram of the strike and dip of a tilted fault block, with an east facing escarpment. This configuration is a simplified version of the kind of faulting that produced the Sierra Nevada.

**FIGURE 14.30**
The steep fault scarp at Santa Elena Canyon, Texas-Mexico. The Rio Grande has cut a canyon into the uplifted and tilted fault block. The left wall of the canyon is in Mexico and the right wall is in the United States.

scarp facing eastward. In Big Bend National Park, Texas, the fault block that forms the walls of Santa Elena Canyon is an excellent example of a fault scarp. Other than the 500-meter-deep canyon that the Rio Grande has cut, the fault block is modified so little by degradation that it preserves much of its blocklike shape ( Fig. 14.30). In the southwestern United States, the Colorado Plateau steps down to the Great Basin by a series of fault scarps that face westward in southern Utah and northern Arizona.

Major uplift of faulted mountain ranges can have a strong impact on other physical systems, and an excellent example is the Sierra Nevada. As the mountains rose, stream erosion accelerated because of the increase in slope. Precipitation on the windward side of the Sierra increased because of orographic lifting. The steep lee side of the tilted fault block became more arid than before because it was situated in the rain shadow of the Sierra. Increased precipitation and lower temperatures at higher elevations changed the climate of the uplifted range significantly, and climate change influenced the vegetation, soils, and animal life. Soils have also been affected by increased runoff and erosion. The uplift of the Sierra has extended over several million years in a gradual but episodic sequence of faulting. The Sierra Nevada Range is continuing to rise rapidly, in a geologic sense—perhaps an average of a centimeter per year. Gradational agents (water, ice, wind, and gravity) have worked to level the land as uplift progressed. The Sierra Nevada, like most high mountain ranges, are much different today than they once were, for they have been altered and etched by glaciation, stream erosion, and downslope gravitational movement of rock material. These processes have carved and shaped valleys in the Sierran fault block, leaving spectacular canyons and mountain peaks.

Faults frequently develop in parallel sets. Where the block between two parallel faults is elevated above the land to either side because it has been pushed up or because the land to either side has dropped down, the raised portion is called a **horst** ( Fig. 14.31). The central European Highlands north of the Alps contain many horsts, as does the Basin and Range region of the southwestern United States. The Basin and Range region extends eastward from California to Utah and southward from Oregon to New Mexico, including all of Nevada. The great Ruwenzori Range of East Africa is a horst, as is the Sinai Peninsula between the fault troughs in the Gulfs of Suez and Aqaba (see again Fig. 13.31).

The opposite of a horst is a **graben,** a depression between two facing fault scarps (see Fig. 14.31) that may occur when the block between two faults drops down or when the land to either side of the faults is uplifted. Classic examples of grabens are Death Valley, California (Fig. 14.32) and the middle Rhine Valley in Germany. **Rift valleys** are lowlands that are fault related and much longer than they are wide. Examples of rift valleys include the Rio Grande rift of New Mexico and Colorado, the Great Rift Valley of East Africa, and the Dead Sea rift valley where that body of water lies at an elevation some 390 meters (1280 ft) below the Mediterranean Sea, which is only 64 kilometers (40 mi) away. Rift valleys also run along the centers of oceanic ridges.

Faulting and folding are important processes of landform development, especially mountain building. Immense regions may be affected by crustal shattering along faults, producing a mosaic of horsts, grabens, and tilted fault blocks. An outstanding example is the Basin and Range region of the western United States, named for the many north–south trending, down-dropped basins and uplifted mountain ranges. Faulting and folding also tend to occur in the same regions and sometimes also in the vicinity of volcanism, the other important mountain-building process. These processes are actively shaping Earth's surface today, as attested to by recent volcanic eruptions and earthquake activity.

**FIGURE 14.31**
Horsts and grabens are blocks of earth material that are bounded by roughly parallel vertical faults. An uplifted block between two faults is a horst, and a graben is a down-dropped block.

## Earthquakes

**Earthquakes,** evidence of present-day tectonic movement, are vibrations that occur when accumulating stress in the crust is suddenly released by displacement along a fault. The shock waves of an earthquake signify the energy released by the sudden, lurching movement of crustal blocks past one another. This energy generates the internal seismic waves that were discussed in Chapter 13 as helpful in understanding Earth's interior, but earthquake waves can also have a great impact on the surface. An earthquake's surface waves, which pass along the crustal exterior or emerge on the surface from

**FIGURE 14.32**
Death Valley, California, is a classic example of a topographic graben. The valley floor is 86 meters (282 ft) below sea level—the lowest elevation in North America.
**What is the difference between a graben and a horst?**

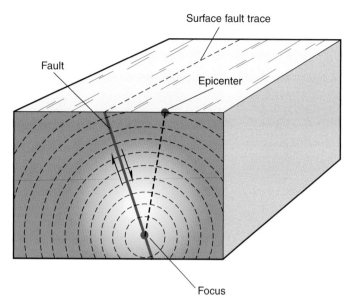

**FIGURE 14.33**
The center of energy release for an earthquake is the focus—the location where the fault broke—which is typically at some depth beneath the surface. The epicenter is the location on the surface directly above the focus.
**Why is the epicenter not located where the fault crosses the Earth's surface?**

below, cause the damage and subsequent loss of life that we associate with major tremors. The subsurface location where an earthquake originates is the **focus,** which may be located anywhere from near the surface to a depth of 700 kilometers (435 mi). The earthquake **epicenter** is the point on Earth's surface that lies directly above the focus and is the location where the strongest shock is normally felt ( Fig. 14.33).

The vast majority of earthquakes are so slight that we cannot feel them, and they produce no injuries or damage. Most earthquakes occur at a focus that is deep enough so that no displacement is visible at the surface. Others may cause mild shaking that rattles a few dishes, while a few are strong enough to topple buildings and break power lines, gas mains, and water pipes. Ground shaking during an earthquake can also trigger landslides. Aftershocks may follow the main earthquake as crustal movement settles. Geophysicists are currently investigating the possibility that foreshocks may alert us to major earthquakes, though evidence is at present inconclusive.

## Measuring the Size of Earthquakes

An earthquake's severity can be expressed in two ways: (1) the size of the event as a physical Earth process and (2) the degree of its impact on humans. These two methods may be related; that is, all other factors being equal, powerful earthquakes should have a greater impact on humans than smaller ones would. However, large tremors may strike in distant, uninhabited places, thus having limited human impact, and smaller quakes may strike densely populated areas, causing much damage and suffering. The two systems, which together measure both the size and the human impact of an earthquake, help scientists and planners understand hazard potential for a locality.

The scale of **earthquake magnitude,** originally developed by Charles F. Richter in 1935, is based on the energy released by an earthquake as recorded on seismographs. Earthquake magnitude is expressed in a number, generally with one decimal place. Every increase of one whole number in magnitude (for example, from 6.0 to 7.0) means the earthquake wave motion is 10 times greater, but the actual energy released may be 30 times greater. The extremely destructive 1906 San Francisco earthquake occurred before the magnitude scale was devised but has been estimated to be the equivalent of a magnitude 8.3. The strongest earthquake in North America (8.6) occurred in Alaska in 1964. The 1989 Loma Prieta earthquake near San Francisco measured 7.1, but it was still responsible for the deaths of more than 60 people. Most were crushed in their automobiles when an elevated freeway collapsed. Many factors other than magnitude affect the damage and loss of human life resulting from an earthquake. Each earthquake has only one magnitude reading that represents the earthquake's size in terms of energy released.

A very different scale of **earthquake intensity** is used to record and understand patterns of damage caused by an earthquake and the degree of its impact on people and their property. This is the **modified Mercalli Scale** of earthquake intensity that uses categories numbered from I to XII (Table 14.1 see page 416) to describe the effects of an earthquake on people and the spatial variation in impacts from the tremor. Roman numerals are used to avoid confusion with earthquake magnitude scales. Although a maximum intensity is determined, an earthquake produces a variety of intensity levels, depending on local conditions. These conditions include distance from the epicenter, how long shaking lasted, the severity of shaking resulting from the local surface materials affected, population density, and building construction in the affected area. After an earthquake, observers gather information about Mercalli intensity levels by noting the damage and talking to residents about their experiences. The intensity levels of responses are spatially mapped so that geographic patterns of damage and shaking intensity can be analyzed. Understanding the spatial patterns of damage and ground response helps us plan and prepare for future tremors.

## Earthquake Hazards

The location of an earthquake epicenter relative to population density, construction type, and the stability of earth materials on which buildings are constructed greatly influences its impact on humans. A strong tremor (magnitude 7.5) that struck the sparsely settled Mojave Desert of California in 1992 caused little loss of life. Yet a magnitude 6.9 quake in

# Geography's Environmental Science Perspective
## Mapping the Distribution of Earthquake Intensity

When an earthquake strikes a populated area, one of the first reports of geoscience information released is the magnitude of the tremor. Magnitude is a numerical expression of an earthquake's size at its source in terms of energy released. In this sense, earthquakes can be compared to explosions: for example, a magnitude 4.0 earthquake releases energy equivalent to exploding 1000 tons of TNT. Because the scale is logarithmic, a 6.9 magnitude is the equivalent of 22.7 million tons of TNT.

Magnitude is an index of an earthquake's power. Larger earthquakes have the potential to cause more damage and human suffering than smaller ones do, but the situation is far more complex than that. A moderate earthquake in a densely populated area may cause great injury and damage, yet a very large earthquake in an isolated region may not affect humans at all. Many factors relating to physical geography can influence an earthquake's impact on people and their built environment. In general, the farther a location is from the earthquake epicenter, the less the effect of shaking, but this generalization does not always apply. An earthquake in 1985 caused great damage in Mexico City, including the complete collapse of buildings, yet it was centered 385 kilometers (240 mi) away.

The Mercalli Scale of earthquake intensity (I–XII) was devised to measure the impact of a tremor on people, their homes, buildings, bridges, and other elements of human habitation. Although every earthquake has only one magnitude, intensity can vary greatly from place to place, so a range of intensities will be encountered. The impact of a single earthquake on a region varies spatially, and the patterns of Mercalli intensity can be mapped. Earthquake intensity maps use lines of equal shaking and earthquake damage, called *isoseismals,* expressed in Mercalli intensity levels. Patterns of isoseismals are useful in assessing what local conditions contributed to spatial variations in

(a) The location of different earth materials during the 1906 San Francisco earthquake.

## TABLE 14.1
### Modified Mercalli Intensity Scale

| | |
|---|---|
| I. | Not felt except by a very few under especially favorable conditions. |
| II. | Felt only by a few persons at rest, especially on upper floors of buildings. |
| III. | Felt quite noticeably by persons indoors, especially on upper floors of buildings. Many people do not recognize it as an earthquake. Standing motor cars may rock slightly. Vibrations similar to the passing of a truck. Duration estimated. |
| IV. | Felt indoors by many, outdoors by few during the day. At night, some awakened. Dishes, windows, doors disturbed; walls make cracking sound. Sensation like heavy truck striking building. Standing motor cars rocked noticeably. |
| V. | Felt by nearly everyone; many awakened. Some dishes, windows broken. Unstable objects overturned. Pendulum clocks may stop. |
| VI. | Felt by all, many frightened. Some heavy furniture moved; a few instances of fallen plaster. Damage slight. |

greater or lesser shaking and impact. Earthquake intensity factors vary geographically according to soil or bedrock conditions, population density, construction type and quality, and topography. Areas on unconsolidated earth materials, with poor construction, or having dense populations tend to suffer more from shaking and tend to experience greater damage.

San Francisco was struck by a major earthquake in 1906 that resulted in an estimated 3000 deaths, many injuries and the destruction of buildings by ground shaking and a fire caused by damage to electrical and gas lines. Neither the magnitude or intensity scales existed in 1906. Follow-up studies, however, suggest that the earthquake magnitude was about 8.0, and the spatial distributions of Mercalli intensity have also been mapped and show great variations in shaking and damage caused by the earthquake. These geographic patterns of isoseismals were analyzed to explain why certain areas suffered more than others did. Areas of bedrock were shown to have experienced lower intensities (less damage) in comparison to areas of unconsolidated earth materials. Much of the worst damage occurred on artificially filled lands along the bayfront and on areas that had been stream valleys but were covered over with loose earth materials in order to construct buildings on the land. In some cases, buildings on one side of a street were destroyed, but on the other side buildings suffered little damage.

Analyzing where intensities were higher or lower than expected, given their locations, helps us understand the reasons for spatial variations in earthquake hazard. The geographic patterns of Mercalli intensity that are generated by small tremors help in planning for larger earthquakes in the same area. The patterns of shaking and isoseismals are likely to be similar for a larger earthquake with the same epicenter. The area affected, however, would be more extensive and levels of intensity would increase for each isoseismal.

(b) Geographic patterns of Mercalli intensity caused by the 1906 earthquake. The areas of intensity VIII+ in the northeast quarter of the city are on artificial landfill.

| | |
|---|---|
| VII. | Damage negligible in buildings of good design and construction; slight to moderate in well-built ordinary structures; considerable damage in poorly built or badly designed structures; some chimneys broken. |
| VIII. | Damage slight in specially designed structures; considerable damage in ordinary substantial buildings with partial collapse. Damage great in poorly built structures. Fall of chimneys, factory stacks, columns, monuments, walls. Heavy furniture overturned. |
| IX. | Damage considerable in specially designed structures; well-designed frame structures thrown out of plumb. Damage great in substantial buildings, with partial collapse. Buildings shifted off foundations. |
| X. | Some well-built wooden structures destroyed; most masonry and frame structures destroyed with foundations. Rails bent. |
| XI. | Few, if any (masonry) structures remain standing. Bridges destroyed. Rails bent greatly. |
| XII. | Damage total. Lines of sight and level are distorted. Objects thrown into the air. |

Source: Abridged from *The Severity of an Earthquake: A U.S. Geological Survey General Interest Publication.* U.S. Government Printing Office: 1989–288–913.

1988 in Armenia (in the former Soviet Union) destroyed two cities, with a death toll of 25,000. The effects of an 8.1 magnitude earthquake with an epicenter 240 miles from Mexico City in 1985 caused a death toll of more than 9000 and extensive damage to many buildings, including high-rise sections of the city. Loosely deposited earth materials tend to shake violently in comparison to areas of solid bedrock. Mexico's densely populated capital city is built on an ancient lake bed made of soft sediments that shook greatly, causing much destruction even though the earthquake was centered in a distant mountain region.

A Los Angeles area earthquake in 1994 killed about 60 people, collapsed freeways, and caused commuter traffic jams for months afterward ( Fig. 14.34a). This earthquake had a magnitude of 6.7, with the epicenter at Northridge in the San Fernando Valley, and was not considered the "big one" that southern Californians have been warned is coming. The $30 billion cost from this earthquake was the greatest in U.S. history because 24,000 buildings were damaged. Residents of the Pacific Northwest have also been warned about an impending "big one" as the North American plate slips over the subducting Juan de Fuca plate. In February 2001, a 6.8 magnitude earthquake struck, damaging buildings, roads, and bridges in a region that includes Seattle and many nearby communities.

Major earthquake tragedies have occurred in Peru (1970, magnitude 7.8), where 65,000 lives were lost when enormous avalanches obliterated entire mountain villages, and in Iran (1990, magnitude 7.7), where at least 50,000 died in remote areas. In January 1995, a 7.2 magnitude quake struck the densely populated port city of Kobe, Japan ( Fig. 14.34b). More than 5000 people were killed and more than 50,000 buildings were destroyed, with property damage estimated at about $100 billion. In terms of economic cost, this was the most expensive natural disaster in history.

Although most major earthquakes are related to known faults and occur in or near mountainous regions, the one most widely felt in North America was not. It was one of a series of tremors that occurred during 1811 and 1812, centered near New Madrid, Missouri, and was felt from Canada to the Gulf of Mexico and from the Rocky Mountains to the Atlantic Ocean. Fortunately, the region was not densely settled at that time in history. Because of that huge earthquake, St. Louis is considered to be at risk from a major earthquake, as shown on the seismic-risk map for the conterminous United States ( Fig. 14.35). Although not common in these regions, recent earthquakes have occurred in New England, New York, and the Mississippi Valley. Probably no area on Earth could be called entirely "earthquake safe."

Volcanic and tectonic processes are natural functioning parts of the Earth system. Many regions of active earthquake or volcanic activity are incredibly scenic and offer attractive environments in which to live, so it is not surprising that some of these hazard-prone areas are densely populated. It is essential that residents in areas where volcanic, tectonic, or other potentially hazardous natural processes are active, and the governments responsible for those regions, make - detailed preparations for coping with disasters before they occur.

(a)

(b)

**FIGURE 14.34**

Earthquake damage along the Pacific rim. (a) The San Fernando Valley earthquake occurred on January 17, 1994. The epicenter was below Northridge, California, and the quake killed about 60 people in the Los Angeles area. (b) Exactly 1 year later, on January 17, 1995, an earthquake struck Kobe, Japan. The deadly tremor killed more than 5000 people and destroyed more than 50,000 buildings.

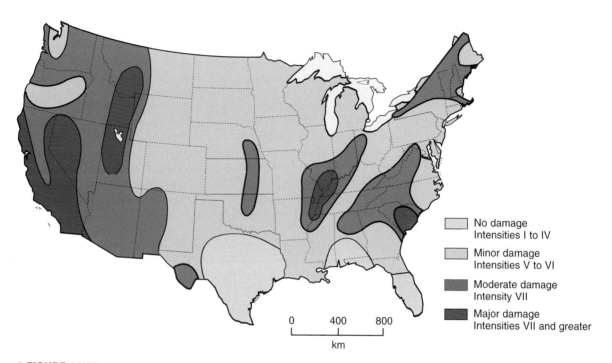

**FIGURE 14.35**
Earthquake hazard potential in the conterminous United States. This map shows a very generalized determination of the greatest expectable earthquake intensities for the 48 states shown.
**What is the earthquake hazard potential for where you live, and what does that level of intensity mean according to the Mercalli scale in Table 14.1?**

# Relationships between Rock Structure and Topography

The landform features that result from tectonic activity vary from microscopic fractures to major mountain ranges. However, few of these remain in original form because of the leveling action of degradation and aggradation. It is important to note the difference between surface topography (*landform features*) and the structural arrangement of the rocks (*rock structure*), as shown in Figure 14.36. For instance, the limb of an anticline may form a surface ridge, but it can also form a valley because of erosional activity on weak or broken rocks. Some mountain tops in the Alps are the erosional remnants of synclines, which are structural downfolds.

As another example, Nashville, Tennessee, occupies a topographic valley, yet it is sited in the remains of a structural dome (a circular, domal, anticline). An important distinction is that the terms *mountain* and *ridge* refer to surface landforms (topography), whereas the terms *anticline* and *normal fault* signify rock structures, which may or may not be visible at the surface. It is also important to remember that the topographic variation on Earth's surface has resulted mostly from three major factors: the interaction among forces that act to create relief, the gradational processes that operate to shape landforms and reduce relief, and the relative strength or resistance of different rock types to degradational processes. Much consideration will be given to these factors in the following chapters that deal primarily with landforms.

**FIGURE 14.36**
Relationships between structure and topography in folded rocks. Not all upfolds form mountains, nor do all downfolds form valleys. The bedrock structure and its configuration may or may not be directly reflected in the terrain. (a) The structure is an anticline (left) and a syncline (right), but the surface is only one of low relief. (b) Here, the center of a downfold, a syncline, has resisted erosion to form a mountain peak. (c) A valley has been eroded into the crest of an anticline.

# Define & Recall

| | | |
|---|---|---|
| geomorphology | shield volcano | syncline |
| igneous processes | cinder cone | recumbent fold |
| tectonic processes | composite cone | overthrust |
| gradation processes | plug dome | faulting |
| relief | caldera | fault |
| landform | intrusion | dip-slip fault |
| degradation | pluton | normal fault |
| aggradation | stock | reverse fault |
| punctuated equilibrium | batholith | thrust fault |
| volcanism | laccolith | strike-slip fault |
| volcano | sill | fault scarp (fault escarpment) |
| plutonism | dike | tilted fault block |
| pyroclastic material | volcanic neck | horst |
| tephra | structure | graben |
| volcanic ash | dip | rift valley |
| lava flow | strike | earthquake |
| pahoehoe | compression | focus |
| aa | tension | epicenter |
| fissure flow | crustal warping | earthquake magnitude |
| flood basalt | folding | earthquake intensity |
| basalt plateau | anticline | modified Mercalli Scale |

## Discuss & Review

1. What are the four basic types of volcanoes? Give an example of each.
2. Which is the most dangerous type of volcano? Explain why.
3. How do intrusive processes and volcanism differ from each other?
4. What is a batholith? What is the difference between a sill and a dike?
5. What are the major differences between folding and faulting? What causes these types of crustal activity?
6. Draw a diagram of folding, showing anticlines, synclines, and overthrusts.
7. What causes an earthquake?
8. What is the relationship between the focus and the epicenter of an earthquake?
9. What is the major difference between volcanism and solid tectonic activity? What structural and landform features are the results of each of these two types of processes?
10. How do earthquake magnitude and earthquake intensity differ? Why are there two systems for evaluating earthquakes?

## Consider & Respond

1. Name the tectonic or volcanic activity that would be most responsible for the following:
   a. Sierra Nevada
   b. Cascades
   c. Basin and Range region
   d. Ridge and Valley section of the Appalachians
2. Name several areas in the United States that are highly susceptible to natural hazards from earthquakes or volcanic activity.
3. List five countries bordering the Pacific Ocean that exhibit evidence of major volcanic activity.
4. Name a few countries that also face tectonic hazards but that are not in the Pacific region.
5. Can you recall any recent earthquakes or volcanic eruptions in the news? If so, where did these occur?
6. Assume that you are a regional planner for an urban area in the western United States. What hazards must you plan for if the region has active fault zones and volcanoes?
7. What recommendations would you make as far as land use and settlement patterns are concerned to lessen the danger from these tectonic hazards?

# Volcanic Landforms
## Map Interpretation

### The Map

The Menan Buttes map area is located on the upper Snake River Plain in eastern Idaho. The Snake River Plain is a region of recent lava flows that extends across southern Idaho. It is part of the vast volcanic Columbia Plateau, which covers more than 520,000 square kilometers (200,000 sq mi) of the northwestern United States. The lava originated from fissure eruptions that spread vast amounts of fluid basaltic lava across the landscape, accumulating to a thickness of several thousand feet. Most of the Snake River Plain has an elevation between 900 and 1500 meters (3000 and 5000 ft). Rising above the basalt plain are numerous volcanic peaks, including Menan Buttes and Craters of the Moon. The Snake River flows westward across the region.

The Snake River Plain has a semiarid, or steppe, climate. Because of the moderately high elevation of the plain, temperatures are cooler than in nearby lowlands, with an average annual temperature of about 10°C (50°F). Precipitation is between 25 and 50 centimeters (10 and 20 in.). It is unpredictable from year to year, and there are thunderstorms during the summer months. The low rainfall total is mainly a result of a rain-shadow effect. Moisture-producing storms from the Pacific Ocean are blocked from reaching the region by the Cascade Range and the Idaho Batholith section of the northern Rockies. The upper Snake River Plain and the Menan Buttes area are left with dry, adiabatically warmed air descending from the lee slopes of the mountains. The vegetation cover in the area is sparse, mainly characterized by sagebrush and bunch grasses.

### Interpreting the Map

1. What type of volcano are the Menan Buttes? On what landform characteristics is your decision based?
2. Define a butte. Are these landforms really buttes?
3. What is the local relief of the northern Menan Butte? What is the depth of the crater of each butte?
4. Do you think that these volcanoes are active at present? What evidence from the map and aerial photograph indicates activity or a period of inactivity?
5. Sketch an east–west profile across the center of northern Menan Butte from the railroad tracks to the channel of Henry's Fork. Is this profile typical of a volcanic summit and crater? What is the horizontal distance of the profile?
6. Slope ratio can be calculated by dividing the relief by the horizontal distance. For example, a 3000-foot-high mountain slope with a horizontal distance of 9000 feet would have a slope ratio of 1:3. What is the slope ratio for the western slope of northern Menan Butte from the crater ridge down to the railroad tracks at the foot of the butte?
7. This is a shaded relief topographic map and differs from most of the other topographic maps in this book. What is the major advantage of this shaded relief mapping technique? Are there any disadvantages, compared with the regular contour topographic maps?
8. Compare the southern Menan Butte on the map with the vertical aerial photograph on this page. Why would it be useful to have both a map and an aerial photograph when studying landforms? What is the chief advantage of each?

Vertical aerial photograph of one of the Menan Buttes, Idaho

Opposite:
Menan Buttes, Idaho
Scale 1:24,000
Contour interval = 10 ft
U.S. Geological Survey

# CHAPTER 15

Weathered rock slabs on a granite dome in Enchanted Rock State Natural Area, Texas.   J. Petersen

# Gradation, Weathering, and Mass Wasting

**PHYSICAL Geography Now™**  This icon, appearing throughout the book, indicates an opportunity to explore interactive tutorials, animations, or practice problems at now.brookscole.com/gabler8

# CHAPTER PREVIEW

**Landforms that are created by volcanism, tectonism, and other processes are then shaped and modified by gradational processes.**

- What evidence in Earth's crust and on its surface supports this statement?
- What does the statement suggest concerning the appearance and distribution of past and future landforms?

**The theoretical end product of gradation is a land surface reduced to low elevation and low relief.**

- What Earth processes operate in opposition to this end result?
- What are the sources of the energy that produces gradation?

**Weathering and mass wasting are initial steps in the overall process of gradation.**

- How are they important in shaping Earth's surface features?
- Why are they initial steps?
- How are they related to erosion and deposition?

**Sedimentation, volcanism, faulting, and folding produce rocks and structures that vary in their resistance and susceptibility to gradational processes.**

- Why is this so?
- What factors cause rocks to vary in this manner?
- How are these processes related to joints and fractures?

**The rates at which rocks weather are not equal because different rocks have varying resistance and different climatic conditions influence the type and intensity of weathering.**

- In what ways does climate influence weathering?
- How do the combined effects of differential weathering and erosion influence the shape and development of landforms?

**Despite spectacular mass wasting events, the imperceptible downhill motion of soil and regolith, called creep, is probably the gravity process with the greatest worldwide impact.**

- How can this be so?
- What factors facilitate creep?

---

The previous two chapters have summarized theories, materials, and structures associated with the processes that create topographic irregularities on Earth's surface. Now we begin to examine gradation, which acts to simultaneously modify, reduce, and perhaps eventually remove these irregularities. *Gradation* involves rock weathering and gravitational processes (discussed in this chapter) in association with erosion and deposition. *Degradation* operates to wear down areas that are steep or have high relief by eroding materials that are then transported and deposited in areas of lower elevation. *Aggradation* processes also construct landform features by building depositional landforms such as deltas (Chapter 17), alluvial fans and sand dunes (Chapter 18), and glacial moraines (Chapter 19).

The removal of rock material from one location and its deposition elsewhere are accomplished by transporting agents, the most important of which are water, glaciers, gravity, and wind. The following two chapters discuss how water underground and on the surface, particularly in streams and rivers, acts as a gradational agent. Chapter 18 covers the development of landforms in arid environments and examines the wind as a factor in erosion and deposition. Chapter 19 focuses on glaciers, glacial processes, and glacial ice as an agent of landform development, and Chapter 21 discusses landform processes in coastal zones. Throughout the study of landform modification, the effects of these processes in relation to humans, other animals, and plants, will also be noted.

## Gradation and Tectonism

The landforms and elevation differences on Earth's surface develop through the interaction of two sets of opposing forces and processes—those that tend to raise the land surface (■ Fig. 15.1a) and those that operate to wear it down (■ Fig. 15.1b). Ultimately, if uplift ceases, the land surface may be reduced to a level and form beyond which further modification by erosion and deposition will be minimal (■ Fig. 15.1c). This is, however, a simplified conceptual model of what may actually occur, because local environmental conditions worldwide vary so extensively.

Gradation includes the removal of loose material, its transportation to another location, and its deposition. **Erosion**—the removal of rock materials—and the **transportation** of those eroded materials act together to lower the land elevation. **Deposition** fills in depressions or raises the land surface by the accumulation of sediments or other rock materials. The combined result of erosion and deposition is a gradual reduction of irregularities in elevation, particularly if uplift ceases or operates at a rate that is low compared to the rates of degradation (see Fig. 15.1c).

Degradation and aggradation are ongoing processes, and their agents are constantly at work wherever there are differences in elevation. The processes that build up Earth's surface create new landforms and increase relief that will be further modified by gradation. Topographic relief is a function of recency and amount of uplift, rates of degradation, and how well the rocks and landforms resist the processes that work to break them down and remove them. Areas of hard rocks and recent uplift tend to be mountainous, with high elevation and high relief. Areas with soft rocks and regions that have been tectonically inactive for a long time in terms of Earth history tend to have low relief and low elevation.

Degradational processes, however, can also produce rugged landscapes. For example, the original surface of a plateau (or a shield volcano) tends to be relatively smooth although its elevation may have been raised thousands of meters. But degradation, working through its primary agent—running water—gradually begins to erode valleys separated by higher land, creating an area of mountains or hills. Eventually, as running water wears away more and more of the uplifted land surfaces, the irregularities originally created by water erosion may also be reduced.

Gravity and solar radiation are the ultimate sources of energy for gradation. Gravity causes water and glacial ice to flow downslope, eroding, transporting, and depositing materials. Acting virtually alone however, gravity is also a major force that moves rock and weathered debris downslope to lower elevations. This process of removal, transportation, and deposition directly by gravity is called **mass wasting**, or

(a) Tectonic disturbance

(b) Gradation begins

(c) Nearly complete gradation

■ **FIGURE 15.1**
The opposing processes of uplift and gradation: (a) tectonic uplift; (b) gradation begins; (c) nearly complete gradation.
**What is the ultimate effect of the gradation process?**

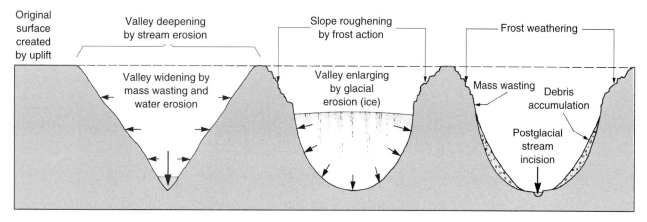

**FIGURE 15.2**
The scenery of most high mountain regions has been formed by at least three separate phases of gradational development: stream cutting and valley formation during tectonic uplift; glacial enlargement of former stream valleys, with intense frost weathering above the level of the ice; and postglacial weathering and mass wasting with recent stream cutting.
**How does the profile of a valley change at each phase?**

mass movement. Toward the end of this chapter, we consider some common processes of mass movement, how they shape the landscape, and their potential impacts on humans.

Weathering, mass wasting, and the various erosional agents typically do not work alone in shaping and developing a landform. More often they combine, simultaneously or sequentially, to modify and level the landscape. For example, both the northern Rockies and the Sierra Nevada were produced by uplift, but much of the spectacular terrain seen there today was produced by glacial activity, running water, and mass wasting that have sculpted mountains and valleys and, to varying degrees, continue to modify the landscape ( Figs. 15.2 and 15.3). However, to simplify the explanation of how gradational forces and processes operate and to recognize and understand their resultant landforms, each of these processes and their associated land features will be discussed separately.

## Weathering

**Weathering** includes various processes that cause rocks at or near the surface to disintegrate or decompose. Weathering loosens rocks and breaks them down into fragments small enough to be moved downslope by erosional or gravitational processes. Some weathered rocks, however, remain in place for thousands or millions of years, gradually becoming part of the soil mantle. The processes involved in rock weathering do not include the movement of weathered rock particles; they only break rocks into smaller fragments that can be more readily removed and transported by erosion or mass wasting, to be deposited elsewhere.

The definition of *weathering* is the breakdown of surface (or near-surface) rock materials *in place,* which involves no movement of the resulting fragments ( Fig. 15.4). It is important to note, however, that little or no erosion or mass wasting will occur without initial preparation by rock weathering.

Weathering is not a single process but includes many processes that fall into two basic categories. **Physical,** or **mechanical, weathering** *disintegrates* rocks, breaking them into fragments without altering their chemical composition. **Chemical weathering** *decomposes* rock through a variety of chemical reactions that cause minerals to weaken and decay. Both types of rock weathering take place at or near the surface, though evidence of weathering has been found as deep as 185 meters (600 ft) underground. Variations in the depths to which weathering occurs, as well as differences in the types and rates of weathering, depend primarily on four factors: (1) the structure and composition of the rocks, particularly rock strength and resistance to weathering processes; (2) climate, particularly temperature and humidity regimes (thus, the term *weathering*); (3) the land surface configuration, its slope and relief; and (4) the type and density of vegetative cover. Relationships among climatic variables and the rates, types, and spatial distributions of weathering processes are of particular interest to geographers ( Fig. 15.5). In most environments, either physical or chemical weathering processes may dominate, but disintegration and decomposition generally work together.

## Physical Weathering

The mechanical disintegration of rocks by physical weathering is important to gradation processes in two ways. First, smaller rock particles are more easily removed and transported

**FIGURE 15.3**
An aerial view of mountain summits in the northern Rockies. The mountains in the foreground are the White Cloud Peaks in the Sawtooth Range of Idaho.
**Can you identify evidence of the three phases shown in Figure 15.2?**

**FIGURE 15.4**
A greatly weathered boulder. This boulder, which was once hard and solid rock, is being broken down by weathering into clays and mineral fragments.
**Why haven't the fragments been washed away to form sediments?**

by one of the gradational processes. Second, the breakup of a large rock into smaller ones encourages chemical weathering because it increases a rock's surface area, thus exposing more of the rock to possible chemical decomposition.

Joints and fractures that develop in igneous, sedimentary, and metamorphic rocks are zones of weakness that expose more surface area, facilitating weathering on and between these exposed rock surfaces. Chemical and physical weathering both proceed faster along joint planes or in zones where the rocks have been broken by stress. Jointing can be found in any solid rock that has been subjected to crustal stresses ( Fig. 15.6a). The processes of volcanism, tectonism, mountain building, and rock formation produce fractures in the rocks that can be exploited by weathering, mass wasting, and erosion to shape landforms. Landscapes and landforms that develop by weathering in fractures, combined with erosion, are strongly influenced by the spatial pattern of joints in the rock. Joints divide rocks

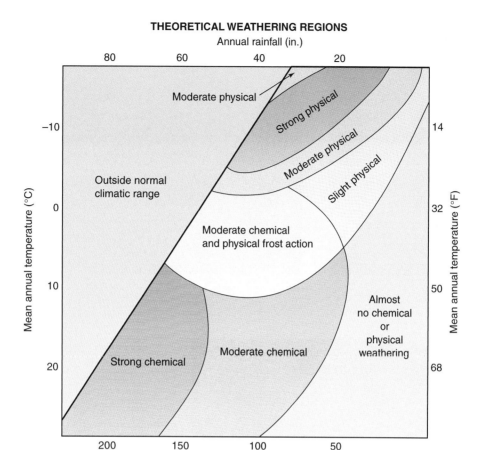

**FIGURE 15.5**

This diagram of weathering regions shows relationships between climate and weathering processes. Physical weathering is most active where temperature and rainfall are both low. Chemical weathering is most active in regions of high temperature and rainfall. Most world regions experience a combination of both physical and chemical weathering.

**In what weathering region would an area that had an annual mean temperature of 5°C and annual rainfall of 100 centimeters be located?**

into many different configurations, most commonly resembling blocks or columns, which will be apparent in a landscape that exposes weathered and eroded bedrock that is fractured into these shapes ( ▌Fig. 15.6b and c).

Plants, animals, and humans also contribute to the disintegration of rocks. Plants begin to grow in narrow cracks provided by rock joints, and their growing roots exert pressure on the rocks, sometimes enough to fracture the rock, causing the joint to open further ( ▌Fig. 15.7). Digging in the ground by burrowing animals, such as gophers and ground squirrels, causes rocks to weaken by fragmentation and disintegration. Certain human activities also encourage physical weathering. To greatly varying degrees, human activities involving, for example, bulldozers, off-road vehicles, bombs, mining, quarrying, and even hikers shatter rocks into smaller fragments.

In climates where the temperature passes between freezing and thawing, water freezing in fractures and joints is an important means of physical weathering. When water freezes, it expands in volume by 9%, exerting tremendous outward force. Drivers who fail to fill their radiators with antifreeze in preparation for winter or for a ski trip to the mountains find this an expensive result of a natural process. In cold-winter areas, water pipes in homes may burst if they are not sufficiently insulated to protect them from freezing.

Similarly, when water seeps into cracks in exposed rock and freezes, the expanding ice forms a wedge that can split the rock apart ( ▌Fig. 15.8). This process, called **frost wedging,** is most intense where cycles of daytime thawing and nighttime freezes are frequent—for example, in middle- to high-latitude regions and at high elevations. This weathering

(a)  (b)

■ FIGURE 15.6
Different landscapes formed by weathering and erosion on fractured rocks. (a) Massive jointing influences the location and shape of canyons in and around Zion National Park, Utah. To get an idea of scale, note the smoke from a wildfire. (b) Weathered granite forms small hills of blocky jointed rock east of the Sierra Nevada in California. (c) Rock spires formed from weathering and erosion of column like jointing in Bryce Canyon National Park, Utah.

process is even attacking the "ageless" 18-meter-high (60-ft) granite faces of presidents on Mount Rushmore. Freeze–thaw cycles have enlarged fractures in the granite since the sculptured faces were completed in 1941 (■ Fig. 15.9). This is a reminder that even solid granite will not last forever when exposed to weathering processes.

Frost wedging is a very important weathering process in most high-latitude regions, but it is most active in high mountains at or near the tree line. Frost wedging breaks up exposed bedrock at these higher elevations (■ Fig. 15.10), and if fragments break loose on a steep slope, gravity causes them to fall or tumble downslope. Broken rock fragments that accumulate at the base of steep slopes or cliffs, called **talus**, often produce a landform called a **talus cone**, named after its distinctive shape (■ Fig. 15.11). Any process that causes rocks to break, fall, and accumulate at the base of a slope can produce talus, but where frost wedging occurs, talus cones are particularly conspicuous. Frost wedging is the weathering process that causes the rocks to break, but the downslope falling, rolling, and accumulation under the influence of gravity cause deposition of talus.

In a process that is similar to frost wedging, mechanical disintegration also results from the growth of salt crystals. This process occurs most commonly in arid or rocky coastal regions, particularly in sandstone or in fractured rocks that can absorb the salty water. Between periods of wetting (rain or surf), the rock dries out and capillary moisture moves salt water outward to the rock surface. Evaporation of the water leaves saline deposits behind, and salt crystals grow, wedging the rock apart. As this process is repeated over time in desert areas, it can create multitudes of niches and shallow caves in large sandstone cliffs and overhangs in other rocks. **Salt weathering,** or **salt wedging,** is also a major factor in the breaking down of rocks in the splash zone along rocky seacoasts and can cause road damage where salt is used to melt ice on frozen roads.

Caves and cliff overhangs created by salt wedging and related weathering processes along with mass wasting provided

**FIGURE 15.7**
Physical weathering: A growing tree has broken up a sidewalk and retaining wall made of concrete, just as trees and other plants can break up natural rock into smaller fragments.
**Besides trees, what other organisms cause organic weathering?**

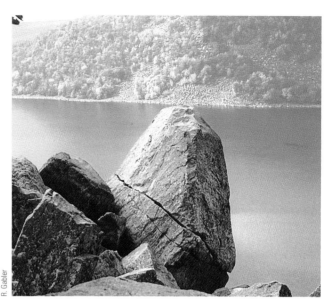

**FIGURE 15.8**
Frost wedging of a rock at Devil's Lake, Wisconsin. As water freezes in rock joints, it expands, wedging further into the rock.
**What is the reason that frost wedging can split such a large rock?**

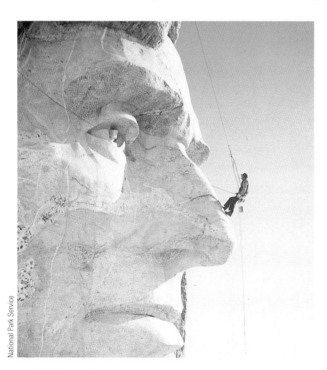

**FIGURE 15.9**
Even the granite face of President Lincoln at Mount Rushmore, South Dakota, is aging as a result of freeze–thaw cycles and frost wedging.
**Will he need a facelift someday?**

**FIGURE 15.10**
Water expands when it freezes into ice. The freezing of water in rock fractures creates a force of expansion great enough to break up rocks.
**How could this process of frost wedging affect the public in areas where freezes are common in winter?**

important living spaces for prehistoric Native American communities in the American Southwest—for example, at Mesa Verde, Colorado, and Canyon de Chelly, Arizona (Fig. 15.12).

These Native American cliff-dwelling villages generally faced toward the south, so they were shaded when the sun was high in summer and warmed by the sun's lower-angle rays in winter.

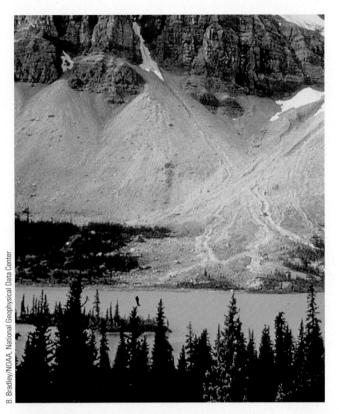

**FIGURE 15.11**
Talus cones in the Canadian Rockies formed by weathering and rockfalls at the base of steep slopes.
**Why does this rock debris form a cone shape?**

**FIGURE 15.12**
Ruins of prehistoric Native American dwellings built into the cliffs at Montezuma Castle National Monument, Arizona.
**Can you think of reasons why people chose these cliffs for their dwellings?**

Granular disintegration, another type of physical weathering, can result from various processes that break rocks down into their constituent mineral grains or particles, especially when the rocks consist of coarse-grained sediments or mineral crystals. Granite, composed of quartz, feldspar, and mica, is one example. These minerals expand with heat and contract with cold at different rates, producing stresses that may cause tiny fractures that can be exploited by various weathering processes. Granular disintegration is a common result of chemical weathering of the weaker minerals in granite and other coarse-grained rocks. When the weaker minerals decompose, the more resistant ones remain as weathered fragments.

It was once believed that extreme diurnal temperature changes in deserts were a primary cause of rock disintegration in those arid regions. Laboratory studies now refute this. Fire, however, can be an effective weathering agent. Because rock is a poor conductor of heat, fire expands only the rock's outer shell, causing it to fracture. In addition, fracturing is compounded by steam pressure formed from fire heating of water in the rocks. Thus, bedrock and boulders exposed to forest and brush fires often undergo cracking and accelerated granular disintegration.

**Exfoliation** refers to the *spalling,* or breaking off, of curved rock shells roughly parallel or concentric to the surface. Exfoliation is especially common in granite and was once thought to be caused by temperature changes. However, this process may actually be related to both chemical and physical weathering processes. Exfoliation that produces thin sheets or spalls a few millimeters or so thick is associated with hydration, a chemical weathering process. Exfoliation of massive rock sheets that are centimeters or meters thick is related to the unloading of pressure on rocks that were once deeply buried. When rocks that were once deep underground (often several kilometers) become exposed at the surface by uplift and erosion, pressure on the rocks is lessened, and they can minutely expand. As they do so, fractures develop, creating concentric shells of rock that resemble the layers of an onion called **exfoliation sheets,** which can be paper thin or meters thick ( Fig. 15.13a). This characteristic fracturing is especially apparent in massive intrusive rocks, such as granite, that crystallized under very high pressures. As rock layers are peeled away from large, solid rock masses by further weathering and erosion, immense dome-shaped hills or mountains, called **exfoliation domes,** are formed. Some of the world's best-known exfoliation domes are Stone Mountain in Georgia ( Fig. 15.13b), Half Dome in Yosemite National Park, and Sugar Loaf Mountain, which overlooks Rio de Janeiro, Brazil. Enchanted Rock and other nearby granitic hills in central Texas are also excellent examples of granite exfoliation domes (see the chapter-opening photograph).

## Chemical Weathering

Chemical weathering, or **decomposition,** prepares rocks for erosion and transportation in three ways. First, chemical alteration forms new minerals that are softer and less resistant to erosion. Second, chemical weathering creates new mineral substances—through chemical reactions, often involving water—that are greater in volume than the original mineral material. This volume increase causes expansion that fractures rocks, thereby accelerating both physical and chemical weathering by increasing the surface area exposed and by weakening the rock mass. Third, chemical weathering may dissolve minerals in a chemically active water solution (weak acid), making them easy to remove and transport. As more and more minerals are removed, the number and size of pore spaces and other openings in the rock are increased, allowing the rate of weathering to increase.

Water is an important catalyst in most chemical weathering processes. Rates of chemical weathering are intensified when water is present, so chemical weathering operates most effectively and rapidly in humid climates. Arid climates generally have enough moisture to allow some chemical weathering to take place, but the rounded boulders that show exfoliation and granular disintegration, common in arid regions of granite rocks, developed much of their present form in earlier times of wetter climates ( Fig. 15.14). The tendency of rock weathering to remove edges and corners, rounding them off to form spheroidally weathered boulders, is called **spheroidal weathering.** Although spheroidal weathering is very common, the term does not refer to a specific weathering process but is the result of a combination of weathering processes. Rounded spherical boulders tend to resist further weathering because there are no more sharp, narrow corners or edges for weathering to attack and a sphere exposes the least amount of surface area for a given volume of rock.

Chemical reactions intensify and occur more rapidly at high temperatures. Consequently, humid tropical regions are subject to intense chemical weathering in comparison to places with cooler temperatures. Subarctic and polar climates are subject to minor chemical weathering processes, but the effects of physical weathering are much more visible in cold regions. Thus, in hot, humid regions, such as the tropical rainforest, savanna, and monsoon climates, chemical weathering is more significant than physical weathering in affecting the topography. The landforms and rocks of warm, humid climates (and also in humid middle-latitude regions) display the importance of chemical weathering in their generally rounded shapes. In contrast, the landforms and rocks of drier and cooler regions, where physical weathering dominates, tend to be sharp, angular, or jagged.

The four principal processes of chemical weathering are hydration and hydrolysis, oxidation, carbonation, and solution. The most important catalysts and reactive agents performing

(a)

(b)

**FIGURE 15.13**

(a) Exfoliation sheets in the Sierra Nevada of California resemble layers in an onion. (b) Stone Mountain, in Georgia, is a huge granite exfoliation dome. Compare the form of Stone Mountain in this photo to the maps in Fig. 2.22.

**Why is granite so susceptible to exfoliation?**

# Geography's Physical Science Perspective

## Expanding and Contracting Soils

A form of weathering that spans both chemical and physical processes is the concept of *shrink and swell*. This process refers to soils that expand and contract with the addition and removal of water.

Soils consist of both organic and inorganic constituents. The inorganic components are of two types. First are silts and sands, derived from parent materials that have not been appreciably altered by chemical weathering. Second are the clay colloids, derived through alteration of the original material. These clay colloids (particles with very small mass but with a high surface area) are tiny, platelike crystals. They are important because they influence the cation-exchange capacity (CEC) of the soil. Cations are positively charged atoms, such as potassium, calcium, and magnesium ($K^+$, $Ca^{2+}$, and $Mg^{2+}$). These are termed *metallic cations* to differentiate them from the nonmetallic cations, such as hydrogen ($H^+$). A high CEC means that clay colloids are well supplied with cations that can be exchanged for different cations in a soil solution. It is from this soil solution that plants acquire cations and nutrients for healthy growth. In a high-CEC situation, the $H^+$ in water ($H_2O$) can also enter and exit the crystal structure of clay minerals.

The capacity of the colloids to exchange cations is a function of their chemical composition and their crystal structure. Clay minerals have layered, "sandwichlike" structures. Their CEC depends on the clay mineral type and how much that type has been weathered. *Illite* and *vermiculite* are two clay minerals associated with slightly weathered soils; *montmorillonite* is an important component of moderately weathered soils; and *kaolinite* is characteristic of highly weathered soils. Of these four basic clay minerals, it is montmorillonite that expands and contracts the most. Water molecules can actually enter and exit the layered crystalline structure and cause the soil to expand when wet and contract when dry. This contraction and expansion (or shrinking and swelling) of these clay minerals can apply physical pressure on surrounding soil particles and rock surfaces, causing them to physically weather at a faster rate. According to some sources, the expansion in volume can vary from a very small percentage to more than 100%. In most cases, however, the expansion in volume is less than 50%.

The addition and removal of water from clay mineral structures is a chemical weathering process known as *hydration*. However when the clay minerals expand and contract in volume, they can then exert the forces associated with physical processes on surrounding earth materials. The end result of chemical and physical processes working in combination is a new process, shrink and swell—a weathering agent of considerable significance where water is present in clay-bearing soils.

Wetting of clays causes an expansion in volume that can shift and crack streets, roads, and building foundations. This process in expansive clay-rich soils causes an estimated $7 billion damage a year to structures in the United States. In areas subject to shrink and swell of soil, special construction techniques can be used to mitigate the problem.

The concrete driveway slab here has been heaved upward by expansive soils, also causing damage to the adjacent railing and retaining wall.

Foundation cracks like this one, caused by expansive soils, can cause serious damage to buildings that is expensive to repair.

**FIGURE 15.14**
Block disintegration of granite rocks, central Texas. The blocks of rock became rounded, or "spheroidal," as angles were attacked by both physical and chemical weathering.
**Can you see where the original joints were located?**

the most common oxides, as these mineral compounds are called, is basically rust, derived from the combination of iron and oxygen (Fig. 15.15). Rust is composed of two iron oxides: the minerals hematite (red) and limonite (yellow). The rusty stains visible on many rocks are often a combination of hematite and limonite, which are softer and more easily removed from rocks than the original iron-bearing minerals from which they were formed. The presence of water is very important in the oxidation weathering of minerals.

**Solution** Though it does not actually involve chemical change, **solution** is an important process of chemical weathering. Solution is also sometimes referred to as *dissolution* because it involves dissolving minerals in water. Dissolved minerals are easily removed and transported by groundwater, soil water, or surface water flow.

Some minerals, such as calcium carbonate (lime), sodium chloride (salt), and calcium sulfate (gypsum), form rocks that are soluble in neutral or acidic waters. Certain mineral deposits, called **evaporites**, are readily soluble because they originally precipitated out of solution as water became saturated with them, as evaporative water loss increased the concentration of minerals

these processes are water, oxygen, and carbon dioxide, all of which are common in soil, rock, precipitation, groundwater, and air.

**Hydration and Hydrolysis** Although both processes involve the addition of water to a substance, hydration and hydrolysis are distinctly different. In **hydration**, water molecules attach to the crystalline structure of a mineral without chemical change. Hydration causes a mineral to expand, which causes wedging that can result in either granular disintegration or exfoliation of thin sheets. Hydration also weakens minerals, making them more susceptible to other chemical weathering processes—particularly oxidation and hydrolysis—as well as to physical weathering.

**Hydrolysis** involves a chemical change as the hydrogen and oxygen in water ($H_2O$) disassociate and combine with another substance. Many common rock-forming minerals are susceptible to hydrolysis, particularly the silicate minerals that form igneous rocks. Hydrolysis, like hydration, also causes mineral grains to expand, creating minute fractures in the rock. Hydrolysis is not limited to rocks that are exposed at the surface; in tropical humid climates, it can decay rock to a depth of 30 meters (100 ft) or more.

**Oxidation** The chemical union of oxygen with another substance to form a new product is called **oxidation.** Chemical compounds formed by oxidation are usually softer than the original substances and have a greater volume. One of

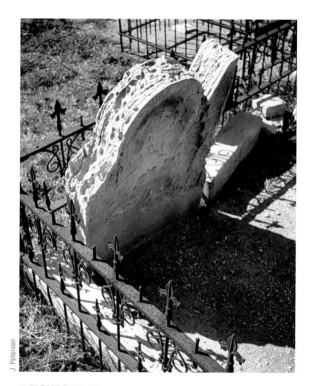

**FIGURE 15.15**
Weathering of limestone tombstones and an iron fence. This tombstone was not made of rock that was very resistant to weathering processes.
**The reddish fence rails result from what kind of chemical weathering?**

left behind. Many of the mineral compounds produced by chemical weathering are soluble. Even silica (the major constituent of glass) can eventually become soluble, particularly in the soils of hot, wet climates. Removal of silica is a major feature of the laterization process (see Chapter 12).

Most minerals that are insoluble or only slightly soluble in pure water dissolve more readily in slightly acidic water. Rain and soil water absorb carbon dioxide from the atmosphere and from decaying vegetation in the soil, typically forming a weak solution of carbonic acid. This mildly acidic water is capable of dissolving a wide variety of minerals, notably calcite (calcium carbonate), the mineral that forms limestone (❚ Fig. 15.16). When acted on by carbonic acid, calcium carbonate forms calcium bicarbonate, a salt that is water-soluble. Thus, the solution of limestone involves both **carbonation** (creation of calcium bicarbonate) and solution. In humid regions, the leaching (or removal) of salts such as calcium bicarbonate can severely weaken rocks by greatly increasing the size of pore openings.

Acid rain and acid fogs, caused by air pollutants dissolved into atmospheric moisture, have accelerated chemical weathering in recent decades. Because many of the world's great monuments and sculptures are made of limestone (calcium carbonate) or marble (calcite), there is a growing concern about weathering damage to these treasures. The Parthenon in Greece, the Taj Mahal in India, and the Great Sphinx in Egypt are examples of structures made of rock where chemical solution and salt buildup are damaging and rotting away monument surfaces.

Carbonic acid is not the only acid active in chemical weathering by solution. Derived primarily from decaying organic matter, other acids present in groundwater are capable of dissolving minerals. Although most organic weathering takes place below the soil surface, it can also affect exposed rock. Lichens and mosses that grow on rock surfaces secrete acid substances that assist rock disintegration and thus produce nutrients for these simple life-forms as they colonize bare rock. Lichens and mosses also absorb precipitation, increasing the frequency of wetting, thus accelerating the potential for weathering away the rock surfaces on which they live.

# Differential Weathering and Erosion

Wherever a variety of rock types occur in a landscape, some will be relatively resistant to weathering, and others will be weak, easily altered, broken down, and removed by erosion. The relative resistance of a rock to weathering is a function of rock type and its nature (for example, folded, fractured, faulted, layered, massive), along with the climatic conditions that it is exposed to over time. Because certain rock types weather more rapidly than others and increased weathering greatly facilitates erosional removal of weathered fragments, areas of diverse rock types undergo **differential weathering and erosion.**

A rock that is strong under certain environmental conditions, compared to other rocks at the same locality, may be weak relative to the same kinds of rocks in a different environmental setting. Rocks that are resistant in a climate dominated by chemical weathering may be weak where physical weathering processes dominate, and vice versa. Quartzite is a good example. It is chemically nearly inert and harder than steel, but it is brittle and can be fractured by physical weathering. In humid regions, limestone is highly susceptible to carbonation and solution, but under arid conditions, limestones tend to be resistant. Granite outcrops in an arid or semiarid region resist weathering. However, the minerals in granite are susceptible to alteration by oxidation, hydration, and hydrolysis, particularly in regions with warm, humid conditions. Accordingly, granitic areas are often covered by a deeply weathered regolith when they have been exposed to a tropical humid environment.

Shale is chemically inert but mechanically weak, susceptible to hydration, which converts it back to clay, its original material. Sandstone is only as strong as its cement, which varies from soluble calcite to inert and resistant silica. In general, the more massive the rock (the fewer the joints and bedding planes), the more resistant it is to all types of weathering.

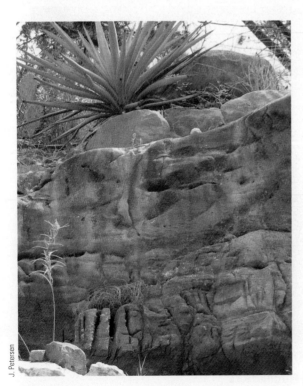

❚ **FIGURE 15.16**
Weathered limestone shows the effects of solution. Exposed limestone quickly discolors and becomes pitted by solution in humid climates.
**Why is limestone so intensely weathered under these climatic conditions?**

## Topography Related to Differential Weathering and Erosion

In the previous chapter, we learned that not all upfolds form mountains and not all downfolds exist in valleys. In general, rocks that are resistant to weathering and erosion tend to stand above their surroundings. It would be hard to imagine a great mountain made of soft, weak rocks because it would be too easily removed by erosion to exist for very long. Nearly anywhere high relief and steep slopes exist, the rocks that form these landform characteristics are resistant to weathering and erosion. Variations in resistance to rock weathering exert a strong and often highly visible influence on the appearance of landforms and landscapes. Differential weathering and erosion operate to make layers in rock structure more detectable unless a deep mantle of soil or weathered regolith covers the area, as is common in tropical, temperate, and middle-latitude regions with humid climates ( Fig. 15.17).

An outstanding example of how differential weathering and erosion can expose rock structure and enhance its expression in the landscape is the scenery at Arizona's Grand Canyon ( Fig. 15.18). In the arid climate of that region, limestone is resistant, as are sandstones and conglomerates, but shale is relatively weak. Resistant rocks are necessary to maintain steep or vertical cliffs. Thus, the stair-stepped walls of the Grand Canyon have cliffs and ledges composed of limestone, sandstone, or conglomerate, separated by gentler slopes of shale. At the canyon base, ancient, resistant metamorphic rocks have produced a steep-walled inner gorge.

Rock structures such as varying strata, joints, folds, and fault scarps tend to be more prominent and obvious in landscapes of arid and semiarid environments. In these dry environments, chemical weathering is minimal, so the slopes are not covered under a significant mantle of soil or weathered rocks. Resistant rocks stand out in the topography as cliffs, ridges, or mountains, while weaker rocks undergo greater weathering and erosion to form gentler slopes, valleys, and subdued hills ( Fig. 15.19).

Differential weathering and erosion can also create varying relief, such as hills and valleys, within an area where uplift

### FIGURE 15.17
Weathering, mass wasting, and erosion work together to make differences in rock structure and strength visible in the landscape. Layers of varying thickness and resistance result in a distinctive array of cliffs and slopes.

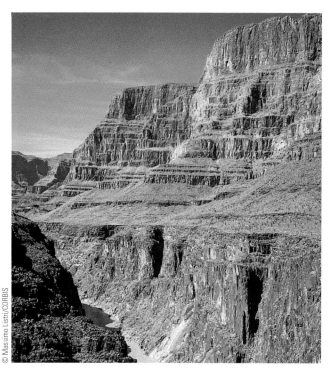

### FIGURE 15.18
The Grand Canyon of the Colorado River in Arizona is a classic example of differential weathering and erosion in an arid climate. This cross section shows the relationship between rock type and surface form in the canyon.
**What surface features do the more resistant rocks, such as sandstone, form?**

and subsequent erosion have affected an entire region. Again, it is important to understand that the processes that work to wear down landform features operate at differential rates. Hard rocks resist degradation and tend to form the higher places in a landscape—hills, ridges, and mountains. Valleys tend to form where the rocks are weaker and subject to intense degradation. Through this process of preferential removal of nonresistant rocks by degradation, the resistant bedrock structures exposed at the surface can be etched into the landscape, increasing the local relief.

An excellent regional example of differential weathering and erosion can be seen in the Appalachian Ridge and Valley region of the eastern United States ( Fig. 15.20). The rock structure here consists of sandstone, conglomerate, shale, and limestone folded into anticlines and synclines. These folds have been eroded so that the edges of steeply dipping rock layers are exposed as prominent ridges. In this humid climate region, forested ridges composed of resistant sandstones and conglomerates stand up to 700 meters (2000 ft) above agricultural lowlands that have been excavated by weathering and erosion out of weaker shales and soluble limestones.

The term *differential weathering and erosion* is often used as a single expression because, as weathering breaks up rocks at differential rates, it combines with erosion—the removal of weathered fragments—to produce distinctive landforms and

(a) Arid and semiarid regions

(b) Humid regions

**FIGURE 15.19**
(a) The bedrock structure of bare, angular hillsides typical in arid and semiarid environments tends to be well exposed, displaying differences in resistance to weathering and erosion of rock layers. (b) Humid region hillsides, in comparison, tend to be more rounded, covered with a mantle of soil or weathered bedrock, and more affected by a cover of vegetation.
**Which form is closer to the hillsides where you live?**

landscapes. Weathering and erosion (as well as the transportation and deposition that follow) are closely linked processes in the development of landforms. Erosion removes loose regolith exposing the fresh or less-weathered bedrock underneath to weathering processes at the surface.

## The Importance of Weathering

The fragmented materials prepared by chemical and physical weathering are not only available to be transported—either by direct movement by gravity or as part of the sediments carried by rivers, glaciers, or wind—but they also provide raw materials for soil development. Regolith formed by weathering is the major source of inorganic soil constituents, without which most vegetation could not grow. Weathered rock material is also a major source of nutrients in the oceans as rivers carry sediments to deposit on or near the coastlines.

Chemical weathering plays an important role in breaking down rocks into their mineral components and in creating new compounds. In this way, valuable ores of iron, aluminum, tin, manganese, and uranium may be concentrated over long periods of time. This effect usually occurs under humid, tropical conditions as part of the laterization process (see Chapter 12). Such secondary ores are to be distinguished from primary metallic ores, which are concentrated by magmatic processes.

**FIGURE 15.20**
A satellite image of Pennsylvania's Ridge and Valley section of the Appalachians clearly shows the effects of weathering and erosion on folded rock layers of different resistance. Resistant rocks form ridges, and weaker rocks form valleys. The pattern of anticlines and synclines was formed by an ancient continental collision that occurred before the formation of the Atlantic Ocean. The Susquehanna River cuts across the ridges and flowed across the region prior to the development of the present ridge and valley landscape.
**Can you see how the topography of the Ridge and Valley section influences human settlement patterns?**

## Mass Wasting

*Mass wasting,* also called *mass movement,* is a collective term for the downslope transport of surface materials in direct response to gravity. Everywhere on the planet's surface, gravity pulls objects toward Earth's center. On sloping surfaces, this force encourages downslope movement of loose or weak materials.

Mass wasting occurs in a wide variety of ways and at many scales. A single rock rolling and tumbling downhill is a form of this gravitational transfer of materials ( Fig. 15.21), as is an entire hillside sliding hundreds or thousands of feet downslope, burying homes, cars, and trees. Some mass movements are catastrophic in their scale and produce instantaneous violence ( Fig. 15.22). However, other types of mass wasting operate so slowly that they are imperceptible by direct observation. In the case of gradual mass movement, tilted telephone poles, gravestones, fence posts, retaining walls, trees, or unfortunate damage

**FIGURE 15.28**
Snow avalanches like this one in Alaska can block roads, knock down trees, carry rocks and tree trunks downslope, and damage structures. Note the heavy-packed nature of the snow, partly a result of pressure at impact. Many people erroneously believe that avalanche snow deposits are light and powdery

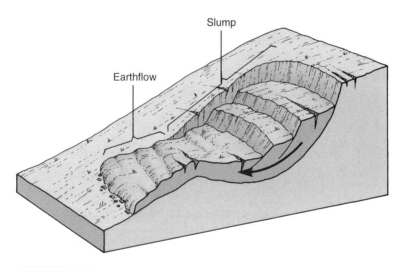

**FIGURE 15.29**
Diagram of a slump and earthflow.
**What is the difference between these two types of mass movement?**

conspicuous phenomena during exceptionally wet winters in California where they frequently damage expensive hillside homes. Slumps often change into **earthflows** in a downslope direction. Earthflows are slow-moving masses of soil and/or rock fragments.

**Debris Flow and Mudflow** The movement of water and rock debris in a mix of various proportions is called a **debris flow** or **mudflow.** Debris flows and mudflows are fast-moving masses that combine tumbling and mixing movements of rock and soil. The main difference in composition between debris and mud is the water content. Mudflows tend to be water-saturated and generally move faster and farther than debris flows. In general, debris flows and mudflows follow valleys although they can also flow as sheets downslope.

Mudflows and debris flows are the most fluid of all mass movements. They result from torrential rainfalls on steep, poorly vegetated slopes, typically in regions with distinct wet and dry seasons, such as coastal California, Oregon, and Washington ( Fig. 15.30). The rains flush weathered debris into canyons where it is mixed with floodwaters. The result is a torrent of mud, gravel, boulders, plant debris, and water that can knock out bridges and destroy buildings and homes. Such torrents frequently occur in wet winters following dry summers in which fires have destroyed hillside vegetation and left the soil vulnerable to removal by water runoff.

Serious mudflow hazards exist in many regions with active volcanoes. Here, steep slopes may be covered with hundreds of feet of loose volcanic ash. During eruptions, emitted steam, cooling and falling as rain, saturates the ash, sending down dangerous and fast-moving volcanic mudflows, known as **lahars.** Of particular concern are high volcanic peaks capped with glaciers and snowfields. Should an eruption melt the ice, rapid and catastrophic lahars rush down the mountains with little warning and bury entire valleys and towns. In the United States, there is a concern over some of the high Cascade volcanoes in the Pacific Northwest. Mounts Rainier, Baker, Hood, and Shasta all have the conditions in place, including nearby populated areas, for potentially disastrous mudflows to occur.

## Slow Mass Movement

Despite spectacular examples of the force and visual impact of rapid mass movements on the land, slow mass movements, especially creep, have a far greater cumulative effect on Earth's surface. **Creep** is the slow downhill movement of soil and regolith. Creep is so gradual that it is visually imperceptible to the observer. Yet creep is the most widespread and persistent form of mass wasting because it affects nearly all slopes where there are weathered materials and soil available for movement.

For the most part, creep does not produce distinctive landforms. Instead, it gradually moves surface materials downward toward the base of sloping land where they accumulate and can be carried away by one of the erosional agents, usually running water. Sometimes the material that creeps downslope is not removed but serves instead to fill in low areas and level the bottom of slopes.

Several factors facilitate creep; in most locations, these factors act in combination to move both regolith and soil slowly downhill. When soil water freezes, it expands with great force,

# Geography's Environmental Science Perspective
## Hazard Potential in the Cascades

With the violent eruption of Mount St. Helens in 1980, the majestic volcanic mountains located in northwest North America known as the Cascades served notice that they contain natural hazards of great significance. One of these hazards is a massive volcanic mudflow known as a *lahar*—the Indonesian word for such a traumatic event. These rapid flows are a slurry mix of rock debris, mud, glacial ice, and water. In 1985 such a volcanic mudflow killed 23,000 people in the Colombian town of Armero, located in a valley 50 kilometers (30 mi) from the Andean volcanic summit of Nevado del Ruiz. The lahar destroyed and then buried the town. A major eruption is not necessary to set off such a deadly event. It can be triggered by a minor eruption, which heats the ice and snow at the summit, or an earthquake, which destabilizes the steep, volcanic rock-, ash-, and ice-covered slopes.

Partly due to its lahar potential, Mount Rainier in the Cascades was targeted by scientists as one of several high-risk world volcanoes for intensive research under the International Decade for Natural Disaster Reduction (IDNDR). Mount Rainier was chosen because it is potentially more dangerous than any other Cascade volcano for several reasons. First, it is the highest and steepest of the peaks. Second, due to corrosive volcanic gases seeping from its active summit area, the rocks have been weakened, and the unstable clay-rich debris could slip. Third, Rainier is second only to Mount St. Helens in earthquake activity, and one of these could jar loose massive amounts of material. Fourth, there is more water stored in Rainier's summit ice cap, its dozen major glaciers, and deep winter snow cover than in the ice and snow of all the other Cascade volcanoes combined. Last, adding to the danger is the rapid growth of the urban area at the base of its western slopes.

Rainier also has a history of past lahar activity. About 5700 years ago a lahar known as the Osceola mudflow removed the summit of Rainier. Scientists have noted that the summit was probably 620 meters (2000 ft) higher prior to the top sliding off the steep volcano. This lahar changed river courses and buried river deltas and valleys with as much as 31

One of the lahars associated with the eruption of Mount St. Helens gave a clear demonstration of the destructive force of these volcanic mudflows as they flowed down river channels. Huge logs from lumbering operations along the Toutle River were swept downstream. Note the large trucks for scale.

Lahars also buried some areas under a concretelike mass consisting mainly of volcanic ash and water. Note the depth of burial on this garage.

meters (100 ft) of material. The flow reached as far as 100 kilometers (62 mi) from the summit and covered an area estimated at 505 square kilometers (195 sq mi).

During the 10-year IDNDR program, scientists used remote sensing, field mapping, and volcanic gas sampling to learn more about Rainier. They hoped to improve their ability to monitor the mountain, predict eruptions, and give ample warning of the potential for a castastrophic lahar roaring downslope toward the populated Puget Sound lowlands. The volcano's hazardous history should play a role in future land-use planning and in disaster preparation. As one scientist commented, "I think a lot of people don't realize the hazard potential that Mount Rainier has. They just look up and see a beautiful mountain."

**FIGURE 15.30**
A 1995 debris flow in La Conchita, California, destroyed several homes. Steep slopes consisting of weak unstable sediments failed during a period of heavy rainfall. This photo was taken in that year. Unfortunately, a similar weather situation triggered massive movement again in 2005, damaging 36 homes and killing 10 people.

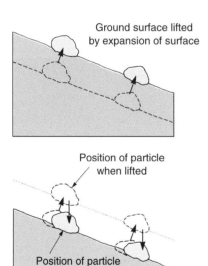

**FIGURE 15.31**
The relationship between creep and the expansion and contraction of soil on a slope. This diagram shows how freeze–thaw or wet–dry cycles produce a mechanism that moves particles downslope.
**Are there places near where you live that show evidence of creep?**

pushing soil particles upward; when thawing occurs, the soil contracts. Soil particles sink back down, but gravity causes them to be displaced slightly downslope of their original position, as shown in Figure 15.31. Cycles of freezing and thawing cause repeated lifting and downslope sinking of soil and rock particles, resulting in downslope creep of the surface materials.

The rate of movement is very slow, usually less than a few centimeters per year.

Alternating cycles of wetting and drying can also cause expansion and contraction of soil or regolith to produce a net downslope movement. This effect occurs because expansion tends to be greater on the downslope side, in response to the ever-present pull of gravity. Likewise, there is less contraction on the downslope side.

Small burrowing animals such as gophers and ground squirrels are effective soil movers. The tunnels that they dig help push soil downhill, and the excavated material that they bring to the surface also tends to fall downslope. Plant roots also push soils outward and in a downward direction on a slope. Even the trampling of soil and regolith by humans or grazing animals tends to push surface material downhill. Every step up or down a steep slope or even across it pushes some surface material downslope to a slightly lower position. Rates of soil creep are greater near the surface and diminish in the subsurface because wet–dry and freeze–thaw cycles are more frequent near the surface and friction increases with depth. Therefore, telephone poles, fence posts, and other human structures (Fig. 15.32a) and even trees (Fig. 15.32b), all of which are anchored at a level below the surface, exhibit a downslope tilt when affected by the downward movement of creep.

**Solifluction** is the relatively slow downslope movement of water-saturated soil and/or regolith. Solifluction, which means "soil flow," is most common in high-latitude or high-elevation tundra regions where there is a layer of permafrost beneath the surface. During summer when temperatures are higher, the top few centimeters or meters of the ground thaw. The frozen permafrost layer beneath the surface prevents downward percolation of meltwater. The upper, unfrozen soil layer then becomes water saturated, creating a soggy mass that sags slowly downslope in response to gravity until the next surface freeze arrives. The annual movement of soil may amount to only several centimeters. Solifluction is encouraged by the sparseness of vegetation found in tundra regions. Plant roots on gentle slopes act to retard mass movements, but where permafrost exists, roots are very shallow. The upper portion of the soil that freezes and thaws annually and that moves when saturated is called the *active layer*.

Evidence of solifluction can be seen in many tundra landscapes where irregular lobes of soil produce hummocky terrain or mounds (Fig. 15.33). Slopes affected by solifluction typically exhibit fractures and lobes formed by compression or tension during downslope movement. The general effect of both creep and solifluction is to produce rounded hillcrests and a landscape free of sharp angular features—the subdued landscapes usually associated with humid climates.

# Weathering, Mass Wasting, and the Landscape

In this chapter, we have concentrated on weathering and mass movement. Although neither weathering nor the slower forms of mass movement attract much human attention, they are important processes in shaping a landscape. They can create distinctive landforms, and every slope reflects the local nature of weathering and mass wasting processes. The influence that rock structure has on a landscape is a function of the resistance of that structure's rocks and minerals, the nature and intensity of weathering, and the way that weathered regolith is affected by gravity. Steep slopes reflect resistant rocks and slow weathering, and gentle slopes result from rapid weathering or weak rocks (which can also develop from weathering). Rounded slope forms suggest the dominance of chemical weathering and slow but active mass movement (creep or solifluction); angular slopes tend to be associated with physical weathering and mass movements in the form of rockfalls, rockslides, and debris flows. There are, however, exceptions to these general relationships. In the following chapters, we consider landforms produced by other dynamic but more generally visible and observable agents of gradation: running water, groundwater, wind, glacial ice, and coastal waves.

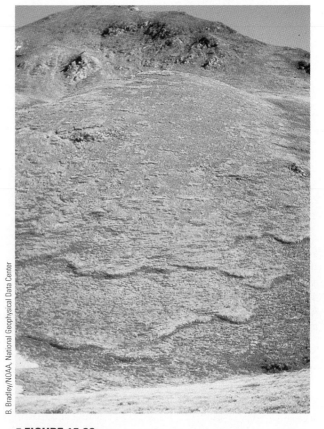

**FIGURE 15.32**
(a) Diagram of common visible landscape effects of creep on natural and human features. (b) Trees attempt to grow vertically, but their trunks become bent if surface creep is occurring.
**What other constructed features might be damaged by creep?**

**FIGURE 15.33**
Solifluction flow near Suslositna, Alaska.
**Solifluction is commonly found in what climate region?**

## Define & Recall

erosion
transportation
deposition
mass wasting (mass movement)
weathering
physical (mechanical) weathering
chemical weathering
frost wedging
talus
talus cone
salt weathering (salt wedging)
exfoliation
exfoliation sheet

exfoliation dome
decomposition
spheroidal weathering
hydration
hydrolysis
oxidation
solution
evaporite
carbonation
differential weathering and erosion
angle of repose
debris
mud

landslide
rockslide
rockfall
avalanche
slump
earthflow
debris flow
mudflow
lahar
creep
solifluction

## Discuss & Review

1. How does weathering differ from the transporting agents of gradation?
2. How does physical weathering encourage chemical weathering in rock?
3. How are the joints and fractures in a rock related to the rate at which weathering takes place?
4. What are several ways in which expansion and contraction can affect the weathering of rock?
5. Why is chemical weathering more rapid in humid climates than in more arid climates?
6. In what ways does chemical decomposition prepare rock for gradation? Give examples of oxidation, hydration, and solution.
7. Compare the ways in which hydrolysis and carbonation work below Earth's surface.
8. What types of rocks best resist all types of weathering? How visible are they in the landscape, compared to less resistant rocks?
9. What are the impacts of differential weathering and erosion on shaping landforms?
10. What two factors encourage mass movement the most? How do they work together?
11. What factors facilitate creep?
12. Describe the general effects of solifluction on the landscape. With what climates is solifluction usually associated?
13. What conditions encourage landslides and rockslides?
14. State the primary differences between a lahar and a debris flow. What causes the development of each?
15. Suggest ways in which weathering and mass movement affect human lives.

## Consider & Respond

1. Based on Figure 15.5, in what weathering region would your local area be classified?
2. Based on Figure 15.5, in what weathering regions would the following geographic regions be located?
   a. Brazil's Amazon Basin
   b. The North Slope of Alaska
   c. The summit of Pike's Peak, Colorado
   d. The Mojave Desert of southern California
   e. The Appalachian Mountains of Pennsylvania
3. What would you recommend as a solution to prevent the loss of valuable historical monuments to weathering processes?
4. If you were an urban planner in a city with numerous steep slopes, what major hazards would you have to plan for? What recommendations would you make to lessen these dangers to the community?

# CHAPTER 16

The work of groundwater can produce limestone caverns of great beauty. © David Hiser/Getty Images

# Underground Water and Karst Landforms

**Physical Geography Now™** This icon, appearing throughout the book, indicates an opportunity to explore interactive tutorials, animations, or practice problems at now.brookscole.com/gabler8

# CHAPTER PREVIEW

**Water is an essential natural resource for life on Earth and the most distinctive feature of our planet's environment.**

- What are the peculiar properties of water that make it so important?
- What role does water play in making Earth a unique planet?

**Groundwater and surface water are closely interrelated components of the hydrosphere, a major Earth subsystem.**

- How are groundwater and surface water related?
- What is the movement of water through the hydrosphere called?
- What are the sources of energy that drives the water movement in the system?

**The depth (location) of the water table, water quality, and water quantity are important factors in the availability of groundwater.**

- How does the water table reflect interactions among precipitation, evaporation, and transpiration?
- How does the water table affect surface landforms and water features?
- What human activities may affect the water table and groundwater quality?

**The processes of dissolution, transportation, and deposition of soluble rock by groundwater and surface water produce distinctive landform features that characterize karst topography.**

- How are these processes involved in the development of karst topography?
- What are some of the distinctive surface and subsurface features that result from these processes?

**Karst landscapes are found primarily in regions where limestone bedrock lies at or near the surface.**

- Why is this so?
- What areas of the United States and of the rest of the world have karst landscapes?

Fresh water is a precious but limited natural resource. Today there is much concern about this natural resource. Many populated regions have limited supplies of fresh water, yet it is abundant in lightly populated or uninhabited areas such as the tundra and tropical rainforest regions. Of the fresh water on Earth, 70% is frozen as ice and snow in the polar regions and thus generally unavailable for human use. When most people think of fresh water, they typically envision rivers and lakes, both of which are important sources, but they represent less than 1% of this resource. The remaining fresh water, nearly 30%, is found and stored beneath the surface as **groundwater.** Groundwater resources represent 90% of the fresh water that is readily available for human use ( Fig. 16.1).

In the last 100 years, worldwide usage of fresh water has grown at a rate that is two times the rate of population growth. The United Nations has declared 2005–2015 to be the *Decade for Action: Water for Life* to promote development, conservation and wise use of water resources.

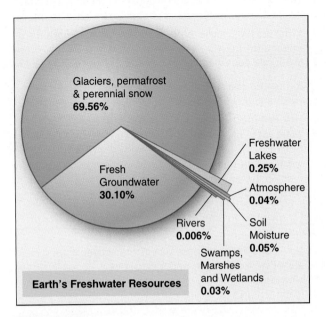

**FIGURE 16.1**
The global supply of fresh water. The available freshwater resources on Earth are limited, as 97% of Earth's water is salt water in the oceans (see again Fig. 6.2). A majority of the fresh water (about 70%) is stored as glacier ice mainly in remote polar regions. Groundwater is an extremely important source of fresh water for human use.
**How can we work to conserve our freshwater resources?**

In this chapter, we consider the significance of groundwater for human life, especially as it affects the environment as well as our domestic, agricultural, and industrial water supplies. Also, in keeping with our continuing examination of processes that shape the land, we focus on the landforms and processes associated with groundwater. Although of lesser importance on a worldwide basis when compared with streams, groundwater has a significant impact on landforms in certain areas.

# Occurrence and Supply of Groundwater

By far the largest proportion of groundwater is originally derived from atmospheric processes—precipitation that infiltrates underground to *recharge* the groundwater supply. Groundwater is brought to the surface by seeps, springs, and wells, and it contributes significantly to standing water bodies, such as lakes, and also to running water in streams.

Some groundwater resources that are being used today are irreplaceable because they accumulated during wetter times in geologic history. A small portion of groundwater is so deep beneath Earth's surface that it has probably never been part of the hydrologic cycle. Other groundwater, because of changes in Earth's surface, has been locked out of the hydrologic cycle for a long period of time. This water is trapped in sediment layers that were deposited by ancient rivers or seas. Future changes in the lithosphere could release these trapped groundwaters and return them to the hydrologic cycle. Through volcanic activity, these waters could be released in the form of *geothermal energy* (steam and hot water). The most obvious evidence of this activity occurs in hot springs and geysers.

## Groundwater Zones and the Water Table

Groundwater includes all subsurface water including that contained within the soil, within the loose regolith, and in bedrock. Under conditions of modest precipitation and good drainage, water infiltrating into the ground first passes through a level where both air and water fill pore spaces within soil and rock (Fig. 16.2). This level is called the **zone of aeration** because the soil and rocks contain air. Further movement downward brings the water to a second level, called the **zone of saturation,** where all the openings are completely filled with water. The **water table** is a surface that marks the upper limit of the zone of saturation. Water tables do not remain at a fixed depth below the land surface. The water table in a particular area fluctuates with the quantity of recent precipitation, loss by outflow to the surface, and the amount of removal by pumping. After heavy precipitation or snowmelt, a water table will rise. Because the depth to a water table generally reflects the precipitation amount for a given location (minus evaporation and other losses), they generally lie closer to the surface in humid regions and tend to be deep underground in arid regions.

Organized by their depth and water content, three distinct groundwater zones exist in humid regions (see Fig. 16.2). The lowest zone is where the subsurface rocks are always saturated. The upper zone is almost never saturated. Between these two is a zone that is saturated under conditions of ample precipitation (and infiltration), but not saturated under conditions of low precipitation. The water table fluctuates through this middle zone that alternates between unsaturated and saturated conditions. Obviously, a well or spring originating within a permanently saturated zone will always bear water, but one originating in an intermediate zone of fluctuation will run dry if the water table falls below it.

In some desert regions, there is no saturated zone at all because, soon after rainstorms, water evaporates at or just below the surface. In many arid and semiarid regions, if considerable groundwater is present, it may be very old, having accumulated during a past period of greater precipitation. Groundwater extracted from wells in these regions is not replaced from the atmosphere under current conditions of aridity, and as pumping continues, the water table will continue to fall. Pumping ancient water from the subsurface faster than it is being replenished has been called **water mining** because these groundwater resources have limited supplies that will not last indefinitely.

Despite their name, water tables are typically not level but tend to follow the general contours of the land surface, being higher under hills or other high topography and lower beneath valleys or depressions. Affected by gravitational force, water tends to seek its own level, so the groundwater under higher land surfaces generally flows downslope to lower elevations, as would a stream on the surface. Thus, at a

### FIGURE 16.2
**Environmental Systems: The Groundwater System.** The groundwater system is a subsystem of the hydrologic cycle. The major input for groundwater is precipitation or snowmelt that percolates into the ground. Infiltration from streams and lakes also contributes to the groundwater system. The major output (loss) of groundwater occurs by release at the surface through wells or in springs.

The groundwater system consists of three major parts: the zone of aeration, the zone of saturation, and the water table. Water percolates downward through the zone of aeration, where air fills most of the openings between soil and rock particles.

Eventually a depth is reached where all openings are saturated with water. This is the zone of saturation. The water table is a surface where the zones of aeration and saturation meet. Water table depths respond to changes in infiltration and outflow. During dry seasons or years, the water table drops, and during wet seasons or years, the water table rises.

More than 50% of the U.S. population receives their drinking water from groundwater sources. In the arid western United States, groundwater is also the major source of irrigation water, and overpumping of wells has caused the water table to drop drastically in this region. Groundwater is generally more difficult to pollute than surface water, but it is also harder to clean up. Some major sources of groundwater pollution are leaking septic systems, animal feed lots, leaking underground storage tanks, and seepage related to disposal of toxic materials.

given location, the water table is usually closer to the surface under low places than under high places.

In humid regions of low relief, the water table may be so high that it intersects the ground surface, producing lakes, ponds, or marshes such as those common in New England and along the Gulf Coast from Louisiana to Florida. Where the landscape is one of hills and narrow valleys, the lowest points on the water table are related to the elevation of valley floors. Some streams flow in valleys that have been eroded below the elevation of the water table. In this instance, groundwater will flow downslope along the water table's gradient and into the stream. This *effluent* condition where groundwater is flowing into a stream helps keep a stream flowing between rains or during dry seasons ( Fig. 16.3a).

Many streams in semiarid and arid regions flow only seasonally or immediately following a significant rainfall. In semiarid regions, the water table typically lies beneath the streambed during dry periods and rises to intersect the streambed during wet periods. These streams are fed by groundwater only during wet seasons, and they lose water by seepage into the subsurface during dry *influent* periods ( Fig. 16.3b). In most desert regions, this influent flow of water is continuous when a stream is flowing because the groundwater table is deep beneath the surface, so no groundwater is available to feed

**FIGURE 16.3**
The relationship between the water table and stream flow. (a) In humid regions, groundwater flows into major stream channels all year, providing streams with continuous flow. This is known as a gaining stream (effluent condition). (b) In semiarid regions or regions with a long dry season, the water table may drop below the streambed, so the stream dries up until the next wet period. (c) In desert regions, the absence of a water table or a water table at great depth can mean that stream flows only occur immediately after a significant rainfall, and flow diminishes downstream because of losses by seepage. This is known as a losing stream (influent) condition.

streams. Under these conditions, surface water only flows during and immediately after rains. Much of this downstream flow is eventually lost by seepage into the dry ground under the stream channel ( Fig. 16.3c).

## Factors Affecting the Distribution of Groundwater

The quantity, quality, and availability of groundwater in an area depend on a variety of factors. Most fundamental is the amount of precipitation that falls in a given location and in the areas that drain into it. Second is the rate of evaporation. Third is the ability of the ground surface to allow water to infiltrate into the groundwater system. A fourth factor is the amount and type of vegetation cover. Although dense vegetation transpires great amounts of moisture back to the atmosphere, it also inhibits rapid runoff of rainfall, encourages infiltration of water into the ground, and lowers evaporation rates by providing shade. Thus, the overall effect of forests in humid regions is to increase the supply of groundwater.

Two additional factors that affect the availability of groundwater are the porosity and permeability of the soil and rocks ( Fig. 16.4). **Porosity** refers to the amount or proportion of space between the particles that make up a soil or rock. Thus, loose gravels and sands, consisting of coarse particles that do not completely interlock, are exceedingly porous and can hold large amounts of water. In general, sedimentary rocks made of coarse fragments (sands and gravels) also have good porosity.

Rocks that are composed of interlocking crystals (rocks such as granite) have virtually no pore space and can hold little water within the rock itself. These crystalline rocks, however, may hold water within joints, which allow the passage of groundwater rather freely. The **permeability** of a material—its ability to allow passage of water through it—is related to

**FIGURE 16.4**
The relationship between porosity and permeability in two different sedimentary rocks. (a) Jointed sandstone is porous and permeable. (b) Unjointed clay is nonporous and impermeable. (c) Jointed rock that is nonporous gains permeability through its fractures.
**Which of these three rock types is the ideal aquifer?**

# Career Vision

**Carmen Yelle, Environmental Geographer/Water Quality, ADEM**
Bachelor of Science, Environmental Science, Emphasizing Physical Geography, University of Alabama

My early interests were in meteorology, which is under the general field of physical geography. I decided to major in geography when I took my first physical geography course. The processes and natural cycles of Earth sparked my interest. I continued taking physical geography courses, but in my junior year, when an environmental science degree was offered, I switched. Most of the required classes for this degree were also in physical geography.

I work for the Alabama Department of Environmental Management (ADEM). This is the state agency that administers major federal environmental laws, including the Clean Water Act, the Clean Air Act, and the Safe Drinking Water Act, as well as federal laws concerning hazardous and solid wastes. I am in the Office of Education and Outreach in the Nonpoint Source (NPS) Unit. Our department manages Environmental Protection Agency grants.

We operate under the Clean Water Act to help protect listed streams from further degradation and to improve their water quality. I also participate in educational activities to inform the general public about nonpoint source pollution (NPS) and its effect on water quality.

In college, I learned about the dynamics of streams and rivers, gaining knowledge that is very important in my career. Knowing how to read topographic maps is a geographic skill that I apply to my work. When dealing with NPS, you have to understand drainage patterns in a watershed and be able to identify them on a topographic map in order to figure out where the pollution is coming from. I have to understand the concepts of physical geography to determine how pollutants enter the stream and what can be done to alleviate the problem.

For students interested in a career like mine, I recommend being involved in an internship. I interned for the Natural Resource Conservation Service, a part of the U.S. Department of Agriculture. They have a volunteer program called the "Earth Team." This experience helped me directly apply what I was learning in class. Other students interned for the U.S. Geological Survey. There are many other agencies that are interested in taking on geography interns.

Physical geography illustrates how Earth operates as a system of interconnected parts and helps you understand relationships between water, air, and land as well as how each is affected by the others. Physical geography is an important course, not only for geography majors but also for those in other science fields, because it focuses on the big picture of Earth's environments.

---

the number and size of spaces within the rock, which may be large pore spaces, joints, bedding planes, faults, or even caverns (see Fig. 16.4). Thus, jointed granite can be described as permeable, even though the rock itself is not porous. Lavas such as basalt may contain much pore space formed by gas bubbles frozen into the rock, but typically these holes are not interconnected, so the rock itself is porous but not permeable. In volcanic areas, groundwater permeability is typically provided by numerous fractures. *Porosity* and *permeability* are not synonymous qualities. A coarse-grained sandstone, large pored and jointed, will typically be porous and permeable. In contrast, a finely porous, unjointed clay may contain significant amounts of water yet also be impermeable because the water clings to the tiny particles by surface tension. Porosity affects the potential amount of groundwater storage (by providing available spaces to the water). Permeability affects the rates and volumes of groundwater movement. Both of these factors affect the availability of groundwater for wells and springs.

An **aquifer** (from Latin: *aqua*, water; *ferre*, to carry) is a sequence of porous and permeable layers of rock or sediments that acts as a storage medium and transmitter of water ( ❙ Fig. 16.5). Although any rock material that is porous and permeable can serve as an aquifer, most aquifers that supply water for human use are sandstones, limestones, or deposits of loose, coarse sediments such as sand and gravel. A rock layer that is relatively impermeable, such as slate or shale, restricts the passage of water and limits its storage and therefore is called an **aquiclude** (from Latin: *aqua*, water; *claudere*, to close off).

Sometimes an aquifer will exist between two aquicludes. In this case, water flows in the aquifer much as it would in a water pipe or hose. Water can pass through the aquifer but does not escape outward through the aquicludes. Furthermore, soil water percolating downward may be prevented from reaching

**FIGURE 16.5**
An aquifer is a natural underground storage medium for groundwater. A perched water table can develop where impermeable rock strata exist above and below an aquifer. In this example, the perched water table is underlain by a dry zone between impermeable beds of shale and clay. Below the dry zone is the regional water table, the surface of the zone of saturation through which water flows toward the nearby river.
**Is a perched water table a reliable source of groundwater?**

the zone of saturation by an aquiclude. An accumulation of groundwater above an aquiclude is called a **perched water table** (see Fig. 16.5). Careless drilling can puncture the aquiclude supporting a perched water table so that the water drains farther down into the subsurface. The well must then be deepened to reach the true water table.

**Springs** are outflows of groundwater to the surface. They are related to many causes—landform configuration, bedrock structure, level of the water table, and the relative position of various types of aquicludes. Springs may occur along a valley wall where a stream or river has cut through the land to a level lower than a perched water table. An impermeable layer of rock prevents further downward percolation of groundwater, forcing the water to move horizontally until it reaches an outlet on the land surface. Spring flows are perennial if the water table always remains at a level above the outlet of a spring; otherwise a spring is intermittent, flowing only when the water table is at a high enough level to feed water to the outlet.

PHYSICAL
Geography Now™ Log on to Physical GeographyNow and select this chapter to work through a Geography Interactive activity on "Permeability" (click Groundwater → Tapping the Ground → Permeability tab).

## Availability of Groundwater

Groundwater is a vital resource to many areas of the United States and most of the world. In fact, half of the U.S. population derives its drinking water from groundwater. In some states, such as Florida, more than 90% of the drinking water is from groundwater. Today, more than two thirds of the groundwater used in the United States is for irrigation. One of the largest aquifers is the Ogallala Aquifer that underlies the Great Plains from west Texas northward to South Dakota. The Ogallala Aquifer supplies more than 30% of the irrigation groundwater in the United States ( Fig 16.6). Today there is much concern about this limited resource. The Ogallala region is semiarid, and much of the water withdrawn from the aquifer accumulated thousands of years ago.

Groundwater plays a major role in supporting many wetlands and forming shallow lakes in land depressions. Water bodies and wetlands, fed by groundwater, are critical habitats for thousands of resident and migratory birds. Adequate groundwater flow is vital for the survival of the Everglades in southern Florida. This "river of grass," its great variety of birds, and many other animals are totally dependent on the continued southward movement of groundwater flow through the region.

## AVAILABILITY OF GROUNDWATER

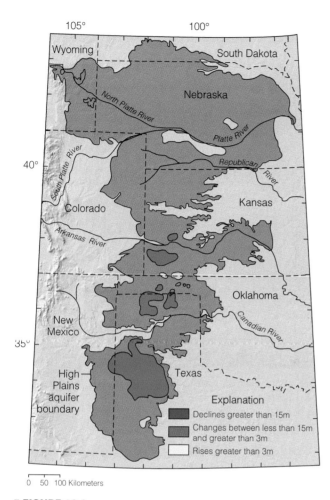

**FIGURE 16.6**
The Ogallala Aquifer supplies water to a wide, semiarid area of the High Plains. Said to be the largest freshwater aquifer in the world, much of the water in the Ogallala accumulated during wetter times thousands of years ago. From J.B. Weeks, et al., *U.S. Geological Survey Professional Paper 1400-A, 1988.*

**Why do you think the drop in water supply has been greatest in the southern part of the aquifer?**

## Wells

**Wells** are artificial openings dug or drilled below the water table. Water is extracted from wells by lifting devices ranging from simple rope-drawn water buckets to pumps powered by gasoline, electricity, or wind. In shallow wells, the supply of water often depends on fluctuations in the water table. Deeper wells that penetrate into lower aquifers and beneath the zone of water table fluctuation provide more reliable sources of water and are less affected by seasonal periods of drought.

In areas where there are many wells or a limited groundwater supply, groundwater removal may exceed the intake of water that replenishes the supply (a process called *groundwater recharge*). In many areas that are irrigated from wells, water tables have fallen below the depth of the original wells (■ Fig. 16.7). Progressively deeper wells must be dug (or the old ones extended) in order to reach the supply of water. In the Ganges River valley of northern India, the development of deep modern wells to replace shallow hand-dug wells has increased pumping capacity but has resulted in significantly lower water tables. In the High Plains of Texas, Oklahoma, and Nebraska, the drawdown of the Ogallala Aquifer, mainly for the irrigation of crops, is of serious concern.

In certain environments, particularly where high groundwater demand has led to heavy pumping, sinking of the land, called *subsidence,* can occur as the water pressure is reduced. Mexico City, Venice, Italy, and the Central Valley of California, among many other places, have subsidence problems. In southern California, groundwater has been artificially replaced by diverting rivers over permeable deposits. This process is known as *artificial recharge.*

Because most groundwater filters down through many layers of soil and rock before it reaches an aquifer, it is free of sediment but often carries a large mineral load dissolved from the materials through which it passes. Groundwater is thus said to be *hard* in comparison with *soft* rainwater. Moreover, just as increases in population, urbanization, and industrialization have resulted in the pollution of some of our surface waters, they have also resulted in the pollution of some of our groundwater supplies. A recent danger to groundwater supply has been from toxic waste seepage. In coastal regions, where groundwater pumping has lowered the water table, saline water from the ocean seeps in and replaces the fresh water because the pressure is no longer sufficient to hold back the salt water. This problem with saltwater replacement has occurred in many localities, notably in southern Florida, Long Island (New York), and Israel.

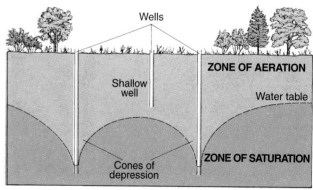

**ACTIVE FIGURE 16.7**
Watch this Active Figure at http://now.brookscole.com/gabler8.
Cones of depression in the water table caused by pumping water from wells. In areas with many wells, adjacent cones of depression intersect, lowering the water table and causing shallow wells to go dry.
**In extreme cases, what other condition may occur?**

# Geography's Physical Science Perspective

## Monitoring Groundwater Flow

It is relatively easy to measure and trace water as it flows on the land surface, particularly when it flows in a stream channel. However, when water infiltrates into the subsurface, the volume, direction, and rate of flow must be studied by using specialized techniques that were designed for monitoring groundwater. In limestone aquifers, groundwater flow rates are relatively high, and the large voids through which water flows do little to filter out pollutants. For these reasons, knowing how long it takes groundwater to flow from one place to another and where it is going as it flows underground are important environmental factors.

A common method of assessing, mapping, and tracing groundwater flow in limestone aquifers involves introducing a nontoxic dye into a location where there are high rates of groundwater flow. This process, called dye tracing, has many applications including cave mapping, tracing contaminated groundwater, locating sewer leaks, mapping pipelines, and a multitude of others.

Fluorescent dyes, which glow under an ultraviolet light, are the most successful and frequently used water tracers. Fluorescein, a green fluorescent dye first developed in 1871, was used a few years later to trace sinking portions of the Danube River in Germany. Today, there are many different fluorescent dyes, but only about ten of them are considered safe for use in groundwater.

Initially, dye tracing was strictly a qualitative science. Dye was injected into a location and the nearby area would be monitored, waiting for traces of the dye to appear in water in wells, springs, or streams, at another location. Later, it was discovered that charcoal grains absorb certain fluorescent dyes even when they are greatly diluted in groundwater. This discovery

Green fluorescein dye is introduced into water that will infiltrate into the groundwater table. The dye helps in the process of tracing the direction and rate of groundwater flow.

allowed for packets of charcoal (sometimes called "bugs") to be left unattended in springs or streams waiting for the dye cloud to pass. The charcoal packets could be analyzed for the presence of the dye. Unbleached cotton can also be used as a monitoring device because it also absorbs the dye as water containing minute traces of the dye flows past. These approaches meant that it would no longer be necessary to have people continually stationed at numerous emergence points to see if the dye would appear.

Originally human vision was relied on to detect presence or absence of the dye, to determine if the dyed water had passed a monitoring point. Further, because the dye was diluted in the groundwater, there were limitations on the dye-concentration levels that were visible. Ultraviolet lights, which made the cotton or carbon grains glow if they contained the dye, helped in this process. A device called a fluorometer, which quantifies the presence of the dye, was a great breakthrough in fluorescence analysis, allowing quantification of dye concentration (dilution) down to 1 part per billion (1 ppb). Since the mid-1980s, new technology has been available and has further refined fluorescence analysis and dye-tracing technologies. Identification of a particular groundwater dye and its concentration now can be accomplished with greater speed and accuracy. Dye tracing has been improved over the years and remains a widely used and very effective tool in groundwater studies.

## Artesian Wells

In an *artesian system,* water under pressure can flow upward to a level above the local water table. In **artesian wells,** water rises to the surface and flows out under its own pressure, without pumping. Certain conditions are required for an artesian water flow ( Fig. 16.8). First, a permeable aquifer such as sandstone or limestone must be exposed at the surface in an area of high recharge by precipitation or infiltration. This aquifer must receive water from the surface, incline downward often hundreds of meters below the surface, and

**ACTIVE FIGURE 16.8**
Watch this Active Figure at http://now.brookscole.com/gabler8.
Conditions that produce an artesian system. The Dakota sandstone, an aquifer that averages 30 meters (100 ft) in thickness, transmits water from the Black Hills to more than 320 kilometers (200 miles) eastward beneath South Dakota.
**What is unique about artesian wells?**

must be confined between impermeable layers that prevent escape of the water, except to the artesian springs or wells. These conditions cause the aquifer to act as a pipe that conducts water through the subsurface. The water in the "pipe" is under pressure from the water above it (closer to the area of intake at the surface). As a consequence of this pressure, water will flow toward any available outlet. If that outlet happens to be a well drilled through the impervious layer and into the aquifer, the water will rise in the well, sometimes gushing out at the surface. The height to which the water rises depends on the amount of pressure exerted on the water. Pressure in turn depends on the quantity of water in the aquifer (more water, more pressure), on the angle of incline (steeper slope, more pressure), and on the number of other outlets, usually wells, available to the water (more outlets, less pressure).

Sandstone exposed at the surface in Colorado and South Dakota transmits artesian water eastward to wells as far as 320 kilometers (200 mi) away (see Fig. 16.8). Other well-known artesian systems are found in Olympia, Washington; the western Sahara; and eastern Australia's Great Artesian Basin, which is the largest artesian formation in the world. The word *artesian* is derived from the Artois region of France, where the first known free-flowing well was dug in the Middle Ages.

# Landform Development by Solution

In areas where the bedrock is soluble in water, groundwater is an important agent in shaping landform features both at the surface and underground. Groundwater is a vital ingredient in subsurface chemical weathering processes, and, like surface water, groundwater dissolves, removes, transports, and deposits materials.

The principal mechanical effect of groundwater is to encourage mass movement by lubricating weathered material and soil, producing slumps, debris flows, mudflows, and landslides. Through chemical activity, groundwater contributes to many other and sometimes unique processes of landform development. Through the removal of rock materials by solution and the deposition of those materials elsewhere, groundwater is an effective land-shaping agent, especially in areas where limestone is present. Limestone can be dissolved by acidic groundwater or surface water, and wherever water can act on any rock type that is significantly soluble in water a distinctive landscape will develop. In many *karst* areas, surface outcrops of limestone are pitted and pockmarked by chemical solution, especially along joints, forming large, flat, furrowed exposures of limestone pavements (Fig. 16.9).

## Karst Landscapes and Landforms

Overwhelmingly the most common soluble rock is limestone. Landform features created by dissolution and precipitation (redeposition) by surface water or groundwater are found in many parts of the world. The eastern Mediterranean region in particular exhibits solution features in limestone on a large scale. These are most clearly developed on the Karst Plateau along Croatia's scenic Dalmatian Coast. Landforms developed by solution, most commonly formed in limestone, are called **karst** landforms after this classic locality. Karst regions are located in Mexico's Yucatán

Peninsula, the larger Caribbean islands, central France, southern China, Laos and many areas of the United States (Fig. 16.10).

The development of a classic karst landscape, in which solution has been the dominant process of land formation, requires several special circumstances. A humid climate with ample precipitation is most conducive to karst development. In arid climates, karst features are typically absent or are not well developed. However, some arid regions have karst features that originated during a period of geologic time when the climate was much wetter than it is today.

Another important factor in the development of karst landforms is the active movement of groundwater so that water, saturated with dissolved calcium carbonate, flows away to be quickly replaced by unsaturated water. Vigorous movement of groundwater occurs when an outlet at a low level is available, such as a deeply cut stream valley or a tectonic depression. In general, the greater the permeability, the faster groundwater will flow.

Because seepage of water into the subsurface tends to be concentrated where joints intersect, these intersections are subject to accelerated solution. Such concentrated solution can cause the development of circular depressions, called **sinkholes,** that are prominent features of many karst landscapes (Fig. 16.11a and b). There are two dominant types

**FIGURE 16.9**
Solution of limestone is most intensive in fracture zones where the dissolved minerals from the rock are removed by surface water infiltrating to the subsurface. This landscape shows a limestone "pavement" where intersecting joints have been widened by dissoluton.

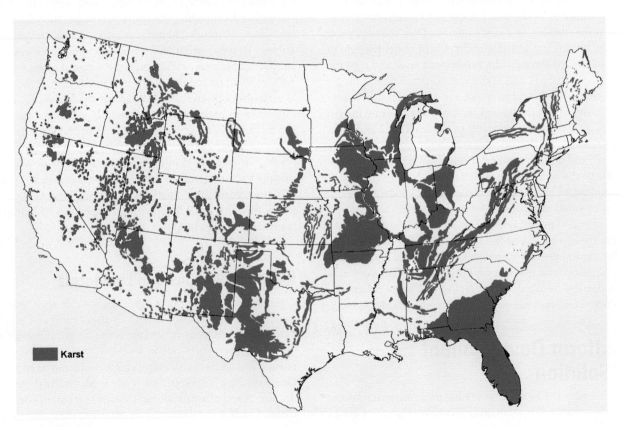

**FIGURE 16.10**
Map of limestone areas in the conterminous United States, where karst landform development exists to varying degrees depending on climate and local bedrock conditions.
**Where is the nearest karst area to where you live?**

### FIGURE 16.11

Idealized development of karst topography. (a) Solution at joint intersections in limestone encourages sinkhole development. Caverns form by groundwater solution along fracture patterns and between bedding planes. Cavern ceilings may collapse, causing larger and deeper sinkholes. Surface streams may disappear into sinkholes to join the groundwater flow. (b) Eventually, the limestone landscape may be divided by merging sinkholes and longer karst valleys, called uvalas. (c) Finally, limestone remnants—called haystack hills, or hums—are isolated above an exposed surface of insoluble rocks.

**Overall, what has been the general effect on the landscape of groundwater solution?**

## Geography's Environmental Science Perspective
### Sinkhole Formation

The sudden collapse of the ground, caused by the creation of a sinkhole, is a problem in many areas of the United States that have soluble rocks at the surface, or hidden beneath a cover of soil or regolith. Ground collapse, whether it is gradual or nearly instantaneous, can cause great difficulties on any land surface that is affected by this process. An example is shown in the accompanying photographs of a road collapse near Bowling Green, Kentucky.

Sinkholes can develop through the solution of underlying limestone bedrock where the water table is above the contact between bedrock and overlying loose rock and soil (regolith). As solution proceeds, groundwater flow washes away some of the regolith along its boundary with karst bedrock, creating a subsurface void. When the surface can no longer support the weight of loose materials above this void, the regolith slumps down and in, creating a sinkhole at the surface. When the surface materials suddenly collapse into the subsurface voids, there can be serious damage to property.

Earth scientists recognize two types of collapse processes that form sinkholes: regolith collapse and bedrock collapse. Sudden regolith collapse can occur in regions with a water table that ordinarily is high, during drought periods or at times when the water table is lowered by excessive pumping from wells. These collapses are caused when the regolith arches covering the openings in the limestone are no longer supported by groundwater buoyancy. Adding additional weight to regolith by constructing roads and buildings or by impounding of water at the surface increases the likelihood that regolith arches will suddenly collapse.

The creation of a sinkhole by bedrock collapse associated with an enlarging underground cavern is rarer than regolith collapse, but this process can also produce a significant environmental hazard. As underground passages are enlarged vertically in a cavern, the roofs above the passages become thinner and weaker, eventually collapsing to form a sinkhole of sizeable proportions. Physical geographers who have examined the Dishman Lane sinkhole report evidence of cave roof collapse in this case. It is likely that most of the roof collapsed thousands of years ago leaving a regolith arch at the surface, but the fact that limestone blocks were found in the rubble indicated that part of the cave roof collapsed in this event. The resulting sinkhole affected over an acre of surface area and caused considerable property damage.

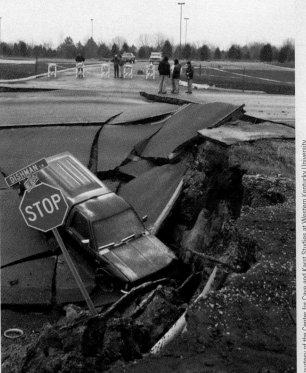

Sinkhole collapse near Bowling Green, Kentucky, caused severe road damage in an area larger than a football field. Luckily, no one was hurt during the rush-hour collapse although several vehicles were damaged. Fixing the problem required completely filling in the sinkhole with rock and repaving the road at a cost of more than $1 million.

of sinkholes identified by their differing formation processes. If the depressions are due primarily to the solution process at or near the surface and the removal of dissolved rock by water infiltrating downward into the subsurface, the depressions are called **solution sinkholes** ( Fig. 16.12a). If the depressions are caused by the caving-in of the land surface above voids created by solution in bedrock below, the depressions are termed **collapse sinkholes** ( Fig. 16.12b).

The two processes, collapse and solution, cooperate to create most sinkholes in soluble rocks. Whether the depressions are termed *collapse* or *solution sinkholes* depends on which of these two processes were *dominant* in their formation. Collapse and solution sinkholes often occur together in a region, and based on their form, they may be either difficult or easy to distinguish from one another. Solution sinkholes tend to be funnel shaped, and collapse sinkholes tend to be steep walled, but these shapes vary greatly.

Sudden collapse of sinkholes is a significant natural hazard that can cause severe property damage and human injury. These rapidly forming sinkholes can be caused by excessive groundwater withdrawal for human use, or they may occur during drought periods. Either of these conditions will lower the water table, causing a loss of buoyant support for the ground above, followed by collapse. Rapid sinkhole collapse has damaged roads and railroads and has even swallowed buildings ( Fig. 16.13).

Many karst regions have very few surface streams that continually flow because the high permeability of exposed fractured bedrock encourages infiltration of surface water underground to the water table. Water, slowly circulating below the surface, flowing along joints and between bedding planes, also dissolves limestone, creating **caves** and **caverns** as a system of connected passageways within the soluble bedrock (see Fig. 16.11a and b).

Surface water seeping into fractures in soluble bedrock widens the fractures by solution. These widened avenues for water flow increase the downward permeability and accelerate infiltration into the groundwater system. In some cases, streams may flow on less permeable bedrock upstream to encounter a highly permeable rock downstream where it loses its surface flow by infiltration (see Fig. 16.9). These are called **disappearing streams** because

**FIGURE 16.12**
The formation of sinkholes. (a) Solution sinkholes develop where surface water funneling into the subsurface dissolves bedrock to create a closed depression in the landscape. (b) Collapse sinkholes form when either the bedrock or regolith above subsurface voids created by solution collapses due to the removal of support.
**Which type of sinkhole poses the greatest potential threat to houses and roads?**

**FIGURE 16.13**
This large sinkhole formed in Winter Park, Florida, when a dropping water table caused the surface to collapse into an underground cavern system during a severe drought in May 1981.
**What human activities might contribute to such hazards?**

(a)

(b)

**FIGURE 16.14**
Disappearing streams. (a) In areas of karst topography, streams may disappear into a solution-widened hole in the ground, flow underground in a cave system, and perhaps emerge elsewhere as a major spring. (b) A topographic map shows a disappearing stream and a swallow hole. Note the hachured contours that indicate the closed depression into which the stream is flowing.

they "vanish" from the surface as the water flows into the subsurface (Fig 16.14a).

In many well-developed karst landscapes, a complex underground drainage system all but replaces the normal surface patterns of water flow. The landscape may consist of many large valleys that contain no streams. Surface streams originally excavated the valleys until the river exposed and broke through into a cavern system. River water was then diverted to underground paths as the stream disappeared into a hole in the cave roof called a **swallow hole** (Fig. 16.14b). This process is characteristic of the Mammoth Cave area in Kentucky. Some of the "lost rivers" may reemerge as springs where they have cut down to impervious beds below the limestone.

Sinkholes, also called **dolines** in some regions and *cenotés* in Mexico, may enlarge and merge over time to form larger karst depressions. As sinkholes, potholes, and dolines coalesce, the larger depressions that develop are often linearly arranged along former underground water courses, forming **valley sinks,** or **uvalas** (see Fig. 16.11b). The terms *doline* and *uvala* are derived from Slavic languages used in the former Yugoslavia.

After intense and long-term karst development, especially in wet tropical conditions, only limestone remnants are left standing above insoluble rock below. These remnants are usually in the form of small, steep-sided, and cave-riddled karst hills called **haystack hills,** or **hums** (see Fig. 16.11b and c). Examples of this landscape are found in Puerto Rico, Cuba, and Jamaica. They have been described as "egg box" landscapes because an aerial view of the numerous sinkholes and hums resembles the surface of an egg box (Fig. 16.15). If the limestone hills are particularly high and steep sided, the landscape is called *tower karst*. Spectacular examples of tower karst landscapes are found in southern China and Southeast Asia (Fig. 16.16).

## Limestone Caverns and Cave Features

Caverns created by the solution of limestone are the most spectacular and best-known karst landforms. Groundwater, sometimes flowing much like an underground stream, carves out networks of caverns. Should the water table drop to the floor of the cave or lower, typically resulting from climatic change or tectonic uplift, the caves later become filled with air. Interaction between the cave air and mineral saturated water will then begin to decorate the cave ceiling, walls, and floor with depositional forms.

Examples of limestone caverns in the United States are Carlsbad Caverns in New Mexico, Mammoth and Colossal Caves in Kentucky, and Luray and Shenandoah Caverns in Virginia. In fact, 34 states have caverns that are open to the public. Some are quite extensive with rooms more than 30 meters (100 ft) high and with kilometers of connecting passageways. Every year, millions of visitors marvel at the intriguing variety of forms, colors, and passageways that they see and experience on tours of limestone caverns. The vast majority of these features in limestone caverns are related to solution and deposition by groundwater.

LANDFORM DEVELOPMENT BY SOLUTION   461

**FIGURE 16.15**
A tropical karst landscape with hums and intergrown sinkholes. Intense solution in wet tropical environments can form a maze of intergrown sinkholes.

The nature of fracturing that exists in soluble bedrock exerts a strong influence on cavern development in karst regions. Groundwater solution widens the space between opposing surfaces of joints, faults, and bedding planes to produce passageways. The relationship between caverns and fracture distributions is evident on cave maps that show linear and parallel patterns of cave passageways (Fig. 16.17). Not all caves contain actively flowing water, but all caverns formed by solution show some evidence of previous water flow, such as deposits of clay and silt on the cavern floor.

Depositional forms in caves (**speleothems**) develop in a great variety of textures and shapes, and many are both delicate and ornate. Speleothems develop when previously dissolved minerals, particularly calcite, precipitate out of

**FIGURE 16.16**
Karst towers (hums) of Guilin, a limestone region in southern China.
**At one time in the geologic past, could this region have looked much like the landscape in Fig. 16.15?**

**FIGURE 16.17**
The Crystal Grottoes in Washington County, Maryland. This cave shows the influence of fractures through their geometric arrangement and spacing of passageways. Karst caverns develop by following zones of weakness, and groundwater flow widens fractures to develop a cave system.
**Why are limestone fractures so susceptible to widening by dissolution?**

groundwater, to produce some of the most beautiful and intricate forms found in nature. The dripping water leaves behind a deposit of calcium carbonate called *travertine*, or *dripstone*. As these travertine deposits grow downward, they form icicle-like spikes called **stalactites** that hang from the ceiling (Fig. 16.18). Calcium saturated water dripping onto the floor of a cavern builds up similar but more massive structures called **stalagmites.** Stalactites and stalagmites often meet and continue growing to form columns or pillars (Fig. 16.19).

Most people believe that evaporation, leaving deposits of lime behind, was the dominant process that produced the spectacular speleothems seen in caves, but that is not the case. Caves that foster active development of speleothems typically have air that is fully saturated, with a relative humidity near 100%, so evaporation is minimal. Where water drips from the ceiling, previously dissolved carbon dioxide gas is vented from the water into the cave air. This degassing makes the water less acidic and less able to hold the lime in solution, so the lime is deposited, eventually enough to form a stalactite, stalagmite, or other depositional cave feature.

Limestone caverns vary greatly in size, shape, and interior character. Some are well decorated with speleothems, and others are not. Many have several levels and are almost spongelike in the pattern of their passageways, whereas others are linear in pattern. The variations in cavern form often indicate differences in mode of origin. Most large caverns appear to have developed either at the water table, where the rate of solution is most rapid, or below the water table in the zone of saturation. A subsequent decline in the water table level, caused by the incision of surface streams, climatic change, or tectonic uplift, replaces the cavern water with air, allowing speleothem formation to begin. Underground rivers deepen some air-filled caverns, and collapse of their ceilings enlarges them upward. Many small caves appear to have been formed entirely above the water table by water percolating downward through the zone of aeration.

Cavern development is a complex process, involving such variables as rock structure, groundwater chemistry, and hydrology, as well as the regional tectonic and erosional history. The science of **speleology** (cavern studies) is

**FIGURE 16.18**
Stalactites form very slowly in places where water drips from the cave ceiling and deposits some of its dissolved minerals, typically lime. These stalactites in the Lehman Caves, Great Basin National Park, Nevada, are hollow and may grow into long, narrow, and very delicate tubes called soda straws.
**Why does evaporation of water tend to be only a minor process in the formation of stalactites?**

**FIGURE 16.19**
Cavern features (speleothems) of dripstone in Carlsbad Caverns National Park, New Mexico.
**How do you explain the presence of this huge cavern in a desert climate?**

particularly challenging. Our knowledge of caves has come from explorations as deep as hundreds of meters underground by "spelunkers" who have made scientific observations while crawling through mud and water and even over bat droppings in dark narrow passages ( Fig. 16.20). Recent cave exploration and mapping of the water-filled caves below the surface in Florida and other karst regions have involved scuba diving. Cave diving in fully submerged, totally dark, and confined passageways, which may contain dangerous currents, is an extremely risky operation.

## Geothermal Water

Hot springs and geysers are both intriguing groundwater-related phenomena. Water heated at its source is referred to as **geothermal water**. Hot or boiling groundwater was probably heated by contact with hot solidified rocks below the surface and is typically accompanied by steam. Geothermal waters flowing or bubbling out fairly continuously form a **hot spring**. Water flow that is intermittent and somewhat eruptive, expelling water and steam vertically out of the ground, as at Old Faithful in Yellowstone National Park, produces a **geyser** ( Fig. 16.21). The word *geyser* is an Icelandic term for the steam eruptions that are so common on that volcanic island. Geysers appear to erupt when the steam pressure below reaches a critical level and forces the column of superheated water and steam out of the fissure in an explosive manner.

Most hot springs and geysers contain significant amounts of minerals in solution, which become deposited in various forms, often as terraces or cones around the vent or spring. These calcareous travertine and siliceous geyserite deposits often accumulate in colorful and impressive forms ( Fig. 16.22).

Geothermal activity is usually associated with areas of tectonic and volcanic activity, especially along plate boundaries and over hot spots. Geothermal energy has been used to produce electricity in areas such as California, Mexico, New Zealand, Italy, and Iceland. The best geothermal water for harnessing energy is not only very hot for generating steam but also "clean"—relatively free of dissolved minerals that can clog pipes and generating equipment.

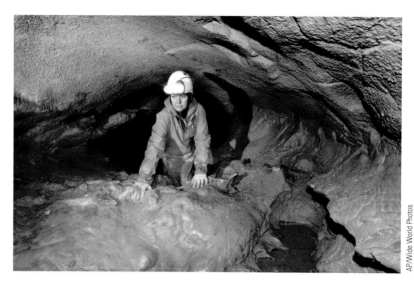

**FIGURE 16.20**
A spelunker explores the Lost River Cave System near Orleans, Indiana. Caving, also called spelunking, can be exciting but is potentially hazardous, so it requires skill, good equipment, and proper judgement of conditions.
**What are some of the potential hazards of caving?**

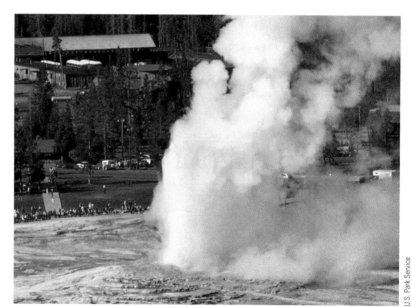

**FIGURE 16.21**
One of the world's most famous geysers is Old Faithful in Yellowstone National Park, Wyoming.

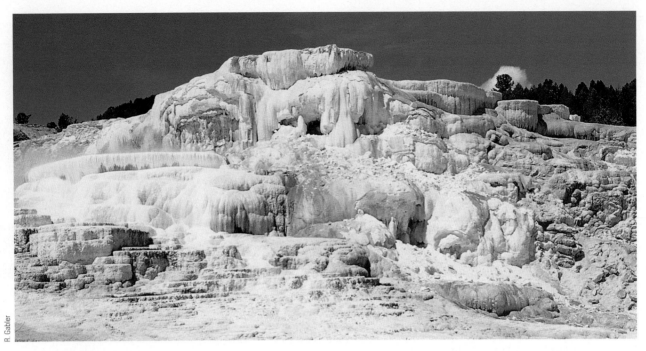

**FIGURE 16.22**
Hot spring calcareous deposits in Yellowstone National Park, Wyoming.
**Explain how these unique deposits are formed.**

## Define & Recall

| | | |
|---|---|---|
| groundwater | spring | doline |
| zone of aeration | well | valley sink (uvala) |
| zone of saturation | artesian well | haystack hill (hum) |
| water table | karst | speleothem |
| water mining | sinkhole | stalactite |
| porosity | solution sinkhole | stalagmite |
| permeability | collapse sinkhole | speleology |
| aquifer | cave (cavern) | geothermal water |
| aquiclude | disappearing stream | hot spring |
| perched water table | swallow hole | geyser |

## Discuss & Review

1. What is the water table? How is it related to climate?
2. What is porosity? What is permeability? How are these two characteristics related to groundwater?
3. Define and describe the conditions necessary for an aquifer.
4. Explain how an artesian well works.
5. Define karst. What conditions are necessary for a region to have karst landforms?
6. Describe a sinkhole. What are the processes of formation involved?
7. Explain how haystack hills are formed. What type of climate would be most conducive to their formation?
8. Describe how caverns form and name some common cavern features.
9. Explain how limestone is deposited by groundwater to form a stalactite in a humid cave.
10. What is a geyser? What is the energy source for a geyser? Where are geysers most often located?

## Consider & Respond

1. If you were a water resources geographer and had to plan for the development of groundwater resources for a community, what would be your major considerations?
2. Describe the major landform features in a region of karst topography. Over time, what changes might be expected in the landscape?
3. What are some human impacts and environmental problems related to the use of groundwater?

# Map Interpretation
## Karst Topography

### The Map

The Interlachen area is in northern central Florida. Florida's peninsula is the emerged portion of a gentle anticline called the Peninsular Arch. The region is underlain by thousands of feet of marine limestones and shales. This great thickness of marine sediments originated in the Mesozoic Era when Florida was a marine basin. As the arch rose, Florida became a shallow shelf and eventually became elevated above sea level.

Although outsiders think of Florida mainly as a state with magnificent beaches and warm winter weather from its humid subtropical climate, it is also a state with hundreds of lakes dotting its center. The lake region is formed on the Ocala Uplift, a gentle arch of limestone that reaches to 46 meters (150 ft) above sea level. Lake Okeechobee is the largest of these lakes and has an average depth of less than 4.5 meters (15 ft). Most of the lakes, such as those in the Interlachen map area, are much smaller.

Florida's central lake region is an ideal area for studying karst topography. Both the surface and the subsurface features express the geomorphic effects of groundwater. Extensive cavern systems exist beneath the surface. Much of the state's runoff is channeled through huge aquifers, and groundwater springs are quite common.

### Interpreting the Map

1. This area would be classified as what type of major landform surface? What is its average elevation?
2. What contour interval is used on this map? Why do you think the cartographers chose this interval?
3. On what type of bedrock is the map area situated? Do you think the climate has any influence on the landforms in the area? Explain.
4. What landform features on the map indicate that this is a karst region?
5. What are the round, steep depressions called? Why do lakes occupy some of the depressions?
6. Locate Clubhouse Lake on the full-page map (scale 1 : 62,500) and the smaller map (1 : 24,000). What is the elevation of Clubhouse Lake? What is its maximum width?
7. What is the area north of Lake Grandin?
8. What is the approximate elevation of the water table? (*Note:* You can determine this from the elevation of the lakes' water surface.)
9. Underground, the water flows through an aquifer. Define an *aquifer* and list the characteristics an aquifer must have. What is the general direction of groundwater flow in the aquifer underlying the Interlachen area?
10. Because much of central Florida is rapidly urbanizing, what problems and hazards do you anticipate in this karst area?

Putnam Hall, Florida Scale 1 : 24,000 Contour interval = 10 ft
U.S. Geological Survey

Opposite:
Interlachen, Florida
Scale 1 : 62,500
Contour interval = 10 ft
U.S. Geological Survey

# CHAPTER 17

The impact of flowing water on the landscape is evident at Vernal Falls in Yosemite National Park, California. Matt Ebiner

# Fluvial Processes and Landforms

**PHYSICAL Geography Now™** This icon, appearing throughout the book, indicates an opportunity to explore interactive tutorials, animations, or practice problems at now.brookscole.com/gabler8

# CHAPTER PREVIEW

**Running water, as sheet wash and stream flow, is the most important single agent of erosion and does more work than all of the other gradational agents combined.**

- What does this suggest to us about Earth's landforms?
- How is this significant to humans?

**Perennial streams flow every day of the year even if it has been several weeks since the last rainstorm.**

- What does this indicate to you about the source of water in perennial streams?
- How will the stream flow differ near the end of a large storm compared to after several weeks without rain?

**Moderate increases in the volume, velocity, and turbulence of stream flow can cause major increases in the sediment-carrying capacity of a stream.**

- What does this imply about stream erosion and deposition?
- What does it suggest about a stream at flood stage?

**The upper, middle, and lower courses of a stream that flows from the mountains to the ocean differ significantly in terms of the dominant gradational processes and landform features.**

- What features would you expect to see in the upper, middle, and lower courses of such a stream?
- How would the gradational processes differ in the three different positions along the stream?

**The level of the average stream rises to the brink of flooding roughly once every 1.5 years.**

- What is the geomorphic significance of this to the floodplain, the gently sloping, low-relief land commonly found adjacent to streams?
- What are the pros and cons of people artificially adding to the height of the natural levees that lie adjacent to many stream channels?

**Streams are natural systems that convey inputs, throughputs, and outputs of water, sediment, and energy from source to mouth.**

- How might a stream adjust to a lowering of sea level at its mouth?
- What changes might a stream channel undergo if a large fire stripped the vegetation off of a large portion of its drainage basin?

Flowing water is more influential in shaping the surface form of our planet than any other gradational process, primarily because of the sheer number of streams on Earth. Through both erosion and deposition, water flowing downslope over the land surface, particularly when concentrated in channels, modifies existing landforms and creates others. With few exceptions, nearly every region of Earth's land surface exhibits at least some topography that has been shaped by the power of flowing water, and many regions exhibit considerable evidence of stream gradation. Flowing water is the main gradational agent in both humid and arid environments. Polar lands buried under thick, perennial ice sheets represent one major exception to Earth's extensive areas of stream-dominated topography.

The study of flowing water as a gradational process, together with the study of the resulting landforms, is termed **fluvial geomorphology** (from Latin: *fluvius,* river). Fluvial geomorphology includes gradation by both channelized and un-channelized flow moving downslope by the force of gravity.

**Stream** is the general term for natural, channelized flow. In the Earth sciences, the term *stream* pertains to water flowing in a channel of any size, although in general usage we may describe large streams as rivers and use local terms, such as creek, brook, run, draw, and bayou, for smaller streams. The land between adjacent channels in a stream-dominated landscape is referred to as the **interfluve** (from Latin: *inter,* between; *fluvius,* river).

Because of the common and widespread occurrence of stream systems and their key role in providing fresh water for people and our agricultural, industrial, and commercial activities, a substantial portion of the world's population lives in close proximity to streams. This makes understanding stream processes, landforms, and hazards fundamental for maintaining human safety and quality of life.

Most streams occasionally expand out of the confines of their channel. Although these **floods** typically last only a few days at most, they reveal the tremendous—and often very dangerous—gradational power potential of flowing water.

The long-term effects of stream flow, whether dominated by erosion or deposition, are sometimes also quite dramatic. Two prime examples in the United States that illustrate the effectiveness of flowing water in creating landforms are the Black Canyon of the Gunnison River (Fig. 17.1a), carved by long-term river erosion into the Rocky Mountains, and the Mississippi River delta (Fig. 17.1b) where fluvial deposition is building new land into the Gulf of Mexico.

## Surface Runoff

Liquid water flowing over the surface of Earth—that is, **surface runoff**—can originate as ice and snow melt or as outflow from springs, but most runoff originates from direct precipitation. When precipitation strikes the ground, several factors interact to determine whether surface runoff will occur. Basically, runoff is generated when the amount, duration, and/or rate of precipitation exceed the ability of

(a)

(b)

**FIGURE 17.1**
(a) Long-term fluvial erosion by the Gunnison River, with the aid of weathering and mass wasting, has carved the Black Canyon of the Gunnison through a part of the Rocky Mountains in Colorado. The Canyon is 829 meters (2,722 ft) deep, 80 kilometers (50 mi) long, and very narrow in width, with resistant rocks comprising the steep walls. The Gunnison River flows beyond the canyon westward into the Colorado River toward the Gulf of California. (b) Through fluvial deposition, the Mississippi River has been building its delta outward into the Gulf of Mexico for thousands of years. Where the river enters the Gulf, the slowing current deposits large amounts of muddy sediment that came from the river's drainage basin. This is a false-color composite, digital image. Muddy water appears light blue, and clearer, deeper water is dark blue.

the ground to soak up the moisture. The process of water soaking into the ground is called **infiltration**, and the amount of water the soil and surface sediments can hold is the **infiltration capacity.** A portion of the infiltrated water will seep down to lower positions and reach the zone of saturation beneath the water table, while much of the rest will eventually return to the atmosphere by evaporation from the soil or by transpiration from plants. When more precipitation falls than can be infiltrated into the ground, the excess water will flow downslope by the force of gravity as surface runoff.

Various factors act individually or together to either enhance or inhibit the generation of surface runoff. Greater infiltration to the subsurface, and therefore less runoff, occurs under conditions of permeable surface materials, deeply weathered sediments and soils, gentle slopes, dry initial soil conditions, and a dense cover of vegetation. **Interception** of precipitation by vegetation allows greater infiltration by slowing down the rate of delivery of precipitation to the ground. Vegetation also enhances infiltration when it takes up soil water and returns it to the atmosphere through transpiration. Given the same precipitation event, surface materials of low permeability and limited weathering, thin soils, steep slopes, preexisting soil moisture, and sparse vegetation each contribute to increasing runoff by decreasing infiltration ( Fig. 17.2). Human activities can impact many of these variables, and in some places the generation of runoff has been greatly modified by urbanization, mining, logging, and agriculture.

Once surface runoff forms, it first starts to flow downslope as a thin sheet of unchannelized water, known as **sheet wash,** or unconcentrated flow. Because of gravity, however, after a short distance, the sheet wash will begin to move preferentially into any preexisting swales or depressions in the terrain. This concentration of flow leads to the formation of tiny channels, called **rills,** or somewhat larger channels, called **gullies.** Rills are on the order of a couple of inches deep and a couple of inches wide, whereas gully depth and width may approach as much as a couple of feet. Water does not flow in rills and gullies all the time but only during and shortly after a precipitation (or snowmelt) event. Channels that are empty of water much of the time like this are described as having **ephemeral flow.** As these small, ephemeral channels continue downslope, rills join to form slightly larger rills, which may join to make gullies. In humid climates, following these successively larger ephemeral channels downslope will eventually lead us to a point where we first encounter **perennial flow**. Perennial streams flow all year, though not always with the same volume or at the same velocity. Most arid region streams flow on an ephemeral basis although some may have **intermittent flow,** which lasts for a couple of months in response to an annual rainy season or spring snowmelt. Because of this contrast in flow duration, and other differences between arid and humid region streams, a full discussion of arid region stream systems appears in the separate chapter on arid region landforms, Chapter 18.

Perennial streams flow throughout the year even if it has been several weeks since a precipitation event. In most cases, this is possible only because the perennial streams continue to receive direct inflow of groundwater (Chapter 16) regardless of the date of most recent precipitation. Slow-moving groundwater seeps directly into the stream through the channel bottom and sides at and below the level of the water surface as **base flow.** Except in rare instances, it takes a humid climate to generate sufficient base flow to maintain a perennial stream between precipitation events.

## The Stream System

Most flowing water becomes quickly channelized into streams as it is pulled downhill by the force of gravity. Continuing downslope, streams form organized channel systems in which small perennial channels join to make larger perennial channels, and larger perennial channels join to create even bigger streams. Smaller streams that contribute their water and sediment load to a larger one in this way are **tributaries** of the larger channel, which is called the **trunk stream** ( Fig. 17.3).

**FIGURE 17.2**
The amount of runoff that occurs is a function of several factors, including the intensity and duration of a rainstorm. Surface features, however, are also important. Any factor that increases infiltration and evapotranspiration will reduce the amount of water available to run off, and vice versa. Deep soil, dense vegetation, fractured bedrock, and gentle slopes tend to reduce runoff. Thin or absent soils, thin vegetation, and steep slopes tend to increase runoff.

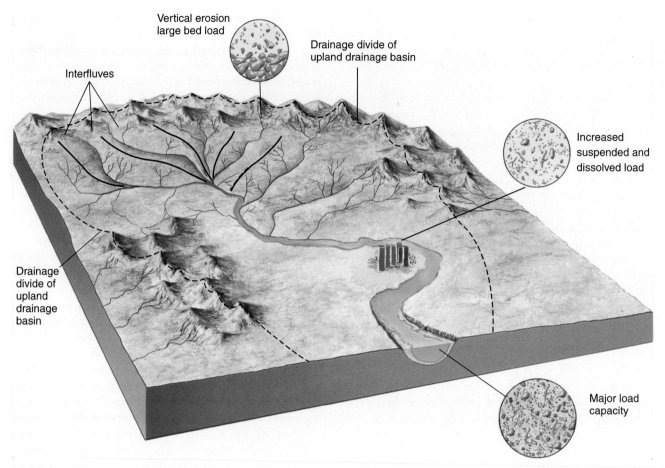

### FIGURE 17.3

**Environmental Systems: The Hydrologic System—Streams.** The stream, or surface runoff, system is a subsystem of the hydrologic cycle. Its major water input is from precipitation. However, groundwater may also contribute to the stream system, particularly in regions of humid climate. The major water output for most stream systems returns water into the ocean. Output or loss of water also occurs by evaporation back to the atmosphere and by infiltration into the groundwater system. Stream systems are divided into natural regions known as drainage basins (also called watersheds), separated from each other by divides; interfluves are the land areas between stream channels in the same drainage basin. Drainage basins are fundamental natural regions of critical importance to life in both humid and arid regions.

Streams are complex systems of moving water that involve energy transfers and the transport of a variety of surface materials. Energy enters the system with precipitation. The runoff flows downslope increasing the amount of energy available to the stream for cutting and eroding channels.

Materials transported by streams, known as load, enter the stream system by erosion and mass movement, particularly in the headwaters of a drainage basin. Much of the surface material eroded there consists of large particles, including boulders. Coarse material is carried along the channel bottom as bed load. As the number and size of tributaries increase downstream, the amount of load carried by the stream generally increases dramatically. This is especially true for finer materials suspended in water (suspended load) and dissolved minerals (dissolved load). Load leaves the stream system when carried and deposited in the sea at the river mouth. Streams also deposit sediment adjacent to their channels as they overflow their banks during floods. Human activities can change the amount of load available for stream systems by building dams, altering land with construction projects, overgrazing, and clearing forests. These activities may also affect water quality downstream, where communities may depend on the stream system for their water supply.

## Drainage Basins

Each individual stream occupies its own **drainage basin** (also known as **watershed,** or **catchment**), the expanse of land from which it receives runoff. **Drainage area** refers to the measured extent of a drainage basin, and is typically expressed in square kilometers or square miles. Because the runoff from a tributary's drainage basin is delivered by the tributary to the trunk stream, the tributary's drainage basin also constitutes part of the drainage basin of the trunk stream. In this way, small tributary basins are nested within, or *subbasins* of, a succession of larger and larger trunk stream drainage basins. Large river systems drain extensive watersheds that consist of numerous inset subbasins.

Drainage basins are open systems that involve inputs and outputs of water, sediment, and energy. Knowing the boundaries of a drainage basin and its component subbasins is critical to properly managing the water resources of a watershed. For example, pollution discovered in a river generally comes from a source within its drainage basin, entering the stream system either at the point where the pollutant was first detected or at a location upstream from that site. This knowledge helps us track, detect, and correct sources of pollution.

The **drainage divide** represents the outside perimeter of a drainage basin and thus also the boundary between it and adjacent basins ( Fig. 17.4). The drainage divide follows the crest of the interfluve between two adjacent drainage basins. In some places, this crest is a definite ridge, but the higher land that constitutes the divide is not always ridge shaped, nor is it necessarily much higher than the rest of the interfluve. Surface runoff generated on one side of the divide flows toward the channel in one drainage basin, while runoff on the other side travels in a very different direction toward the channel in the adjacent drainage basin. The *Continental Divide* separates North America into a western region where most runoff flows to the Pacific Ocean and an eastern region where runoff flows to the Atlantic Ocean. The Continental Divide generally follows the crest line of high ridges in the Rocky Mountains, but in some locations the highest point between the two huge basins lies along the crest of gently sloping high plains.

A stream, like the Mississippi River, that has a very large number of tributaries and encompasses several levels of nested subbasins, will have some major differences from a small creek that has no perennial tributaries and lies high up in the drainage basin near the divide. Knowing where each stream lies in the hierarchical order of tributaries helps earth scientists make more meaningful comparisons among different streams. The importance of quantitatively describing a stream or drainage basin's position in the hierarchy was realized long ago. In the 1940s, a hydrologist, Robert Horton, first proposed a system for determining this **stream order.** The stream-ordering system in most common use today is a modified version of what Horton suggested. In this system, *first-order streams* have no perennial tributaries. Although they are generally the smallest channels in the drainage basin, first-order channels can be mapped on large-scale topographic maps. Most first-order streams lie high up in the drainage basin near the drainage divide, the **source** area of the stream system. Two first-order streams must meet in order to form a *second-order stream,* which is larger than each of the first-order streams. It takes the intersection of two second-order channels to make a *third-order stream* regardless of how many first-order streams might independently join the second-order channels. The ordering system continues in this way, requiring two streams of a given order to combine to create a stream of the next higher order. The order of a drainage basin derives from the largest stream order found within it ( Fig. 17.5). For example, the Mississippi is a tenth-order drainage basin because the Mississippi River is a tenth-order stream. Stream ordering allows us to compare various attributes of streams quantitatively by relative size, which helps us better understand how stream systems work. Among other things, comparing streams on the basis of order has shown that typically as stream order increases, basin area, channel length, channel size, and amount of flow also increase.

Water moves through a stream system via higher- and higher-order channels as gravity pulls it downslope toward

**FIGURE 17.4**
An aerial photograph of a small stream channel network, basin, and divide. Every arm of the stream occupies its own subbasin (separated by divides) that combines with others to form the basin of the main stream channel.
**Can you mentally trace the outline of the main drainage basin shown here?**

# Geography's Spatial Science Perspective
## Watersheds as Critical Natural Regions

Perhaps the most environmentally logical division of Earth's land surface is the watershed, the drainage basin for a stream system. Humans have a vital responsibility to monitor, maintain, and protect the quality of freshwater resources. Surface water from watershed sources provides much of the potable water resources for the world's population. As the population grows, the need for fresh water increases, as does the human impact on these resources.

In recent years, many government agencies have designated watersheds as critical regions for environmental management and have worked to map the drainage divides that form the boundaries for these areas. There are several sound reasons for this spatial strategy. Watersheds are not only natural regions but also function as spatially coherent environmental systems.

Virtually all of Earth's land surface forms part of a watershed. Along with the channel network that occupies these basins, watersheds form well-integrated natural systems that involve water quality, soil, rock, terrain, vegetation, wildlife, domesticated animals, and humans.

The environmental processes and physical components of a watershed are strongly interrelated. Problems in one part of the system are likely to cause problems elsewhere.

Complex biotic habitats exist in these basins, and they are greatly affected by the environmental quality of the watershed. Increasing human populations and land-use intensities in previously natural drainage basins place pressures on these habitats and the water quality in the stream system. Monitoring and managing watersheds requires an interdisciplinary effort that considers aspects of all four of the world's major spheres.

Watersheds share the hierarchy of stream numbers and can be subdivided into smaller subbasins for local studies and management. Watersheds of large river systems can also be studied at a broad, regional scale. Water quality can be monitored for the whole system or for any subbasin. Streams form a network of flow in one direction (downstream), and problems with pollution can be located by tracing them upstream.

Watersheds are clearly defined, well-integrated, natural regions of critical importance to life on Earth, and they make a logical spatial division for environmental management. Unfortunately, rarely do political and administrative regions coincide with the divide that defines the edge of a watershed. A stream system may flow through many counties, cities, states, and governmentally administered lands and through more than one country. Each of these jurisdictions may have very different needs, goals, and strategies for using and managing their part of a watershed and often a variety of strategies results in direct conflict. Yet cooperative management of the watershed system as a whole is the best approach. The U.S. Environmental Protection Agency advocates and urges this management approach, and many cooperative river basin authorities have been established to encourage a united effort to protect a shared watershed. This watershed-oriented management strategy, based on a fundamental natural region, is an important step in protecting our freshwater resources.

The Mississippi River watershed covers an extensive region of the United States.

**FIGURE 17.5**
The hierarchy of stream ordering is illustrated by the channels of this fourth-order watershed. Note that when two streams of a given order meet, they join to form a segment of the next higher order. Order does not change unless two streams of the same order flow together. When the stream order changes, the channel is considered a new segment.
**What is the highest-order stream in a selected drainage basin called?**

its downstream end, or **mouth.** The mouth of most humid region, perennial stream systems lies at sea level where the channel system finally ends and the stream water is delivered to the ocean. Drainage basins with channel systems that convey water to the ocean have **exterior drainage.** Many arid region streams have **interior drainage** because they do not have enough flow to reach the ocean but terminate instead in local or regional areas of low elevation. Every stream has a **base level,** the elevation below which it cannot flow. Sea level is the *ultimate base level* for virtually all stream gradation. Streams with exterior drainage reach ultimate base level; the low point of flow by a stream with interior drainage is referred to as a *regional base level.* In some drainage basins, a very resistant rock layer located somewhere upstream from the river mouth can act as *temporary base level,* temporarily controlling the lowest elevation of the flow upstream from it until the stream is finally able to cut down through it (Fig. 17.6).

## Drainage Density and Pattern

Each organized system of stream tributaries exhibits spatial characteristics that provide important information about the nature of the drainage basin. The extent of channelization can be represented by measuring **drainage density** ($D_d$),

where $D_d = L/A_d$, the total length of all channels ($L$) divided by the area of the drainage basin ($A_d$). Drainage density indicates how dissected the landscape is by channels; thus, it reflects both the tendency of the drainage basin to generate surface runoff and the erodibility of the surface materials (Fig. 17.7). Regions with high drainage densities will promote considerable runoff, have limited infiltration, and have at least moderately erodible surface materials. In addition to the factors noted previously that promote runoff and reduce infiltration—impermeable sediments, thin soils, steep slopes, sparse vegetation—the ideal climate for high drainage densities is one that is semiarid. Humid climates encourage extensive vegetation cover, which promotes infiltration through interception and reduces channel formation by holding soils and surface sediment in place. In arid climates, although the vegetation cover is sparse, there is generally insufficient precipitation to create enough runoff to carve many channels. Semiarid climates have enough precipitation input to produce overland flow, but not enough to support an extensive vegetative cover. The easily eroded Dakota Badlands, which occur in a steppe climate, have an extremely high drainage density of over 125 (125 km of channel per 1 sq km of land), whereas very resistant granite hills in a humid climate may have a drainage density of only 5 (5 km of stream channel for every square kilometer of basin area). Another way to understand the concept of drainage density is to think about what would happen on a hillside in a humid climate if the natural vegetation were burned off in a fire. Erosion would rapidly cut gullies, creating more channels than previously existed there. In other words, the drainage density would increase. We could use the quantitative measure of $D_d = L/A_d$ to determine precisely the change and to monitor the change over time.

When viewed from the air or on maps, the tributaries of various stream systems may also form distinct **drainage patterns** (also called **stream patterns**). Two primary factors that influence drainage patterns are the bedrock structure and the nature of the land surface. A *dendritic* (from Greek: *dendros,* tree) stream pattern (Fig. 17.8a), see page 477, is an irregular branching pattern with tributaries joining larger streams at acute angles (less than 90°). A dendritic stream pattern is the most common type, in part because water flow in this pattern of channel network is highly efficient. Dendritic patterns form where the underlying rock structure does not strongly control the position of stream channels. Hence, dendritic patterns tend to develop in areas where the rocks have a roughly equal resistance to weathering and erosion. In contrast, a *trellis* stream pattern consists of long, parallel streams linked by short, right-angled segments (Fig. 17.8b). Trellis drainage is usually evidence of folding where parallel outcrops of erodible rocks form valleys between more resistant ridges, as in the Ridge and Valley region of the Appalachians. A *radial* pattern develops where streams flow away from a common high point on cone- or dome-shaped geologic structures such as volcanoes and domal uplifts (Fig. 17.8c).

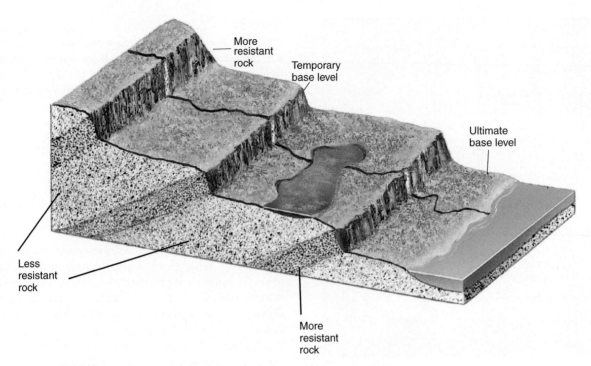

**FIGURE 17.6**
The lowest point to which a stream can flow is its base level. Stream water travels downslope by the force of gravity until it can flow no lower due to factors of topography, climate, or both. Most humid region streams have sufficient flow to make it all the way to an ocean basin, which is essentially a megascale topographic low. Thus, sea level represents ultimate base level for all of Earth's streams. In arid climates, many streams lose so much water by evaporation to the atmosphere and infiltration into the channel bed that they cannot flow to the sea. The lowest point they can reach is instead a regional base level, a topographic basin on the continent. A temporary base level is formed when a rock unit lying in the pathway of a stream is significantly more resistant than the rock upstream from it. The stream will not be able to cut into the less resistant rock any faster than it can cut into the resistant rock of the temporary base level.

The opposite pattern is *centripetal*, with the streams converging on a central area as in an arid region basin of interior drainage (Fig. 17.8d). *Rectangular* patterns occur where streams follow sets of intersecting fractures to produce a blocky network of straight channels with right-angle bends (Fig. 17.8e).

Most drainage patterns that occupy a large region require tens of thousands to millions of years to become well established. In regions that were covered by extensive glacier ice until about 10,000 years ago, streams flow on low-gradient terrain left by receding glaciers, wandering between marshes and small lakes in a chaotic pattern called *deranged* drainage (Fig. 17.8f). These streams have yet to establish a "normal" drainage pattern because, given the low gradient, there has not been enough time since deglaciation for a stream network to become well developed.

Many streams follow the "grain" of the topography or bedrock structure, eroding valleys in weaker rocks and flowing between divides formed on resistant rocks. Many examples also exist, however, of streams that flow across—that is, transverse to—the structure, cutting a gap or canyon through mountains or ridges. These **transverse streams** can be puzzling, giving rise to questions as to how a stream can cut a

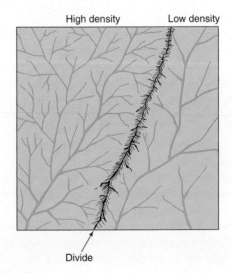

**FIGURE 17.7**
Drainage density—the length of channels per unit area—varies according to several environmental factors. For example, everything else being equal, highly erodible and impermeable rocks tend to have higher drainage density than areas dominated by resistant or permeable rocks. Slope and vegetation cover can also affect drainage density.
**What kind of drainage density would you expect in an area of steep slopes and sparse vegetation cover?**

### FIGURE 17.8
Drainage patterns often reflect bedrock structure. (a) A dendritic pattern is found where rocks have uniform resistance to weathering and erosion. (b) A trellis pattern indicates parallel valleys of weak rock between ridges of resistant rock. (c) A radial pattern results from multiple channels trending away from the top of a domed upland or volcano. (d) A centripetal pattern shows multiple channels flowing inward toward the center of a structural lowland or basin. (e) A rectangular pattern indicates linear joint patterns in the bedrock structure. (f) A deranged pattern typically results following the retreat of continental ice sheets; it is characterized by a chaotic arrangement of channels connecting small lakes and marshes.

gorge through a mountain range or how a stream can move from one side of a mountain range to the other. Such streams are probably either *antecedent* or *superimposed*.

Antecedent streams existed before the formation of the mountains that they flow through, maintaining their courses by cutting an ever-deepening canyon as gradual mountain building took place across their paths. The Columbia River Gorge through the Cascade Range in Washington and Oregon and many of the great canyons in the Rocky Mountain region, such as Royal Gorge and the Black Canyon of the Gunnison, both in Colorado, and Flaming Gorge in Utah/Wyoming, probably originated as antecedent streams ( Fig. 17.9a). Other rivers, including some in the central Appalachians, have cut gaps through mountains in a very different manner. These streams originated on earlier rock strata, since stripped away by erosion, so that the streams have been superimposed on the rocks beneath ( Fig. 17.9b). This sequence would explain why, in many instances, the rivers flow *across* folded rock structure, creating water gaps through mountain ridges. Examples include the Cumberland Gap and the gap formed by the Susquehanna River in Pennsylvania, both of which were important travel routes for the first American settlers crossing the Appalachians.

## Stream Discharge

The amount of water flowing in a stream not only depends on the impact of recent weather patterns but also on such drainage basin factors as its size, relief, climate, vegetation, rock types, and land-use history. Stream flow varies considerably from time to time and place to place. Most streams experience occasional brief periods when the amount of flow exceeds the ability of the channel to contain it, resulting in the flooding of channel-adjacent land areas.

Just as it was important to develop the technique of stream ordering to indicate quantitatively a channel's place in the hierarchy of tributaries, it is also crucial to be able to describe quantitatively the amount of flow being conveyed through a stream channel. **Stream discharge** ($Q$) is the volume of water ($V$) flowing past a given cross section of the channel per unit time ($t$): $Q = V/t$. Discharge is most commonly expressed in units of cubic meters per second ($m^3/s$)

**FIGURE 17.9**
How transverse streams form valleys and canyons across mountains. (a) An antecedent stream and its valley. A stream maintains its course by cutting a valley through a mountain as it is uplifted gradually over much time. (b) A superimposed stream and its valley. This stream has uncovered and excavated ancient structures that were buried beneath the surface. As the stream erodes the landscape downward, it cuts across and through the ancient structures.

or cubic feet per second (ft$^3$/s). A cross section is essentially a thin slice extending from one stream bank straight across the channel to the other stream bank and oriented perpendicular to the channel. If a drainage basin experiences a rainfall event that produces significant runoff, the volume of water ($V$) reaching the channel will increase. Notice from the discharge equation that this increase in volume ($V$) will cause an increase in stream discharge ($Q$). It is important to collect and analyze discharge data for several reasons. For example, it can be used to compare the amount of flow carried in different streams, at different sites along a single stream, or at different times at a single cross section. Discharge data indicate the size of a stream and, in times of excessive flow, provide an index of flood severity. In general, as the size of the drainage basin increases, so do the length of the river and its discharge. Among the major rivers of the world, the Amazon has by far the greatest discharge and the largest drainage basin; the Mississippi River system is ranked fourth in terms of discharge (Table 17.1).

The volume of water rushing through a cross section of a stream per second is extremely difficult to measure directly.

In reality, discharge is determined not by measuring $Q = V/t$ directly but by using the fact that discharge ($Q$) is also equal to the area of the cross section ($A$) times the average stream velocity ($v$). This equation, $Q = Av$, can also be stated as $Q = wdv$ because the cross-sectional area ($A$) is equivalent to channel width ($w$) times channel depth ($d$), two factors that are relatively easy to measure in the field (Fig. 17.10). Notice that, since cross-sectional area ($A$) is measured in square meters or square feet and average velocity is measured in meters per second or feet per second, solving the equation $Q = Av$ yields discharge values in units of volume per unit time (cubic meters per second or cubic feet per second). This analysis of the measurement units should convince you that volume per unit time is indeed equivalent to cross-sectional area times average velocity—that stream discharge is *both* volume per unit time and cross-sectional area times average velocity, $Q = V/t = Av$.

As is true for any equation, a change on one side of the discharge equation must be accompanied by a change on the other. If discharge increases because a rainstorm delivers a large volume of water to the stream, that increase in

**TABLE 17.1**
**Ten Largest Rivers of the World**

|  | Length | | Area of Drainage Basin | | Discharge | |
|---|---|---|---|---|---|---|
|  | km | mi | sq km (×1000) | sq mi | 1000 m³/s | 1000 cfs |
| Amazon | 6276 | 3900 | 6133 | 2368 | 112–140 | 4000–5000 |
| Congo (Zaire) | 4666 | 2900 | 4014 | 1550 | 39.2 | 1400 |
| Chang Jiang (Yangtze) | 5793 | 3600 | 1942 | 750 | 21.5 | 770 |
| Mississippi–Missouri | 6260 | 3890 | 3222 | 1244 | 17.4 | 620 |
| Yenisei | 4506 | 2800 | 2590 | 1000 | 17.2 | 615 |
| Lena | 4280 | 2660 | 2424 | 936 | 15.3 | 547 |
| Paraná | 2414 | 1500 | 2305 | 890 | 14.7 | 526 |
| Ob | 5150 | 3200 | 2484 | 959 | 12.3 | 441 |
| Amur | 4666 | 2900 | 1844 | 712 | 9.5 | 338 |
| Nile | 6695 | 4160 | 2978 | 1150 | 2.8 | 100 |

*Source:* Adapted from Morisawa, *Streams: Their Dynamics and Morphology.* New York, McGraw-Hill Book Company, 1968

**FIGURE 17.10**
These university students are measuring a stream's velocity, depth, and width in order to find discharge, the volume flowing in a stream per unit time.

(flow depth, $d$) rises, most streams also experience an increase in width ($w$) because channels generally flare out a bit as they rise up toward their banks. Stream channels continually adjust their cross-sectional area and/or flow velocity in response to changes in flow volume. Understanding the relationships among the factors involved in discharge is very important for understanding how streams work.

## Stream Energy

When people have more energy, we can accomplish more, or do more work, than when we have less energy. The same is true for streams and the other geomorphic gradational agents. The ability of a stream to erode and transport sediment—that is, to perform geomorphic work—depends on its available energy. Picking up pieces of rock and moving them require the stream to have **kinetic energy,** the energy of motion. When a stream has more kinetic energy, it can pick up and move more clasts (rock particles) and heavier clasts than when it has less energy. The kinetic energy ($E_k$) equation, $E_k = 1/2mv^2$, shows that the amount of energy a stream has depends on its mass ($m$), but especially on its *velocity* of flow ($v$) because kinetic energy varies with

volume ($V$) will occupy a larger cross-sectional area ($A$) or flow through the cross section at a faster rate ($v$), and it usually does some of both. In other words, as you probably already know, a large rainfall event will cause the level of a stream to rise and the water to flow faster. As the water level velocity squared ($v^2$). Thus, stream velocity is a critical factor in determining the amount of gradation accomplished by a stream. As we saw in the previous section, when stream discharge (Q) varies due to changes in runoff, so does stream velocity ($v$) because $Q = V/t = wdv$.

In addition to variations in flow discharge, another major way to alter stream velocity, and thereby stream energy, is through a change in the slope, or gradient, of the stream. **Stream gradient** is the drop in stream elevation over a given downstream distance and is typically expressed in meters per kilometer (m/km) or feet per mile (ft/mi). Everything else being equal, water flows faster down channels with steeper slopes and slows down over gentler slopes. Steeper channels typically occupy locations that are farther upstream and higher in the drainage basin, as well as places where the stream flows over rock types that are more resistant to erosion. Gentler stream gradients tend to occur closer to the mouth of a stream and where the stream crosses easily eroded rock types. As water moves through a channel system, its ability to erode and transport sediment—that is, its energy—is continuously changing as variations in stream gradient and discharge cause changes in flow velocity.

Different factors work to decrease the energy of a stream. Friction along the bottom and sides of a channel and even between the stream surface and the atmosphere slows down the stream velocity and therefore contributes to a decrease in stream energy. A channel with numerous substantial irregularities, such as those due to large rocks and vegetation sticking up into the flow, has a great amount of **channel roughness**, which causes a considerable decrease in stream energy due to the resulting frictional effects. Smooth, bedrock channels without these irregularities have lower channel roughness and a smaller loss in stream energy due to frictional effects. In all cases, however, friction along the bottom and sides of a channel slows down the flow and slows it down the most at the channel boundaries so that the maximum velocity usually occurs somewhat below the stream surface at the deepest part of the cross section. In addition to the external friction at the channel boundary, streams lose energy because of internal friction in the flow related to eddies, currents, and the interaction among water molecules. About 95% of a stream's energy is consumed in overcoming all types of frictional effects. Only the remaining energy, probably less than 5%, is available for eroding and transporting sediment.

Stream gradients are usually steepest at the headwaters and in new tributaries and diminish in the downstream direction ( Fig. 17.11). Discharge, on the other hand, increases downstream in humid regions as the size of the contributing drainage basin increases. As the water originally conveyed in numerous small channels high in the drainage basin is collected in the downstream direction into a shrinking number of larger and larger channels, flow efficiency improves and frictional resistance decreases. As a result, flow velocity tends to be higher in downstream parts of a channel system than over the steep gradients in the headwaters. The ability of the stream to carry sediment, therefore, may be equivalent or even greater downstream than upstream.

The sediment being transported by a stream is called the **stream load.** Carrying sediment is a major part of the geomorphic work accomplished by a stream. In order from smallest to largest, the size of sediment that a stream may transport includes clay, silt, sand, granules, pebbles, cobbles, and boulders. Sand marks the boundary between fine-grained (small) clasts and coarse-grained (large) clasts. *Gravel* is a general term for any sediment larger in size than sand. The maximum size of rock particles that a stream is able to transport, referred to as the **stream competence** (measured as particle diameter in centimeters or inches), and the total amount of load being moved by a stream, termed **stream capacity** (measured as weight per unit time), depend on available stream energy and thus on the flow velocity. As a stream flows along, its velocity is always changing, reflecting the constant variations in stream discharge, stream gradient, and frictional resistance factors. As a result, the size and amount of load that the stream can carry is also changing constantly. If the material is available when flow velocity increases, the stream will pick up from its own channel bed larger clasts and more load. When its velocity decreases and the stream can no longer transport such large clasts and so much material, it drops the larger clasts onto its bed, depositing them there until a new increase in velocity provides the energy to entrain and transport it. Thus, the particular gradational process being carried out by a stream at any moment, whether it is

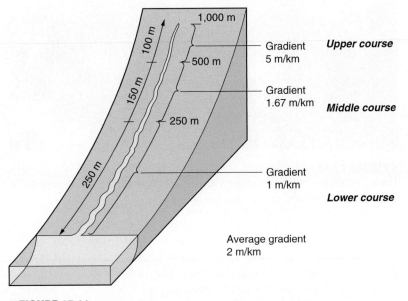

**FIGURE 17.11**
How stream gradient changes from source to mouth for the ideal stream

## Career Vision

**Andy Wohlsperger, Flood Mapping Analyst, Remote Sensing/GIS Specialist, Watershed Concepts**
Bachelor's Degree, University of Texas–Austin

After 2 years of coursework at the University of Texas at El Paso, Andy transferred to the University of Texas at Austin. While at the University of Texas, he worked for a professor creating layouts for the display of marketing data for several research studies. After graduation, Andy worked for an aerial photography/topographic mapping company as a stereo plotter operator. Here he gained most of his aerial photo interpretation skills and knowledge of working with and creating digital elevation models (DEM) and digital terrain models (DTM). Seven months after graduation, Watershed Concepts, a major engineering firm, hired him, where he works as GIS staff. He helps produce Digital Flood Insurance Rate Maps and Flood Recovery Maps. Andy fits into his company's mission because of his knowledge and training in GIS and remote sensing, and his general understanding of geographic concepts.

I decided to major in geography mainly because I've always been very interested in my surrounding world, especially nature, politics, economics, and world affairs. Geography is such a broad field that whatever you are interested in you will find a niche. I'm also very interested in new technologies; therefore, GIS and remote sensing attracted me right away.

*What did you learn in your physical geography education (particularly GIS) that is now important to you in your career?* Everything! I've probably used every concept I learned in my GIS, remote sensing, and cartography courses (projections, layouts, resolution, buffers, databases, intersecting, plotting, data display, the list goes on . . . ).

*What sort of academic preparation would you recommend to students who are interested in a career similar to yours?* Take all the techniques courses you can, pertaining to geographic information systems, remote sensing, basic cartography, digital cartography, physical geography, and so on. Make friends with the computer and the software applications; you will need to be able to use them to succeed.

*Why do you think that physical geography is an important course for those who are majoring in other fields?* A physical geography course answers a lot of questions. Why is the sky blue? Why are deserts where they are? How would global warming affect us? El Niño? La Niña? These are all issues that we are exposed to every day in our natural surrounding or that are addressed in the news. Therefore, we should all have a basic understanding of these issues. My wife just recently told me that her physical geography class was the most beneficial class in her college career when it comes to day-to-day events and understanding our world.

---

eroding or depositing, is determined by the complex of factors that control its energy.

Both stream capacity and stream competence can increase in response to a relatively small increase in velocity. A stream that doubles its velocity during a flood may increase its amount of sediment load six to eight times. The boulders seen in many mountain streams arrived there during some past flood that greatly increased stream competence; they will be moved again when a flow of similar magnitude occurs. In fact, rivers do most of their heavy earth-moving work during short periods of flood ( ▌Fig. 17.12).

Streams have no influence over the amount of water entering the channel system, nor can they change the type of rocks over which they flow. Streams do, however, have some control over channel size, shape, and gradient. For example, when a stream undergoes a decrease in energy so that it deposits some of its sediment, the deposit raises the channel bottom and locally creates a steeper gradient downchannel from the deposit. With continued deposition, the location

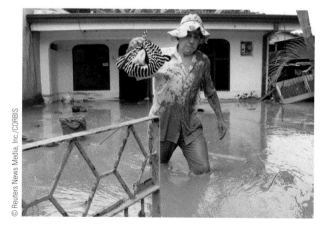

**▌FIGURE 17.12**
It is not just the power of flowing water that causes damage in a flood. A flooding stream typically has a high suspended load, can transport large heavy materials, and may carry floating debris that can damage homes and other structures, as in this image from the village of Orosi de Cartago, Costa Rica, 35 kilometers (22 mi) southeast of the capital, San Jose.

will eventually attain a slope steep enough to cause a sufficient increase in velocity so that the flow can again entrain and carry that deposited sediment. Streams tend to adjust their channel properties so that they can move the sediment supplied to them. *Dynamic equilibrium* is maintained by adjustments among channel slope, shape, and roughness; amount of load eroded, carried, or deposited; and the velocity and discharge of stream flow. A **graded stream** has just the velocity and discharge necessary to transport the load eroded from the drainage basin.

## Fluvial Gradational Processes

### Erosion by Streams

*Fluvial erosion* is the removal of rock material by flowing water. Fluvial erosion may take the form of the chemical removal of ions from rocks or the physical removal of rock fragments (clasts). Physical removal of rock fragments includes breaking off new pieces of bedrock from the channel bed or sides and moving them as well as picking up and removing preexisting clasts that were temporarily resting on the channel bottom. Breaking off new pieces of bedrock proceeds very slowly where highly resistant rock types are found.

Erosion is simply the removal of rock material; erosion of sediments from the bottom of a stream channel does not necessarily mean that the channel will occupy a lower position in the landscape. If the eroded rock fragments are replaced by the deposition of other fragments transported in from upstream, there will be no lowering of the channel bottom. Such lowering, or channel incision, occurs only when there is *net erosion* compared to deposition. Net erosion results in the lowering of the affected part of the landscape and is termed *degradation*. Net deposition of sediments results in a building up, or *aggradation*, of the landscape.

One way that streams erode occurs when stream water chemically dissolves rock material and then transports the ions away in the flow. This fluvial erosion process, called **corrosion** (or *solution*, or *dissolution*), has a limited effect on many rocks but can be significant in certain rock types, such as limestone.

**Hydraulic action** refers to the physical, as opposed to chemical, process of stream water alone removing pieces of rock. As stream water flows downslope by the force of gravity, it exerts stress on the streambed. Whether this stress results in entrainment and removal of a preexisting clast currently resting on the channel bottom, or even the breaking off of a new piece of bedrock from the channel, depends on several factors including the volume of water, flow velocity, flow depth, stream gradient, friction with the streambed, the strength and size of the rocks over which the stream flows, and the degree of stream turbulence. **Turbulence** is chaotic flow that mixes and churns the water, often with a significant upward component, that greatly increases the rate of erosion as well as the load-carrying capacity of the stream. Turbulence is controlled by channel roughness and the gradient over which the stream is flowing. A rough channel bottom increases the intensity of turbulent flow. Likewise, even a small increase in velocity caused by a steeper gradient can result in a significant increase in turbulence. Turbulent currents contribute to erosion by hydraulic action when they wedge under or pound away at rock slabs and loose fragments on the channel bed and sides, dislodging clasts that are then carried away in the current. **Plunge pools** at the base of waterfalls and in rapids reveal the power of turbulence-enhanced hydraulic action where it is directed toward a localized point ( Fig. 17.13).

As soon as a stream begins carrying rock fragments as load, it can start to erode by **abrasion,** a process even more powerful than hydraulic action. As rock particles bounce, scrape, and drag along the bottom and sides of a stream channel, they break off additional rock fragments. Because solid rock particles are denser than water, the impact of having clastic load thrown against the channel bottom and sides by the current is much more effective than the impact of water alone. Under certain conditions, stream abrasion makes distinctive round depressions called **potholes** in the rock of a bedrock streambed ( Fig. 17.14). Potholes generally origi-

**FIGURE 17.13**
Plunge pool at the base of Vernal Falls in Yosemite National Park, California.
**Why is a deep plunge pool found at the base of most waterfalls?**

nate in special circumstances, such as below waterfalls or swirling rapids, or at points of structural weakness, which include joint intersections in the streambed. Potholes range in diameter and depth from a few centimeters to many meters. If you peer into a pothole, you can often see one or more round stones at the bottom. These are the *grinders*. Swirling whirlpool movements of the stream water cause such stones to grind the bedrock and enlarge the pothole while finer sediments are carried away in the current.

In a process related to abrasion, as rock fragments moving as load are transported downstream, they are gradually reduced in size, and their shape changes from angular to rounded. This wear and tear experienced by sediments as they tumble and bounce against one another and against the stream channel is called **attrition.** Attrition explains why gravels found in streambeds are rounded and why the load carried in the lower reaches of most large rivers is composed primarily of fine-grained sediments and dissolved minerals.

Stream erosion widens and lengthens stream channels and the valleys they occupy. Lengthening occurs primarily at the source through **headward erosion,** accomplished partly by surface runoff flowing into a stream and partly by springs undermining the slope. The lengthening of a river's course in an upstream direction is particularly important where erosional gullies are rapidly dissecting agricultural land. Such gullying may be counteracted by soil conservation practices to reduce erosional soil loss. Channel lengthening and thus a decrease in stream gradient also occurs if a stream becomes more sinuous.

■ **FIGURE 17.14**
These potholes were cut into a streambed of solid rock during times of high flow. Potholes result from stream abrasion and the swirling currents in a fast-flowing river.
**Explain how a pothole is carved into bedrock on the stream bottom.**

## Stream Transportation

A stream directly erodes some of the sediment that it transports, and most chemical sediments are delivered to the channel in base flow, but a far greater proportion of its load is delivered to the stream channel by surface runoff and mass movement. Regardless of the sediment source, streams transport their load in several ways (■ Fig. 17.15). Some minerals are dissolved in the water and are thus carried in *solution*. The finest solid particles are carried in *suspension,* buoyed by vertical turbulence. Such small grains can remain suspended in the water column for long periods, as long as the force of upward turbulence is stronger than the downward settling tendency of

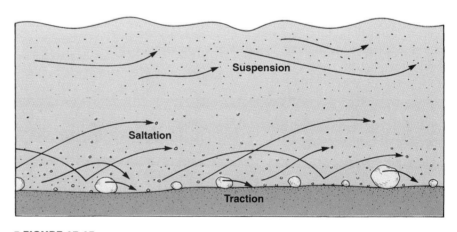

■ **FIGURE 17.15**
Transport of solid load in a stream. Clay and silt particles are carried in suspension. Sand typically travels by suspension and saltation. The largest (heaviest) particles move by traction.
**What is the difference between traction and saltation?**

the particles. Some grains too large and heavy to be carried in suspension bounce along the channel bottom in a process known as **saltation** (from French: *sauter,* to jump). Particles that are too large and heavy to move by saltation may slide and roll along the channel bottom in the transportation process of **traction.**

There are three main types of stream load. Ions of rock material held in solution constitute the **dissolved load. Suspended load** consists of the small clastic particles being moved in suspension. Larger particles that saltate or move in traction along the streambed comprise the **bed load.** The total amount of load that a stream carries is expressed in terms of the weight of the transported material per unit time.

The relative proportion of each load type present in a given stream varies with such drainage basin characteristics as climate, vegetative cover, slope, rock type, and the infiltration capacities and permeabilities of the rock and soil types. In general, dissolved loads will be larger than average in basins with high amounts of infiltration and base flow, and therefore limited surface runoff, because slow-moving groundwater that feeds the base flow acquires ions from the rocks through which it moves. Humid regions experience considerable weathering, which produces much fine-grained sediment, and thus humid region streams tend to have a large amount of suspended load. Rivers that are carrying a high suspended load look characteristically muddy (❚ Fig. 17.16). The Huang He in northern China, known as the "Yellow River" because of the color of its silty suspended load, carries a huge amount of sediment in suspension, with more than 1 million tons of suspended load per year. Compared to the "muddy" Mississippi River, the Huang He transports five times the suspended sediment load with only one fifth the discharge. Streams dominated by bed load tend to occur in arid regions because of the limited weathering rate in arid climates. Limited weathering leaves considerable coarse-grained sediment in the landscape available for transportation by the stream system.

## Stream Deposition

Because the capacity and competence of a stream to carry material depend primarily on flow velocity, a decrease in velocity will cause a stream to reduce its load through deposition. Velocity decreases over time when flow subsides—for example, after the impact of a storm—but it also varies from place to place along the stream. Shallow parts of a channel that in cross section lie far from the deepest and fastest flow typically experience low-flow velocity and become sites of recurring deposition. The resulting accumulation of sediment, like what forms on the inside of a channel bend, is referred to as a **bar.** Sediment also collects in locations where velocity falls due to an abrupt reduction in stream gradient, where the river current meets the standing body of water at its mouth, and on the land adjacent to the stream channel during floods.

**Alluvium** is the general name given to fluvial deposits, regardless of the type or size of material. Alluvium is usually recognized by the characteristic sorting and/or rounding of sediments that streams perform. A stream sorts particles by size, transporting the sizes that it can and depositing larger ones. As velocity fluctuates due to changes in discharge, channel gradient, and roughness, particle sizes that can be picked up, transported, and deposited vary accordingly (❚ Fig. 17.17). The alluvium deposited by a stream with fluctuating velocity will exhibit alternating layers of coarser and finer sediment.

When streams leave the confines of their channels during floods, the channel cross-sectional width is suddenly enlarged so much that the velocity of flow must slow down to counterbalance it ($Q = wdv$). The resulting decrease in stream competence and capacity cause deposition of sediment on the flooded land adjacent to the channel. This sedimentation is greatest right next to the channel where aggradation constructs channel-bounding ridges known as **natural levees,** but some alluvium will be left behind wherever load settled out of the receding flood waters.

**Floodplains** constitute the often-extensive, low-gradient land areas composed of alluvium that lie adjacent to many stream channels (❚ Fig. 17.18). Floodplains are aptly named because they are inundated during floods and because they are at least partially composed of *vertical accretion deposits,* the sediment that settles out of slowing and standing floodwater. Most floodplains also contain *lateral accretion deposits.* These are generally channel bar deposits that get left behind as a channel gradually shifts its

**❚ FIGURE 17.16**
Some rivers carry a tremendous load of sediment in suspension and show it in their muddy appearance, as in this aerial view of the Mississippi River in Louisiana. Much of the load carried by the Mississippi River is brought to it by its major tributaries.
**What are some of the major tributaries that enter the Mississippi River?**

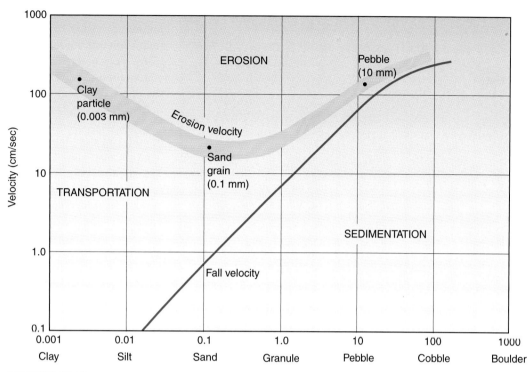

**FIGURE 17.17**
This graph shows the relationship between stream flow velocity and the ability to erode or transport material of varying sizes (inability to erode or transport particles of a particular size or larger will result in deposition). Note that small pebbles (particles with a diameter of 10 mm, for example) need a high stream flow velocity to be moved because of their size and weight. The fine silts and clays (smaller than 0.05 mm) also need high velocities for movement because they stick together cohesively. Sand-sized particles (between 0.05 and 2.0 mm) are relatively easily eroded and transported, compared to clays or gravel (a mix of any particles larger than 2.0 mm).

**FIGURE 17.18**
During floods, low areas adjacent to the river are inundated with sediment-laden water that flows over the banks to deposit alluvium, mainly silts and clays, on the floodplain. This is the Missouri River floodplain at Jefferson City, Missouri, during the 1993 Midwest flood.

**What would the river floodwaters leave behind in flooded homes after the water recedes?**

position in a sideways fashion (laterally) across the floodplain ( Fig. 17.19).

## Channel Patterns

Three principal types of stream channel have traditionally been recognized when considering the form of a given channel segment in map view. Although *straight channels* may exist for short distances under natural circumstances, especially along fault zones, joints, or steep gradients, most channels with parallel, linear banks are artificial features that were totally or partially constructed by people.

If a stream has a high proportion of bed load in relation to its discharge, it deposits much of its load as sand and gravel bars in the streambed. These obstructions break the stream into strands that interweave, separate, and rejoin, giving a braided appearance to the channel, and indeed such a pattern is called **braided** ( Fig. 17.20). This channel pattern may develop wherever the coarse-sediment input into a stream is extremely high owing to banks of loose

**FIGURE 17.19**
Sediment deposited in a bar on the inside of a channel bend can become part of the floodplain alluvium if the stream migrates away leaving the bar deposits behind, as occurred with these lateral accretion deposits. In this photo, remnants of winter ice fill swales between ridges that mark successive crests of the laterally accreted bar deposits.

sand and gravel or the presence of a nearly unlimited coarse bed load, such as that found downstream from glaciers and also in many desert areas. Braided streams are common on the Great Plains (for example, the Platte River), in the desert Southwest, in Alaska, and in Canada's Yukon.

The most common channel pattern in humid climates displays broad, sweeping bends in map view. Over time, these sinuous, **meandering channels** also wander from side to side across their low-gradient floodplains widening the valley by lateral erosion on the outside of meander bends and leaving behind lateral accretion (bar) deposits on the inside of meander bends ( Fig. 17.21). These streams and their floodplains have a higher proportion of fine-grained sediment and thus greater bank cohesion than the typical braided stream.

## Land Sculpture by Streams

One way to understand the variety of landform features resulting from fluvial processes is to examine the course of an idealized river as it flows from headwaters in the mountains to its mouth at the ocean. The gradient of this river would diminish continually downstream as it flows from its source toward base level. In nature, exceptions exist to this idealized profile because some streams flow entirely over a low gradient ( Fig. 17.22a), while other streams, particularly small ones on mountainous coasts, flow down a steep gradient all the way to the standing body of water at their mouth ( Fig. 17.22b). Rather than having a smoothly decreasing slope from headwaters to mouth, we would expect real streams to have some irregularities in this **longitudinal profile,** the stream gradient from source to mouth (see again Fig. 17.11). The following discussion subdivides the ideal stream course into upper, middle, and lower river sections flowing over steep, moderate, and low gradients, respectively. Fluvial erosion processes dominate the steep upper course, whereas deposition predominates in the lower course. The middle course displays important elements of both fluvial erosion and deposition.

**FIGURE 17.20**
The braided stream channel of the Brahmaputra River in Tibet, viewed from the International Space Station. Stream braiding results from an abundant bed load of coarse sediment that obstructs flow and separates the main stream into numerous strands. Braided channels typically occur when a stream has low discharge compared to the amount of bed load.
**What does the common occurrence of braided channels just downstream from glaciers tell you about the sediment transported and deposited by moving ice?**

valley, gorge, or ravine as the stream channel in the bottom of the valley cuts deeply into the land. Little if any floodplain is present, and the valley walls typically slope directly to the edge of the stream channel. Steep valley sides encourage mass movement of rock material directly into the flowing stream. Valleys of this type, dominated by the downcutting activity of the stream, are often called **V-shaped valleys** because with their steep slopes they attain the form of a letter "V" (Fig. 17.23).

The effects of *differential erosion* can be significant in the upper course where rivers cut through rock layers of varying resistance. Typically, rivers flowing over resistant rock have a steeper gradient than where they encounter weaker rock. This steep gradient gives the stream flow more energy, which the stream needs to erode the resistant rock. Rapids and waterfalls may mark the location of resistant materials in a stream's upper course. Where rocks are particularly resistant to weathering and erosion, valleys will be narrow, steep-sided gorges or canyons; where rocks are less resistant, valleys tend to be more spacious.

Many streams spill from lake to lake in their upper courses, either over open land (like the Niagara River at Niagara Falls, between Lake Erie and Lake Ontario) or through gorges. In either case, the lakes may eventually be eliminated if stream erosion lowers their outlets enough or if fluvial sediment deposited at the inlet points fills the lakes.

## Features of the Middle Course

In the middle section of the ideal longitudinal profile, the stream flows over a moderate gradient and on a moderately smooth channel bed. Here the river valley includes a floodplain, but remaining ridges beyond the floodplain still form definite valley walls. The stream lies closer to its base level, flows over a gentler gradient, and thus directs less energy toward vertical erosion than in its upper course. The stream still has considerable energy, however, due to the downstream increase in flow volume and reduction in bed friction. The river now uses much of its available energy for transporting the considerable load that it has accumulated and toward lateral erosion of the channel sides. The stream displays a definite meandering channel pattern with its sinuous bends that wander over time across the valley floor. The stream erodes a **cut bank** on the outside of meander loops, where the channel is deep and centrifugal force accelerates stream velocity. The cut bank is a steep slope, and slumping may occur there particularly when there is a rapid fall in water level. Slumping on the outside of meander bends contributes to the effect of lateral erosion by the stream and adds load to the stream. In the low velocity and shallow flow on the inside of the meander bends, the stream deposits a **point bar** (Fig. 17.24). Erosion on the outside and deposition on the inside of river meander bends results

**FIGURE 17.21**
Over time, a meandering (sinuous) stream channel, like the Blackstone River of Massachusetts shown here, may swing back and forth across its valley. Where the outside of a meander bend impinges on the valley side walls, stream erosion can undercut the wall and with the assistance of mass wasting contribute to valley widening.

## Features of the Upper Course

At the headwaters in the upper course of a river, the stream generally flows in contact with bedrock. Over the steep gradient high above its base level, the stream works to erode vertically downward by hydraulic action and abrasion. Erosion in the upper course generally creates a steep-sided

### FIGURE 17.22
Not all streams fit the generalized pattern of characteristics for upper, middle, and lower stream segments. (a) The Mississippi River, here a relatively small stream, meanders on a low gradient near its headwaters in Minnesota, far upstream from its mouth. (b) A stream plunges down a steep gradient on the island of Hawaii.

### FIGURE 17.23
Where the upper course of a stream lies in a mountainous region, its valley typically has a characteristic "V" shape near the headwaters. Such a stream flows in a steep-walled valley, with rapids and waterfalls, as shown here in Yellowstone Canyon, Wyoming.

**How does the gradient of the Yellowstone River compare with that of the stretch of the Mississippi River shown in Fig. 17.22a?**

(a)

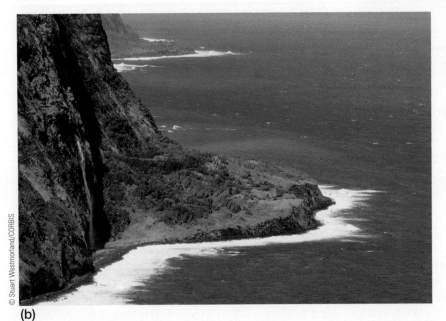

(b)

in the sideways displacement, or **lateral migration,** of meanders. This helps increase the area of the gently sloping floodplain when cut banks impact the confining valley walls. Tributaries flowing into a larger stream also aid in widening the valley through which the trunk stream flows. Though flooding of the valley floor is always a potential hazard, the richness of floodplain soils offers an irresistible lure for farmers.

## Features of the Lower Course

The minimal gradient and close proximity to base level along the ideal lower river course make downcutting virtually impossible. Stream energy, now derived almost exclusively from the higher discharge rather than the downslope pull of gravity, leads to considerable lateral shifting of the river channel. The river meanders around helping to create a large depositional, rather than erosional, plain (see Map Interpretation: Fluvial Landforms). The lower floodplain of a major river is much wider than the width of its meander belt and shows evidence of many changes in course (Fig. 17.25). The stream migrates laterally through its own previously deposited sediment in a channel composed exclusively of alluvium. During floods, these extensive floodplains, or **alluvial plains,** become inundated with sediment-laden water that contributes vertical accretion deposits to the large natural levees and to the already thick alluvial valley fill of the floodplain in general. Natural levees along the Mississippi River rise up to 5 meters (16 ft) above the rest of the floodplain.

A common landform in this deposition-dominated environment provides evidence of the meandering of a river over time. Especially during floods, **meander cut-offs** occur when a stream seeks a shorter, steeper, and straighter path; breaches through the levees; and leaves a former meander bend isolated from the new channel position. If the cut-off meander remains filled with water, which is common, it forms an **oxbow lake** (Fig. 17.26).

Sometimes humans attempt to control streams by building up levees artificially in order to keep the river in its channel. During times of reduced discharge, however, when a river has less energy, deposition occurs in the channel. Thus, in an artificially constrained channel, a river can sometimes raise the level of its channel bed. In some instances, as in China's Huang He and the Yuba River in northern California, the stream channel may be raised above the surrounding floodplains. Flooding presents a very serious danger in this situation because much of the floodplain may actually lie below the level of the river. Unfortunately, when floodwaters eventually overtop or breach the levees, they may be even more extensive and destructive than they would have been in the natural case.

The presence of levees—both natural and artificial—can prevent tributaries in the lower course from joining the main stream. Smaller streams are forced to flow parallel to the main river until a convenient junction can be found. These parallel tributaries are called **yazoo**

**FIGURE 17.24**

Characteristics of a meandering river channel. Note that water flowing in a channel has a tendency to flow downstream in a helical, or "corkscrew," fashion, which moves water against one side of the channel and then to the opposite side. The up-and-down motion of the water contributes to the processes of erosion and deposition.

(a)

(b)

(c)

**FIGURE 17.25**
This colorized radar image shows part of the Mississippi River floodplain along the Arkansas–Louisiana–Mississippi state lines. Images like this help us assess flood potential and learn much about the geomorphic history of the river and its floodplain. The colors are used to enhance landscape features such as water bodies (dark), field patterns, and forested areas (green). Note how the river has changed its channel position many times, leaving oxbow lakes and meander scars on the floodplain.

**streams,** named after the Yazoo River, which parallels the Mississippi River for more than 160 kilometers (100 mi) until it finally joins the larger river near Vicksburg, Mississippi.

## Deltas

Where a stream flows into a standing body of water, such as a lake or the ocean, the flow is no longer confined in a channel. The current expands in width, causing a reduction in flow velocity and thus a decrease in load competence and capacity. If the stream is carrying much load, the sediment will begin to settle out, with larger particles deposited first, closer to the river mouth, and smaller particles deposited farther out in the water body. With continued aggradation, a distinctive landform, called a **delta** because the map view shapes of some resemble the Greek letter delta (Δ), may be constructed ( ▌Fig. 17.27a). Deltas form at the interface between fluvial systems and coastal environments of lakes or the ocean and therefore originate in part from fluvial and in part from coastal processes. Deltas have a subaqueous (underwater) coastal component, called the **prodelta,** and a fluvial part, the **delta plain,** that exists at, to slightly above, the lake level or sea level. In general, deltas can form only at river mouths where the fluvial sediment supply is high, where the underwater topography does not drop too sharply, and where waves, currents, and tides cannot transport away all the sediments delivered by the river. Although these circumstances exist at the mouths of many rivers, not all rivers have deltas.

Delta construction may be a slow, ongoing process. A river channel that approaches its base level at a large standing body of water typically has a very low gradient. Lacking the ability to incise its channel below base level, the stream may divide into two channels and may do so multiple times, to convey its water and load to the lake or ocean. These multiple channels flowing out from the main stream are called **distributaries,** are typical features of the delta plain, and help direct flow and sediment toward the lake or ocean. Natural levees accumulate along the banks of these distributary channels. Continued deposition and delta formation can extend the delta plain and create new land far out from the original shoreline. Rich alluvial deposits and the abundance of moisture allow vegetation to quickly become established on these fertile deposits and further secure the delta's position. Delta plains, such as those of the Mekong, Indus, and Ganges Rivers, form important agricultural areas that feed the dense populations of many parts of Asia.

Where it flows into the Gulf of Mexico, the Mississippi River has constructed a type of delta called a *bird's-foot delta* ( ▌Fig. 17.27b). Bird's-foot deltas form in settings where the influence of the fluvial system far exceeds the ability of waves, currents, and tides of the standing water body to rework the deltaic sediment into coastal landforms or to transport it away. Natural levee crests along numerous distributaries remain intact slightly above sea level and extend far out into the receiving water body. Occasional changes in the distributary channel system occur when a major new distributary is cut that siphons flow away from a previous one, causing the center of deposition to switch to a new location far from its previous center. The appearance in map view of the natural levees extending toward the present and former depositional centers leaves the delta resembling a bird's foot.

Different types of deltas are found in other kinds of settings. An *arcuate delta,* like that of the Nile River, projects to a

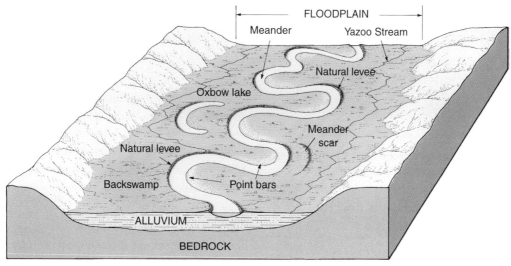

**ACTIVE FIGURE 17.26**

Watch this Active Figure at http://now.brookscol.com gabler8.

Features of a large floodplain common in the lower courses of major rivers. Low marshy or swampy parts of the floodplain, generally at the water table, are called backswamps.

**What is the origin of an oxbow lake?**

(a)

(b)

**FIGURE 17.27**

Satellite views of two different types of deltas. (a) The Nile River delta at the edge of the Mediterranean Sea is an arcuate delta, displaying the classic triangular arc shape. Waves and currents smooth out irregularities along the seaward edge of the delta. (b) The unusual shape of the Mississippi River delta resembles a bird's foot. Waves, currents, and tide in the Gulf of Mexico do little to change the visible shape of this type of delta.

**Why are the shapes of some deltas controlled more by fluvial processes whereas the shapes of others are strongly influenced by coastal processes?**

limited extent into the receiving water body, but the smoother, more regular seaward edge of this kind of delta shows greater reworking of the fluvial deposits by waves and currents than in the case of the bird's-foot delta. *Cuspate deltas,* like the São Francisco in Brazil, form where strong coastal processes push the sediments back toward the mainland and rework it into beach ridges on either side of the river mouth.

## Base Level Changes and Tectonism

A change in elevation along a stream's longitudinal profile will cause an increase or decrease in the stream's gradient and thus will impact the stream's type and amount of geomorphic work. Elevation changes can occur in the drainage basin due to tectonic uplift or depression. Base level changes for basins of exterior drainage result principally from climate change. Sea level drops in response to large-scale growth of glaciers and rises with substantial glacier shrinking. Tectonic uplift or a drop in base level give the stream a steeper gradient and increased energy for erosion and transportation. The landscape and its stream are then said to be **rejuvenated** because the stream uses its renewed energy to incise its channel to the new base level. Valleys with waterfalls or rapids may develop as these channels are deepened by erosion. Tectonic depression of the drainage basin or a rise in sea level reduces the stream's gradient and energy, enhancing deposition.

If new uplift occurs gradually in an area where stream meanders have formed, these meanders may become **entrenched** as the stream deepens its valley ( Fig. 17.28). Now, instead of eroding the land laterally, with meanders migrating across an alluvial plain, the rejuvenated stream's primary activity is vertical incision.

It is important to note that virtually all rivers reaching the sea incised their valleys during the Pleistocene in response to the lowering of sea level associated with continental glaciation. The accumulation of water on the land in the form of glacial ice caused sea level to drop as much as 120 meters (400 ft) and lowered the base level for streams of exterior drainage. Consequently, near their mouths, the streams eroded deep valleys for their channels. Subsequent melting of the glaciers again elevated base level. This base level rise caused the streams to deposit their sediment loads, filling valleys with sediment, as the streams adjusted their channels to a second new base level. These consecutive changes in base level produced broad, flat, alluvial floodplains above buried valleys cut far below today's sea level.

While a drop in base level causes downcutting and a rise causes deposition, an upset of stream equilibrium resulting from sizable increases or decreases in discharge or load can have similar effects on the landscape. Research has shown that variations in base level, tectonic movements, and changes in stream equilibrium can each cause downcutting in stream valleys, so the valley is slightly deepened and remnants of the older, higher valley floor are preserved in "stair-stepped" banks along the walls of the valley. These remnants of previous valley floors are **stream terraces.** Multiple terraces are a consequence of successive periods of downcutting and deposition ( Fig. 17.29). Stream terraces provide a great deal of evidence about the geomorphic history of the river and its surrounding region.

## Stream Hazards

Although there are many benefits to living near streams, settlement along a river has its risks, particularly in the form of floods. Variability of stream flow constitutes the greatest problem for life along rivers and is also an impediment to their use. Stream channels can generally contain the maximum flows that are estimated to occur once every year or two. The maximum flows that are probable over longer periods of 5, 10, 100, or 1000 years overflow the channel and inundate the surrounding land, sometimes with disastrous results ( Fig. 17.30). Similarly, exceptionally low flows may produce crises in water supply and bring river transportation to a halt.

The U.S. Geological Survey maintains more than 6000 gaging stations for the measurement of stream discharge in the United States ( Fig. 17.31). Many of these gaging stations operate on solar energy, measure stream discharge in an automated fashion, and beam their data to a satellite that relays them to a receiving

**FIGURE 17.28**
A spectacular example of an entrenched meander from the Colorado River.

■ **FIGURE 17.34**
The multipurpose Lookout Point Dam on the Middle Fork of the Willamette River, Oregon.
**What are some of the functions that multipurpose dams serve?**

■ **FIGURE 17.35**
Many of the world's lakes are of glacial origin; shown here is Stanley Lake in the Sawtooth Range of Idaho.
**Why might this lake be considered a temporary feature?**

instructive examples of the damage that can be done to a large, complex natural system by human misuse over a short period of time.

## Quantitative Fluvial Geomorphology

Quantitative methods are important in studying virtually all aspects of the Earth system, and they are used by climatologists, meteorologists, biogeographers, soil scientists, and hydrologists as well as geomorphologists. The importance of quantitative methods to the objective analysis of fluvial systems in particular may be discerned, at least in part, from the material presented in this chapter. Streams are complicated, dynamic systems with input, throughput, and output of energy and matter that depend on numerous, often-interrelated variables. Geomorphologists routinely measure stream channel, drainage basin, and flow properties and analyze them using statistical methods and the principles of fluid mechanics so that they can describe, compare, monitor, predict, and learn more about streams and the geomorphic work that they perform. Drainage area, stream order, drainage density, stream discharge, stream velocity, channel width, and channel depth are just some of the numeric data that are collected in the field, on topographic maps, from digital elevation models, or from remotely sensed imagery to facilitate the study of stream systems. Extensive efforts are made to gather and analyze numeric stream data because of the widespread occurrence of streams and their great importance to human existence. Future quantitative studies will not only help us better understand the origins and formational processes of landforms and landscapes, but they will also help us better predict water supplies and flood hazards, estimate soil erosion, and trace sources of pollution.

# Define & Recall

fluvial geomorphology
stream
interfluve
flood
surface runoff
infiltration
infiltration capacity
interception
sheet wash
rill
gully
ephemeral flow
perennial flow
intermittent flow
base flow
tributary
trunk stream
drainage basin (watershed, catchment)
drainage area
drainage divide
stream order
source
mouth
exterior drainage
interior drainage

base level
drainage density
drainage pattern (stream pattern)
transverse stream
stream discharge
kinetic energy
stream gradient
channel roughness
stream load
stream competence
stream capacity
graded stream
corrosion
hydraulic action
turbulence
plunge pool
abrasion
pothole
attrition
headward erosion
saltation
traction
dissolved load
suspended load
bed load

bar
alluvium
natural levee
floodplain
braided channel
meandering channel
longitudinal profile
V-shaped valley
cut bank
point bar
lateral migration
alluvial plain
meander cut-off
oxbow lake
yazoo stream
delta
prodelta
delta plain
distributary
rejuvenated stream
entrenched stream
stream terrace
stream hydrograph
reservoir

# Discuss & Review

1. On a worldwide basis, what is the most important geomorphic agent operating to shape the landscape of the continents? Why is that agent so important?
2. What is the difference between a stream and a river or creek?
3. What is the relationship between infiltration and surface runoff? What are some of the factors that enhance infiltration? What are some of the factors that enhance surface runoff?
4. Consider a fifth-order drainage basin. What difference would you expect between the first- and fifth-order channels in terms of overall number of channels? How would a typical first-order channel differ from the fifth-order channel in terms of length, discharge, velocity, and gradient? What other factors would differ between the two channels?
5. What factors affect the discharge of a stream? What are the two very different equations that can be used to represent stream discharge?
6. Why do drainage basins in semiarid climates tend to have greater drainage density than those in humid climates?
7. Name two major ways in which groundwater influences streams.
8. Which is the most effective fluvial erosion process? Why is it so effective?
9. What are the differences among the fluvial transportation processes? Which moves the largest particles?
10. How does a stream sort alluvium? Explain the relationship of this sorting to velocity changes in a stream.
11. Describe the development of natural levees.
12. What are stream terraces? How are they formed? How did changes in sea level during the Pleistocene cause stream terraces to form on land?
13. Why is it important to monitor stream gaging stations and the stream flow data that they provide?
14. What is the relationship between rivers and lakes, and how do the two types of features interact?

## Consider & Respond

1. How do streams represent the concept of dynamic equilibrium? What are some examples of negative feedback in a stream system?
2. In the study of a drainage basin, what types of geographic observations and quantitative data might prove useful in planning for flood control and water supply?
3. How does urbanization affect runoff, discharge, and flood potential in a drainage basin?

# Map Interpretation
## Fluvial Landforms

### The Map

Campti is in northwestern Louisiana on the Gulf Coastal Plain. This coastal region stretches from northern Florida to the Texas–Mexico border, and extends inland for more than 200 miles in some areas. Elevations on the Gulf Coastal Plain gradually increase from sea level at the shoreline to several hundred feet far inland. The region is underlain by gently dipping sedimentary rock layers. The surface material includes marine sediments and alluvial deposits from rivers that cross the region, especially those of the Mississippi drainage system. This is a land of meanders, natural levees, and bayous.

The Red River's headwaters are in the semiarid plains of the Texas Panhandle, but it flows eastward toward an increasingly more humid climate. About midcourse, the Red River enters a humid subtropical climate region, which supports rich farmland and dense forests. The river flows into the Mississippi in southern Louisiana, about 161 kilometers (100 mi) downstream of the map area. The Red River is the southernmost major tributary of the Mississippi. Louisiana has mild winters with hot, humid summers.

Annual rainfall totals for Campti average about 127 centimeters (50 in.). The warm waters of the Gulf of Mexico supply vast amounts of atmospheric energy and moisture, producing a high frequency of thunderstorms, tornadoes, and, on occasion, hurricanes that strike the Gulf Coast.

### Interpreting the Map

1. How would you describe the general topography of the Campti map area? What is the local relief? What is the elevation of the banks of the Red River at Campti?
2. What kind of landform is the low-relief surface that the river is flowing on? Is this mainly an erosional or depositional landform?
3. In what general direction does the Red River flow? Is the direction of flow easy or hard to determine just from the map area? Why?
4. Does the Red River have a gentle or steep gradient? Why is it difficult to determine the river gradient from the map area?
5. What is the origin of Smith Island? Adjacent to Smith Island is Old River; what is this feature called?
6. Explain the stippled brown areas in the meanders south of Campti. Are these areas on the inside or the outside of the meander bend? Explain why.
7. How would you describe the features labeled a "bayou"?
8. Is this map area more typical of the upper, middle, or lower course of a river?
9. Although this is not a tectonically active region, what do you think would happen to this area if it were uplifted? What would be the major geomorphic process change?

Campti, Louisiana
Scale 1 : 62,500
Contour interval = 20 ft
U.S. Geological Survey

U.S. Department of Agriculture

# CHAPTER 18

These vertical rock cliffs adjacent to a flat, dry lake bed in western Utah represent just one example of the striking beauty that is commonly seen in desert landscapes.   D. Sack, Ohio University

# Arid Landforms and Eolian Processes

**Physical Geography Now™** This icon, appearing throughout the book, indicates an opportunity to explore interactive tutorials, animations, or practice problems at now.brookscole.com/gabler8

## CHAPTER PREVIEW

In most deserts, running water only operates occasionally, so it may seem odd that water rather than wind is the chief agent of gradation in dry regions.

- How can this be?
- What features in arid regions are produced predominantly by water action?

A variety of landforms in many deserts today could not have been created under present conditions of aridity; they are evidence of landscapes formed under earlier, wetter climate conditions.

- What is some landform evidence of climate change in deserts?
- How might the most recent cooler, wetter period experienced by midlatitude deserts be associated with glaciation?

Stream channels in desert areas are often quite different from those in humid regions.

- What are some of these differences?
- What combination of factors causes these differences?
- What are the most common landforms of fluvial erosion and deposition in arid environments?

Landforms of fluvial deposition on desert plains and along the base of mountains in desert areas are attractive locations for human settlement.

- What are some example locations of this in the United States?
- What combination of factors causes this to be true?

Over most land surfaces, the wind exerts little influence on landform development, but under certain conditions, it can be an important agent of gradation.

- Why is wind so limited as a major agent in landform development?
- Under what conditions and in what environments is the wind an important agent of gradation?

Surprisingly, there can be a large number of lakes in some arid regions.

- How do these lakes differ from those in humid regions?
- Under what circumstances do these lakes form?

The wind deposits different kinds of sand dunes, but it also deposits loess.

- How does loess differ from sand dunes, and how do the wind processes involved in depositing loess differ from those involved in creating sand dunes?
- Why has deposition of loess had a positive impact on humans in a number of major world regions?

Relatively barren of vegetation, arid regions display distinctive scenery that is unlike that of most other environments. Hill slopes in humid regions tend to be rounded and mantled in soil. Desert mountains and hill slopes are generally angular, with extensive bedrock exposures, separated by low-relief areas either filled in by sedimentation or controlled by resistant rock layers. Many desert regions in North America hold a mythical fascination for the people who live there as well as for ecotourists, two groups of people who especially appreciate arid landscapes. A great number of movies have used striking desert scenery of all kinds as a location for filming. Desert landscapes often display, in stark beauty, the colors, characteristics, and structure of the rocks that make up the area. The desert's barrenness reveals considerable evidence about landforms and geomorphic processes that would be much more difficult to observe in humid environments, due to the cover of soil and vegetation. Much of our understanding of how landscapes and landforms develop in a wide variety of environments has come from important studies and scientific explorations conducted in desert regions.

Although the wind plays an important role in arid region geomorphology, most desert landforms are produced by the action of water. The effects of wind erosion are mainly confined to removing fine, dust-sized (silt and clay) particles from desert regions and to dislodging loose rock fragments of sand-sized materials in deserts. Still, we tend to associate arid environments with eolian (wind) geomorphic processes because some desert areas display notable accumulations of wind-deposited sediment, usually in the form of sand dunes. Because of sparse vegetation and other environmental characteristics, eolian gradation reaches its optimum in arid environments. However, because gases have much lower densities than liquids, even in deserts the geomorphic work of the wind is outmatched by fluvial geomorphic processes. We should also understand that eolian processes and landforms are not confined to arid regions; they are also conspicuous in many coastal areas and anywhere that loose sediments are frequently exposed to winds strong enough to move them.

## Surface Runoff in the Desert

Landforms, rather than vegetation, typically dominate desert scenery. The precipitation and evaporation regimes of an arid climate result in a sparse cover of vegetation and, because many weathering processes require water, relatively low rates of weathering. With low weathering rates, insufficient vegetation to break the force of raindrop impacts, and a lack of extensive plant root networks to help hold rock fragments in place, a blanket of moisture-retentive soil cannot accumulate on slopes. Soils tend to be thin, rocky, and discontinuous. This absence of a continuous vegetative and soil cover gives desert landforms their unique character. Under these surface conditions of very limited interception and low permeability, much of the rain that falls in the desert quickly becomes surface runoff available to perform fluvial gradational work. With little vegetative cover, any grains of rock that have been loosened by weathering may be swept away in surface runoff produced by the next storm. Although desert landscapes strongly reflect a deficiency of water, the effects of running water are widely evident on slopes as well as in valley bottoms ( ❙ Fig. 18.1). Where vegetation is sparse, running water, when it is available, is extremely effective in shaping the land.

Desert climates characteristically receive small amounts of precipitation and are subjected to high rates of potential evapotranspiration. In exceptional circumstances, years may pass without any rain in certain desert areas. Most desert locations, however, receive some precipitation each year although the frequency and amount are highly unpredictable. Rains that do fall are often brief and limited in their spatial coverage, but they can also be quite intense. While times of rainfall are short, unreliable, and difficult to predict, potential evapotranspiration remains high throughout the year in most arid regions. The most important impact of rain on landform development in deserts is that when rainfall does occur, much of it falls on impermeable surfaces,

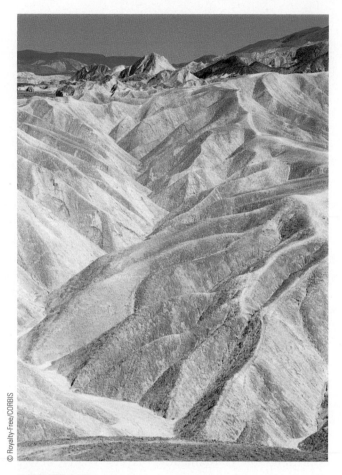

❙ **FIGURE 18.1**
In almost all deserts, even in the most arid locations, the effects of erosion and deposition by running water are prominent in the landscape. This is Death Valley, California.
**Why do you think there is a high drainage density here?**

producing intense runoff, generating flash floods, and operating as a powerful agent of erosion.

The importance of water as a gradational agent in arid regions is related to not only the climate of those areas today but also past climates. Paleogeographic evidence indicates that most deserts have not always had the arid climates that exist today. Geomorphologists studying arid regions have found certain landforms that are incompatible with the present climate and attribute these to the work of water under earlier, wetter climates. A great majority of desert areas were wetter in the past, most recently during the Pleistocene Epoch. While glaciers were advancing in high latitudes and in mountain regions during the Pleistocene, precipitation was also greater than it is today in the basins, valleys, and plains of the middle and subtropical latitudes where deserts are now found. Cooler temperatures for these regions meant that they experienced lower evaporation rates. In many of today's deserts, evidence of past wet periods includes deposits and wave-cut shorelines of now-extinct lakes ( ❙ Fig. 18.2) and immense canyons occupied by streams that today are too small to have eroded such large valleys.

Running water is a highly effective agent of landform development in deserts even though it operates only occasionally. In most desert regions, running water is active just during and shortly after rainstorms. Desert streams, therefore, are typically *ephemeral channels,* containing water only for brief intervals. Ephemeral stream channels are exposed and dry the rest of the time. In contrast to *perennial channels,* which flow all year and are typical of humid environments, ephemeral streams do not receive seepage from groundwater to sustain them between episodes of surface runoff. Ephemeral streams instead generally lose water to the groundwater system through infiltration into the channel bed. Because of the low weathering rates in desert environments, most arid region streams receive an abundance of coarse sediment that they must transport as bed load. As a result, *braided channels,* in which multiple threads of flow split and rejoin around temporary deposits of coarse-grained sediments, are common in deserts ( Fig. 18.3).

Unlike the typical situation for humid region streams, many desert streams undergo a downstream decrease, rather than increase, in discharge. A discharge decrease occurs because continued losses of stream water by infiltration into the gravelly channel bed are accompanied downstream by increasing evaporation due to warmer temperatures at the lower elevations. As a result of the diminishing discharge, many desert streams terminate before reaching the ocean. The same mountains that contribute to aridity through the rain-shadow effect can effectively block desert streams from flowing to the sea. Without sufficient discharge to reach ultimate base level, desert streams terminate in depressions in the continental interior where they commonly form shallow, ephemeral lakes. Ephemeral lakes evaporate and disappear and then reappear when rain provides another episode of adequate inflow. During the cooler and wetter times of the Pleistocene Epoch, many closed basins in now-arid regions were filled with considerable amounts of water that in some cases formed large perennial freshwater lakes instead of the shallow ephemeral lakes that they contain today.

Where surface runoff drains into closed desert basins, sea level does not govern erosional base level as it does for streams that flow into the ocean and thereby attain *exterior*

■ FIGURE 18.2
Many desert basins in Nevada, Utah, and California have remnant shorelines that were created by wave action from lakes they contained during the Pleistocene. The linear feature extending across this hill slope in Utah is a shoreline from one of these ancient lakes.
**What can we learn about climatic change from studying these relict lake features?**

■ FIGURE 18.3
This braided stream in Canyon de Chelly National Monument near Chinle, Arizona, splits and rejoins multiple times as it works to carry extensive bed load of coarse sand.
**Why do you think the number and position of the multiple channels can change rapidly?**

*drainage.* Desert drainage basins characterized by streams that terminate in interior depressions are known as basins of *interior drainage* ( Fig. 18.4); such streams are controlled by a **regional base level** instead of ultimate base level. When sedimentation raises the elevation of the desert basin floor located at the stream's terminus, the stream's base level rises, which causes a decrease in the stream's slope, velocity, and energy. If tectonic activity lowers the basin floor, the regional base level is depressed, which may lead to rejuvenation of the desert stream. Tectonism has even created some desert basins of interior drainage with floors below sea level, as in Death Valley, California; the Dead Sea Basin in the Middle East; the

■ FIGURE 18.4
The Sierra Nevada (upper part of photo) poses a topographic barrier to streams on its rain-shadow side (lower part of photo) and did so even during the wetter times of the Pleistocene so that few flowed to the sea. The other streams filled depressions to form lakes, most of which are completely dry today. This image, oriented with north to the right, shows the bed of Owens Lake (large white area), which shrank because of climate change and was desiccated when its waters were diverted to urban areas in southern California. A small amount of moisture (elongated dark zones) occupied part of the dry lake bed when the photo was taken.

Turfan Basin in western China; and Australia's Lake Eyre (see Map Interpretation: Desert Landforms).

Many streams found in deserts originate in nearby humid regions or in cooler, wetter mountain areas adjacent to the desert. Even these, however, rarely have sufficient discharge to sustain flow across a large desert (■ Fig. 18.5). With few tributaries and virtually no inflow from groundwater, stream water losses to evaporation and underground seepage are not replenished. In most cases, the flow dwindles and finally disappears. The Humboldt River in Nevada is an outstanding example; after rising in the mountains of central Nevada and flowing 465 kilometers (290 mi), the river disappears into the Humboldt Basin, a closed depression. Only a few large rivers that originate in humid uplands have sufficient volume to survive the long journey across hundreds of kilometers of desert to the sea (■ Fig. 18.6). Called **exotic streams,** the rivers that successfully traverse the desert erode toward a base level governed by sea level and provide drainage that is external to the arid region. Classic examples of exotic streams include the Nile (Egypt and Sudan), Tigris-Euphrates (Iraq), Indus (Pakistan), Murray (Australia), and Colorado (United States and Mexico) Rivers.

# Water as a Gradational Agent in Arid Lands

When rain falls in the desert, sheets of water run down unprotected slopes, picking up and moving sediment. Dry channels quickly change to flooding streams. The material removed by runoff and surface streams is transported, just as in humid lands, until flow velocity decreases sufficiently for deposition to occur. Eventually these streams disappear when seepage and evaporation losses exceed their discharge. Huge amounts of sediment can be deposited along the way as the stream loses volume and velocity. The processes of erosion, transportation, and deposition by running water are essentially the same in both arid and humid lands. However, the resulting

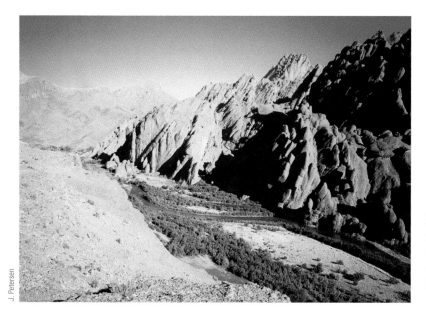

# FIGURE 18.5
A stream flows through a deep gorge in the Atlas Mountains of Morocco. This is the arid, rain-shadow side of the mountains, facing the Sahara to the east. This stream loses water by infiltration and evaporation and disappears into the Sahara. Note the steeply dipping, folded rocks of the Atlas and the thin line of vegetation along the stream channel.
**Was the gorge eroded by the stream with this amount of flow? If not, what factors might have produced more discharge to erode the deep canyon?**

landforms differ because of the sporadic nature of desert runoff, the lack of vegetation to protect surface materials against rapid erosion, and the common occurrence of streams that do not reach the sea.

## Arid Region Landforms of Fluvial Erosion

Among the most common desert landforms created by surface runoff and erosion are the channels of ephemeral streams. Known as **washes** or **arroyos** in the southwestern United States, **barrancas** in Latin America, and **wadis** in North Africa and Southwest Asia, these channels usually form where rushing surface waters cut into unconsolidated alluvium ( Fig. 18.7). These typically gravelly, braided channels are prone to flash floods, which makes them potentially very dangerous sites. Though it may sound strange, many people have drowned in the desert during flash floods.

Where weak, easily erodible clays and shales underlie steep slopes, rapid runoff produces a dense network of barren ridges dissected by a maze of steep, dry gullies and ravines. Early fur trappers in the Dakotas called such areas "bad lands to cross" ( Fig. 18.8). The phrase stuck, and those regions are still called the Badlands, while that type of rugged, barren, and highly dissected terrain is termed **badlands** topography. Badlands topography has an extremely high *drainage density*, defined as the length of stream channels per unit area of the drainage basin. Besides the Dakotas, extensive badlands can be seen in Death Valley National Park, California; Big Bend National Park, Texas; and southern Alberta, Canada. Badlands generally do not form naturally in humid climates because the vegetation there slows runoff and erosion, leading to lower drainage densities. Removing the vegetation from clay or shale areas by overgrazing, mining, or logging, however, can cause badlands topography to develop even in humid environments.

A **plateau** is an extensive, elevated region with a fairly flat top surface. Plateaus are generally dominated by a structure of horizontal rock layers. Many striking plateaus exist in the deserts and semiarid regions of the world. An excellent example in the United States is the Colorado Plateau, centered on the Four Corners area of Arizona, Colorado, New Mexico, and Utah. In tectonically uplifted desert plateau regions such as this, streams and their tributaries

# FIGURE 18.6
A false-color satellite image of the Nile River meandering across the Sahara in Egypt. The dark-red irrigated croplands contrast with the barren desert terrain. The Nile is an exotic stream. Its headwaters are in the wetter climates of the Ethiopian Highlands and lakes in the east African rift zone, which support its northward flow across the Sahara to the Mediterranean Sea.

respond to uplift by cutting narrow, steep-sided canyons. Where the canyon walls consist of horizontal layers of alternating resistant and erodible rocks, differential weathering and erosion exert a strong influence on the canyon walls. Canyons in these areas tend to have stair-stepped walls, with near-vertical cliffs marking the resistant layers (ordinarily sandstone, limestone, or basalt) and weaker rocks (often shales) forming the slopes. The distinctive walls of the Grand Canyon have this appearance, which exposes the structure of horizontal rock layers of varying thickness and resistance (▮ Fig. 18.9). The rim of the Grand Canyon is a flat-topped cliff made of a **caprock,** a term that refers to a resistant horizontal layer that forms (caps) the top of a landform.

Caprocks top plateaus and constitute canyon rims, but they also form the summits of other, smaller kinds of flat-topped landforms that, although they are found in many climate regions, are most characteristic of deserts. Weathering and erosion will eventually reduce the extent of a caprock until only flat-topped, steep-sided **mesas** remain (*mesa* means "table" in Spanish). A mesa has a smaller surface

▮ **FIGURE 18.7**
This dry streambed, or wash, has a channel bed of coarse alluvium and conveys water only during and slightly after a rainstorm.
**Why would this desert stream channel have a high risk for flash floods?**

▮ **FIGURE 18.8**
The Badlands of South Dakota. Impermeable clays that lack a soil cover produce rapid runoff, leading to intensive gully erosion and a high drainage density.
**Was this rugged terrain named appropriately?**

area than a plateau and is roughly as broad across as it is tall. Mesas are relatively common landscape features in the Colorado Plateau region. Through additional erosion of the caprock from all sides, a mesa may be reduced to a **butte,** which is a similar, flat-topped erosional remnant but with a smaller surface area than a mesa ( Fig. 18.10). Mesas and buttes in a landscape are generally evidence that uplift occurred in the past and that weathering and erosion have been extensive since that time. Variations in the form of the slope extending down the sides of buttes, mesas, and plateaus are related to the height of the cliff at the top, which is controlled by the thickness of the caprock in comparison to the size of landform feature. Monument Valley, in the Navajo Tribal Reservation on the Utah–Arizona state line, is an exquisite example of such a landscape formed with a caprock that is particularly thick, contributing to the distinctive scenery ( Fig. 18.11). Many famous western movies have been filmed in Monument Valley and in nearby areas of the Colorado Plateau because of the striking, colorful, and photogenic desert landscape.

Sheet wash and gully development generally accomplish extensive erosion of mountain and hill slopes fringing a desert basin or plain. Particularly in desert regions with exterior drainage or a sizable trunk stream on the basin or plain, this fluvial action, aided by weathering, can lead to the gradual erosional retreat of bedrock slopes. This retreat of the steep mountain front can leave behind a more gently sloping surface of eroded bedrock, called a **pediment** ( Fig. 18.12). Characteristically in desert areas, there tends to be a sharp break in slope between the base of steep hills or mountains, which rise at angles of 20 to 30 degrees or steeper, and the gentle pediment, whose slope is usually only 2 to 7 degrees. Resistant knobs of the bedrock comprising the pediment may project up above the surface on some pediments. These resistant knobs are referred to as **inselbergs** (from German: *insel,* island; *berg,* mountain).

Geomorphologists do not agree on exactly how pediments are formed, perhaps because different processes may be responsible for their formation in various world regions. However, there is general agreement that most pediments are erosional surfaces created or partially created by the action of running water. In some areas, weathering, perhaps when the climate was wetter in the past, may also have played a strong role in the development of pediments.

**FIGURE 18.9**
The earliest European American explorers of the Grand Canyon took along an artist to record the geomorphology of the canyon, here beautifully shown in great detail. Aridity creates an environment where the bare rocks are exposed to our view; differential weathering and erosion give the stair-stepped quality to the walls of the Grand Canyon.

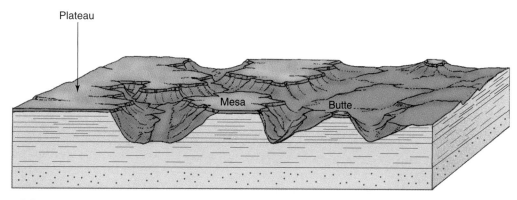

**FIGURE 18.10**
Diagram of landforms developed through weathering and erosion in an area of horizontal rock layers with a resistant caprock (such as the Colorado Plateau of Arizona, Colorado, New Mexico, and Utah)

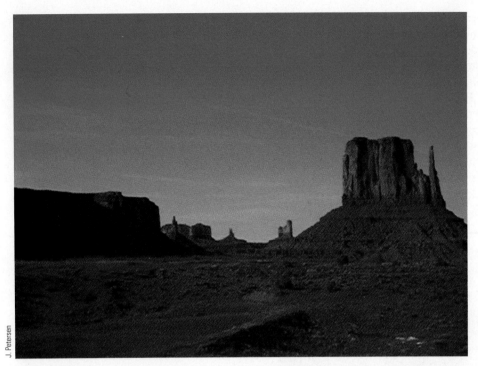

**FIGURE 18.11**
Monument Valley, Arizona, with prominent buttes and mesas. The caprock here is particularly thick and represents a rock layer that once covered the entire region. The buttes and mesas are erosional remnants of that layer.
**Compare this photo to the diagram in Figure 18.10. How were these landforms produced?**

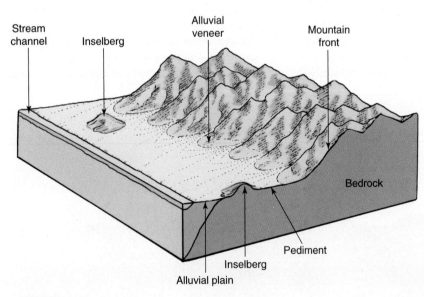

**FIGURE 18.12**
Pediments are erosion surfaces cut into bedrock beyond the present mountain front in arid regions. Pediments most commonly occur in desert areas with exterior drainage that can remove a portion of the erosion products. At many locations, pediment surfaces are covered with a thin veneer of alluvium, and their gently sloping surface may be interrupted by resistant knobs of bedrock, called inselbergs, that stick up above the pediment surface.

# Arid Region Landforms of Fluvial Deposition

Deposition is as important as erosion in creating landform features in arid regions, and in many areas sedimentation by water does as much to level the land as does erosion. Many desert areas have wide expanses of alluvium deposited either in closed basins or at the base of mountains by streams as they lose water in the arid environment. As the flow of a stream diminishes, so does its *capacity*—the amount of load it can transport. Most landforms developed by fluvial deposition in arid lands are not exclusive to desert regions but are particularly common and visible in dry environments due to their thin soil and sparse vegetative cover.

**Alluvial Fans** Where streams, particularly ephemeral (sporadic) or *intermittent* (seasonal) ones, flow out of narrow canyons and onto open plains, their channels may flare out to become wide and shallow. Because stream discharge ($Q$) equals channel cross-sectional width ($w$) and depth ($d$) multiplied times flow velocity ($v$) ($Q = wdv$), the increase in width causes a decrease in flow velocity, reducing stream competence (maximum size of load) and stream capacity (maximum amount of load) Discharge also decreases as water seeps from the channel into coarse alluvium below. As a result, most of the sediment load carried by such streams is deposited along the base of the highland.

Upstream of the mouth of the canyon, the channel is constrained by bedrock valley walls, but as it flows out of the canyon, the channel is free to not only widen but also shift its position laterally. Sediments are deposited beyond the canyon mouth and build up the plain area there. Eventually this aggradation results in the channel shifting laterally to deposit sediment in and build up an adjacent area and perhaps subdividing into smaller channels. The canyon mouth serves as a pivotal point anchoring the channel as it swings back and forth over the plain downstream from the canyon, leaving alluvium

**FIGURE 18.13**
Alluvial fans are constructed at the base of a mountain range. Where streams come out of confined canyons, they are free to widen, which causes a decrease in stream velocity, capacity, and competence. The apex of a fan lies at the mouth of the canyon or wash. Fans are particularly common landforms in the arid Basin and Range region of the western United States. For scale, note that a road runs across the lower portion, or toe, of this alluvial fan in Death Valley, California.
**How do alluvial fans differ from pediments?**

behind. This creates a fan-shaped depositional landform, called an **alluvial fan,** in which deposition takes place radially away from that pivotal point, or **fan apex** ( Fig. 18.13).

An important characteristic of an alluvial fan is the sorting of sediment that typically occurs on its surface. Coarse sediments, like boulders and cobbles, are deposited near the fan apex where the stream first undergoes a decrease in competence and capacity as it emerges from the confinement of the canyon. In part because of the large size of the clasts deposited there, the slope of an alluvial fan is steepest at its apex and gradually diminishes, along with grain size, with increasing distance downstream from the canyon mouth. In areas where the uplands generate debris flows rather than stream flows, **debris flow fans** or even mixed debris and alluvial fans are constructed instead of purely alluvial fans. Debris flow fans tend to be steeper than alluvial fans and do not display the same degree of downslope sorting shown by the fluvial counterpart.

Although they can be found in mountainous areas of almost any climate, alluvial fans are particularly common where ephemeral or intermittent streams laden with coarse sediment flow out of a mountainous region onto desert plains or into arid interior basins. In the western United States, alluvial fans are a major landform feature in landscapes consisting of fault-block mountains and basins, as in the Great Basin of California, Nevada, and Utah ( Fig. 18.14). Here streams laden with sediment periodically rush from canyons cut into uplifted fault-block mountains and deposit their load in the adjacent desert basins. Everything else being equal, fans associated with larger drainage basins within the uplifted fault-block mountains tend to have greater area and be less steep than fans developed from streams draining smaller upland drainage basins.

Large, conspicuous alluvial fans develop in environmental settings like the Great Basin for several reasons. First, highland areas in desert regions are subject to intense erosion, primarily because of the low density of vegetative cover, steep slopes, and the orographically intensified downpours that can occur over mountains. In addition, streams in arid regions typically carry a greater concentration of sediment load (in comparison to the discharge) than comparable streams in more humid regions. As the streams flow from confined mountain canyons into desert basins, they deposit most of their coarse sediment near the canyon mouth. Flowing into the desert basin, their width increases, their depth and velocity decrease, and their volumes are significantly reduced through infiltration into the alluvial channel bed. Not far from the canyon, the stream itself may disappear, or it may occasionally reach the desert basin floor where it deposits its remaining load, the silts and clays. Extensive alluvial fans are not as common in humid as in arid regions because most highland streams in moist climates are perennial and have sufficient flow to continue across adjacent lowlands.

Along the bases of mountains in arid regions, adjacent alluvial fans may become so large that they join together along their sides to form a continuous ramplike slope of alluvium called a **bajada** ( Fig. 18.15). A bajada consists of adjacent alluvial fans that have coalesced to form an "apron" of alluvium along the mountain base. Where extensive fans coalesce over very wide areas, they form a **piedmont alluvial plain,** like the area surrounding Phoenix, Arizona ( Fig. 18.16).

Piedmont alluvial plains generally have rich soils and the potential to be transformed into productive agricultural lands. The major obstacle is inadequate water supply to grow crops in an arid environment. In many world regions, arid alluvial plains are irrigated with water diverted from mountain areas or obtained from reservoirs on exotic streams. The alluvial farmlands near Phoenix are a good example, producing citrus fruits, dates, cotton, alfalfa, and vegetables.

Where a veneer of alluvium has been deposited on a pediment, the land surface may closely resemble a water-deposited alluvial fan. In some situations, it may not be possible

**FIGURE 18.14**
Map of the Great Basin of the western United States showing major lakes that existed during glacial times. More than 100 lakes formed in the fault-block basins of this region during the Pleistocene Epoch.
**Why are there only a few remnant lakes in this region today?**

to determine the existence of an underlying pediment without either excavating through the surface alluvium or finding the erosional pediment surface exposed in the walls of washes or gullies. In locations where no extensive pediment exists, alluvial deposits beneath the fans can be tens or even hundreds of meters thick. In contrast, the layer of alluvium overlying a pediment is only a relatively thin layer, no more than a few meters deep and sometimes much less.

**FIGURE 18.15**
A bajada is formed when a series of alluvial fans coalsece, forming a continuous alluvial slope along the front of an eroding mountain range. This example is from a site in Utah within the Great Basin.
**Why would a series of alluvial fans have a tendency to eventually join to form a bajada?**

**FIGURE 18.16**
A large desert alluvial plain in Arizona that is extensively urbanized.

**Playas** Desert basins of interior drainage surrounded by mountains are sometimes called **bolsons.** Most bolsons were formed as faulting created basins between uplifted mountains. The lowest part of most bolsons is occupied by a landform called a **playa** (in Spanish: *playa,* beach or shore), which is the fine-grained bed of an ephemeral lake. Large rainfall (or snowmelt) events or wet seasons occasionally cover the playa with a very shallow body of water, called a **playa lake.** Direct precipitation onto the playa, inflow from surface runoff, or discharge from the groundwater zone can contribute water to the playa lake. The playa lake may persist for a day or two or for several weeks ( Fig. 18.17) Wind blowing over the playa lake moves the shallow water, along with its suspended and dissolved load, around on the playa surface. This helps to fill in any low spots on the playa and contributes to making playas one of the flattest of all landforms on Earth. Playa lakes lose most of their water by evaporation to the desert air.

Although playas are very flat, considerable variation exists in the nature of playa surfaces. Playas that receive most of their water from surface runoff typically have a smooth clay surface, baked hard by the desert sun when it is dry but extremely sticky and slippery when wet. In contrast, playas that receive much of their water from groundwater may be damp most of the time and encrusted with salt mineral deposits crystallizing out of the evaporating groundwater. Playas with salt crusts comprised of carbonates, sulfates, and chlorides (for example, calcite, gypsum, and rock salt, respectively) were formed when lakes that occupied now-desert basins during the Pleistocene Epoch desiccated due to climate change. Saline playas are also known as **salt flats,** or **salinas** ( Fig. 18.18).

Playas are useful in several ways. For one, companies mine the rich deposits of evaporite minerals, including such important industrial chemicals as potassium chloride, sodium chloride, sodium nitrate, and borates, that have been deposited in some playa beds. Also, the extensive, flat surfaces of some playas make them suitable as racetracks and airstrips. Utah's famous Bonneville Salt Flats mark the bed of an extinct Pleistocene lake. The western portion of the Salt Flats, where world land-speed records are set, still floods to a depth of 30–60 centimeters (1–2 ft) in the cool, wet, winter season. The hard, flat playa surface at Edwards Air Force Base, in California's Mojave Desert, has served for several decades as a landing site for military aircraft and, in recent decades, for the space shuttle. These landings have occasionally been disrupted due to flooding of the playa.

**FIGURE 18.17**
Badwater, in Death Valley, California. The playa surface, seen just behind the edge of the playa lake water, is the lowest elevation in North America at 86 meters (282 ft) below sea level. Badwater, the small lake, is fed by groundwater that flows down from the surrounding mountains into alluvial deposits under the surface of the basin. The water quickly evaporates in this extremely arid environment.
**Why was this small lake called Badwater?**

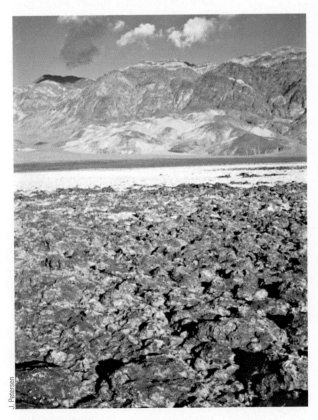

**FIGURE 18.18**
This view of Death Valley shows deposits of salt left behind by evaporation primarily of groundwater derived from the surrounding mountains. Evaporation is accumulating salts in the mud in the foreground, creating the playa microtopography known as puffy ground.

# Wind as a Gradational Agent

On a worldwide basis, wind is less effective than running water, waves, groundwater, moving ice, or mass movement in accomplishing geomorphic work. Under certain circumstances, however, wind can be a significant agent in the modification of topography. Landforms—whether in the desert or elsewhere—that are created by wind are called **eolian** (or **aeolian**) landforms, after Aeolus, the god of winds in classical Greek mythology (Fig. 18.19). The three principal conditions necessary for wind to become an effective gradational agent are a sparse vegetative cover, the presence of dry, loose materials at the surface, and a wind velocity that is high enough to pick up and move those surface materials. These three conditions occur most widely in arid regions and on beaches although they are also found on or adjacent to exposed lake beds, areas of recent alluvial or glacial deposition, newly plowed fields, and overgrazed lands.

A dense vegetative cover reduces wind velocity near the surface by providing frictional resistance, prevents wind from being directed against the land surface, and holds materials in place with its root network. Without such a protective cover, fine-grained and sufficiently dry surface materials are subject to removal by strong gusts of wind. If surface particles are damp, they tend to adhere together in wind-resistant aggregates due to increased cohesion provided by the water. The arid conditions of deserts therefore make those regions most susceptible to wind erosion.

Eolian processes have many things in common with fluvial processes because air and water are both fluids. Some important contrasts also exist, however, due to fundamental differences between gases and liquids. For example, rock-forming materials cannot dissolve in air, as some can in water; thus, air does not erode by corrosion or move load in solution. Otherwise, the wind detaches and transports rock fragments in ways comparable to flowing water, but it does so with less overall effectiveness because air has a much lower density than water. Another difference is that, compared to streams, the wind has fewer lateral or vertical limitations on movement. As a result, the dissemination of material by the wind can be more widespread and unpredictable than that by streams.

A principal similarity between the gradational properties of wind and running water is that flow velocity controls their competence—that is, the size of particles each can pick up and carry. However, because of its low density, the

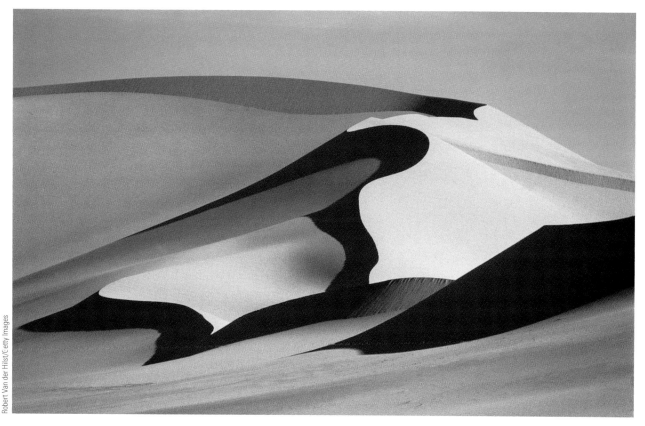

**FIGURE 18.19**
This view of sand dunes in Africa's Namib Desert shows how eolian processsess can create a stunningly beautiful landscape.
**What makes the processes that formed these dunes so different from those that form alluvial fans?**

competence of moving air is generally limited to rock fragments that are sand sized or smaller. Wind erosion selectively entrains small particles, leaving behind the coarser and heavier particles that it is incapable of lifting. Like water-laid sediments, wind deposits are stratified according to changes in its velocity although within a much narrower range of grain sizes than occurs with most alluvium.

## Wind Erosion and Transportation

Strong winds blow frequently in arid regions, whipping up loose surface materials and transporting them within turbulent air currents. Wind erodes surface materials by two main processes. The first of these is **deflation,** which is similar to the hydraulic force of running water. Deflation occurs when wind blowing fast enough or with enough turbulence over an area of loose sediment is able to pick up and remove small fragments of rock. The finest particles transported by winds, clays and silts, are moved in **suspension,** buoyed by vertical currents ( Fig. 18.20). Such particles essentially comprise a fine dust that will remain in suspension as long as the strength of upward air currents exceeds the tendency of particles to settle out to the ground due to the force of gravity. The sediments carried in suspension by the wind make up its suspended load. If the wind velocity surpasses 16 kilometers (10 mi) per hour, surface sand grains can be put into motion. As with fluvial transportation, particles that are too large to be carried in eolian suspension are bounced along the ground as part of the bed load in the transportation process of *saltation*. When particles moving in eolian saltation, which are typically sand sized, bounce on the ground, they generally dislodge other particles that are then added to the wind's suspended or saltating load and push forward along the ground surface, in a process called **surface creep,** even larger sand grains too heavy to be lifted into the air. Grains with low saltation trajectories become organized into small wave forms termed **ripples.** The velocity required for the wind to pick up and start to move a grain is greater than that required to keep it moving, and this is primarily due to such surface factors as roughness and cohesion.

The second way that the wind erodes is by *abrasion*. Once the wind obtains some load, the impact of those wind-driven solid particles is more effective than the wind alone in dislodging and entraining other grains or for breaking off new fragments of rock. This process is analogous to abrasion that occurs with stream flow and the other geomorphic agents, but eolian abrasion operates on a much more limited

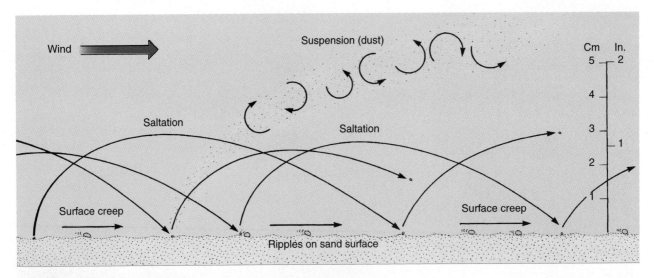

**FIGURE 18.20**
Wind moves sediment in the transportation processes of suspension, saltation, and surface creep. Some of the bed load forms ripples, which can be seen moving forward when the wind is strong.
**Why are grains sized larger than sand not generally moved by the wind?**

**FIGURE 18.21**
Dust storms occur when strong winds mobilize, erode, and suspend silt particles, picked up in areas of barren soil or alluvium. This is a dust storm in Texas during the "Dust Bowl" era of the 1930s.
**Can you suggest a continent that might be a source of major dust storms today?**

scale. Most eolian abrasion is quite literally sandblasting, and quartz sand, which is common in many desert areas, can be a very effective abrasive agent in eolian processes. Yet sand grains are typically the largest size of clast that the wind can move, and they rarely are lifted higher than 1 meter (3 ft) above the surface. Consequently, the effect of this natural sandblast is limited to a zone close to ground level.

Where loose dust particles exist on the land surface, they will be picked up and carried in suspension by strong winds. The result is a thick, dark, swiftly moving dust cloud swirling over the land. **Dust storms** (Fig. 18.21) can be so severe that visibility drops to nearly zero and almost all sunlight is blocked. They can also be highly destructive, removing layers of surface materials and depositing them elsewhere, sometimes in thick, choking new layers, all within a matter of a few hours. The infamous Dust Bowl era of the 1930s particularly impacted the southern Great Plains of the United States in this way when devastating dust storms were brought about by years of drought and poor agricultural practices. **Sandstorms** may occur in areas where sand is abundant at the surface. Because sand grains are larger and heavier than dust particles, most sandstorms are confined to a low level near the surface. Evidence of the restricted height of desert sandstorms can be seen on vehicles that have traveled through the desert, as well as on fence posts, utility poles, and other structures. After a sandstorm, the pitting, gouging, and abrading effects of natural sandblast are more damaging to the lower portions of vehicles and other objects that were subjected to the abrasion.

Erosion by deflation can produce shallow depressions in a barren surface of unconsolidated materials. These depressions, which can vary in diameter from a few centimeters to a few kilometers, are called **deflation hollows**. Deflation hollows tend to collect rainwater and may hold water for a time, depending on permeability and evaporation rates. Often deflation hollows form at sites that were already exhibiting a slight depression or where vegetation cover has been breached by overgrazing, fire, and other means.

Deflation may be one of several factors that help to produce **desert pavement** (**reg** in North Africa, **gibber** in Australia), a close-fitting mosaic of gravel-sized stones that overlies a deposit of mostly finer grained sediments. Desert pavement is common in many arid regions, particularly in parts of the Sahara, interior Australia, the Gobi in central Asia, and the American Southwest. If deflation selectively removes the smaller particles (clays, silts, and possibly sands) from a desert surface of mixed particle sizes, the gravel-sized clasts left behind can form a concentration of stones at the surface overlying the mixed grain sizes below ( Fig. 18.22). Sheet wash (unchannelized running water) may contribute to the formation of desert pavement by selectively eroding only the fine-grained clasts from an area of mixed grain sizes. Research has also shown that pavements can form by eolian deposition, rather than erosion, of the fine-grained sediments found beneath the stony surface layer. Regardless of its origin, desert pavement is important for the protection it affords the finer material below the surface layer of gravel. Pavement formation stabilizes desert surfaces by preventing continuous erosion. Unfortunately, off-road recreational vehicles can disrupt the surface stones and disturb this stability, thus damaging desert ecological systems.

Like deflation, eolian abrasion is also responsible for creating interesting desert landform features. Where the land surface is exposed bedrock, wind abrasion can polish, groove, or pit the rock surface and in some cases produces **ventifacts,** which are individual wind-fashioned rocks. A ventifact is a rock fragment that has been trimmed back to a smooth slope on one or more sides by sandblast. Because of frictional effects at the surface, the ability of the wind to erode by abrasion increases with increasing distance from the ground surface, at least up to a certain height. Thus, abrasion carves the windward side of a rock into a smooth, sloping surface, or face. Ventifacts subjected to multiple sand-transporting wind directions have multiple faces, called facets, which meet along sharp edges ( Fig. 18.23). Although not extremely common, ventifacts are plentiful in local areas where wind and surface rock conditions are ideal for their formation.

Another feature often attributed to wind abrasion is the pedestaled, or balanced, rock—commonly and incorrectly thought to form where eolian abrasion attacks the base of an individual rock so that the larger top part appears balanced on a thinner pedestal below. Actually, such forms result from various weathering processes in the damper environment at the base of an exposed rock and are not typically related to eolian abrasion ( Fig. 18.24).

Rates of eolian erosion in arid regions often reflect the strength of the materials exposed. Where eolian abrasion affects rocks of varying resistance, differential erosion etches away softer rocks faster than the more resistant rocks. Even in desert locations of extensive soft rock, such as shale, or semiconsolidated sediments, like ancient lake deposits, abrasion may not act in a uniform fashion over the entire exposure. A **yardang** is a wind-sculpted remnant ridge, often of easily eroded rock or semilithified sediments, left behind after the surrounding material has been eroded by abrasion ( Fig. 18.25), perhaps with deflation assisting in removal of fine-grained fragments. Everything else being equal, abrasion and deflation are most effective where rocks are soft or weak.

## Wind Deposition

All materials transported by the wind are deposited somewhere, generally in a distinctive manner that is related to characteristics of the wind as well as the nature and grain size of the deposits. Coarser, sand-sized material is often deposited in

**FIGURE 18.22**
Some desert pavement may be created when the fine-grained fraction of a deposit of mixed grain sizes is removed by the wind or by sheet wash, leaving a lag of stones at the surface.
**Is desert pavement a surface indestructible to human activities? Why?**

**FIGURE 18.23**
This approximate true-color image of a ventifact on Mars was taken by the Mars Exploration Rover *Spirit*. With high wind speeds, plenty of loose surface materials, and lack of vegetation cover, the rocks on the surface of Mars display strong evidence of abrasion by the wind.

drifts in the shape of hills, mounds, or ridges, called **sand dunes.** Finer-grained sediment, such as silt, can be transported in suspension long distances from its source area before blanketing and sometimes modifying the existing topography as a deposit called **loess** ( Fig. 18.26).

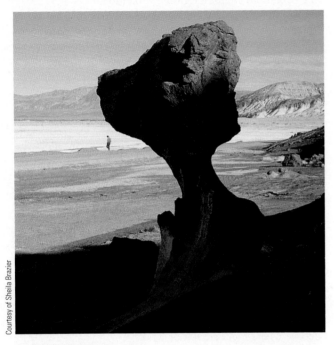

■ **FIGURE 18.24**
Mushroom Rock in Death Valley, California, is a pedestal rock caused by desert weathering processes.
**What would account for such an unusual shape?**

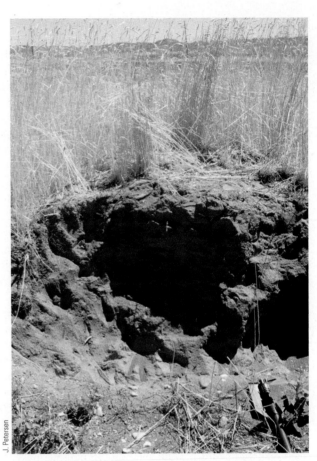

■ **FIGURE 18.26**
Loess, seen here near Ogden, Utah, is wind-deposited silt.

■ **FIGURE 18.25**
Eolian erosion can leave behind yardangs, aerodynamically shaped ridges like this one in the Kharga Depression of Egypt.

**Sand Dunes** To many people, the word *desert* evokes the image of endless sand dunes, blinding sandstorms, a blazing sun, mirages, and an occasional palm oasis. Although these features do exist, particularly in Arabia and North Africa, most of the areas of the world's deserts have rocky or gravelly surfaces, scrubby vegetation, and few or no sand dunes. Nevertheless, sand dunes are certainly the most spectacular features of wind deposition, whether they occur as seemingly endless dune regions, called **sand seas** (or **ergs**), as small dune fields, or as sandy ridges behind a beach ( Fig. 18.27).

Dune topography is highly variable. For instance, dunes in the great sand seas of the Sahara and Arabia look like rolling ocean waves. Others have aerodynamic crescent forms. Eolian sand deposits can also form **sand sheets,**

## Geography's Environmental Science Perspective
### Invasions by Sand Dunes

This chapter discusses the various types and movements of sand dunes within arid regions of Earth. Because sand dunes can play a role in preserving wildlife species and protecting human development, protecting dunes against environmental degradation is important. However, there is also a counterargument to the view of dunes as beneficial. Sand dunes can and do invade delicate environments and human communities, and this invasion of massive amounts of sand can be devastating.

For sand dunes to actively move, three conditions are necessary: a source of sand, winds that are strong enough to move the sand, and a lack of stabilizing vegetation. Depending on the relative balance of these three factors, different types of dunes may result. For example, parabolic dunes can form with a moderate amount of sand but with some vegetation present; longitudinal dunes can exist with minimal sand but strong winds; transverse dunes need abundant sand and little or no vegetation. If sand supply is minimal, winds are weak, and/or there is abundant vegetation, no dunes will form.

In trying to predict if sand dunes will be active or stabilized, research in the Colorado Plateau has shown that comparing precipitation ($P$) and potential evapotranspiration ($PE$) rates in a region often holds the answer. As discussed in Chapter 9, arid regions are deficient in moisture overall; that is, the ratio of $P$ to $PE$ is less than 1. The question then becomes, Is there enough moisture to support stabilizing vegetation? In regions where the $P/PE$ is between 0.3 and 0.5, there is still enough moisture to support sagebrush and grasses at lower elevations and piñon pine and juniper at higher elevations. Thus, at this time, the Colorado Plateau region, despite its abundance of sand and wind capable of moving sand, has few actively migrating dunes.

Dune migration is more fully quantified by comparing the $P/PE$ ratio with $W$, the percentage of time wind is capable of moving sand. The ratio of $W$ to the $P/PE$ value is referred to as the *dune mobility index*. Between the two factors of this index, $W$ is not as critical as $P/PE$; thus, a decrease in moisture is the factor that may activate dune mobility in the future. Dunes are therefore more mobile in extremely dry regions and arid regions undergoing severe drought.

There are multiple examples of areas where sand dune migration threatens human development. In areas of Saudi Arabia, teams of workers have dug into shifting dunes in attempts to alter their movement as they invade precious desert oases. These invaluable water sources are vital to desert survival. In southern Peru, dunes from the northern reaches of the Atacama Desert invade cities that border the region. In the Colorado Plateau region, the Navajo and Hopi Indian Reservations lie downwind from large areas of sand dunes. Activation of these sand dunes in the future may disrupt their housing, grazing for their sheep and cattle, and their farming practices. Only the continual monitoring of moisture conditions to identify regions of severe drought can provide warning that massive deposits of sand may again be on the move.

---

with no dune formation at all. Research has shown that the specific type of dune that forms depends on the amount of sand available, the strength and direction of sand-transporting winds, and the amount of vegetative cover. As wind-carrying sand encounters surface obstacles or topographic obstructions that decrease its velocity, sand is deposited and piles up in drifts. These sand piles interfere with the wind regime and the sand-transporting capabilities of the wind, so the dunes grow larger until equilibrium is reached between dune size and the ability of winds to feed sand to the dune.

Sand dunes may be classified as either *active* or *stabilized* (Fig. 18.28). Active dunes change their shape or advance downwind as a result of wind action. Dunes may change their shape with variations in wind direction and/or wind strength.

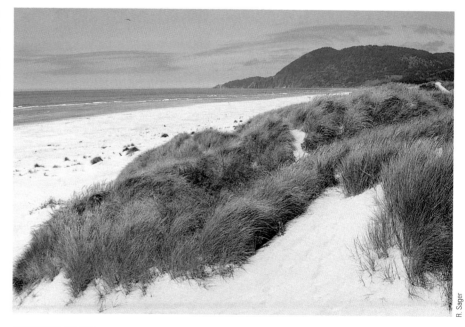

**FIGURE 18.27**
Dunes along the Oregon coast.
**Why are coastlines such good locations for dune formation?**

**FIGURE 18.28**
Active and stabilized dunes. (a) Active dunes usually have sharp crests. The gentle back slopes face upwind, while the downwind advancing slip faces are steep. Sediments transported across the dune, some of which may have been eroded from the back slope, are deposited on the slip face. The slip face is at the angle of repose for dry sand, as shown by these active dunes. (b) If plants can establish themselves in a dune area, they bring up moisture from beneath the dunes, which stabilizes the dune movement. Stabilized dunes tend to have more rounded crests.

**Why are active dunes sharper while stabilized dunes are more rounded in form?**

Dunes travel forward as the wind erodes sand from their upwind (windward) slope, depositing it on their downwind (leeward) side. Sand entrained on the upwind slope moves by saltation and surface creep up to the dune crest and over and onto the steep leeward slope, which is the **slip face.** The slip face of a dune lies at the angle of repose for dry, loose sand. The angle of repose (about 35 degrees for sand) is the steepest slope that a pile of dry, loose material can maintain without experiencing slipping or sliding down the slope. When wind direction and velocity are relatively constant, a dune can move forward while maintaining its general form by this downwind transfer of sediment ( Fig. 18.29). The speed at which active dunes move downwind varies greatly, but as with many processes, the movement is episodic; the dune advances only when the wind is strong enough to move sand from the upwind to the downwind side. Because of their greater height and especially their greater mass of sand, large dunes travel more slowly than smaller dunes, which may migrate up to 40 meters (130 ft) a year. During sandstorms, a dune may migrate more than 1 meter (3 ft) in a single day. Some dunes are affected by seasonal wind reversals so that they do not advance, but the crest at the top moves back and forth annually under the influence of seasonally opposing winds.

Stabilized dunes maintain their shape and position over time. Vegetative cover normally stabilizes dunes. If the vegetation cover becomes breached on a stabilized dune, perhaps due to the effects of range animals or off-road vehicles, the wind can then remove some of the sand, creating a **blowout.** In places where plants, including trees, lie in the path of an advancing dune, the sand cover may move over and smother the vegetation. Where invading dunes and blowing sands are a problem, attempts are frequently made to plant grasses or other vegetation to stabilize the dunes, halting their advance. Vegetation can stabilize a sand dune if plants can gain a foothold and send roots down to moisture deep within the dune ( Fig. 18.30). This task is difficult for most plants because the sand itself offers little in the way of nutrients and because of its high permeability and limited moisture.

One extensive area of stabilized dunes in North America is the Sand Hills of Nebraska. This region features large dunes that formed during a drier period between glacial advances in the Pleistocene Epoch. These impressive dunes are now stabilized by a cover of grasses ( Fig. 18.31). Similar stabilized dunes are found along the southern edge of the Sahara, where the desert has clearly extended farther toward the equator in the recent geologic past. Both locations involve changes in climate that affected sand supply, wind patterns, and moisture availability (see Map Interpretation: Eolian Landforms).

## Types of Sand Dunes

Sand dunes are classified according to their shape and their relationship to the wind direction. The different types are also related to the amount of available sand, which affects not only the dune size but also its shape.

**Barchans** are one kind of crescent-shaped dune (Fig. 18.32a). The two arms of the crescent, called the dune's horns, point downwind (Fig. 18.33). The main body of the crescent lies on the upwind side of the dune. From the desert floor at its upwind edge, the dune rises as a gentle slope up which the sand moves until it reaches the highest point, or crest, of the dune and, just beyond that, the slip face at the angle of repose. The slip face is oriented perpendicular to the barchan's arms. The arms extend downwind beyond the location of the slip face. Barchans form in areas of minimal sand supply where winds are strong enough to move sand downwind in a single prevailing direction. They may be most common in smaller desert basins surrounded by highlands where they tend to form near the downslope, sandy edge (toe) of alluvial fans, and adjacent to small playas. Although they form as isolated dunes, barchans often appear in small groups, called barchan fields.

**Parabolic dunes** are similar to barchans in that they are also crescentic dunes, but their orientation is reversed from that of barchans (Fig. 18.32b). Here, the arms of the crescent tend to be stabilized by vegetation, long, and pointing upwind, trailing behind the unvegetated main body and crest of the dune, rather than extending downwind from it. The main body of a parabolic dune points

### FIGURE 18.29
Active dunes move downwind. The wind transports sand from the dune's upwind side, up its back slope toward the sharp dune crest. The sand then slides down the steeper slip face on the downwind (leeward) side of the dune, causing the dune to advance, or migrate. Sand supply, dune size, and wind speed and duration are major factors controlling the speed of dune advance.
**Why does the inside of the migrating dune consist of former slip faces?**

### FIGURE 18.30
A sand dune advances into a vegetated area on Padre Island, Texas. In coastal dune regions, landscape change is common as dunes move inland and are subsequently invaded by plants.
**Explain how plants can stabilize dunes.**

### FIGURE 18.31
The rolling grazing lands of the Sand Hills in central Nebraska were once a major region of active sand dunes.
**Why are these dunes no longer active today?**

**FIGURE 18.32**
Five principal sand dune types: (a) barchan, (b) parabolic, (c) transverse, (d) longitudinal, and (e) star. The wind direction is shown in all figures, as indicated by arrows.
**What factors play a role in which type of dune will be found in a region?**

downwind, and the slip face along its downwind edge has a convex shape when viewed from above. Parabolic dunes commonly occur just inland of beaches and along the margin of active dune areas in deserts.

**Transverse dunes** are created where winds blow from a constant direction and the supply of sand is abundant ( Fig. 18.32c). The upward slope of a transverse dune ridge is gentle like that of the barchans, while the steeper downwind slip face is at the angle of repose. In the downwind direction, transverse dunes form ridge after ridge separated from each other by low swales in a repeating wavelike fashion. Each dune ridge is laterally extensive perpendicular to the sand-transporting wind direction. The dune ridges, slip faces, and interdune swales trend perpendicular to the direction of prevailing winds, hence the name *transverse*. Abundant sand supply may derive from such sources as widespread exposure of easily eroded sandstone bedrock, sandy alluvium deposited by exotic streams or during wetter climates in the Pleistocene Epoch, or from sandy deltas and beaches left in the landscape after the desiccation of ancient lakes.

**Longitudinal dunes** are long dunes aligned parallel to the average wind direction ( Fig. 18.32d). There is no consistent distinction between the back slopes and slip faces of

**FIGURE 18.33**
A small barchan in southern Utah.

**FIGURE 18.34**
A satellite view of longitudinal dunes in the Sahara. The area in the photograph is approximately 160 kilometers (100 mi) across.
**Estimate the length of the dunes in this satellite image.**

these dunes, and their summits may be either rounded or sharp. Strong winds are important to the formation of most longitudinal dunes, which do not migrate but instead elongate in the downwind direction. Small longitudinal dunes, such as those found in North America, may simply represent the long trailing ridges of breached parabolic dunes or a sand streak extending from a somewhat isolated source of sand. Much higher and considerably longer than these are the impressive longitudinal dunes that cross vast areas of the flatter, more open desert topography of North Africa, the Kalahari, the Arabian peninsula, and interior Australia ( Fig. 18.34). These develop under bidirectional wind regimes, where the two major sand-transporting wind directions come from the same quadrant, such as a northwesterly and southwesterly wind. A type of large longitudinal dune called a **seif** (from the Arabic for "sword") is found in the deserts of Arabia and North Africa. Seif dunes are huge, sinuous rather than linear, sharp-crested dunes, sometimes hundreds of kilometers long, whose troughs are almost free of sand. They may reach 180 meters (600 ft) in height.

**Star dunes** are large, widely spaced, pyramid-shaped dunes in which ridges of sand radiate out from a peaklike center to resemble a star in map view ( Fig. 18.32e). These dunes are most common in areas where there is a great quantity of sand, changing wind directions, and an extremely hot and dry climate. Star dunes are stationary, but ridges and slip faces shift orientation with wind variations.

**Dune Protection** To many who visit the desert or beaches, dunes are one of nature's most beautiful landforms. However, dune areas are also very attractive sites for recreation, and they are particularly exciting for drivers of ORVs (off-road vehicles). Although dunes appear to be indestructible and rapidly changing environments that do not damage easily, this is far from the truth. Dunes are fragile environments with easily impacted ecologies. Because dune regions are the result of an environmental balance between moving dunes and the plants trying to stabilize them, the environmental equilibrium is easily upset. Many of the most spectacular dune areas in the United States have special protection in national parks, national monuments, or national seashores, such as White Sands, New Mexico; Great Sand Dunes, Colorado; Indiana Dunes, Indiana; and Cape Cod, Massachusetts. Many dune areas, however, do not have special protection, and environmental degradation is a constant threat.

There are many practical reasons for dune preservation. In coastal zones, dunes play an important role in coastal protection and are sometimes the last defense of coastal communities from storm waves ( ▌ Fig. 18.35). They are particularly important along the low-lying Gulf and Atlantic coasts of the United States where occasional hurricanes or "nor'easters" batter coastlines and erode beaches in front of the dunes. In nations such as the Netherlands, coastal dunes are extremely important because the land behind them is below sea level, thus a breach through the dunes could mean disaster. Coastal dune regions also are critical wildlife habitats, especially for many bird species.

**Loess Deposits** The wind can carry in suspension dust-sized particles of clay and silt, removed by deflation, for hundreds or thousands of kilometers before depositing them. Eventually these particles settle out to form a tan or grayish blanket of loess that may cover or bury the existing topography over widespread areas. These deposits vary in thickness from a few centimeters or less to more than 100 meters (330 ft). In northern China downwind from the Gobi Desert, the loess is 30–90 meters (100–300 ft) thick ( ▌ Fig. 18.36). Loess may originate from deserts, other sparsely vegetated surfaces, or river floodplains. The widespread loess deposits of the American Midwest and Europe were derived from extensive glacial and meltwater deposits of retreating glaciers. As winds blew across the barren glaciated regions or glacial meltwater areas, they picked up a large load of fine sediment and deposited it as loess in downwind regions.

Certain interesting characteristics of loess affect the landscape where it forms the surface material. For example, though fine and dusty to the touch, because of its high cohesion, particularly when damp, loess maintains vertical walls when cut through naturally by a stream or artificially by a road. Sometimes slumping will occur on these steep faces. Slumping gives a steplike profile to many loess bluffs. In addition, loess is easily eroded by either the wind or running water because of its fine texture and unconsolidated character. As a result, loess-covered regions that are unprotected by vegetation often become gullied. Where loess covers hills, gully erosion and slumping are conspicuous processes. A particularly severe erosion problem is responsible for the recent collapse of high loess bluffs along the Mississippi River at Vicksburg, Mississippi ( ▌ Fig. 18.37).

Because of its high calcium carbonate content and unleached characteristics, loess is the parent material for many of Earth's most fertile agricultural soils. Extensive loess deposits are found in northern China, the Pampas of Argentina, the North European Plain, Ukraine, and Kazakhstan. In the United States, the midwestern plains, the Mississippi Valley, and the Palouse region of eastern Washington are underlain by rich loess soils ( ▌ Fig. 18.38). All these areas are extremely productive grain-farming regions.

## Landscape Development in Deserts

Landscape development in arid climates is comparable in many ways to that in humid climates, but not in all ways. Weathering and mass movement processes operate, and fluvial processes predominate in both environments but at different rates and with the tendency to produce some different landform and landscape features in the two contrasting environments. Arid environments experience the added regionally or locally important effect of eolian processes, which are not very common in humid settings beyond the coastal zone. The major differences in the results of gradation in arid climates, as compared to humid climates, are caused by the great expanses of exposed bedrock, a lack of continuous water flow, and a more active role of the wind in arid regions.

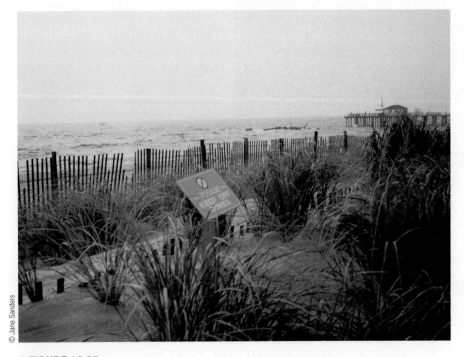

**▌ FIGURE 18.35**
A sign directs visitors on how to help protect this dune area along the New Jersey coast.
**Why do you think some dune areas need to be protected from human activities such as "dune buggies" and other recreational vehicles?**

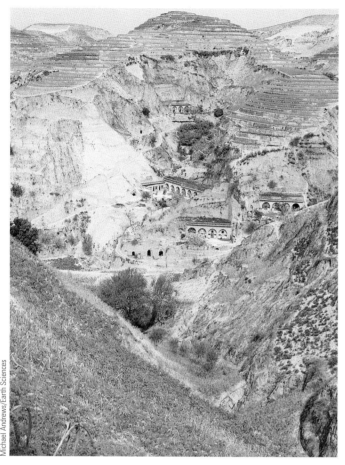

■ FIGURE 18.36
A steep gully eroded into the thick loess deposits of northern China.
**Where is the origin of these loess deposits?**

■ FIGURE 18.37
The steep and unstable loess bluffs of Vicksburg, Mississippi.
**Why might the instability of loess cliffs be a problem?**

Some desert landscapes, such as most of those in the American Southwest, are found in regions of considerable topographic relief, whereas others, like much of the Sahara and interior Australia, occupy huge expanses of open terrain with few mountains. An excellent example of desert landscape gradation in a region of considerable structural relief is found in the Basin and Range region of western North America. The region extends from west Texas and northwestern Mexico to eastern Oregon. It includes all of Nevada and large portions of New Mexico, Arizona, Utah, and eastern California. Here, more than 200 mountain ranges, with basins between them, dominate the topography. The Great Basin—a large subregion characterized by interior drainage, numerous alternating mountain ranges and basins, active tectonism, and centered on Nevada—occupies much of the central and northern part of the Basin and Range.

Fault-block mountains in the Basin and Range region rise thousands of meters above the desert basins, and many form continuous ranges ( ■ Fig. 18.39). These high ranges, such as the Guadalupe Mountains, Sandia Mountains, Warner Mountains, and the Panamint Range, to name just a few, encourage orographic rainfall. Fluvial erosion dissects the mountain blocks to carve canyons between peaks and cut washes between interfluves. Where active tectonism continues so that the uplift of mountain ranges matches or exceeds the rate of their erosion, as in the Great Basin, fluvial deposition constructs alluvial fans extending from canyon mouths outward toward the basin floor. In many basins, the alluvial fans have coalesced to create a bajada. In these tectonically active basins with interior drainage, playas often occupy the lowest part of the basin, beyond the toe of the alluvial fans ( ■ Fig. 18.40a). Although the mountain ranges create considerable roughness, the wind can remove sandy alluvium from the toe areas of alluvial fans, sand-sized aggregates of playa sediments, and, in some cases, sandy sediment from beaches left behind by ancient perennial lakes. These sediments may be transported relatively short distances before being deposited in local dune fields.

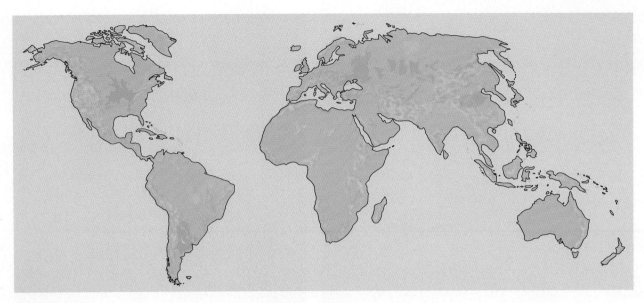

**FIGURE 18.38**
Map of the major loess regions of the world. Most loess deposits are peripheral to deserts and recently glaciated regions.

**FIGURE 18.39**
A false-color satellite image of Death Valley, California, which is a desert bolson in the Basin and Range region. Death Valley lies between the Amargosa Range to the northeast and the Panamint Range to the southwest. Alluvial fans, which constitute much of the red areas of this image, dominate the edges of the valley.
**What do you think the white areas are in the center of the valley?**

Pediments lie along the base of some mountains, particularly in the tectonically less active areas south of the Great Basin, where much of the terrain is open to exterior drainage. In these areas, mountains are being lowered and the pediments extended. Resistant inselbergs remain on some of the pediments. Sandy alluvium along streams and sandy deposits left by ancient rivers and deltas provide sediment to be reworked by the wind into occasional dune fields. The landscapes of the Mojave Desert in California and parts of the Sonoran Desert in Arizona have localities where uplift along faults has been inactive long enough for the landscape to be dominated by extensive desert plains interrupted by a few isolated inselbergs as reminders of earlier, tectonically active, mountainous landscapes (Fig. 18.40b).

The geologic structures and geomorphic processes found in desert areas are, for the most part, the same as those found in humid regions. It is variations in the effects and rates of these processes that make the desert landscape distinctive. Although fault-block mountains and fault-block basins dominate the geologic structure of immense regions of the American West and other arid locations around the world, it is important to know that desert landscapes are as varied as those of other climatic environments. Deserts exist at localities where the landscape has developed in nearly every imaginable geologic setting, including volcanoes, ash deposits and lava flows, folded rocks

**FIGURE 18.40**
(a) Alluvial fans and playas tend to occur in mountain and basin deserts of active tectonism, which typically have interior drainage. (b) Pediments and inselbergs are common in desert areas that have been tectonically stable since a distant period of mountain formation. Development of exterior drainage helps transport sediment eroded from the mountains out of the basin.

Sahara, common landforms include inselbergs surrounded by extensive desert plains (Fig. 18.41), deflation hollows, playas, washes, the channels of ancient streams, and the beds of shallow, ancient lakes. With limited terrain roughness, areas of large longitudinal dunes can develop in these settings.

Arid landscapes and eolian landscapes in any climatic environment can be beautiful and stark, but those who are unfamiliar with these environments too often misunderstand them. Deserts are not wastelands, and dunes are not merely piles of sand. It is true that most desert and eolian areas have little to offer in terms of directly exploitable economic value, but their unique characteristics and scenic beauty fully qualify them for preservation and protection. The austere, angular character of their landforms, the fragility of their environments, and the opportunities they provide for learning about how certain Earth systems operate continue to attract those who seek to understand and experience arid and eolian environments (Fig. 18.42). One only needs to count the number of national parks, national monuments, and other scenic attractions in the arid southwestern United States and in regions of sand dunes to find ample support for their survival. Deserts and dune localities are places with a beauty all their own and are areas worthy of appreciation and appropriate environmental protection.

**FIGURE 18.41**
Ayers Rock is a striking sandstone inselberg, an erosional remnant, rising above the arid, flat interior of the Australian Outback. The Olgas, in the background, formed as another erosional remnant divided into many bedrock hills by the widening of joints.
**Explain how inselbergs form.**

forming ridges and valleys, horizontal strata, and exposures of massive intrusive rocks. Where arid climates occur in expansive regions of largely open, low-relief terrain, as in the ancient and geologically stable deserts of Australia and the

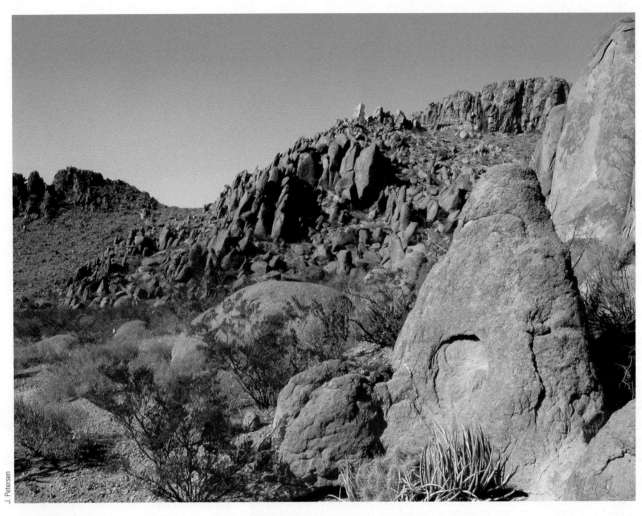

**FIGURE 18.42**
This photograph of the Grapevine Hills in Big Bend National Park, Texas, illustrates the dramatic beauty of desert landscapes. The arid climate of the Chihuahuan Desert has produced an intriguing region featuring fractured granitic hills and boulders.
**What aspects of this environment make this an attractive landscape?**

# Define & Recall

regional base level
exotic stream
wash (arroyo, barranca, wadi)
badlands
plateau
caprock
mesa
butte
pediment

inselberg
alluvial fan
fan apex
debris flow fan
bajada
piedmont alluvial plain
bolson
playa
playa lake

salt flats (salinas)
eolian (aeolian)
deflation
suspension
surface creep
ripple
dust storm
sandstorm
deflation hollow

desert pavement (reg, gibber)
ventifact
yardang
sand dune
loess

sand sea (erg)
sand sheet
slip face
blowout
barchan

parabolic dune
transverse dune
longitudinal dune
seif
star dune

## Discuss & Review

1. What are eolian landforms? What gradational agent most significantly affects the desert landscape?
2. How do climate and vegetation affect landforms in the desert?
3. How do basins of interior drainage differ from basins of exterior drainage?
4. How does an exotic stream differ from an ephemeral stream? Give three examples of exotic streams.
5. How does the formation of a mesa or butte differ from the formation of a pediment?
6. Describe the main characteristics of an alluvial fan.
7. Why do temporary lakes form in bolsons? How are such lakes related to playas?
8. What are the major differences between deflation and eolian abrasion?
9. Why are most sandstorms confined to a low height? How do they differ from dust storms?
10. Describe the formation of a ventifact.
11. What is the difference between a barchan and a transverse dune?
12. What is loess? Where are some major regions in the world where loess deposits are located? What is an important economic activity related to loess in many regions?

## Consider & Respond

1. How are eolian erosion and transportation processes similar to fluvial erosion and transportation processes, and how do they differ? What are the main reasons for these similarities and differences? Is eolian deposition similar to fluvial deposition? If so, what are some of the similarities?
2. What physical geographic factors contribute to create so many basins of interior drainage in arid regions? What are some ways in which base level changes can occur in basins of interior drainage?
3. How does stream gradation in arid regions differ from stream gradation in humid regions? What are the most common erosional and depositional landforms associated with stream gradation in arid regions?
4. Name several types of arid landforms commonly found in tectonically active desert regions, such as the American Great Basin. Name several types of arid landforms commonly found in deserts, like interior Australia, that have been tectonically stable for a very long time.
5. Deserts are considered fragile environments. What are some ways in which desert landforms fit or contribute to this designation?
6. Why are there so many national parks and monuments in the arid parts of the American Southwest? What landform characteristics and other factors of desert landscapes draw people to these locations?

# Map Interpretation
## Desert Landforms

### The Map

The map shows a section of Death Valley National Park, California. This is part of the Basin and Range region, characterized by fault-block mountains (ranges) separated by down-faulted valleys (basins). As the rugged ranges erode, the basins fill with debris carried by infrequent flash floods.

The mountain block that forms the western slope of Death Valley (in the map area) is the Panamint Range. The highest summit in that range, Telescope Peak, reaches an elevation of 3368 meters (11,049 ft). The Amargosa Range forms the eastern boundary. Death Valley's lowest elevation is −86 meters below mean sea level (−282 ft).

The present climate of the Basin and Range is arid, except for high-elevation mountains that receive more precipitation. With an average annual rainfall of 3.5 centimeters (1.7 in.) and a potential evaporation that may exceed 380 centimeters (150 in.), Death Valley has the most extreme aridity in the region. Summer temperatures commonly exceed 40°C (104°F), and the record maximum is 57°C (134°F). In winter, much of the Basin and Range region has freezing nighttime temperatures.

During the Pleistocene, lakes several hundred feet deep filled numerous basins in the region. Death Valley was occupied by Lake Manly, and traces of this 183-meter (550-ft) deep lake's shoreline can be seen from the valley floor. This is a classic site for studying desert landforms.

### Interpreting the Map

*Note: In answering these questions it will be helpful to refer to both the figures and text on pages 510–514.*

1. Based on the general location of Death Valley, why is the area so arid?
2. Describe the general topography of the map area.
3. What is the lowest elevation on the map? What is significant about the elevation of Death Valley?
4. What specific type of arid landform is the blue-striped feature located in the depression? Why are the edges of this feature shown with blue-dashed lines?
5. What types of surface materials compose the feature in Question 4? Explain how this feature was formed.
6. What specific type of landform is indicated on the map by the curved contours at the base of the mountains? What do the blue _ _ _ _ . _ _ _ _ . . _ _ _ lines crossing the contours represent?
7. What types of surface materials compose the feature in Question 6? Explain how this feature was formed.
8. Note that the large, curved landforms at the mountain base coalesce. What is the specific landform name for such broad alluvial features?
9. What evidence from the map indicates that this is an interior drainage basin?
10. Sketch a general east–west profile from the benchmark (elevation −270 feet) at Devil's Speedway to the 2389-foot benchmark on the mountain straight to the west. Label the following landforms: mountain front, pediment, alluvial fan, and basin.

A digital terrain model shows the topography of Death Valley. Figure 18.39 presents an alternate view with a topographic profile (black foreground) at about 3× vertical exaggeration.

Opposite:
Furnace Creek, California
Scale 1 : 62,500
Contour interval = 80 ft
U.S. Geological Survey

# CHAPTER PREVIEW

**A glacier is an excellent example of an open system.**
- Why is a glacier considered an open system?
- What materials enter and leave the system?
- How does energy move through the system?

**The two major forms of glaciation are similar in many respects, yet they leave their own distinct impacts on the landscape and originate in different environmental settings.**
- What are the two forms?
- In what environmental settings do they originate?

**Much of the stark grandeur and scenic beauty of mountain regions is a result of alpine glaciation.**
- What characteristics of most high mountains have resulted from alpine glaciation?
- How has the human use of mountains been influenced by alpine glaciation?

**The ice budget of a glacier is determined by two factors: the annual accumulation of snow and ice and the annual ablation (loss of snow and ice by melting, evaporation, and sublimation).**
- What happens to the glacier when accumulation exceeds ablation?
- When ablation exceeds accumulation?
- What happens when dynamic equilibrium exists in a glacier?

**The most visible effects of continental glaciation in North America are from the Wisconsinan, the last stage of Pleistocene glaciation.**
- How did Pleistocene glaciation change the physical appearance of North America?
- Why are the effects of Wisconsinan glaciation so prominent in the physical landscape?

**Erosional and depositional landforms produced by continental glaciation exhibit major differences.**
- What are the major differences?
- How did they develop through glaciation?
- How has human use been affected by these landscapes?

Glaciers play several important roles in the Earth system. They are excellent climate indicators because certain environmental conditions are required for glaciers to exist and glaciers respond visibly to climate variation. Glaciers become established, expand, contract, and disappear in response to changes in climate. Their long-term storage of fresh water as ice has a tremendous impact on the hydrologic cycle and the oceans, and the accumulation of ice by glaciers provides a record of past climates that can be studied in ice cores. Where glaciers once existed or where they were once much larger than they are today, much evidence can be found concerning past climatic conditions.

The processes of erosion and deposition by glaciers, whether ongoing or in the past, leave a distinctive stamp on a landscape. Some of the most beautiful and rugged terrain in the world exists in mountainous and highland regions that have been sculpted by glaciers. Virtually every high mountain region in the world displays glacial landscapes, including the Alps, the Rocky Mountains, the Himalayas, and the Andes. Glaciers also carved coastal fjords in Norway, Chile, New Zealand, and Alaska. Rugged mountain peaks rising high above lake-filled valleys or narrow, deep-sea lanes create the ultimate in scenic appeal for many people. Glaciers, masses of moving ice, have transformed the appearance of high mountains, as well as large portions of continents, into distinctive landscapes. The flowing ice of glaciers is an effective and spectacular geomorphic agent on major portions of Earth's surface.

## Glacier Formation and the Hydrologic Cycle

**Glaciers** are masses of flowing ice that have accumulated on land in areas where the annual snowfall exceeds the yearly loss of snow by melting ( Fig. 19.1). Snow falls as hexagonal ice crystals that form flakes of intricate beauty and variety. Once snow accumulates on the land, it becomes transformed by compaction into a mass of smaller, rounded grains. As the air space around them is reduced by compression and melting, the snow grains become denser. Through melting, refreezing, and pressure caused by the increasing weight from burial under newer snowfalls, the snow further compacts into a crystalline granular stage between flakes and ice known as **firn**. After additional time, small firn granules become larger by recrystallizing into interlocked ice crystals through pressure, partial melting, and refreezing. When the ice is deep enough, the pressure from burial under many layers of snow and firn will cause the ice below to become plastic and flow outward or downward away from the area of greatest snow and ice accumulation.

Glaciers are open systems with inputs of snow entering, being stored as snow and ice, and leaving the system, mainly as meltwater. A glacial system is controlled by two basic climatic conditions: precipitation in the form of snow and freezing temperatures. First, to establish a glacier, there must be sufficient snowfall to exceed the annual loss through melting, evaporation, and **sublimation** (direct change from solid ice to gaseous water vapor). **Calving** ( Fig. 19.2), another form of glacial ice loss, occurs when a glacier loses masses of ice that break away as icebergs into an ocean or other water body. Mountains along middle-latitude coastlines and high mountains near the equator can support glaciers because of heavy orographic snowfalls, despite intense sunshine and warm climates in the surrounding lowlands. Yet some very cold polar regions in subarctic Alaska and Siberia and a few valleys in Antarctica have no glaciers because the climate is too dry.

A second climatic condition that affects a glacier is temperature. Summer temperatures must not be high for too long, or all of the snow accumulation from the previous winter will melt. Surplus snowfall is essential because it allows the pressure from years of accumulated snow layers to transform older buried snow into firn and ice. When the ice

**FIGURE 19.1**
Glaciers form in environments that receive more snowfall in a year than will melt. This mountaineer considers negotiating a crevasse on a snow-covered glacier on the mountain of Huascaran in the Peruvian Cordillera Blanca Range.

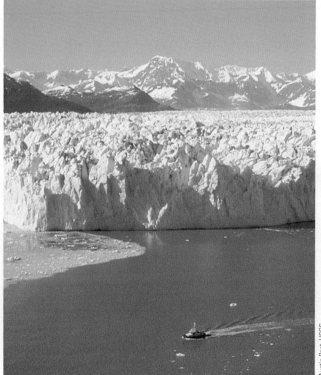

**FIGURE 19.2**
Glaciers that flow into the sea or a deep lake have icebergs break off in a process called calving. This is the Columbia Glacier in Alaska, flowing into Prince William Sound.

## Geography's Physical Science Perspective
### Glacial Ice Is Blue!

When we make ice in our freezers, clear colorless water turns to relatively clear ice cubes. The ice cubes may contain some white crystalline forms and air bubbles, but in general the ice is clear. In nature, however, the process of making ice is very different from that of an ice maker in a refrigerator-freezer. As snow falls in colder, higher latitudes and altitudes, it forms a layer of snow (known as snow pack) on the surface. Each successive snowfall forms another layer as it piles onto the previous snow pack. The weight of the successive layers of snow creates pressure that compresses the older layers beneath. Through time, the layers of snow become layers of solid ice. While compaction is occurring, there may be periods of partial melting and refreezing, which may speed the process along.

Snow is white in color because the hexagonal crystalline structure of the snowflakes reflects all wavelengths of visible light with equal intensity, thus appearing white to the human eye. However, when the snow strata in a glacier are compacted through the years (perhaps hundreds or thousands of years) and under great pressure, the ice becomes denser. Basically, under this pressure, more ice crystals are squeezed into the same volume. The dense glacial ice reflects a high level of the shorter wavelengths of visible light (the blue end of the color spectrum). Therefore, the denser the ice, the more blue it appears. However, one must be careful not to assume that deeper blue layers in a glacier are the oldest layers. Again, melting and refreezing may affect the density of the ice layer, and the packing of a wet snow as opposed to drier snow layers can affect the density of a particular layer. Nevertheless, what is certain when looking at massive ice accumulations in nature—such as glaciers, ice caps (smaller and more localized), and ice sheets (greater and regional in extent)—is that shades of the color blue will appear.

This iceberg in Antarctica displays very old layers of glacial ice.

reaches a depth of about 30 meters (100 ft), a pressure threshold is reached that changes ice into a plastic, flowing solid, forming a glacier. Glaciers are sometimes classified by temperature, either as more active *temperate glaciers* or as colder, slower-flowing *polar glaciers*.

Glaciers are an important part of Earth's hydrologic cycle and are second only to the oceans in the total amount of water they contain. About 2.25% of Earth's water is currently frozen in glaciers. Two percent may be a deceiving figure, however, because about 70% of the world's *fresh* water is locked up as ice in glaciers, the vast majority in Antarctica and Greenland. The total amount of ice is even more impressive if we estimate the water that would be released if all of the world's glaciers were to melt. Sea level would rise about 65 meters (215 ft). This would change the geography of the planet considerably. In contrast, if another ice age were to occur, sea level would drop drastically (see Geography's Spatial Science Perspective, Climate Change and Its Impact on Coastlines, in Chapter 8). During the last major glacial advance, sea level dropped about 120 meters (400 ft).

Unlike the water in a stream system, which returns rapidly to the sea or atmosphere (unless it is stored in lakes or as groundwater), the snow that becomes glacier ice is stored for the long term in a much more slowly flowing system. Glaciers may store water as ice for hundreds or even hundreds of thousands of years before it is released as meltwater into the liquid part of the hydrologic system. Yet, glacial ice is not stagnant. It moves slowly but with tremendous energy across the land. Glaciers reshape the landscape by engulfing, pushing, dragging, transporting, and finally depositing rock debris, often in places far from its original location. Long after glaciers recede from a landscape, glacial landforms remain as a reminder of the energy of a glacial system and as evidence of past climates.

During most of Earth history, glaciers did not exist on the planet. However, when a period of time occurs during which significant areas of the middle latitudes are covered by glaciers, we call it an *ice age*. Today, glaciers cover about 10% of Earth's land surface. Present-day glaciers are found in Antarctica, in Greenland, and at high latitudes and high elevations on all continents except Australia. During recent Earth history, from about 2.4 million years ago to about

10,000 years before the present, glaciers periodically covered nearly a third of Earth's land area. In the much more distant geologic past, other ice ages have occurred.

## Types of Glaciers

There are two major types of glaciers, alpine glaciers and continental ice sheets. **Alpine glaciers** are fed by ice and snow in mountain areas, usually occupying valleys originally initiated by stream erosion. Those that are confined by the rock walls of the valley they occupy may also be called **valley glaciers** ( Fig. 19.3). A related glacial type, called a **piedmont glacier,** forms when a glacier flows beyond its valley over flatter land and spreads out in a broad lobe shape because the bedrock walls no longer confine it (see the chapter-opening photo). Some alpine glaciers, however, do not reach the valleys below the zone of high peaks. Instead, they occupy distinctive steep-sided and amphitheater-like depressions eroded by ice flow at the head of a valley, called **cirques,** and are known as **cirque glaciers** ( Fig. 19.4). Cirque glaciers are the smallest type of glacier, and most occupy cirques that formed when they were occupied by a larger alpine glacier.

Alpine glaciers create the characteristic rugged scenery of high mountains in most world regions. Today glaciers can be found in the Rockies, Sierra Nevada, Cascades, the Olympic Mountains, the Coast Ranges, and in numerous Alaskan ranges of North America. They also exist in the Andes, New Zealand, the Alps, the Himalayas, the Pamirs, and other high Asian mountain ranges. Small glaciers are even found on tropical mountains in New Guinea and in East Africa on Mounts Kenya and Kilimanjaro. The largest alpine glaciers in existence today are found in Alaska and the Himalayas, where some reach lengths of more than 100 kilometers (62 mi).

The second and largest type of glacier, far more extensive than the valley glacier, is the **continental ice sheet** ( Fig. 19.5). At one time, continental ice sheets covered as much as 30% of the Earth's land area. They still blanket Greenland and Antarctica to a depth of 3 kilometers

### FIGURE 19.3
A system of valley glaciers flow together to form the larger Susitna Glacier in Alaska. Mt. Hayes, the highest mountain in the Alaska Range, is the peak in the background.
**How are valley glaciers similar to rivers?**

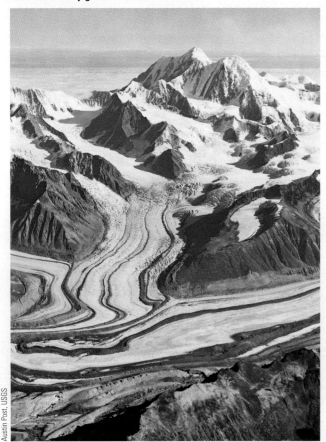

### FIGURE 19.4
A cirque glacier in Alaska. A bergschrund, or great crevasse, can be seen along the upper margin of the glacier.
**What does such a large crevasse indicate about this glacier?**

**FIGURE 19.5**
Continental ice sheet. This terrain model of Antarctica shows the domelike form of an ice sheet. Ice flows away from the center of the continent, near the South Pole, where the pressure is greatest, because the ice is thickest there.
**These glaciers flow in a roughly radial pattern. Remembering radial drainage of streams, what does this mean?**

(2 mi) or more, and smaller glaciers called **ice caps** are present on some Arctic islands and on Iceland. In contrast to alpine glaciers, continental ice sheets are mainly unconfined and even flow over higher portions of the land. Ice sheets and ice caps are not elongated, and they flow radially outward in all directions from their source area where ice thickness and pressures are greatest.

## Features of an Alpine Glacier

Glaciers can be divided into two functional parts: a **zone of accumulation** and a **zone of ablation** (Fig. 19.6). The upper portion of a glacier, where snowfall exceeds **ablation** (ice loss through melting, evaporation, and sublimation), is the zone of accumulation. The lower elevation portion of an alpine glacier, where ablation exceeds snowfall, is the zone of ablation. Under the force of gravity or pressure from ice thickness, glacial ice flows from the snowy, cooler zone of accumulation to the warmer, less snowy zone of ablation. Downslope movement is seen in the development of a great crack, or crevasse, known as a **bergschrund,** around the head of an alpine glacier (see Fig. 19.6). It shows that the ice mass is pulling away from the confining rock walls of the cirque, moving outward and downward. The end of a glacier—its **terminus,** or **snout**—marks the farthest extent of the glacier. If the terminus reaches the sea, it will calve icebergs.

The **equilibrium line,** which separates the two zones of a glacier, is a balance point between snowfall and ablation (Fig. 19.7). The equilibrium line represents the ablation–accumulation balance line averaged over a record of years and is different from the snow line. The snow line is the elevation where snow cover begins at a certain point in time. A snow line changes position in response to the seasons and the weather, as well as after every snowfall. The equilibrium line is the elevation line where snow exists on the ground all year long and tends to coincide with the late-summer snow line. Several factors influence the location of the equilibrium line. The interaction between latitude and elevation, both of which affect temperature, is an important factor. On mountains near the equator, the equilibrium line is at very high elevations but becomes lower with higher latitudes until it exists at sea level in the polar regions.

Equally important is the amount of snowfall received during the winter. In general, with colder temperatures and greater snowfalls, the equilibrium line will go down in elevation and retreat to higher elevations if the climate warms. Other factors causing variations in the equilibrium-line elevation include the amount of insolation. A shady mountain slope will have a lower equilibrium line than one that receives more insolation. Wind is another factor because it produces snowdrifts on the lee side of mountain ranges. In the middle latitudes of the Northern Hemisphere, the equilibrium line is lower on the north (shaded) and east (leeward) slopes of mountains. Consequently, the most significant glacier development is on these north-facing and east-facing slopes (Fig. 19.8).

Some alpine glaciers are caused primarily not by snowfall on the glacier itself but by the accumulation of snow

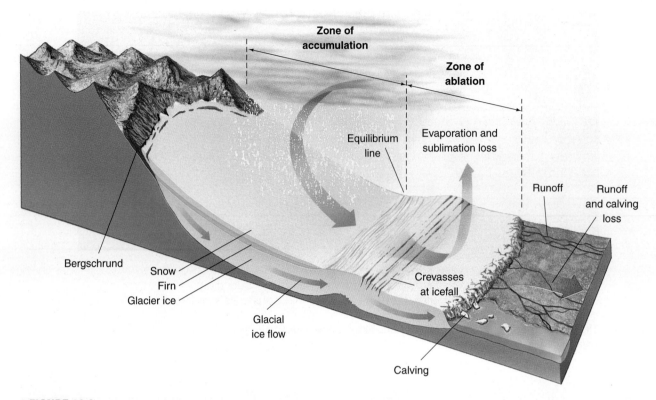

**FIGURE 19.6**
**Environmental Systems: The Hydrologic System—Glaciers.** Glacial systems are controlled by the input of snowfall (accumulation) and the loss of snow and ice (ablation) resulting from evaporation or melting. Most of the glacial growth is in the zone of accumulation where temperatures permit freezing much of the year and annual snowfall exceeds melting. Over the years in the accumulation zone, buried under newer snowfall, snow changes into granular firn and much later transforms into glacial ice. When the ice reaches a threshold depth, the deeply buried ice will deform plastically and flow downslope.

Most loss from a glacier is in the zone of ablation—the area below the equilibrium line where annual melting exceeds snowfall. Here the glacier loses mass from sublimation from the glacier's surface, from runoff of meltwater, and by calving icebergs at the terminus if the glacier flows into a body of water deep enough to float them.

A glacier is in equilibrium if a balance is achieved between inputs of snow and ice and outputs by ablation. If accumulation exceeds ablation for several years, the glacier's snout will advance; if ablation exceeds accumulation, the glacier will lose mass and recede. At present, most of the world's glaciers are receding. Many scientists are concerned about the potential effects of global warming on glacial systems, especially on the continental ice sheets of Antarctica and Greenland. Increasing world temperatures can shift glacial systems toward a new equilibrium, at a smaller size and mass, with more rapid ice loss through melting and calving. These changes in turn could affect world sea level and could cause local increases in flood hazards.

blowing, drifting, and avalanching onto the glacier's surface. The Colorado Rockies and the Ural Mountains in Russia provide good examples of alpine glaciers formed by accumulations of snowdrift.

## Equilibrium and the Glacial Budget

When the snout of a glacier neither advances, nor retreats, the glacier is said to be in a state of equilibrium; that is, a balance in the system has been achieved between accumulation and ablation of ice and snow. As long as equilibrium is maintained, the glacier's snout will remain in the same location although the glacial ice continues to flow forward, like the stairs of an escalator.

Let's assume that exceedingly heavy amounts of snow are received for several years and the glacier's equilibrium is upset. Under the pressure of increased snow accumulation, more ice will be produced, and the snout will advance until it reaches a new point of equilibrium where receipt of ice and snow equals wastage. A deficit of snow in the glacial budget will cause the snout to recede, by melting back faster than it is flowing, until equilibrium between accumulation

FEATURES OF AN ALPINE GLACIER    541

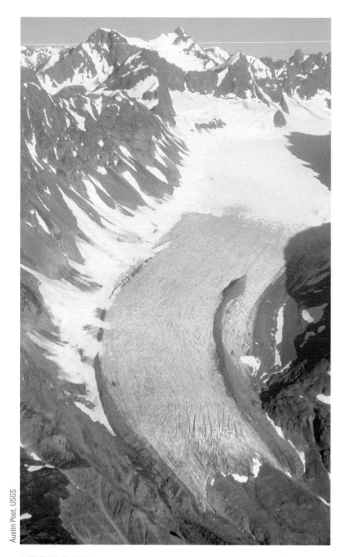

**FIGURE 19.7**
A valley glacier on Alaska's Kenai Peninsula. A clearly defined equilibrium line separates the white upper accumulation zone from the blue-toned lower ablation zone.
**Explain the significance of the equilibrium line on a glacier.**

**FIGURE 19.8**
This topographic map of part of Rocky Mountain National Park in Colorado illustrates the impact of slope aspect on glaciation. In the Northern Hemisphere at middle to high latitudes, slopes facing north or northeast tend to be shaded from the sun. Thus, more snow and ice can accumulate there, and glaciation tends to be more extensive. More snow melts from south-facing slopes because they receive more direct sunlight. (See Chapter 3.)
**Are there differences of any kind between north- and south-facing slopes where you live?**

and wastage is again achieved. An increase or decrease in temperature that occurs over several years can also cause an advance or a retreat of the glacier.

It is important to understand that even when a glacier is retreating, the ice continues to flow forward (or outward in ice sheets and ice caps). Forward movement stops only if the ice becomes too thin to maintain plastic flow. In fact, if ice flow ceases, a mass of ice is no longer considered a glacier. Glaciers can fluctuate from year to year, but most changes in the position of the terminus are seen by monitoring glaciers over the long term ( Fig. 19.9). From about 1890 to the present, most Northern Hemisphere glaciers have been receding. This overall retreat may be an indication of global warming, yet each individual glacier has its own balance. For example, in 1986, Alaska's Hubbard Glacier advanced so rapidly that it cut off

Russell Fjord from the ocean, trapping many seals and porpoises. Yet just a few hundred miles away, the giant Columbia Glacier is rapidly receding and calving increased numbers of icebergs into Prince William Sound. Scientists and the Coast Guard are concerned because of the increased hazard from icebergs in oil tanker lanes from the Trans-Alaska Pipeline.

## How Does a Glacier Flow?

The mechanics by which a glacier flows are complex. Most ice movement in a valley glacier, however, is related to downslope gravitational pull, and all glaciers flow in response to ice pressure. Glacier flow occurs through a combination of processes, one of which is *basal slip,* the sliding of ice over a bedrock floor ( Fig. 19.10). Meltwater, which reduces

**FIGURE 19.9**
The Jacobshavn Glacier, Greenland's largest outlet glacier, has been generally retreating since the beginning of measurement and monitoring by scientists in 1850. The colored lines mark the former position of the glacier's terminus.
**Why is the rapid receding of this glacier of particular concern to scientists?**

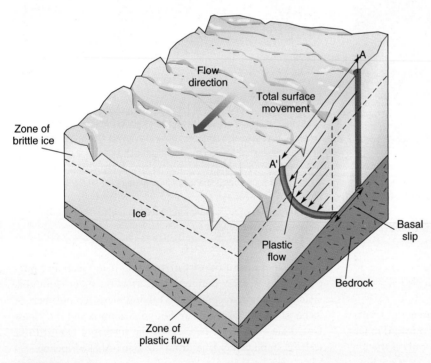

**FIGURE 19.10**
Most of the ice movement in a glacier is caused by ice that deforms under pressure and flows as a plastic solid. A lesser part of glacial motion results from ice sliding on bedrock at its base. This diagram shows that the upper part of the ice moves faster than the lower ice that is in friction with the bed. Plastic flow is also cumulative upward, so the velocity is greatest on the glacier's surface.
**Does the ice on the top of the glacier flow or just ride along on the ice below?**

friction through lubrication and hydrostatic pressure, and steep slopes are both involved in this kind of glacial movement. Basal slip is important in middle-latitude glaciers, particularly during the summer when much of the glacier is near its melting point and meltwater is available. During winter and in polar climate glaciers with little meltwater available, the glacier may freeze onto the bedrock, and basal slip may be prevented.

*Internal plastic deformation* is the dominant process by which the ice in a glacier flows. At depth, the tremendous weight of ice causes ice crystals to arrange themselves in parallel layers that glide over each other, much like a deck of cards. This type of flow occurs when a threshold pressure is exceeded, causing the ice to flow through plastic deformation. This generally occurs only under about 30 meters (100 ft) of ice in an area within a glacier and extending to its base, known as the *zone of plastic flow*.

The upper surface of a glacier is brittle ice that does not experience plastic deformation. Movement of the ice in this upper *zone of brittle ice* occurs by fracturing and faulting, basically breaking up the ice as the glacier rides along on the flowing ice below. Here the ice fractures and cracks as a solid, breakable material. These ice cracks, called **crevasses,** are common wherever a glacier becomes stretched, experiencing tensional stress, particularly where it flows over a break in slope ( Fig. 19.11). A glacier flowing over a very steep slope forms an **icefall.** Here, intersecting crevasses break the ice into a morass of unstable ice blocks that are riding on relatively rapid flowing ice below in the zone of plastic flow. These are extremely dangerous areas for mountain climbers and scientists who venture onto the ice ( Fig. 19.12). Locations where the snout of a glacier meets seawater can also be dangerous because large waves can be created as huge blocks of ice calve off and topple into the water.

The speed of glacial flow varies from imperceptible fractions of a centimeter per day to as much as 30 meters (100 ft) per day. Also, the flow of an individual glacier varies from time to time with changes in the dynamic equilibrium and from place to

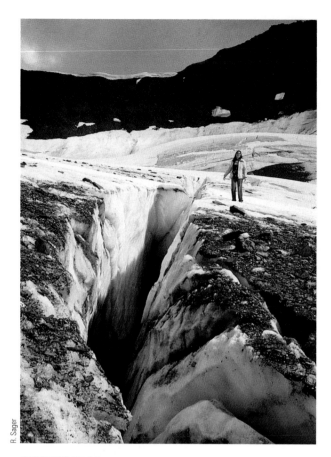

**FIGURE 19.11**
A large crevasse on the Aleyska Glacier, Alaska.
**Why does the upper part of the glacier break into crevasses?**

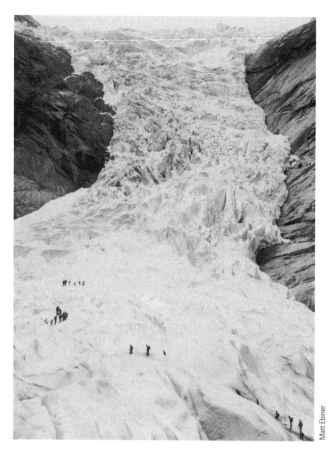

**FIGURE 19.12**
Icefalls are the glacial equivalent of rapids or waterfalls in a river and are riddled with crevasses that form unstable ice blocks. Although glacial ice flows much more slowly than running water, these are still the most rapidly moving and changing part of a glacier. Icefalls are one of the most treacherous parts of a high mountain climb because they are made of huge ice blocks that can shift at any time.

place because of variations in the gradient over which it flows or differences in the friction encountered with adjacent rock. Glaciers flow more rapidly where the slopes are steep, where the ice is thickest, and where temperatures are warmest. For example, the Nisqually Glacier, on the steep slopes of Washington's Mt. Rainier, flows 38 centimeters (16 in.) per day in summer. As a general rule, temperate alpine glaciers flow much faster than the cold polar Antarctic or Greenland ice sheets.

The velocity of ice flow also varies depending on its location in a glacier. In the middle of the upper surface where friction is least and motion accumulates from plastic flow of layers below, the speed of flow is greatest. Friction with surfaces on the sides and bottom of a glacier slows the rate of flow.

Sometimes a glacier's velocity will increase by many times its normal rate, causing the glacier to advance hundreds of meters per year. The reasons for such enormous *glacial surges,* as these velocity increases are called, are not completely clear although lubrication of a glacier's bed by pockets of meltwater explains some of them.

## Glaciers as Agents of Gradation

As a glacier moves over the land, whether as an ice sheet or an alpine glacier, it scrapes along the land surface, picking up and carrying boulders and rock fragments. This erosive process of lifting and incorporating rocks and soil into a glacier is called **glacial plucking.** Weathering, particularly the freezing of water in bedrock joints and fractures, breaks rock fragments loose, encouraging plucking.

Glaciers also drag rock fragments along at their base. Abrasive rock fragments traveling on the bed of a glacier become tools of erosion as they scrape and gouge stationary bedrock. The **striations** (gouges, grooves, and scratches) produced by such **glacial abrasion,** as this erosion process is called, provide evidence of previous glaciation in areas that are devoid of glaciers today ( Fig. 19.13). Striations in the bedrock indicate the direction of flow long after a glacier has disappeared.

Bedrock obstructions subjected to intense glacial abrasion are typically smoother and more rounded than those produced by plucking. The glacially produced landform developed on such rock surfaces is a **roche moutonnée**—a bedrock hill, or knob, that is smoothly rounded on the upstream side most subject to abrasion, with plucking evident on the downstream side (Fig. 19.14a and b).

The ice volume and velocity of a glacier do not directly determine the particle sizes it can erode and transport, as is the case with running water. Plucking and abrasion provide glaciers with a load of rock fragments of all sizes, from the finest-ground rock flour to giant boulders. The sorting of sediment by size that occurs in running water also does not occur with flowing ice. Much of the load is concentrated in areas where a glacier is in contact with a land surface—the source of most of a glacier's load. Additional rock material, particularly on alpine glaciers, is transported on a glacier's surface, having been deposited by processes associated with erosion or mass wasting. Ice wedging on steep slopes above a glacier encourages rockfalls that deposit large quantities of rock debris along the valley sides. Where valley glaciers join, these rock materials cause the characteristic dark stripes on the surface of many alpine glaciers (Fig. 19.15).

Glacial erosion is especially aggressive in removing nonresistant rock to produce depressions. The bottom ice currents

(a)

(b)

■ **FIGURE 19.13**
Glacial abrasion produces smooth rock surfaces with scratches and grooves parallel to the direction of ice movement. Note that the ice flowed across the tops of columnar jointed basalts with their characteristic hexagonal joint pattern.
**Can you see evidence of the direction of ice flow from this photograph in Devils Postpile National Monument, California?**

■ **FIGURE 19.14**
(a) The typically asymmetric form produced by glacial erosion of a bedrock hill, caused by abrasion upstream and plucking downstream. Arrows indicate the direction of ice flow. (b) An example of this kind of glacial landform in Yosemite National Park. These bedrock knobs or hills, are known as roche moutonnées, or sheep-backed rocks.
**Why would crevasses form above this feature?**

may move obstructions even if the ice has to move slightly uphill to drag material out of the depression. After the ice has melted, previously glaciated surfaces may have many bedrock depressions often filled with water to form lakes.

## Erosional Landforms Produced by Alpine Glaciation

Glacial erosion works headward even as the glacier flows downslope. The headward erosion of a glacier produces a valley head, called a *cirque headwall*, that is shaped like a steep-sided bowl or amphitheater. Glacial undercutting of rock basins. During glaciation, crevasses will be present at the steps as well as icefalls if the break in slope is steep. The result is a type of topography on the valley floor called a "glacial stairway." When the ice recedes, rockbound lakes may fill the basins, often looking like beads connected by a glacial stream flowing down the glacial trough. Such lake chains are called **paternoster lakes.**

Most large valley glaciers have tributary glaciers that merge into them. These tributary glaciers, like the main ice stream, also carve U-shaped channels. However, because these tributaries have less ice volume than the main glacier, they also have lower rates of erosion and less ability to erode their channels. As a result, their troughs are smaller and not as deep as those of the main glacier. Nevertheless, during peak glacial phases, the ice surface of the smaller tributary glacier flows in at the surface level of the larger glacier. Not until the two glaciers begin to wane does the difference in height between their trough floors become apparent. The higher trough, typically no longer occupied by a tributary glacier, is called a **hanging valley.** A stream that flows down such a channel will drop down to the lower glacial valley by a high waterfall or a series of steep rapids. Yosemite Falls and Bridalveil Falls in Yosemite National Park are excellent examples of waterfalls cascading out of hanging valleys. Yosemite Valley itself is a classic example of a glacial trough. A hypothetical sequence in the development of alpine glacial landforms is illustrated in Figure 19.17.

Landscapes eroded by alpine glaciers show a sharp contrast between the glacial troughs scoured smooth by ice flow and the jagged peaks above the former ice levels. The rugged quality of these upper surfaces is caused primarily by mechanical weathering above the ice surface and by glacial undercutting ( Fig. 19.18). Rugged alpine glacial terrains are found in the mountains of Alaska, in California's Sierra Nevada, and in the Rockies (see Map Interpretation: Alpine Glaciation).

Glaciers flowing into the sea calve icebergs. Today, iceberg calving is an ongoing process along the coasts of British Columbia, southern Alaska, Chile, Greenland, and Antarctica. During the Pleistocene, it was characteristic of Scotland, Norway, Iceland, and New Zealand. As a coastal glacier recedes landward, the ocean invades the abandoned glacial trough, creating a deep, narrow inlet of the sea called a **fjord** ( Fig. 19.19). Unlike streams, which can erode only to base level, glaciers can erode below sea level, so the sea can flow into deep glacial channels after the ice has melted. Many fjords formed during times of large-scale Pleistocene glaciation when sea level was lower and were later submerged as sea level rose with melting of the glaciers. Most of the deep, narrow channels of Washington's Puget Sound were carved into bedrock by glacial erosion and later invaded by the sea.

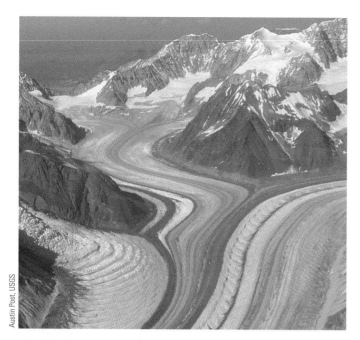

**FIGURE 19.15**
Where valley glaciers join, the eroded materials from glacial erosion, weathering, and mass movement form characteristic stripes as shown here in the Alaska Range.
**Explain how these stripes can be found in the middle of a valley glacier.**

walls above the ice level and of the bedrock below, combined with mechanical weathering and mass movement, increases the size and depth of the bowl, forming a cirque. In fact, at many mountain wilderness and recreation areas, cirques are formally or informally referred to as "bowls." Lakes that form in a cirque depression after the ice melts are called **tarns,** or cirque lakes ( Fig. 19.16a).

Often two or more cirques develop on adjoining sides of a mountain. As the cirques or valleys of two adjacent glaciers enlarge, the bedrock ridge between them will be shaped into a jagged, sawtooth-shaped spine of rock, called an **arête** ( Fig. 19.16b). Where three or more cirques meet at a mountain summit, they form a characteristic pyramid-like peak called a **horn** ( Fig. 19.16c). The Matterhorn in the Swiss Alps is the world's classic example. A **col** is a pass formed by the headward erosion of two cirques that have intersected to produce a low saddle in a high-mountain ridge, or arête.

Unlike streams, which initially erode V-shaped valleys, glaciers erode characteristically steep-sided U-shaped valleys called **glacial troughs** ( Fig. 19.16d). In addition, a glacier's tendency to flow straight ahead rather than to meander causes it to straighten out the preexisting valley that it occupies.

By eroding weaker rocks on the valley floor, alpine glaciers typically create a sequence of rock steps and excavated

**FIGURE 19.16**
Erosional landforms produced by alpine glaciation. (a) A cirque and tarn (lake) formed by headward erosion of a glacier in the Sierra Nevada of California. (b) Jagged sawtooth spines of rock, such as these in Sierra Nevada, are called arêtes. (c) The Matterhorn in the Swiss Alps is a classic example of a horn. Horns are formed when several glaciers cut headward into a mountain peak. (d) Glaciers carve steep-sided U-shaped valleys called glacial troughs. Little Cottonwood Canyon, East of Salt Lake City in the Wasatch Range of the Rocky Mountains, is a classic example of a glacial trough.

## Depositional Landforms Produced by Alpine Glaciation

A glacier can carry debris on its surface, frozen into its interior, and dragged along at its base. As mentioned previously, the load glaciers can transport and deposit includes boulders, rocks, and fragments plucked by glaciers from their channel sides and floor, as well as smaller fragments and particles produced by abrasion. Rockfalls from steep sidewalls also supply a glacier with debris. Glacial deposits can include a mix of huge chunks of bedrock, pebbles, cobbles, sand, fine rock flour, layers of pollen, dead trees and other plants, soil, and volcanic ash.

All glacial deposits are included within the general term **drift,** whether they are unsorted and unstratified ice deposits or orderly deposits of meltwater streams issuing from the glacier. To differentiate these two types of deposits, the term **till** is applied to unsorted drift laid down by ice ( Fig. 19.20). **Glaciofluvial deposits** refer to better-sorted and stratified sediments deposited by meltwater.

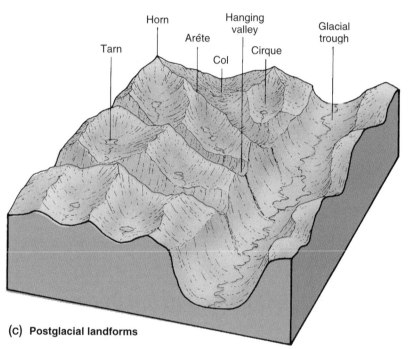

**FIGURE 19.17**
The development of glacial landforms in an alpine region: (a) Mountain topography prior to glaciation, (b) maximum valley glaciation, and (c) major postglacial landforms.
**How have the valley profiles changed from preglacial to postglacial times?**

Glaciers deposit a portion of their load when their capacity is reduced. Landforms constructed from glacial deposits, typically ridges formed along the margins of glaciers, are called **moraines.** Till deposits along the side margins of a glacier form depositional ridges called **lateral moraines** ( Fig. 19.21a). Where two tributary glaciers flow together, their conjoining lateral moraines merge downstream to form a **medial moraine** in the center of the main glacier. At the snout of a glacier, debris carried forward by the "conveyor belt" of ice or pushed ahead of the glacier is deposited in a jumbled heap of rocks and fine material, forming a curved depositional ridge called an **end moraine** ( Fig. 19.21b). End moraines that mark the farthest advance of a glacier are referred to as **terminal moraines.** End moraines deposited as a consequence of a temporary pause by a receding glacier, followed by a stabilization of the ice front prior to further recession, are called **recessional moraines.** A retreating glacier also deposits a great deal of till on the floor of its channels, as the ice melts away and leaves its load of drift behind. The hummocky landscape created by these glacial till deposits is called **ground moraine.**

Braided meltwater streams, laden with sediment, commonly issue from the glacier terminus. The sediment, called **glacial outwash,** is deposited beyond the terminal moraine, with larger rocks and debris deposited first, followed downstream by progressively finer particles. Often resembling an alluvial fan confined by valley walls, this

**FIGURE 19.18**
The alpine glacial topography of the north-central Chugach Mountains, Alaska. This scene includes cirques, arêtes, horns, moraines, a moraine-dammed lake, and other landforms of alpine glaciation.
**What features of glaciation do you recognize in this photograph?**

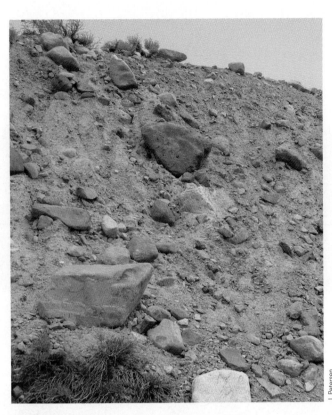

**FIGURE 19.20**
Glacial till, here deposited by an alpine glacier in the Sierra Nevada, consists of an unsorted, unstratified, rather jumbled mass of boulders, cobbles, sand, and silt.
**Why does till have these disorganized characteristics?**

**FIGURE 19.19**
Dobbin Bay, a fjord on Ellesmere Island in the Canadian Arctic, seen on an enhanced satellite image. The main glacier is calving many icebergs in this summer scene, a time of minimal cover by snow and ice.
**What kinds of valley glaciers can you recognize here?**

depositional form left by braided streams is called a **valley train.** Valleys in glaciated regions may be filled to depths of a few hundred meters by outwash or deposits in moraine-dammed lakes, producing extremely flat valley floors (Fig. 19.22).

# Continental Ice Sheets

In terms of their size, shape, and direction of ice flow, continental ice sheets are very different from alpine glaciers. However, all glaciers share certain glacial characteristics and processes, and much of what we have discussed about alpine glaciers also applies to continental ice sheets. The gradational activities of the two types of glaciers are primarily different in scale, attributable to the enormous disparity in size between the two (Fig. 19.23).

## Existing Ice Sheets

Glacial ice currently covers about 10% of Earth's land area. However, in area and ice mass, alpine glaciers are insignificant compared to the ice sheets of Greenland and Antarctica, which account for 96% of the area covered by glaciers today. Glaciers resembling those of Greenland and Antarctica but on a much smaller scale, called ice caps, are also present in Iceland, on the arctic islands of Canada and Russia, in Alaska, and in the Canadian Rockies.

**FIGURE 19.21**
(a) Moraines clearly mark the former position and extent of a glacier that has receded. This is on the Kenai Peninsula in Alaska. (b) The glacial deposits of an end moraine form a ridge that rims the position of this glacier's terminus, although the large valley downstream shows that the glacier was once much larger. Outwash from meltwater forms a valley train in the foreground. This glacier is in Pakistan.
**What can we learn from studying moraines?**

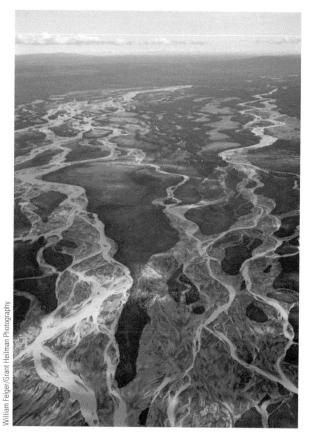

**FIGURE 19.22**
Meltwaters from glaciers in the Chugach Mountains of Alaska formed these braided streams from deposits of glacial outwash.
**What is the source of all this material?**

The Greenland ice sheet covers the world's largest island with a glacier that is more than 3000 meters (2 mi) thick in the center. The only land exposed in Greenland is a narrow, mountainous coastal strip (Fig. 19.24). Where the ice reaches the sea, it usually does so through fjords. These ice flows to the sea resemble alpine glaciers and are called **outlet glaciers.** The action of waves and tides breaks off huge ice masses that float away. The resulting **icebergs** are a hazard to vessels in the North Atlantic shipping lanes south of Greenland. Tragic maritime disasters, such as the sinking of the *Titanic,* have been caused by collisions with these huge irregular chunks of ice, which are nine tenths submerged and thus mostly invisible to ships. Today, with iceberg tracking by radar, satellites, and the ships and aircraft of the International Ice Patrol, these sea disasters are minimized.

The Antarctic ice sheet (Figure 19.25) covers some 13 million square kilometers (5 million sq mi), almost 7.5 times the area of the Greenland ice sheet. As in Greenland, little land is exposed in Antarctica, and the weight of the 4500-meter- (nearly 3 mi) thick ice in some interior areas has depressed the land well below sea level. Where the ice reaches the sea, it floats in enormous flat-topped plates called **ice shelves.** These ice shelves are the source of icebergs in Antarctic waters, which do not have the irregular shape of Greenland's icebergs. Because they do not float into heavily used shipping lanes, these huge tabular-shaped Antarctic icebergs are not as much of a hazard to navigation (Fig. 19.26). They do, however, add to the problem of access to Antarctica for scientists. The huge wall of the ice

**FIGURE 19.23**
The tabular edge of the continental ice sheet that covers almost all of Antarctica. In March 2000, the breakup from the Ross Ice Shelf, which extends out into the ocean on Antarctica's coast, formed an iceberg of almost unbelievable size. Called "B-15," this tabular iceberg was 300 kilometers (188 mi) long and 40 kilometers (25 mi) in width. This is 12,000 square kilometers, or about 4630 square miles, larger than the island of Jamaica. Since it formed, it has broken into several (still huge) sections and continues to shift position along the coast of Antarctica.

shelf itself, the massive, broken-up sea ice, and the extreme climate combine to make Antarctica inaccessible to all but the hardiest individuals and equipment. This icy continent serves, however, as a natural laboratory for scientists from many countries to study Antarctic glaciology, climatology, and ecology.

## Pleistocene Ice Sheets

The Pleistocene Epoch was a time of great climate change that began about 2.4 million years ago and ended around 10,000 years before the present. There were a great number of glacial fluctuations during the Pleistocene, marked by numerous major advances and retreats of continental ice sheets over large portions of the world's landmasses. When the Pleistocene glaciers advanced, ice sheets spread outward from centers in Canada, Scandinavia, and eastern Siberia, and glaciers also expanded in high-mountain ranges. At their maximum extent, glaciers covered nearly a third of Earth's land surface (Fig. 19.27). At the same time, sea ice expanded equatorward. In the Northern Hemisphere, sea ice was present along coasts as far south as Delaware in North America and Spain in Europe. Between each glacial advance, a warmer time called an *interglacial* occurred, during which the enormous continental ice sheets and sea ice retreated and almost completely disappeared. An examination of glacial deposits has determined that within each major glacial advance, many minor

**FIGURE 19.24**
The Greenland ice sheet. (a) Except for the mountainous edges, the Greenland ice sheet almost completely covers the world's largest island. Ice thickness is more than 3000 meters (10,000 ft) and depresses the bedrock below sea level. In this satellite view, several outlet glaciers flow seaward from the ice sheet to the east coast of the island. (b) Map showing the extent of the Greenland ice sheet.

(a)

(b)

**FIGURE 19.25**
The Antarctic ice sheet. The world's largest ice sheet covers an area larger than the United States and Mexico and has a thickness more than 4500 meters (14,000 ft). This satellite mosaic image covers the whole South Polar continent. Notice that most of Antarctica is ice covered (white and blue on the image). The only rocky areas (darker areas on the image) are the Antarctic Peninsula and the Transantarctic Mountains. Large ice shelves flow to the coastline, the largest being the Ross Ice Shelf. The image is oriented with the Greenwich Meridian at the top.

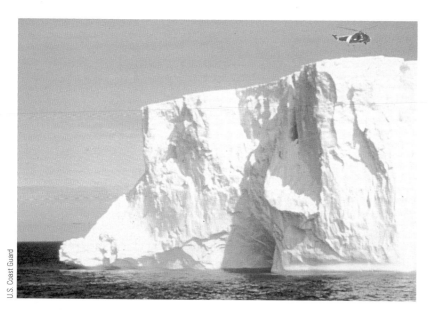

**FIGURE 19.26**
A closeup view of the edge of a tabular Antarctic iceberg.
**What portion of the iceberg is hidden below the sea surface?**

retreats and advances reflected smaller changes in global temperature and precipitation (see the discussion of climate change in Chapter 8).

The geomorphologic effects of the last major glacial advance, known as the Wisconsinan stage, which ended about 10,000 years ago, are the most visible in landscapes today. The glacial landforms created during the Wisconsinan stage are relatively recent and have not been destroyed to any great extent by erosion. Consequently, we can derive a fairly clear picture of the extent and actions of the ice sheets at that time.

Major glacial advances occurred in North America and Eurasia. Continental ice sheets in North America extended as far south as the Missouri and Ohio Rivers and covered nearly all of Canada and much of the northern Great Plains, the Midwest, and the northeastern United States. In New England, the ice was thick enough to overrun the highest mountains, including Mount Washington, which has an elevation of 2063 meters (6288 ft). The ice was more than 2000 meters (6500 ft) thick in the Great Lakes region. In Europe, glaciers spread over what is now most of Great Britain, Ireland, Scandinavia, northern Germany, Poland, and western Russia. In much of Siberia and interior Alaska, it was too cold and dry (as it is today) to generate the massive snow and ice accumulations that occurred farther south in North America and northwestern Europe. During each advance of the ice sheets, alpine glaciers were much more numerous, extensive, and massive in highland areas than they are today.

Where did the water locked up in all the ice and snow come from? Its original source was the oceans. During the periods of glacial advance, there was a general lowering of sea level, exposing large portions of the continental shelf and forming land bridges across the present-day North, Bering, and Java Seas. The most recent melting and glacial retreat raised the oceans a similar amount—about 120 meters (400 ft). Evidence for this rise in sea level can be seen along many coastlines around the world.

# Movement of Continental Ice Sheets

It is a popular misconception that the continental ice sheets originated at the poles and spread toward the equator. Actually, the great centers of Pleistocene glacial accumulation

## Geography's Spatial Science Perspective
### The Driftless Area—A Natural Region

The Driftless Area, a region in the midwestern United States, is located mainly in Wisconsin but also occupies sections of the adjacent states of Iowa, Minnesota, and Illinois. This region was not glaciated by the last Pleistocene ice sheets that extended over the rest of the northern Midwest, so it is free of glacial deposits from that time. As discussed in this chapter, *drift* is a term for glacial deposits, so the regional name is appropriate and linked to its geomorphic history. There is some debate about whether earlier Pleistocene glaciations covered the region, but the most recent glacial advance did not override this locality.

The landscape of this region is very different from the muted terrain and low rounded hills of adjoining glaciated areas. Narrow stream valleys, steep bluffs, caves, sinkholes, and loess-covered hills produce a scenic landscape with many landforms that would not have survived erosion by a massive glacier.

In some parts of the United States, lobes of the ice sheet extended as much as 300 miles farther south than the latitude of the Driftless Area. Why this region was glacier free has to do with the topography directly to the north. A highland of resistant rock, called the Superior Upland, caused the front of the glacier to split into two lobes, diverting the southward-flowing ice around this topographic obstruction.

The diverging lobes flowed into two valleys that today form Lake Superior and Lake Michigan. These two lowlands were oriented in directions that channeled the glacial lobes away from the Driftless Area. Although not overridden by these glaciers, the region did receive outwash deposits and a cover of wind-deposited loess, both derived from the surrounding glaciers and their sediments.

A map of the Midwest shows that the Driftless Area was virtually surrounded by the advancing Pleistocene ice sheets yet was not glaciated.

The Driftless Area has a unique landscape because of its isolation, an unglaciated "peninsula" almost completely surrounded by extensive glaciated regions. The rocks, soils, terrain, relief, vegetation, and habitats contrast strongly, with the nearby drift-covered areas and provide a glimpse of what the preglacial landscapes of this midwestern region may have been like. The region also provides distinctive habitats for flora and fauna that do not exist in the glaciated terrain. Unique and unusual aspects of the topography and ecology in this region attract ecotourists and offer countless opportunities for scientific study. For these reasons, there are many protected natural lands in the Driftless Area, an excellent example of a natural region defined by its physical geographic characteristics.

---

(aside from highland areas, Antarctica, and Greenland) were in the upper-middle latitudes, in the vicinity of Hudson Bay in Canada, on the Scandinavian Shield, and in eastern Siberia.

Continental ice sheets flow radially outward in all directions from a central zone of accumulation where the ice is thickest and the pressure is greatest ( Fig. 19.28). As with valley glaciers, the flow direction of advancing ice sheets was also determined by the path of least resistance, found in preexisting valleys and belts of softer rock. Both ice sheets and ice caps are shaped somewhat like a convex lens in cross section, thicker in the center and thinning toward the edges. Thus, the radial expansion is analogous to the spreading of pancake batter in a frying pan. If enough batter is poured into the center of the pan, it will eventually spread outward to cover the entire bottom of the pan. Like all glaciers, an ice sheet flows outward from a zone of accumulation to a zone of ablation. Also like all glaciers, ice sheets advance and recede by responding to

MOVEMENT OF CONTINENTAL ICE SHEETS   553

**FIGURE 19.27**
This map shows the maximum extent of glacial ice coverage in the Northern Hemisphere during the Pleistocene. Glaciers up to several thousand meters thick covered much of North America and Eurasia.
**Why were some very cold areas, such as portions of interior Alaska and Siberia, ice free during this time?**

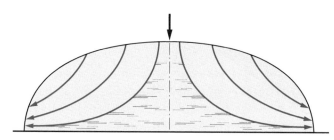

**FIGURE 19.28**
This diagram shows how ice flows in an ice sheet, downward from the center where the glacier is thickest and outward.
**How is this manner of ice flow different from and similar to that of a valley glacier?**

changes in temperature and snowfall. Eventually a glacier will continually retreat and finally disappear if ablation continues to exceed accumulation.

## Ice Sheets and Erosional Landforms

Ice sheets erode the land through processes that are similar to those of alpine glaciers but on a much larger scale. As a result, erosional landforms created by ice sheets are far more extensive than those formed by alpine glaciation, stretching over millions of square kilometers of North America, Scandinavia, and Russia. As ice sheets flowed over the land, they gouged Earth's surface, enlarging valleys that already existed, scouring out rock basins, and smoothing off existing hills. The ice sheets removed most of the soil and then eroded the bedrock below.

**FIGURE 19.29**
Seen from the space shuttle, this section of the Canadian Shield has been deeply eroded and scoured by ice sheets and exposes the remains of an ancient meteor crater, gouged out to make a circular lake (reservoir). Although the overall relief is low, note how erosion has exploited zones of weakness in fractures to make linear valleys and lakes.

Today, these **ice-scoured plains** are areas of low, rounded hills, lake-filled depressions, and wide exposures of bedrock (Fig. 19.29). Ice sheets cover and totally disrupt the former stream patterns. Because the last glaciation was so recent in terms of landscape development, new drainage systems have not had time to form a well-integrated system of stream channels. Ice-scoured plains are characterized by thousands of lakes, marshes, and areas of *muskeg* (poorly drained areas, grown over with vegetation, that form in cold climates). The major characteristics of glacially eroded landscapes, such as those found in Canada and Finland, are the great expanses of exposed gouged bedrock and standing water.

## Ice Sheets and Depositional Landforms

The sheer disparity in scale makes the deposits and depositional landforms created by ice sheets different in comparison to those of alpine glaciers. Though terminal and recessional moraines, ground moraines, and glaciofluvial deposits are produced by both glacier types, they form significantly larger features in the landscapes left by retreating ice sheets ( ▌ Fig. 19.30 and Map Interpretation: Continental Glaciation).

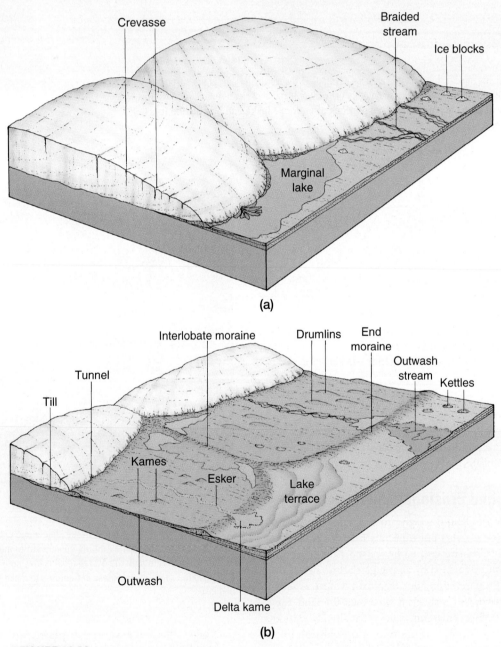

▌ **FIGURE 19.30**
Landscape alteration by ice sheet deposition. (a) Features associated with ice stagnation at the edge of the ice sheet. (b) Landforms resulting from further modification by glacial meltwater as the ice sheet retreats.
**What is the direct cause of most of these landscape features?**

**FIGURE 19.31**
The hilly topography of an end moraine deposited by a continental ice sheet. This aerial photo is of an end moraine on the Waterville Plateau in eastern Washington.
**How do you account for the bumpier terrain at the left of the photo compared to the smoother surface of the plain at the right? Refer to Figure 19.30 (b) to help explain your answer.**

**Terminal and Recessional Moraines** Terminal and recessional moraines left by Pleistocene ice sheets form belts of low hills and ridges deposited on the land in glaciated areas. These features are rarely more than 60 meters (200 ft) high ( Fig. 19.31), despite the massive size of the continental glaciers. The last major Pleistocene glacial advance through New England left a terminal moraine running the length of New York's Long Island and formed the offshore islands of Martha's Vineyard and Nantucket, Massachusetts. Recessional moraines formed both Cape Cod and Lake Michigan's rounded southern end. End moraines are usually arc shaped and convex toward the direction of ice flow. Their pattern and placement indicate that the ice sheets did not maintain an even front but spread out in tongue-shaped lobes channeled by the underlying terrain ( Fig. 19.32). The positions of terminal and recessional moraines provide more evidence than simply the direction of ice flow. Examining the character of deposited materials helps us detect the sequence of advances and retreats of each successive ice sheet.

**Till Plains** In the zone of ice-sheet deposition, massive amounts of unsorted glacial till accumulated, often to depths of 30 meters (100 ft) or more. Because of the uneven nature of deposition, the topographic configuration of land covered by till varies from place to place. In some areas, the till is too thin to hide the original contours of the land, and in other regions, thick deposits of till form broad, rolling plains of low relief. Small hills and slight depressions, some filled with water, characterize most **till plains,** reflecting the uneven glacial deposition. Some of the best agricultural land of the United States is found on the gently rolling till plains of Illinois and Iowa. The young, dark-colored, grassland soils (mostly mollisols) that developed on the till are extremely fertile.

**Outwash Plains** Beyond belts of hills formed by terminal and recessional moraines are areas formed by meltwater deposits called **outwash plains.** These are extensive, relatively smooth plains covered with sorted deposits transported by meltwater from the ice sheets. Outwash plains, which may cover hundreds of square kilometers, are analogous to the valley trains of alpine glaciers.

Pits, called **kettles,** which were formed when blocks of ice were buried in the glacial deposits, mark some outwash plains, till plains, and moraines. Eventually, where the blocks of ice melted, surface depressions remained. Although kettles have formed in deposits left by alpine glaciers, the vast majority of kettles form lakes in landscapes glaciated by ice sheets. For example, most of Minnesota's famous 10,000 lakes are **kettle lakes** ( Fig. 19.33).

**Drumlins** A **drumlin** is a streamlined hill, often about 0.5 kilometer (0.3 mi) in length and less than 50 meters (160 ft) high, molded in glacial drift on the till plains ( Fig. 19.34a). Drumlins are oddly shaped, resembling half an egg or the convex side of a teaspoon, and are usually found in swarms, with as many as a hundred or more clustered together. Glaciologists do not yet understand exactly how drumlins were formed. The most conspicuous feature is their elongated, streamlined shape, which follows the direction of ice flow. Thus, the geometry of a drumlin is the reverse of that of roches moutonnées. The broad, steep slope of a drumlin faces the direction from which the ice advanced, while gently sloping, narrower tail its points in the direction of ice flow. Drumlins are well developed in Canada, Ireland, and in the states of New York and Wisconsin. Boston's Bunker Hill, a drumlin, is one of America's best-known historical sites.

**Eskers** An **esker** is a narrow and typically winding ridge composed of glaciofluvial sands and gravels ( Fig. 19.34b). Some eskers are as long as 200 kilometers (130 mi), but usually they do not exceed several kilometers. It is believed that most eskers were formed by meltwater streams flowing in ice tunnels at the base of ice sheets. Eskers are prime sources of gravel and sand for the construction industry. Being natural

**FIGURE 19.32**
Glacial deposits in the Great Lakes region.
**Why do the many end moraines have such a curved pattern?**

embankments, they are frequently used in marshy, glaciated landscapes as highway and railroad beds. Eskers are especially well developed in Finland, Sweden, and Russia.

**Kames** Roughly conical hills composed of sorted glaciofluvial deposits are called **kames.** They are presumed to have formed when sediments accumulated in glacial ice pits, in crevasses, and among jumbles of detached ice blocks. Like eskers, kames are excellent sources for mining sand and gravel and are especially common in New England. **Kame terraces** are landforms resulting from accumulations of glaciofluvial sand and gravel along the margins of ice lobes that melted away in valleys of hilly regions. Examples of kame terraces can be seen in New England and New York.

**Erratics** Boulders scattered in and on the surface of glacial deposits or on glacially scoured bedrock are called **erratics** (Fig. 19.35) because the rock they consist of differs from the local bedrock. Erratics are found in association with alpine glaciers, but they are probably best known when deposited by ice sheets, which can move much larger rocks much father from their source. The source regions of erratics usually can be identified by the rock type, providing evidence of the direction of ice flow. In Illinois, glacially deposited erratics have come from source regions as far away as Canada.

Before the scientific proof of the existence of ice ages was confirmed, many hypotheses were developed to explain the existence of erratics. Among these explanations was one

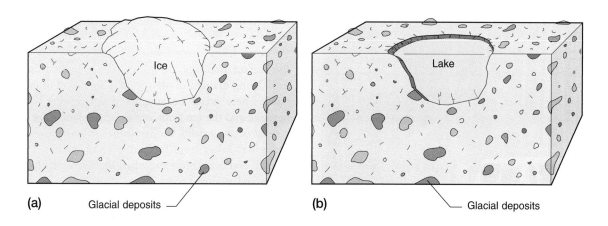

### FIGURE 19.33
Kettle lakes form (a) when huge blocks of ice, (b) melt away after burial or partial burial in glacial deposits. (c) This kettle, near the headwaters of the Thelon River in the Northwest Territories, Canada, formed in an area of ground moraine deposits, as can be seen by the hummocky topography.

### FIGURE 19.34
Drumlins and eskers. (a) A drumlin in Montana near the Canadian border. Drumlins are streamlined hills elongated in the direction of ice flow. (b) An esker near Albert Lea, Minnesota. Eskers are depositional ridges formed by meltwater deposits in a tunnel under the ice.
**What economic importance do eskers have?**

# Map Interpretation
## Alpine Glaciation

### The Map

The Chief Mountain, Montana, map area is located in Glacier National Park, which adjoins Canada's Waterton Lakes National Park located across the international border. Alpine glaciation has produced much of the spectacular scenery of this region of the northern Rocky Mountains.

Most of the glaciated landscape was produced during the Pleistocene. Today, glaciers still exist in Glacier National Park, but because of rapid melting in recent decades, it is estimated that they may be gone within 70 years. Prior to glaciation, this region was severely faulted and folded during the formation of the northern Rocky Mountains.

The oblique aerial photo clearly shows the rugged terrain of this map area. The steep slopes, glaciers, lakes, and sharp arête ridges are obvious landform characteristics of alpine glaciated regions. Temperature is a primary control of the highland climate of this region. Elevation is another important factor.

As you would expect, the rapid decrease in temperature with increasing elevation results in a variety of microclimates within alpine regions. Exposure is also an important highland climate control. West-facing slopes receive the warm afternoon sun, whereas east-facing slopes are sunlit only in the cool of the early morning.

### Interpreting the Map

1. What is the approximate local relief depicted by this map? How does the scale of this topographic map compare to those previously shown in the book (is this map of larger or smaller scale?)
2. Examining the topographic map and the aerial photograph of this mountain region, do you think most of the glaciated landscape was produced by erosion or deposition?
3. Locate Grinnell, Swiftcurrent, and Sperry Glaciers. What type of glaciers are they?
4. What evidence indicates that the glaciers were once larger and extended farther down the valleys?
5. Note that most of the existing glaciers are located to the northeast of the mountain summits. Explain this orientation. At what elevation are the glaciers found?
6. What types of glacial landform are the following features?
    a. The features occupied by Gunsite, Iceberg, and Ipasha Lakes
    b. The feature occupied by McDonald Lake (in the southwest corner of the map) and Lake Josephine
    c. Mt. Gould and Mt. Wilbur
    d. The series of lakes that occupy the valley of Swiftcurrent Creek
7. Along the high ridges runs a dashed line labeled "Continental Divide." What is its significance?
8. If you were to hike southeast from Auto Camp on McDonald Creek up Avalanche Creek to the base of Sperry Glacier, how far would you travel, and how much elevation change would you encounter?

An oblique aerial photo of Glacier National Park.

Opposite:
Chief Mountain, Montana
Scale 1 : 125,000
Contour interval = 100 ft
U.S. Geological Survey

# Map Interpretation
## Continental Glaciation

### The Map

The Jackson, Michigan, map area is located in the Great Lakes section of the Central Lowlands. Massive continental ice sheets covered this region during the Pleistocene. As the glaciers melted and the ice sheets retreated, a totally new terrain, vastly different from that of preglacial times, was exposed. Today, evidence of glaciation is found throughout the region. The most obvious glacially produced landforms are the thousands of lakes (including the Great Lakes), the knobby terrain, and moraine ridges. Moraines left by advancing and retreating tongue-shaped ice lobes that extended generally southward from the main continental ice sheet also influenced the shapes of the Great Lakes.

The ice sheet, its deposits of sediment, and its meltwaters also created other, smaller-scale features. The glaciers left a jumbled mosaic of deposits from boulders to sand, silt, and clay that has produced a hilly and hummocky terrain. The overall relief is low, in part because of glacial erosion. Many landform features of continental glaciation are well illustrated on the Jackson, Michigan, 7.5 minute quadrangle map.

This region has a humid continental climate. The summers are mild and pleasant. Excessively warm and humid air seldom reaches this area for more than a few days at a time. Instead, cool but pleasant evening temperatures tend to be the rule in summer. Winters, however, are long and often harsh. Snow can be abundant and on the ground continuously for many weeks or even months at a time.

The annual temperature range is significant, and precipitation occurs year-round but is provided primarily by storms moving along polar fronts.

### Interpreting the Map

1. Describe the general topography of the map area.
2. What is the local relief? Why is it so difficult to find the exact highest and lowest points on this map?
3. Does the topography of this region indicate glacial erosion or glacial deposition?
4. Does the area appear to be well drained? What are the three main hydrographic features that indicate the drainage conditions?
5. What type of glacial landform is Blue Ridge? What are its dimensions (length and average height)?
6. How is a feature such as Blue Ridge formed? What economic value might it have?
7. What is the majority of the surface material in the map area? What is the term for this type of surface cover?
8. What probably caused the many small depressions and rounded lakes? What are these depressions called?
9. Describe how the topography of the Jackson, Michigan, area appears on the satellite image.
10. What is the advantage of having satellite image and topographic map coverage of an area of study?

Landsat image of the region near Jackson, Michigan

Michigan Department of Natural Resources

Opposite:
Jackson, Michigan
Scale 1 : 62,500
Contour interval = 10 ft
U.S. Geological Survey

# CHAPTER 20

The Atlantic Ocean, the Caribbean Sea, and the Gulf of Mexico merge together in this scene. The bright blue–green colors in the water indicate dense colonies of plankton. Jacques Descloitres, MODIS Land Rapid Response Team, NASA/FSFC

# The Global Ocean

**PHYSICAL**
**Geography Now™** This icon, appearing throughout the book, indicates an opportunity to explore interactive tutorials, animations, or practice problems at now.brookscole.com/gabler8

Having a knowledge of the oceans and seas is of major importance in physical geography. Oceans cover most of Earth and are an essential component in the planet's physical systems. Their impact on weather and climate is vital. The oceans are the world's major collector of solar radiation because they continually absorb, store, and distribute much of Earth's energy. The oceans are a fundamental part of the hydrologic cycle, and in the global circulation system, they are a major control of weather and climate. The oceans are a vital producer of the world's oxygen and an important absorber of carbon dioxide. The seafloor contains two thirds of Earth's surface topography, and its submarine geologic processes are still being discovered. Finally, the oceans are a critical source for food, minerals, and energy resources.

# CHAPTER PREVIEW

**The ocean covers more than 70% of Earth's surface, yet we have just begun to investigate many unanswered questions about its nature.**
- What does the concept of a global ocean mean to you personally?
- What are the five major divisions of the global ocean?

**There are significant surface and depth differences in the salinity, temperature, density, and pressure of seawater.**
- What are the most important differences?
- Why does salinity vary in different geographic locations?
- What effects and limitations does water pressure have on humans?

**Contrary to early ideas about the seafloor, there are topographic irregularities on the ocean floor comparable to those on the continents.**
- What did early scientists believe the ocean floor looked like? Why?
- What are the major features associated with the continental margins and the deep-ocean floor?

**Islands are landmasses that are smaller than continents and surrounded by water.**
- What are the three major types of islands?
- How are they formed?
- What are coral reefs?

**The surface movements of the global ocean have major impacts on humans and their activities.**
- What are the three most important ocean movements?
- How do they impact humans?

# Introduction to the Oceans

The oceans contain more than 97% of Earth's water and cover 71% of Earth's surface. In fact, our planet should probably be called Oceanus or Hydro because water dominates its surface and it is the only planet in the solar system with liquid water at the surface. Yet our knowledge of the oceans is not very extensive because they are difficult to explore. The oceans are vast and dark at depth, and the pressures existing at ocean depths present a hostile environment to humans. Only since the British expedition aboard HMS *Challenger*, from 1872 to 1876, have scientists systematically begun to explore the ocean's complexity. Oceanography was not generally recognized as an important science until the 20th century. Most of our knowledge of the ocean has been developed with rather recent technologies, such as sonar (using sound waves to determine ocean floor topography), deep-diving submersibles, underwater photography and videography, deep-ocean drilling ships, environmental satellites, and the global positioning system (GPS).

The oceans actually form a single, large, continuous body of water. The "global ocean" is divided into five individual oceans that are distinguished largely on the basis of their geographic locations although they do differ somewhat in certain marine characteristics. In 2000 the International Hydrographic Organization declared that the ocean surrounding Antarctica (poleward of 60°S) should be considered a separate Southern Ocean. The rationale for this decision is based on oceanographic characteristics, particularly the Antarctic circumpolar current, which flows eastward around the continent.

The five oceans are the Pacific, Atlantic, Indian, Southern, and Arctic Oceans. The approximate area of the Pacific Ocean ( Fig. 20.1) is 156 million square kilometers (60 million sq mi). The Atlantic is about half that size, or 77 million square kilometers (30 million sq mi), and the Indian Ocean is about 69 million square kilometers (27 million sq mi). The Southern Ocean is 20 million square kilometers (8 million sq miles), and the Arctic Ocean is 14 million square kilometers (5 million sq mi). In comparison, the continent of North America is 23 million square kilometers (9 million sq mi) in area. In fact, the total land area of Earth—149.5 million square kilometers (57.5 million sq mi)—is smaller than the Pacific Ocean, which is the Earth's largest geographic feature.

The oceans also vary greatly in depth. The Pacific Ocean has an average depth of 4200 meters (13,800 ft), and the Atlantic and Indian Oceans have average depths of 3900 meters (12,800 ft). The average depth of the Southern Ocean is 4000–5000 meters (13,100–16,400 ft), and the Arctic Ocean is the shallowest averaging only 930 meters (3100 ft). However, these figures are somewhat misleading because the ocean floors are not just flat plains but have mountains, trenches, and basins that vary considerably in depth. The Pacific reaches a maximum depth of more than 11,000 meters (36,000 ft).

*Seas* are saltwater bodies that are smaller than oceans and somewhat enclosed by land. Unlike lakes, which can be

**FIGURE 20.1**
This view shows the vast extent of the Pacific Ocean.
**Why has Earth been referred to as the "ocean planet"?**

of either fresh water or saltwater, seas interchange water with the oceans. Examples of major seas include the Mediterranean, Baltic, Bering, Arabian, Caribbean, Coral, North, Black, Yellow, and Red Seas. Some large semienclosed areas of the oceans are also called gulfs or bays but are actually seas as well. Examples of these include the Gulf of Mexico, Gulf of Alaska, Gulf of California, and Bay of Bengal. Large salty lakes, such as the Aral, Caspian, and Dead "Seas," are incorrectly named because they are landlocked.

## Characteristics of Ocean Waters

The world ocean is 96.5% water by weight, but seawater is a dilute solution with on average 3.5%, or 35 parts per thousand (ppt), dissolved solids. Of these dissolved minerals, the most common substance by far is sodium chloride, or common table salt ( Fig. 20.2). Other important constituents of seawater are magnesium, sulfur, calcium, and potassium. The oceans also contain the dissolved atmospheric gases, especially oxygen and great amounts of carbon dioxide.

Ocean waters contain virtually every element found on Earth. Most, however, are trace elements in concentrations of a few parts per billion (ppb). For example, there are about 40 pounds of gold and 200 pounds of lead per cubic mile of ocean water. We now extract salt and magnesium from the sea, and the future extraction of other substances, including hydrogen for energy, may prove worthwhile.

The measurement of all dissolved solids in seawater is referred to as **salinity**, and it varies throughout Earth's oceans and seas. Although, as already mentioned, the average salinity is 3.5% (35 ppt), this can vary in the open ocean from 3.2 to 3.8% (32–38 ppt). The salinity variation is even greater in enclosed or partially enclosed seas.

Several factors affect salinity—primarily, amounts of precipitation, inflow from rivers, and rates of evaporation. In humid regions where precipitation is high, fresh water tends to dilute the seawater and reduce salinity. Arid and semiarid regions have a higher salinity because precipitation is low, few rivers exist, and evaporation is high.

Ocean temperatures also vary and together with salinity affect density; these factors are also affected by latitudinal climate zones (Fig. 20.3). Density is the mass per unit volume measured in kilograms per cubic meter (or pounds per cubic foot). The maximum density of fresh water occurs at about 4°C (39°F) and is 1000 kilograms per cubic meter (62.4 lb/cu ft). An increase in temperature results in a decrease in density. Salinity and density are directly proportional—the greater the salinity, the denser the water. The salinity in seawater raises water density to about 1025 kilograms per cubic meter (64 lb/cu ft). Density affects circulation because density differences cause gravitational displacement of water. Warm water is less dense than cold, so it tends to float above colder, denser water. Cold or highly saline water near the surface sinks and is replaced by warmer or less saline water. These changes produce patterns of surface-water movement that vary with the seasons. Density differences also cause deep-ocean currents (*thermohaline* currents) that circulate the ocean's deepest waters.

## Surface Variations

Salinity varies in response to precipitation and evaporation rates at different latitudes (see again Fig 20.3). Salinity is highest in subtropical high pressure regions near 30°N and S because of low precipitation and high evaporation rates. Salinity decreases toward the equator because of abundant rainfall, heavy stream flow, and lower evaporation rates due to increased cloud cover and humidity. For example, salinity is only 2% (20 ppt) near the equatorial coast of Brazil because of the diluting influence of the Amazon River as well as abundant rainfall and low evaporation.

Low salinity is also found in the polar regions because of extremely low evaporation and the summer freshwater inflow from melting snow and glacial ice. Moving equatorward, salinity increases in the middle latitudes with their more temperate ocean waters and greater evaporation, especially during the summer.

The highest marine salinities are found in semienclosed seas in hot, arid regions where the evaporation rates are very high and there is no stream flow. For example, the Red Sea, between the Sahara and the Arabian Deserts, has a salinity of more than 4% (40 ppt). Desert basin lakes have the world's

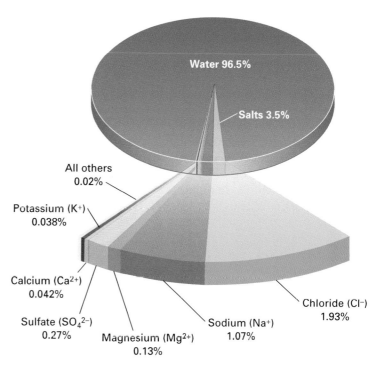

**FIGURE 20.2**
The average salinity of seawater in the global oceans is 3.5%. Six chemical ions make up most of the salts in ocean water.
**What factors could cause seawater to increase its salinity in a certain area? What could cause salinity to decrease?**

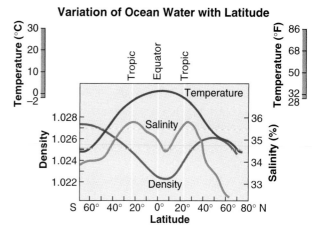

**FIGURE 20.3**
The temperature, salinity, and density of seawater at the ocean surface tend to be related to latitude.
**What latitudinal climatic factors are directly related to the variation shown on this graph?**

highest salinity; the Dead Sea is so saline—23.8% (238 ppt)—that most aquatic organisms cannot survive (hence its name). The Great Salt Lake in Utah has a maximum salinity of 22% (220 ppt).

## Variations with Depth

The vertical temperature distribution in the oceans results from solar heating that is concentrated in the surface waters. Some of this heat is mixed into deeper water by conduction and currents. Oceans tend to have three vertical temperature zones: (1) a surface layer of generally warmer water, (2) a transition zone of rapidly cooling waters with increasing depth, and (3) the uniformly cold (average 3.5°C/37°F) waters of the deep ocean. The change from the heated surface layer to cold deep water occurs in a well-defined transition zone called the **thermocline** ( ▌ Fig. 20.4). The thermocline is most apparent in the tropics and during summer in middle latitudes when surface waters are most strongly heated. In polar waters, there may be no thermocline because seawater temperatures are often as cold at the surface as they are at great depth.

Pressure increases rapidly with depth ( ▌ Fig. 20.5). Water is more than 800 times denser than air, and water pressure increases by 1 atmosphere (14.7 pounds per square inch) for every 10 meters (33 ft) of water depth. At a depth of 300 meters (1000 ft), water pressure is 445 pounds per square inch. At 600 meters (2000 ft), it is 892 pounds per square inch, and at 1800 meters (6000 ft), it is 2685 pounds per square inch. At the greatest ocean depths in the Pacific's Mariana Trench, pressure exceeds 16,000 pounds per square inch. Dangerous water pressure will strain air-filled human organs such as eardrums and lungs, even during shallow descents into the ocean. For this reason, the deepest scuba dives have been only to about 100 meters (330 ft). However, deep-diving submarines and research submersibles can descend thousands of meters while maintaining a normal atmospheric pressure for humans inside ( ▌ Fig. 20.6).

## The Ocean Floor

It was not until the 20th century that scientists were able to explore the ocean floor and discover details of its submarine topography. The development of sonar (SOund Navigation And Ranging) has helped enormously, as have specialized research ships designed for deep-sea drilling, such as the *Glomar Challenger,* and deep-diving vehicles like *Alvin* (see Fig. 20.6).

At present, we know more about the surface of the moon than we do about our own seabed. However, recent innovations are increasing our knowledge and ability to explore deep-sea terrain. Undersea research laboratories allow

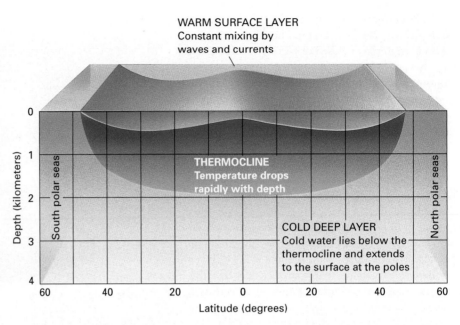

▌ **FIGURE 20.4**
Ocean waters near the surface are warmed by absorbed solar radiation. The thermocline is a boundary at depth (except in the polar regions) where the water below rapidly gets much colder.
**Why does the thermocline change in depth from the equator to the polar regions?**

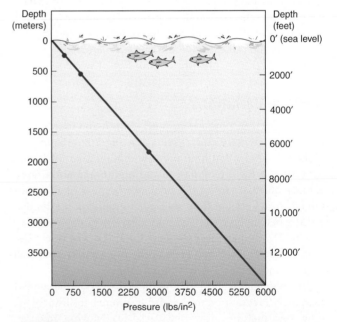

▌ **FIGURE 20.5**
Water pressure in the ocean increases directly with greater depth.

scuba-diving scientists to stay in shallower reaches of the ocean for days at a time. New side-scanning sonar can record swaths of seafloor topography several kilometers wide with great accuracy and can produce three-dimensional bathymetric contour maps. Camera and video systems can image the deepest sections of the ocean floor. Though submersibles such as *Alvin* will continue to take scientists to the ocean floor, robotics are the way of the future for ocean exploration. Remote-controlled submersibles such as *Argo-Jason,* which helped locate HMS *Titanic* and other famous sunken vessels, are the safest and most efficient deep-ocean exploration vehicles.

Before the most recent explorations of the sea, it was believed that most of the ocean floor consisted of flat plains. We now know that beneath the ocean waters there are landforms greater in size and extent than those found on the continents. Mountains, basins, plains, volcanoes, escarpments, canyons, and trenches are all present beneath the ocean ( ▌Fig. 20.7).

In the 1960s, Bruce Heezen and Marie Tharp produced the first global map of the ocean floor ( ▌Fig. 20.8). Since then, mapping of the ocean floor's topography, known as **bathymetry,** has advanced rapidly, with thousands of echo soundings by ships and the use of computer-imaging techniques to produce detailed bathymetric maps. Amazingly, the seafloor can also be mapped from satellites, based on the measurement of sea level. Variations in gravity cause sea level to bulge slightly over higher areas such as submarine mountains and to dip over deep marine trenches.

Much of the shallow seafloor is geologically a part of the continents. As a result, the ocean basins have two major topographic subdivisions: (1) the continental margins consisting of the continental shelf and continental slope, which are part of the continental crust, and (2) the deep-ocean floor, which is formed by oceanic crust (see Fig. 20.8).

## Features of the Continental Margins

**Continental Shelf** The **continental shelf** consists of the edges of continents that slope gently from sea level to depths of about 200 meters (650 ft) below sea level. The continental

▌ **FIGURE 20.6**
The deep-diving submersible *Alvin* has taken scientists to the ocean bottom to study its characteristics. Inside *Alvin,* crew members are under normal atmospheric pressure, while outside the submersible is under tremendous water pressure.
**Does all undersea exploration require manned submersibles? Why or why not?**

shelves vary a great deal in width. In some places, they are almost nonexistent; in others, they are as wide as 1300 kilometers (800 mi). Generally, where broad continental plains slope to meet the ocean (passive margins), the continental shelf is wide. Where mountains are found along the continental edge (active margins), the shelf is narrow. These extremes are found on the two coasts of the United States as seen in Figure 20.9. Beyond the Atlantic and Gulf coasts, the continental shelf extends outward as much as 480 kilometers (300 mi). On the mountainous Pacific Coast, the continental shelf is narrow.

The continental shelves are the "treasure chests of the sea." The amount of sunlight that they receive and the supply of nutrients washed into these waters by rivers support vast populations of marine plants and animals. Thus, 90% of the

▌ **FIGURE 20.7**
Major features of the ocean floor are shown in this profile of the North Atlantic Ocean from the East Coast of the United States to the Mid-Atlantic Ridge.
**What is the deepest part of this section of the North Atlantic Ocean?**

■ FIGURE 20.8
The "Floor of the Ocean" map by Heezen and Tharp. This map was produced in the 1960s and was the first detailed submarine relief map of the world ocean floor. It still serves well for a general view of the major ocean-floor features.
**Locate the deepest point in the global ocean.**

# THE OCEAN FLOOR

**FIGURE 20.9**
The continental shelves are wide on the East Coast of North America, far from the plate boundary in midocean. In contrast, on the tectonically active West Coast continental shelves are much more narrow.
**Why are the continental shelves so different in width on opposite sides of North America?**

fish we eat are caught in the shallow waters above the continental shelf. Virtually all lobster, crab, shrimp, and shellfish live in these rich waters. Offshore reserves of gas and oil are stored in the shelf sediments, as are gravels (for construction) and minerals. Presently, about 25% of the world's petroleum is produced from offshore oil rigs.

The topography of a continental shelf is not as smooth as we might expect from the effects of constant wave motion, settling of sediments (sand, silt, and clay) brought by rivers, and global sea-level changes. Although it is usually a relatively smooth, gently sloping plain, a shelf often has ridges, depressions, hills, valleys, and canyons. Some of the higher features break the water surface as islands, such as New York's Long Island and Massachusetts's Martha's Vineyard. During periods of lower sea level during the Pleistocene when much water was stored on the land as glacial ice, the continental shelves formed land bridges connecting Alaska to Siberia and Great Britain to continental Europe. The teeth and bones of Pleistocene land mammals are occasionally brought up in fishing nets from continental shelves previously exposed as land areas. The balance between water in the oceans and glacial ice on land is important in determining the location of the coastline. If global warming accelerates and causes rapid glacial melting, many low-lying coastal areas will become submerged parts of the continental shelves.

**Submarine canyons** are probably the most striking features of the continental margin. These canyons are steep-sided, erosional valleys cut into the edge of the shelf and slope. They resemble canyons cut by rivers on land ( Fig. 20.10), and some are larger than the Grand Canyon. V-shaped in cross section, these submarine canyons are often found opposite the mouths of major rivers, such as the Congo or Hudson. This distribution has led to the hypothesis that rivers helped in forming the canyons. Yet some submarine canyons are not located off present-day or ancient river mouths. The most widely accepted explanation is that they were carved by **turbidity currents**—periodic submarine flows of sediment and water (a slurry) that move from the continental shelves down to the deeper ocean floor. These gravity flows of dense sediment-laden water have a high capacity for eroding into the seabed. The upper portions of the canyons may have formed by surface erosion during the Pleistocene when lower sea levels exposed much of the present-day continental shelves. The lower portions of some submarine canyons are cut thousands of meters below sea level and have fans of sediment at their mouths on the ocean floor.

**Continental Slope** Marking the outer edge of the continental shelf is a relatively steep drop, usually 3000–3600 meters (10,000–12,000 ft) to the ocean floor ( Fig. 20.11). The **continental slope** forms the true boundary between the continents and the ocean basins. The continental slope begins at depths of 120–180 meters (400–600 ft).

The continental slope is not an extremely steep incline but descends at an angle of about 15 degrees. However, this is much steeper than the surface of the adjacent continental shelf,

Some ocean-floor areas at the base of the continental slope form a gently sloping surface, known as the **continental rise** (see Fig. 20.11). These continental rises are well developed along the margins of the Indian and Atlantic Oceans, especially off major river mouths, such as in the Bay of Bengal. However, they are almost nonexistent along the Pacific margins where trenches are located. The continental rise consists of muddy sediments from turbidity currents and landslide deposits from the shelf and slope. Deep-ocean currents may distribute these materials along the base of the continental slopes and out onto the deep-ocean floor.

## Features of the Deep-Ocean Floor

The deep-ocean floor lies at an average depth of about 4000 meters (12,000 ft). Until recently it was believed to be a relatively smooth submarine plain. Now, with sophisticated bathymetric maps, we know that the topography of the deep-ocean floor is as irregular as that of the continents. The ridges and depressions of the ocean floor often rival and surpass the relief of continental landforms. In contrast, large areas of the deep ocean are blanketed by thousands of meters of marine sediments forming virtually featureless plains.

The two most impressive submarine features of the deep-ocean floor are oceanic ridges and trenches. Other significant terrain features of the deep ocean are abyssal plains, seamounts, and guyots.

### Oceanic Ridges
The **oceanic ridges** (or **midocean ridges**) form an interconnected network of submarine volcanic mountains that extends through all four oceans (see Figs. 20.7 and 20.8). The best known of these, the Mid-Atlantic Ridge (Fig. 20.12), extends southward from the Arctic Ocean to Iceland. It then trends southward, midway between the Americas and Africa, almost to Antarctica before it turns eastward south of Africa and extends into the Indian Ocean. The Mid-Indian Ridge forms an inverted "Y" with one arm extending into the Red Sea (see Fig. 20.8). Its other arm links the Mid-Atlantic Ridge to the Pacific Ridge and on to the East Pacific Rise that extends across the South Pacific and then northward into the Gulf of California, separating the Baja California peninsula from mainland Mexico. This continuous global network of submarine mountains is 64,000 kilometers (40,000 mi) long and

**FIGURE 20.10**
Monterey Canyon, off the central California coast, is a submarine canyon that is deeper than the Grand Canyon of the Colorado River. This image was created from high-resolution satellite imagery and high-tech digital lithography. The area is now designated the Monterey Bay National Marine Sanctuary.
**What is the probable origin of submarine canyons such as this one?**

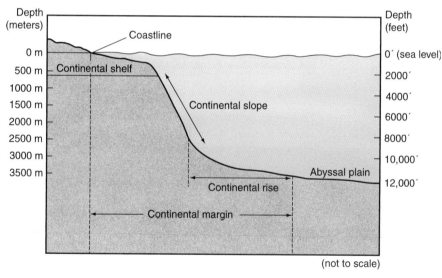

**FIGURE 20.11**
A general profile of the major features of the continental margin.
**Where is the true boundary between the continents and the deep-ocean basins?**

which slopes only about 1 or 2 degrees. What is most characteristic of a continental slope is its great descent, usually some 3600 meters (12,000 ft), but sometimes as much as 9000 meters (30,000 ft), to the ocean floor or into depths of the trenches. There is evidence of massive submarine mudslides or slump flows off the edges of some continental slopes.

**FIGURE 20.12**
Relief map of the floor of the North Atlantic Ocean, showing such features as the Mid-Atlantic Ridge, abyssal plains, numerous seamounts, and the continental shelves and slopes of surrounding continents.
**What is the most obvious submarine topographic feature in the North Atlantic Ocean?**

**FIGURE 20.13**
Hydrothermal vents on the midocean ridges are called "black smokers."

averages about 1600 kilometers (1000 mi) wide. The oceanic ridges rise an average of 1500–3000 meters (5000–10,000 ft) above the ocean floor.

In some areas, the highest peaks of the midocean ridges rise above sea level to form islands. Iceland, the Azores, and Ascension Island are all high volcanic peaks of the Mid-Atlantic Ridge. The Azores rise more than 8100 meters (27,000 ft) above the ocean floor.

The topography of oceanic ridges is very rugged, and the whole range consists of volcanic rocks, chiefly basalts forming "pillow lavas." Pillow lavas are globular formations formed by lava cooling rapidly under seawater. Numerous fractures and faults add to the complexity of the oceanic ridges. Running through the middle of the oceanic ridges is a central rift valley that may reach several kilometers in width. It is volcanically active and the center of earthquake activity. Most dramatic are the undersea volcanoes called "black smokers" (❙ Fig. 20.13) that spew out hydrothermal fluids (hot water mixed with iron and zinc sulfides). Fracture zones that cut across the oceanic ridges offset the rift valleys.

These fracture zones are actually transform faults where sections of tectonic plates slide past each other, causing the rift valleys to have a zigzag pattern.

The Deep Sea Drilling Project (DSDP) operated the drill ship *Glomar Challenger* between 1968 and 1983. *Glomar Challenger* drilled more than 1000 cores in all the major oceans. The most remarkable discovery was the confirmation of seafloor spreading. The addition of new oceanic crust rock along the midocean ridges is pushing older parts of the crust apart. The Ocean Drilling Program (ODP) continues to explore the ocean bottom with a newer drill ship.

**Trenches** Trenches, the deepest parts of the ocean, are long, narrow, arc shaped, steep sided, and aptly named. Most trenches are located close to and along active oceanic margins. Trenches are found adjacent to zones that have a concentration of volcanic and earthquake activity and are most common on the seaward (convex) side of curving **island arcs** formed by volcanoes, such as the Aleutian Islands.

Most trenches, including the deepest ones, are found around the Pacific "Ring of Fire." Challenger Deep, in the North Pacific's Mariana Trench, is the deepest known part of the ocean, reaching 10,915 meters (35,810 ft) below sea level. In 1960 Jacques Piccard and Donald Walsh descended in the bathyscaphe *Trieste* to the bottom of the Mariana Trench. Placed in that trench, Mount Everest would still have a mile of seawater above its summit. Other major trenches in the Pacific are the Aleutian, Kuril, Japan, Philippine, Tonga, and Peru–Chile Trenches ( ∎ Fig. 20.14). The deepest part of the Atlantic is the Puerto Rico Trench, at 8648 meters (28,374 ft), and the Java Trench, at 7125 meters (23,377 ft), is the Indian Ocean's deepest point (see Fig. 20.8).

As with the oceanic ridges, the trenches play a major role in Earth's geologic evolution. Oceanic plates are descending below continental plates and are being recycled to Earth's interior at the trenches. Ocean-floor drilling has shown that no ocean-floor rock is older than about 200 million years. The youngest oceanic crust is being created at oceanic ridges, while the oldest oceanic crustal rock is being destroyed in the trenches by subduction. Thus, oceanic crust is completely recycled over a period of a 2–3 hundred million years.

> **PHYSICAL**
> **Geography Now™** Log on to Physical GeographyNow and select this chapter to work through a Geography Interactive activity on "Triple Junctions and Seafloor Study" (click Plate Tectonics → Triple Junctions and Seafloor Study).

**Abyssal Plains** Abyssal plains are vast submarine plains of very low relief. They lie at depths of 3000–6000 meters (10,000–20,000 ft) and cover about 40% of the ocean floor. Most abyssal plains are covered by layers of sediments that bury the relief of the oceanic crust. Much of this sedimentary blanket consists of fine brown and red clays, contributed by wind deposition and volcanic eruptions. A significant portion also consists of the remains of microscopic marine organisms. Such organic marine deposits are known as **ooze.** Britain's famous white chalk cliffs of Dover are uplifted ancient ocean-floor ooze deposits ( ∎ Fig. 20.15). In regions of slow deposition, manganese nodules form on the abyssal plains. In the future, these potato-sized concentrations of manganese, iron, copper, and nickel may be mined.

∎ **FIGURE 20.14**
The floor of the western Pacific Ocean, showing the major trenches off Asia. Note the great depths of the Japan and Mariana trenches.
**What caused the formation of these trenches?**

∎ **FIGURE 20.15**
The Seven Sisters section of the white chalk cliffs of Dover along the English Channel.
**What is the origin of the cliffs of Dover?**

**FIGURE 20.16**
Hawaii, the largest and youngest Hawaiian island, was formed over a "hot spot." (a) Two large shield volcanoes, Mauna Loa and Mauna Kea, dominate the island. Mauna Loa and Kilauea on its eastern slope are still very active. Loihi seamount, an active submarine volcano, is forming to the south and may be above sea level in 50,000 years. (b) A robotic submersible gathers data on Loihi's solidified lava at a depth of 980 meters (3200 ft). The yellow areas are iron-rich minerals spewed from the volcanic vents.

**What is the difference between an oceanic island and a seamount?**

**Seamounts and Guyots** One of the most common features of submarine topography is the **seamount**, a submarine volcanic mountain. Seamounts are relatively isolated volcanic mountains or groups of mountains, usually with a height above the ocean floor of 900 meters (3000 ft) or more. Typically steep sided with small summits, seamounts are volcanic peaks that grow from the deep-ocean floor.

**Guyots** are flat-topped seamounts. The origin of their flat tops at such depths is not certain. Research indicates that they are volcanoes whose summits have been planed off by wave erosion. Later they subsided to their present depths, possibly during lateral movement of the seafloor away from the oceanic ridges and "hot spots."

## Islands and Coral Reefs

There are three basic types of islands: continental, oceanic, and coral. **Continental islands** are usually found on the continental shelves. They are geologically part of the continent but are separated from it because of global sea-level changes or regional tectonic activity. The world's largest islands, such as Greenland, New Guinea, Borneo, and Great Britain, are continental. Smaller continental islands include New York's Long Island, California's Channel Islands, and Vancouver Island off the west coast of Canada. The barrier islands along the Gulf and Atlantic coasts of the United States are also continental. A few large continental islands, such as New Zealand and Madagascar, are isolated "continental fragments" that separated from continents millions of years ago.

**Oceanic islands** are volcanoes that rise from the deep-ocean floor and are not geologically related to the continents.

Most oceanic islands, such as the Aleutians, Tonga, and the Marianas, occur in island arcs along the edges of the trenches. Others, like Iceland and the Azores, are peaks of oceanic ridges rising above sea level. Many oceanic islands occur in chains, such as the Hawaiian Islands. The oceanic crust sliding over a stationary "hot spot" in the mantle causes these island chains. In the future, the Hawaiian Islands will move northwestward with the Pacific plate and slowly submerge to become seamounts. A new volcanic island, named Loihi, will form to the southeast ( Fig. 20.16). Evidence of the plate motion is indicated by the fact that the youngest islands of the Hawaiian chain, Hawaii and Maui, are to the southeast, while the older islands, such as Kauai and Midway, are located to the northwest.

**Coral reefs** are shallow, wave-resistant structures formed by an accumulation of remains of tiny sea animals that secrete a limy skeleton of calcium carbonate. Many other organisms, including algae, sponges, and mollusks, add material to the reef structure. Reef corals need special conditions to grow—clear and well-aerated water, water temperatures above 20°C (68°F), plenty of sunlight, and normal salinity. These conditions are found in the shallow waters of tropical regions such as Hawaii, the West Indies, Indonesia, the Red Sea, and the coast of Queensland in Australia. Today, coastal water pollution, dredging, souvenir coral collecting, and possibly global warming threaten the survival of many coral reefs.

A **fringing reef** is a coral reef attached to the coast ( Fig. 20.17a). Fringing reefs tend to be wider where there is more wave action that brings a continuous supply of well-aerated water and additional nutrients for increased coral growth. They are usually absent where there is a river mouth because the corals cannot grow where the waters are laden with sediment or where river water lowers the salinity of the marine environment.

Sometimes coral forms a **barrier reef**, which lies offshore, separated from the land by a shallow lagoon ( Fig. 20.17b). Most barrier reefs form in association with slowly subsiding oceanic islands, growing at a pace that keeps them above sea level. Other barrier reefs, such as Australia's Great Barrier Reef, the Florida Keys, and the Bahamas, were formed on continental shelves and grew upward as sea level rose after the Pleistocene ice age waned. The world's largest organic structure, the Great Barrier Reef of Australia, is more than 1930 kilometers (120 mi) long.

An **atoll** is a ring of coral reefs encircling a central lagoon that has no inner volcanic island ( Fig. 20.17c). Figure 20.18 illustrates the manner in which atolls develop. As a volcanic island subsides, the fringing reef grows upward keeping pace with the seafloor subsidence, becoming a barrier reef and finally an atoll. Charles Darwin proposed this explanation of atoll formation in the 1830s. Drilling evidence indicates that there has been as much as 1200 meters (4000 ft) of subsidence and an equal amount of reef development in the past 60 million years.

Atolls pose severe challenges as environments for human habitation. First, they have a low elevation above sea level and provide no defense against huge storm waves that can inundate the entire atoll, drowning all its inhabitants. Possible future sea-level rise from global warming would also threaten these low islands. Second, there is little fresh water available on the porous coral limestone surface. Third, little vegetation can survive in the lime-rich rock and soil of the atoll islands. The coconut palm is an exception, and coconuts were vital to the survival of early inhabitants of the atoll islands in Polynesia and Micronesia.

## Tides and Waves

Oceans exhibit three major forms of water movement: currents, tides, and waves. First are the major ocean currents—the global flows of surface water pushed by the major wind belts. Because ocean currents and winds play such a significant role in the global circulation system and its effect on climates, they were discussed in Chapter 5. Second are the tides—the periodic rise and fall of sea level caused by the gravitational forces of the moon, sun, and Earth. Third are waves—oscillatory motions on the ocean surface that are generated mainly by the force of winds. Coastlines (to be discussed in Chapter 21) are primarily a product of the work of waves, especially storm waves. The tides allow the waves to break closer to or farther from shore and produce local coastal currents that move sediments and circulate water in bays and estuaries.

### Tides

The periodic rise and fall of ocean waters in response to gravitational forces are called the **tides.** For the waters to rise in one place as a high tide, they must fall in another part of the ocean, resulting in a low tide there. The gravitational pull of the moon and sun and the force produced by rotation of the Earth–moon system are the major causes of the tides ( Fig. 20.19). Tides are complex phenomena; to explain them, we must first consider the relationship between Earth and the moon.

Because the moon revolves around Earth every 29.5 days (one "moonth," or month), one might assume that the center of its revolution would be at the center of Earth. However, both the moon and Earth actually revolve around a common center of gravity that lies within Earth, always on the side of Earth that faces the moon. As the Earth–moon system rotates around this axis, centrifugal force causes Earth and the moon to tend to fly away from each other. However, this tendency is overcome because the centrifugal force is balanced by the gravitational attraction between Earth and the moon. Centrifugal force causes Earth's ocean to bulge outward on the side opposite the moon. This effect is one of the external forces that raises the ocean surface enough to produce a high tide. The ocean

**FIGURE 20.17**
The major types of coral reefs are evident in the Society Island chain of French Polynesia. (a) Moorea is a rugged young oceanic island with a fringing reef. (b) The older island of Bora Bora is subsiding and now has a barrier reef around it. (c) Tetiaroa Atoll has no surface evidence of its former volcanic core.

TIDES AND WAVES 581

(a) Fringing reef

(b) Barrier reef

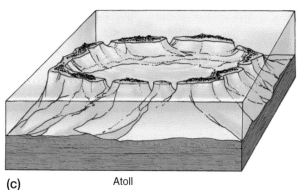
(c) Atoll

**PHYSICAL**
**Geography Now™ ■ ACTIVE FIGURE 20.18**
Watch this Active Figure at http://now.brookscole.com/gabler8.

The theory of coral reef development around oceanic islands was proposed by Charles Darwin. Three successive reef forms develop from island subsidence and coral reef building. (a) First, a fringing reef forms along the shore. (b) Later, as the island erodes and subsides, a barrier reef forms. (c) Further subsidence causes the coral to build upward while the volcanic core of the island is completely submerged below a central lagoon, forming an atoll.
**Explain why the island subsides while the coral grows upward.**

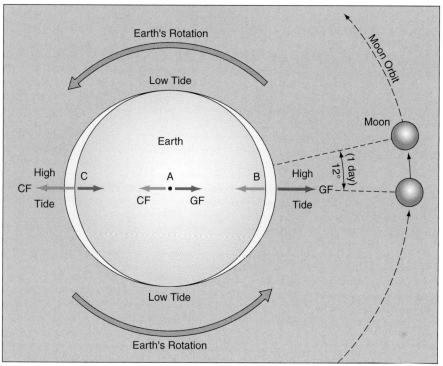

A. Gravitational force (GF) and centrifugal force (CF) are equal. Thus separation between Earth and moon remains constant.

B. Gravitational force exceeds centrifugal force, causing ocean water to be pulled toward moon.

C. Centrifugal force exceeds gravitational force, causing ocean water to be forced outward away from moon.

**PHYSICAL**
**Geography Now™ ■ ACTIVE FIGURE 20.19**
Watch this Active Figure at http://now.brookscole.com/gabler8.

The tides are a response to the moon's gravitational attraction (periodically reinforced or opposed by the sun), which pulls a bulge of water toward it while the centrifugal force of rotation of the Earth–moon system forces an opposing mass of water to be flung outward on the opposite side of Earth. Earth rotates through these two bulges each day. Actually, a "tidal" day is 24 hours and 50 minutes long.
**Explain why a tidal day is slightly longer than an Earth day. How many high tides and low tides are there during each tidal day?**

waters on the side of Earth facing the moon respond to the moon's gravitational force by bulging toward the moon. This bulge of water is directly opposite the bulge produced by centrifugal force, and Earth rotates through two high tides daily, 180° of longitude apart, separated by about 12 hours (half a day, because 180° is half of Earth's circumference). Between these two tidal bulges, the waters recede as they are pulled toward the areas of high tide. Accordingly, there are two low tides midway (90° of longitude) between the high tides.

Earth's 24-hour rotation together with the moon's daily movement of 12° eastward along its monthly 360° orbit around Earth means that theoretically coastlines will experience two high tides and two low tides every 24 hours, 50 minutes (the length of a tidal day). The time between the two high tides, called the **tidal interval,** thus averages 12 hours, 25 minutes. However, this ideal tidal pattern does not occur everywhere, as seen in Figure 20.20. The most common tidal pattern, however, approaches the ideal of two high tides and two lows in a tidal day. This *semidiurnal* tidal regime is characteristic along the Atlantic coast of the United States. In a few seas that have restricted access to the open ocean, such as the Gulf of Mexico, tidal patterns of only one high and one low tide occur during a tidal day. This type of tide, called *diurnal,* is not very common. A third type of tidal pattern consists of two high tides of unequal height and two low tides, one lower than the other. The waters of the Pacific coast of the United States exhibit this *mixed pattern*.

Tidal regimes are extremely complex and variable, even over short distances. We have described the tides as though they were bulges of water through which Earth rotates. Actually, this "wave of equilibrium" theory does not account for many variations in tidal regimes. A somewhat different theory explains the tides as an oscillatory movement. This "stationary-wave theory" describes the tides as a back-and-forth sloshing of water, like the waves produced if a tray of water is tipped first one way and then the other. A circular motion, caused by Earth's rotation, is superimposed on this up-and-down or rocking movement.

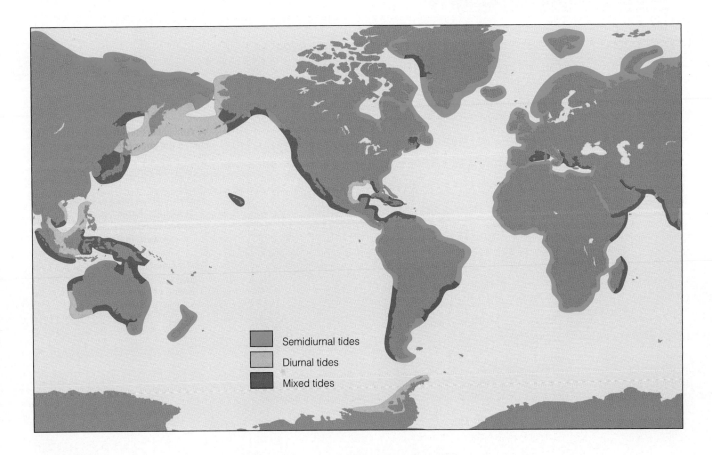

**FIGURE 20.20**
A map of world tidal patterns shows the geographic distribution of diurnal, semidiurnal, and mixed tides.
**What is the tidal pattern on the coastal area nearest where you live or where you like to visit?**

# CHAPTER PREVIEW

**The ever-changing coastal regions of Earth are greatly affected by human activities and should be protected from destructive uses and increasing pollution.**

- Why are the coasts such dynamic environments?
- Why are the coastal zones the most polluted ocean environments?

**Waves are the dominant agent of erosion and deposition in the narrow coastal zones where the land and water meet.**

- What are the major coastal landforms produced by wave erosion?
- What are the major coastal landforms produced by deposition?
- What would be the ultimate result of wave action on shorelines if wave processes continued without interruption?

**Beaches are the most visible evidence of coastal deposition, and they reflect a balance between input and removal of material by waves and currents.**

- Of what types of materials are beaches made?
- How do coastal currents affect the formation of major coastal landforms?

**Barrier islands exist along much of the Atlantic and Gulf coasts of the United States.**

- What are the main physical characteristics of barrier islands?
- What natural processes and environmental changes can affect these islands in ways that are detrimental to human occupation on barrier islands?

**Coasts are dynamic and complex systems, and they are hard to classify because of their great variety and changeable nature.**

- Based on global plate tectonics, what are the two major types of coastlines?
- What are the differences between coastlines of emergence and coastlines of submergence?

---

The most continually changeable environment in the world is the coastal zone. This is a boundary zone where the hydrosphere, lithosphere, biosphere, and atmosphere meet. Coastlines are constantly affected by interactions among Earth's physical spheres. The coastlines of the world offer a tremendous variety of landforms and weather patterns, as well as marine and coastal organisms. The coastal waters of the oceans are not only dynamic from a geomorphic viewpoint but also are biologically dynamic, with a great variety of ecological niches for organisms to occupy.

The world's coastal regions attract more tourism than any other natural environment. Places such as the Mediterranean Riviera, Hawaii, Florida, southern California, and Australia compete for the hundreds of millions of tourists who head for the seashore on their vacations each year. The coastal zones, however, include some of our most polluted waters. They are heavily affected by a multitude of human activities, including rapid urban development, port operations, offshore oil production, tourism, and agricultural runoff. The coasts need protection from these human impacts, and by understanding the natural processes that operate there, we can work toward solving future problems.

Natural hazards also cause problems that require understanding in order to live safely along a coast. At this interface between land and water, people may experience the pleasures of warm sun and gentle sea breezes on many occasions, but they must also face the prospect of dangerous winds and powerful storms, depending on the location, season, and weather conditions. Low-lying coasts are subject to flooding, storm surges, and, in certain places, tsunamis. High-relief coasts are susceptible to landslides. Most coastal settlements deal with many issues concerning tides and waves and the movement of sediment related to these processes. In this chapter, we examine both the dynamic forces that shape our coastal zones and the magnificent landforms produced by these forces.

## Coastal Landforms and Processes

Waves and currents are the major geomorphic processes associated with large bodies of water—oceans, seas, and major lakes. Their effects are felt in a narrow, dynamic zone where land, air, and water meet, called the **shoreline.** Landform changes are continuous both landward and seaward of the shoreline. The position of the shoreline fluctuates with tides, storms, and long-term rises and falls in sea (or lake) levels, as well as with tectonic movements. The **coastal zone** includes the dynamic region on land as well as areas currently submerged under water through which the shoreline boundary fluctuates.

Waves—like streams, glaciers, and wind—erode, move, and deposit materials. By removing, transporting, and depositing material, waves continually work the narrow strip of land with which they are in contact. Although large storm waves (and tsunamis) generate the most effective erosive force, all waves that reach the coastal zone do some work in shaping the land ( ▌Fig. 21.1). Changing tidal levels allow the waves to extend their impact over a wider zone, and global sea-level changes do the same in a much longer time frame.

As waves pound the coast, their general effect is to straighten and smooth the shoreline. Peninsulas or headlands, which extend farther into the water than other parts of the land, are gradually cut back by wave action. Bays and inlets are gradually filled in by coastal deposition ( ▌Fig. 21.2).

▌**FIGURE 21.1**
Large breakers pounding the shore at Point Lobos, California.
**What is the origin of such large breakers?**

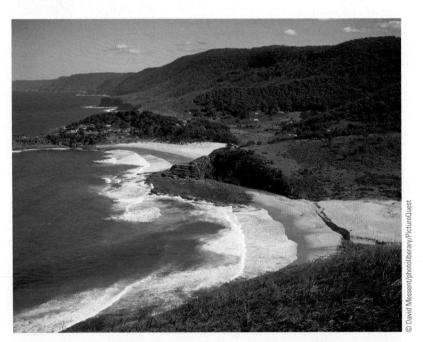

▌**FIGURE 21.2**
Wave action along an irregular coast tends to straighten the shoreline by eroding protrusions and filling inlets with sediment.
**Why does this process occur?**

### The Breaking of Waves

As long as the water is deep, meaning that it has a depth that is more than half the wavelength, waves of oscillation can roll along without disturbing the bottom. In such deep water, the ocean (or lake or sea) floor will remain unaffected. When the water depth is about half the wavelength, however, the waves begin to "feel" the bottom and are slowed

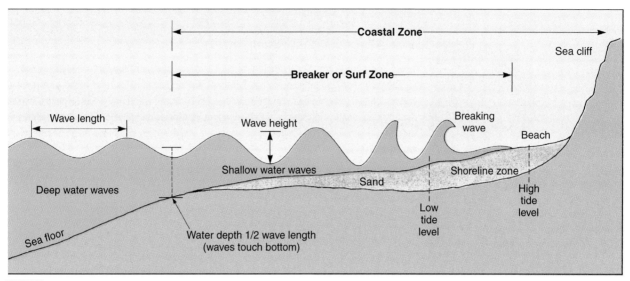

**PHYSICAL Geography Now™ ACTIVE FIGURE 21.3**
Watch this Active Figure at http://now.brookscole.com/gabler8.
Waves begin to "feel" bottom when the water depth becomes half the distance between wave crests. Then the wave velocity and wavelength decrease while the wave height and steepness increase until breaking occurs.
**Why don't the waves break in deeper water?**

down by friction with sand and rock (Fig. 21.3). As these shallow water waves (waves of translation) slow down, their wavelength decreases and both wave height and steepness increase. Eventually wave height reaches a point of instability, and the waves curl and collapse, producing *surf,* or *breakers.* If the waters out from shore are shallow, the surf will break some distance offshore. If, however, water remains deep adjacent to the coast, the surf is forced to break against the land. The tidal range also influences how far from shore the waves will break.

If a wave crashes directly against solid material, the impact can be sufficient to break the material apart, whether it is rock or a built structure. The direct impact of storm waves erodes coastal cliffs and headlands. More commonly, the waves break before striking the land directly, and the water surges forward as foaming **swash,** which picks up sand and carries it onto a beach. Gravity eventually overcomes the waning momentum of the water, which finally reverses the flow direction and drains back down the beach slope as **backwash,** taking some sand with it. The material returned to the water in the backwash is flung back onto the shore during the breaking of the next wave.

Waves are a powerful, continuous, and relentless erosive force. When waves are large, chunks of rock and sand serve as abrasive tools similar to those of a river's bed load. Abrasion is an important factor in wave erosion, just as it is in fluvial erosion. The power of waves, combined with the buoyancy of water, enables large storm waves to carry rocks and even move boulders. During storms, large pieces of rock or coral may be thrown against cliffs. The force of impact has a "cannonball effect" that breaks the rock surface. Wave erosion of the coastal zone is also assisted by solution, as is the case with surface water and groundwater. There is also the power of hydraulic action from the sheer physical force of the pounding water and the explosive effect of the air compressed between breaking waves and cliff faces. The growth of salt crystals from evaporating ocean spray (salt weathering) also detaches mineral grains from cliffs and rocks, helping to push back the land.

## Wave Refraction

**Wave refraction** is the bending of waves as they approach a shore. An important consequence of wave refraction is that wave energy becomes concentrated along certain parts of the shoreline and is greatly reduced in others. To understand how this happens, imagine an irregular shoreline of bays and headlands (Fig. 21.4). Offshore waves are essentially parallel to each other and may approach the shore either directly or obliquely. However, as they approach land, the waves reach shallow water in some places sooner than in others. The shallow waters off headlands will be reached by a wave before it reaches the shallow waters in a bay. As a result, the part of a wave that approaches a headland will be slowed down and forced to break before the part that is approaching the bay. Because one part of the wave is slowed down before another part, the wave is bent, or refracted.

Consequently, wave energy is directed more toward headlands and less toward bays and coastal indentations. This concentration of wave energy at headlands makes wave erosion more intense there. The debris produced by wave erosion at coastal headlands and cliffs joins with sediment brought to the coast by streams and is transported to areas where there is less wave energy, accumulating in the bays between headlands. This effect of wave refraction is the reason why waves tend to

## Career Vision

**Aaron Young, Coastal Mapping Analyst, Computer Modeling of Harbors**
**Bachelor of Arts, Geography, University of Florida**

Courtesy of Aaron Young

When I was in the army as an artillery scout, I would occupy a hilltop for 3–4 days, looking for people and drawing the terrain out. It was an amazing experience to sit in the same spot where you can't see anyone else for 2–3 days. I knew that I wanted to be involved in a career that involved observing aspects of the environment. Many other physical geographers have the same experience. You can look at a landscape and see different things than other people would see.

Today, I work creating virtual three-dimensional (3-D) harbors, computer visualizations for boating and shipping. We use satellite imagery, digital terrain elevation data (DTAT), bathymetry, and other ancillary data. We geoposition the imagery, getting it lined up to a coordinate system, and map data to the outer bounds of the harbor. We also decide what amount of detail should be presented in each area of a harbor. If you're coming into a harbor on a ship, you want a high level of detail in the outer areas but may not need such a high level of detail farther inland.

These three-dimensional models must be very accurate and precise. We include data about buildings and cultural features. Image-to-image change detection is used to identify shoreline movement and new manmade features and to determine areas of potential change. We then create a virtual 3-D environmental model that can simulate the movements of a ship that is very similar to being at the port. Through these computer models, mariners can "sail" into port.

We have also created digital elevations models using NASA imagery of Mars. A great aspect of that project was that, in analyzing the digital terrain, we might have been the first humans to physically see certain terrain features at such a high resolution on the Martian surface.

A good foundation would be to first learn the scientific method. Take an object or an issue or a concept, look at it from every single angle, and try to root out the core truth. What has also been helpful on a daily basis was to learn the physical characteristics of landforms. A solid background in photogrammetry is very helpful. A solid base in core GIS concepts is critical for a continued successful career in this field.

Another concern is the observation that our physical landscape is going through huge anthropogenic transformations. It's somewhat disturbing to be told about, in the third person, the horrible atrocities to our environment; it's quite another to see the destruction firsthand through imagery.

Everything has a connection with everything else. Every theme, whether it is political, social, or environmental, has a geographic component to it. If one is not a physical geographer and has no plans on being one, he or she would still benefit from the physical geographer's perspective when becoming aware that we all live and have an impact on our Earth.

reduce irregularities in shorelines. The concentration of erosion on headlands wears them back, while beach and land-sediment deposition in bays fills them seaward.

## Erosional Coastal Landforms

As waves erode the coastal zone, they create distinctive landforms ( Fig. 21.5; also see Map Interpretation: Active-Margin Coastlines). As is the case with differential weathering and water erosion inland, rock resistance to wave erosion is a major factor in the formation and development of coastal landforms. **Sea cliffs** are created where waves pound directly against steeply sloping land or resistant headlands. The erosive power of waves gradually undercuts the steep shoreline, creating coastal cliffs ( Fig. 21.6a). Where the rocks are well jointed but cohesive, wave erosion may create **sea caves** along lines of weakness ( Fig. 21.6b). **Sea arches** result where two caves meet from each side of a headland ( Fig. 21.6c). When the top of an arch

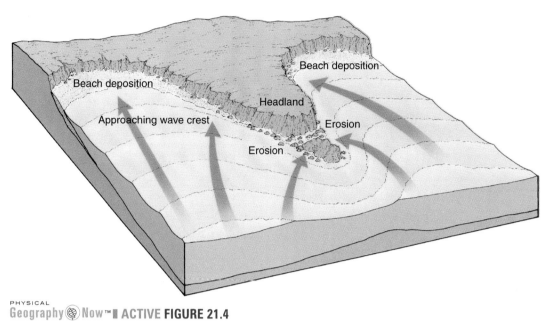

**PHYSICAL Geography Now™ ACTIVE FIGURE 21.4**
Watch this Active Figure at http://now.brookscole.com/gabler8.
Wave refraction causes wave energy to be concentrated on headlands, eroding them back, while in bays, wave deposition causes beaches to grow seaward.
**How will this coastline change over a long period of time?**

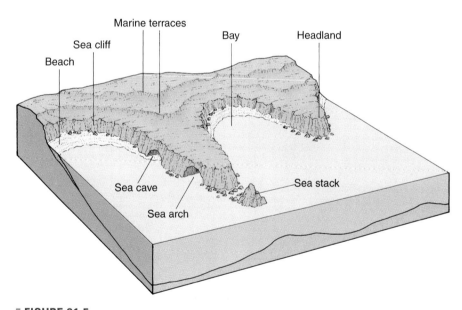

**FIGURE 21.5**
Diagram of the major coastal erosional landforms associated with wave activity.

collapses or a sea cliff retreats and a resistant pillar is left standing, the remaining landform is called a **sea stack** (Fig. 21.6d).

Waves frequently erode notches into a sea cliff. Wave-cut notches are particularly common and well developed where the sea cliffs are made of limestone. Seawater dissolves the limestone just as groundwater does on land, creating notches or lateral indentations at the water level. Notches are also cut into other rock types by abrasion and are an important factor in the erosional retreat of sea cliffs. Rates of coastal erosion are controlled by the interaction between wave energy and rock type. Coastal erosion is greatly accelerated during severe storms.

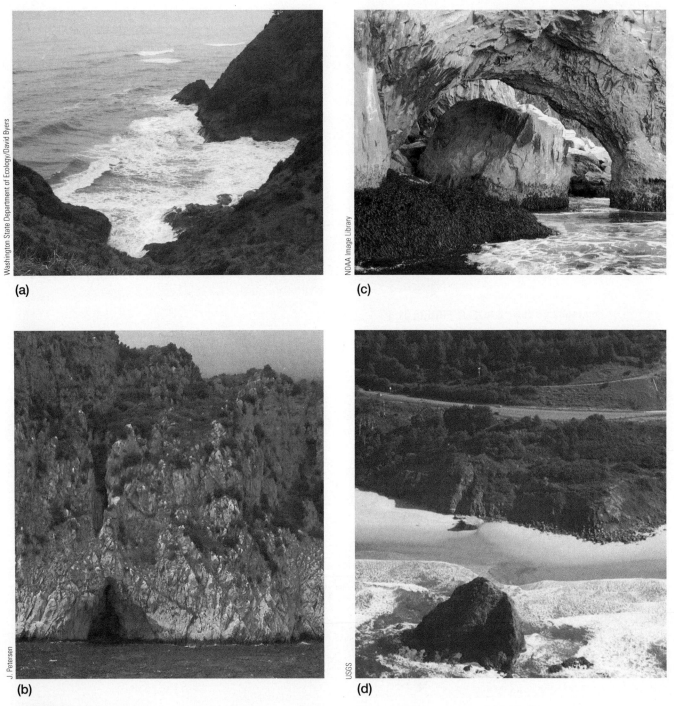

### FIGURE 21.6
Examples of the major coastal landforms shown in Figure 21.5. (a) Rugged sea cliffs along the wave-eroded and uplifted Washington coastline. (b) Sea caves in the steep limestone sea cliffs on Italy's Amalfi Coast on the Mediterranean Sea. (c) Sea arches, such as this one in Alaska, form as sea cliffs on opposite sides of a headland are eroded completely through. (d) A sea stack, such as this one off the Oregon Coast, forms when sea cliffs retreat, leaving a resistant pillar of rock standing above the waves.

The presence of a wave-cut cliff implies erosion and removal of a large mass of material. The cliff is the vertical portion of a notch cut into the land. The horizontal portion of the notch is below water level and consists of a broad abrasion platform or **wave-cut bench** (Fig. 21.7). Rock material eroded from sea cliffs and wave-cut benches comes to rest in the beaches that occupy the coastal indentations. Backwash and coastal currents then transport some

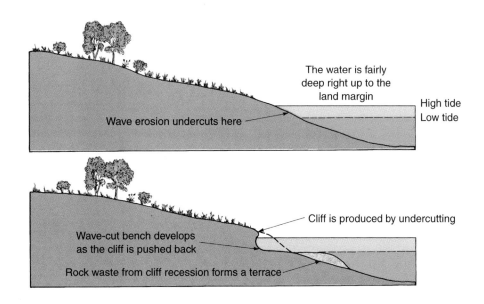

sedimentation, like fluvial deposition, is organized by size and weight with the finest particles being deposited farthest from shore. As these deposits accumulate, a **wave-built terrace** is created just seaward of the wave-cut bench. If tectonic activity uplifts these wave-cut benches and wave-built terraces above sea level, they are called **marine terraces** ( Fig. 21.8). Successive periods of uplift may create a coastal topography that resembles a series of steps. Each step represents a period of time that a terrace was at sea level. Some areas, such as the Palos Verdes peninsula near Los Angeles, may have more than ten marine terraces, each representing an episode of uplift, separated by periods of terrace cutting and building.

**FIGURE 21.7**
(a) Diagram of the origin of sea cliffs, wave-cut benches, and wave-built terraces.
(b) A wave-cut bench beveled across solid bedrock by wave erosion on a central California beach.
**What is the difference between a wave-cut bench and a wave-built terrace?**

# Beaches and Coastal Deposition

**Beaches** are the most visible evidence of coastal deposition. They reflect an equilibrium state in a dynamic coastal system between input of material by swash and removal by backwash. Though sand-sized sediments are the most common beach material, not all beaches are made of sand. Actually, beach deposits may consist of a

of the eroded material seaward. Because turbulence, important in transporting debris, is lower seaward, coarse debris accumulates and is battered into smaller particles close to the land. Finer particles move out into the sea. Thus, marine

## Geography's Environmental Science Perspective
### Beach Building

Miami Beach, Florida, is a classic example of how modern engineering can protect a coastline from the ravages of a menacing sea. Miami Beach is a popular holiday destination for millions of annual visitors, drawn to the magnificent beaches. However, in the past, sand removal by natural wave erosion placed the beaches in jeopardy.

Waves at Miami Beach relentlessly pounded against the sea wall, and by 1970 the beaches had mostly been eroded away. With the beaches gone, shoreline condominiums and hotels were threatened with serious damage from storm waves, and the hotels were half empty as tourists traveled elsewhere.

The U.S. Army Corps of Engineers (the federal agency responsible for the development of inland and coastal waterways) chose Miami Beach to experiment with a new technology for protecting shorelines. Referred to as *beach building*, the technology involves transferring millions of cubic yards of sand from offshore to replenish existing beaches or build new ones where beaches have eroded away. Huge barge-mounted dredges dig sand from the sea bottom near the shore, and the sand is pumped as a slurry through massive movable tubes to be deposited on a beach.

The initial project cost $72 million, but it was so successful that within 2 years Miami Beach again had a sand beach 90 meters (300 ft) wide and 16 km (10 mi) long. With the return of the beach, the number of visitors increased to 21 million annually, triple the number prior to beach building.

Today, beach building has become the accepted way to respond to beach erosion. Nearly $1.7 billion was spent on beach restoration during the last decade of the 20th century.

Visitors and residents of Miami Beach benefit from an intensive program of beach building that imported sand to maintain the beach.

**FIGURE 21.8**
An uplifted marine terrace on the California coast, the former sea cliff is inland, just beyond the highway.
**How does a marine terrace form?**

wide range of sediments, from boulders to large cobbles to fine silt (Fig. 21.9). Granite, basalt, shale, conglomerate, and coral all result in beaches that differ in type, color, and texture.

Where wave energy is high, particles are larger and beaches will be steeper than where only fine material is present and wave energy is low. Because the wave height generated by storms is greater in winter than in summer in the middle latitudes, winter beaches are generally narrower, steeper, and composed of coarser material than are summer beaches. Winter storm waves are more erosive and destructive, and smaller summer waves tend to be more depositional and constructive. On the Pacific coast of the United

According to the Corps of Engineers, these projects are not meant primarily to provide beaches for recreational purposes but to protect buildings on or near the beach. In the Corps's opinion, the true value of beach building can best be measured in savings from storm damage. For example, it has been estimated that beach replenishment along the coast of Miami–Dade County prevented property losses of $24 million during Hurricane Andrew in 1992.

When beaches are rebuilt or widened to protect against storms, continued erosion is inevitable. In deciding how much sand must be pumped in, it is important to determine the beach width necessary to protect shoreline property from storms. This is usually about 30 meters (100 ft). Then an amount of additional "sacrificial" sand is added to produce a beach twice the width of the permanent beach. In about 7 years, the sacrificial sand will be lost to wave erosion. If rebuilding is repeated each time the sacrificial sand is removed, however, the permanent beach will remain in place, and the coastal zone will be protected. An estimated $5.5 billion has already been committed to the continuous rebuilding of existing projects over the next 50 years.

Is the battle with the environment at the beachfront worth the price we are paying? Many people are not so sure, but most environmental scientists respond to the question with a resounding no, for they are concerned with a price that is not measured solely in appropriated dollars. This price is paid in destroyed natural beach environments, reduced offshore water quality, eliminated or displaced species, and repeated damage to food chains for coastal wildlife each time a beach is rebuilt. Clearly, beach building is a mixed blessing.

Seagulls enjoy the swash on this sand-replenished beach in Alameda, California, on the San Francisco Bay.

States, summer beaches are generally temporary accumulations of sand deposited over winter beach materials (Fig. 21.10a). Beach deposits, particularly sand and pebbles, may be removed entirely by destructive winter storm waves (Fig 21.10b). On the Atlantic and Gulf coasts of the United States, the late summer to early fall hurricane season is a time when beach erosion can be most severe.

## Coastal Currents

A **longshore current** is the movement of water in the surf zone, parallel to the coast, that is generated by obliquely approaching waves (Fig. 21.11). The movement of beach sediment carried in longshore currents is known as *longshore drift*. When waves break onto a beach at an angle to the shoreline, the swash pushes material ahead of it toward shore. The waves and consequent swash strike the shore at an angle, but the backwash, responding to gravity, moves directly downslope perpendicular to the shoreline. As a result, beach material is pushed diagonally up the beach by swash but not returned to its original position by backwash. Continuously repeated over time, this zigzag movement causes the mass transport of tons of sediment along the shore and beach.

**Rip currents** (Fig. 21.12) are strong, seaward-moving currents found near the shore and are caused by the channeled return of water and sediments from large waves that have broken against the land. Rip currents (also known as *rip tides*) can be dangerous because the rapidly moving water can pull swimmers out to sea. Rip currents are often visible as streaks of foamy, turbid water flowing perpendicular to the shore and directly offshore.

Beach systems are in equilibrium when income and outgo of sand are in balance. An increase in the size of a beach may be accomplished by building an obstruction to the longshore current. This will prevent sand removal while sand input remains the same. The obstruction is accomplished by constructing a **jetty,** or **groin,** usually a concrete or rock wall built perpendicular to the beach. Of course, any obstruction starves the adjacent

**FIGURE 21.9**
Beaches are the most visible evidence of wave deposition and may be made of any material deposited by waves. (a) Drake's Beach, north of San Francisco, California, in Point Reyes National Seashore is a sandy beach. (b) A boulder beach on Mt. Desert Island, Acadia National Park, Maine (note the person for scale). (c) Fine-grained, light-colored sand beach common on tropical islands with coral reefs, such as Palmyra Atoll in the South Pacific. (d) A black-sand beach on the island of Tahiti, formed from wave deposits of volcanic material.

**FIGURE 21.10**
Because of seasonal variations in wave energy, the differences in a beach from summer and winter can be striking. (a) Waves in summer tend to be mild and deposit sand on the beach. (b) Winter waves, associated with higher winds and storms, remove the sand, leaving boulders and bare bedrock in the beach area.

**FIGURE 21.11**
Waves that approach a beach at an angle tend to push beach sediment downwave, along the shoreline.
**What role does swash play in this process of transporting sand?**

**FIGURE 21.12**
Rip currents move water seaward from a beach. Here, the current can be seen moving offshore, opposite to the wave direction.
**Why are these currents a hazard to swimmers?**

down-current beach area of sand, which now has little or no sediment input to a beach but still has the usual rate of removal ( Fig. 21.13). Beach deposition is often engineered to keep harbors free of sediment or to encourage recreational beach growth. When human activity upsets the natural sand supply, by damming rivers or building jetties to slow beach migration, the beach's temporary nature may be accelerated. In Florida and New Jersey, hundreds of millions of dollars have been spent to replenish the sand beaches. The beaches not only serve obvious recreational needs but also help protect coastal settlements from storm waves.

## Depositional Coastal Landforms

Wave erosion of nonresistant rock materials produces a great deal of sediment and can construct a wide variety of coastal depositional landforms ( Fig. 21.14a). Rivers flowing into the sea bring additional sediment that is added to the coastal system. Material not carried out into deeper waters or deposited on a wave-built terrace is transported along the shore from areas of intense wave action to more protected areas where deposition can take place. **Spits** are depositional landforms, made of sediments from longshore drift, that extend beyond the shore of coastal protrusions, indentations, or islands ( Fig. 21.14b and the chapter-opening photo). Where a shoreline is relatively straight except for a bay, the material carried by longshore currents may be deposited as a spit that continues along the shoreline at the mouth of the bay. If two spits grow together from opposite sides of a bay or if a single spit grows completely across a bay, the result is called a **baymouth bar** ( Fig. 21.14c).

**FIGURE 21.13**
Longshore currents carry beach sediments parallel along the coast. A jetty traps sand on one side and can help keep a harbor entrance open but may also starve beaches on the other side. This jetty is at Ocean City, Maryland.
**What direction does the longshore current move in this aerial photo?**

The formation of a baymouth bar changes the bay into a protected lagoon. The salinity in the lagoon will now vary from that of the open sea, depending on river inflow and evaporation. Salinity levels affect marine organisms living in the lagoon. A river that flows into the waters of a bay protected from wave erosion by a spit or bar will build a delta. Eventually, the entire lagoon may be transformed into a salt marsh or coastal wetland by river deposits on one side trapped behind wave deposits on the other. A **tombolo** is formed if wave deposition builds up enough beach material to connect an offshore island or a sea stack to the mainland ( Fig. 21.14d).

**FIGURE 21.14**
(a) Diagram of the major coastal landforms formed from deposition by waves and longshore currents. (b) A spit is a beach connected at one end to the coast, as illustrated by this example on the Oregon coast. (c) A baymouth bar is connected to the coast at each end of a bay, as this one at Big Sur, California. (d) A tombolo is formed when wave-deposited materials connect a nearby island with the mainland, as shown by Point Sur on the California coast.

# Geography's Physical Science Perspective
## Beach Erosion and Protection

If you go to the same coastal area year after year, you may have noticed that some beaches undergo significant changes through time. At times, sand and other loose materials are eroded away and expanses of beach have simply disappeared. At other times, there may be an enlargement of the beach by *accretion* as the natural processes of wind, waves, and currents deposit new materials. Beaches are among the most dynamic landscapes on Earth because their form and area are altered, sometimes changing at an alarming rate. As discussed in this chapter and Chapter 20, several natural forces affect beaches daily, seasonally, and in response to storm events. The most important processes that cause beach change are *waves, tides/tidal currents,* and *ocean currents*. In most locations, waves are the dominant force affecting beaches because their actions are continual as long as the winds generate them. Tides, with their corresponding currents of ebb and flow, also affect most oceanic coastlines. The offshore flow of ocean currents also can have a significant impact on coasts and the beaches along them.

In addition to these three major processes, there are also periodic and drastic changes caused by hurricanes and other sea storms as well as tsunamis. The coastal *surges* and waves of these natural events can bring about rapid change in a coastline. Yet, even without experiencing a major onshore surge of the size generated by these processes, beaches are continually being affected by a continuous regime of waves, tides, and currents. As the coastal population continues to grow, so have the efforts to protect beachfront property with a variety of methods. Some are discussed in this chapter; others are discussed below.

Protecting the sediments that make a beach can be accomplished using a number of structures in addition to the groins, or jetties, described in this chapter. There are also *seawalls* (massive structures built along the shore to protect the beach against erosion) and *offshore breakwaters* (structures built out in the

Groins on the Atlantic shoreline in Norfolk, Virginia, capture sand to maintain the beach.

Boulders were brought in and offshore breakwaters were built to protect the beach at Sand Island, Oahu, Hawaii.

water but parallel to the shore, which break the energy of the incoming waves). In addition, *rock revetments* (layers of flat stones laid upon the shore to act as armor against erosive wave energy) may also protect the beach sands.

In many parts of the world, annual change in beaches is a normal and natural process that occurs seasonally, with erosive, high-energy waves in winter and deposition during summer. Every beach has a distinctive character that results from the interaction of coastal processes, sediment availability, and the local environment. Measuring and monitoring the shoreline processes that affect a beach help us understand and identify potential problem areas. Being aware of the natural processes and materials involved in a beach system is the best strategy for deciding if artificial beach protection would be beneficial and, if so, what approach would be most effective.

**FIGURE 21.15**
Barrier islands, such as this one near Pamlico Sound on the North Carolina coast, form from wave deposition along coasts with gentle slopes and an adequate sediment supply.
**Name several other barrier islands along the U.S. Atlantic and Gulf coasts.**

On gently sloping coasts or along coastlines where waves are forced to break at considerable distance from shore, churning motion of the water builds submerged **offshore bars** running parallel to the coast. As these bars grow and emerge above sea level, they form barrier beaches and possibly larger **barrier islands** (Fig. 21.15; also see Map Interpretation: Passive-Margin Coastlines). The origin of barrier islands is somewhat controversial, but most coastal experts agree that rising sea level since the Pleistocene has played a major role in their formation. They also agree that barrier islands are temporary landforms that migrate over long periods of time (Fig. 21.16) and may change drastically during severe storms, especially hurricanes. Most barrier islands have three distinct zones: (1) a sandy beach, (2) a sand dune zone, and (3) a shallow lagoon between the barrier island and the mainland (Fig. 21.17). Barrier islands are the most common type of beach-deposited landform on low-relief coastlines. Barrier islands dominate the Atlantic and Gulf coasts from New York to Texas. Some excellent examples of long barrier islands are Fire Island (New York), Cape Hatteras (North Carolina), Cape Canaveral and Miami Beach (Florida), and Padre Island (Texas).

**FIGURE 21.16**
Historic maps show us that barrier islands shift their shapes and positions over time, with major changes coming during storm events. This pair of photos shows hurricane-generated damage to homes built along the shore on a barrier island. Compare the before and after photos.
**What is the solution to problems caused by this natural process?**

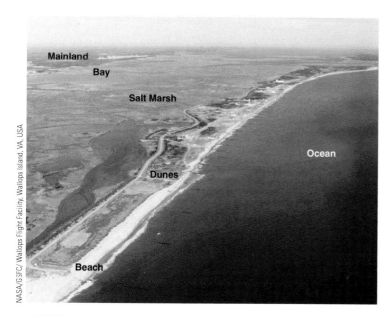

**FIGURE 21.17**
A typical barrier island consists of three major sections: a beach, dunes, and salt marsh.

## Types of Coasts

Coasts are dynamic and complex systems that are influenced by tectonics, global sea-level change, storms, and marine and continental geomorphic processes. Because of this complexity, there is no universally accepted classification system for coasts. However, there are a few coastal classifications that do contribute to an understanding of such complex systems.

On a global scale, coastal classification is based on plate tectonic relationships. Here, there are two major coastal types: passive-margin coasts and active-margin coasts. **Passive-margin coasts** are well represented by the coastal regions of continents along the Atlantic Ocean (Fig 21.18). Most major tectonic activity is in the center of the ocean along the Mid-Atlantic Ridge, whereas the coastal areas are tectonically passive, with little mountain-building or volcanic activity. Generally, passive-margin coasts have low relief with broad coastal plains and wide continental shelves. Most have been modified by marine deposition and some subsidence. The East Coast of the United States is a good example of a passive-margin coast (see Map Interpretation: Passive-Margin Coastlines). There are some exceptions among the younger passive-margin coasts, such as those of the Red Sea and the Gulf of California, that do not show this low-relief pattern.

**Active-margin coasts** are best represented by coastal regions along the Pacific Ocean (Fig. 21.19). Here, most tectonic activity occurs around the ocean margins because the "Pacific Ring of Fire" is formed along active subduction and transform plate boundaries. Active-margin coasts are usually characterized by high relief with narrow coastal plains, narrow continental shelves, earthquake activity, and volcanism. These coasts tend to be erosional and spectacular in nature, with much less time within Earth history for the development of marine or continental depositional features. The West Coast of the United States is an excellent example of an active-margin coast (see Map Interpretation: Active-Margin Coastlines).

On a regional scale, coasts may be classified as coastlines of emergence or coastlines of submergence. **Coastlines of emergence** occur where the water level has been lowered or the land has risen in the coastal zone. In either case, land emerges from an area that was once below sea level. Coastal landforms dominated by wave action, such as marine terraces, sea cliffs, sea stacks, and beaches, are found above the level of the present shoreline. The position above the present shoreline of wave-created landforms serves as evidence of emergence. Coastlines of emergence were probably common during the glacial phases of the Pleistocene, prior to 12,000 years ago, because sea level lowered about 120 meters (400 ft). Features of emergence are best developed along active-margin coasts such as those of California, Oregon, and

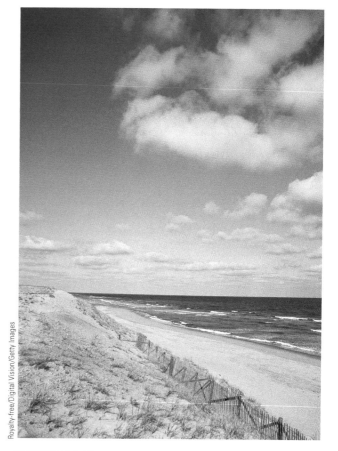

**FIGURE 21.18**
A sandy passive-margin coast on the Atlantic Ocean at Marconi Beach, Cape Cod National Seashore, Massachusetts.

Washington where marine terraces are found as much as 370 meters (1200 ft) above sea level (Fig. 21.20). Less spectacular emergent coastlines are located in areas where isostatic rebound has raised the land after the retreat of the continental ice sheets, such as around the Baltic Sea and Hudson Bay.

As the Pleistocene ice sheets retreated, global sea level rose, creating **coastlines of submergence.** These coasts have many features of the coastal zone that were formerly above sea level but are now submerged. Coastlines of submergence may also occur where tectonic forces have lowered the level of the land, as in San Francisco Bay. Great thickness of river deposits and compaction of alluvial sediments, as along the Louisiana coast, may also cause coastal submergence. The features of a new coastline of submergence are related to the character of coastal lands prior to submergence. Plains, for instance, will produce a far more regular shoreline than will a mountainous region or an area of hills and valleys. When areas of low relief with soft sedimentary rocks are submerged, barrier islands form with shallow bays and lagoons behind them. The classic example of this type of submerging coastline is the Atlantic and Gulf coasts of the United States.

Two special types of submerged coastlines are ria and fjord coasts. **Rias** are created where river valleys are "drowned" by a rise in sea level or a sinking of the coastal area (Fig. 21.21). These irregular coastlines are formed when valleys become narrow bays and the ridges form peninsulas. The Aegean coast of Greece and Turkey is an outstanding example of a ria coastline. *Fjords,* which are drowned glacial valleys, form scenically spectacular shorelines (Fig. 21.22). A fjord shoreline is highly irregular, with deep, steep-sided arms of the sea penetrating far inland in troughs originally deepened by glaciers. Tributary streams cascade down fjord walls that may be several thousand feet high. Fjord coastlines are found in Norway (where the term originated), Chile, New Zealand, Greenland, and Alaska. Canada, however, has more fjords than any other coastal nation. In many fjords, the glaciers have retreated far inland, but some, especially in Greenland and Alaska, have "tidewater" glaciers that calve icebergs into the cold fjord waters.

Some coastlines, such as those formed by coral reefs and river deltas, cannot be classified as either submerging or emerging. Actually, most shorelines show evidence of more than one type of development largely because the land elevation and the level of the seas have changed many times during geologic history. For this reason, features of both submerged and emerged shorelines characterize many coastlines.

Because both continental and marine geomorphic processes shape coastlines, another regional classification system recognizes two types of coasts: primary and secondary coast-

**FIGURE 21.19**
A rugged active-margin coastline at Point Lobos, California, has experienced much tectonic unrest that includes general uplift in addition to great lateral movement along the San Andreas Fault.

**FIGURE 21.20**
An emergent coastline, Cape Blanco, Oregon. The flat surface on which the lighthouse is built is a marine terrace formed by wave erosion and deposition prior to tectonic uplift. The elevation of the marine terrace is about 61 meters (200 ft) above sea level.

lines. **Primary coastlines** are formed mainly by land erosion and deposition processes. Such formation is due to tectonic activity or sea-level changes that cause the shoreline to change its level too rapidly for the marine processes to shape the coast. The major types of primary coastlines (with examples) are drowned river valleys (Delaware and Chesapeake Bays); glacial erosion coasts (southeastern Alaska, British Columbia, and

TYPES OF COASTS    609

### FIGURE 21.21
Submergent coasts like the Chesapeake Bay region are characterized by rias, or drowned river valleys, that developed as sea level rose at the end of the Pleistocene. The white edges on the barrier islands are beaches.

Puget Sound in Washington, and from Maine to Newfoundland); glacial deposition coasts (Cape Cod, Massachusetts [Fig. 21.23], and the north shore of New York's Long Island); river deltas (Mississippi Delta, Louisiana); volcanic coasts (Hawaii); and faulted coasts (California).

**Secondary coastlines** are those formed mainly by marine geomorphic agents, especially waves, and by marine organisms. Such features as sea cliffs, arches, stacks, and sea caves dominate marine erosional coasts (Oregon). Marine depositional coasts have such features as barrier islands, spits, and bars (North Carolina). An example of coasts built by marine organisms is the coral reef (Fig. 21.24). The Florida Keys were built by coral growth. Mangrove trees and salt-marsh grasses also trap sediments to build new land areas in shallow coastal waters.

Regardless of the classification type, coastlines are one of the most spectacular and dynamic environments on Earth (Fig. 21.25). Shorelines are a meeting place of all Earth's spheres—hydrosphere, lithosphere, atmosphere, and biosphere.

### FIGURE 21.22
Fjords, like this one in Greenland, are glacial valleys that were drowned by the sea following the Pleistocene epoch as the glaciers receded and sea levels rose.

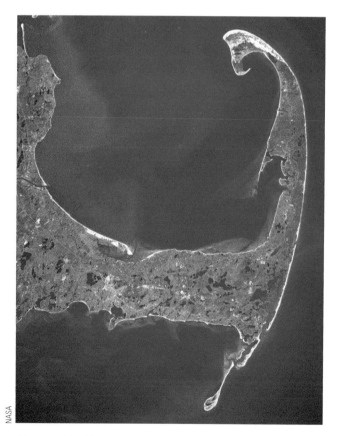

### FIGURE 21.23
Cape Cod is an example of a primary shoreline of glacial deposits being modified by waves and currents.
**How has wave action modified the moraine that originally formed Cape Cod?**

**FIGURE 21.24**
Corals building the coastline at Vieques National Wildlife Refuge, Puerto Rico, provide an example of a secondary shoreline.

**FIGURE 21.25**
The shorelines of the world are complex environments where many elements of the hydrosphere, lithosphere, atmosphere, and biosphere interact.

# Define & Recall

shoreline
coastal zone
swash
backwash
wave refraction
sea cliff
sea cave
sea arch
sea stack
wave-cut bench

wave-built terrace
marine terrace
beach
longshore current
rip current
jetty (groin)
spit
baymouth bar
tombolo
offshore bar

barrier island
passive-margin coast
active-margin coast
coastline of emergence
coastline of submergence
ria
primary coastline
secondary coastline

# Discuss & Review

1. What is the difference between a shoreline and the coastal zone?
2. How do ocean waves change when they enter shallow coastal waters? What is the main factor causing this change in the waves?
3. What is wave refraction? How is wave refraction related to the shape of the coastline?
4. Describe how sea cliffs form. Name three other coastal erosional landforms that form in sea cliff areas.
5. Describe a marine terrace. What does it indicate?
6. What are the differences between longshore currents and rip currents?
7. Explain the relationship between the sorting of coastal marine sediment and the development of beaches. Why do beaches change seasonally?
8. Why do baymouth bars develop? What effect do they have on bays?
9. What is the difference between an offshore bar and a barrier island? Name several examples of barrier islands. Where are most barrier islands located in the United States?
10. What are the major differences between active-margin coastlines and passive-margin coastlines? Why do they look so different?
11. What would be the major changes in the world's coastal zones if sea level were to rise?

## Consider & Respond

1. Explain why the coastal regions of the world are considered to include the most dynamic systems of Earth's environments.
2. What are some of the major ways humans change coastlines? Explain how this human activity may interfere with natural coastal processes.
3. Assume you are the geographer in charge of coastal zone management for a rapidly growing coastal community on the Atlantic or Gulf coast of the United States. Describe the major coastal landforms. What natural hazards would you have to plan for to protect the community?
4. Assume you are a geographer in charge of coastal zone management for a rapidly growing coastal community on the Pacific coast of the United States. Describe the major coastal landforms. What natural hazards would you have to plan for to protect the community?

# Map Interpretation
## Active-Margin Coastlines

### The Map

Point Reyes National Seashore is north of San Francisco on the rugged California coast. Point Reyes consists of resistant bedrock that is being eroded by the forces of the sea. Marine life is abundant, including many marine birds and marine mammals (note Sea Lion Cove below Point Reyes). A hilly area, Punta de Los Reyes, separates Drake's Bay and Point Reyes from Tomales Bay (visible on the aerial photograph as a ribbon of water across the top, but barely showing in the northeast corner of the map).

The oblique aerial photo of Point Reyes was taken from a high-altitude NASA aircraft. Color infrared film makes vegetation appear red. The view is looking northeast. Trending across the upper portion of the photograph is the San Andreas Fault, which forms the linear Tomales Bay. The San Andreas Fault also separates the Pacific plate, on which Point Reyes is located, from the North American plate, which underlies the area east of the fault. Point Reyes is moving northwest along this fault. Emergent coastal features and recent tectonic activity characterize active-margin coastlines.

Point Reyes has a Mediterranean climate, influenced by the cool offshore California current. Here, the sea is not hospitable for swimming because of the uncomfortably cool seawater, large surf, and dangerous rip currents. The coastal location and onshore westerly winds create a truly temperate climate. Although very hot and very cold temperatures are rarely experienced, this is one of the windiest and foggiest coastlines in the United States. It is an area of rugged natural beauty with wind-sculpted trees, grasslands, rocky sea cliffs, and long sandy beaches.

### Interpreting the Map

1. Which area of the coast is most exposed to wave erosion? What features indicate this type of high-energy activity?
2. Which area of the map is under the influence of strong longshore currents? What is the general direction of flow, and what coastal feature would indicate this flow?
3. Looking into the future, what may happen to Drakes Estero? What would Limantour Spit become?
4. Which area of the map appears to have strong wind activity? What would indicate this? What do you think the prevailing wind direction is?
5. Locate examples of the following coastal erosional landforms:
    a. Headland
    b. Sea stack
    c. Sea cliffs
6. Note that the bays and *esteros* (Spanish for estuary) have mud bottoms. Because the creeks and streams are very small in the region, what is the probable source and cause for the movement of the mud?
7. Note the offshore bathymetric contours (blue isolines). Which has a steeper gradient, the Pacific coast (west side) or Drakes Bay (east side)? Why do you think there is such a great difference?

High-altitude oblique aerial photo of Point Reyes.

Opposite:
Point Reyes, California
Scale: 1 : 62,500
Contour interval = 80 ft
U.S. Geological Survey

# Map Interpretation
## Passive-Margin Coastlines

### The Map

Eastport, New York, is located on the south shore of Long Island, 70 miles east of New York City. Long Island is part of the Atlantic Coastal Plain, which extends from Cape Cod, Massachusetts, to Florida. Water bodies such as Delaware Bay, Chesapeake Bay, and Long Island Sound embay much of the Atlantic Coastal Plain in the eastern United States. The south shore of Long Island has low relief, and its coastal location moderates its humid continental climate.

Although this is a coastal region, its recent glacial history has played an important role in the landforms that exist today. Two east–west trending glacial terminal moraines deposited during the Pleistocene glacial advances formed Long Island. Between the coast and the south moraine is a sandy glacial outwash plain that forms the higher elevations at the northern part of the map area. As the glaciers melted, sea level rose, submerging a lowland that now forms Long Island Sound. This water body separates Long Island from the mainland of the Atlantic Coastal Plain.

The Atlantic and Gulf coasts of the United States have nearly 300 barrier islands with a combined length of over 2500 kilometers (1600 mi). New York, especially the coastal zone of Long Island's south shore, has 15 barrier islands with a total length of over 240 kilometers (150 mi). Coastal barrier islands protect the mainland from storm waves, and they form coastal wetlands on their landward side, which are a critical habitat for fish, shellfish, and birds.

### Interpreting the Map

1. What is the long, narrow feature called that forms the straight shoreline labeled Westhampton Beach?
2. What is the highest elevation on the linear coastal feature? What do you think forms the highest portions of this feature?
3. What forms the outer seaward portion of the landform in Question 1?
4. Behind the linear feature is a water body. Is it deep or shallow? Is it a high-energy or low-energy environment? What is a water body such as this called?
5. Is the Eastport coastline erosional or depositional? What features support your answer? Is this a coastline of submergence or emergence?
6. Note Beaverdam Creek in the upper middle of the map area. Does it have a steep or gentle gradient?
7. Based on its gradient, does Beaverdam Creek flow seaward continuously? If not, what would influence its flow?
8. What kind of topographic feature exists as indicated by the map symbols in the area surrounding Oneck?
9. How do you think geomorphic processes will likely modify this map area in the future? Which area will probably be modified the most? Explain.
10. What type of natural hazard is this coastal area of Long Island most susceptible to? Why?

Satellite image of Long Island.

Opposite:
Eastport, New York
Scale: 1 : 24,000
Contour interval = 10 ft
U.S. Geological Survey

**gap** an area within the territory occupied by a plant community when the climax vegetation has been destroyed or damaged by some natural process, such as a hurricane, forest fire, or landslide.

**General Circulation Model (GCM)** complex computer simulations based on the relationships of selected variables within the Earth system that are used in attempts to predict future climates.

**generalist** species that can survive on a wide range of food supplies.

**genetic classification** classification process based on the causes, theory, or origins of phenomena; generally ignoring their statistical, physical, or observable characteristics.

**geocoding** the process and reference system used to tie map locations to field locations using a grid system.

**Geographic Information Systems (GIS)** complex computer programs that combine the features of automated (computer) cartography and database management to produce new data to solve spatial problems.

**geography** study of Earth phenomena; includes an analysis of distributional patterns and interrelationships among these phenomena.

**geomorphology** the study of the origin and development of landforms.

**geostationary orbit** an orbit that synchronizes a satellite's position and speed with Earth rotation so that it continually images the same location.

**geostrophic winds** upper-level winds in which the Coriolis effect and pressure gradient are balanced, resulting in a wind flowing parallel to the isobars.

**geothermal water** groundwater that has been heated by contact with hot rock in the subsurface.

**geyser** a fountain of hot groundwater that spews above the surface.

**giant planets** the four largest planets—Jupiter, Saturn, Uranus, and Neptune.

**gibber** Australian term for an extensive desert plain covered with pebble or cobble-sized rocks.

**glacial outwash** the fluvial deposits derived from glacial meltwater streams.

**glacial plucking** erosive pulling away of rock material underneath a glacier by glacial ice flowing away from a bedrock obstruction.

**glacial trough** a u-shaped valley carved by glacial erosion.

**glacier** a mass of ice that is flowing as a plastic solid.

**glaciofluvial deposit** sorted glacial drift deposited by meltwater.

**glaciolacustrine deposit** sorted glacial drift deposited by meltwater in lakes associated with the margins of glaciers.

**glaze (freezing rain)** translucent coating of ice that develops when rain strikes a freezing surface.

**gleization** soil-forming process of poorly drained areas in cold, wet climates. The resulting soils have a heavy surface layer of humus with a water-saturated clay horizon directly beneath.

**Global Positioning System (GPS)** GPS uses satellites and computers to compute positions anywhere on Earth to within a few centimeters of their true location.

**gnomonic projection** planar projection with greatly distorted land and water areas; valuable for navigation because all great circles on the projection appear as straight lines.

**graben** depressed landform or crustal trough that develops when the crust between two parallel faults is lowered relative to blocks on either side.

**gradational processes** processes that derive their energy indirectly from the sun and directly from Earth gravitation and serve to wear down, fill in, and level off Earth's surface.

**graded stream** stream where slope and channel size provide velocity just sufficient to transport the load supplied by the drainage basin; a theoretical balanced state averaged over a period of many years.

**gradient** a term for slope often used to describe the angle of a streambed.

**granite** a coarse-grained intrusive igneous rock generally associated with continental crust.

**graphic (bar) scale** a ruler-like device placed on maps for making direct measurements in ground distances.

**gravel** a general term for sediment sizes larger than sand.

**gravitation** the attractive force one body has for another. The force increases as the mass of the bodies increases, and the distance between them decreases.

**gravitational water** meteoric water that passes through the soil under the influence of gravitation.

**gravity** the mutual attraction of bodies or particles.

**great circle** any circle formed by a full circumference of the globe; the plane of a great circle passes through the center of the globe.

**greenhouse effect** warming of the atmosphere that occurs because short-wave solar radiation heats the planet's surface, but the loss of long-wave heat radiation is hindered by the release of gases associated with human activity (e.g., CO2).

**Greenwich mean time (GMT)** time at zero degrees longitude used as the base time for Earth's 24 time zones; also called Universal Time or Zulu Time.

**ground moraine** glacial till deposited on Earth's surface beneath a melting glacier.

**ground-inversion fog** *see* radiation fog.

**groundwater (underground water)** all subsurface water, especially in the zone of saturation.

**gullies** trench-like channels eroded by running water.

**guyot** flat-topped seamount thought to be formed by the slow subsidence of a volcanic island.

**gyre** broad circular patterns of major surface ocean currents produced by large subtropical high pressure systems.

**habitat** location within an ecosystem occupied by a particular organism.

**hail** form of precipitation consisting of pellets or balls of ice with a concentric layered structure usually associated with the strong convection of cumulonimbus clouds.

**hamada** desert plains covered with boulders or featuring large expanses of exposed bedrock.

**hanging valley** tributary trough that enters a main glaciated valley at a level high above the valley floor.

**hardpan** dense, compacted, clay-rich layer occasionally found in the subsoil (B horizon) that is an end product of excessive illuviation.

**Haystack hill (hum)** hills that remain after considerable erosion of a karst landscape.

**headward erosion** gullying and valley cutting that extends a stream channel in an upstream direction.

**heat** the total kinetic energy of all the atoms that make up a substance.

**heat energy budget** relationship between solar energy input, storage, and output within the Earth system.

**heat island** mass of warmer air overlying urban areas.

**hemisphere** half of a sphere; for example, the northern or southern half of Earth divided by the equator or the eastern and western half divided by 2 meridians, the 0° and 180° meridians.

**herbivore** an animal that eats only living plant material.

**heterotroph** organism that is incapable of producing its own food and that must survive by consuming other organisms.

**high** See anticyclone.

**highland climates** a general climate classification for regions of high, yet varying, elevations.

**Holocene** the most recent time interval of warm, relatively stable climate that began with the retreat of major glaciers about 10,000 years ago.

**horizon** the visual boundary between Earth and sky.

**horn** pyramid-like peak created where three or more expanding cirques meet at a mountain summit.

**horst** raised landform that develops when the crust between two parallel faults is uplifted relative to blocks on either side.

**hot spot** a mass of hot molten rock material at a fixed location beneath a lithospheric plate.

**human geography** specialization in the systematic study of geography that focuses on the location, distribution, and spatial interaction of human (cultural) phenomena.

**humidity** amount of water vapor in an air mass at a given time.

**humus** organic matter found in the surface soil layers that is in various stages of decomposition as a result of bacterial action.

**hurricane** severe tropical cyclone of great size with nearly concentric isobars. Its torrential rains and high-velocity winds create unusually high seas and extensive coastal flooding; also called willy-willies, tropical cyclones, baguios, and typhoons.

**hydration** attachment of water molecules to molecules of other elements or compounds without chemical change.

**hydraulic action** the physical processes associated with erosion by running water.

**hydrologic cycle** circulation of water within the Earth system, from evaporation to condensation, precipitation, runoff, storage, and re-evaporation back into the atmosphere.

**hydrolysis** union of water with other substances involving chemical change and the formation of new compounds.

**hydrosphere** major Earth subsystem consisting of the waters of Earth, including oceans, ice, freshwater bodies, groundwater, and water within the atmosphere and biomass.

**hygroscopic water** water in the soil that adheres to mineral particles.

**ice age** period of Earth history when much of Earth's surface was covered with massive continental glaciers. The most recent ice age is referred to as the Pleistocene Epoch.

**ice cap** small ice sheet found in highland areas that usually covers all but the highest mountain peaks.

**ice fall** portion of a glacier moving over and down a steep slope, creating a rigid white cascade, criss-crossed with deep crevasses.

**ice sheet** mass of glacial ice thousands of feet thick that is of continental proportions and covers all but the highest points of land. The sheet usually flows from one or more areas of accumulation outward in all directions.

**ice shelf** large flat-topped plate of ice from the Antarctic ice cap, which overlies Antarctic waters and is a source of icebergs.

**iceberg** free-floating mass of glacier broken off by melting, tidal, and wave action.

**Icelandic Low** center of low atmospheric pressure located in the north Atlantic, especially persistent in winter.

**ice-marginal lake** temporary lake formed by the disruption of meltwater drainage by deposition along a glacial margin, usually in the area of an end moraine.

**ice-scoured plain** a broad area of low relief, deeply eroded by a glacier.

**igneous rock** one of the three major rock types: formed from the cooling and solidification of molten Earth material.

**illuviation** deposition of fine soil components in the subsoil (*B* horizon) by gravitational water.

**infiltration** water seeping downward into the soil or other surface materials.

**infiltration capacity** the greatest amount of infiltrated water that a surface material can hold.

**inner core** the innermost portion of the Earth's core that forms the center of the Earth and is considered to be a dense hot solid mass of iron.

**inputs** energy and material entering an Earth system.

**inselberg** remnant residual hill rising above an arid or semiarid plain; produced by stream erosion of a former mountainous area.

**insolation** incoming solar radiation, i.e., energy received from the sun.

**instability** condition of air when it is warmer than the surrounding atmosphere and is buoyant with a tendency to rise; the lapse rate of the surrounding atmosphere is greater than that of *unstable* air.

**interfluve** the land between two tributary streams.

**interglacial** warmer period between glacial advances, during which continental ice sheets and many valley glaciers retreat and disappear or are greatly reduced in size.

**interior drainage pattern** a closed topographic depression that water flows down into, but has no outlet, or outflowing stream.

**intermittent stream** stream that flows part of the time, usually only during, and shortly after, a rainy period.

**International Date Line** line roughly along the 180-degree meridian, where each day begins and ends; it is always a day later west of the line than east of the line.

**Intertropical Convergence Zone (ITCZ)** zone of low pressure and calms along the equator, where air carried by the trade winds from both sides of the equator converges and is forced to rise.

**intrusion** see pluton.

**intrusive rock** igneous rock that was cooled and crystallized beneath the Earth's surface.

**inversion** see temperature inversion.

**isarithm** line on a map that connects all points of the same numerical value, such as isotherms, isobars, and isobaths.

**island arc** curved row of volcanic islands along a deep oceanic trench; found near tectonic plate boundaries where subduction occurs.

**isobar** line drawn on a map to connect all points with the same atmospheric pressure.

**isoline** a line on a map that represents equal values of some numerical measurement such as lines of equal temperature or elevation contours.

**isostasy** theory which holds that Earth's crust *floats* in hydrostatic equilibrium in the denser plastic layer of the mantle.

**isotherm** line drawn on a map to connect all points with the same temperature.

**jet stream** high-velocity upper-air current with speeds of 120–640 kph (75–250 mph).

**jetty** artificial structure extending into a body of water; built to protect a harbor, inlet, or beach by modifying action of waves or currents.

**joints** cracks or systems of cracks revealing lines of weakness in bedrock.

**jungle** dense tangle of trees and vines in areas where sunlight reaches the ground surface (not a true rainforest).

**kame** conical hill composed of sorted glaciofluvial deposits; presumed to have formed in contact with glacial ice when sediment accumulated in ice pits, crevasses, and among jumbles of detached ice blocks.

**kame terraces** landform resulting from accumulation of glaciofluvial sand and gravel along the margin of a glacier occupying a valley in an area of hilly relief.

**karst** unique landforms developed as a result of the dissolving of limestone by groundwater.

**katabatic wind** downslope flow of cold, dense air that has accumulated in a high mountain valley or over an elevated plateau or ice cap.

**Kelvin scale** temperature scale developed by Lord Kelvin, equal to Celsius scale plus 273; no temperature can drop below absolute zero, or 0 degrees Kelvin.

**kettle** depression formed by the melting of an ice block buried in glacial deposits left by a retreating glacier.

**kettle hole** water-filled pit formed by the melting of a remnant ice block left buried in drift after the retreat of a glacier.

**kettle lake** a small lake or pond occupying a kettle hole.

**Köppen system** climate classification based on monthly and annual averages of temperature and precipitation; boundaries between climate classes are designed so that climate types coincide with vegetation regions.

**La Niña** cold sea-surface temperature anomaly in the Equatorial Pacific (opposite of El Niño).

**laccolith** massive igneous intrusion that bows overlying rock layers upwards in a domal fashion as it forces its way toward the surface.

**lahar** rapid form of mass movement involving mudflows from volcanic materials.

**lamination planes** very thin layers in rock.

**land breeze** air flow at night from the land toward the sea, caused by the movement of air from a zone of higher pressure associated with cooler nighttime temperatures over the land.

**Landsat** a family of U.S. satellites that have been returning digital images since the 1970s.

**landslide** mass of Earth material, including all loose debris and often portions of bedrock, moving as a unit rapidly downslope.

**lapse rate** *see* normal lapse rate.

**latent heat of condensation** energy release in the form of heat, as water is converted from the gaseous (vapor) to the liquid state.

**latent heat of evaporation** amount of heat absorbed by water to evaporate from a surface (i.e., 590 calories/gram of water).

**lateral migration** the sideways shift in the position of a stream channel over time.

**lateral moraine** moraine deposited along the side margin of an alpine glacier or lobe of a continental ice sheet.

**laterite** iron, aluminum, and manganese rich layer in the subsoil (*B* horizon) that can be an end product of laterization in the wet-dry tropics (tropical savanna climate).

**laterization** soil-forming process of hot, wet climates. Oxisols, the typical end product of the process, are characterized by the presence of little or no humus, the removal of soluble and most fine soil components, and the heavy accumulation of iron and aluminum compounds.

**latitude** angular distance (distance measured in degrees) north or south of the equator.

**lava** molten Earth material expelled at the surface from volcanoes or fissures. From this material extrusive igneous rock is formed.

**leaching** removal by gravitational water of soluble inorganic soil components from the surface layers of the soil.

**leeward** located on the side facing away from the wind.

**levee** natural raised alluvial bank along margins of a river on a floodplain; artificial levees may be constructed along river banks for flood control.

**liana** woody vine found in tropical forests that roots in the forest floor but uses trees for support as it grows upward toward available sunshine.

**life-support system** interacting and interdependent units (e.g., oxygen cycle, nitrogen cycle) that together provide an environment within which life can exist.

**light year** the distance light travels in one year—6 trillion miles.

**lightning** visible electrical discharge produced within a thunderstorm.

**lithification** the combined processes of compaction and cementation that transform clastic sediments into sedimentary rocks.

**lithosphere** solid crust of Earth that forms one of the major Earth subsystems. In a more technical definition related to tectonic plate theory, the lithosphere consists of Earth's crust and the uppermost rigid zone of the mantle, which is divided into individual plates that move independently on the plastic material of the asthenosphere.

**lithospheric plates** huge slab-like segments of Earth's exterior including the crust and solid part of the upper mantle.

**Little Ice Age** an especially cold interval of time during the early 14th century that had major impacts on civilizations in the Northern Hemisphere.

**llanos** region of characteristic tropical savanna vegetation in Venezuela, located primarily in the plains of the Orinoco River.

**loam soil** soil with a texture in which none of the three soil grades (sand, silt, or clay) predominate over the others.

**loess** wind-deposited silt; usually transported in dust storms and derived from arid or glaciated regions.

**longitude** angular distance (distance measured in degrees) east or west of the prime meridian.

**longitudinal dune** a linear ridge-like sand dune that is oriented parallel to the prevailing wind direction.

**longshore current** current flowing parallel to the shore within the surf zone, produced by waves breaking at an angle to the shore.

**long-wave radiation** electromagnetic radiation emitted by Earth in the form of waves more than 4.0 micrometers in amplitude, which includes heat reradiated by Earth's surface.

**low** *see* cyclone.

**magma** melt or molten Earth material, situated beneath Earth's surface, from which plutonic and intrusive igneous rock is formed.

**magnetic declination** horizontal angle between geographic north and magnetic north.

**mantle** moderately dense, relatively thick (2885 km/1800 mi) middle layer of Earth's interior that separates the crust from the outer core.

**map projection** any presentation of the spherical Earth on a flat surface.

**maquis** sclerophyllous woodland and plant community, similar to North American chaparral; can be found growing throughout the Mediterranean region.

**marine terrace** horizontal land surface, now above sea level, that was either wave-cut or wave-built by shoreline.

**maritime** relating to weather, climate, or atmospheric conditions in coastal or oceanic areas.

**mass** a measure of the total amount of matter in a body.

**mass wasting (mass movement)** movement of surface materials downslope as a result of Earth gravitation.

**mathematical/statistical model** computer-generated representation of an area or Earth system using statistical data.

**matrix** the dominant area of a mosaic (ecosystem supporting a particular plant community) where the major plant in the community is concentrated.

**meander** a broad, sweeping bend in a river or stream.

**medial moraine** central moraine in a large valley glacier; formed when two smaller valley glaciers come together to form the larger glacier and their interior lateral moraines merge.

**mental map** conceptual model of special significance in geography because it consists of spatial information.

**Mercalli Scale, modified** an earthquake intensity scale with Roman numerals from I–XII used to assess spatial variations in the degree of impact that a tremor generates.

**Mercator projection** mathematically produced, conformal map projection showing true compass bearings as straight lines.

**mercury barometer** instrument measuring atmospheric pressure by balancing it against a column of mercury.

**meridian** one half of a great circle on the globe connecting all points of equal longitude; all meridians connect the North and South Poles.

**mesa** flat-topped erosional remnant of a tableland characteristic of arid regions with flat-lying sediments; typically bordered by steep-sided escarpments and may cover large areas.

**mesopause** upper limit of mesophere, separating it from the thermosphere.

**mesosphere** layer of atmosphere above the stratosphere; characterized by temperatures that decrease regularly with altitude.

**mesothermal climates** climate regions or conditions with hot, warm, or mild summers that do not have any months that average below freezing.

**metamorphic rock** one of the three major rock types; formed from other rock within the crust by change induced by heat and pressure.

**meteor** the luminous phenomenon observed when a small piece of solid matter enters Earth's atmosphere and burns up.

**meteorite** any fragment of a meteor that reaches Earth's surface.

**meteorology** study of the patterns and causes associated with short-term changes in the elements of the atmosphere.

**microclimate** climate associated with a small area at or near Earth's surface; the area may range from a few inches to one mile in size.

**microplate terrane** material added to continents as they collide with smaller areas of distinct geology such as volcanic islands or continent fragments.

**microthermal climates** climate regions or conditions with warm or mild summers that have winter months with temperatures averaging below freezing.

**middle-latitude disturbance** convergence of cold polar and warm subtropical air masses over the middle latitudes.

**millibar** unit of measurement for atmospheric pressure; one millibar equals a force of 1000 dynes per square centimeter; 1013.2 millibars is standard sea level pressure.

**mineral** naturally occurring inorganic substance that possesses fairly definite physical characteristics and unique chemical composition.

**mistral** cold downslope wind in southern France (*see* katabatic wind).

**model** a useful simplification of a more complex reality that permits prediction.

**Mohorovicic discontinuity (Moho)** interface between Earth's crust and the denser mantle.

**monadnock** erosional remnant of more resistant rock on a plain of old age; associated with a theoretical cycle of erosion in humid lands.

**monsoon** seasonal wind that reverses direction during the year in response to a reversal of pressure over a large landmass. The classic monsoons of Southeast Asia blow onshore in response to low pressure over Eurasia in summer and offshore in response to high pressure in winter.

**moraine** unsorted glacial drift deposited beneath and along the margins of a glacier.

**mosaic** a plant community and the ecosystem upon which it is based, viewed as a landscape of interlocking parts by ecologists.

**mountain breeze** air flow downslope from mountains toward valleys during the night.

**mud** a mix of rock and soil with ample water.

**mudflow** downslope movement of mud with mixing and tumbling as it moves.

**multispectral scanning** using a number of energy wavelength bands to create images.

**muskeg** poorly drained vegetation-rich marshes or swamps usually overlying permafrost areas of polar climatic regions.

**natural levee** the banks of a stream or river that have been built up by flood deposits.

**natural resource** any element, material, or organism existing in nature that may be useful to humans.

**natural vegetation** vegetation that has been allowed to develop naturally without obvious interference from or modification by humans.

**navigation** the science of location and finding one's way, position, or direction.

**neap tide** tide of less than average range; occurs at first- and third-quarter moon.

**near-infrared (NIR) film** photographic film that makes pictures using near-infrared light that is not visible to the human eye.

**negative feedback** reaction to initial change in a system that counteracts the initial change and leads to dynamic equilibrium in the system.

**nekton** marine organisms that swim freely in the oceans.

**nimbo** a prefix for cloud types that means rain-producing.

**nimbus** term used in cloud description to indicate precipitation; thus cumulonimbus is a cumulus cloud from which rain is falling.

**normal fault** a vertical fault with the footwall up and the hanging wall down, caused by tensional stress.

**normal lapse rate** decrease in temperature with altitude under normal atmospheric conditions; approximately 6.5°C/1000 m (3.6°F/1000 ft).

**North Atlantic Oscillation** oscillating (see-saw) pressure tendencies between the Azores High and the Icelandic Low.

**North pole** maximum north latitude (90° N), at the point marking the axis of rotation.

**northeast trades** *see* trade winds.

**oblate spheroid** Earth's shape—a slightly flattened sphere.

**obliquity cycle** the change in the tilt of the Earth's axis relative to the plane of the ecliptic over a 41,000 year period.

**occluded front** boundary between a rapidly advancing cold air mass and an uplifted warm air mass cut off from Earth's surface; denotes the last stage of a middle-latitude cyclone.

**ocean current** horizontal movement of ocean water, usually in response to major patterns of atmospheric circulation.

**oceanic crust** the denser (av. 3.0 g/cm3) portion of the Earth's crust that underlies the ocean basins.

**oceanic islands** volcanic islands that rise from the deep ocean floor.

**oceanic ridge (midocean ridge)** linear seismic mountain range that interconnects through all the major oceans; it is where new molten crustal material rises through the oceanic crust.

**oceanic trench (trench)** long, narrow depression on the sea floor usually associated with an island arc. Trenches mark the deepest portions of the oceans and are associated with subduction of oceanic crust.

**omnivore** animal that can feed on both plants and other animals.

**ooze** sedimentary deposit containing the microscopic remains of sea organisms.

**open system** system in which energy and/or materials can freely cross its boundaries.

**organic sedimentary rocks** rocks that consist of deposits of abundant organic material such as carbon from plants (coal), or the shells of sea creatures (some limestones).

**orographic precipitation** precipitation resulting from condensation of water vapor in an air mass that is forced to rise over a mountain range or other raised landform.

**orographic thunderstorm** a thunderstorm produced by the orographic uplift mechanism.

**outcrop** bedrock exposed at Earth's surface with no overlying regolith or soil.

**outer core** the upper portion of the Earth's core; considered to be composed of molten iron liquefied by the Earth's internal heat.

**outlet glacier** a valley glacier that flows outward from a larger glacier, such as an ice sheet.

**outputs** energy and material leaving an Earth system.

**outwash** glacial drift deposited beyond an end moraine by glacial meltwater.

**outwash plain** extensive, relatively smooth plain covered with sorted deposits carried forward by the meltwater from an ice sheet.

**overthrust** faulting that pushes the rocks on one side a considerable distance over the opposite side of the fault; the rocks that have overridden others in this manner.

**oxbow lake** crescent-shaped lake or pond formed on a river floodplain in an abandoned meander channel.

**oxidation** chemical union of oxygen with other elements to form new chemical compounds.

**oxide** a mineral group composed of oxygen combining with other Earth elements, especially metallics.

**oxygen-isotope analysis** a dating method used to reconstruct climate history; it is based on the varying evaporation rates of different oxygen isotopes and the changing ratio between the isotopes revealed in foraminifera fossils.

**ozone** gas with a molecule consisting of three atoms of oxygen, ($O_3$); forms a layer in the upper atmosphere that serves to screen out ultraviolet radiation harmful at Earth's surface.

**Pacific high** persistent cell of high atmospheric pressure located in the subtropics of the North Pacific Ocean.

**pahoehoe** a smooth, ropy surface on a lava flow.

**paleogeography** the study of past geographic environments, based on climatic and geologic evidence.

**paleomagnetism** the historic record of changes in Earth's magnetic field.

**Pangaea** ancient continent that consisted of all of today's continental landmasses.

**parabolic sand dune** a hairpin-shaped sand dune, most common in coastal regions.

**parallel** circle on the globe connecting all points of equal latitude.

**parallelism** tendency of Earth's polar axis to remain parallel to itself at all positions in its orbit around the sun.

**parent material** residual (derived from bedrock directly beneath) or transported (by water, wind, or ice) mineral matter from which soil is formed.

**passive-margin coast** coastal region that is far removed from the volcanism and tectonism associated with plate boundaries.

**patch** a gap or area within a matrix (territory occupied by a dominant plant community) where the dominant vegetation is not supported due to natural causes.

**paternoster lakes** chain of lakes connected by a post-glacial stream occupying the trough of a glaciated mountain valley.

**patterned ground (frost polygons)** polygonal shapes formed on the surface in subarctic and tundra climates, formed by repeated freezing and thawing of soils.

**ped** naturally forming soil aggregate or clump with a distinctive shape that characterizes a soil's structure.

**pediment** gently sloping bedrock surface, usually covered with fluvial gravels, located at the base of a stream-eroded mountain range in an arid region.

**pediplain** desert plain of pediments and alluvial fans; the presumed final erosion stage in an arid region.

**peneplain** theoretical plain of extreme old age; the last stage in a cycle of erosion, reached when a landmass has been reduced to near base level by stream erosion in a humid region.

**perched water table** a minor water table that exists above the regional water table.

**perennial stream** a stream with regular and adequate discharge to flow all year.

**periglacial landscape** The environment and landforms that developed or exists today near the margins of a glacier.

**perihelion** position of Earth at closest distance to sun during each Earth revolution.

**permafrost** permanently frozen layer of subsoil and underlying rock found in subarctic and polar climates where the season is too short for summer thaw to penetrate more than a few feet below ground level.

**permeability** characteristic of soil or bedrock that determines the ease with which water moves through Earth material.

**pH scale** scale from 0 to 14 that describes the acidity or alkalinity of a substance and which is based on a measurement of hydrogen ions; pH values below 7 indicate acidic conditions; pH values above 7 indicate alkaline conditions.

**photosynthesis** the process by which carbohydrates (sugars and starches) are manufactured in plant cells; requires carbon dioxide, water, light, and chlorophyll (the green color in plants).

**physical (mechanical) weathering** the various surface processes that break rocks into smaller fragments without causing chemical changes.

**physical geography** specialization in the systematic study of geography that focuses on the location, distribution, and spatial interaction of physical (environmental) phenomena.

**physical model** three-dimensional representation of all or a portion of Earth's surface.

**phytoplankton** tiny plants, algae and bacteria, that float and drift with currents in water bodies.

**pictorial/graphic model** representation of a portion of Earth's surface by means of maps, photographs, graphs, or diagrams.

**piedmont alluvial plain** a plain created by stream deposits at the base of an upland, such as a mountain, a hilly region, or a plateau.

**piedmont glacier** glacier that forms where a valley glacier flows out of its valley and spreads out at the base of a mountain.

**pixel** the smallest area that can be resolved in a digital image. Pixels, short for "picture element," are much like pieces in a mosaic, fitted together in a grid to make an image.

**plane of the ecliptic** plane of Earth's orbit about the sun and the apparent annual path of the sun along the stars.

**planet** any of the nine largest bodies revolving about the sun, or any similar bodies that may orbit other stars.

**plankton** passively drifting or weakly swimming marine organisms, including both phytoplankton (plants) and zooplankton (animals).

**plant community** variety of individual plants living in harmony with each other and the surrounding physical environment.

**plastic solid** any solid material that changes its shape under stress, and retains that deformed shape after the stress is relieved.

**plate convergence** the collision of tectonic plates that are moving toward each other.

**plate divergence** the separation of tectonic plates as they move away from each other.

**plate tectonics** theory that superseded continental drift and is based on the idea that the lithosphere is composed of a number of segments or *plates* that move independently of one another, at varying speeds, over Earth's surface.

**plateau** an extensive, flat-topped landform or region characterized by relatively high elevation, but low relief.

**playa** dry lake bed in a desert basin.

**playa lake** a temporary lake that forms on a playa from runoff after a rainstorm or during a wet season.

**Pleistocene** the name given to the most recent "ice age" or period of Earth history experiencing cycles of continental glaciation; it commenced approximately 2.4 million years ago.

**plucking** *see* quarrying.

**plug dome** a steep-sided, explosive type of volcano with its central vent or vents plugged by the rapid congealing of its highly acidic lava.

**plunge pool** a depression at the base of a waterfall formed by the impact of cascading water.

**pluton (also intrusion)** a mass of igneous rock that cooled beneath Earth's surface.

**plutonic rock** igneous rock formed by the cooling of magma deep within the Earth's crust.

**plutonism** the processes associated with the formation of rocks from magma cooling deep beneath Earth's surface.

**pluvial** rainy time period, usually pertaining to glacial periods when deserts were wetter than at present.

**podzolization** soil-forming process of humid climates with long cold winter seasons. Spodosols, the typical end product of the process, are characterized by the surface accumulation of raw humus, strong acidity, and the leaching or eluviation of soluble bases and iron and aluminum compounds.

**point bar** a deposit of sand or gravel on the inside of a bend in a stream or river.

**polar climates** climate regions that do not have a warm season and are frozen much or all of the year.

**polar easterlies** easterly surface winds that move out from the polar highs toward the subpolar lows.

**polar front** shifting boundary between cold polar air and warm subtropical air, located within the middle latitudes and strongly influenced by the polar jet stream.

**polar highs** high pressure systems located near the poles where air is settling and diverging.

**polar jet stream** high-velocity air current within the upper air westerlies.

**polar** referring to the North or South Polar regions

**pollution** alteration of the physical, chemical, or biological balance of the environment that has adverse effects on the normal functioning of all life forms, including humans.

**porosity** characteristic of soil or bedrock that relates to the amount of pore space between individual peds or soil and rock particles and which determines the water storage capacity of Earth material.

**positive feedback** reaction to initial change in a system that reinforces the initial change and leads to imbalance in the system.

**potential evapotranspiration** hypothetical rate of evapotranspiration if at all times there is a more than adequate amount of soil water for growing plants.

**pothole** a bedrock depression in a streambed drilled by abrasive rocks swirling in a whirlpool.

**prairie** grassland regions of the middle latitudes. Tall-grass prairie varied from 2 to 10 feet in height and was native to areas of moderate rainfall; short-grass prairie of lesser height remains common in subhumid and semi-arid (steppe) environments.

**precession cycle** changes in the time (date) of the year that perihelion occurs; the date is determined on the basis of a major period 21,000 years in length and a secondary period 19,000 years in length.

**precipitation** water in liquid or solid form that falls from the atmosphere and reaches Earth's surface.

**pressure belts** zones of high or low pressure that tend to circle Earth parallel to the equator in a theoretical model of world atmospheric pressure.

**pressure gradient** rate of change of atmospheric pressure horizontally with distance, measured along a line perpendicular to the isobars on a map of pressure distribution.

**prevailing wind** direction from which the wind for a particular location blows during the greatest proportion of the time.

**primary coastline** coast that has primarily developed its current form and shape from land-based processes (fluvial or glacial, erosion or deposition, tectonism), and is only moderately modified by coastal processes.

**primary productivity** *see* autotrophs, and productivity.

**prime meridian (Greenwich meridian)** half of a great circle that connects the North and South Poles and marks zero degrees longitude. By international agreement the meridian passes through the Royal Observatory at Greenwich, England.

**prodelta** the underwater part of a river delta.

**productivity** rate at which new organic material is created at a particular trophic level. Primary productivity through photosynthesis by autotrophs is at the first trophic level; secondary productivity is by heterotrophs at subsequent trophic levels.

**profile** a graph of changes in height over a linear distance, such as a topographic profile.

**punctuated equilibrium** the concept that many changes in Earth systems occur during episodes of intense activity.

**pyroclastic material (tephra)** solid rock material (cinders, ash, and rock fragments) thrown into the air by a volcanic eruption.

**quarrying (glacial plucking)** process whereby active glaciers break away and carry forward weathered and fractured bedrock.

**radiation** emission of waves that transmit energy through space. (see *also* short-wave radiation and long-wave radiation.)

**radiation fog** fog produced by cooling of air in contact with a cold ground surface.

**rain** falling droplets of liquid water.

**rain shadow** dry, leeward side of a mountain range, resulting from the adiabatic warming of descending air.

**recessional moraine** end moraine deposited behind the terminal moraine, marking pauses in the retreat of a valley glacier or ice sheet.

**recumbent fold** a fold in rock that has been completely overturned.

**reg** desert surface of gravel and pebbles with finer materials removed; common to large areas in the Sahara.

**regional geography** specialization in the systematic study of geography that focuses on the location, distribution, and spatial interaction of phenomena organized within arbitrary areas of Earth space designated as regions.

**regions** areas identified by certain characteristics they contain that make them distinctive and separates them from surrounding areas.

**regolith** weathered surface materials that usually cover bedrock.

**rejuvenated stream** a river or stream that has deepened its channel by erosion because of uplift or lowering of base level.

**relative humidity** ratio between the amount of water vapor in air of a given temperature and the maximum amount of water vapor that the air could hold at that temperature, if saturated; usually expressed as a percentage.

**relative location** location of an object in respect to its position relative to some other object or feature.

**relief** a measurement or expression of the difference between the highest and lowest location in a specified area.

**remote sensing devices** variety of techniques by which information about Earth can be gathered from great heights, typically from very high-flying aircraft or spacecraft.

**remote sensing** mechanical collection of information about the environment from a distance, usually from aircraft or spacecraft, e.g., photography, radar, infrared.

**representative fraction (RF) scale** A map scale presented as a fraction or ratio between the size of a unit on the map to the size of the same unit on the ground, as in 1/24,000 or 1:24,000.

**reverse fault** a vertical fault with the hanging wall up and the footwall down, caused by compressional stress.

**revolution (Earth)** motion of Earth along a path, or orbit, around the sun. One complete revolution requires approximately $365\frac{1}{4}$ days and determines an Earth year.

**rhumb line** line of true compass bearing (heading).

**ria** coastline with many narrow bays mainly due to submerged river valleys.

**ribbon falls** high, narrow waterfalls dropping from a hanging glacial valley.

**rift valley** major lowland that forms in a graben or down-faulted crustal block.

**rills** tiny channels formed by running water.

**rime** ice crystals formed along the windward side of tree branches, airplane wings, etc., under conditions of supercooling.

**rip current** strong, narrow surface current flowing away from shore. It is produced by the return flow of water piled up near shore by incoming waves.

**roche moutonnée** bedrock hill subjected to intense glacial abrasion on its upstream side, with some plucking evident on the downstream side.

**rock** a naturally formed aggregate of minerals or of particles of other rocks.

**rock cycle** a circular sequence of the processes that form the different kinds of rocks.

**rock flour** rock fragments finely ground between the base of a glacier and the underlying bedrock surface.

**rockfall** nearly vertical drop of individual rocks or a small rock mass caused by the pull of gravity on steep slopes.

**rockslide** rapid downslope movement of huge masses of bedrock.

**Rossby waves** horizontal undulations in the flow of the upper air winds of the middle and upper latitudes.

**rotation (Earth)** turning of Earth on its polar axis; one complete rotation requires 24 hours and determines one Earth day.

**runoff** flow of water from the land surface, generally in the form of streams and rivers.

**salinas** *see* salt flats.

**salinity** the amount of dissolved solids in seawater.

**salinization** soil-forming process of low-lying areas in desert regions; the resulting soils are characterized by a high concentration of soluble salts as a result of the evaporation of surface water.

**salt flat** a low-relief deposit of saline minerals, typically in desert regions.

**salt weathering (salt wedging)** rock weathering caused by the growth of salt crystals in tiny rock fractures, common in arid and coastal regions.

**saltation** the transportation by running water or wind of particles too large to be carried in suspension; the particles are bounced along on the surface or streambed by repeated lifting and deposition.

**sand sea** an extensive area covered by sand dunes.

**sand** sediment particles ranging in size from about .05 mm to 2.0 mm.

**sandstorm** strong winds blowing sand along the ground surface.

**Santa Ana** very dry foehn wind occurring in southern California. (see *also* foehn wind.)

**satellite** any body which orbits a larger primary body, for example, the moon orbiting Earth.

**saturation (saturated air)** point at which sufficient cooling has occurred so that an air mass contains the maximum amount of water vapor it can hold. Further cooling produces condensation of excess water vapor.

**savanna** tropical vegetation consisting primarily of coarse grasses, often associated with scattered low-growing trees or patches of bare ground.

**scale** ratio between distance as measured on Earth and the same distance as measured on a map, globe, or other representation of Earth.

**sclerophyllous** vegetation type commonly associated with the Mediterranean climate; characterized by tough surfaces, deep roots, and thick, shiny leaves that resist moisture loss.

**sea breeze** air flow by day from the sea toward the land; caused by the movement of air toward a zone of lower pressure associated with higher daytime temperatures over the land.

**sea stack** mass of rock that forms a small near-shore island, isolated from the shoreline by wave erosion.

**seafloor spreading** movement of oceanic crust in opposite directions away from the midocean ridges, associated with the formation of new crust at the ridges and subduction of old crust at ocean margins.

**seamount** submarine volcanic peak rising from the deep ocean floor.

**secondary coastline** coast that has primarily developed its current form from sea-based processes (erosion or deposition by waves and currents, reef-building).

**secondary productivity** the formation of new organic matter by heterotrophs, consumers of other life forms. (see productivity).

**section** a square parcel of land with an area of one square mile as defined by the U.S. Public Lands Survey System.

**sedimentary rock** one of the three major rock types; formed by the accumulation, compaction, and cementation of fragmented Earth materials, organic remains, or chemical precipitates.

**seif** a term for longitudinal dune used in North Africa and the Middle East.

**seismograph** scientific instrument utilized to read the passage of vibratory earthquake and shock waves.

**selva** characteristic tropical rainforest comprised of multistoried, broad-leaf evergreen trees with significant development of lianas and relatively little undergrowth.

**sextant** navigation instrument used to determine latitude by star and sun positions.

**sheet wash** water that runs off over a surface rather than flowing in a channel.

**shield volcano** gentle-sloped volcano formed by the cooling and accumulation of successive fluid lava flows extruded from a central vent or system of vents.

**shoreline** line of intersection between a water body and a landmass.

**short-wave radiation** radiation energy emitted by the sun in the form of waves of less than 4.0 micrometers (1 micrometer equals one ten-thousandth of a centimeter); includes X rays, gamma rays, ultraviolet rays, and visible light waves.

**Siberian high** intensively developed center of high atmospheric pressure located in northern central Asia in winter.

**side-looking airborne radar (SLAR)** a radar system that is used for making maps of terrain features.

**silicate** the largest mineral group, composed of oxygen and silica and forming most of the Earth's crust.

**sill** igneous intrusion that forms a horizontal rock mass after molten material has been forced between rock layers and subsequently cools.

**silt** sediment particles with a grain size between 0.002 mm and 0.05 mm.

**sinkhole** circular surface depression produced by the dissolving of limestone by groundwater.

**slash-and-burn (shifting) cultivation** also called swidden or shifting cultivation; typical subsistence agriculture of primitive societies in the tropical rainforest. Trees are cut, the smaller residue is burned, and crops are planted between the larger trees or stumps before rapid deterioration of the soil forces a move to a new area.

**sleet** form of precipitation produced when raindrops freeze as they fall through a layer of cold air; may also, locally, refer to a mixture of rain and snow.

**slip face** the steep, downwind side of a sand dune.

**slope aspect** direction a mountain slope faces in respect to the sun's rays.

**slump** mass of soil and regolith that slips or collapses downslope with a backward rotation.

**small circle** any circle that is not a full circumference of the globe. The plane of a small circle does not pass through the center of the globe.

**smog** combination of chemical pollutants and particulate matter in the lower atmosphere, typically over urban industrial areas.

**snow line** elevation in mountain regions above which summer melting is insufficient to prevent the accumulation of permanent snow or ice.

**snow** precipitation in the form of ice crystals.

**soil** a dynamic, natural layer on Earth's surface that is a complex mixture of inorganic minerals, organic materials, microorganisms, water, and air.

**soil grade** classification of soil texture by particle size: clay (less than 0.002 mm), silt (0.002–0.05 mm), and sand (0.05–2.0 mm) are soil grades.

**soil horizon** distinct soil layer characteristic of vertical zonation in soils; horizons are distinguished by their general appearances and their specific chemical and physical properties.

**soil ped** *see* ped.

**soil profile** vertical cross section of a soil that displays the various horizons or soil layers that characterize it; used for classification.

**soil survey** a publication of the United States Soil Survey Division of the Natural Resources Conservation Service which includes maps showing the distribution of soil within a given area, usually a county.

**soil taxonomy** the classification and naming of soils.

**soil texture** the distribution of particle sizes in a soil that give it a distinctive "feel."

**solar constant** rate at which insolation is received just outside Earth's atmosphere on a surface at right angles to the incoming radiation.

**solar energy** *see* insolation.

**solar noon** the time of day when the sun angle is at a maximum above the horizon (zenith).

**solar system** the system of the sun and the planets, their satellites, comets, meteoroids, and other objects revolving around the sun.

**solar wind** streams of hot ions (protons and electrons) traveling outward from the sun.

**solid tectonic processes** those processes that distort the solid Earth crust by bending, folding, warping, or fracturing (faulting).

**solifluction** slow movement or flow of water-saturated soil and regolith downslope due to gravity; causes characteristic lobes on slopes in permafrost areas where only the top few feet of Earth material thaws in summer and drainage is poor.

**solstice** one of two times each year when the position of the noon sun is overhead at its farthest distance from the equator; this occurs when the sun is overhead at the Tropic of Cancer (about June 21) and the Tropic of Capricorn (about December 21).

**solution** dissolving in a fluid such as water, or the liquid containing dissolved material.

**solution sinkhole** a topographic depression formed mainly by the solution and removal of soluble rock, such as limestone.

**sonar** a system that uses sound waves for location and mapping underwater.

**source region** nearly homogeneous surface of land or ocean over which an air mass acquires its temperature and humidity characteristics.

**South pole** maximum south latitude (90° S), at the point marking the axis of rotation.

**southeast trades** *see* trade winds.

**Southern Oscillation** the systematic variation in atmospheric pressure between the eastern and western Pacific Ocean.

**spatial distribution** location and extent of an area or areas where a feature exists.

**spatial interaction** process whereby different phenomena are linked or interconnected, and, as a result, impact one another through Earth space.

**spatial pattern** arrangement of a feature as it is distributed through Earth space.

**spatial science** term used when defining geography as the science that examines phenomena as it is located, distributed, and interacting with other phenomena throughout Earth space.

**specific humidity** mass of water vapor present per unit mass of air, expressed as grams per kilogram of moist air.

**speleology** the science of studying caves.

**speleothem** a cave feature made from the deposition of minerals by groundwater.

**spheroidal weathering** a result of various weathering processes that remove the corners of block-like rocks, creating a rounded rock mass.

**spit** beach feature attached to the mainland and built partially across a bay or inlet by the depositional action of longshore currents.

**spring** any surface outflow of groundwater, generally where the water table intersects the ground surface.

**spring tide** tide that occurs about the time of a new or full moon, the tide that rises highest and falls lowest from mean sea level

**squall line** narrow line of rapidly advancing storm clouds, strong winds, and heavy precipitation; usually develops in front of a fast-moving cold front.

**stability** condition of air when it is cooler than the surrounding atmosphere and resists the tendency to rise; the lapse rate of the surrounding atmosphere is less than that of *stable* air.

**stalactite** An icicle-like deposit of dripstone that hangs from a cave ceiling.

**stalagmite** a vertical deposit of dripstone that sticks up from a cave floor.

**star dune** a star-shaped sand dune formed by rising air currents.

**stationary front** frontal system between air masses of nearly equal strength; produces stagnation over one location for an extended period of time.

**steppe climate** characterized by middle-latitude semiarid vegetation, treeless and dominated by short bunch grasses.

**stock** individual deep-seated igneous mass of limited size; it is often associated with other igneous masses known generally as plutons.

**storm** local atmospheric disturbance often associated with rain, hail, snow, sleet, lightning, or strong winds.

**storm surge** rise in sea level due to wind and reduced air pressure during a hurricane or other severe storm.

**storm track** path frequently traveled by a cyclonic storm as it moves in a generally eastward direction from its point of origin.

**strata (stratification)** distinct layers or beds of sedimentary rock.

**strato** signifies a low-level cloud (i.e., from the surface to 2000 meters in elevation).

**stratopause** upper limit of stratosphere, separating it from the mesosphere.

**stratosphere** layer of atmosphere lying above the troposphere and below the mesosphere, characterized by fairly constant temperatures and ozone concentration.

**stratovolcano** *see* composite cone.

**stratus** uniform layer of low sheet-like clouds, frequently grayish in appearance.

**stream** a body of water that is flowing in a channel (a river is a stream, but generally larger); also a specific reference to a small river.

**stream capacity** how much sediment and dissolved load a stream can carry.

**stream competence** the largest particle size that a stream can carry.

**stream discharge** the volume of water flowing past a point in a stream or river channel in a given unit of time.

**stream gradient** the drop of a streambed in a given distance, generally given in feet per mile or meters per kilometer.

**stream hydrograph** a graph of changes in stream flow over time.

**stream load** amount of material transported by a stream at a given instant; includes bed load, suspended load, and dissolved load.

**stream order** a numerical index of the significance of a channel (or system of channels) in a stream channel network.

**stream terraces** former banks of a stream or river that today are above the level of the stream.

**striations** gouges, grooves, and scratches produced in bedrock by rock fragments and boulders imbedded in a glacier.

**strike** the compass direction taken by a rock stratum or fault plane, which is at right angles to their dip.

**strike-slip fault** a fault with horizontal motion, where movement takes place along the strike of the fault.

**structure** the descriptive physical characteristics and arrangement of bedrock, such as folded, faulted, layered, fractured, massive.

**subduction** process associated with plate tectonic theory whereby an oceanic crustal plate is forced downward into the mantle beneath a lighter continental plate when the two converge.

**submarine canyon** steep-sided erosional valley cut into the continental shelf or continental slope.

**subpolar lows** east/west trending belts or cells of low atmospheric pressure located in the upper middle latitudes.

**subsurface horizon** buried soil layer that possesses specific characteristics essential to the identification of soils in the National Resources Conservation Service System.

**subsystem** separate system operating within the boundaries of a larger Earth system.

**subtropical highs** cells of high atmospheric pressure centered over the eastern portions of the oceans in the vicinity of 30°N and 30°S latitude; source of the westerlies poleward and the trades equatorward.

**subtropical jet stream** high-velocity air current flowing above the sinking air of the subtropical high pressure cells; most prominent in the winter season.

**succession** progression of natural vegetation from one plant community to the next until a final stage of equilibrium has been reached with the natural environment.

**sunspots** temporary, dark, cooler, spots on the solar surface. Their numbers follow an approximate 11-year cycle.

**surface of discontinuity** three-dimensional surface with length, width, and height separating two different air masses; also referred to as a *front*.

**surge (glacial)** sudden shift downslope of glacial ice, possibly caused by a reduction of basal friction with underlying bedrock.

**suspended load** solid particles, held in suspension, that are transported by flowing water.

**swallow hole** a depression where a river or stream disappears, flowing downward into the groundwater system.

**swash** onshore flow of water by waves.

**swell** regular longer-period sea wave traveling a significant distance from the area where it was generated by the wind.

**syncline** trough or downfold in a wave of crustal folding.

**system** group of interacting and interdependent units that together form an organized whole.

**taiga** term used to describe the northern coniferous forest of subarctic regions on the Eurasian landmass.

**taku** cold downslope wind in Alaska. (*See also* katabatic wind.)

**talus (talus cone)** rock debris in a cone-shaped deposit at the base of a steep slope or escarpment; usually a result of frost wedging and individual rockfalls with debris accumulating at the angle of rest.

**tarn** mountain lake in a glacial cirque.

**tectonic processes** processes that derive their energy from within Earth's interior and serve to create landforms by elevating, disrupting, and roughening Earth's surface.

**temperature** degree of heat or cold and its measurement.

**temperature gradient** rate of change of temperature with distance in any direction from a given point; refers to rate of change horizontally; a vertical temperature gradient is referred to as the *lapse rate*.

**temperature inversion** reverse of the normal pattern of vertical distribution of air temperature; in the case of inversion, temperature *increases* rather than decreases with increasing altitude.

**tension** stress resulting from divergent forces, pulling apart.

**tephra** *see* pyroclastic material.

**terminal moraine** end moraine that marks the farthest advance of an alpine glacier or ice sheet.

**terminus (snout)** the lower end of a glacier.

**terra rossa** characteristic calcium-rich (developed over limestone bedrock) red-brown soils of the climate regions surrounding the Mediterranean Sea.

**terrestrial planets** the four closest planets to the sun—Mercury, Venus, Earth, and Mars.

**thematic map** a map designed to present information or data about a specific theme, as in a population distribution map, a map of climate or vegetation.

**thematic mapper (TM)** a family of imaging systems that return images of Earth from Landsat satellites.

**thermal infrared (TIR) scanning** images made with scanning equipment that produces an image of heat differences.

**thermocline** vertical zone of ocean water where there is a sharp change in temperature with depth.

**thermosphere** highest layer of atmosphere extending from the mesopause to outer space.

**Thornthwaite system** climate classification based on moisture availability and of greatest use at the local level; climate types are distinguished by examining and comparing potential and actual evapotranspiration.

**threshold** condition within a system that causes dramatic and often irreversible change for long periods of time to all variables in the system.

**thrust fault** a reverse fault where compression has pushed the hanging wall on one side of the fault over the rocks on the other side.

**thunder** sound produced by the rapidly expanding, heated air along the channel of a lightning discharge.

**thunderstorm** intense convectional storm characterized by thunder and lightning, short in duration and often accompanied by heavy rain, hail, and strong winds.

**tidal current** onshore or offshore flow of water that results from tidal fluctuation.

**tidal interval** the time between successive high tides, or between successive low tides.

**tidal range** elevation distance between water levels at high tide and low tide.

**tide** periodic rise and fall of sea level in response to the gravitational interaction of the moon, sun, and Earth.

**till plain** a broad area of low relief covered by glacial deposits.

**till** unsorted glacial drift, characterized by variation in size of deposit from clay particles to boulders.

**tilted fault block** a crustal block that has been uplifted on one side and down-dropped on the other by faulting.

**tolerance** ability of a species to survive under specific environmental conditions.

**tombolo** wave depositional beach feature connecting an island to the mainland.

**topographic contour line** line on a map connecting points that are the same elevation above mean sea level.

**tornado** small, intense, funnel-shaped cyclonic storm of very low pressure, violent updrafts, and converging winds of enormous velocity.

**trace** less than a measurable amount of rain or snow (i.e., less than 1 mm or 0.01 in.

**trade winds** consistent surface winds blowing in low latitudes from the subtropical highs toward the intertropical convergence zone; labeled northeast trades in the Northern Hemisphere and southeast trades in the Southern Hemisphere.

**transform movement** horizontal sliding of tectonic plates, alongside each other.

**transpiration** transfer of moisture from living plants to the atmosphere by the emission of water vapor, primarily from leaf pores.

**transportation** movement of Earth materials from one site to another as a result of the transporting power of water, wind, or glacial ice.

**transverse dune** a linear ridge-like sand dune that is oriented at right angles to the prevailing wind direction.

**transverse stream** a stream or river that flows across the general orientation or "grain" of the topography, such as mountains or ridges.

**travertine** calcium carbonate (limestone) deposits resulting from the evaporation in caves or caverns and near surface openings of groundwater saturated with lime.

**tree line** elevation in mountain regions above which cold temperatures and wind stress prohibit tree growth.

**tributary** a stream or river that flows into another stream or river.

**trophic level** number of feeding steps that a given organism is removed from the autotrophs (e.g., green plant—first level, herbivore—second level, carnivore—third level, etc.).

**trophic structure** organization of an ecosystem based on the feeding patterns of the organisms that comprise the ecosystem.

**Tropic of Cancer** parallel of latitude at $23\frac{1}{2}°N$; the northern limit to the migration of the sun's vertical rays throughout the year.

**Tropic of Capricorn** parallel of latitude at $23\frac{1}{2}°S$; the southern limit to the migration of the sun's vertical rays throughout the year.

**tropical climates** climate regions that are warm all year.

**tropical easterlies** winds that blow from the east in tropical regions.

**tropical** region on Earth lying between the Tropic of Cancer ($23\frac{1}{2}°N$ latitude), and the Tropic of Capricorn ($23\frac{1}{2}°S$ latitude).

**tropopause** boundary between the troposphere and stratosphere.

**troposphere** lowest layer of the atmosphere, exhibiting a steady decrease in temperature with increasing altitude and containing virtually all atmospheric dust and water vapor.

**trough** elongated area or "belt" of low atmospheric pressure; also glacial trough, a U-shaped valley carved by a glacier.

**trunk stream** the main or largest channel in a stream or river system.

**tsunami** ocean wave produced by submarine earthquake, volcanic eruption, or landslide; not noticeable in deep ocean waters, but building to dangerous heights in shallow waters.

**tundra** high latitude or high altitude environments or climate regions that are not able to support tree growth because the growing season is too cold or too short.

**tundra climate** characterized by treeless vegetation of polar regions and very high mountains, consisting of mosses, lichens, and low-growing shrubs and flowering plants.

**turbidity current** submarine flow of sediment-laden water.

**turbulence** chaotic, mixing, churning flow, typically of air or water.

**typhoon** a tropical cyclone found in the western Pacific, the same as a hurricane.

**U.S. Public Lands Survey System** a method for locating and dividing land, used in much of the Midwest and western United States. This system divides land into six by six mile square *townships* consisting of thirty-six *sections* of land (each one square mile). Sections can also be subdivided into halves, quarter sections, and quarter-quarter-sections.

**unconformity** an interruption in the sequence of deposition mainly due to erosion.

**uniformitarianism** widely accepted theory that Earth's landforms have developed over exceedingly long periods of time as a result of processes that may be observed in the present landscape.

**uplift mechanisms** methods of lifting surface air aloft, they are: orographic, frontal, convergence (cyclonic), and convectional.

**upper air westerlies** system of westerly winds in the upper atmosphere, flowing in latitudes poleward of $20°$.

**upwelling** upward movement of colder, nutrient-rich, subsurface ocean water, replacing surface water that is pushed away from shore by winds.

**urban heat island** *see* heat island.

**valley breeze** air flow upslope from the valleys toward the mountains during the day.

**valley glacier** *see* alpine glacier.

**valley sink (uvala)** a karst depression where a stream flows from the surface into the underground.

**valley train** outwash deposit from glacial meltwater, resembling an alluvial fan confined by valley walls.

**variable** one of a set of objects and/or characteristics of objects, which are interrelated in such a way that they function together as a system.

**varve** a pairing of organic-rich summer sediments and organic poor winter sediments found in exposed lake beds; because each pair represents one year of time, counting varves is useful as a dating technique for recent Earth history.

**veering wind shift** the change in wind direction clockwise around the compass; e.g., east to southeast to south, to southwest, to west, and northwest.

**ventifact** wind-fashioned rock produced by wind abrasion (sandblasting).

**verbal scale** stating the scale of a map using words such as "one inch represents one mile."

**vertical exaggeration** a technique that stretches the height representation of terrain in order to emphasize topographic detail.

**vertical rays** sun's rays that strike Earth's surface at a 90 degree angle.

**visualization** a wide array of computer techniques used to vividly illustrate a place or concept, or the illustration produced by one of these techniques.

**volcanic ash** small fragments of lava (sand to dust sized) thrown into the atmosphere by a volcano.

**volcanic neck** the throat or conduit forming the passageway for molten rock in a volcano.

**volcanism** the upward movement of molten material (magma) and its cooling above Earth's surface.

**v-shaped valley** the typical shape of a stream valley where the gradient is steep.

**warm front** leading edge of a relatively warmer, less dense air mass advancing upon a cooler, denser air mass.

**warping** broad and general uplift or settling of Earth's crust with little or no local distortion.

**wash (arroyo, wadi, barranca)** generally steep-walled channel of an ephemeral stream in an arid region; the streambed is characteristically choked with coarse alluvium.

**water budget** relationship between evaporation, condensation, and storage of water within the Earth system.

**water mining** taking more groundwater out of an aquifer through pumping than is being replaced by natural processes in the same period of time.

**water table** upper limit of the zone of saturation below which all pore spaces are filled with water.

**water vapor** water in its gaseous form.

**wave refraction** bending of waves as they approach a shore, aligning themselves with the bottom contours of the surf zone.

**wave-cut bench** gently sloping surface produced by wave erosion at the base of a sea cliff.

**weather** atmospheric conditions, at a given time, in a specific location.

**weathering** physical (mechanical) fragmentation and chemical decomposition of rocks and minerals in Earth's crust.

**westerlies** surface winds flowing from the polar portions of the subtropical highs, carrying fronts, storms, and variable weather conditions from west to east through the middle latitudes.

**wet adiabatic lapse rate** rate at which a rising mass of air is cooled by expansion when condensation is taking place. The rate varies but averages 5°C/1000 m (3.2°F/1000 ft).

**white frost** a heavy coating of white crystalline frost.

**wind** air in motion from areas of higher pressure to areas of lower pressure; movement is generally horizontal, relative to the ground surface.

**windward** location on the side that faces toward the wind and is therefore exposed or unprotected; usually refers to mountain and island locations.

**xerophytic** vegetation type that has genetically evolved to withstand the extended periods of drought common to arid regions.

**yardang** bedrock features that have been shaped by wind erosion.

**yazoo stream** a stream tributary that flows parallel to the main stream for a considerable distance before joining it.

**zone of ablation** lower portion of a glacier, below the firn line, where melting, evaporation, and sublimation exceed net snow accumulation.

**zone of accumulation** subsoil or *B* horizon of a soil, characterized by deposition or illuviation of soil components by gravitational water; also the upper portion of a glacier, above the firn line, where net snow accumulation exceeds the melting, evaporation, and sublimation of snowfall.

**zone of aeration** upper groundwater zone above the water table where pore spaces may be alternately filled with air or water.

**zone of depletion** top layer, or *A* horizon, of a soil, characterized by the removal of soluble and insoluble soil components through leaching and eluviation by gravitational water.

**zone of saturation** zone immediately below the water table, where all pore spaces in soil and rock are filled with groundwater.

**zone of transition** an area of gradual change from one region to another.

**zooplankton** tiny animals that float and drift with currents in water bodies.

# Index

## A

Aa (viscous lava), 397, 397f
Abiotic ecosystem components, 294
Abrasion, 482
  eolian, 515–516
  glacial, 543
Absolute humidity, 149
Absolute location, 10
Absolute zero, 100
Abyssal plains, 577
Accretion, 384
Acidity, soil, 338
Acid precipitation, 17
Acid rain, 145, 434
Active layer, of soil, 443
Active-margin coasts, 607–608
Actual ET. *See* Actual evapotranspiration
Actual evapotranspiration, 208
Actual temperature, 276
Adiabatic cooling, 159
Adiabatic heating, 159
Adiabatic lapse rate, 160
Advection, 98
Advection fog, 154–155, 267
Aeolus, 514
Aerial photography, 57
Afforestation, 310
African plate, 383
*The African Queen*, 238
African Sahel, 251
Agassiz, Louis, 217
Age of dinosaurs, 387–388
Aggradation, 395, 417, 423, 424, 482
Agriculture
  dangers of steppe climates for, 254
  in highland climates, 288
  in humid continental climates, 274–275
  in humid continental, hot summer climates, 270
  in humid subtropical climates, 263, 264
  in marine west coast climates, 266–267, 268
  in Mediterranean climates, 260, 261, 262
  secondary succession associated with, 304f
  slash-and-burn, 242, 348
  in subarctic climates, 279
  in tropical mountains, 289f
  in tropical rainforest climates, 242
  in tropical savanna regions, 245–246
Air
  effect of horizontal movement on, 101
  flow, cyclones and, 186–187
  soil, 334–335
  stable, 160
  temperature, 99–110
  unstable, 160
Air masses, 176–179
  "aggressor," 179
  classification of, 176
  Continental Arctic, 177
  Continental Polar, 177
  Continental Tropical, 179
  defined, 176
  letter code identification of, 176
  Maritime Polar, 177–178
  Maritime Tropical, 178–179
  modification and stability of, 176
  North American, 176
  source regions of, 176
  types of, 177t
Air pollution, 18–19
Air pressure
  basic systems of, 117–118
  belts, global, 126–127, 129
  decrease of, with increasing altitude, 286
  distribution, mapping, 118
  horizontal variations in, 117
  vertical variations in, 117
  *See also* Atmospheric pressure
Alaska Tsunami Warning Center (ATWC), 586
Albedo, 101
Aleutian Low, 129
Alfisols, 355
Alkalinity, soil, 338
Allen's Rule, 307
Alluvial fans, 510–512, 511
Alluvial plains, 489
Alluvium, 346, 484
Alpine glaciers, 538
  depositional landforms produced by, 546–548
  erosional landforms produced by, 544–546
  features of, 539–548
Altithermal, 219, 226
Altitude
  effects on human body of, 286
  temperature and, 107
Altocumulus clouds, 158
Altostratus clouds, 158
*Alvin* (submersible), 327, 570, 571
Amazon rainforest, 241. *See also* Tropical rainforest climates
American Meteorological Society (AMS), 132
Analemma, 83
Anderson, Roxie, 317
Andesite, 366
Andisols, 351
Angle of inclination, 77
Angle of repose, 437
Angular velocity, 76
Animals
  arboreal, of tropical rainforest, 316
  contributing to rock disintegration, 427
  in desert climates, 251, 306
  distinctions between rainforest and monsoon climates, 243
  ecosystems and, 15
  in humid subtropical climates, 264
  response to seasonal changes of, 84
  as soil movers, 443
  in steppe climates, 254
  in subarctic climates, 279
  in tropical rainforest climates, 240
  of tundra, 325
Annual lag of temperature, 109–110
Annual march of temperature, 109–110
Annual temperature range, 238
Antarctica
  dryness of, 536
  glaciers in, 537
  ice sheet in, 549
  ozone depletion over, 284
  strategic and scientific importance of, 284
  temperature at, 282
Antarctic Circle, 82
  pattern of day and night in, 279
Antarctic Circumpolar Drift, 138
Antarctic zone, 84
Antecedent streams, 477
Anticlines, 407, 417
Anticyclones, 117, 129
  cyclones, and winds, 122
  general movement of, 182
  nature, size, and appearance on maps, 181–182
  polar, 278, 282
  weather resulting from influence of, 188
  *See also* High pressure systems
Aphelion, 77
*Apollo* spacecraft, 74–75
Apparent temperature, 276, 277
Aqualfs, 355
Aquicludes, 451–452
Aquifers, 451, 452f
  permeable, 454–455
Arctic Circle, 81
  pattern of day and night in, 279
Arctic Ocean
  depth of, 568
  size of, 568
Arctic zone, 84
Areas, 42
  locating, 38
  on maps, 50f
Arête, 545
*Argo-Jason* (submersible), 571
Argon, 88–89
Arid climate regions, 246–254, 247f, 248t
  distinctive scenery of, 503
  water as gradational agent in, 506–513
  wind as gradational agent in, 514
Arid climates, 211, 214–215
Aridisols, 351, 354
Arroyos, 507

643

Artesian wells, 454–455
Artificial recharge, 453
Asteroids, 69
Asthenosphere, 372
Atlantic Ocean
    "chemosynthetic" vent communities in, 327
    coastal regions of continents along, 607
    depth of, 568
    floor formation of, 380
    midocean ridges in, 378
    runoff flows to, 473
    size of, 568
Atmosphere, 13, 16, 87
    changes in Earth's, 222–225
    characteristics of, 88–99
    chemical composition, system of layering by, 94–95
    colors of, 97
    composition of, 88–89
    effects of, on solar radiation, 95
    elements of, as ingredients of weather and climate, 111
    environmental issues related to, 89–91
    friction within, 88
    functional characteristics, system of layering, 94
    gases in, 88–89, 89t, 91
    general circulation of Earth's, 130f
    heat energy budget in, 99
    heating, 97–98
    of Mars, 74
    ozone layer of, 17
    of planets, 72
    solid water in, 89
    sources of moisture in, 151–152
    temperature characteristics, system of layering by, 93–94
    of Venus, 74
    vertical layers of, 92–95
    water in, 147–151
Atmospheric disturbances, 180–193, 195–300. *See also specific types of disturbances*
Atmospheric pressure, 115–116
    belts, seasonal variations in pattern of global, 127, 129
    effect on weather and climate of, 115
Atmospheric pressure *(continued)*
    global pattern of, 127
    variations in, 117
    *See also* Air pressure
Atmospheric scattering, 97
Atolls, 579–580
Atomic mass, 220
Atomic number, 220
Atomic weight, 220
Atoms, 220
Attrition, 483
Aurora Australis, 70
Aurora Borealis, 70
Auroras, 70
Austral spring, 92
Austral winter, 92
Autotrophs, 295
Avalanches, 438, 440
Azimuthal map, 43, 44f
Azimuth system, 49
Azores High, 129, 141

## B
Backing wind shift, 186
Backwash, 595, 598
Bacteria, in soil, 342
Badlands, 507
Bagyos, 199

Bajada, 511
Balance of nature, 24
Bar, 484
Barchans, 521
Barometric pressure, 115–116
Barrancas, 507
Barrier islands, 606, 609
Barrier reefs, 579
Bar scale, on maps, 46
Basal slip, 541–542
Basalt, 365, 366, 373, 376
    at undersea ridges, 378
Basalt plateaus, 397
Base flow, 471
Base level, 475
    changes and tectonism, 492
Base lines, 38
Batholiths, 402–403
Bathymetric charts, 51
Bathymetric contours, 49
Bathymetry, 571, 596
Bathyscaphe, 577
Baymouth bars, 603, 604f
Bays, 568
Beach building, 600–601
Beaches, 599–601, 607
    engineered deposition of, 603
    erosion of, 605
    protection of, 605
Bearing system, 49
Bedding planes, 368
Bed load, 484
Bedrock, 363
    collapse, sinkhole creation due to, 458
    obstructions subjected to glacial abrasion, 544
    soluble, 459
    structure, 404
Benguela Current, 138
Benioff, Hugo, 383
Benioff zones, 383
Benthos, 325, 327
Bergeron Process, 161–162
Bergeron, Tor, 162
Bergmann's Rule, 307
Bergschrund, 539
Bermuda High, 129
Biogeographers, 6, 293, 302
Biogeography, 6, 293
    theory of island, 302–303
Biomass, 297
    rainforest, 316
Biomes, 306
    forest, 312, 316–321
    grassland, 321–323
Biosphere, 13, 16
Biotic associations, 294
Bird's-foot delta, 490
Bjerknes, Jacob, 180
Black smokers, 576
Blizzards, 189–190
Block diagrams, 52
Blowout, 520
Bogart, Humphrey, 238
Bolsons, 513
Bora, 124
Boralfs, 355
Borates, 368
Boreal forests, 278, 320
Boundaries, transitional nature of, 8–9
Braided channels, 485, 486f
    in deserts, 505
Brazil Current, 138
Breakers, 595

Breccia, 367
Broad-leaf deciduous forest, 318–319
Broad-leaf evergreen forest, 319
Buoyancy, principle of, 385
Bush tundra, 325
Buttes, 509

## C
Calcification, 347, 349
Calcretes, 347
Calderas, 401
Caliche, 334, 347
California Current, 106
Calorie, 75
Calving, 536, 608
Campos, 244–245
Canadian High, 129
Canary Current, 138
Canyons, 510–511
    submarine, 574
Capacity, 148
Capacity, stream, 510
Capillary action, 146
Capillary water, 334
Caprocks, 508
Carbonation, 434
Carbon cycle, 89
Carbon dioxide, 88–89
    absorption by oceans, 567
    and greenhouse effect, 90–91
    levels, 222–223
    photosynthesis and, 90–91
Carbonic acid, 145, 434
Careers, related to geography
    biogeographers, 6
    cartographers, 7, 32, 42, 343
    climatologists, 5
    coastal mapping analyst, 596
    environmental geographer, 451
    flood mapping analysts, 481
    geographers, 7, 67
    geomorphologists, 5
    glaciologists, 6
    human geographers, 5
    hydrologists, 6
    meteorological computer training consultants, 185
    meteorologists, 5, 132, 172
    oceanographers, 6
    physical geographers, 4, 14, 17, 67, 317
    soil scientists, 6
    storm spotters, 194
Carnivores, 295
Cartography, 7
    computerization of, 32
    defined, 32
    technological revolution in, 51
    *See also* Mapmaking
*Cassini* spacecraft, 72
Catastrophes, 309
Catastrophism, 374
Catchments, 473
Cation-exchange capacity (CEC), 432
Caverns, 459
    development of, 462
    limestone, 460–463
Caves, 459
    depositional forms in, 461–462
    diving in, 463
    limestone, 460–463
Celsius, Anders, 99
Celsius scale, 99–100, 100f
Cementation, 367

Cenotés, 460
Cenozoic Era, 386
Centigrade scale, 99
Central meridian, 36
Centripetal stream pattern, 476
HMS *Challenger*, 568
Challenger Deep, 33, 576–577
Channel patterns, 485–486
Channel roughness, 480
Chaparral, 261
Charts, bathymetric, 51
Chemical sedimentary rocks, 368
Chemical weathering, 430, 431, 433–434, 436
  carbonation, 434
  hydration and hydrolysis, 433
  oxidation, 433
  principal processes of, 431
  solution, 433–434
Chinooks, 124
Chlorofluorocarbons (CFCs), 25, 27, 91
  minimizing U.S. use of, 26
Chronometer, 37
Cinder cone, 398–399
Circle of illumination, 33, 77f
Circumpolar vortex, 92
Cirque glaciers, 538
Cirque headwall, 544
Cirques, 538
Cirrocumulus clouds, 158
Cirrostratus clouds, 158
Cirrus clouds, 157, 158
Clastic sedimentary rocks, 367
Clay, in soil, 336
Clean Air Act, 451
Clean Water Act, 451
Clear air turbulence (CAT), 191
Cliffs, 435
Climate, 115
  arid, 211, 214–215
  classifying, 206–217
  defined, 110
  descriptions, 205
  desert, 247, 248–252
  empirical classification of, 206
  future, 226–228
  genetic classification of, 206
  highland, 215, 284–289
  humid continental, hot summer, 270–273
  humid continental, mild summer, 273–275
  humid mesothermal, 258–268, 258f
  humid subtropical, 262–264
  humid tropical, 237t
  ice-sheet, 210–211, 281–282
  ingredients of, 111
  marine west coast, 264–268
  Mediterranean, 125, 211, 258–262
  mesothermal, 211, 259t
  microthermal, 211, 269, 271t
  middle-latitude, 258
  oceans and, 567
  of the past, 217–219
  polar, 210–211, 279–284
  regions, 215–216. *See also* Climate regions
  relation between ecosystems and, 306–308
  relation of Earth and sun to, 68
  scale and, 216–217
  soil formation and, 342–344
  steppe, 247, 252–254
  subarctic, 275–279
  trends of past 150,00 years, 227f
  tropical, 209–210
  tropical monsoon, 242–244
  tropical rainforest, 238–242
  tundra, 210, 280–281
  types, distribution of, 209
  weather and, 110–111
  *See also* Climate change; Climate regions
Climate change
  atmospheric gases and, 222–223
  causes of, 220–226
  and impact on coastlines, 225
  landmasses and, 223–225
  new knowledge gained regarding, 217–219
  ocean and, 223
  rates of, 219–220
  volcanic activity and, 222
  *See also* Climate
Climate regions, 215–216
  arid, 246–254
  highland, 284–289
  humid mesothermal, 258–268
  humid microthermal, 268–279
  humid tropical, 236–246
  polar, 279–284
  *See also* Climate
Climatologists, 5
Climatology, 110
Climax community, 304–306
  defined, 305
Climographs, 216, 216f
  of subarctic climate stations, 278f
Cl, O, R, P, T (mnemonic for soil development), 340
Closed systems, 23, 316
  hydrologic cycle, 146
  hydrosphere, 147
Cloud cover, 101
  albedo of, 101–102
Cloud forests, 239–240, 239f
Clouds, 156–160, 156f
  adiabatic heating and cooling, and formation of, 159–160
  classification scheme for, 156
  cloud cover, 101
  color of, 97
  convective currents as cause of, 98
  effect of stability and instability on formation of, 160–161
  forms of, 156–159
  lifting condensation level (LCL) of, 164
  in marine west coast climates, 267
  solar radiation in, 97–98
  thunderstorm, 157, 159
  types of, 158–159f
  *See also specific cloud types*
Coal, 368
Coastal zone, 593
  components of, 594
  currents, 601–603
  depositional landforms, 603, 606
  erosional landforms, 596–599
  human activities affecting, 593–594
  landforms and processes, 594–59
  types of coasts along, 607–609
  wave erosion of, 595
Coastlines of emergence, 607–608
Coastlines of submergence, 608
Coasts, 607–609
  active-margin, 607–608
  climate change and impact on, 225
  glacial deposition, 609
  glacial erosion, 608–609
  passive-margin, 607
Col, 545
Cold currents, 137
Cold fronts, 179
Coleridge, Samuel Taylor, 131
Collapse sinkholes, 459
Collision-coalescence process, 161
Color composite image, 61
Columnar joints, 366
Columns, in Township and Range System, 38–39
Comets, 69
Communities, 304
  "chemosynthetic" vent, 327
  climax, 304–306
  succession, 300–301, 304
Compaction, 367
Compass directions, 49
Composite cones, 399–400
Compression, 405
Compromise projections, 44–45
Computers
  effect on cartography of, 32, 52–53
  weather analysis by, 141, 195
  weather forecasting uses of, 188
Conceptual models, 21
Condensation, 96, 147, 152–156
  clouds, 156
  defined, 152
  factors needed for, 153
  latent heat of, 98, 99, 147–148
  other minor forms of, 155
  *See also specific types of fog*
Condensation nuclei, 153, 163
Conduction, 98, 99
Conformal maps, 41
Conglomerates, 367
Conic projections, 41, 44
Coniferous forests, 319–321
Constructive plate margins, 380
Continental Arctic air masses, 177
Continental collision, 383
Continental crust, 373
Continental Divide, 473
Continental drift, 374–375
  supporting evidence for, 375, 377
  *See also* Plate tectonics
Continental highs, 129
Continental ice sheets, 538–539, 548–551, 548–551
  movement of, 551–560
Continental islands, 578
Continentality, 101, 260
  effect on subarctic climate of, 276–278
  in marine west coast climates, 265
Continental margins, 571–575
  submarine canyons, 574
Continental Polar air masses, 177
Continental rise, 575
Continental shelf, 571, 574
  topography of, 574
Continental shields, 384
Continental slope, 574–575
Continental suturing, 383
Continental Tropical air masses, 179
Continents
  in motion, 374–375, 377
Continuous data, 49
Continuous variables, 50f
Contour interval, 50
Contour lines, 50–51
Controls, 104
Convection, 98, 99, 153, 239, 379
Convectional precipitation, 163, 165–166
Convectional system, 98
Convective current, 98
Convective thunderstorms, 189
Convergence, 239

Convergent wind circulation, 117–118, 130
Cool currents, 137
Coordinate system, 34
Coral reefs, 578–580, 608, 609
Core, Earth's, 371
Coriolis effect, 120
   defined, 120
   gyres and, 136
   and wind, 120
Corridors, in environmental mosaic, 305
Corrosion, 482
Coulees, 559
Creep, 438
   factors facilitating, 441, 443
Cretaceous Period, 386, 387
Crevasses, 542
Cross-bedding, 368
Cross sections, 52
Crustal warping, 405–406
Crust, Earth's, 362, 373
   continental, 373
   folding of, 406–407
   most common elements in, 362t
   oceanic, 373
   warping of, 405–406
Cryoturbation, 351
Cultural geography, 5
Cumulonimbus clouds, 157, 159
   updrafts in, 163
Cumulus clouds, 157–158
Cumulus congestus clouds, 157
Currents
   coastal, 598–599, 602–603
   cold or cool, 137
   ocean, 105–107, 136–138, 223, 569
   thermohaline, 569
   tidal, 584–585
   turbidity, 574, 575
   warm, 136–137
Cuspate deltas, 491
Cut bank, 487
Cyclones, 117, 199
   anticyclones, and winds, 122
   and local weather, 183–186
   middle-latitude, 183f, 184f
   nature, size, and appearance on maps, 181–182
   origin, development, and characteristics of, 182–183
   storm tracks of, 182
   and upper air flow, 186–187
   *See also* Low pressure systems
Cyclonic precipitation, 165, 166, 267
Cylindrical projection, 41

**D**

Daily (diurnal) temperature ranges, 238
Daily march of temperature, 101
Dams, 495–496
Darwin, Charles, 342, 579
Data
   discrete and continuous, 49
   displaying spatial, and information on maps, 49–51
Daylight, 80t
   in Antarctic Cirle, 279
   in Arctic Circle, 81
   duration of, 306
Debris, 438
Debris flow, 438, 441
Debris flow fans, 511
December solstice, 84
Deciduous forests, 318–319
Declination, 376
Declination, of sun, 83

Decomposers, 295
Decomposition, 431
Deep Sea Drilling Project (DSDP), 576
Deep-sea sedimentary record, 218
Deflation, 515–517
Deflation hollows, 516–517
Deforestation, 17, 228, 310
   of rainforests, 241
   related to desertification, 250
Degradation, 394, 417, 423, 424, 482
Delta plain, 490
Deltas, 489–491, 608
   bird's-foot, 490
   construction of, 490
   cuspate, 491
Dendritic stream pattern, 475
Dendrochronology, 221
Deposition, 424
   by glaciers, 535
   stream, 484–485
   wind, 517–524
Deranged drainage, 476
Desert climate, 247
Desert climates, 248–252
   adaptations by plants and animals in, 251–252
   challenges of, to human activities, 251–252
   extremes of, 248–250
Desertification, 17, 250–251
Desertification Convention, 251
Desert pavement, 517
Deserts, 170, 246, 247, 247f, 323–324
   landscape development in, 524–527
   landscapes of, 503
   similarities between steppe regions and, 252–253
   streams of, 505–506
   surface runoff in, 504–506
Desert tundra, 325
Detritivores, 295
Dew, 155
   cooling process producing, 159
Dew point
   defined, 149
   saturation and, 148–149
Dew point temperature, 149
Diaz, Sandra, 132
Differential erosion, 487
Differential weathering and erosion, 434
   topography related to, 435–436
Digital elevation models (DEM), 481
Digital images, 56–57
   resolution of, 57
Digital mapmaking, 52–53
Digital terrain elevation data (DTAT), 596
Digital terrain models (DTM), 52–53, 481
Dikes, 404
Dinosaurs, age of, 387–388
Diorite, 366
Dips, 405
Dip slip, 409
Direction, 43
   compass, 49
   on maps, 46
Disappearing streams, 459–460, 460f
Discrete data, 49
Discrete variables, 50f
Dissolution, 433–434, 482
Dissolved load, 483
Distance, 43
   distortion of, in gnomonic projections, 44f
   measurement in universe, 68
Distributaries, 490
Diurnal tides, 581

Divergent wind circulation, 118, 130
Dobson, G. M. B., 92
Dobson units (du), 92
Doldrums, 131, 244
   humid air of, 238–239
Dolines, 460
Dolomite, 368
Doppler radar, 59
   tornado detection and forecasting using, 191
Drainage areas, 473
Drainage basins, 473, 475
Drainage density, 507
Drainage divides, 473
Drainage patterns, 475
Drainage winds, 124
Drift, 546, 558
Driftless Area, 552
Dripstone, 462
Drizzle, 162
Drought, 171, 226
   defined, 250
   in Mediterranean climates, 260
   -resistant needle-leaf trees, 318
   in savanna climates, 244
   in steppe climates, 254
   in West Africa, 246
Drumlins, 555, 557f
Dry adiabatic lapse rate, 159
Dune mobility index, 519
Dust Bowl, 254
Dust storms, 516
Dynamic equilibrium, 24, 98, 482

**E**

Earth
   –atmosphere system, 115
   axial tilt of, 76f, 77–78
   changes in atmosphere of, 222–225
   changes in orbit of, 221–222
   complexity of energy systems of
   controls of surface temperatures of, 104–108
   core, 371
   crust, 362, 373
   crustal warping of, 405–406
   ever-changing, 11, 13
   first circumnavigation of, 38
   gravity of, 94
   heat energy budget, 90
   heat energy budget at surface of, 98–99
   information on Internet regarding, 6–7
   insolation received by, 78
   as life-support system, 16–19
   lines on, delimiting solar energy, 82–83
   location on, 32
   magnetic field of, 70, 376, 378
   mantle, 372–373
   maps of features of, 41
   and moon, gravitational attraction between, 581
   movements of, 76–78
   planetary structure of, 371–373
   question of life beyond, 72
   radiation budget of, 96f
   revolution of, 77–78
   rotation of, 36, 76–77, 78, 117
   satellite imaging of, 7
   size and shape of, 32–33
   subsystems, 23
   temperature distribution at surface of, 108–109
   viewing the "big picture" of, 3–4
   water, as transportation system for, 146
Earth–moon system, 581

Earth–sun system, 74–78
　movements of Earth, 76–78
　solar energy and atmospheric dynamics, 75–76
Earthflows, 441
Earthquake intensity, 413
Earthquake magnitude scale, 413
Earthquakes, 412–413
　epicenter of, 413
　focus of, 413
　hazards related to, 413, 416–417
　magnitude, 414
　measuring size of, 413
　San Francisco (1906), 409, 415
　shallow, 380
　spatial relationships between plate boundaries, volcanoes, and, 406–407
Earth system, 13–14
　alteration of water in, 95–97
　contribution of digital technologies to understanding of, 61
　dynamic aspects of, 13
　energy source of sun for, 75–76
　equilibrium in, 24–27
　insolation received by, 84
　operating in sudden or extraordinary fashion, 16
　role of glaciers in, 535
Easterly wave, 195
East Pacific Rise, 575
Ebb tide, 584
Eccentricity cycle, 221
Ecological niche, 300
Ecological optimum, 306
Ecology, 14, 294
Ecosystems, 14–16
　artificial, 299
　biotic factors, 309
　characteristics of, 15–16
　climatic factors, 306–308
　damage to Everglades, 19, 20f
　defined, 14, 294
　dynamic aspects of, 15
　effect of deforestation on, 241
　energy flow in, 296–297
　human impact on, 310, 312
　major components of, 294–295
　marine, 325–327
　organization within, 294–300
　productivity in, 298–300
　trophic structure of, 295, 296t
Ecotone, 306
Effective precipitation, 247
Effective temperature, 276, 277
Electromagnetic energy, 75
Electrons, 220
El Niño, 136, 139
　causes of, 223
　and global weather, 141–142
　and the Southern Oscillation, 139–140
Eluviation, 333
Empirical classification, 206
End moraines, 547
Energy
　budget, effect of deforestation on, 241
　caloric, 75
　electromagnetic, 75
　exchanges, 96–97
　flow in ecosystem, 296–297
　geothermal, 448
　kinetic, 479–480
　released into environment, stored, 96f
　solar, 69
　stream, 479–482

sun and its, 69–71
of tsunamis, 586
wave, 595–596
wind, 119
See also Heat energy
ENSO (El Niño/Southern Oscillation), 140, 141
　onset of, 223
Entisols, 350
Environmental geographers, 451
Environmental lapse rate, 160
Environmental Protection Agency (EPA), 255, 474
Environmental science perspective, 14–21
　Amazon rainforest, 241
　beach building, 600–601
　cumulus congestus: an aspiring thunderstorm, 157
　effects of altitude on human body, 286
　harnessing wind, 119
　hazard potential in Cascades, 442
　human–environment equation, 19, 21
　human–environment interactions, 16–17
　invasions by sand dunes, 519
　life-support system, 16–19
　mapping distribution of earthquake intensity, 414–415
　restoring the Everglades, 494
　rogue waves, 585
　sinkhole formation, 458
　solar energy, 69
　theory of island biogeography, 302–303
Environment(s), 14
　alterations of natural, consequences of, 19
　and atmospheric issues relating to, 89–91
　biotic factors, 309–310
　changes in, causing regional changes, 8
　controls, 306–310, 312
　effects of greenhouse warming of Earth's, 91
　effects of human activities on, 16–17
　limiting, of subarctic climate, 278–279
　map illustrations of relationships in, 41
　most continually changeable, 593
　natural catastrophes, 309
　prohibitive, of polar regions, 282
　remote sensing of, 56–57
　slash-and-burn agriculture, 348
　soil and topography, 308–309
　stored energy released into, 96f
　study of past geographic. See Paleogeography
Eolian processes, 504
　similarity of fluvial processes to, 514
Ephemeral channels, 505
Ephemeral flow, 471
Epicenter, earthquake, 413
Epipedons, 350
Epochs, 386
Equal-area maps, 42–43, 43f
Equator, 34
　influence of ITCZ on, 209–210
Equatorial Current, 136
Equatorial low, 126
Equatorial trough, 126, 129
Equidistance, 43
Equilibrium
　attempt to reach, by systems of Earth, 88
　beach systems in, 602–603
　dynamic, 24, 98, 482
　in Earth systems, 24–27
　glacial budget and, 540–541
　punctuated, 395
Equilibrium line, 539
Equinox, 81–82, 83f
Eras, 386
Eratosthenes, 81

Ergs, 518
Erosion, 424
　beach, 605
　coastal landforms caused by, 596–599
　differential, 487
　fluvial, 483
　by glaciers, 535
　gully, 310, 357f
　headward, 483
　by streams, 482–483
　and transportation, wind, 515–517
Erratics, 556, 558
Eruptions, volcanic, 396
Eskers, 555–556, 557f
Mount Etna, 400
Eurasian plate, 381, 383
Evaporation, 96, 148
　of deep lakes, 219
　latent heat of, 98
　in physical weathering, 428
　rate of, 151–152
Evaporites, 433–434
Evapotranspiration, 151
　distribution of actual mean, 151t
　potential, 152
Mount Everest
　air pressure at top of, 117
　height of, 33
Everglades
　damage to, 19, 20f
　restoring, 494
Exfoliation, 430
Exfoliation domes, 430
Exfoliation sheets, 430
Exotic species, 311
Exotic streams, 506
Exposure, 285
Exterior drainage, 475
　in deserts, 505
Extinction, of species, 17, 388
Extratropical disturbances, 180
Extrusive igneous rocks, 365–366

F
Fahrenheit, Daniel, 99
Fahrenheit scale, 99–100, 100f
Falls, 438
Famine, 246
Fan apex, 511
Fault escarpments, 410
Faulting, 408–412
　defined, 408
Faults
　development of, 411
　normal, 409, 417
　reverse, 410
　strike-slip, 410
　thrust, 408, 410
　transform, 383
　vertical displacement along, 409
Fault scarps, 410–411
Feedback, 24
　cycle, 225
Feedback loop, 25
　nature's controlling mechanism–negative, 26f
Fertility, 356
Fertilization, soil, 333
Fetch, 586
Fire ants, 311
Firn, 536
First-order streams, 473
Fissure flows, 397
Fjords, 267, 536, 545, 608

Flood basalts, 397
Flood crest, 493
Floodplains, 484
Floods, 171, 246, 470
    caused by frontal precipitation, 166
    human activities and, 494
    meander cut-offs during, 489
    preparations for future, 494
Flood tide, 584
Flows, 438
Fluorescent dyes, 454
Fluorometer, 454
Fluvial deposition, 484–485
    arid region landforms of, 510–513
Fluvial erosion, 482
    arid region landforms of, 507–509
Fluvial geomorphology, 469–470
    quantitative, 497
Focus, earthquake, 413
Foehn, 124
Fog, 103–104, 153–154
    acid, 434
    advection, 154–155, 267
    cooling process producing, 159
    problems caused by, 154
    radiation, 154
    sea, 155
    temperature-inversion, 154
    upslope, 155
Folding, 406–407
    anticlines, 407
    recumbent folds, 407–408
    synclines, 407
Foliations, 370
Food chain, 295
Food web, 295
Forest biomes, 312, 316–321
Forests
    boreal, 320
    broad-leaf deciduous, 318–319
    broad-leaf evergreen, 319
    coniferous, 319–321
    deciduous, 318–319
    middle-latitude, 318–321
    mixed, 319
    scrub, 318
    thornbush, 318
    tropical, 316–318
Forties, 39
Fossil fuels, 90
Fossils, 218, 377
Fractures, 426–427
Freezing, 96, 146, 427, 440
Freezing nuclei, 162
Freezing rain, 163
Fresh water
    in glaciers, 537
    global supply of, 447f
    growth in usage of, 447
Friagem, 195
Friction
    within atmosphere, 88
    and interaction of pressure gradient and Coriolis effect, 120
    slowing of waves by, 594–595
    waves and, 585
    and wind, 121
Fringing reefs, 579
Frontal precipitation, 163, 166
Frontal thunderstorms, 189
Fronts, 166, 179–180
    cold, 179
    occluded, 180, 183
    stationary, 180
    warm, 179
    weather conflicts occurring at, 273
Frost, 103–104, 266
    damage, in Mediterranean climates, 260
Frostbite, 277
Frost polygons, 279
Frost wedging, 427–428
F-scale, 192. *See also* Fujita Intensity Scale
Fujita Intensity Scale, 192, 192t. *See also* F-scale
Fujita, T. Theodore, 192
Fuller, Buckminster, 17–18
Fusion (thermonuclear) reactions, 69–70

## G

Gabbro, 365, 366
Galactic movement, 76
Galaxy, 68
Galileo, 70, 206
Galleria forest, 317
Gamma rays, 75
Gap creation, in plant succession process, 304, 309
Gases
    in atmosphere, 88–89, 89t
    in atmospheric layer of planets, 72
    changes in atmospheric, 222–223
    greenhouse, 222–223, 228
    ionized, 70
    separation of, 94
Gas planets, 71
Gelisols, 351
General circulation models (GCMs), 226
Generalists, 300
Genetic classification, 206
Geocoding, 55
Geographers
    environmental, 451
    holistic approach to research of, 4
    meaning and significance of *region* to, 8
    subsystem divisions of, 22
Geographic grid, 36
    geometry of, 41, 46
    orientation of, 46
Geographic information system (GIS), 53–54
    data and attribute entry, 55
    four basic subsystems of, 55f
    function of, 54
    registration and display, 55
    in the workplace, 56
Geography, 5
    career opportunities associated with, 4, 5, 6, 7, 14, 17, 32, 42, 67, 132, 172, 185, 191, 317, 451, 481, 596
    cultural, 5
    description of, 4
    environmental science perspective, 14–21, 16–17, 69, 119, 157, 241, 286, 302–303, 348, 414–415, 442, 458, 494, 519, 585, 600–601
    human, 5
    paleomagnetism: evidence of Earth's ancient, 376
    physical. *See* Physical geography
    physical science perspective, 12, 13–14, 60, 81, 97, 164, 191, 220–221, 276–277, 336, 385, 432, 454, 537, 586, 605
    regional, 5
    spatial science perspective, 8–9, 52–53, 92–93, 125, 198, 225, 250–251, 311, 376, 406–407, 474, 552
    study of, 4–7
    subdivisions, 4–5
Geologic cross sections, 52

Geomorphology, 5–6
    fluvial, 469–470
    landforms and, 394–396
    quantitative fluvial, 497
    scientific goals in, 393
Geostationary orbit, 60
    characteristics of, 60
Geostrophic wind, 121
Geosynchronous orbit, 60
Geothermal energy, 448
Geothermal water, 463
Geysers, 463
Giant planets, 71
Gibber, 517
Glacial abrasion, 543
Glacial epochs, 217
Glacial outwash, 547
Glacial plucking, 543
Glacial surges, 543
Glacial troughs, 545
Glaciation, 217
    depositional landforms produced by alpine, 546–548
    erosional landforms produced by alpine, 544–546
    evidence of ancient, 377
    evidence of, in Sahara Desert, 374
Glaciers, 217, 225
    as agents of gradation, 543–544
    defined, 536
    formation of, and hydrologic cycle, 536–538
    fresh water in, 537
    functional parts of, 539
    in hydrologic system, 540f
    location of present-day, 537
    mechanics of flow for, 541–543
    melting of, 91, 219
    role of, in Earth system, 535
    "tidewater," 608
    types of, 538–539
Glaciolacustrine, 558
Glaciologists, 6
Glaze, 163
Gleization, 349–350
Global positioning system (GPS), 39, 568
Global pressure belts, 126–127, 129
    effect of seasons on, 127, 129
    idealized, 127–128
    in January, 129
    in July, 129
    pattern of, 127
Global surface wind systems, 129–134
    conditions within latitudinal zones, 131–132
    effects of seasonal migration, 133
    idealized model of atmospheric circulation, 130–131
    longitudinal differences in winds, 134
Global warming, 17
    effect of deforestation on, 241
Globes, 32
    and great circles, 33–34
*Glomar Challenger,* 570, 576
GMT. *See* Greenwich Mean Time
Gneiss, 370
Gnomonic projections, 44
    distortion of distances in, 44f
Gondwana, 375, 375f, 387
Grabens, 412
Gradation processes, 394, 423
    agents of, 411
    carried out by streams, 480–481
    episodic nature of, 395

fluvial, 482–485
glaciers as agents of, 543–544
sources of energy for, 424
tectonism and, 424–425
water as gradational agent in arid lands, 506–513
wind as gradational agent, 514
Graded streams, 482
Gradient, 51
Grand Canyon, 435, 435f
caprocks in, 508
geomorphology of, 509f
Granite, 365, 366, 370, 373, 430, 434
Granular disintegrations, 430
Graphic scale, on maps, 46
Grassland biomes, 321–323
Grass tundra, 325
Gravel, 480
Gravitation, 72
Gravitational processes, 423
Gravitational water, 333
Gravity, 94, 385
as energy source for gradation, 424
as factor in mass movement, 437
Great circles
globes and, 33–34
on Mercator projection, 44
Great Lakes, 559, 560
Greenhouse effect
carbon dioxide and, 90–91
defined, 90
Greenhouse gas, 222–223, 228
Greenwich Mean Time, 37
Grid cells, 34
Grinders, 483
Groins, 603, 605
Ground moraines, 547
Groundwater
availability of, 452–455
in comparison with rainwater, 453
components of, 448
effluent condition, 449, 450f
factors affecting distribution of, 450–452
flow, monitoring, 454
influent condition, 449–450, 450f
occurrence and supply of, 448–452
as source of fresh water, 447
zones and water table, 448–450
Groundwater recharge, 453
Gulfs, 568
Gulf Stream, 106, 136–137, 138
Gulf Stream Current, 224
Gullies, 471
Gully erosion, 310, 357f
Guyots, 578
Gypsum, 368
Gyres, 136

## H

Habitat, 300
unique aspects of, 305–306
watershed, 474
Hail, 89, 159, 163
Hailstones, 163
Halite, 368
Hanging valley, 546
Hardpan, 333
Hawaii
island chains of, 578
volcanic activity in, 398
Haystack hills, 460
Headward erosion, 483
Heat energy

transfer, processes of, 98
water as, 95–97
See also heat energy budget
Heat energy budget, 90
in atmosphere, 99
at Earth's surface, 98–99
variations in, 99
water budget and its relation to, 147–148
Heezen, Bruce, 571
map of ocean floor by, 572–573f
Hemispheres, 33
Hepburn, Katharine, 238
Herbivores, 295
Heterosphere, 94, 95
Heterotrophs, 295
Hibernation, 84
Highland climates, 215, 284–289
complexity of, 285
controls of, 285
human activities in, 288
mountainous areas in, 287–288
peculiarities of, 287–288
High pressure
belts, 117
cells, 126, 188. See also Anticyclones
ridges of, 135
winds and, 118
High pressure systems
Azores High, 129
Bermuda High, 129
Canadian High, 129
defined, 117
divergent wind circulation in, 117–118. See also Winds
isobars shown on, 118
mountain breeze–valley breeze cycle and, 123–124
Pacific High, 129
polar highs, 127
pressure gradient and, 130
Siberian High, 129
subtropical highs, 126, 131
winds and, 118, 120, 122
See also Anticyclones; Atmospheric pressure
Highs, 117
High tides, 581
Hills, 435
in humid regions, 503
Himalayas, 33
continental suturing and, 383
uplift of, 224
Histosols, 350–351
Holocene Epoch, 218, 226, 227f, 386
Homosphere, 94
Horn, 545
Horsts, 411
Horton, Robert, 473
Hot spots, 384, 578
Hot springs, 463
deposits, 464f
Howard, Luke, 156
Human–environment interactions, 16–17
and possible threshold population, 19, 21
two-way relationship of, 14
See also Human activities
Human activities
affecting coastal zones, 593–594
beaches and, 603
causing wildfires, 125
challenge of desert climates to, 251–252
contributing to rock disintegration, 427
disruption of climate trends due to, 228
effects of altitude on, 286

highland climates and, 288
in humid continental climates, 274–275
in humid subtropical climates, 264
impact on ecosystems of, 310, 312
increasing risk of flooding, 494
influence on microclimates of, 216–217
modification of runoff by, 471
in polar regions, 283–284
streams and, 489
in subarctic climates, 279
temperature and, 107–108
in tropical rainforest climates, 240, 242
and urban heat-island effect, 107, 217
Human geography, 5
Humboldt Current, 138, 140
Humid continental, hot summer climate
internal variations, 270–272
seasonal changes in, 272–273
Humid continental, hot summer climates, 270–273
comparison with mild summer climates, 273
Humid continental, mild summer climates, 273–275
comparison with hot summer climates, 273
human activities in, 274–275
Humidity, 149–151
absolute and specific, 149
in marine west coast climates, 265
rainforests and, 151
relative, 149–151
in tropical rainforest climates, 238
Humid microthermal climates, 269f
Humid subtropical climates, 262–264
comparison with Mediterranean climates, 262–263
human activities in, 264
productive qualities of, 263–264
Hums, 460
Humus, 335, 342
Hurricanes, 194, 196–200
circulation patterns within, 197f
defined, 196
devastating, 200, 589, 601
El Niño and, 141
name assignment of, 197
paths and landfall probability maps for, 198
Huygens probe, 72
Hydration, 432, 433
Hydraulic action, 482
Hydrographs, 493–495
curves, part of, 494
Hydrologic cycle, 146–147, 148f
effect of deforestation on, 241
fog in, 153–154
glacier formation and, 536–538
glaciers and, 540f
groundwater system, as subsystem of, 449f
impact of glaciers on, 535
streams, 472f
Hydrologists, 6
Hydrolysis, 433
Hydrosphere, 13, 16, 146
Hygroscopic water, 334
Hygrothermograph, 150f
Hypotheses, 12
Hypoxia, 286

## I

Ice
color of glacial, 537
glacier, 537
reflective qualities of, 101
Ice ages, 217, 220, 225, 537

Icebergs, 608
Ice caps, 537, 539
Ice crystal process, 161–162
Ice crystals, 89
    atmospheric scattering caused by, 97
    in Bergeron Process, 162
    in clouds, 156
    forming rime, 155
Icefall, 542
Icelandic Low, 129, 141
Ice-marginal lakes, 558–559
Ice-scoured plains, 553–554
Ice-sheet climates, 210–211, 279, 281–282
    animal life in, 284
Ice sheets
    analysis of, 219
    color of, 537
    continental. *See* Continental ice sheets
    and depositional landforms, 554–558
    erosional landforms and, 553–554
    existing, 548–550
    melting, 91
    new knowledge gained regarding, 217
    Pleistocene, 550–551
Ice shelves, 549
Igneous processes, 394
    intrusive activity and intrusions, 401–404
    landforms and, 396–404
Igneous rocks, 364–365
    categories of, 365
    chemical composition of, 365
    extrusive, 365–366
    intrusive, 365–366
Illite, 432
Illuviation, 333
Imaging radar, 59
Inceptisols, 350
Inclination, 376
Indian Ocean
    depth of, 568
    size of, 568
Indiscriminant scattering, 97
Industrialization, 17
Industrial Revolution, 90
Infiltrate, 337
Infiltration, 471
Infiltration capacity, 471
Inner core, 371
Inputs
    of incoming shortwave solar radiation, 98
    into system, 22
Inselbergs, 509
Insolation, 78
    daily effects of, 100–101
    in ice-sheet climates, 282
    variations of, with latitude, 83–84
    *See also* solar energy
Instability, 160
Integrated Ocean Drilling Program, 372–373
Interception, 471
Interfluve, 470
Interglacial, 550
Interior drainage, 475
    in deserts, 505
Intermittent flow, 471
Internal plastic deformation, 542
*International Cloud Atlas,* 157
International Date Line (IDL), 37–38, 38f
International Decade for Natural Disaster Reduction (IDNDR), 442
International Hydrographic Organization, 568
International Meridian Conference (1884), 36
International Space Station, 95
Internet, information regarding Earth on, 6–7

Interrupted projection, 45
Interstices, 146
Intertropical convergence zone (ITCZ), 131, 209–210, 244
    shifting of, 242–243
    unstable conditions along, 238–239
Intrusions, igneous, 401–404
Intrusive igneous rocks, 365–366
Inversions
    surface, 103–104
    temperature, 103
Ionosphere, 94, 95
Ions, 220
Island arcs, 382, 578
Islands
    barrier, 606, 609
    biogeography of, 302–303
    continental, 578
    oceanic, 578
Isobars, 49, 118
Isobaths, 49
Isogonic map, 49f
Isohyets, 49
Isolines, 49
Isoseismals, 414
Isostasy, 385
Isostatic rebound, 385, 560, 608
Isotherms, 49, 108
    in marine west coast climates, 265
Isotopes, 220

## J

Jenny, Hans, 340
Jet streams, 126, 134–136
    advantage to pilots of, 136
Jetties, 603, 605
Joint planes, 426
Joints, 426–427
    in igneous rocks, 366
June solstice, 81, 82, 83–84
Jungle, 240, 242, 316–317
Jupiter
    characteristics of, 72
    composition of, 71
    possibility of life on moons of, 71

## K

Kames, 556
Kame terraces, 556
Kaolinite, 432
Karst landforms, 455–456, 457f, 459–460
    important factor in development of, 456
    limestone caverns and caves, 460–463
Katabatic winds, 124, 282
Kelvin, Lord, 100
Kelvin scale, 100
Kettle lakes, 555, 557f
Kettles, 555, 557f
Kinetic energy, 479–480
Köppen system, 208–209
    advantages and limitations of, 209
    classification, modified, 209–215
    simplified, climate classes, 210t
    world climate map, modified, 212–213f
Köppen, Wladimir, 208–209
Krakatoa, explosion of, 400
Krill, 326
Kuiper Belt, 72
Kuroshio Current, 136–137

## L

Labrador Current, 137–138, 155
Laccoliths, 403
Lagoons, 603
Lahars, 441, 442

Lake effect, 496
    snows, 176
Lakes, 496–497
    glacial, 558–560
    ice-marginal, 558–559
    kettle, 555
    paternoster, 546
    playa, 513
Lambert, Johann, 78
Lambert's Law, 78, 84
Land
    differential heating of, 101
    sculpture by streams, 486–489
    and water distribution, 104–105
Land breeze–sea breeze cycle, 122–123
Land degradation, 310
Landform barriers, 107
Landforms
    created by intrusive activity, 401–404
    dating absolute ages of, 217
    depositional coastal, 603, 606
    depositional, ice sheets and, 554–558
    developed by weathering in fractures, 426
    development by solution, 455–463
    dominating desert scenery, 504
    eolian, 514
    erosional coastal, 596–599, 597f
    erosional, produced by alpine glaciation, 544–546
    of fluvial deposition, arid region, 510–513
    of fluvial erosion, arid region, 507–509
    geomorphology and, 394–396
    ice sheets and erosional, 553–554
    igneous processes and, 396–404
    karst landscapes and, 455–456, 457f, 459–460
    and processes, coastal, 594–59
    produced by alpine glaciation, depositional, 546–548
    response of local winds to, 122
    rocks and development of, 370
    stresses in Earth's crust and development of, 366
    tectonic processes, rock structure, and, 404–412
    variations in, 393–394
Landmasses
    changes in, 223–225
    colliding, 383
    continental crust in, 373
    movement of, 374
    orientation of ancient, 376
Landscapes, 305
    described by relief, 394
    desert, 503
    developed by weathering in fractures, 426
    development of desert, 524–527
    karst, 455–456, 457f, 459–460
    periglacial, 558–560
    rejuvenated, 492
    unique, of Driftless Area, 552
    weathering, mass wasting, and, 444
Landslides, 439
La Niña, 136, 140
    causes of, 223
Lapse rates, 159–160
Large-scale maps, 46, 47f
Lassen Peak, 401
Latent heat, 96
    of condensation, 98, 99
    of evaporation, 98
Latent heat of condensation, 147–148
Lateral accretion deposits, 484–485
Lateral migration, of meanders, 487–488
Lateral moraines, 547
Laterite, 344, 347
Laterization, 347

Latitude, 34
 conditions within zones of, 131–132
 and duration of daylight, 80t
 high, effect on subarctic climate of, 276–278
 and longitude, finding location by, 35f
 measuring, 34–35
 migration of global-scale pressure systems by, 129
 precipitation distribution and zones of, 170
 reference line for measuring, 35
 temperature and, 104
 variations of insolation with, 83–84
Laurasia, 375, 375f, 387
Lava, 362–363, 364–365
 formed by columnar joints, 366
Leaching, 333–334
Leap year, 77
Leeward, 121–122
Legends, map, 45–46
Levees, 489
 natural, 484
 protecting New Orleans, 589
Lianas, 316
Lichens, 434
Life-support system, 16–19
 threats to, 18
Lifting condensation level (LCL), 164
Light
 intensity and quality of, 306
 plant competition for, 309
 speed of, 68
Lightning, 125, 159
 conditions producing, 188
Light-year, 68
Limestone, 368, 370
 aquifers, 454
 caverns and cave features, 460–463
Linear velocity, 76
Lines
 east–west. See Base lines
 on maps, 50f
 north–south. See Principal meridians
Liquids, 89
Lithosphere, 13, 16, 372
 isostasy: balancing Earth's, 385
Lithospheric plates, 377
Litter, forest, 316
Little Ice Age, 226
Llanos, 244–245
Load, 472f, 480
 bed, 484
 dissolved, 483
 suspended, 483–484
 transport of, 483
Loam, 337
Local time, 36
Local winds, 122
Location, 10
 absolute, 10
 on Earth, 32
 finding, by latitude and longitude, 35f
 relative, 10
Loess, 346, 518
 deposits, 524
Longitude
 defined, 36
 differences in winds due to, 134
 finding location by latitude and, 35f
 measuring, 35–36
 and time, 36–37
Longitudinal dunes, 522–523
Longitudinal profile, stream, 486
Longshore current, 601
Longshore drift, 601
Longwave radiation, 98, 99, 222

Long waves, 135
Low pressure
 of Asian continent, 242–243
 belts, 117, 126
 cyclonic activity and, 185. See also Cyclones
 troughs of, 135
 winds and, 118
Low pressure systems, 166, 167
 Aleutian Low, 129
 convergent and divergent circulation in, 117–118. See also Winds
 defined, 117
 in doldrums, 131
 equatorial low, 126
 Icelandic Low, 129
 isobars shown on, 118
 land breeze–sea breeze cycle and, 122–123
 pressure gradient and, 130
 subpolar lows, 126
 subtropical, 170
 See also Atmospheric pressure; Cyclones
Lows, 117
Low tides, 581
Lunar missions, 74–75

## M

Magellan, 38
Magellan spacecraft, 72
Magma, 362–363, 364–365
 types of, 396
Magnetic field, Earth's, 376, 378
Magnetic north, 46
 map representation of, 49f
Mallee, 261
Mantle, Earth's, 372–373
 hot spots in, 384
Mapmaking, 7
 digital, 52–53
 maps and, 32
 technology, 51–56
 See also Cartography
Map projections
 compromise, 44–45
 conic, 41, 44
 cylindrical, 41
 examples of, 43–45
 gnomonic, 44
 interrupted, 45
 maps and, 39–45
 Mercator projection, 42
 planar, 41
 properties of, 41
 Robinson projection, 45f
Maps
 advantages of, 40–41
 air pressure distribution, 118
 azimuthal, 43
 of coastline changes, 225
 conformal, 41
 depiction of cyclones and anticyclones on weather, 181–182, 182f
 direction on, 46
 displaying spatial data and information on, 49–51
 equal-area, 42–43
 human–environmental, 9
 hurricane paths and landfall probability, 198
 isogonic, 49f
 large-scale, 46
 legends on, 45–46
 limitations of, 41
 major frontal symbols used on weather, 181f
 and mapmaking, 32
 and map projections, 39–45
 scale of, 46

 small-scale, 46
 thematic, 49
 titles on, 45
 topographic, 50–51, 52–53
 verbal scale on, 46
 of water-filled caves, 463
 world average annual precipitation, 168–169f
Map view, 52
Maquis, 261, 318
Marble, 370
Mariana Trench, 33, 570, 576–577
Marine influence, 265
Marine terraces, 599, 607, 608
Marine west coast climates, 264–268
 clouds and precipitation, 267
 oceanic influences on, 265–266
 resource potential of, 268
Maritime locations, 101
Maritime Polar air masses, 177–178
Maritime Tropical air masses, 178–179
Mars, 72
 atmosphere of, 74
 images taken of, 74
 possibility of life on, 71
*Mars Global Survey*, 74
Mass movement, 424–425
 gravity and, 437
 rapid, 438–441
 slow, 441, 443
 See also Mass wasting
Mass, of Mercury, 72
Mass wasting, 424–425, 436–437
 kinds of motion, 438
 processes, classification of, 437–438
 speed of movement, 438
 type of earth materials, 438
 weathering, and landscape, 444
 See also Mass movement
Mathematical/statistical models, 21
Matrix, environmental, 305
Mattoral, 261
Meander cut-offs, 489
Meandering channels, 486
Meanders, 486
 entrenched, 492
 lateral migration of, 487–488
Mechanical weathering. See Physical weathering
Medial moraines, 547
Mediterranean climates, 125, 211, 258–262
 comparison with humid subtropical climates, 262–263
 drought in, 260
 major characteristics of, 260
 special adaptations in regions with, 260–262
Mediterranean sclerophyllous woodland, 318
Melting, 96
Mental maps, 21
Mercalli Scale of earthquake intensity, 414
Mercator, Gerhardus, 44
Mercator projection, 42, 42f, 43–44
Mercury barometer, 115–116
 measurement taken by, 116
Mercury, characteristics of, 72
Meridians, 36
 on Mercator projection, 43–44
Mesas, 508
Mesopause, 94
Mesosphere, 94
Mesothermal climates, 211, 259t
Mesozoic Era, 386, 387
Metallic cations, 432
Metamorphic rocks, 364, 369–370
Meteorites, 69
Meteorology, 110, 132
Meteors, 69, 88

Microclimate, 216
Microplate terranes, 384, 386
Microthermal climates, 211, 271t, 275–279
Mid-Atlantic Ridge, 575, 607
Middle-latitude climates, 258
Middle-latitude disturbances, 180
Middle-latitude forests, 318–321
Middle-latitude grasslands, 322–323
Middle-latitude zone, 83–84
Mid-Indian Ridge, 575
Midocean ridges, 378, 383, 575–576
Mie, Gustav, 97
Mie scattering, 97
Migration, seasonal, 133
Milankovitch Cycles, 78
Milankovitch, Milutin, 78, 222
Milky Way Galaxy, 76
Minerals, 362–363
    defined, 362
    in solution, 463
Mississippi River
    load of, 484
    watershed, 474
Mistral, 124
Mixed pattern tides, 581
Models
    computer-generated, 21
    conceptual, 21
    defined, 21
    mathematical/statistical models, 21
    physical, 21
    pictorial/graphic, 21
    regions as spatial, 9
    and systems, 21–27
    uses of, 23
Modified Mercalli Scale of earthquake intensity, 413, 414–415t
Moho, 372
Mohorovicic discontinuity, 372
Mollisols, 355
Monsoon, 124–126
    rainforest, 317–318
Montmorillonite, 432
Moon, 88
    Earth–moon system, 581
    gravitational attraction between Earth and, 581
Moraines, 547
Mosaic, environmental, 305
Mosses, 434
Mountain breeze, 124
Mountain breeze–valley breeze cycle, 123–124
Mountain building, 426
Mountain climates, 287–288
    orographic precipitation in, 166
    winds in, 124
Mountains, 417, 435
    characteristics of weather in, 287–288
    fault-block, 525, 526
    submarine, 575–576
Mouth, stream, 475
Mud, 438
Mudflows, 441
Multiple working hypotheses, 12
Multispectral remote sensing, 59, 61
Multispectral scanners, 59
Muskeg, 280, 554

## N

National Centers for Environmental Predication, 202
National Ocean Service, 51
National Weather Association (NWA), 132
National Weather Service (NWS), 185, 194
Natural levees, 484
Natural resources, 18
    fresh water, as limited, 447
    need for sustainable, 241
    nonrenewable, 19
    soils as critical, 356
Natural Resources Conservation Service (NRCS), 346–347
    soil classification system, 350, 351t
    Soil Survey Division, 350
Natural vegetation, 301
    world map of, 314–315f
Nature, balance of, 24
Navigation, 32
    advent of electronic, 37
    importance of maps to, 41
Nazca plate, 381
Neap tide, 583
Near-infrared (NIR) energy, 57, 61
Negative feedback, 24
    loop, nature's controlling mechanism, 26f
    system, 219–220
Nekton, 325
Neptune
    characteristics of, 72
    composition of, 71
Net erosion, 482
Neutrons, 220
NEXRAD, 191
Nimbostratus clouds, 157, 159
Nitrogen, 88–89
Nonpoint source pollution (NPS), 451
Normal faults, 409, 417
Normal lapse rate, 93, 102–103
North American air masses, 176
    source regions of, 178f
North Atlantic Drift, 106, 137, 224, 265
North Atlantic Oscillation (NAO), 141
Northeast trades, 130
Northern Hemisphere, 34, 80
    temperatures in, 106t
Northern lights, 70
North Pacific Drift, 137
North polar zone, 84
North Pole, as reference point, 34
North Star, 35, 78
    map representation of, 49f
Novak, Thomas G., 185

## O

Oblate spheroid, 33
Obliquity cycle, 221
Obsidian, 366
Occluded fronts, 180
    conditions producing, 183
Ocean currents, 105–107, 136–138
    accentuation of coastal aridity from, 247
    deep, 569
    surface, 223
    and temperature, 139t
Ocean Drilling Program (ODP), 576
Ocean floor, 570–571, 574–578
    abyssal plains, 577
    continental shelf, 571, 574
    continental slope, 574–575
    deep-sea sedimentary record of, 218
    features of, 575–578
    guyots, 578
    map of (Heezen and Tharp), 572–573f
    oceanic ridges, 575–576
    seamounts, 578
    topographic subdivisions, 571
    trenches, 576–577
    See also Ocean(s)
Oceanic crust, 373
Oceanic islands, 578
Oceanic ridges, 378, 380, 383, 575–576
    peaks of, 578
    topography of, 576
Oceanic trenches, 381–382
Oceanographers, 6
Ocean(s)
    changes in, 223
    coastal waters of, 593
    currents, deep-, 569
    difficulty of exploring, 568
    ecosystems of, 325–327
    forms of water movement in, 580–587
    impact of glaciers on, 535
    importance of, 567
    influence of, on marine west coast climate, 265–266
    knowledge of, 568
    moderating effect on temperature of, 122–123
    surface variations, 569
    temperature variation of, 569
    use of robotics for exploration of, 571
    vertical temperature zones of, 570
    waters, characteristics of, 568–569
    See also Ocean floor
Offshore bars, 606
Offshore breakwaters, 605
Ogallala aquifer, 452, 453, 453f
Omnivores, 295
Ooze, 577
Open energy system, 98
Open systems, 15, 23, 294
    glaciers, 536
    radiation budget, 147
    soils functioning as, 332
*Operating Manual for Spaceship Earth* (Fuller), 17–18
Opportunity, 74
Orbit(s)
    of Earth, 76, 77f
    planetary, 72
    of satellites, 60
    variations in Earth's, 221–222
Organic sedimentary rocks, 368
Orographic lifting, 411
Orographic precipitation, 165, 166, 267
Orographic thunderstorms, 189
Osceola mudflow, 442
Outcrops, 363
Outer core, 371
Outlet glaciers, 549
Outputs, 22
    of longwave terrestrial radiation, 98
Outwash plains, 555
Overthrust, 408
Oxbow lakes, 489
Oxidation, 88–89, 296, 433
Oxisols, 356
Oxygen, 88, 88–89
    effect of deforestation on atmospheric, 241
    production by oceans, 567
Oxygen-isotope analysis, 218, 220
    of Antarctic ice cores, 284
Oyashio Current, 137–138
Ozone, 25, 89–91
    depletion, 92–93, 284
Ozone holes, 92–93
Ozone layer, 25, 91
    damage to, 17
Ozonosphere, 94

## P

Pacific High, 129
Pacific Ocean
    "chemosynthetic" vent communities in, 327
    coastal regions along, 607–608
    depth of, 568
    El Niño and, 223

midocean ridges in, 378
runoff flows to, 473
size of, 568
tsunamis in, 586
*See also* El Niño
Pacific plate, 377, 381, 382, 384, 578
Pacific Ridge, 575
Pacific Ring of Fire, 406, 576, 607
Pacific Time Zone, 37
Pacific Tsunami Warning Center (PTWC), 586
Pahoehoe, 397, 397f
Paleoclimatologists, 220
Paleogeography, 386–389
Paleomagnetism, 375, 376
Paleomaps, 389
Palynology, 221
Pampas, 244–245, 264
Pangaea, 374–375, 375f
Parabolic dunes, 521–522
Parallelism, 77
Parallels, 36
Parent material, of soil, 339
residual, 341
transported, 341
Partial pressure of oxygen gas, 286
Particulates, 89
Passive-margin coasts, 607
Patches, within environmental matrix, 305
Paternoster lakes, 545
Patterned ground, 279, 559
Pediments, 509
Pedosphere, 332
Mount Pelée, 401
Perched water table, 452
Perennial channels, 505
Perennial flow, 471
Periglacial landscapes, 559–560
Perihelion, 77
Periodic Table of Elements, 220
Periods, 386
Permafrost, 278–279, 351, 559
Permeability, 337, 450–451
Perspective, 52
Photography, aerial, 57
Photosynthesis, 89
carbon dioxide and, 90–91
factors affecting, 298
Photovoltaic (PV) cells, 69
pH scale, 338, 339f
Physical geographers, 317
study of environmental processes by, 14
technology, tools, and methods of, 6–7
Physical geography, 5–6
career opportunities in, 132
major perspectives in, 7–21
you and, 27–28
Physical models, 21
Physical science perspective, 13–14
basic soil analysis, 336
beach erosion and protection, 605
color of glacial ice, 537
colors of the atmosphere, 97
determining past climates, 220–221
Earth system, 13
effective temperature, 276–277
expanding and contracting soils, 432
isostasy: balancing Earth's lithosphere, 385
lifting condensation level (LCL), 164
major subsystems of Earth, 13–14
monitoring groundwater flow, 454
polar versus geostationary satellite orbits, 60
scientific method, 12
storm chasers and storm spotters, 194
tsunami forecasts and warnings, 586

using sun's rays to measure Earth (2200 years ago), 81
Physical weathering, 425–430
contribution of freezing and thawing to, 427
granular disintegration, 430
Phytoplankton, 325–326
Piccard, Jacques, 577
Pictorial/graphic models, 21
Piedmont alluvial plains, 511
Piedmont glaciers, 538
Pillow lavas, 576
Mount Pinatubo, eruption of, 396, 400
Pioneer, 304
Pioneer community, 304
Pixels, 56–57
Places, characteristics of, 10
Planar projection, 41
Plane of the ecliptic, 77, 78f
seasons and, 80
Planets, 69, 71–74
comparison of, 74t
phenomena common to, 72
in solar system, 73f
*See also specific planets*
Plankton, 325
Plant communities, 301
Plants
associations of, 301
contributing to rock disintegration, 427
essentiality of, 295
*See also* Vegetation
Plan view, 52
Plastic solid, 372
Plateaus, 507–508
Plate convergence, 380–383
Plate divergence, 380, 406
Plate tectonics, 377–384
plate convergence, 380–383
plate divergence, 380
seafloor spreading and convection currents, 378–379
tectonic plate movement, 379–380
transform movement, 383
*See also* Continental drift
Playa lakes, 513
Playas, 513
Pleistocene Epoch, 217, 218, 224, 225, 267, 386, 504, 574, 606, 607
formation of dunes during, 520
ice sheets, 550–551
Plug dome volcanoes, 401
Plunge pools, 482
Pluto, 72
characteristics of, 72
Plutonic rocks, 365
Plutonism, 396
Plutons, 401
Podzolization, 347
Point bar, 487
Points
locating, 38
on maps, 50f
Polar climates, 210–211, 279–284, 280t
characteristics of, 279
human activities in, 283–284
pattern of day and night in, 279
Polar deserts, 282
Polar easterlies, 130
Polar front, 132–133, 180–181, 266
development of wavelike kink along, 182–183
Polar front jet stream, 135, 180–181
Polar front theory, 180
Polar glaciers, 537
Polar highs, 127, 129

Polaris, 35, 78
map representation of, 49f
Polarity reversals, 376, 378
Polar orbit, 60
Polar outbreak, 195
Polar winds, 132
Pollutants, air, 434
Pollution, 17, 18–19
atmospheric transporting of, 92
nonpoint source, 451
of surface waters, 453
worldwide problem of, 19
Pompeii, 395f, 400
Ponds, 496
Popocatepetl, 401
Porosity, 337, 450
Positive feedback, 24
system, 219
Potential ET. *See* Potential evapotranspiration
Potential evapotranspiration, 152, 207–208, 247
in arid climates, 214
in the contiguous 48 states, 152f
Potholes, 482–483
P (primary) waves, 371
Prairies
short-grass, 323
tall-grass, 322–323
Precambrian Period, 386
Precession cycle, 222
Precipitation, 147
acid, 17, 145
in arid climates, 248
components of, 156
convectional, 163, 165–166
convective currents as cause of, 98
cyclonic, 165, 166, 267
in desert climates, 248, 249–250
distinctions between rainforest and monsoon climate, 243
distribution of, 166–170
distribution within latitudinal zones, 170
factors necessary for, 163, 165
frontal, 163, 166
in highlands, 286
horizontal distribution of, 167, 170
in humid subtropical climates, 262–263
in humid tropical climates, 236
in ice-sheet climates, 282
interception of, 471
latitudinal distribution, 171f
major forms of, 162–163
in marine west coast climates, 267
in Mediterranean climates, 258, 260
in microthermal climates, 269
orographic, 165, 166, 267
principal cause of, 164f
processes, 161–162
of savanna climates, 244
in steppe climates, 252
subarctic, 278
in tropical rainforest climates, 238, 239
variability of, 170–172
world map of average annual, 168–169f
Precipitation effectiveness, 247
Pressure, average sea-level, 129f
Pressure gradients, 118
assessment of, 135
and interaction of friction and Coriolis effect, 120
winds and, 118, 120
Pressure systems, basic, 117–118
Prevailing winds, 122
Primary coastlines, 608–609
Primary productivity, 298–300
Primary succession, 301, 304

Prime meridian, 36
Principal meridians, 38
Prodelta, 489
Productivity, in ecosystems, 298–300
 primary, 298–300
 secondary, 298, 300
Profile, 51
 topographic, 52
Protons, 220
Punctuated equilibrium, 395, 395f
Pyroclastics, 366, 396

## Q
Quarter-quarter sections, 39
Quarter sections, 39
Quartzite, 370
Quasi-stationary fronts, 180
Quaternary Period, 386

## R
Radar, 58–59
Radial stream pattern, 475
Radiation, 98
Radiation fog, 154
Radiocarbon dating, 217
Radiometric-dating techniques, 217
Rain, 162
 trace of, 166
Rainbows, 97
Rainforests, 9, 316–317
 climate of, 210
 deforestation of, 241
 humidity and, 151
 large-scale harvesting of, 240
 leaf canopies in, 316
 monsoon, 317–318
 oppressive heat found in, 239–240
Mount Rainier
 Nisqually Glacier on, 543
 risk associated with, 442
Rain shadow, 166
Rainwater, 145
Range, 39
Rayleigh, Lord, 97
Rayleigh scattering, 97
Recessional moraines, 547, 555
Reconstructions, paleographic, 377
Rectangular stream pattern, 476
Recumbent folds, 407–408
Reef corals, 578
Reflection, 101
Regional base level, 475
 in deserts, 505
Regional geography, 5
Regions, 5
 arid climate, 246–254
 climate, 215–216. *See also* Climate regions
 defined, 215
 division into subregions of, 8
 humid tropical climate, 236–246
 map representations of, 49
 natural and environmental, 8–9
Regolith, 340f, 363, 448
 collapse, sinkhole creation due to, 458
 expansion and contraction of, 443
 formed by weathering, 436
 gravity and, 437
Relative humidity, 149–151
Relative location, 10
Relief
 increase of, 394
 reduction of, 395
Remote sensing
 defined, 56
 of environment, 56–57

multispectral, 59, 61
 techniques, specialized, 57
Representative fraction (RF) scale, 46
Reservoirs, 495
Residual parent material, 341
Resolution, 57
Resources
 from oceans, 567
 protection of freshwater, 474
Respiration, 296–297
Reverse faults, 410
Revolution, 76
Rhumb line, 44
Rhyolite, 366
Rias, 608
Richter, Charles F., 413
Ridges, of high pressure, 135
Ridges (surface forms), 417, 435
Rift valleys, 412
 oceanic, 576
Rills, 471
Rime, 155
*Rime of the Ancient Mariner* (Coleridge), 131
Rip currents, 601
Ripples, 515
Rip tides, 602
Rivers. *See* Streams
Robinson projection, 45f
Robotic rovers, 74
Roche moutonnée, 544, 555
Rock cycle, 364–366
Rockfalls, 438, 440
Rock revetments, 605
Rocks, 363–366
 clastic, 368
 creation of sinkholes in soluble, 459
 decomposition and disintegration of, 425
 defined, 362
 formation of, 426
 igneous, 364–365
 jointing in, 366
 metamorphic, 364, 369–370
 relationship between topography and structure of, 417
 sedimentary, 364, 367–369
 tectonic processes, landforms, and structure of, 404–412
Rockslides, 438, 439
Rock weathering, 340, 362, 423
Rocky Mountains, 473
Rogue waves, 585
Rossby waves, 135
Rotation, 76–77
 days and nights related to, 77
 Earth's, 36, 117
 planetary, 72

## S
Safe Drinking Water Act, 451
Saffir–Simpson Hurricane Scale, 197, 199t
Salinas, 513
Salinity, 568–569
 levels in lagoons, 603
 variations in, 569
Salinization, 349
Saltation, 483, 515
Salt flats, 513
Saltwater replacement, 453
Salt weathering, 428–429
Salt wedging, 428–429
San Andreas Fault zone, 383, 406, 409, 410f
Sand dunes, 518–524
 classification of, 519
 invasions by, 519

protection of, 523–524
 types of, 521–523
Sand, in soil, 336–337
Sand seas, 518
Sand sheets, 518–519
Sandstone, 367, 434
 aquifers, 454
Sandstorms, 516
Santa Ana winds, 124, 125
Satellites
 geostationary, 60
 global positioning system (GPS), 39
 imaging of Earth by, 7
 orbits of, 60
 weather, 58, 141, 195
Satellites (in solar system), 69
Saturation, 148
 and dew point, 148–149
Saturn
 characteristics of, 72
 composition of, 71
 possibility of life on moons of, 71
Savannas, 245–246, 321–322
 drought in, 244, 246
 potential for African, 245–246
 transitional features of, 244–245
Scale, of maps, 46
 graphic or bar, 46
 representative fraction (RF), 46
 small, medium, or large, 46
 verbal, 46
Scanners, multispectral, 59
Schist, 370
Scientific method, 12
 steps in, 12
Sclerophyllous vegetation, 260–261, 318
Scrub, 318
Sea arches, 596–597, 609
Sea caves, 596, 609
Sea cliffs, 596, 607, 609
Seafloor spreading, 376
 and convection currents, 378–379
 defined, 378
Sea fog, 155
Sea levels, rising, 91
Seamounts, 578
Seas, 568
Seasonal migration, of wind systems, 133
Seasons, 78, 80–82
 animal response to, 84
 in humid continental, hot summer climates, 272–273
 in Mediterranean climates, 258
 in microthermal climates, 268–269
 in subarctic climate, 276–277
Sea stacks, 597, 597f, 607, 609
Seawalls, 605
Secondary coastlines, 609
Secondary productivity, 298, 300
Secondary succession, 304
Second-order streams, 473
Sections, 39
Sedimentary rocks, 364, 367–369
 chemical, 368
 clastic, 367
 composition of common, 367f
 information gleaned from, 404–405
 organic, 368
Seif, 523
Seismic waves, 371, 412
Seismographs, 371, 413
Selva, 240
Semidiurnal tidal regime, 581
Semipermanent pressure systems, 129

Sensible temperature. *See* Effective temperature
Sextant, 35
Shale, 367, 434
Shape, 41–42
Sheet wash, 471
Shield volcanoes, 397–398
Shifting cultivation, 242
Shock waves, 412
Shoreline, 594
Short-grass prairies, 323
Shortwave radiation, 75, 98
Siberian High, 129
Side-Looking Airborne Radar (SLAR), 59
Silicate minerals, 362
Sills, 404
Silt, in soil, 336
Siltstone, 367
Sinkholes, 456, 459
    collapse, 459
    formation of, 458
    solution, 459
SKYWARN networks, 194
Slack water, 584
Slash-and-burn agriculture, 242, 348
Slate, 370
Sleet, 89, 163
Slides, 438
Slip faces, 520
Slope aspect, 216, 285
    effect on microclimates of, 344
Slumps, 438, 440–441
Small circles, 33
Small-scale maps, 46, 47f
Snout, glacier, 539
Snow, 89, 162
    color of, 537
    reflective qualities of, 101
Snow belts, 176
Snowflakes, patterns of, 162
Snow flurries, 189
Snow line, 287
Snowmelt, 471
Snow pack, 537
Snow showers, 189
Snowstorms, 189
Soil, 331–332
    acidity and alkalinity, 338
    active layer, 443
    air, 334–335
    analysis, basic, 336
    categorization of, 340f
    chaparral as check against erosion of, 261
    characteristics of, 335–338
    classification, 350–356
    color, 335–336
    components, major, 332–335
    as critical natural resource, 356
    development of, 339–340
    ecosystems and, 15
    expansion and contraction of, 432, 443
    fertilization, 333
    formation. *See* Soil formation
    horizons, 340
    in humid continental climates, 275
    in humid continental, hot summer climates, 270
    in humid subtropical climates, 264
    inorganic materials in, 333
    in marine west coast climates, 268
    in Mediterranean climates, 261–262
    organic matter contained in, 335
    peculiar nature of savanna, 321–322
    rainforest, 240, 316
    savanna, 245, 246
    slash-and-burn agriculture and, 348
    structure, 337–338
    texture, 336–337
    and topography, 308–309
    water, 333–334
Soil formation
    climate and, 342–344
    effect of time on, 344–347
    factors affecting, 340–347
    influence of parent material on, 341
    land surface configuration and, 344
    organic activity, 342–342
Soil-forming regimes, 347, 349–350
    calcification, 347, 349
    gleization, 349–350
    laterization, 347
    podzolization, 347
    salinization, 349
Soil grade, 337
Soil horizons, 340
Soil orders, NRCS, 350–356
    Alfisols, 355
    Andisols, 351
    Aridisols, 351, 354
    Entisols, 350
    Gelisols, 351
    global distribution of soils by, 352–353f
    Histosols, 350–351
    Inceptisols, 350
    Mollisols, 355
    Oxisols, 356
    Spodosols, 355–356
    Ultisols, 356
    Vertisols, 354–355
Soil peds, 337
Soil profile, 339–340
Soil scientists, 6
Soil Survey Division of the Natural Resources Conservation Service (NRCS), 350
Soil surveys, 350
Soil taxonomy, 350. *See also* Soil orders, NRCS
Solar constant, 75–76
Solar energy, 69
    and atmospheric dynamics, 75–76
    components of, 75
    and land and water distribution, 104–105
    lines on Earth delimiting, 82–83
    reflective qualities of, 101
    *See also* Insolation
Solar noon, 36
Solar radiation
    in clouds, 97–98
    effects of atmosphere on, 95
    as energy source for gradation, 424
    intensity of, 79t
    involvement in energy exchanges of, 95
Solar system, 68f
    and beyond, 68–74
    defined, 68
    planets in, 73f
    sun as center of, 67–68
Solar wind, 70
Solberg, Halvor, 180
Solifluction, 443, 559
Solstices, 80, 81, 82f
Solution, 482
    coastal erosion assisted by, 595
    landform development by, 455–463
    minerals in, 463
    as process of chemical weathering, 433–434
    sinkhole development from concentrated, 456
    transport of stream load by, 483
Solution sinkholes, 459
Sonar, 59, 568, 570
Soufriere Hills volcano, eruption of, 400–401
Sound waves, 88
Source regions, 176
    of North American air masses, 178f
South American plate, 381
Southeast trades, 130
Southern Hemisphere, 34, 80
Southern lights, 70
Southern Ocean
    depth of, 568
    size of, 568
Southern Oscillation
    El Niño and, 139–140
    relationships defining, 140
South middle-latitude zone, 84
South polar zone, 84
South Pole, 92
    as reference point, 34
South tropical zone, 84
Spacecraft, exploratory, 74
Spaceship Earth, 17
Spalling, 430
Spatial distribution, 10
Spatial interaction, 11
Spatial patterns, 10–11
Spatial science, 4, 8. *See also* Geography
Spatial science perspective, 8–11
    characteristics of places, 10
    climate change and impact on coastlines, 225
    desertification, 250–251
    Driftless Area, a natural region, 552
    ever-changing Earth, 11, 13
    hurricane paths and landfall probability maps, 198
    introduction and diffusion of exotic species, 311
    location, 10
    location of ozone holes, 92–93
    paleomagnetism, 376
    regional concept: natural and environmental regions, 8–9
    relationships between plate boundaries, volcanoes, and earthquakes, 406–407
    Santa Ana winds and fire, 125
    spatial distributions and spatial patterns, 10–11
    spatial interaction, 10
    using vertical exaggeration to portray topography, 52–53
    watersheds as critical natural regions, 474
Species
    diversity on islands, 302–303
    generalist, 300
Specific heat, 104
Specific humidity, 149
Speleology, 462–463
Speleothems, 461–462
Spheroidal weathering, 431
*Spirit*, 74
Spits, 603, 604f, 609
Spodosols, 355–356
Springs, 452
Spring tide, 583
S (secondary) waves, 371
Stability, 160
Stalactites, 462
Stalagmites, 462
Star dunes, 523
Stars, 68
Stationary fronts, 180
Steffen, Amy Jo, 343
Steppe climates, 247
    dangerous appeal of, 253
    drought in, 254
    similarities between desert regions and, 252–253
Stereoscope, 57
Mount St. Helens, eruption of, 136, 399–400, 442

Stock, 402
Storm chasers, 194
Storm spotters, 194
Storm surges, 196, 198, 588–589
    erosion by, 605
Storm tracks, 182
Straight channels, 485
Strata, 368
Stratification, 334, 368
Stratiform clouds, 157
Stratocumulus clouds, 158
Stratopause, 94
Stratosphere, 93
    ice clouds in, 92
Stratospheric ice clouds (SIC), 92
Stratovolcanoes, 399–400
Stratus clouds, 157
Stream capacity, 480
Stream competence, 480
Stream discharge, 477–479
Stream gradients, 480
Stream hydrograph, 493, 496f
Stream load, 480
Stream order, 473
Stream patterns, 475, 475–476
Streams, 471–473, 472f, 475–477
    antecedent or superimposed, 477
    base level changes and tectonism, 492
    base level of, 475
    capacity of, 510
    channel patterns, 485–486
    defined, 470
    deposition, 484–485
    desert, 505–506
    disappearing, 459–460, 460f
    drainage basins, 473, 475
    drainage density and pattern, 475–477
    energy of, 479–482
    ephemeral, 505, 510
    erosion by, 482–483
    exotic, 506
    hazards related to, 492–495. *See also* Floods
    importance of, 495–496
    intermittent, 510
    land sculpture by, 486–489
    lower course features, 488–489
    middle course features, 487–488
    order of, 473
    perennial, 471
    rejuvenated, 492
    stream discharge, 477–479
    transportation and, 483–484
    transverse, 476–477
    trunk, 471
    upper course features, 487
    yazoo, 489
Stream terraces, 492
Striations, 543
Strikes, 405
Strike-slip faults, 410
Subarctic climates, 275–279
    effects of high latitude and continentality on, 276–278
    human activities in, 279
    landscapes of, 279
    limiting environment of, 278–279
Subduction, 380–383
Subduction zones, 381–382, 383
Sublimation, 162, 536
Submarine canyons, 574
Subpolar lows, 126, 129
Subsidence, 453
Subsurface horizons, 350

Subsystems, 22
Subtropical highs, 126, 131
Subtropical jet stream, 135
Succession, 301, 304
    final step in, 305
    predictable steps in, 304–305
    primary, 301, 304
    secondary, 304
Summer solstice, 80, 82
Sun
    angle, duration, and insolation, 78
    at aphelion, 77
    as center of solar system, 67–68
    declination of, 83
    Earth–sun system, 74–78
    and its energy, 69–71
    at perihelion, 77
    size and distance from Earth of, 71
    solar wind produced by, 70
    vertical (direct) rays of, 82
Sunspots, 70
Sun synchronous orbit, 60
Supercontinents, 375f, 387
Supercooled water, 155, 162
Superimposed streams, 477
Surf, 595
Surface creep, 515
Surface inversions, 103–104
Surface of discontinuity, 179. *See also* Fronts
Surface runoff, 470–471
    in deserts, 504–506
Surface weather map, 202, 203
Suspended load, 483–484
Suspension, 515
    transport of stream load by, 483
Swallow holes, 460
Swash, 595
Synclines, 407, 417
Systems, 13
    closed, 23, 146, 147, 316
    Earth, 13–14
    Earth processes as, 25
    inputs and outputs of, 22
    models and, 21–27
    open, 15, 23, 147, 294, 332, 536
    satellite, 60
    wetland, 19
    workings of, 22–23
Systems theory, 22

## T

Taiga, 278, 320
Taku, 124
Tall-grass prairies, 322–323
Talus, 428
Talus cone, 428
Technology
    digital, 61
    modern mapping, 51–56
    photovoltaic (PV), 69
    for protecting shorelines, 600–601
    radar, 191–193
    satellite, 7, 39, 195
    for tracking tropical storms, 198
Tectonic processes, 394
    gradation and, 424–425
    rock structure, and landforms, 404–412
Tectonism, 426
    base level changes and, 492
    desert, 505–506
Temperate glaciers, 537
Temperate oceanic climate. *See* Marine west coast climate

Temperature
    adiabatic changes in, 159–160
    affecting rate of evaporation, 151–152
    annual lag of, 109–110
    annual march of, 109–110
    in arid climates, 248
    controls of Earth's surface, 104–108
    daily march of, 101
    defined, 99
    in desert climates, 248–249
    distinctions between rainforest and monsoon climate, 243
    distribution at Earth's surface, 108–109
    effect of cloud cover on, 101
    effects on soil of, 343–344
    global rise in, 91
    gradients, 108, 109
    and heat, 99
    in highlands, 287
    in humid continental, hot summer climates, 270
    in humid subtropical climates, 263
    in humid tropical climates, 236, 236f
    in ice-sheet climates, 282
    inversions, 103
    in marine west coast climates, 265
    in Mediterranean climates, 258, 260
    moderating effect of ocean on, 122–123
    in Northern Hemisphere, 106t
    ocean currents and, 139t
    scales, 99–100
    shifts in, 91
    short-term variations in, 100–102
    in steppe climate, 253
    in subarctic climate, 276–278
    in tropical rainforest climates, 238
    in tundra climate, 280
    variation of ocean, 569
    vertical distribution of, 102–104
Temperature gradients, 108, 109
Temperature-inversion fog, 154
Temperature inversions, 103, 247
Temporary base level, 475
Tension, 405
Tephra, 396
Terminal moraines, 547, 555
Terminus, glacier, 539
Terranes, 384, 386
Terrestrial ecosystems
    abiotic component in, 294
    classification of, 312, 313f
    major types of, 306
    soil and topography in, 308–309
Tharp, Marie, 571
    map of ocean floor by, 572–573f
Thawing, 427, 440
Thematic maps, 49
Thermal expansion, 90, 91
Thermal infrared (TIR) images, 58
Thermocline, 570
Thermohaline currents, 569
Thermometers, 206
Thermosphere, 94, 95
Third-order streams, 473
Thornbush, 318
Thornthwaite, C. Warren, 207
Thornthwaite system, 206–208
Threshold, 24
Thrust faults, 408, 410
Thunder, 188
Thunderstorms, 157, 188–189, 272
    convective, 189
    cross-sectional view, 188f

frontal, 189
orographic, 189
Tidal bores, 584, 585
Tidal currents, 583–584
Tidal interval, 582
Tidal ranges, 583, 583f
Tidal waves. *See* Tsunamis
Tides, 579–585
    defined, 579
    high and low, 582
    spring and neap, 583
    theories of, 582
    tidal interval between two high, 582
Tide tables, 583
*Tierra caliente*, 288
*Tierra helada*, 288
*Tierra templada*, 288
Tiers, in Township and Range System, 38
Till plains, 555
Tilted fault blocks, 410
Time
    local, 36
    longitude and, 36–37
    soil formation and, 344–347
    zones, 36
Timescale, geologic, 386
Time zones, 36
    world, 37f
Titan, 72
*Titania*, 549, 571
Titles, map, 45
Tolerance, organism's range of, 306
Tombolo, 603, 604f
Topoclimates, 216
Topographic contour lines, 50
Topographic maps, 50–51
    features shown on, 51
    using vertical exaggeration in, 52–53
Topographic profile, 52
Topography
    of continental shelf, 574
    Earth's, contained in ocean seafloor, 567
    karst, 457f
    of oceanic ridges, 576
    related to differential weathering and erosion, 435–436
    relationship between rock structure and, 417
    soil and, 308–309
    use of vertical exaggeration to portray, 52–53
Tornado Alley, 191
Tornadoes, 159, 190, 190f, 191–193
    defined, 192
    use of Doppler radar for detection of, 192
Torricelli, Evangelista, 115–116
Tower karst, 460
Township and Range System, 38–39
Townships, 39
Toxic waste, 453
Trace of rain, 166
Traction, 483
Trade winds, 130, 131, 136
Transform faults, 383
Transform movement, 383
Transhumance, 288
Transpiration, 151
Transportation, 424
    stream, 483–484
    wind erosion and, 515–517
Transported parent material, 341
Transverse dunes, 522
Transverse streams, 476–477
Travertine, 462
Tree line, 287

Trellis stream pattern, 475
Trenches, 576–577
Tributaries, 471
*Trieste*, 577
Trophic level, 295
Trophic structure, 295, 296t
    nutrient cycles, 295–296
Tropical climates, 209–210
Tropical cycles, 193
Tropical cyclones. *See* Hurricanes
Tropical easterlies, 131, 136
Tropical forests, 316–318
Tropical monsoon climates, 242–244
    distinctions between tropical rainforest climates and, 243
    effects of seasonal change, 244
Tropical rainforest climates, 238–242
    cloud forests, 239–240
    constant heat and humidity of, 238–239
Tropical rainforest climates (*continued*)
    delicate balance of, 240
    distinctions between tropical monsoon climates and, 243
    human activities in, 240, 242
Tropical rainforests, 316–317
Tropical savanna climates, 244–246
    drought in, 244, 246
Tropical savanna grasslands, 321–322
Tropical storms, 197
    technology for tracking, 198
Tropical zone, 83–84
Tropic of Cancer, 83, 127
Tropic of Capricorn, 83, 127
Tropopause, 93
Troposphere, 93
    upper, 134
Troughs, 126
    of low pressure, 135
True north, 46
Trunk streams, 471
Tsunamis, 585, 587–588
    erosion by, 605
    from explosion of Krakatoa, 400
    forecasts and warnings for, 586
    Hilo, Hawaii (1946), 588
    Indian Ocean (2004), 14, 16, 587
    period and wavelength of, 586
Tundra, 280–281
    arctic and alpine, 324–325
Tundra climate, 210, 279, 280–281, 559
    characteristics of, 280–281
Turbidity currents, 574, 575
Turbulence, 482
Typhoons, 193, 198, 199

## U

Udalfs, 355
Ultimate base level, 475
Ultisols, 356
Ultraviolet rays, 75
Ultraviolet (UV) radiation, 25, 87, 89–90, 91
    in mountain climates, 287
Unconformity, 368
Uniformitarianism, 374, 389
United Nations Environment Program (UNEP), 251
Universal Time. *See* Greenwich Mean Time
Universe, measurement of distance in, 68
Updrafts, 157
    formation of hail in, 163
Uplift mechanisms, 163, 165
Upper air easterlies, 134
Upper air westerlies, 134

Upslope fog, 155
Upwelling, 138, 327
Uranus
    characteristics of, 72
    composition of, 71
    rotation of, 72
Urban heat island, 107, 217
Urbanization, 228
U.S. Army Corps of Engineers, 600–601
U.S. Geological Survey (USGS), 46, 225, 492
U.S. Public Lands Survey System, 38–39
Ustalfs, 355
UTC. *See* Greenwich Mean Time
Uvalas, 460

## V

Valley breeze, 124
Valley glaciers, 538
Valleys, 435, 492
    drowned river, 608
    in glaciated regions, 548
    hanging, 545
    V-shaped, 487
Valley sinks, 460
Valley train, 548
Variables, 13
    continuous, 50f
    discrete, 50f
Vegetation
    characteristic subarctic, 278
    contributing to soil composition, 342
    distinctions between rainforest and monsoon climates, 243
    in highlands, 287
    in humid subtropical climates, 263–264
    interception of precipitation by, 471
    in marine west coast climates, 268
    in Mediterranean climates, 258, 260, 262
    in monsoon forests, 318
    natural, 301
    rainforest, 240, 316
    savanna, 244–245
    sclerophyllous, 260–261, 318
    as source of water vapor, 151
    in steppe climates, 252
    in tropical mountains, 289f
    in tropical rainforest climates, 238
    tundra, 324–325
    use of carbon dioxide by, 90–91
    and wildfires, 125
    xerophytic, 251–252, 323–324
Ventifacts, 517
Venus, 73, 75
    atmosphere of, 74
    rotation of, 72
Verbal scale, on maps, 46
Vermiculite, 432
Vernal equinox, 82
Vertical accretion deposits, 484
Vertical (direct) rays, 82
Vertical exaggeration
    in digital terrain models, 52–53
    using, to portray topography, 52–53
Vertisols, 354–355
Mount Vesuvius, 395f
    eruption of, 400
Vietnam War, 238
Visualization, 53
Vitamin D, 91
Volcanic ash, 396, 401
Volcanic neck, 404
Volcanism, 396, 426
Volcanoes, 224, 396

cinder cone, 398–399
composite cones, 399–400
eruptions and, 396
newly forming undersea, 384
plug dome, 401
shield, 397–398
spatial relationships between plate boundaries, earthquakes, and, 406–407
types of, 397–401, 398f
undersea, 576
V-shaped valleys, 487

## W

Wadis, 507
Walker, Sir Gilbert, 139–140
Walsh, Donald, 577
Warm currents, 136–137
Warm fronts, 179
Washes, 507
Waste, toxic, 453
Water
  in atmosphere, 147–151
  capillary, 334
  as catalyst in chemical weathering processes, 431
  density of, 570
  differential heating of, 101
  distribution of land and, 104–105
  in ecosystems, availability of, 306
  geothermal, 463
  as gradational agent in arid lands, 506–513
  as heat energy, 95–97
  hygroscopic, 334
  importance of surface, 495–497
  mining, 448
  properties of, 145–146
  soil, 333–334
  supercooled, 155, 162
  transparency of, 104
  as transportation system for Earth, 146
  as universal solvent, 145
Water budget, 147
  fog in, 153–154
  and its relation to heat budget, 147–148
Water mining, 448
Water resource managers, 6
Watersheds, 473
  as critical natural regions, 474
Water table, 333
  defined, 448
  groundwater zones and, 448–450
  relationship between stream flow and, 450f
Water vapor, 89, 95
  evapotranspiration and, 151
  and relative humidity, 149–150
  and specific humidity, 149
  vegetation as source of, 151

Wave-built terraces, 599
Wave crest, 586
Wave-cut benches, 598
Wavelength, 586
Wave period, 586
Wave refraction, 595–596, 597f
Waves, 584–589
  breaking of, 594–595
  catastrophic, 587–589
  erosion by, 597, 605
  refraction of, 595–596
  rogue, 585
Waves of oscillation, 586
Wave trough, 586
Weak tropical disturbances, 193–195
Weather, 87, 115
  changes in, 88
  climate and, 110–111
  cyclones and local, 183–186
  defined, 110
  El Niño and global, 141–142
  general characteristic of mountain, 287–288
  ingredients of, 111
  and latent heat of condensation and power of severe, 98
  oceans and, 567
  systems, shifts in, 91
Weather forecasting, 187–188
  importance of computers to, 188
Weathering, 394
  chemical, 431, 433–434
  defined, 425
  factors involved in, 425
  importance of, 436
  mass wasting, and landscape, 444
  physical, 425–430
  salt, 428–429
  spheroidal, 431
Weather radar, 59
Weather satellites, 141
  use of TIR by, 58
Weather Service Doppler Radar, 193
Wedges, 559
Wegener, Alfred, 374–375, 377
Wells, 453
  artesian, 454–455
Westerlies, 130, 131–132, 133, 170
West Wind Drift, 138
Wet adiabatic lapse rate, 159
Wetland systems, 19
  draining of, 310
  role of groundwater in supporting, 452
White frost, 155
Whiteout, 189
Whiteouts, 282
Wild and Scenic Rivers Act, 495
Wildfires, 125

Wind chill chart, 277
Wind power, 119
Winds
  air pressure gradients and, 118, 120
  Chinooks and other warming, 124
  convergent and divergent circulation, 117–118
  Coriolis effect and, 120
  cyclones, anticyclones and, 122
  defined, 118
  deposition by, 517–524
  erosion and transportation, 515–517
  friction and, 121
  geostrophic, 121
  global surface wind systems, 129–134
  as gradational agent, 514
  longitudinal differences in, 134
  monsoon, 124–126
  movement of, 105–106
  polar, 282
  power from, 119
  role of, in arid region geomorphology, 504
  seasonal migration of, 133
  subglobal surface-wind systems, 122–126
  terminology pertaining to, 121–122
  upper air, 134–138
  waves initiated by, 585. *See also* Waves
  and wildfires, 125
Windward, 121
Winter solstice, 82
Wisconsin stage, 551
Wohlsperger, Andy, 481
World Meteorological Organization, 157

## X

Xeralfs, 355
Xerophytes, 251–252, 323–324
X-rays, 75

## Y

Yardang, 517
Yazoo streams, 489
Yelle, Carmen, 451
Young, Aaron, 596

## Z

Zone of ablation, 539
Zone of accumulation, 539
Zone of aeration, 448
Zone of brittle ice, 542
Zone of plastic flow, 542
Zone of saturation, 448
Zones of transition, 215, 243
Zooplankton, 325
Zulu Time. *See* Greenwich Mean Time

**World Map of Population Density**